STAINLESS STEELS '87

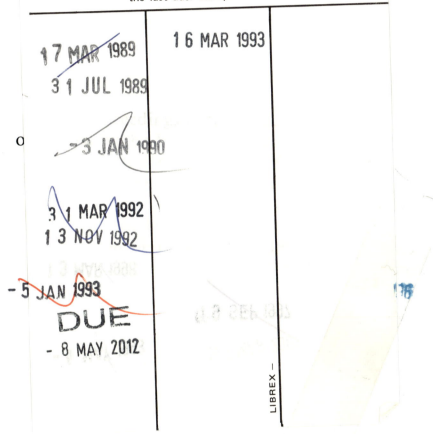

THE INSTITUTE OF METALS
1988

Book 426
published in May 1988 by

The Institute of Metals
1 Carlton House Terrace, London SW1Y 5DB
and
The Institute of Metals, N American Publications Center,
Old Post Road, Brookfield, VT 05036, USA

ISBN 0 901462 42 X

British Library Cataloguing in Publication Data

Stainless Steels '87 *(Conference : York)*
Stainless steels 87
1 Stainless steel
I Title II Institute of Metals (1985)
669'.142

ISBN 0-901462-42-X

Library of Congress Cataloging in Publication Data

Stainless steels '87
1 Steel, stainless — congresses
I Institute of Metals
TA479.575678 1988 620.1'788-841

ISBN 0-901462-42-X

Compiled from typescripts and illustrations provided by the authors

Printed and made in England by Whitstable Litho Printers Ltd

CONTENTS

SESSION III
CORROSION

SESSION IV
CORROSION

SESSION V
MECHANICAL PROPERTIES

SESSION VI
ELEVATED TEMPERATURE EFFECTS

POSTERS

FOREWORD

The success of the Stainless Steels 1984 Conference held in Göteborg, Sweden, made it inevitable that serious consideration would be given to a second international conference on this subject.

The timing of 'follow-up' conferences is clearly critical in that the Organizing Committee has to balance the need to maintain continuity with a duty to ensure that sufficient technical progress has been made to justify a further conference.

It was against such a background that the Organizing Committee commenced its deliberations in late 1986 and quickly came to the conclusion to proceed with 'Stainless Steels '87', choosing the University of York for the venue.

The original intention was to broaden the scope of the proposed second Conference to include steelmaking and fabrication processes in addition to those topics highlighted at Göteborg. This approach met with only limited success, however, and by far the majority of abstracts submitted were concerned with progress in the fields of corrosion, mechanical properties, elevated temperature effects and welding technology.

The overwhelming response to the call for papers justified the initial decision to go ahead but did present the Committee with the daunting and unenviable task of selecting papers. Unfortunately, despite extending the Conference to a full three days, some papers could not be admitted and it was decided to offer unsuccessful authors the opportunity to participate in a poster display. A number accepted the proposal and as a result a total of 65 contributions and keynote papers were accommodated.

This work reviewed developments since the Göteborg Conference and clearly illustrated the continuing international interest in the metallurgy of stainless steels. All six sessions were well attended and my overall impression was that the Conference served a useful role in providing both a forum for a critical assessment of progress and an opportunity to publish this volume of proceedings which represents the current state of the art.

On behalf of the Organizing Committee I wish to thank the keynote speakers, the Session Chairmen, the authors and participants and, last but not least, the personnel of The Institute of Metals, all of whom contributed to the success of the Conference.

I would also like to add my personal thanks to Mr D. B. Bray, Master Cutler of The Company of Cutlers in Hallamshire and Chairman of BSC Stainless, who despite an extremely heavy schedule agreed to be guest speaker at the Conference Dinner.

J R BLANK
Operations Director
Stocksbridge Engineering Steels
Chairman of Organizing Committee

SESSION I
PROCESSING

Influence of steelmaking and primary processing factors on availability and properties of stainless steels

H Everson and M A Clarke

The authors are with Stockbridge Engineering Steels, Stocksbridge, Sheffield.

SYNOPSIS

Steelmaking and Primary Processing factors often influence compositional and microstructural features of stainless steels and, as such have a significant bearing on applications and properties together with a corresponding influence on availability and cost.

The paper will summarise key features of the process route that have a major influence, including historical aspects.

Each of the main types of stainless steel; austenitic, ferritic, martensitic, precipitation hardening and duplex has its own processing peculiarities which affect overall production features with a corresponding effect on end user availability and applications.

Modern methods of manufacture such as AOD, VOD and continuous casting together with remelting techniques have allowed significant improvements in basic steel quality and properties to be made in addition to reducing overall cost.

Introduction

Many questions are asked of steelmakers regarding the processing, availability and properties of stainless steels and similar alloys.

The processing performance and the ease with which properties can be obtained has an obvious bearing on the availability and cost of a particular alloy.

Reference to the proceedings of several previous conferences indicates that there have been brief mentions of these relationships; there have also been a limited number of papers describing various elements of the process route involved in making stainless steels. However, the past 10 years have seen many changes in the industry, many facilities have closed, new facilities have been built or existing facilities modernised with, in some cases, new processes dovetailing into older equipment at other stages of the process route: therefore it is opportune to describe essentials of the route.

Additionally, many engineers and metallurgists concerned with the fabrication and application of stainless steel who have not had the opportunity of seeing for themselves will have learned their stainless production metallurgy several years ago - often this aspect only gets a brief mention in lectures and standard text books - therefore it will be useful to explain some of the limitations of the processes involved in order that the user has a better appreciation of these limits.

Historical factors have played their part and where appropriate mention is made of this aspect.

As it is easier to illustrate points using specific examples the production of billet and primary rolled bar at Stocksbridge Engineering Steels is used as a framework but other production routes are drawn on where necessary.

Raw Materials

Like many other interesting and successful discoveries eg. America and penicillin, stainless steels were discovered by someone stumbling across them by accident whilst looking for something else.

In England Brearley(1) was looking for an improved gun barrel steel. In Germany Strauss(2) (3) (4) (5) et al were experimenting with high nickel alloys when they discovered martensitic and austenitic types respectively.

Certainly Brearley was fortunate in the raw materials at his disposal - low carbon steel and high carbon ferrochrome gave him a composition around 0.3%C and 12%Cr. Earlier observations(6) (7) by other workers had been misled in their studies of the effects of chromium on iron alloys by the influence of carbon and had not obtained results worth developing. The earliest stainless steels were made by adding the appropriate alloys either as ferro-alloys or metal to melts of mild steel.

On the introduction of austenitic steel manufacture to England a manufacturer's publication of around 1925 claims "The chromium is added to the molten steel in the form of pure ferro chromium, and the nickel is added in the form of pure nickel,[8] a very expensive method.

The major raw material in todays processing is, of course, scrap- internally arising scrap from steelworks processes, customers' process scrap and scrap bought in from scrap merchants, in a wide variety of forms and of varying degrees of purity.

Effective scrap control is of paramount importance since the steelmaker aims to maximise the amount of chromium and other alloying elements that can be obtained from the scrap without introducing any undesirable contaminant. This is particularly important in preventing contamination by elements that can't be removed during steelmaking eg. when working to low cobalt specification for nuclear purposes or low lead requirements. At Stocksbridge scrap is carefully segregated and each wagon load sampled and analysed. Where necessary mixed scraps may have several representative samples taken to make a laboratory melt which gives an average analysis. Increasing use is made of portable computerised analysis equipment which is now available, in the sorting and classification of scrap.

Regular analysis checks have to be made on all alloying additions as changes in source of supply can lead to fluctuations in trace elements eg. cobalt and arsenic in nickel supplies.

Careful monitoring of other steelmaking materials such as limestone and fluorspar is also necessary since occasions have arisen where contamination of the melt from these sources could have had disastrous effects if undetected. One example of this is lead pick-up from certain supplies of fluorspar.

The effects of increased lead on the elevated temperature ductility of a 12% Cr Ni MO V steel - intended for elevated temperature use - are shown in [9] Fig 1.

Early developments in the production of stainless steels using fossil-fuelled furnaces gave excessive carbon pick up. Refractory crucibles used at that time also contained carbon which added to the problem and electric arc melting was the preferred method.

Brearley's first stainless type with about 0.3%C and 13%Cr could be made quite easily in the arc furnace. As lower carbon ferrochrome became available the so-called "Stainless Irons" were developed which had a .10%C and 13%Cr base composition.

In Hadfield's early work he used crucible melting and Fe Cr containing 6%C and 66%Cr therefore the resultant 11% and 15% chromium alloys contained 1.27% and 1.79% Carbon respectively. Both alloys as tested by Hadfield showed poor corrosion resistance and the latter alloy was too high in carbon to satisfactorily hot work. Were it not for these raw material and processing limitations it is reasonable to assume that stainless steels would have been discovered 20 years earlier.

In parallel with the 13% Cr types the austenitic 18% Cr 9% Ni types were developed. As with the 13% Cr type manufacturers found a readily attainable level around 0.10% C which served well for many applications and kept research workers busy for a number of years in developing ways to avoid the classic carbide precipitation or "weld decay" problems associated with this carbon level.

Primary Melting

The arc furnace is still the principal method used for melting stainless steel; larger and larger units were used increasing in size up to about 15 tonnes in 1939, 50/60 tonnes in 1953 - units of up to 150 tonnes capacity are now used. However, such is the variety of stainless manufacture that the range in sizes of arc furnaces of current manufacturers is around 7 to 150 tonnes.

As a melting unit for scrap the developed high energy input arc furnace is ideal but as a refining unit for stainless steels it has its limitations.

Methods of reducing carbon level to typical levels of 0.05/0.06 using oxygen lancing were developed and further extended by the availability of lower carbon ferrochromium enabling the "L" grades with 0.03% C max to be produced in bulk.[10]

The methods were not without their disadvantages. Oxygen blowing led to very high bath temperatures in excess of 2000°C which gave refractory wear problems - once a low carbon was achieved the subsequent refining to desulphurise gave rise to increased carbon due to pick-up from electrodes and from the constituents used in the desulphurising slag.

Effectively one was often faced with a choice; low carbon or low sulphur, achieving both could prove difficult. The strongly oxidising conditions were also inefficient with regard to chromium recovery.

The effects of carbon on stainless steels were highlighted at an early stage of their development.

In parallel with problems of carbon removal, methods of chemical analysis were slow in the first 40 years of stainless production, thus whilst waiting for a carbon result the molten bath in the arc furnace was absorbing more carbon.

The introduction of the high-frequency furnace for stainless manufacture in the 1920s was one method of avoiding carbon pick up but facilities for carbon and sulphur removal were limited and whilst this method could make steels with low carbon and sulphur contents, expensive low carbon and sulphur feedstocks were usually required.

Despite the limitations of the various raw materials and processes available, many stainless types were developed in the first 30 to 40 years of commercial stainless production; martensitic, ferritic austenitic and duplex types were developed. The original 12%Cr and 18%Cr 9%Ni types however remained the most common grades. Stabilisation of the carbon by titanium (11) or niobium was generally accepted as the best method of avoiding weld decay, although tungsten (12) and vanadium (11) were also used.

Nevertheless the principal brake on the large-scale production of low carbon types was the lack of an economic process to achieve very low carbon levels.

BS970 (1955) did not contain specifications for low carbon steels ie. 0.03% maximum carbon.

The development of the AOD and VOD processes unlocked the door to the economic production of the low carbon, low sulphur steels. The processes are well enough described elsewhere (13) (14) (15) (16) (17) but their impact on the development of new kinds of stainless cannot be over-emphasised. Parallel developments in rapid chemical analysis, particularly of C N and S also aided developments.

The higher oxygen potentials being achieved in the AOD/VOD process during the blow do not aid the removal of phosphorous which can impair the corrosion resistance of the stainless steel. Although oxidation occurs this is in equilibrium with the chromium in the slag and the economic advantages of the AOD/VOD are lost if the slag (higher in chromium) is taken off. On reduction the phosphorus pentoxide and chromic oxide are both converted back into the melt. To achieve low phosphorous levels the steelmaker has to resort to low phosphorous scraps although research is taking place to achieve a more economic hot process.

In considering the major categories of stainless steels modern steelmaking methods are able to produce the complete S range. Restrictions on steelmaking of low carbon steels are removed, therefore availability is limited by other aspects of processing such as hot workability or cold manipulation characteristics.

Casting

Having achieved a melt of the correct composition unless castings or metal powder are being made, the steel is cast either by continuous casting or into ingots.

Continuous Casting

In 1981/1982 SES installed and commissioned a VAD unit and a 4-strand billet continuous casting machine. Several papers have been published (18) describing these plants and with the tradition of stainless steelmaking at Stocksbridge it was inevitable stainless grades would be produced for rerolling, forging and seamless tube production.

A modest programme was followed, introducing casting techniques for the production of 5Cr and the 9Cr steels via the concast plant and continuous tube mills. The success of these products enabled SES to move to 13Cr martensitic steels and 304/316 austenitic steels, the latter being made to the traditional arc furnace process.

Although the 300 series steels can readily be continuously cast, care is necessary on the casting of the 13Cr martensitic steels to avoid too-rapid cooling and cracking problems, as is careful post-casting heat treatment to enable the as-cast product to be handled without breaking. Similarly for the austenitic steels the composition balance must be carefully controlled to achieve the optimum solidification mode[19] and care is needed in the selection and use of mould fluxes and the setting up of casting conditions.

Seamless tubes have been successfully produced from concast stainless steels using both the traditional tube-making methods; extrusion and via the continuous mills.

In Fig 2 the as-cast structure is shown.

Processing via the extrusion route involves machining the surface, boring out the centre and contouring the leading end. During reheating for extrusion, recrystallisation of the heavily worked surface and bore take place thus avoiding oxide grain boundary penetration down the columnar grains - a reason for surface break-up during hot working. Figs 3 & 4 show these features and Fig 5 shows the final tube.

Although there are obvious processing, yield and quality advantages of using continuously cast feedstock via the extrusion route, further cost reductions can be achieved when tube conversion is via a continuous mill using the black as-cast feedstock without the need to remove the surface or bore the centre out. Initial results both in the UK and Europe are very encouraging and this process will no doubt become the standard route for the future.

Ingot Casting

Modern ingot casting techniques generally use indirect teeming, with or without shrouding; mould fluxes are used to improve surface quality and insulating head tiles and exothermic powders are used to minimise pipe and segregation.

Ingot Preparation and Cooling

Until the full introduction of the improved casting methods about 30 years ago, ingot surfaces were poor and frequently required dressing overall by grinding or planing if defects in the wrought product were to be avoided. By their nature, stainless steels are scale resistant therefore the high scaling rate that was relied upon to remove ingot defects from the surface of carbon and low alloy steels does not apply. It is still common practice for very highly alloyed stainless ingots to be overall surface-dressed prior to rolling to prevent cracking on rolling. Dressing of ingots made at SES is now required only at top and bottom ingot positions. However, on certain grades high titanium and high aluminium levels can give rise to poor surface despite the use of mould additives.

The cooling out of ingots after casting differs markedly with the various types of stainless. Austenitic and duplex types can be satisfactorily air cooled. Most martensitic types have to be cooled slowly and in many cases annealed, unless they are charged immediately after stripping to a reheating furnace to prevent stress cracking on cooling down or subsequent reheating.

Fully ferritic low interstitial grades are extremely brittle in the as-cast state with impact transition temperatures in excess of 200°C and in the authors' experience these types can not be cooled out, either in air or slowly in an insulated pit, and subsequently reheated without an extremely high incidence of cracking. The only satisfactory way of producing fully ferritic stainless is to charge the ingots to a reheating furnace whilst still hot, keeping the ingot above the impact transition temperature.

There was considerable interest in the fully ferritic types during the 1970s as the AOD/VOD processes offered an apparently easy method of producing cheap low interstitial nickel-free ferritics. Whilst the fully ferritic types did find applications, particularly in thin sheet form, the very brittle nature of coarse-grained high chromium ferrite prevented their use for thick sections. The necessity to direct hot charge ingots reduced planning flexibility thus providing a restraint on availability. Problems of a similar nature are found on continuous casting fully ferritic steels with low interstitials.

Quasi-ferritic steels eg 430 types with around 0.05% carbon are not quite as difficult as the fully ferritic types but poor impact resistance in thick sections has limited the applications of this potentially attractive composition.

Reheating & Primary Hot Rolling

The most economic method of producing wrought stainless products from as cast ingots or concast feedstock is hot rolling.

Several features are important in determining the hot working characteristics of stainless steels and in broad terms they are composition related.

1. Hot Ductility

Ductility over normal hot working temperature range - between 900°C and 1300°C-should be high. As a general rule material showing over 40% Reduction of Diameter (R of D) on a high strain rate tensile test over at least a 300°C temperature range (eg 950 to 1250°C) has satisfactory ductility for the majority of primary rolling operations using a single reheating cycle.

It is possible to roll materials with a ductility level that falls to as low as 20% R of D at the lower end of the rolling temperature range using a double reheating cycle, provided a proportion of the ductility curve is above 40% R of D and a temperature range of 200°C is covered within these limits.

Within a grade there is cast-to-cast scatter but differences between grades are quite marked. Typical values are shown in Fig 6.

The poor ductility of high alloy austenitics and high carbon martensitics when compared with standard 304 type and the high ductility of ferritic stainless, particularly at low temperatures, is important in considering hot working characteristics.

2. Hot Strength

Hot strength is another key factor in determining the primary rolling characteristics. Very high hot strength requires high deformation loads which in turn can lead to mill overloads or mill failure.

The normal method of determining hot strength is the hot torsion test and typical curves are given in Fig 7.

Fully ferritic high chromium steels are weak at hot working temperatures making them prone to damage and difficult to manipulate into mill guides as they are not very well self supporting. For this reason fully ferritic grades are preferably rolled from lower soaking temperatures eg 1050°C instead of about 1250°C used for standard martensitic and austenitic grades. The lower rolling temperature also assists in achieving a finer grain, giving added shock resistance when cold.
Increasing the alloy content of martensitic and austenitic grades generally increases the hot strength. Molybdenum, Nitrogen, Vanadium, Nickel and Tungsten increase hot strength significantly. The effect of nitrogen is particularly marked as can be seen by comparing the torque curve for 316N with that of 316.

Alloys with a high hot strength and limited ductility range can not be satisfactorily rolled from large ingots and limited deformations per pass are necessary. However, once the material is reduced to a small billet or bloom it is often possible to re-roll the product to smaller sizes.

Compositional Factors

Composition balance is an important factor. Austenitic stainless steels with a delta ferrite content of greater than about 6% can have high ductility as measured by the hot tensile test but very poor hot workability, showing severe cracking on hot rolling or forging. The formation of a ferrite network around the austenite grains at hot working temperatures gives a thin film of relatively weak phase around a relatively stiff phase. The tendency is for the thin weak phase to try and absorb the majority of the deformation eventually exceeding its ductility limit and causing surface break-up of the material.
As the proportion of ferrite increases towards the levels of duplex stainless, ie. around 40 to 60%, the hot workability improves as the ferrite is no longer just a thin film around the austenite.

Several calculation methods for the estimation of austenite/ferrite balance have been published including those of Schaeffler [18] and Pryce and Andrews [19]. Processing conditions and experience vary from plant to plant and manufacturers devise their own variations of these formulae to suit their particular process. For example in the case of the common 304 and 316 grades it has been found that as-rolled size is a factor and the following formulae are used.

$$Cr^1 = \% Cr + 3 (\% Si) + \% Mo$$

$$Ni^1 = \% Ni + 0.5 (\% Mn) + 21 (\% C) + 11.5 (\% N)$$

In the case of 304 the % delta ferrite $\pm$ 3% (94% confidence limit) is given by

$$\% \text{ ferrite} = \frac{Cr^1 - (14.74 - (0.004 \times \text{dia mm})) - (.451 \times Ni^1)}{3}$$

For 316 the % delta ferrite $\pm$ 2.5% for 94% confidence

$$= \frac{Cr^1 - (13-(.0075 \times \text{dia mm})) - (0.6 \times Ni^1)}{0.3}$$

No doubt other steelmakers have devised modifications of previously published formulae based on their own processing experience.

Low ferrite levels (2%) are often specified by seamless tube manufacturers to prevent bore splitting on hot working.

Ferrite can also cause problems on martensitic stainless steels. The 17%Cr 4%NiCuNb precipitation hardening steel (AMS 5643) is an example. Compositions towards the ferritic end of the range are difficult to hot work due to splitting on ferrite. At the austenitic end of the composition range full transformation to martensite can not be readily achieved on final heat treatment therefore despite a wide specified chemical analysis range the steelmaker has to work to a very restricted chemical analysis range to satisfy both processing and mechanical property criteria.

Sulphur levels are also important. The high sulphur level of free machining types gives rise to severe end splitting on both primary and secondary hot rolling which causes excessive yield loss. Even quite small changes in sulphur level can have a significant effect on the hot ductility of 22%Cr 5%Ni duplex stainless steel as shown in Fig 8.

Various "additive" elements are frequently used as a means of improving hot workability. Boron is the most widely used for austenitic grades with a typical level of 0.0015 to 0.004% although Ti, Zr, Ca and Mg are also used. See Fig 9 and Fig 10.

The chemical and microstructural refinement that takes place during consumable electrode remelting (ESR and VAR) can also improve hot ductility levels, see Fig 11.

By way of illustration the typical process route for the hot rolling of standard austenitic stainless steel ingots to an as-rolled size of 180mm is as follows:-

Ingots - uphill teemed, wide end up
typical weight 4.6 tonnes
Air cool after casting.

Preheating furnace - temperature raised to 800-900°C at approx 100°C per hour.

Soaking Pit - Setting 1300-1320°C 4 to 5 hours.

Initial Rolling - 40" reversing mill
600mmsq ingot reduced to 225mm sq using 50mm draftings.

Discard at bloom shear - Up to 6% top end 1.5% bottom end.

Intermediate rolling - 42" reversing mill
225mm sq bloom reduced to 195mm sq.

Finish rolling - 32" reversing mill roll to 180mm dia using a round cornered square - oval - round sequence

Post Rolling and Finishing Operations

After rolling austenitic, duplex and ferritic stainless are normally cooled in air. Fully ferritic steels may be transferred hot to a surface grinding machine when the temperature has fallen to below 350°C and the billets hot ground to prevent cracking.

Martensitic stainlesses may be slow cooled but may be cooled in air to room temperature then annealed immediately to prevent cracking.

Stainless steels are normally fully surface ground or turned to remove surface defects prior to further hot working, because unlike C & A steels the scaling during reheating for rework is very low and any small defects are not removed.

The surface quality of continuously cast material can be suitable for further hot processing without overall surface dressing.

In the case of austenitic grades the finishing temperature is generally between 1000 and 1050°C. With the normal air cool from rolling the product is effectively auto solution treating, being free from carbide network.

However, for contractual/specification reasons material not intended for further hot working has to be solution treated. Obviously it is not necessary to solution treat material intended for subsequent hot working.

Steel Cleanness

Modern stainless steelmaking methods together with shrouded teeming techniques, selected refractory holloware give very good cleanness. Very low sulphur levels lead to reduced MnS content, secondary refining with its associated argon/nitrogen purging promotes inclusion flotation and allows effective de-slagging.

Silicon deoxidation can give rise to the formation of plastic silicates. Aluminium deoxidation eliminates the formation of silicates.

The production of clean steels is particularly important for stainless steels because of the improvements which can be obtained in corrosion properties and polishability. As a simple example a comparison of the Huey test on air melt and Electro Slag Remelted steel shows that typically airmelted stainless 304L steel will begin to show increasing weight loss after 4 or 5 periods, whereas the same steel processed via the ESR process shows no "take off" even after 8 to 10 periods.

Remelted Steels

As has been stated, modern melting and casting techniques can give far better levels of cleanness than was previously possible. Nevertheless applications arise where even higher standards of cleanness are required and a remelting process is specified. In addition to improving cleanness the two common remelting processes ESR and VAR give refinement of the ingot structure plus some impurity removal which lead to enhanced ductility and toughness in the finished product. This is particularly true of transverse properties, as illustrated in Table I [20].

Not all steels are suitable for both re-melting processes. For example the ESR process will remove titanium therefore titanium bearing steels are difficult to produce by the ESR route. Titanium is not lost during the VAR process therefore VAR is the most suitable technique for titanium bearing stainless steels. Conversely the VAR process removes nitrogen, which is retained during the ESR process thus ESR is the most suitable process for nitrogen bearing stainless steels, although special "pressure melting" techniques are being developed for high nitrogen steels via VAR methods.

Machinability

The very low sulphur levels and high standard of cleanness inherent in AOD and VOD produced steels has resulted in generally poorer machinability. This is particularly true of the standard 304/316 type austenitics where high cleanness levels can cause problems on high speed automatics due to the swarf coming off in long continuous spirals which cannot be removed and tend to get tangled in the tooling.

An obvious answer for non critical applications is to use a free machining grade of the 303 type which contains 0.2% sulphur. However, such a high sulphur level causes a marked loss of corrosion resistance - even standard processing operations such as passivating and descaling of 303 type can show severe pitting along the sulphides.

This has led several manufacturers to develop improved machinability grades such as IM 304 and IM 316 which, whilst maintaining the sulphur level within the 0.030% max typical of most national specifications have a controlled composition optimised to improve machinability. These improvements can be seen in Figs 12 & 13 which show results of standard and treated steels when subjected to various machinability tests.

Conclusions

Primary processing characteristics will continue to be the most influential factor in determining the availability of stainless steels. Only if an alloy can be produced into the full range of product forms eg bar strip plate and tube, will it achieve ready acceptance. Advances in steelmaking techniques over the last 20 years have encouraged the development of a wide range of very low carbon steels with enhanced corrosion resistance. However, other aspects of primary processing such as hot workability or lack of toughness in the as-cast or finished product have frequently limited the widespread adoption of steel compositions with potentially attractive properties.

Nevertheless such are the increasingly demanding needs of modern industry that steelmakers and fabrications will continue to develop new compositions, with all that that entails in developing manufacturing techniques.

References

1. H. Brearley Canadian Patent No 164 622 Aug 31st 1915
2. Clement Pasel British Patent 13414 (1913)
3. Clement Pasel British Patent 13415 (1913)
4. B Strauss Zeit F Angerwandt Chemie, 1914 27 P 633 Strauss and Manners
5. Die Hochlegresten Chromnickelstable als Nichrostende Stahle, Kruppoche Monalshefte, Aug 1920
6. R A Hadfield JISI 1892 II 48
7. P Monnartz Metallurgie Val V111 1911 p161
8. Thos. Firth & Sons Ltd "The Development of Staybrite Steel"
9. Orr J. Swinden Labs.
10. D C Hilty et al J.I.S.I. 1955 180 116
11. F Krupp AG British Patent Application 22875 July 25th 1929.
12. WH Hatfield and H Green. British Patent 316, 964 Aug 1st 1929.
13. M Schmidt et al Stahl Eisen 1968 88 (4) 153
14. H Baner et al Stahl Eisen 1970 90 (14) 153
15. J M Saccamo et al J Met 1969 21 (2) 59
16. Al Hodge Ironmaking and Steelmaking 1977 4 (2) 81.
17. J C C Leach et al Metals Society conference proceedings "Secondary Steelmaking" London 5-6 May 1977.
18. I G Davies et al "Continuous Casting '85" Institute of Metals London 1985. Ironmaking and Steelmaking 1986 Vol 13 No 1.
19. Deneuville and Wadier "Continuous Casting '85" Institute of Metals London 1985, Paper 26.
20. AL Schaeffer Brit. Weld. J. 1960, 7, 27.
21. L Pryce and K W Andrews JISI 1960, 195, 415 - 417
22. H Everson and F R Milner BNES Conference "Ferrite Steels for Fast Reactor Steam Generators" London June 1977, Paper 69.

TABLE 1

12CrNiMoV Hardened and Tempered							
		0.2% PS N/mm^2	UTS N/mm^2	E1%	RA%	Izod Impact Joules	
ESR ingot As Cast	Longitudinal	755	980	15.4	58	81	81
	Transverse	750	992	15.4	55	70	76
Air Melt Rolled Billet	Longitudinal	787	995	16.4	57	85	81
	Transverse	781	998	15.0	47	31	34

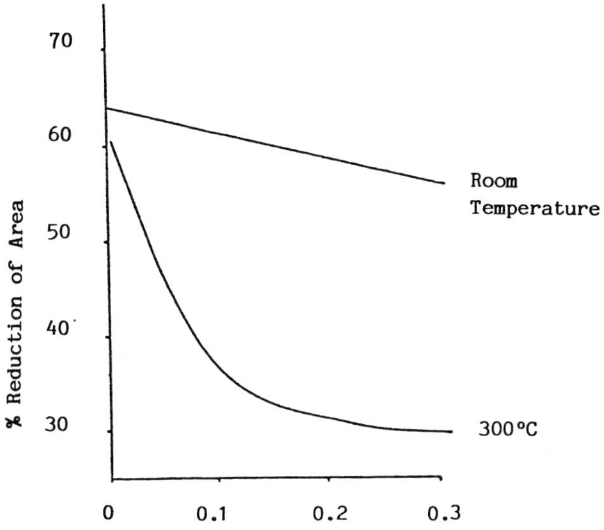

1 Effect of lead on tensile ductility of Jethete M152

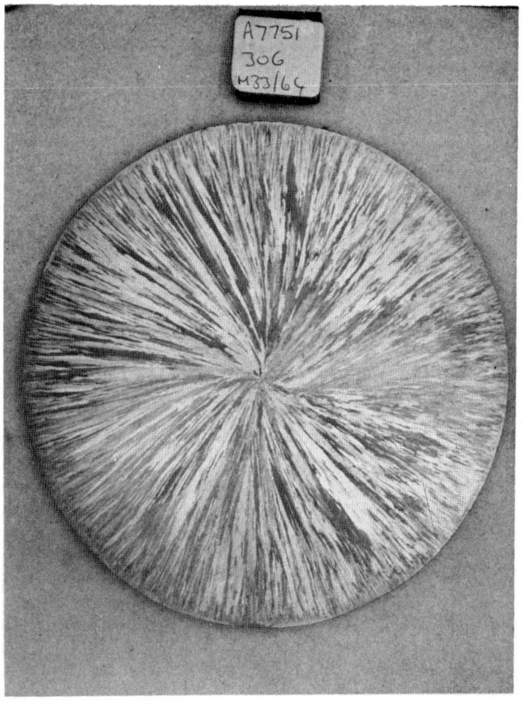

2 175mm diameter type 304 as-cast

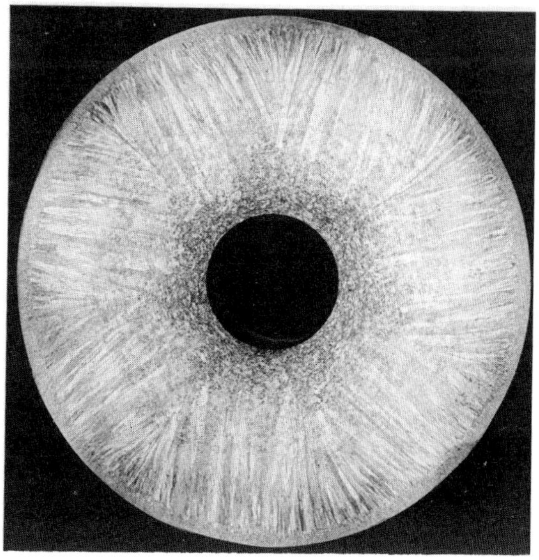

3 "Lifted use" showing lightly worked surface and central regions with recrystallised structure

4 Cross section through sample shown in Figure 3

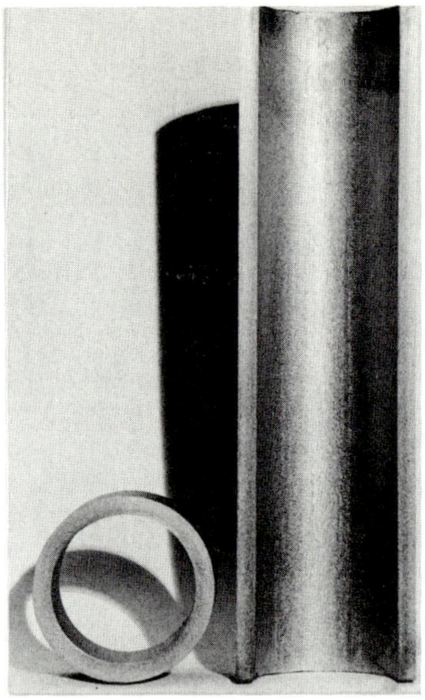

5 Final tube - SF304

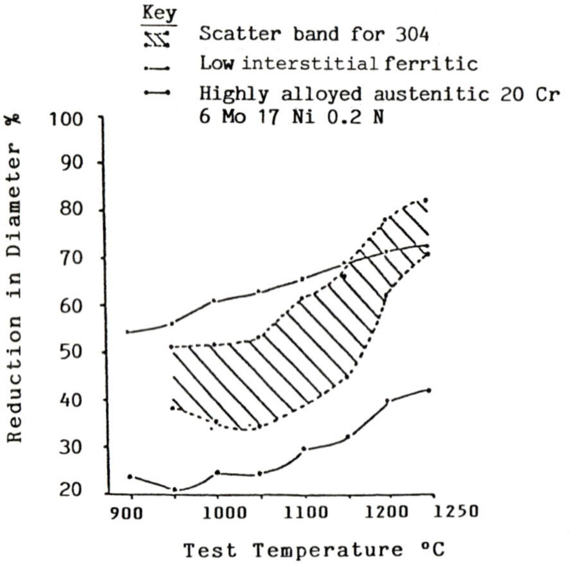

6 Comparison of hot ductility

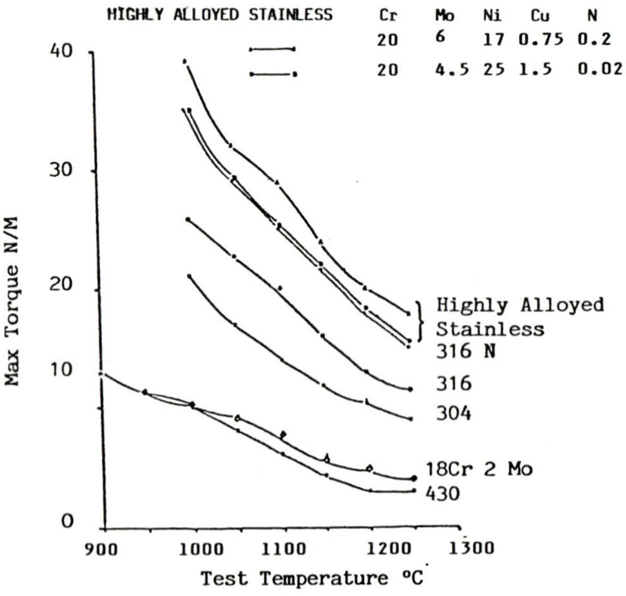

7 Comparison of hot strength of several types of stainless steels as assessed by torsion test

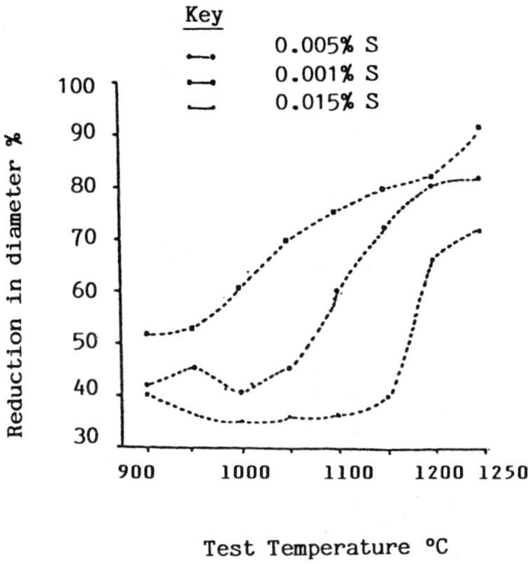

8 Effect of sulphur on hot ductility of 22%Cr5%Ni duplex stainless steel (soaking temperature 1250°C)

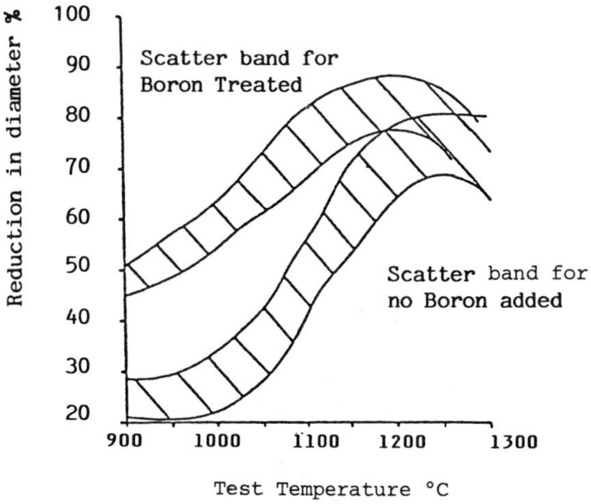

9 Effect of boron on hot ductility of 32%Ni20%Cr alloy - commercial casts

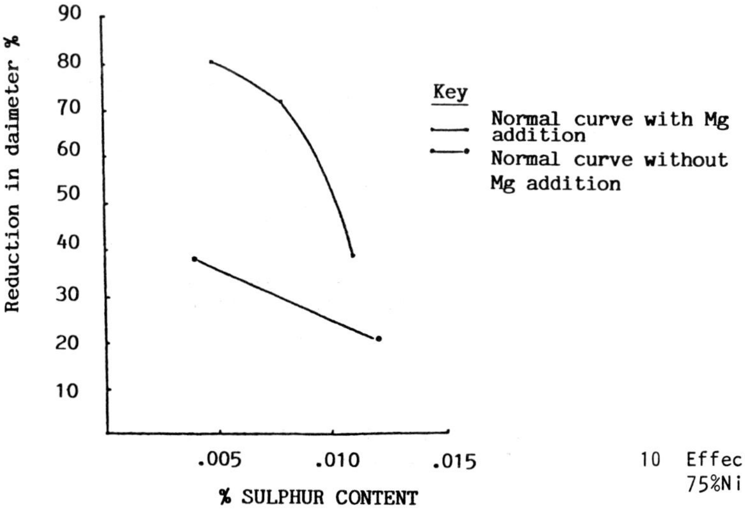

10 Effect of sulphur and magnesium on ESS 600 75%Ni15%Cr8%Fe

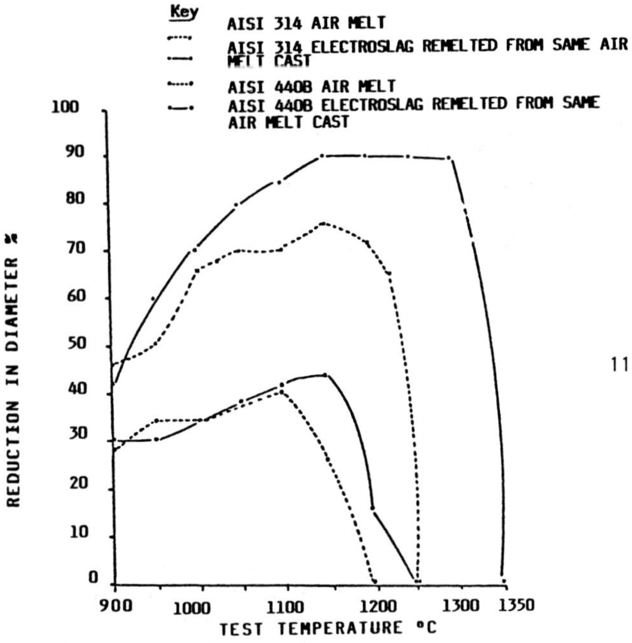

11 Effect of electroslag refining on hot ductility of an austenitic heat resisting steel AISI 314 and a martensitic bearing steel AISI 440B

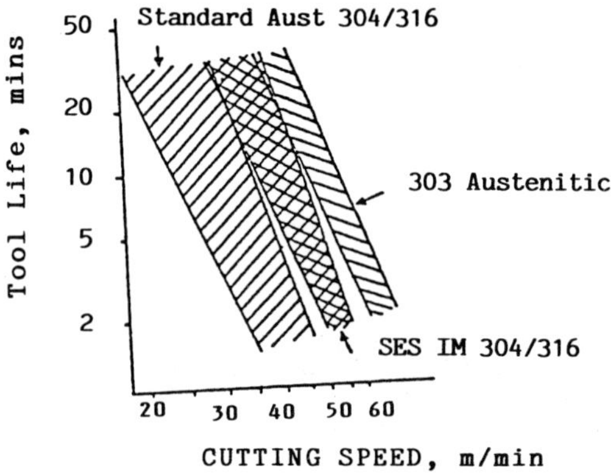

12 Tool steel: M2, feed rate: 0.125mm/REV, depth of cut: 1.25mm

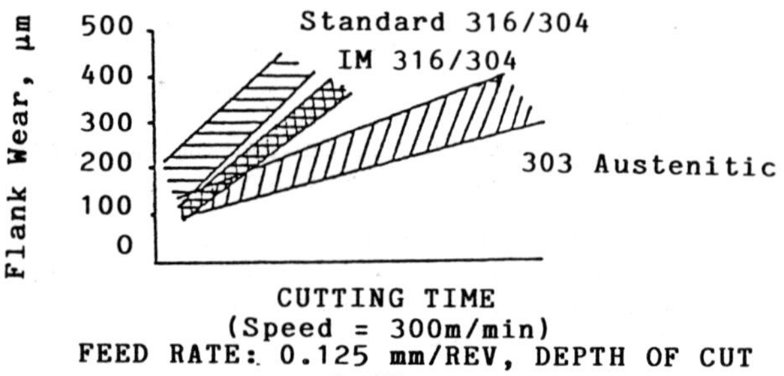

13 Coated carbide tools

Application of 'Horicast' for continuous casting of stainless and high alloy steels

D A Preshaw

D. A. Preshaw is with Davy McKee HORICAST in Sheffield. Davy McKee is one of the United Kingdom's leading companies in the plant and processing engineering field.

SYNOPSIS

Horizontal continuous casting of steels is rapidly gaining in importance as more plants are brought on stream. One of the initial horizontal casting processes is that of HORICAST by Davy McKee. This paper describes some of the more important fundamentals necessary for the application of HORICAST in the field of stainless and high-alloy steel casting.

1 INTRODUCTION

The Davy McKee horizontal continuous casting (H.C.C.) HORICAST process was originally envisaged and developed as a simple low-cost means of enjoying the benefits of continuous casting as an alternative to vertical and curved-mould machines. The changing market conditions, increases in competitiveness between steelmakers, and improvements in performance and product quality using vertical continuous casters (V.C.C.), has led to the continuing development of the Davy McKee Horicast process and its application in the field of stainless and high-alloy steel casting.

These developments began with the commissioning in 1983 of a Horicast for round blooms and billets at the Ohgishima plant of the Keihin Works of N.K.K. in Japan. Significant differences in the properties of stainless steels, when compared with the carbon and low-alloy steels for which Horicast had originally been developed, presented new and specific problems. Solutions to those problems had to be found if the process was to become industrially acceptable. This short paper discusses the development which has culminated in the sale of three twin-strand Horicast machines to important European special-steel producers during the last two years.

2 IMPORTANT FEATURES

Although there have been many articles published which give detailed explanations of the basic principles and features of the Horicast Process (1, 2, 3, 4) it is necessary to understand the more important principles before considering the application to the casting of stainless steels.

The three main features which characterise Horicast and differentiate between H.C.C. and V.C.C. are:

i) The tundish and mould are joined together in a configuration which does not allow molten metal penetration.

ii) The mould is fixed and does not oscillate as on V.C.C. machines.

iii) The solidifying strand is withdrawn in a cyclic manner comprising a pull-pause-push combination.

Let us consider each of these areas.

i) Mould-Tundish Joint

A schematic representation of a section through this area is shown in Figure 1.

Steel is teemed from a bottom-poured casting ladle into the tundish and controlled with a ferrostatic head of between 800mm and 1000mm under normal casting conditions. The liquid steel flows through the feed tube and into the water-cooled copper mould where it begins to solidify in a controlled fashion against the mould wall and break ring. It is the control of this initial solidification which determines the surface quality of the cast billet; reliability of the casting operation and consistency in production.

It is the choice of material, design and configuration and the methods of application which determine the successful performance in a novel situation. Prerequisites which determined the choice of materials for the various components in the joint were:

a) Feed Tube

i) Resistance to erosion by the type of steel to be cast.

ii) Low thermal conductivity (necessary to prevent premature freezing and skulling).

iii) Low coefficient of thermal expansion.

iv) Good thermal shock resistance (cracking or collapse of this component during casting would result in an immediate termination of the cast).

v) Resistance to thermal creep/deformation under load.

vi) Acceptable cost/performance ratio.

b) Break Ring

i) Maximum erosion resistance to the grades of steel being cast (not more than 1mm per hour).

ii) Ability to be manufactured to very close tolerances similar to machined components.

iii) Excellent thermal shock resistance under severe temperature gradients and thermal cycling conditions, associated with low thermal expansion.

iv) Good mechanical strength.

v) Impervious to the effects of water (condensed water vapour).

vi) Acceptable cost/performance ratio.

c) Mould Tube

i) Sufficiently high thermal conductivity to produce a solidification rate in the cast billet which would sustain a commercially acceptable casting rate.

ii) High temperature hardness and creep resistance.

iii) Ability to be fabricated into the required configuration using normally available techniques.

iv) Multiple life performance.

Whilst we have identified the most important properties of the components comprising the joint between the tundish and the mould it is also necessary to understand the mode of solidification at the junction of the mould and break ring. This can be explained using the sequence of events shown schematically in Figure 2.

At the beginning of the withdrawal cycle the skin which was formed during the previous complete cycle can be seen as having solidified against the break ring and the mould wall (Figure 2, Part A).

During the next cycle the strand is withdrawn so that molten steel can flow into the gap between the break ring and the previously solidified skin and begins to solidify against the mould wall (Figure 2, Part B).

Upon completion of the pull-phase of the cycle the solidifying skin is characterised by a weak point which is the position of most recent solidification. Strand movement is stopped to allow consolidation of the skin at this weak point (Figure 2, Part C).

After a short period the strand is pushed back in a reverse direction to compensate for contraction while further skin growth takes place (Figure 2, Part D). This completes the cycle which is then repeated at a frequency of between 80 and 120 cycles per minute. This cyclic withdrawal pattern produces two characteristic surface marks on the surface of the cast strand, for each withdrawal cycle. They have regrettably become known as the "cold shut" and the "hot tear".

ii) Fixed Mould

Because of the constant high ferrostatic head in the tundish a bulging force is exerted on the solidifying shell which keeps the shell in contact with the mould wall until the shell has sufficient strength to oppose the bulging force and can move away due to contraction.

This not only produces high rates of shell growth and permits high casting speeds but leads to an increased thermal load on the mould tube and associated water-cooling system. Excessive mould surface temperatures must be prevented if there is not to be premature failure of the mould due to creep or cracking. However, rapid shell growth does produce quickly a strand with a strong shell with the benefit that relatively short moulds can be used.

The mould is the heart of any continuous casting process since this is where the solidification is initiated. On V.C.C. machines the solidifying strand is withdrawn from the mould continuously at a uniform speed. The mould oscillates up and down at a controlled frequency and stroke, in such a manner that during the down stroke the mould travels slightly faster than the strand. This action strips the solidifying shell from the mould wall whilst instantaneously keeping the shell in compression, thus reducing the tendency for hot-tearing and rupturing of the shell.

This feature is known as compression release or negative strip and is applied to all industrial V.C.C. operations. The mould surface is continuously lubricated using special oils or powder fluxes which further improves the casting operation. Neither mould oscillation nor continuous lubrication can be applied to the Horicast mould which means that an alternative approach has to be adopted. The solution has been to adopt a cyclic withdrawal system.

iii) Cyclic Withdrawal Pattern

The pull-pause-push configuration of the withdrawal cycle which was mentioned in the explanation of shell growth above is shown in the velocity-time diagram in Figure 3.

The withdrawal of the strand is effected by a twin-strand pinch roll with all four rolls driven through a precision gear box from an axial-piston hydraulic motor. This is carried out under closed-loop control using a high-speed digital computer with pressure and velocity feedback.

A hydraulic system is used in preference to an electric-drive system because of its ability to be accurately controlled in a high-torque low-speed application. This is particularly important when trying to control push-back distances of less than 0.4mm.

The appropriate characteristic during each phase of the cycle — acceleration, velocity, deceleration, position — must be precisely controlled if problems with caster operation and product quality are to be avoided.

Details of the hydraulic withdrawal mechanism and its various modes of control are covered in a number of Davy McKee patents. (5, 6, 7)

3 DEVELOPMENT OF CASTING OF STAINLESS AND SUPER ALLOYS

Although Davy McKee had identified significant differences between the casting conditions required for stainless steels as compared with carbon steels during pilot plant operations at Wiggin Steel and Alloys in 1979 — 80, (8) no industrial-scale casting operations had been carried out on stainless steel prior to 1983. At this time Davy McKee's Japanese partner in the exploitation of H.C.C. (9) began casting stainless steels on their newly commissioned Horicast at their Keihin Works. (10)

The problems of casting stainless alloys were immediately apparent when using the technology which had been developed to cope with the properties of carbon and low-alloy steels.

The specific differences in chemical and physical properties which affected the castability of the stainless alloys were as follows:

i) Chemical reactivity with the break ring ceramic.

ii) "Stickiness" factor.

iii) Fluidity of molten steel (and surface tension).

iv) Wide freezing range.

v) Increased heat capacity.

vi) Low thermal conductivity.

vii) High hot strength.

viii) Surface scale characteristics.

These differences resulted in changes to the established Horicast technology, particularly in the areas of break ring, mould, withdrawal pattern and the application of strand electromagnetic stirring. Modifications which were carried out and improvements in results obtained are discussed for each of these areas below.

4 BREAK RING DEVELOPMENT

Typical temperature gradients and stress distribution patterns which are set up during casting are shown in Figures 4 and 5. The induced stresses can be accommodated by use of a suitable refractory ceramic material and by application of compressive forces as explained in our patents. (11,12) Reaction-bonded silicon nitride had been used originally for carbon steels and this had been further improved by the addition of approximately 10% of boron nitride.

However this material proved to be totally unacceptable for casting stainless alloys due to the very high rate of chemical erosion. Pure boron nitride seemed to be a suitable material from the point of view of erosion and thermal shock resistance, but was subject to moisture attack and was thought to be unacceptably expensive. A recent innovation had been the development of Sialon materials (silicon aluminium oxy-nitride) which appeared to have most of the properties required for an improved break ring.

Immersed stirring tests in molten 304 stainless steel using hot-pressed Yttrium-stabilised Sialon (Syalon™) showed a dramatic improvement in erosion resistance which was sufficiently encouraging to justify the development of these materials for break ring production.

Several grades of Beta Sialon containing between 10 — 20% boron nitride were produced and used for casting a wide range of stainless alloys. The improvements obtained are shown in Figure 6.

Two grades of Sialon-based material are now used in break rings for casting the whole range of stainless steels.

5 MOULD MODIFICATIONS

The cold-shut mark mentioned previously can result in a cracking defect at every cycle on the billet surface if there is failure to weld at the triple point of solidification between the molten steel, mould and break ring, due to excessive cooling. The influence of various factors upon this triple point temperature was investigated using a two-dimensional finite-difference model. Results showed quite clearly that withdrawal cycle frequency had the biggest influence and that casting should be carried out at a frequency of 150 cpm.

A comparison of the depth of the cold-shut crack for carbon and low-alloy steel for different withdrawal cycle frequencies was made and the results presented in Figure 7. It can be seen that the cold shut disappears at a frequency of 150cpm for carbon steel but it remains at a depth of almost one millimetre for type 304 stainless at the same cycle frequency. To increase the frequency beyond 150cpm was considered to be impracticable because of the increased demands upon the withdrawal machine and control system.

Researchers at N.K.K. decided to modify the geometry of the mould in the area of the triple point in an attempt to overcome this problem.

This new design of mould (which is currently the subject of worldwide patent applications) increases the temperature of the solidifying skin at the triple point, changing the surface characteristics and microstructure and reducing the depth of the cold shut (Figure 8).

Other improvements which were made to the moulds in order to improve the casting performance on stainless alloys were:

a) In order to reduce the tendency of the skin towards sticking, and to cope with the more abrasive effects of the solidified shell; a precipitation-hardening Be-Cu alloy mould without any surface plating of Cr or Ni was developed. The surface is prelubricated with a specially developed high-temperature lubricant.

b) Compound mould tapers are used to ensure uniform solidification which is a prerequisite of the maintenance of good billet shape and freedom from cracking.

c) The mould tube was split into two parts along its length in order to reduce fabrication and refurbishing costs and to improve the overall mould-life performance in terms of cost per tonne of steel cast.

6 WITHDRAWAL CYCLE

Because of the differences in specific heat capacity and thermal conductivity between carbon and stainless steels, the latter produces a thinner shell with an increased temperature gradient across the thickness of the skin. This factor makes stainless alloys more susceptible to changes in casting conditions, such as superheat and break ring performance. This means that stainless steels are normally cast at speeds 20 – 25% lower than for plain carbon and low-alloy grades, using a different withdrawal cycle pattern.

The importance of the correct cycle pattern can be explained best by reference to Figure 9. The pull speed (Vc) and the pull time (T_1) determine the length of the withdrawal stroke, which is usually in the range of 10 – 20mm, and also affect the overall casting speed.

The pause time (T_2) determines the integrity of the billet surface at the hot-tear mark; too long and the billet skin will tear due to thermal contraction; too short and the skin will have insufficient time to become thick enough to resist the push-back force applied during the next part of the cycle and skin buckling will occur. The push-back time (T_3) is kept to a minimum in order to keep the overall cycle time within the range required for the type of steel being cast. However it must be of a sufficient duration to permit consolidation of the solidifying skin to a point where the shell can resist tearing during the subsequent withdrawal stroke.

7 ELECTROMAGNETIC STIRRING

Results of the effects of electromagnetic strand stirring are presented in detail elsewhere.

8 RESULTS

The following results were obtained from the casting trials undertaken by NKK on the Keihin Works Horicast.(13)

Figure 10 shows the results of secondary dendrite-arm spacing measurements on type 304 stainless. The spacing in the lower half of the billet is smaller than that for the upper half which can be attributed to asymmetrical cooling due to natural convection and cooling by the support rolls.

The very fine chill-structure at the surface of the billet results in rapid homogenisation during hot-rolling. In an "as cast" sample of type 304 billet the delta ferrite content is less than 2% at a distance of 20mm from the surface and increases to approximately 4% at the centre.

The rapid cooling achieved in H.C.C. suppresses the tendency of delta ferrite formation (Figure 11).

When compared with plain carbon and low-alloy steels the stainless grades produce almost no scale during the continuous casting process or during reheating prior to subsequent hot-rolling or forging. Since there is little chance to "scale off" any surface defects, the surface condition requirements placed upon the casting operation are quite severe. 115mm-square billets were rolled into 30mm-diameter bars and 7mm-diameter wire rod without any surface conditioning and checked for surface defect levels (Figure 12).

A comparison between measured mechanical properties and theoretical calculated values on the relationship between tensile strength and composition, ferrite content and austenite grain size was made on rolled wire rod. Both values agreed with each other very well, indicating that there is no difference between Horicast and ingot-route billets.

9 CONCLUSIONS

9.1 High-quality alloy, stainless and superalloy steels have been produced on a commercial basis using the HORICAST process.

9.2 Except for products for the most severe applications (eg. cold heading and upset forging) there is generally no need for surface conditioning.

9.3 The new mould and break ring technology has enabled stable casting to be achieved and produced a remarkable improvement in surface quality.

9.4 Four Horicast plants for stainless steel are now in commercial operation at:-

a)	Keihin Works of N.K.K.	-	Japan
b)	Olarra S.A. Bilbao	-	Spain
c)	C3F Ondaine (Usinor)	-	France

A fourth machine will begin operation at the Panteg Works of B.S.C. during the Autumn of 1988.

10 REFERENCES

1. Development of the Horicast process for casting steel billets - Shearn, Marsh and Toothill. - 38th Electric Furnace Conference Chicago 1980.
2. Horizontal continuous casting of steel billets. D. Toothill. Solidification technology conference - Warwick University September 1980.
3. Development and metallurgical analysis of H.C.C. - Miyashita et al I & S.M. August 1981.
4. H.C.C. Process for high alloy stainless tubular goods - Komori et al I & S.M. January 1985.
5. U.K. Patent 1380582 Gamble & Marsh - Davy International.
6. British Patent Application 8126051 and 8130484. Toothill - Davy Loewy.
7. British Patent Application Toothill - Davy Loewy.
8. Continuous Casting of Stainless and High Alloy Steels International Continuous Casting Conference London 1982. D. Toothill & J. Marsh.
9. Steel Times International - Continuous Casting Supplement March 1987. D. Toothill.
10. Development and industrialisation of Horicast. T. Koyano and M. Ito. 4th I.S.C. Continuous Casting London May 1982.
11. European Patent 0022373 - Apparatus connecting tundish and mould.
12. British Patent 1,337,971 - Improvements in continuous casting apparatus.
13. Development of H.C.C. for high grade steel. Hanmyo et al. N.K.K. Technical report 116 February 1987.

1 SECTION THROUGH MOULD TO TUNDISH JOINT

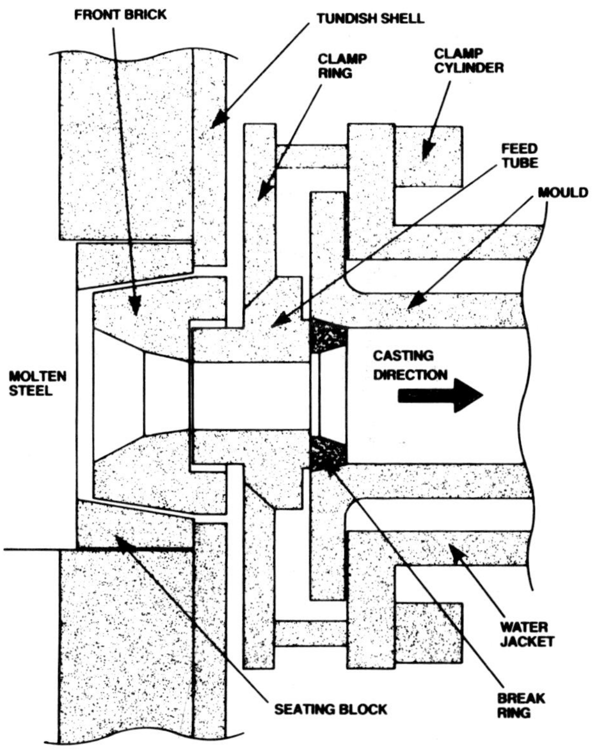

2 SHELL FORMATION

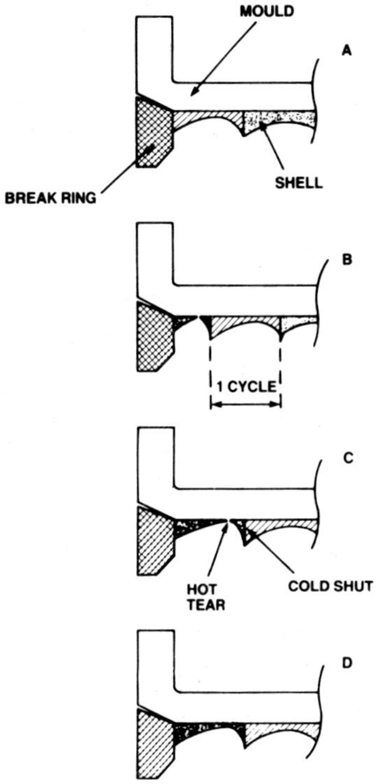

3 WITHDRAWAL VELOCITY – TIME DIAGRAM

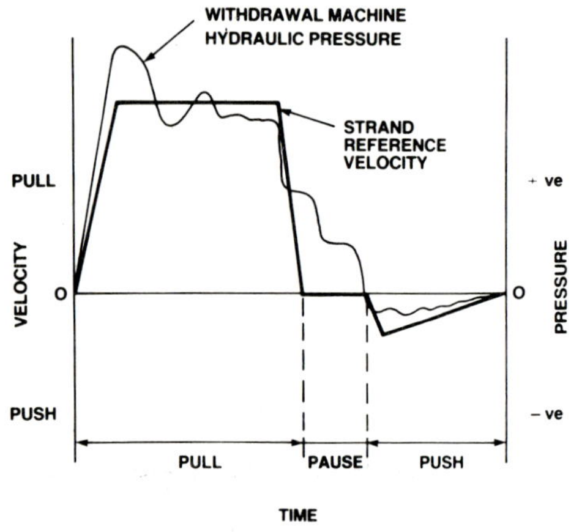

4 TEMPERATURE DISTRIBUTION IN BREAK RING AFTER 50 SECS.

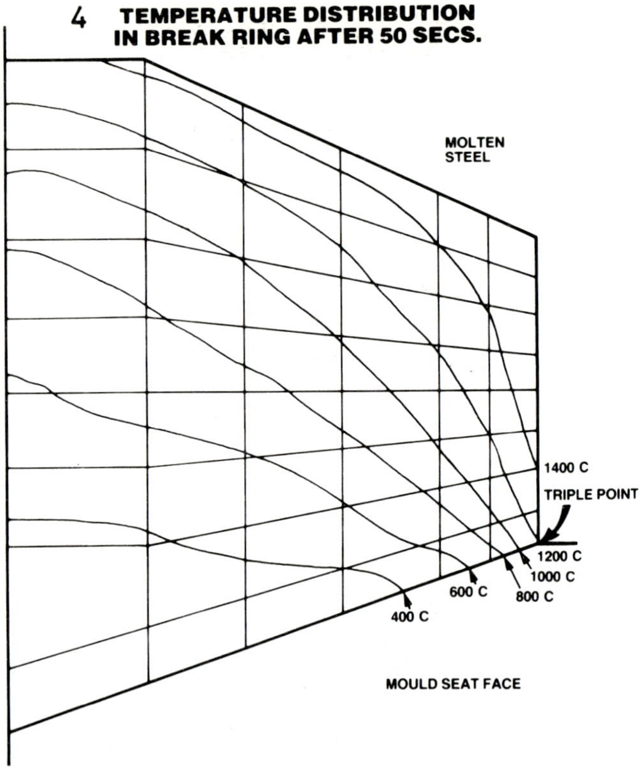

5 BREAK RING STRESS DISTRIBUTION PATTERN

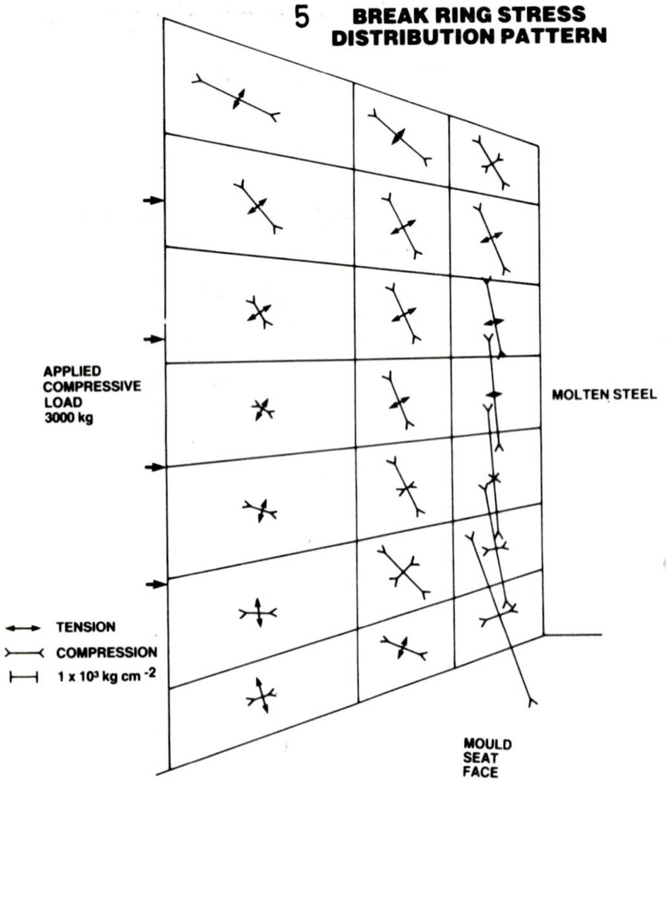

6 EFFECT OF STEEL GRADE ON EROSION OF BREAK RING

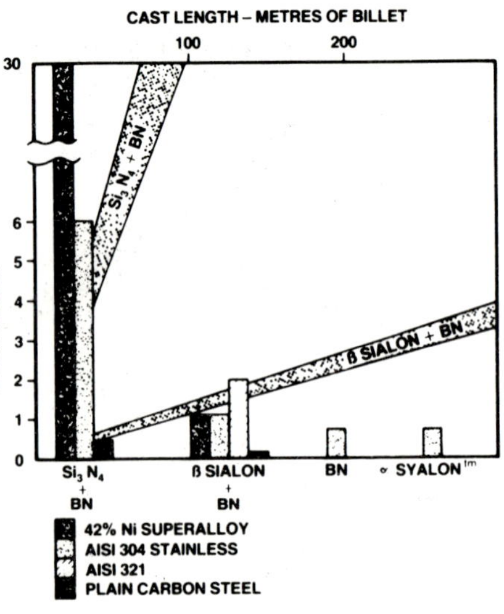

7 EFFECT OF CYCLE FREQUENCY ON COLD SHUT DEPTH

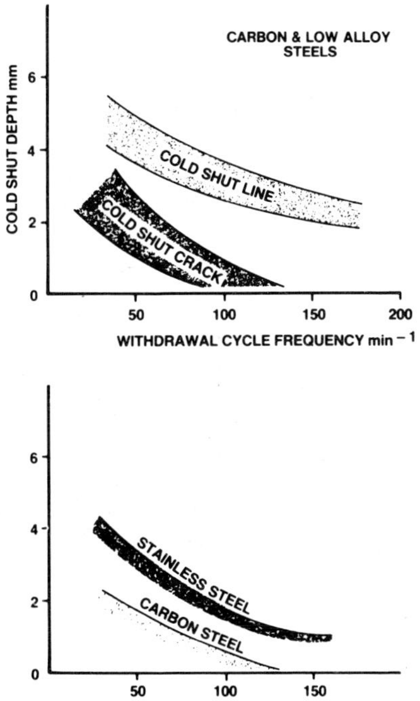

8 EFFECT OF MOULD SHAPE ON COLD SHUT DEPTH

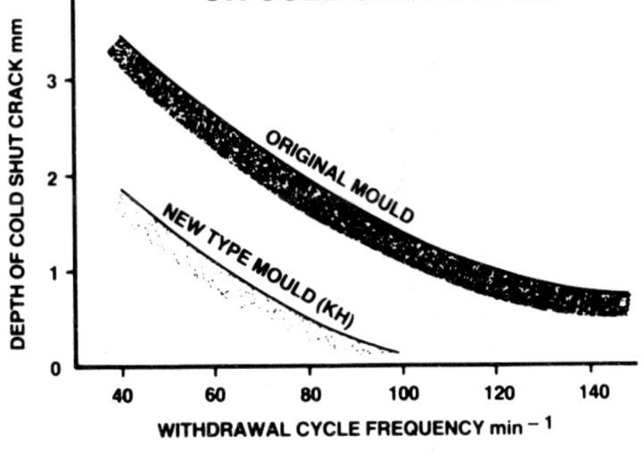

9 WITHDRAWAL CYCLE PATTERN

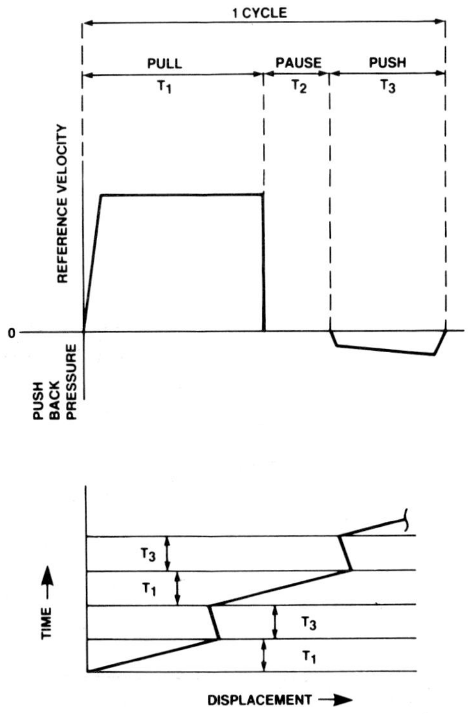

10 DISTRIBUTION OF SECONDARY DENDRITE ARM SPACING

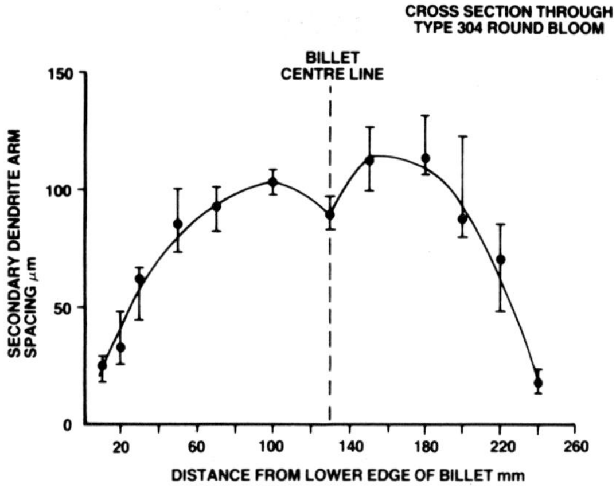

11 DELTA FERRITE DISTRIBUTION

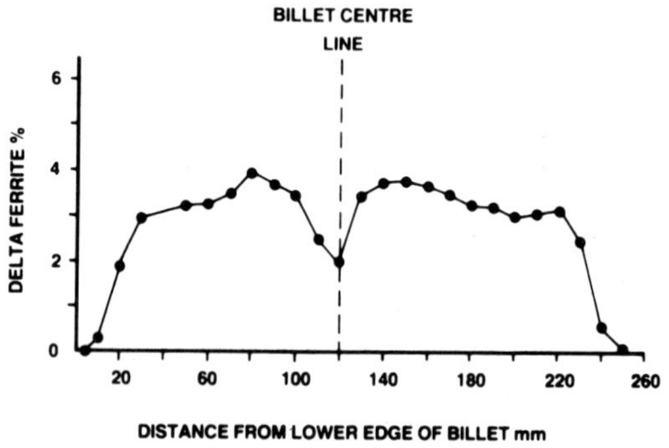

12 ROLLED BAR SURFACE QUALITY STAINLESS ALLOYS

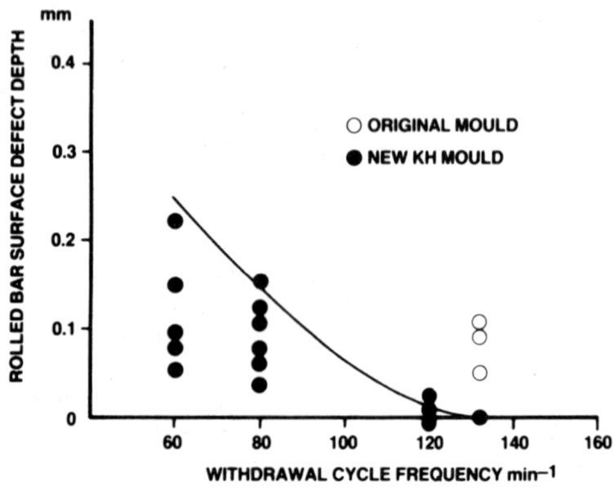

☐

Hot extrusion of special sections in stainless steel

J Middleton

Mr. Middleton is Technical Director with Osborn Steel Extrusions Ltd., an Aurora Company.

SYNOPSIS

This paper shows how the hot extrusion process has developed from its origins in the 18th Century as a method of producing lead pipes. It has since refined lubricants, tooling materials and die design to enable complex solid and hollow shapes to be produced in a wide range of ferrous alloys, including stainless steels.

The paper discusses criteria to be considered when designing stainless steel sections for extrusion, and it concludes by illustrating (by means of several industrial applications) the economic benefits offered by extrusion.

THE ORIGINS OF HOT EXTRUSION

The design of the first press for the extrusion of metal is usually attributed to a hydraulic engineer, Joseph Bramah, who in 1797 was granted a patent for a device for the manufacture of lead (and other soft metal) pipes using molten lead as a feedstock. Thomas Burr, a Shrewsbury plumber by trade, saw the potential of this process and by 1820 was extruding lead pipes from solidified lead billets.

Over the next 100 years the process was developed extensively for the manufacture of non-ferrous solid and hollow sections and it is for the extrusion of the non-ferrous alloys that extrusion is best known, UK production of aluminium and its alloy alone amounting to some 150,000 tonnes in 1986 compared with around 15,000 tonnes of ferrous alloy extrusion.

LUBRICATION

By 1925 in Germany and France, steel tubes were being extruded but at a high cost in terms of tool life and surface finish of the product; the problem being the inability of the lubricants which were being used to prevent metal-to-metal contact between the billet and tooling. Solid, liquid and gaseous lubricants were tried. Solid lubricants such as mica have high coefficients of friction and resulted in high extrusion pressures beyond the range of most presses at that time. Liquid and gaseous lubricants, many of which contained carbon, reduced friction and hence extrusion pressure, but were lost very early in the extrusion and caused carburisation when used for lubricating stainless steels. The ideal lubricant therefore had to flow uniformly during the complete extrusion cycle; not react with the extrusion billet material at extrusion temperatures of 1000 - 1250°C and provide a mechanical and thermal barrier between billets at these temperatures and tooling which was typically at 350 - 550°C.

In 1947 two Frenchmen, Labertaille and Sejournet, after working on this problem using mixtures of salts, discovered that ordinary window glass at these extrusion temperatures possessed just the correct properties so that when hot steel billets were coated, the softened glass formed an insulating and lubricating film between the extruded product and the tooling (Fig. 1).

It is common practice nowadays to use several types of glass for lubricating different alloys, each type of glass having a closely controlled composition and size to ensure optimum lubricating properties at the chosen extrusion temperature. Figure 2 shows the main constituents of glasses used for extrusion lubrication. For lubrication at the extrusion/die interface, a viscosity of 10^3-10^4 poise is considered desirable, whilst at the billet/liner interface where chilling is a greater problem, a viscosity of 10^1 - 10^2 poise is chosen to ensure that friction is minimised, thereby avoiding shearing at the billet surface. For stainless steel therefore, being extruded at 1150 - 1200°C, a soda, lime 70% silica glass is chosen for die lubrication and a high borax/low silica glass chosen for billet/liner lubrication. For extrusion at temperatures in excess of 1200°C, borosilicate glasses typically 70% SiO_3, 15% B_2O_3 are used at the die face.

Ferrous extrusion in contrast to non-ferrous extrusion is performed rapidly to minimise heat loss, maintain optimum lubrication conditions, maximise tool life and minimise extrusion pressure. On the Osborn Press at Bradford, a 100kg billet is typically extruded to a 6–7-metre length of bar in less than 5 seconds - undergoing a reduction in area of typically 85 - 95%. The glass lubricant at the die must therefore be able to soften rapidly but at a controlled rate to ensure that the whole of the extruded product is adequately lubricated. This rate of softening is controlled by careful selections of glass frit size where lubricating pads are used or by selection of fibre diameter where glass fibre is used - too coarse a frit or fibre and the glass does not soften and acts as an abrasive on the tooling; too fine a frit and the pad is too dense, again resulting in tool wear (Fig. 3). The optimum conditions are obtained using a mixture of frit sizes which, when formed into a pad, produce an open structure giving the desired softening properties.

TOOLING MATERIALS AND METHODS OF MANUFACTURE

Extrusion dies are manufactured by one of two methods, casting or machining. The method chosen by the extruder tends to determine the type of die material and die coatings which are used.

Cast dies are usually produced in close proximity to the extrusion press in materials such as H10A and stellite (TABLE 1). They may be coated with zirconia or similar refractory material and are generally used only once before being remelted and recast.

The use of cast dies necessitates the manufacture of large numbers of dies in order to maintain continuity of production. Material costs however are low since this can be equated with the irrecoverable losses associated with the melting and casting process. This technique also reduces inspection costs (where one die per extrusion is used), reduces the cost of die stocks since only sufficient dies necessary to maintain continuity are required, and enables lower alloy die materials to be used.

Set against these advantages are the capital cost of the foundry to produce the dies, the cost of patterns for small production quantities and lead times which may be up to 2 weeks.

The alternative method of manufacture involves the machining of dies from solid, usually wrought, blanks which may be cut from bar or produced as forged blanks. The materials used are generally hot work tool steels H10A, H13 or H21 types (TABLE 1). The higher grade alloys are used when complex shapes are required since these alloys offer a higher hot strength, although other creep-resistant nickel and stellite alloys have been used with varying degrees of success. Dies are generally used many times, and are inspected after each extrusion for signs of wear. When worn beyond predetermined limits, the worn areas on the dies are refurbished by weld-depositing various hard-facing alloys which are then ground or spark-eroded back to the original form. In this way, in excess of one hundred extrusions can be produced from one die with several intermediate refurbishments. The use of machined wrought dies therefore reduces the number of dies required for a given production programme considerably when compared with the use of cast dies, although die stocks tend to be large. No pattern costs are involved so no penalty is incurred in the production of small production quantities and modern low cost CAD/CAM systems linked to CNC milling and EDM machines can produce finished dies within 24 hours enabling flexibility to be maintained both in terms of meeting varying production requirements and being able to react quickly to changing customer schedules.

COATINGS

A variety of ceramic and metallic coatings have been used in an effort to improve die life of both cast and wrought dies: in the former case, longer extruded lengths and closer tolerances have been sought whilst for wrought dies more extrusions per die has been an additional requirement.

To select a suitable coating, it is important to appreciate how an extrusion die fails. This is normally by movement (creep) of die material from the front face and approach radii of the die into the die aperture, thus producing a build-up of die material on or near the land of the die which effects the dimensions of the section. This movement is caused by high temperature creep of the surface material of the die and therefore the solutions have sought (i) to thermally insulate the die from the extruded product using ceramic materials such as zirconia with appropriate intermediate substrate materials, zirconia having the added advantage of being resistant to abrasion, or (ii) to apply a surface coating which, at the temperatures reached during extrusion, has a higher strength than that of the die material. The depth of this layer is obviously important; too thin and the die material will still "creep" rendering the coating ineffective, too much and the cost and time taken for its application becomes prohibitive. It follows therefore for the extrusion of stainless steel that techniques such as CVD, nitriding and salt-bath diffusion of carbides, all of which produce very thin layers and have been used with great success in the extrusion of non-ferrous alloys, are not suitable for application to dies for stainless steel extrusion.

Following extensive trials at Osborn Steel Extrusions we have identified a range of Ni-and Co-based alloys for use in refurbishment of dies, such alloys having a particular use dependent on the severity of wear experienced in a particular feature of the die.

DIE DESIGN

This aspect of steel extrusion has received far less attention than for aluminium extrusion where detailed design criteria have been established to ensure close dimensional control and minimum distortion of the extruded product and to minimise extrusion load. Some of these principles have been applied successfully to steel extrusion but refinements such as "tuning" the dies by making small alterations to approach angles and land lengths, are less relevant to steel extrusion where die lives are relatively short (typically only 1 or 2% of the life of an aluminium die).

Most extrusion presses for steel extrusion have interchangeable containers of varying diameter, the press at Osborn Steel Extrusions having four of 105mm, 130mm, 153mm and 180mm in diameter - thus providing a range of maximum specific pressures at the die face ranging in our case from 1250MPa on the 105mm container to 650MPa on the 180mm container.

The selection of a die face suitable for the extrusion of a particular section depends upon:

(i) The circumscribing circle of the section; this should be less than 80% of the diameter of die face to ensure adequate lubrication of the extruded section

and

(ii) the pressure required for the extrusion of the section in the alloy specified; this calculated pressure must obviously be less than the maximum specific pressure available on the chosen container.

All dies used by Osborn Steel Extrusions Ltd are of the type shown in Fig. 4, ie flat-faced with a radiused approach and short land. It has been found that for lubricated steel extrusion, conical entry dies do not afford a significant reduction in extrusion pressure (cf 20-30% reduction for non-lubricated extrusions) and provided that lubrication is maintained, the dead metal zone adjacent to the die face is avoided.

The entry radius into the land is a nominal 10mm but is varied according to the need to control rate of metal flow through the die. Angled "lead-ins" to re-entrant and thin portions of the die are also used to improve lubrication and metal flow in these difficult areas.

The entry radius leads on to a parallel land portion of the die which is generally 6mm long; this is the portion of the die which controls section size and is cleaned and checked for size using a template after each extrusion.

The position of the aperture on the die face affects the degree of bow and twist on the extruded product. As a general rule, the centre of area of the section is placed on the centre of the die face where the rate of flow of metal is greater. Positioning the aperture in this way reduces any twisting moment on the section as it is extruded. Unbalanced sections– that is those possessing portions of greatly varying perimeter to area ratios – are positioned so that the portion with the greatest ratio, ie that portion where metal flow will be most restricted, is placed close to the centre of the die where rate of flow is greater with the portion with the lowest ratio being placed closer to the edge of the die. In this way and by further influencing flow by changes to approach radii and angles, the rate of metal flow is adjusted so that it is approximately uniform throughout the die aperture thereby minimising bow and twist. In addition to die design, longitudinal and radial temperature gradients induced in the billets by induction heating can also be used to significantly affect rates of metal flow across the die face. Multi-aperture dies may be used for steel extrusion to reduce the pressure required for extrusion by effectively reducing the extrusion ratio as well as increasing the extruded weight per cycle. Obviously it is no longer possible to place the centre of gravity of each section on the centre of the die and this invariably results in considerable twist in the extruded product. To ensure uniform flow through each aperture, all are arranged symmetrically around the centre of the die with the portions with the lowest P/A ratio positioned closest to the edge of the die for the same reasons as stated above.

RAW MATERIAL

OSE extrude a wide range of austenitic, martensitic and some duplex stainless steels including free cutting and stabilised grades. The raw material is purchased in the form of rolled or forged bar which has been turned or preferably peeled to produce a bright product with a surface finish of better than 3.2 micrometers. Billets are heated for extrusion by N_2-purged induction heaters and so are scale-free when transferred to the press for extrusion. Furthermore, surface finish is important because minor surface imperfections are not scaled off during heating and are carried through as surface marks or defects on the extruded product, since as the process is lubricated the billet surface is also extruded and in effect becomes the surface of the extruded product. A typical defect caused by heavy turning marks in the billet is shown in Fig. 5. These "chevron" marks can, in severe cases, lead to hot-tearing at external corners.

The analysis of austenitic stainless billet material is controlled to keep levels of delta ferrite at less than 4% to improve hot workability, particularly important where sections have features such as thin webs and sharp external corners which are prone to "tearing". Every bar is stamped with the cast number by the raw material supplier and this identification is transferred at every process to ensure complete traceability of each batch of material despatched, as demanded by many of our customers.

THE MANUFACTURING PROCESS

Figure 6 shows a schematic representation of the process route at Osborn Steel Extrusions showing key operations for the production of a section in austenitic stainless steel.

Billet cutting

The incoming bar, which is supplied bright in approximately 6-metre lengths is cut on band saws to a weight determined by the finished product length required. Billet weights of between 20-90 kgs are commonly processed and current modifications to the press will increase this to 120 kgs by the end of 1987.

Billet radiusing

The billet face which will be in contact with the die prior to extrusion is radiused to provide a lead-in to the die which reduces peak pressure at the start of extrusion and improves the surface finish at the front of the extruded bar.

Billet heating

Mains frequency induction heaters, each rated at 500 Kw, are used to raise the billet to the required extrusion temperatures. Heating times vary according to billet diameter, weight and material type but would be typically 7 minutes for a 90 Kg billet of austenitic stainless steel. Three single-billet induction heaters are used in sequence at Osborn Steel Extrusions, giving production rates of approximately 25/hour.

When compared with other billet heating methods induction heating offers:

(1) High productivity as billets are heated individually enabling reheating temperatures to be changed for different alloys without delay.

(2) High-quality products made possible by accurate temperature control and by the virtual absence of scaling or decarburisation on the reheated billet.

(3) Maximum economy, since fully heated billets are ready within under 10 minutes of start-up and heaters can be switched off during production delays.

(4) The ability to control rates of metal flow through the die aperture by generating longitudinal and radial temperature gradients in the billet.

Lubrication and extrusion

The type of lubricants used have been discussed in detail earlier, a low-viscosity glass being used to lubricate between the billet and extrusion liner and a higher viscosity glass being used to lubricate the extrusion itself.

After extrusion, which is generally completed within 5 seconds, the bar may be air cooled or, in the case of austenitic stainless products, water quenched to provide a fully solution treated product. Water quenching has the added benefit of cracking the adherent lubricating glass film from the extruded section thus promoting rapid descaling on subsequent acid pickling.

Stretch-straightening and detwisting

We have seen how careful die design and adequate lubrication will minimise the degree of bow and twist which occurs during extrusion. However, in order to produce a bar which is commercially satisfactory, most of our stainless product is cold straightened on a Fielding 320 ton stretch detwisting machine. The bar is gripped at each end by shaped jaws, bow is corrected by elongating the bar by 2-4% and twist is removed by rotating one of the gripping heads. In this way straightness and twist can be controlled to within 1 : 500.

Descaling

Residual glass and scale are removed by immersing the austenitic stainless sections in a hot nitric/hydrofluoric acid descaling bath The hydrofluoric acid is essential to remove the adherent glass lubricant. Martensitic stainless steels are shotblasted to remove heat treatment scale and residual glass lubricant.

Inspection

A full dimensional inspection, with the aid of a shadowgraph projector if necessary, is performed on each batch of extrusions to ensure the profile conforms to drawing tolerances prior to despatch.

DESIGN CRITERIA FOR AUSTENITIC STAINLESS SECTIONS

As we have seen earlier, the greater resistance to hot deformation and the higher hot working temperatures necessary for the successful extrusion of ferrous alloys compared with non-ferrous alloys led to the development of new lubricants and die materials capable of withstanding these more severe conditions. Figure 7 shows a comparison of extrusion temperatures and pressures required for the extrusion of both ferrous and non-ferrous materials and demonstrates the clear dividing line at around 1000°C above which all ferrous extrusion is performed. It is also clear that even at these higher temperatures, stainless steels have a higher resistance to deformation and therefore require greater extrusion pressures than non-ferrous materials.

It is the resistance to deformation (ρ) which is a major factor in determining how a particular section and alloy is to be extruded and the limitations which must be imposed on the complexity of the extruded profile.

The following points should be borne in mind when designing steel sections for extrusion:

Avoid thin webs and sharp corners

Figure 8(a) shows a profile with sharp corners and a web thickness of under 4 mm. Metal flow through the aperture in the die is severely restricted (particularly for the stiffer stainless steels) preventing the production of a satisfactory section. By thickening the web and increasing the corner radii, an extruded section such as that shown in Fig. 8 (b) can be achieved.

Avoid deep slots

The tongue on the die used to produce the section shown in Fig. 9 (a) will be subjected to high temperatures and pressures during extrusion, causing softening and rapid wear. To minimise temperature increases in these areas of the die a ratio of width thickness of the tongue (and hence slot on the extruded section) should be of the order of 1 : 1.

Profiled hollow sections

For hollow sections, a further design limitation must be considered when a profiled bore and profiled outer surface are required. Whilst this arrangement is possible using an appropriately shaped mandrel and die, the alignment between the inner and outer profile is more difficult to achieve consistently.

Dimensional tolerances

These are similar to those achieved by other hot metal forming processes, they are dependent on the alloy being extruded, the complexity of the profile and the length of the extruded bar.

Typical tolerances for a stainless steel would be:

Nominal Dimension	Tolerance Band
<25 mm	0.75 mm
≥25 <50 mm	1.00 mm
≥50 <75 mm	1.25 mm
≥75	1.50 mm

Minimum Radii

Internal - 3.0 mm
External - 1.5 mm

APPLICATIONS FOR EXTRUDED STEEL PROFILES

The extrusion process is used to produce both solid and hollow sections. The production of seamless austenitic stainless tube represents by far the bulk of extruded steel tonnage in the world, modern tube presses being capable of extruding and finishing 100,000 tpa. The market for solid and hollow shapes (Fig. 10) is the one in which Osborn Steel Extrusions has concentrated its efforts over the last 25 years, and it is this market which will now be discussed.

Osborn Steel Extrusions have during this period supplied extruded sections for many and diverse applications ranging from sections used on lawnmowers through to sections for nuclear fuel cans. The list of industries supplied is impressive and includes:

- Aerospace
- Agriculture
- Automotive
- Business machines
- Defence
- Mechanical engineering
- Mining
- Power generation, both nuclear and fossil fuelled
- Process control instrumentation

Shapes can obviously be manufactured by a number of hot and cold forming processes, in particular:

- Rolling
- Forging
- Casting
- Fabrication
- Machining from solid

It is therefore with these processes that extrusion usually competes.

So what advantages can be offered to the designer and manufacturer to make extrusion his first choice at the expense of these competing processes?

To demonstrate the reasons and hence advantages, the author has selected a number of examples of sections which were either designed as extrusions from their initial concept or have been changed to extrusions from alternative processes.

The reasons for choosing an extrusion of course vary but can be grouped under the following general headings:

- Elimination of welding
- Elimination or reduction of machining
- Saving on material cost
- No other viable method of manufacture

In addition to these, further general advantages are offered by the extrusion process:

Lower tooling costs - Typically less than £2000

Versatile tooling - Initial prototype sections can if necessary be modified by altering existing tooling, normally at a fraction of the original die cost.

Small production runs - As few as 2 or 3 extrusions can be manufactured for trial purposes without undue disruption to the extrusion programme. Once designs have been finalised and trials completed, minimum production quantities of 1000 Kgs are preferred.

Short lead times - Using billet material held in stock, new sections can usually be supplied within 6 weeks of accepting the order.

The first example, Fig. 11 from the aerospace industry, is chosen to illustrate how the use of an extruded steel section can produce savings on both raw material costs and machining time for the end user. In the example, the weight of the extruded section is 7.92 Kg/metre, whilst that of the equivalent flat rolled bar is 14.25 Kg/metre, an advantage in favour of the extrusion of almost 2 : 1. This advantage in weight of material saved more than offsets the higher cost per kilo of the extruded section, making it more than competitive at this stage. Then comes the added bonus to the end user, in that as much less metal has to be removed, substantial savings in machining time can be expected.

With the high and ever-increasing cost of materials used in the aerospace industry the benefits of more efficient material utilisation offered by the extrusion process are expected to be exploited more widely by engineering industries in this field.

Figures 12(a) and (b) illustrate clearly how the steel extruder working in close co-operation with the engineer can significantly reduce manufacturing costs whilst maintaining the integrity of the designer's original concept. In the example, which comes from the field of heat transfer engineering, the original design called for a tube, a machined spyder and non-metallic rods. For the design to achieve the desired efficiency, a close fit was required between the tube, spyder and rods. In Fig. 12(b), we see how one of the particular advantages offered by the steel extruder, the ability to produce relatively long lengths of internally profiled tube, was incorporated into the design. The one-piece extrusion replaces both the tube and the spyder, and the rods are replaced by one rod of larger diameter, thereby channelling the flow of fluid into the region adjacent to the tube wall and achieving required heat transfer.

The final example is taken from the metal refining industry and is chosen to show how extrusion can assist the designer by taking a relatively simple, often standard shape and developing the profile to meet specific design criteria through a series of limited extrusion and site trials.

Osborn Steel Extrusions supply the client with straightened cut lengths approximately 1.4 m long in 304 stainless steel. A cathode plate is then welded perpendicular to the table of the 'I' section.

In service these cathode plates must hang vertically supported by the 'I' sections so the original concept was to cold draw the top and bottom faces of the extruded section to achieve the required tolerance (Fig. 13(a)). This cold work resulted however in some distortion on subsequent welding and so the design was modified to omit and drawing operation and to incorporate a radiused face (Fig. 13(b)) onto which the cathode plate was welded, the plate acting in effect as a keel allowing the face to always hang vertically.

An additional requirement at this stage was to ensure that any liquor splashed onto the 'I' section during service would not collect and result in premature corrosion failure so a further simple die modification to produce a sloping shoulder to assist drainage was incorporated.

Development work of this type is only possible through the ability of the extrusion press at O.S.E. to process small quantities economically using induction heating and the ability to modify tooling at a relatively low cost compared with other hot forming processes.

Whilst it is not possible to give exact savings that have accrued from the use of the sections shown in the preceding examples, it can be said that these are significant and in one of the cases it is known that at least 50% of the original cost has been saved by the use of an extrusion.

The author hopes that this paper has provided a better insight into the design possibilities offered by the steel extrusion process.

The versatility of the process, achieved by the method of heating and lubrication, has thus been illustrated. This versatility has enabled a wide range of ferrous and many superalloys to be extruded into both solid and hollow profiles.

The diversity of products available has provided, in many cases, a cost-effective alternative to more traditional metal forming processes by solving current manufacturing problems, reducing machining times and, especially where raw material costs are high, reducing wastage.

Acknowledgements

The author wishes to thank colleagues at Osborn Steel Extrusions Ltd who have assisted in the preparation of this paper.

TABLE 1 COMPOSITION AND HOT TENSILE STRENGTH OF TYPICAL DIE MATERIALS

Alloy	Analysis							Rm		MPa	
	C	CR	W	Mo	Ni	Co	V	20°	200°	500°	650°C
BH 10A	.35	3.0	-	2.75	-	3.0	0.45	1515	1340	1160	585
BH 12	.35	5.0	1.4	1.6	-	-	0.25	1575	1400	1150	585
BH 13	.39	5.0	-	1.4	-	-	1.0	1745	1590	1280	600
BH 21	.30	3.0	9.0	-	-	-	0.3	1600	1475	1175	670
STELLITE 4	1.0	33.0	14.0	-	-	Bal	-	849	-	-	770
STELLITE 6	1.0	26.0	5.0	-	-	Bal	-	896	-	-	680
STELLITE 21	0.2	27.0	-	6.0	2.0	Bal	-	694	-	-	740

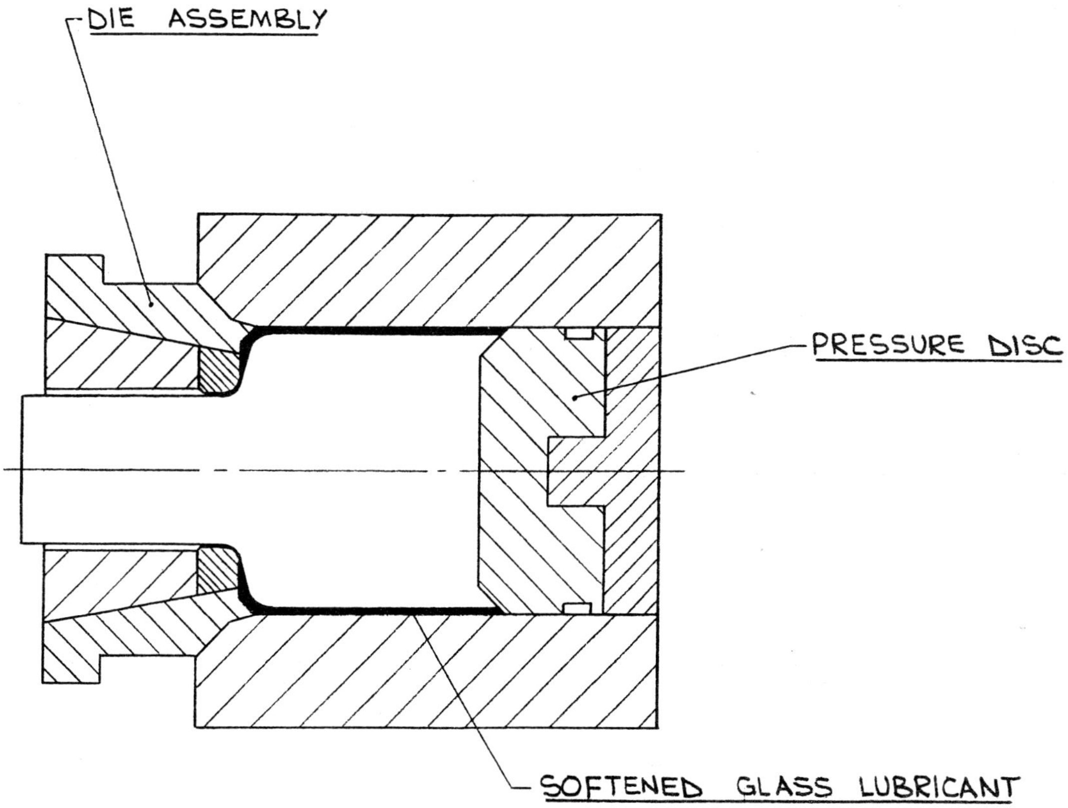

1 Cross-section showing the softened glass lubricating film

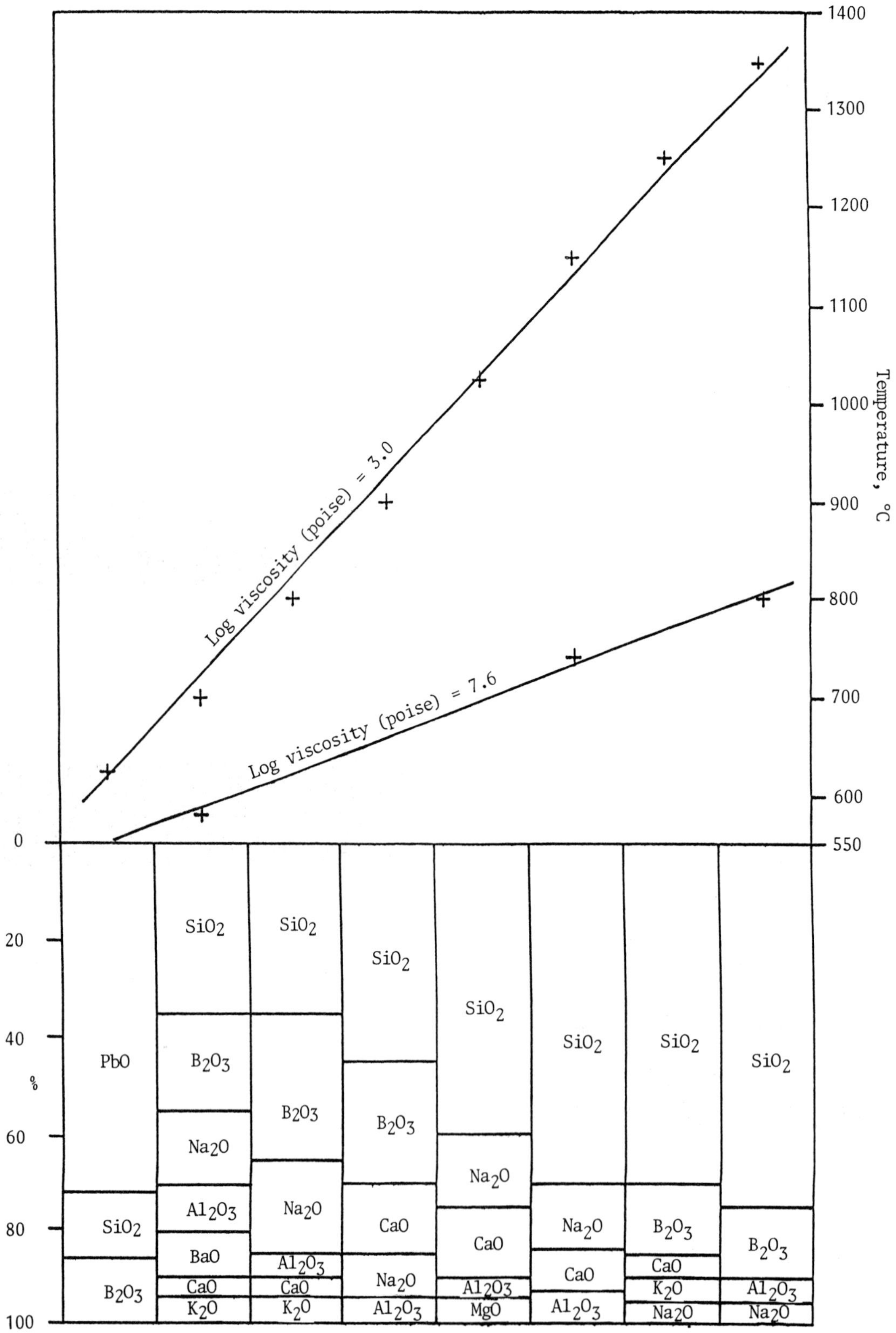

2 Relationship between glass composition, extrusion temperature and viscosity

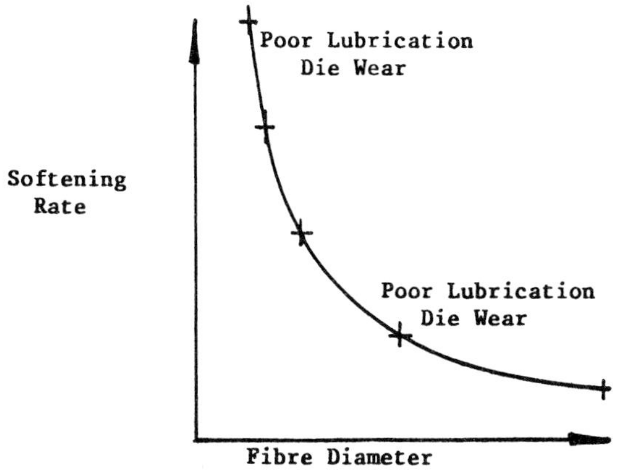

3 Effect of glass fibre diameter on softening rate and abrasive die wear

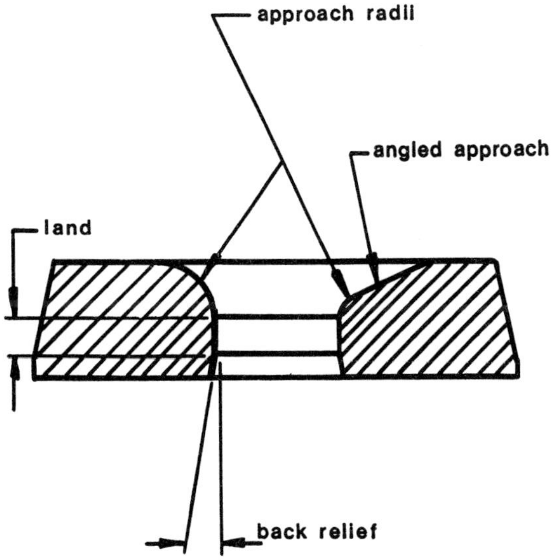

4 Flat faced extrusion die

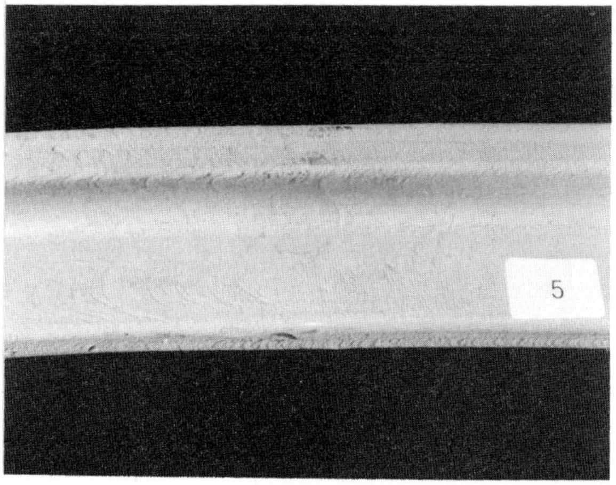

5 Inferior extruded surface caused by machining marks on billet surface

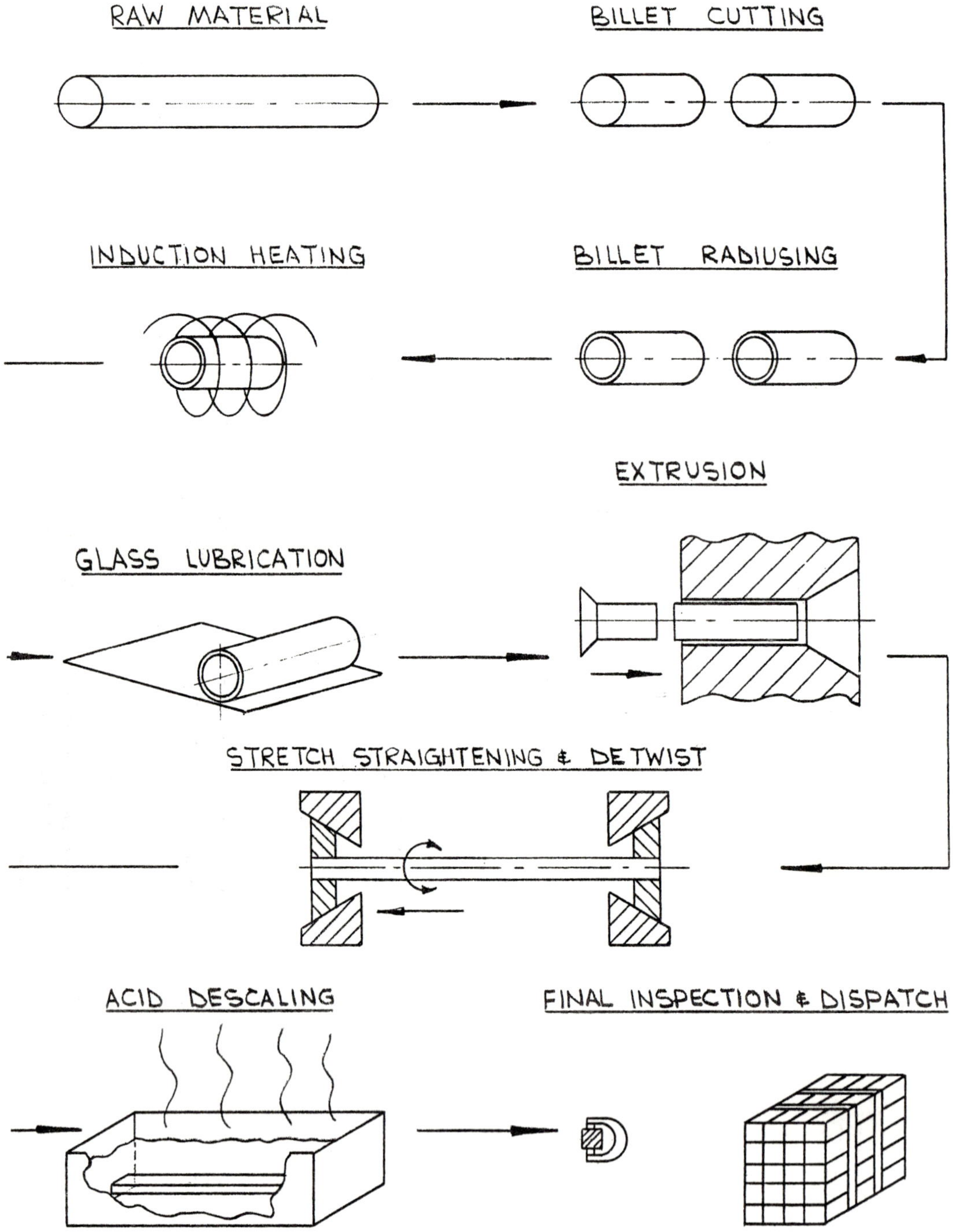

6 The manufacturing route at Osborn Steel Extrusions for an austenitic stainless steel

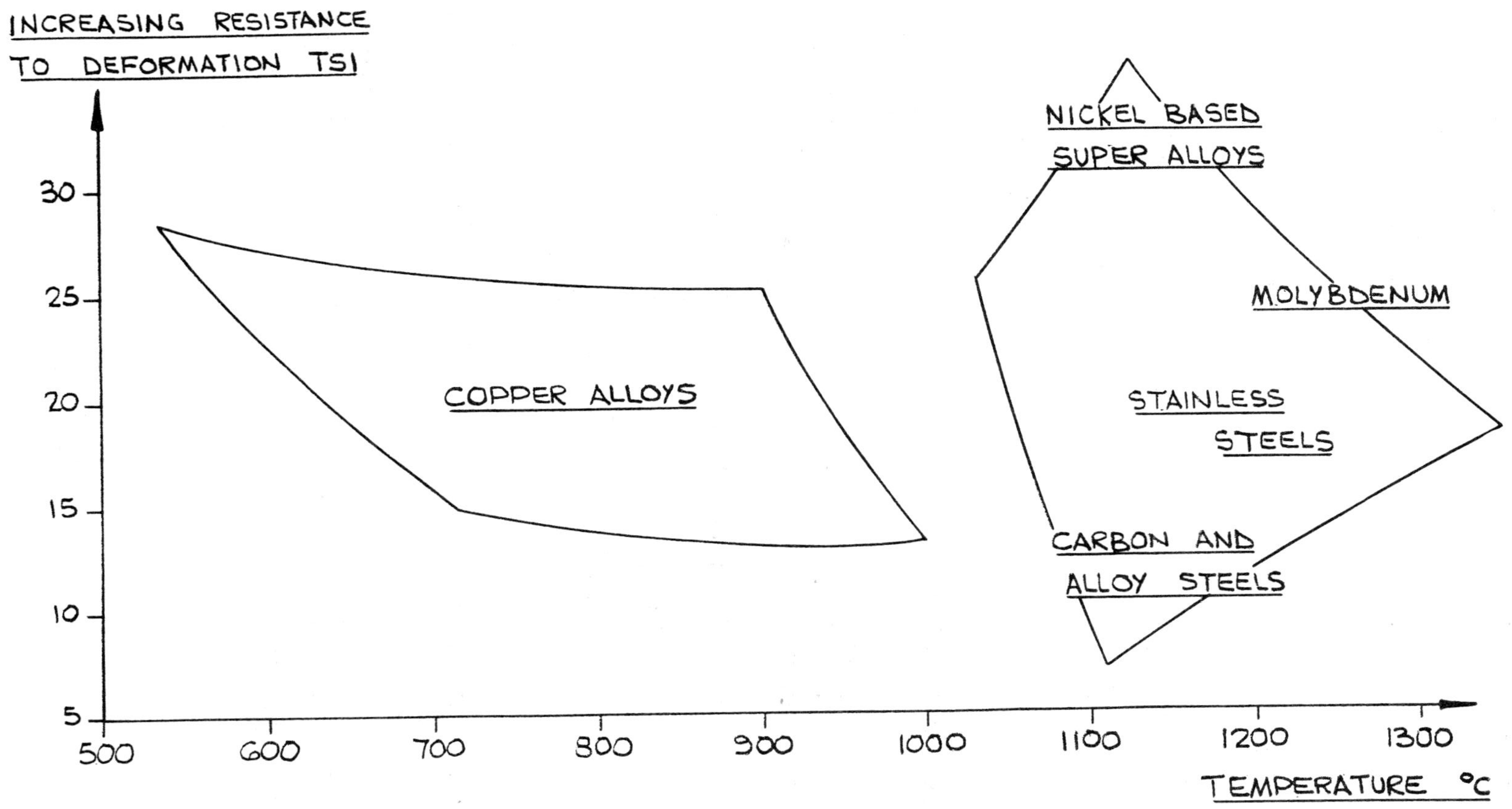

7 Resistance to deformation vs. extrusion temperature for ferrous and non-ferrous alloys

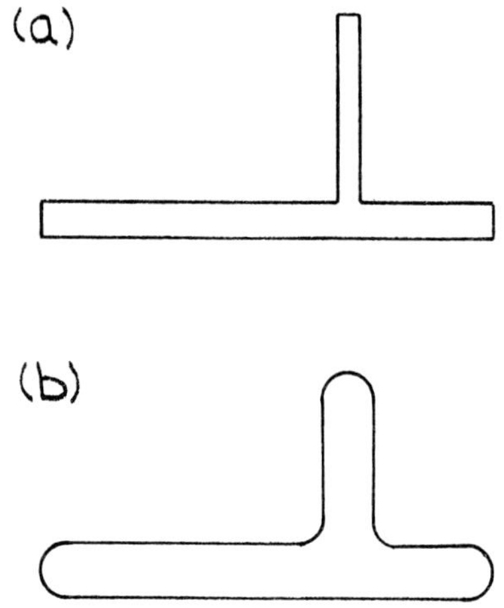

8 Designing a section: avoid thin sections and sharp corners (a)

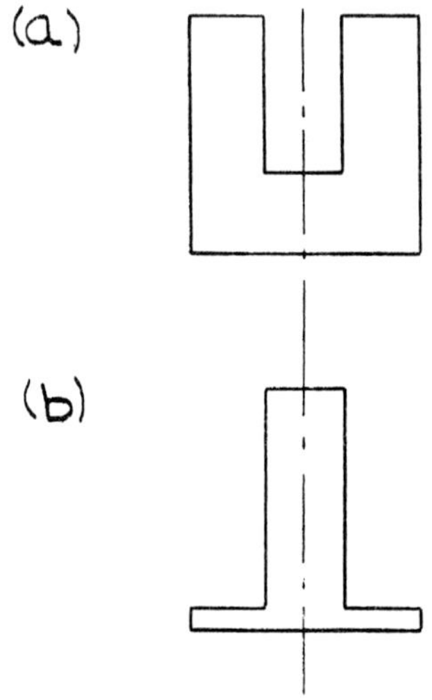

9 Designing a section: avoid deep slots (a) and thin extensions (b)

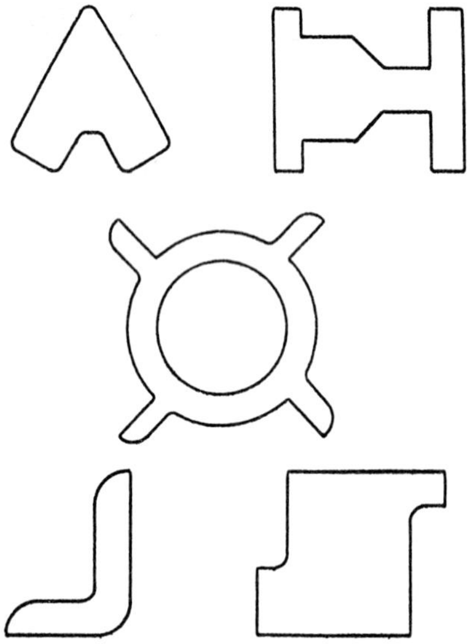

10 Typical solid and hollow extruded sections

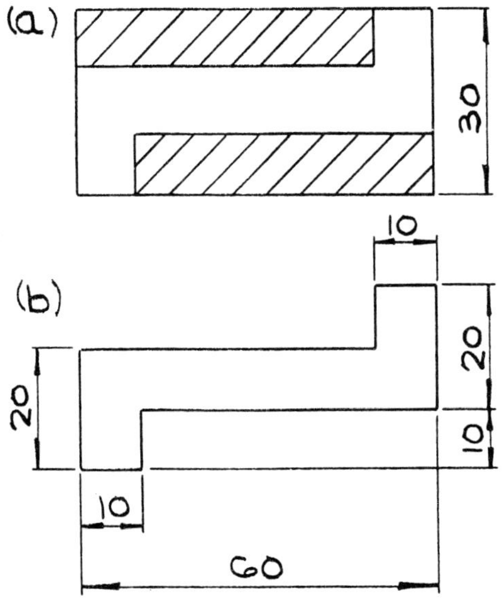

11 Applications for extrusions: savings on raw material and machining

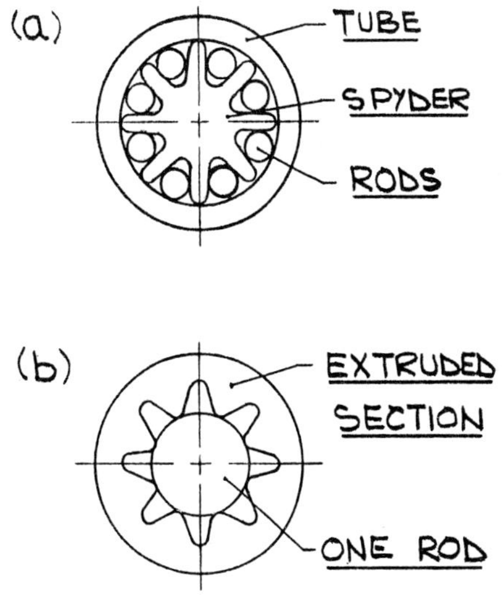

12 Applications for extrusions: internally finned tube

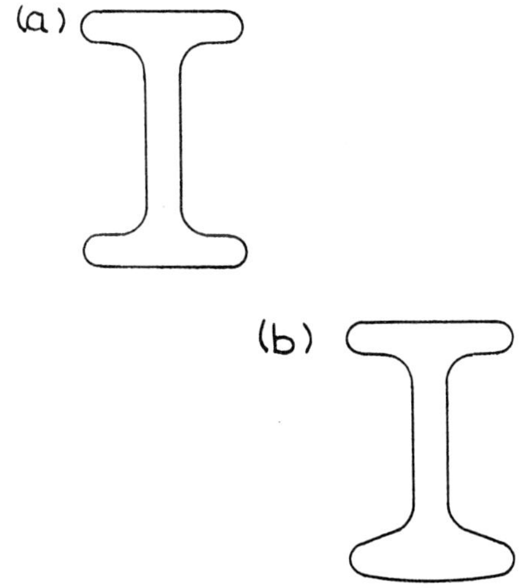

13 Applications for extrusions: custom-designed 'I' section: (a) design concept; (b) final design achieved by joint client/extruder liaison

Duplex stainless steels produced by hot isostatic pressing: opportunity to improve materials and develop new materials

M Lindenmo

The author is with ASEA POWDERMET AB, Surahammar, Sweden.

ABSTRACT

Duplex stainless steels produced by Hot Isostatic Pressing (HIP) of gas-atomized powder have equal and in some respects better properties than those of comparable conventionally made steels. Alloys with compositions which make them very difficult to produce by conventional methods, may be produced by the P/M-HIP technique. An example is the highly nitrogen-alloyed duplex stainless steel APM 2389, which combines very high strength with high corrosion resistance. The P/M-HIP technique also enables complex products to be directly formed to the Near Net Shape (NNS) of the end product. Other advantages of the technique are short lead times and new and/or integrated design solutions.

1. INTRODUCTION

This paper presents a comparison between conventionally produced and P/M-HIP-produced duplex stainless steels. The production process of P/M-HIP steels and some general advantages of the technique are also described.

The P/M-HIP duplex steel has always a homogeneous, isotropic and fine-grained structure which can engender a higher ductility and toughness compared to corresponding conventional products. Fig. 1 illustrates the production of P/M-HIP steels at ASEA POWDERMET. The production starts with melting and alloying in a 2.5-ton high-frequency furnace. The melt is tapped into a ladle, refined and, by inert gas jets, atomized into small spherical particles. The powder is poured into capsules, which can be given very complex forms. The capsules retain their form during compaction, which enables Near Net Shape (NNS) production of a wide range of sizes of products. After evacuation the capsules are compacted in the Hot Isostatic Press (HIP). Presses for products with a diameter of 1100 mm and a height of 3100 mm are today available. The compaction results in a fully dense material.

2. A COMPARISON BETWEEN CONVENTIONALLY AND P/M-HIP PRODUCED DUPLEX STAINLESS STEELS

The comparison will be confined to products with large thickness of material eg heavy "forgings".

2.1 Microstructure

The very high cooling rate during solidification of the powder (appr. 1000°C/s) in combination with the Hot Isostatic Pressing (HIP) results in a fully dense, homogeneous, fine-grained and isotropic structure, which is independent of the size of the P/M-HIP product. In heavy forgings the microstructure can be coarse, segregated and anisotropic. This is to some extent illustrated in Fig. 2, which shows the microstructure of a conventional heavy forging of steel SS2324 (AISI 329) and of a corresponding P/M-HIP-made product. Due to the atomization with nitrogen, the P/M-HIP steel has a somewhat higher nitrogen content compared to the conventional steel. The higher nitrogen content engenders an increase of the amount of austenite.

2.2 Strength and ductility

Tensile tests of material from two conventional forgings (weight appr. 450 kg) and from one P/M-HIP-made object of the same size, form and material were carried out. The test material was taken from different locations and in different directions. The results are shown in Table 1. Based on this and many similar investigations it can be concluded that duplex stainless steels made by P/M-HIP have the same yield strength and a higher ductility compared to conventionally made steels of the same type. Possible explanations are the more homogeneous and fine grained structure of the P/M-HIP steel.

2.3 Impact toughness

Fig. 3 shows the results of impact testing of a conventional forging and of a corresponding P/M-HIP-made object. In the thinner walls, where the deformation during forging was relatively large and where there may exist a favourable orientation of austenite, the conventional forging has higher impact toughness than the P/M-HIP steel. In the thicker central part the impact toughness of the forging is lower than that of the P/M-HIP steel. Contrary to the conventional forging the P/M-HIP object has homogeneous and isotropic properties. Similar results have been obtained for heavy components of type AISI 329 duplex steel.

2.4 Fatigue properties

A heavy forging and a similar P/M-HIP-made object were fatigue-tested and the results are given in Table 2. The conventional forging was made of electroslag remelted steel, which has a low amount of non-metallic inclusions. The fatigue strength of the P/M-HIP steel is higher. This is probably dependent on the higher tensile strength of the P/M-HIP material, since the ratio between fatigue and tensile strength is nearly the same for both materials.

Based on these tests it is not possible to draw any safe conclusions, but it is likely that the fatigue strength of P/M-HIP-made duplex stainless steels is at least as high as that of conventional ESR-steels of the same type (Rm = appr. 675 MPa, large thickness of material). This conclusion is confirmed by results from martensitic 12% Cr steels,(1) where the production via the P/M-HIP technique results in a material with higher fatigue strength compared to conventional material.

2.5 Corrosion properties

One should expect no difference in corrosion properties between conventionally and P/M-HIP-made stainless steels of the same composition, unless the conventional steel exhibits a high degree of segregation. P/M-based duplex stainless steels have been found to have slightly higher pitting potentials compared to conventional steels.(2) This was explained by the more homogeneous and fine-grained structure of the P/M-based materials. Table 3 shows the pitting potential of SS 2324 (conventional and P/M-HIP) and of NU44LN (conventional and hot extruded P/M-based). The P/M-based materials have higher pitting potentials. In the case of SS 2324 this may partly be explained by a somewhat higher N-content in the P/M-HIP steel. In the case of NU44LN the two materials have practically the same composition. Other investigators (3, 4) have obtained similar results for highly alloyed austenitic stainless steels.

2.6 Non-destructive testing

Ultrasonic testing for inner defects may be required for security reasons when producing critical components. Conventional heavy forgings of duplex stainless steels can be very difficult to test due to coarse microstructure, which increases the scattering of the sound. This is not the case with P/M-HIP duplex steels, which independently of dimensions have fine-grained microstructure, and therefore are possible to test with ultrasonics.

3. A NEW HIGH-STRENGTH DUPLEX STAINLESS STEEL

Nitrogen has a positive effect on yield strength in different types of stainless steels (5, 6). By increasing the N-content to appr. 0.3% it is possible to obtain a duplex stainless steel with the yield strength min. 600 MPa. The high N-content causes however a substantial decrease in the hot-workability of conventional ingot-based material, and the material yield is therefore very poor. By producing the steel via the P/M-HIP technology these problems are eliminated. The duplex stainless steel is called APM 2389 and was developed in order to combine high strength with high corrosion resistance.

3.1 Composition and microstructure of APM 2389

The nominal composition of APM 2389 is shown in Table 4. The microstructure consists of finely distributed austenite in a ferritic matrix, (Fig. 4). The amount of austenite is appr. 40%. The annealing is carried out at somewhat higher temperatures than normally used for duplex stainless steels.

3.2 Mechanical properties

Typical mechanical properties of APM 2389 (products with weights ~400 kg and wall thicknesses ~80 mm) are listed in table 5. The yield strength is much higher compared to the ordinary duplex stainless steels (eg. Table 1). The fatigue strength of APM 2389 is higher than 500 MPa, which is a large improvement compared to standardized duplex steels (eg. Table 2).

3.3 Corrosion properties

The high Cr, Mo- and also N-content of APM 2389 result in a good corrosion resistance. The resistance against pitting and crevice corrosion in chloride-containing solutions have been investigated for a number of steels.(7, 8) Fig. 5 shows the test results. The results indicate that the pitting resistance of APM 2389 is higher than of the well-known duplex steel SAF 2205 and in the same range as the highly alloyed austenitic steel 904L. This can also be expected from the Pitting Resistance Equivalent according to Fig. 5. A low activation-pH in chloride-containing solutions is normally related to a high crevice corrosion resistance. According to Fig. 5 the activation-pH of APM 2389 is lower (better) compared to SAF 2205 and 904L and in the same range as Sanicro 28.

3.4 Practical applications of APM 2389

APM 2389 is suitable for applications with high demands on both corrosion resistance and mechanical strength (static and/or dynamic). Up until May 1987 about 70 components in APM 2389, each weighing 300 to 600 kg, have been commercially produced.

4. PRODUCTION OF P/M-HIP STEELS

The overall flow sheet of the production of P/M-HIP steels at ASEA POWDERMET is given in Fig. 6. The process is essentially the same as used for the production of P/M High Speed Steels (eg. ASP-steels of Kloster Speedsteel) and for P/M Superalloys for the aerospace industry.

4.1 Melting and refining

After melting in a 2.5-ton high-frequency furnace the melt is tapped into a ladle of ASEA CALIDUS type, which is situated on top of the atomization unit. Here refining, heating, stirring, vacuum treatment, sampling and temperature control may be performed. Thus precise conditions can be obtained before the start and maintained during the atomization.

4.2 Atomization

The melt is cast through nozzles in the bottom of the ladle directly into the atomization unit. In the Horizontal Gas Atomization unit (HORGA) the molten steel is broken up by horizontal gas jets. Nitrogen or Argon is used depending on the steel. The atomized melt rapidly solidifies into small spherical particles. Typical cooling rates during gas-atomization of superalloys are 10^2 - 10^5 °C/s.(9) The solidified powder is collected in the bottom of the atomization unit, sieved (normally <500 µm) and poured into sealed containers.

Every powder grain has the same composition as the melt. The structure within the powder grains is very fine (Fig. 7). Typical for gas-atomized powder is the spherical form (Fig. 7) and low oxygen content (< 200 ppm O).(10) This powder is the prerequisite for the isotropic, fine-grained and homogeneous structure of the P/M-HIP product. The powder used for sintered P/M parts has irregular shape and high oxygen content (≤ 3000 ppm O), and is mainly produced by direct reduction or water atomization.(11, 12) Gas-atomized powder is also used for different spray deposition methods and extrusion of tubes.(13)

4.3 Encapsulation

The capsules are normally made from sheet segments, which are preformed, welded together and leakage-tested. The capsules are filled with powder under controlled conditions. Since the filled powder has approximately 70% of the density of the fully dense material, there is a linear shrinkage of about 10% during the compaction. The capsules retain their form during compaction. This enables the material to be formed to Near Net Shape (NNS) of the end product.

Very complex forms are possible to produce. This is illustrated by Fig. 8, which shows the design of the capsule for a hollow cylinder with internal undercut. This object would not be possible to produce by conventional forging methods. A photo of a complex valve body weighing about 200 kg, is also shown in Fig. 8. The capsule becomes during compaction an integrated part of the object and is normally removed eg. by machining or pickling.

Capsules, not made of steel sheet, are also possible to use. Capsules similar to investment casting molds are used in the Ceramic Mold Process (Crucible).(14) A fluid die is used in the Rapid Omnidirectional Compaction process (ROC), which furthermore does not involve HIPing.(15) A sintered preform without capsule is used in the Ceracon process.(16)

4.4 Hot Isostatic Pressing

The compaction of the capsules takes place in a Hot Isostatic Press (HIP). Argon gas is used as pressure medium. Typical pressures are 100-150 MPa, typical temperatures 1100-1200°C. During HIPing the space between the powder particles disappears completely, and the result is a 100% dense body. Plastic flow (yield), creep and diffusion are the active mechanisms. This is illustrated by HIP-maps, which can be theoretically calculated, Fig. 9 .(17) In connection with the densification a diffusion bonding is established between the powder particles. Densification and diffusion bonding require normally about 1 hour holding at full temperature and pressure.

Fig. 10 shows the principles of a Quintus Hot Isostatic Press. Objects with diameters of about 1200 mm and heights of about 3100 mm are possible to produce in the HIP of ASEA POWDERMET. Today most HIPs in the world are used for defect-healing of investment castings for the aerospace industry. Compaction of powder is the second greatest application for HIPs.

Other methods for compaction of powder into 100%-dense objects have already been mentioned; Extrusion,(15) Rapid Omnidirectional compaction (ROC).(17)

4.5 Post-treatment

The products are normally delivered after heat treatment, machining and control. Quality control is also performed throughout the entire process. Hot working is used when the size or shape make HIPing impossible.

5. COMPETITIVE FACTORS

The "dimensionless" powder stock in combination with "simple" sheet-forming tools gives the P/M-HIP process flexibility and short production times. This gives the customer short lead times, which in turn means lower tied-up capital in inventory. The P/M-HIP technique has also a small-batch economy.

Compared to conventional forging and casting techniques the P/M-HIP technique can lower the overall costs of a component. Important factors include:
- Better material yield
- Improved machinability
- Less downtime
- Fewer rejects
- Lower tied-up capital due to short lead times
- Lower tooling and prototype costs

Compared to the conventional techniques a higher value of the P/M-HIP product can be achieved by means of:
- Improved materials properties
- New materials
- Compound/Composite materials

- New design solutions
- Integrated design solutions

6. CONCLUSIONS AND FUTURE AIMS

P/M-HIP-produced duplex stainless steels have compared to conventional steels of the same type:

* higher ductility and impact toughness (especially in products with large thickness of material)

* homogeneous and isotropic properties with very small influence of materials thickness

* on the whole equal yield-strength, fatigue strength and corrosions properties

The P/M-HIP technique also enables the production of a new duplex stainless steel with very high yield-strength, APM 2389.

The products can be formed to Near Net Shape of the end product utilizing capsules of steel sheet. Other advantages of the technique are flexibility, small-batch economy, lower tied-up capital in inventory and new design solutions.

The company ASEA POWDERMET was formed in 1984 in order to exploit the possibilities of the P/M-HIP technique. The company believes that the P/M-HIP steels are now standing on the threshold of a broad application within the engineering industry.

7. REFERENCES

1. Berglin L., "HIP-stål", 27-28 Nov 1986, STF-ingenjörsutbildning Nr 61394, Stockholm, 1986.
2. Brandrup-Wognsen H., in "Modern Developments in Powder Metallurgy", 1984 Int. Powder Conf. June 17-22, Toronto, 1984.
3. Andersson T., in "Stainless Steels 84", Göteborg 3-4 Sept 1984, The Institute of Metals, London, 1985, p. 485-493
4. Guntz G., et. al., ibid., p. 494-502
5. Nilsson J-O., Thorvaldsson T., Scand. J. Metallurgy 15 (1986) p. 83-89
6. Harzenmoser M.A., Uggowitzer P.J., in "Ergebnisse der Werkstoff- Forschung, Band 1" Schweizerische Akademie der Werkstoffwissenschaften, 1987, p. 219-247
7. Simpson S.P., Private Communication, Sulzer Brothers Limited, Wintherthur, Switzerland
8. Schläpfer H.W., Weber J., Austenitic-Ferritic Duplex Steels, Internal report, Sulzer Brothers limited, Winterthur, Switzerland
9. Moll J.H., et.al. in "Powder Metallurgy Applications, Advantages and Limitations" ASM, Ohio, 1983, p. 247-298
10. Ambs HD., et. al. in "Metals Handbook, Ninth edition, Volume 7, Powder Metallurgy", ASM, Ohio, 1984, p. 100-104
11. Lenel F.V., ibid p. 79-82
12. Klar E., ibid, p. 83-86
13. Tornberg C., in "1984 ASM Int. Conf. on New Developments in Stainless Steels Technology", Detroit, 17-20 Sept. 1984, ASM 8410-013
14. Price P.E., Kohler S.P., in "Metals Handbook Ninth edition, Volume 7, Powder Metallurgy" ASM, OHIO, 1984, p. 419-443
15. Kelto, C.A., ibid, p. 542-546
16. Ferguson B.L., Smith O.P., ibid, p. 537-541
17. Helle A.S., et. al., Acta Metall., 33, (1985) 12, p. 2163-2174

Table 1

Tensile testing of an object weighing appr. 450 kg and made of duplex stainless steel (C max 0,04, Cr23, Ni5, Mo2,7, N0,12). Comparison between conventionally and P/M-HIP produced objects.

Position/ direction	$R_{p0,2}$ (N/mm^2)			R_m (N/mm^2)			A_5 (%)		
	HIP	Conv.		HIP	Conv.		HIP	Conv	
1/tangential	500	500	490	730	690	690	39	33	37
2/tangential	490	-	510	730	-	690	38	-	33
2/axial	480	510	-	720	690	-	39	31	-
3/tangential	490	510	480	720	680	690	39	31	30
4/tangential	480	490	-	710	670	-	39	32	-
4/radial	500	500	480	730	670	680	41	34	34
5/radial	520	-	500	730	-	690	38	-	33

Table 2

Fatigue testing (pulsating, R=O, $\sigma u = \sigma max$) of conventional forging of SS 2324 (Appr. AISI 329) and of similar P/M-HIP made object.

Material	Fatigue strength σu (10^7) (N/mm^2)	Tensile strength R_m (N/mm^2)	$\frac{\sigma u\ (10^7)}{R_m}$
P/M-HIP	453	687	0,66
Conv. Electro-slag remelted	411	651	0,63

Table 3

Pitting potential (1M NaCl, 20mV/min) of conventionally and P/M-made duplex stainless steels. SS 2324 (appr. AISI 329) NU44LN (appr. SS 2324 with <0,03 C, 0,15 N).

Steel	Production method	Pitting potential	Standard deviation	Test temperature
2324	P/M-HIP	485 mV SCE	30mV SCE	40°C
	Conv. forging	192 "	38 "	"
44LN	P/M-Hot extruded	522 "	85 "	60°C
	Conv. Hot rolled	343 "	86 "	"

Table 4

Nominal composition of APM 2389.

C max	Si	Mn	S max	P max	Cr	Ni	Mo	N
0,07	0,5	0,5	0,030	0,030	26	4,5	2,5	0,3

Table 5
Typical mechanical properties of quench-annealed APM 2389. Wall thickness of products appr. 80 mm.

$R_{p0.2}$ N/mm^2	R_m N/mm^2	A_5 %	KU J	σ_u* N/mm^2
630-670	800-850	27-30	40-50	500-520

*R = 0, $\sigma_u = \sigma_{max}$

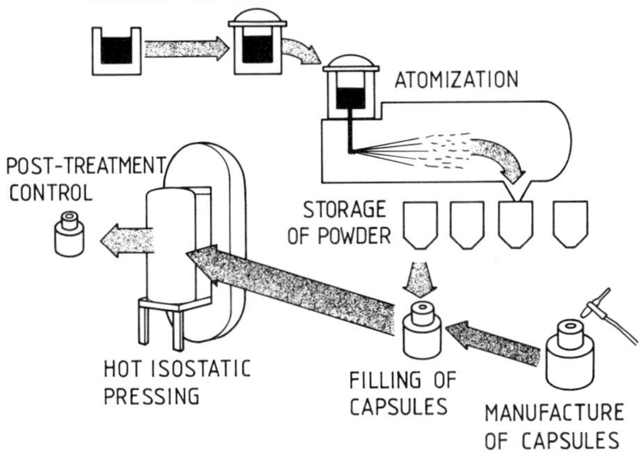

1. **Production of a Near Net Shape object by the P/M-HIP process**

Steel: C max0.04, Cr23, Ni5, Mo2.7, N 0.12 Weight appr. 500 kg

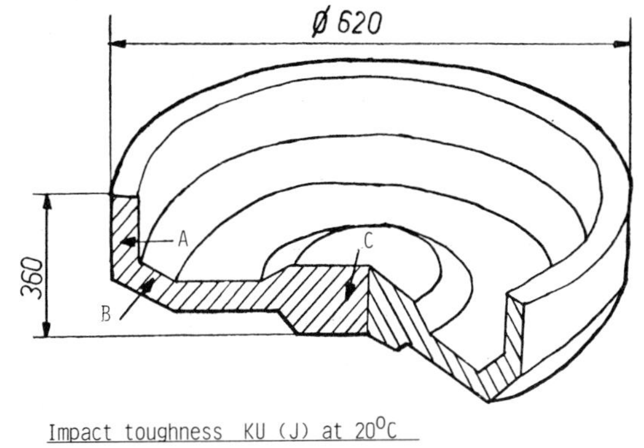

Impact toughness KU (J) at 20°C

Location/ direction	P/M-HIP	Conventional
A/tangential	81	95
B/tangential	82	95
C/radial	81	70
C/axial	82	55

3. **Impact toughness of a conventional heavy forging and a corresponding P/M-HIP-made object exhibiting homogeneous and isotropic properties**

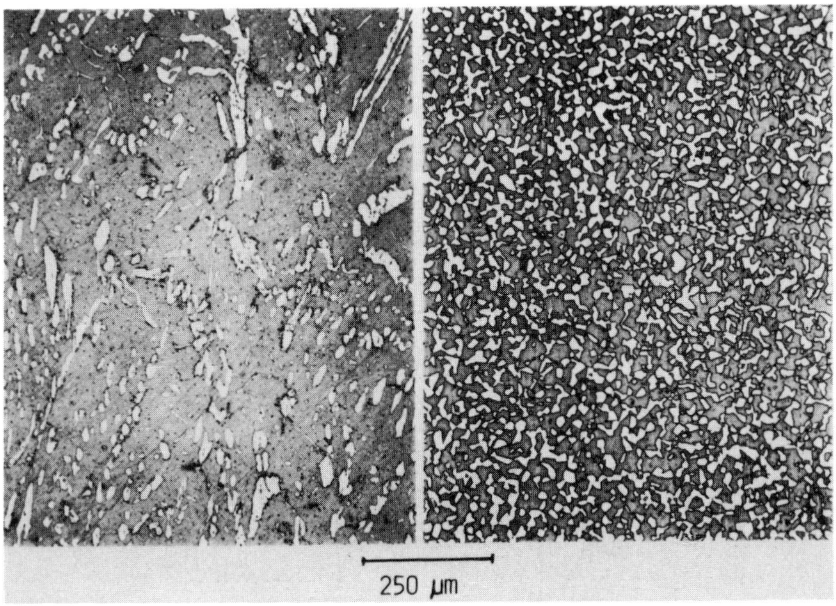

2. **Illustrating the homogeneous and isotropic microstructure of P/M-HIP-produced duplex stainless steel SS 2324 (AISI 329)**

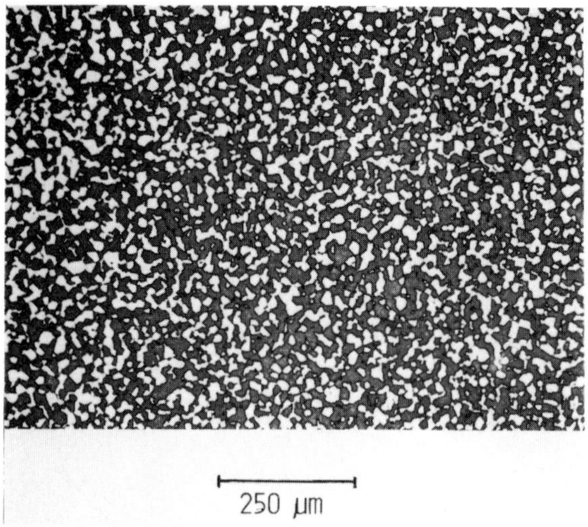

4. **Microstructure of the highly nitrogen-alloyed duplex steel APM 2389**

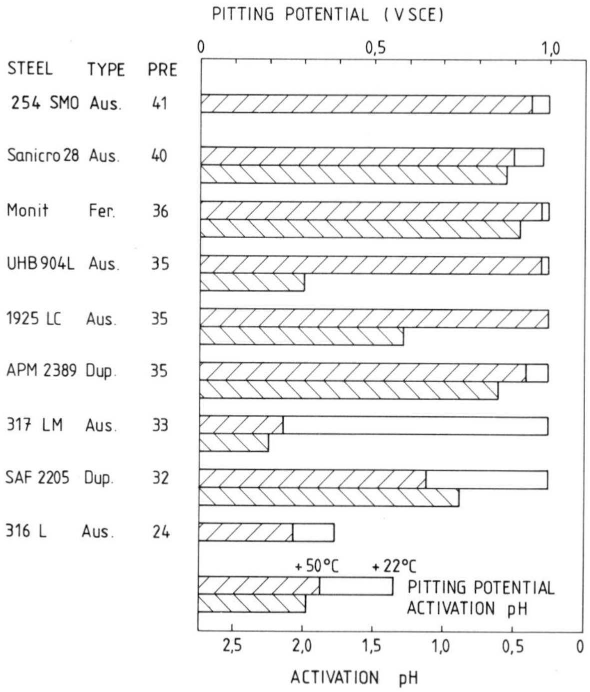

5. **Pitting potential (synthetic seawater, 10 mV/min) and activation pH (2M NaCl+ HCl 25°, 10mV/min i_p> 10µA/cm²) of different stainless steels PRE = % Cr + 3 • % Mo (7)**

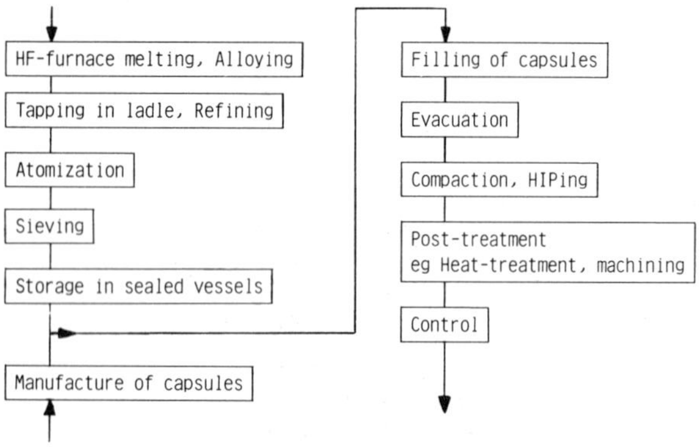

6. **Flow sheet of the production of P/M-HIP steels at ASEA POWDERMET**

7. **Gas-atomized duplex stainless steel powder: a), b) SEM, c) cross-section of powder (annealed in order to develop a duplex structure)**

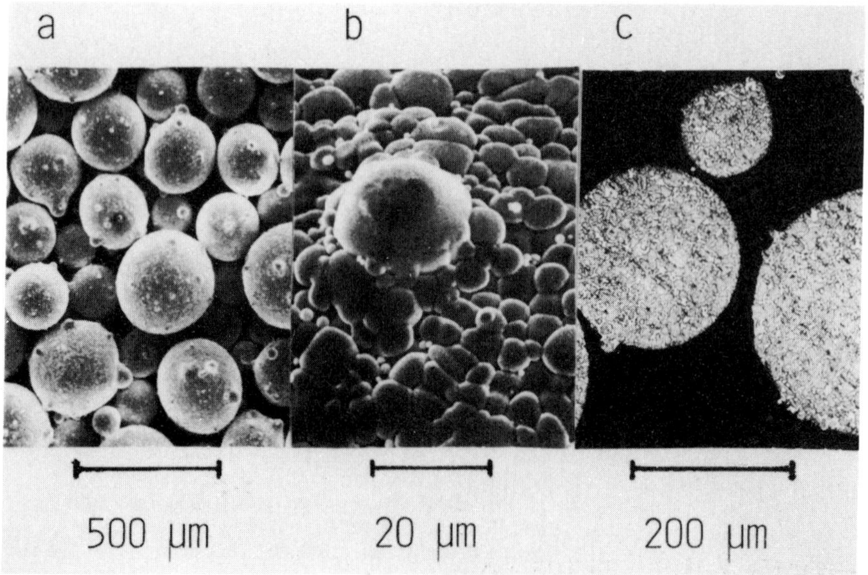

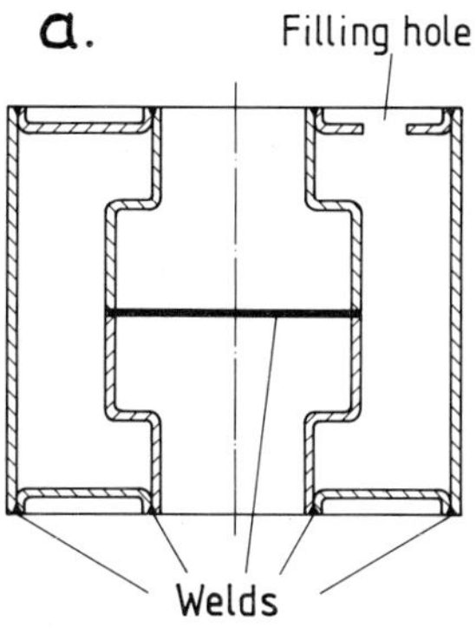

8. a) Design of a capsule for a complex cylinder
b) Valve body weighing 225 kg

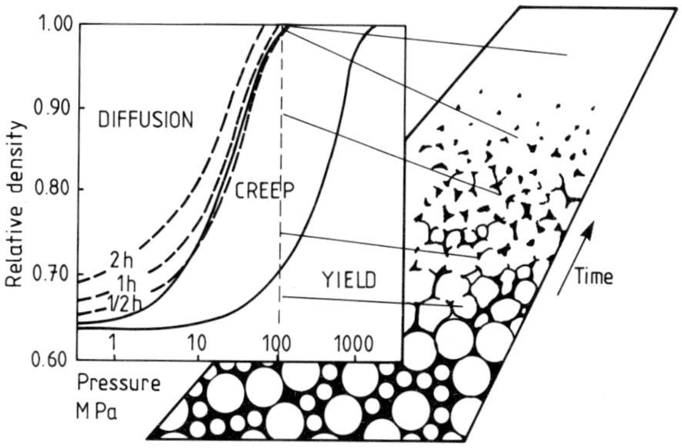

9. HIP-map showing the different mechanisms during densification of high-speed steel powder (diam. 100 μm) at 1200°C (17)

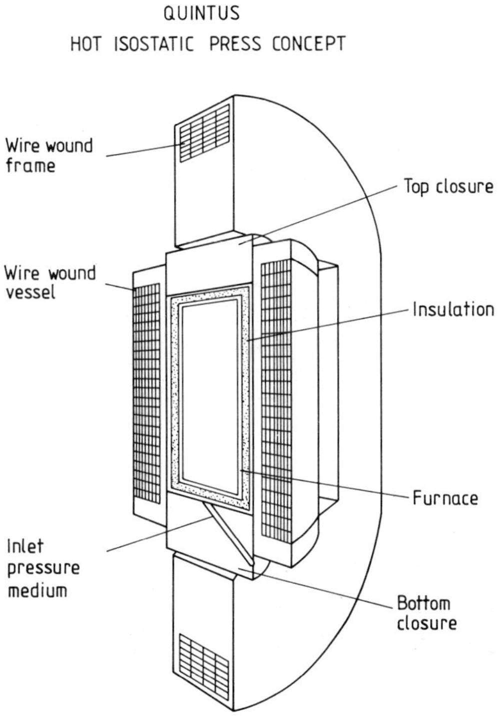

10. Design of a ASEA-QUINTUS Hot Isostatic Press

In-line heat treatment of austenitic and ferritic stainless steel wire

G Baralis, M Farinet and F Guglielmi

G. B. and M. F. are with DeltaCogne, Aosta, Italy; F. G. is with C.S.M., Rome, Italy.

SYNOPSIS

Hot rolling of austenitic and ferritic stainless steel wire is usually followed by cooling and heat treatment of the coils (quench annealing for austenitic and recrystallization annealing for ferritic steel).
In-line heat treatment allows energy savings and shortening of cycle time, and leads to properties fully equivalent to those obtained with the conventional heat treatment.
Furthermore in the considered rolling mill in-line cycle times are very similar for austenitic and ferritic steels, so a single continuous furnace can handle both steel families.
Austenitic stainless steel wire (AISI 302 - 304 - 304L - 316) is coiled at about 1000-1035°C; the coils can be charged into the furnace at 940-1000°C, for reheating up to 1070-1100°C in 30 min with subsequent water cooling.
Ferritic stainless steel wire (AISI 430) is coiled at about 800-840°C; the coils can be charged into the furnace at a somewhat lower temperature, then be kept at 750-800°C for 30 min and water cooled.
In the case of austenitic steels a shortened cycle giving a wire with good cold drawing properties has been devised.

1. INTRODUCTION

Hot rolling of austenitic and ferritic stainless steel wire is usually followed by cooling and heat treatment of the coils. According to the present practice, coils of austenitic stainless steel wire are held at 1050-1100°C for 1-2 hours and water cooled (quench annealing), coils of ferritic stainless steel wire are held at about 800°C for 1-2 hours and water cooled (recrystallization annealing).

Immediate ("in-line") heat treatment of the hot rolled coils in a continuous furnace allows energy savings (recovery of rolling heat) and simplifies operating cycles.

The aim of the present paper is to describe the results of the experimental study carried out on an industrial rolling mill in order to define the operating conditions for in-line heat treatment of austenitic and ferritic stainless steel wire coils.

2. MATERIALS

The study has been carried out on the following steel types :

Austenitic :	AISI	302	(EU X10CrNi189)
	"	304	(EU X6CrNi1810)
	"	304L	(EU X3CrNi1810)
	"	316	(EU X6CrNiMo17122)
Ferritic	AISI	430	(EU X8Cr17)

The average chemical composition of the above steel types is reported in Table 1.

Wire samples were taken from 21 industrial heats of austenitic and 5 industrial heats of ferritic steel.

All heats were melted in an electric arc furnace and were argon-oxygen decarburized (AOD); some of them have been cast in 3600 kg ingots, others were continuously cast (240 x 180 mm strands).

The cast material was rolled to square billets (110 mm side). After inspection and grinding the billets were hot rolled to wire.

3. RESULTS

3.1 AUSTENITIC STAINLESS STEELS

Wire samples of austenitic stainless steels (diameters ranging from 5,5 to 11 mm) were taken after different thermal cycles :

- water cooling immediately before or immediately after coiling
- air cooling after coiling
- industrial quench annealing of the coils (1 hour at 1080°C - water cooling)

3.1.1 Mechanical properties

Average mechanical characteristics of the wire samples are illustrated in Table 2. More details have been reported elsewhere. (1) From Table 2 it can be seen that mechanical characteristics are little affected by the rate of cooling. More important effects are brought about by quench annealing, i.e. lowering of ultimate (UTS) and yield strength ($YS_{0.2}$), increase of tensile elongation (El_5) of strain-hardening exponent (n) and of grain size.

Optical and electron microscope observations showed that both rapidly and slowly cooled hot rolled samples present no evidence of work hardening; therefore the differences in mechanical properties can be ascribed mostly to grain size.

3.1.2 Resistance to intergranular corrosion

In the case of austenitic steels rolling temperature is high enough to avoid carbide precipitation during rolling and subsequent hot coiling. Consequently rapid cooling of the coiled wire (e.g. water cooling or air – blast cooling) should avoid carbide precipitation and allow high resistance to intergranular corrosion.

No traces of grain boundary carbides were detected in rapidly cooled austenitic stainless steel wire by means of transmission electron microscope observations. Tests carried out in accordance with the ASM standard A 262-77a: practice C for 302, 304 and 304L steels (boiling nitric acid : Huey test) and practice D (nitric-hydrofluoric acid test) for 316 steel showed excellent resistance to intergranular corrosion (corrosion rate lower than 0,5 mm/year for 302, 304 and 304L; corrosion ratio lower than 1.5 for 316).

Quite similar behaviour was observed in the industrially quench annealed steel.

Therefore in-line quench annealing of the coils should only allow the grain growth necessary for softening the steel, when required.

For this reason two types of in-line heat treatment were devised for austenitic stainless steel :

- simplified treatment : the wire is rapidly cooled immediately after coiling
- substitutive heat treatment : the coils are held in a hot environment until the grain grows up to the required size (usually ASTM number 4-6), then they are rapidly cooled (usually by plunging in water).

3.1.3 Simplified treatment

Industrial and laboratory tests were carried out in order to verify the properties of the rapidly cooled coiled wire.

During the industrial tests, several 400 kg coils of 304 and 316 stainless steel (wire diameter 6 and 9.5 mm) were rapidly cooled by plunging in water in an experimental apparatus.

Each coil was separated into two halves : one half was subjected to the usual treatment of quench annealing. Both halves were then cold drawn under identical operating conditions in an industrial drawing plant, until a reduction of section of 80% was reached (6 to 8 drawing steps). No appreciable difference in drawing behaviour was found between rapidly cooled and quench annealed steel. Tensile tests carried out on samples taken after each drawing step gave the results exemplified in Fig. 1. It can be seen that the differences in UTS and $YS_{0.2}$ stay approximately constant irrespective of drawing reduction. Reduction of area at rupture (R of A) shows a slight advantage for the quench annealed steel.

Laboratory tests were used for evaluating cold headability of rapidly cooled steels through the measurement of the Cold Work Hardening Factor (CWHF)(2,3) and of the Tozawa index. (3,4)

The CWHF evaluates the energy per unit volume absorbed by a cylindrical specimen, whose height equals two times the diameter, when it is cold compressed to a final height equalling 37.5% of the initial one.

The Tozawa test evaluates the ability of the steel to withstand heavy plastic deformation by compression before cracking. In this test a cylindrical specimen (height = 1.5 times the diameter) is compressed between two cylindrical tools (Fig. 2) so as to produce a highly uneven deformation. Tozawa index is the ratio between the initial height (h_o) and the height (h_1) at which the first crack appears.

Wire samples of 7.5 to 11 mm diameter, cooled with different rates, were compared with quench annealed samples (Tables 3 and 4). The quench annealed wire showed superior cold headability (lower energy absorption and higher deformation to rupture) while cooling rate had little effect on these properties.

In conclusion austenitic stainless steel wire rapidly cooled after hot rolling is suitable when high cold drawability and good intergranular corrosion resistance are required in connection with high YS (e.g. production of springs). Quench annealing is necessary to improve cold headability (e.g. production of bolts).

3.1.4 Substitutive heat treatment

In order to evaluate temperature and time required for grain growth, samples of 304, 304L and 316 stainless steel wire (diameter 5.5 and 9.5 mm) were heated in a salt - bath furnace at different temperatures (1000-1050-1100°C) for different times (5,10,15,30 minutes) and then cooled by dipping in water. The observed grain growth is reported in Figs. 3 and 4. A simultaneous decrease of UTS has been observed (Fig. 5).

It can be seen that the grain grows quite rapidly in the first 5 minutes, then the growth gradually slows down. In order to obtain a grain size corresponding to ASTM range 4 - 6, a holding time of 10 min at 1100°C seems to be suitable. This holding time is a little higher than those found by other authors ,(5) but this fact can be ascribed to the different starting grain size (ASTM 10-12 in the place of 7-9).

In order to substantiate the above results, a set of hot torsion tests was run. The testpieces were machined from steel bars 12.5 mm square drawn from the the rolling mill halfway through rolling process by means of flying shears. The testpieces were subjected to 11 deformation steps in the hot-torsion machine, in order to simulate hot rolling of the wire from 12.5 mm square to 5.5 mm round section. Two deformation temperatures were tested : 1000 and 1050°C. The deformation rate was 3.6 s^{-1}, that is lower than the rolling deformation rate (about 100 s^{-1}). After deformation the following heat treatments were carried out before air cooling :

- 1 s or 5 - 15 min holding at the deformation temperature or heating from the deformation temperature up to 1100°C in 5 min then holding for 0 s or 5 - 10 min.

Grain sizes measured on torsioned and heat treated specimens are in good agreement with those observed in comparable conditions of heat treatment on the salt-bath annealed samples of rolled wire.

Obviously thermal inertia of the coils makes the above cycles not transferable to industrial production. In our case, on the base of heat exchange models used by furnace designers, we took into account the heating cycle shown in Fig. 6 : this cycle is considered realistic in the case of 500 kg coils of 5.5 mm diameter wire. It was assumed that the coils cool from 1000-1035°C (coiling temperature) to 940°C on the surface and to 1000°C in the inner part before they were introduced in the furnace.

In order to verify the effectiveness of the above cycle a number of samples of 304, 304L and 316 wire with diameters from 5.5 to 9.5 mm were heat treated in a programmable temperature laboratory furnace.

Four types of temperature cycles were tested, i.e. the cycles reported in Fig. 6 (surface and inner part of the coils) and the same cycles with 15°C reduction of temperature. That takes into account the possibility of temperature errors during the industrial operation.

Samples were examined after 8,14,22 and 28 minutes of each cycle : the measured grain sizes are reported in Fig. 7. It can be seen that after 22 min nearly all measured grain sizes fall into the preferred range (ASTM 4-6). At the end of the heating cycle all the samples fall into the preferred range.

Tension tests, carried out on some of the samples heat treated according to the above cycles, gave results (UTS, YS, El, R of A) in full agreement with the values found on the industrially quench annealed steel. It can then be concluded that the proposed heat treating cycle gives results fully equivalent to industrial quench annealing.

3.2 FERRITIC STAINLESS STEEL

Wire samples of ferritic stainless steel (diameters 5.5-6 mm) were taken after the following thermal cycles :

- air cooling of the coils
- industrial heat treatment of the coils (recrystallization annealing for 2 hours at 800°C and water cooling).

3.2.1 Mechanical properties

Typical mechanical characteristics of the wire are illustrated in Table 5. It can be seen that air cooled samples show a wide scatter of properties. Recrystallization annealing reduces UTS and $YS_{0.2}$, increases El_5 , R of A and n and reduces data scatter; grain size is nearly unaffected.

3.2.2 Resistance to intergranular corrosion

In the case of ferritic stainless steel wire, resistance to intergranular corrosion was measured in accordance with the ASTM Standard A 763-83 practice X (boiling ferric sulphate-sulphuric acid test : Streicher test).

Metallographic examinations showed that when the corrosion rate is lower than 10 mm/year intergranular corrosion is absent.

Ferritic steel from air cooled coils showed very scattered corrosion rates, ranging from very high to quite low values. Recrystallized wire showed corrosion rates consistently lower than 10 mm/year.

3.2.3 Heat treatment

Rolling temperature of 430 steel is near to or higher than 900°C. In these conditions ferrite partially transforms to austenite; during cooling after rolling austenite undergoes martensitic transformation. Moreover cooling leads to a very rapid chromium - carbide precipitation on grain boundaries, that can hardly be prevented by rapid cooling.

Recrystallization annealing has a double aim :

- to transform martensite into ferrite (softening of the steel)
- to allow chromium diffusion to the depleted zones near the grain boundaries, in order to restore the resistance to intergranular corrosion.

During the last rolling passes wire temperature is near to the lower limit for austenite formation and this explains the scattering of characteristics of the air cooled wire. Generally the presence of some austenite is preferred, in order to avoid abnormal grain growth in the subsequent recrystallization annealing.

It goes without saying that in this case rapid cooling of the steel is not an acceptable choice; a substitutive heat treatment (recrystallization annealing carried out without cooling of the hot coils) should meet the following requirements :

- to transform austenite directly into ferrite
- to allow carbides precipitation on and chromium rediffusion to the grain boundaries.

3.2.4 Substitutive heat treatment

Because of martensitic transformation and carbides precipitation, samples taken from air cooled coils cannot be used to simulate substitutive heat treatment.

Three ways have been devised to carry out the tests :

- heating wire samples to 900 or 950°C to give rise to austenite formation, then cooling them to 760°C or 800°C and holding there for different times;
- simulation of the rolling deformation and subsequent treatment on a hot torsion machine;
- taking hot wire samples from the rolling mill by means of flying shears, introducing them immediately into a hot furnace and holding at 750 or 800 or 850°C for different times (10,15,30,60 min).

In all cases after heat treatment the samples were air cooled or water cooled.

The results obtained through the first way are shown in Figs. 8 and 9 : after holding 15 min at 760°C the softening of the steel is completed (Fig. 8). Resistance to intergranular corrosion is very good after holding 15 min at 800°C, if the samples are water cooled at the end of the treatment (Fig. 9). Air cooling gave good results only if the samples were initially heated at 900°C (temperature nearer to the final rolling one).

In all cases heating at 900-950°C without deformation produced an appreciable growth of ferritic grain and austenite islands sizes, that could affect transformation rates and corrosion resistance.

The testpieces for the hot torsion tests (second way as above) were machined as already described in the case of austenitic steels (see above). The testpieces were heated to 950°C (15 min holding time), were deformed at 930°C or 880°C and recrystallized at 800°C for 15 or 60 min. Reference tests were carried out with cooling to room temperature before recrystallization. Vickers hardness measurements carried out on the external layer of the treated testpieces showed the same evolution of mechanical strength already reported in Fig. 8.

Metallographic observations confirmed that the structure of the external layer of the testpieces is similar to the one of the industrially produced steel wire both before and after recrystallization annealing (Figs. 10, 11, 12, 13, 14).

Obviously, because of disuniform deformation, torsion testpieces are not suitable for corrosion tests. Therefore extensive use of the third way described above was made to test resistance of the wire to intergranular corrosion.

Holding at 750 and 800°C gave good corrosion resistance (corrosion rate lower than 10 mm/year), irrespective of holding time (10 min minimum) and type of cooling (air or water, Fig. 15), though water cooling gave somewhat lower corrosion rates. Corrosion rates decreased slowly with increasing holding time.

Holding at 850°C gave erratic results : corrosion rates were generally higher than in the previous cases , and sometimes marked intergranular corrosion was evident.

The structure of the heat-treated samples is very similar to the structure of industrially recrystallization annealed steel (Figs. 11 and 16).

It has been concluded that holding the coils at 750-800°C for 30 min followed by cooling in water are safe conditions for in-line heat treatment.

In the considered rolling mill 430 steel is usually coiled at about 900°C. The most convenient way to reduce the temperature for in-line heat treatment has been found to be water-spray cooling of the wire before coiling. Experiments with 6.5 mm-diameter wire showed that coiling can be carried out at 800-830°C without any inconvenience.

4. CONCLUSIONS

Austenitic and ferritic stainless steels can be heat treated in-line. The obtained product has properties quite similar to the properties of the equivalent batch heat treated steel.

In the case of the considered rolling mill, optimized treatment times are 30 min , so a single continuous furnace can be taken into account for the heat treatment of both austenitic and ferritic stainless steels.

Alternatively austenitic stainless steel wire coils can be water cooled immediately after hot coiling : the wire so produced shows very good intergranular corrosion resistance, excellent drawability, slightly impaired cold headability and increased yield and rupture strength. Its use is suggested for springs production.

ACKNOWLEDGEMENTS

This work has been carried out with the financial aid of the ECSC (Convention n° 7210 EA/413).

Thanks are due to the Direction of DeltaCogne spa for the permission to publish this work.

REFERENCES

1) F. Guglielmi, G. Baralis, G. Lanfranco: ECSC Research Agreement N° 7210-EA/413 Final Report, 1985

2) F.K. Bloom, G.N. Goller, P.G. Mabus: Trans. Am. Soc. Metals 39 , 843-64, 1947

3) M. Civera, V. Faccenda, B. Ligia : CSM (Rome) Report n° 4790R, Feb. 1984

4) Y. Tozawa : Recommended method in Japan for testing for cold upsettability - ICFG 12th Plenary Meeting (Stuttgart, 5-6 September 1975)

5) W. Murata, J. Tominaga, W. Shinada, N. Sakao : Transactions I.S.I.J. 23 , B 99, 1983

Table 1   Average percent composition of the different stainless steel types

Steel Type	C	S	P	Si	Mn	Cr	Ni	Mo
302	0,07	<0,015	<0,030	0,40	0,90	18,0	8,5	<0,5
304	0,05	<0,020	<0,030	0,40	1,50	18,0	9,5	<0,4
304L	0,03	<0,015	<0,030	0,40	1,80	18,0	10,5	<0,4
316	0,05	<0,015	<0,030	0,40	1,80	17,0	12,0	2,25
430	0,04	<0,020	<0,030	0,40	0,40	17,0	<0,40	<0,20

Table 2 - Average mechanical properties and grain size of austenitic steels in different heat treatment conditions

Steel type	Heat treatment condition	wire diameter mm	UTS MPa	$YS_{0.2}$ MPa	El_5 %	R of A %	n	Grain size ASTM number
302	air cooled	5.5	700	320	62	76	0.41	-
	quench annealed	5.5	590	225	76	79	0.53	5
304	air cooled	5.5	694	300	67	78	0.43	11
	id. id.	8.0	656	290	59	75	0.41	11
	water cooled	5.5	701	332	64	78	0.40	11
	id. id.	9.5	610	260	58	76	0.42	10.5
	quench annealed	5.5	596	227	78	81	0.51	5
	id. id.	8.0	560	234	73	79	0.46	6
304L	air cooled	5.5	637	290	67	79	0.43	11.5
	id. id.	7.5	624	281	57	74	0.42	10.5
	id. id.	9.5	634	310	52	75	0.39	11
	id. id.	11.0	620	285	51	72	0.40	11.5
	quench annealed	5.5	562	193	79	79	0.51	5.5
	id. id.	7.5	522	188	70	78	0.52	5
	id. id.	9.5	517	195	73	80	0.51	5
	id. id.	11.0	497	180	69	76	0.50	5
316	air cooled	5.5	657	323	65	78	0.40	11
	id. id.	11.0	653	327	50	70	0.37	11.5
	water cooled	5.5	668	333	63	75	0.38	11.5
	quench annealed	5.5	560	222	79	81	0.44	5
	id. id.	11.0	534	226	66	77	0.42	5.5

Table 3 – Cold Work Hardening Factor – Average values

Steel type	Heat treatment condition	Wire diameter mm	CWHF $\frac{ft.pounds}{in^3} \cdot 100$
304	air cooled	8.0	104.2
	water cooled	9.5	97.1
	quench annealed	8.0	94.1
304L	air cooled	7.5	99.4
	id. id.	9.5	97.9
	id. id.	11.0	95.8
	quench annealed	7.5	91.6
	id. id.	9.5	86.9
	id. id.	11.0	85.2
316	air cooled	11.0	96.6
	quench annealed	11.0	88.7

Table 4 – Tozawa index – Average values

Steel type	Heat treatment condition	Wire diameter mm	Tozawa index
304	air cooled	8.0	4.0
	water cooled	9.5	4.2
	quench annealed	8.0	5.5
304L	air cooled	7.5	3.3
	id. id.	9.5	3.8
	id. id.	11.0	3.0
	quench annealed	7.5	5.0
	id. id.	9.5	6.3
	id. id.	11.0	5.5
316	air cooled	11.0	4.3
	quench annealed	11.0	6.4

Table 5 – Steel 430. Range of mechanical properties and grain size in different heat treatment conditions. Wire diameter 5,5–6 mm.

Heat treatment	UTS MPa	$YS_{0.2}$ MPa	El_5 %	R of A %	n	Grain size ASTM number
As rolled and air cooled	474 728	283 460	44 16	83 72	0.25 0.12	10 12
Recrystallization annealed	450 539	270 305	40 44	83 79	0.23 0.25	10 10

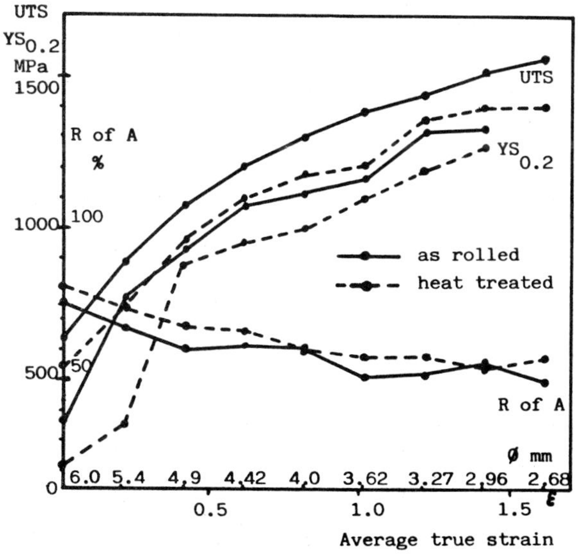

Fig. 1 – Steel 316. Mechanical properties vs. reduction in cross section after cold drawing. Initial ø = 6 mm
as rolled = coil cooled by plunging in water
heat treated = quench annealed (1050°C – 2 hours)

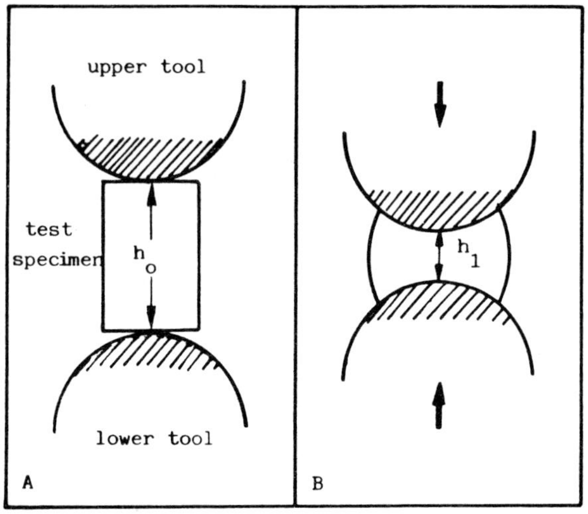

Fig. 2 - Outline of Tozawa test
Dimensions of test specimen :
ø from 5 to 20 mm
ho from 7.5 to 30 mm
Tools : cylindrical dies : ø from 10 to 40 mm
Compression rate : 1.5 mm/min
A : Before compression
B : After compression

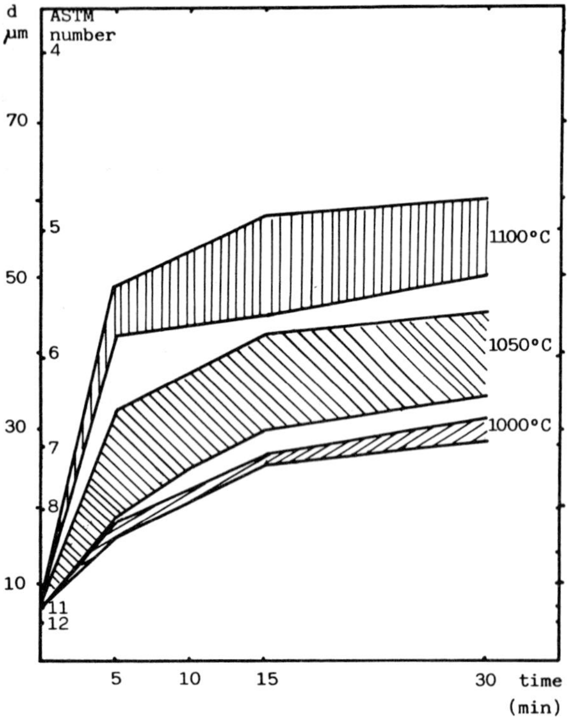

Fig. 3 - Steels 304 and 304L.
Grain growth vs. holding time at various temperatures
d : average grain diameter (Snyder & Graff)

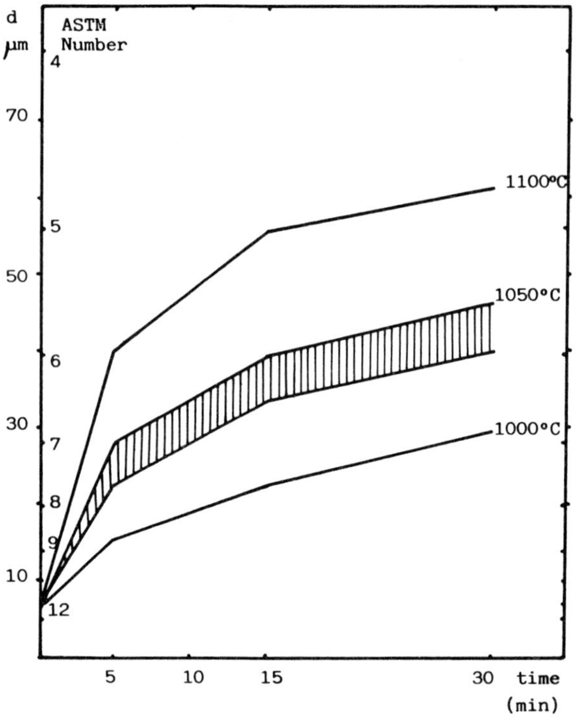

Fig. 4 - Steel 316. Grain growth vs. holding time at various temperatures
d : average grain diameter (Snyder & Graff)

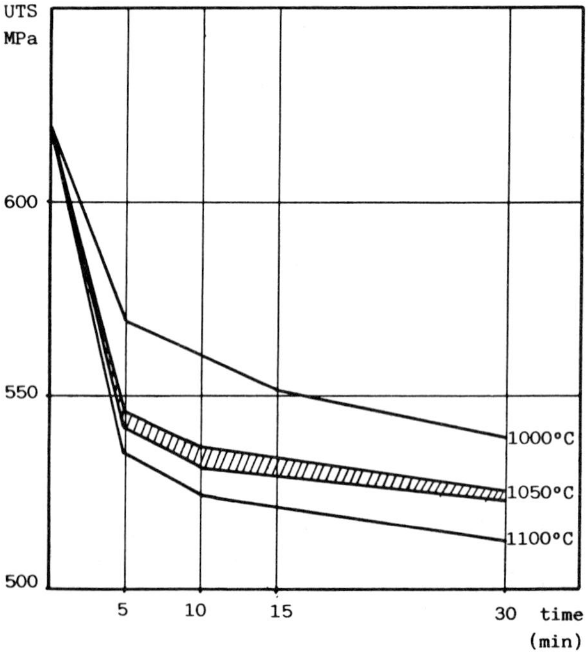

Fig. 5 - Steel 304. Effect of holding time at various temperatures on UTS.

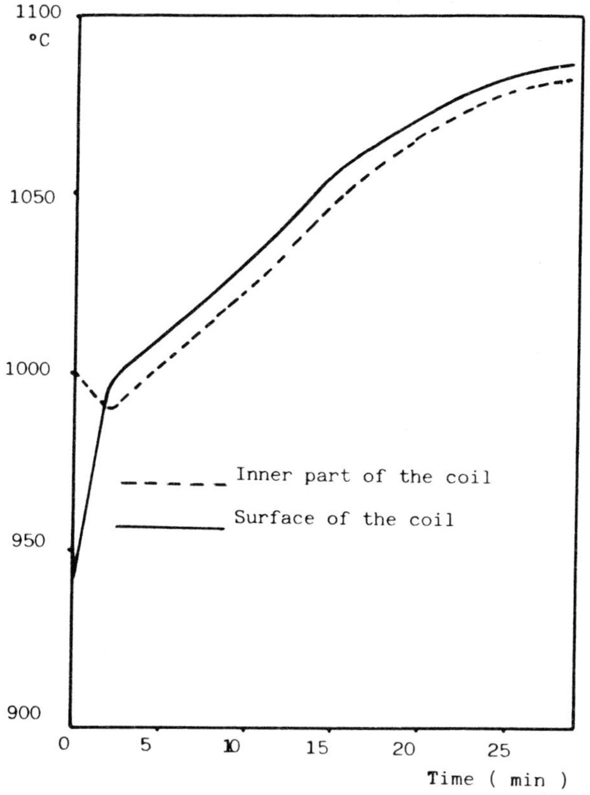

Fig. 6 - Heat cycles used to simulate in-line quench annealing of austenitic stainless steels.

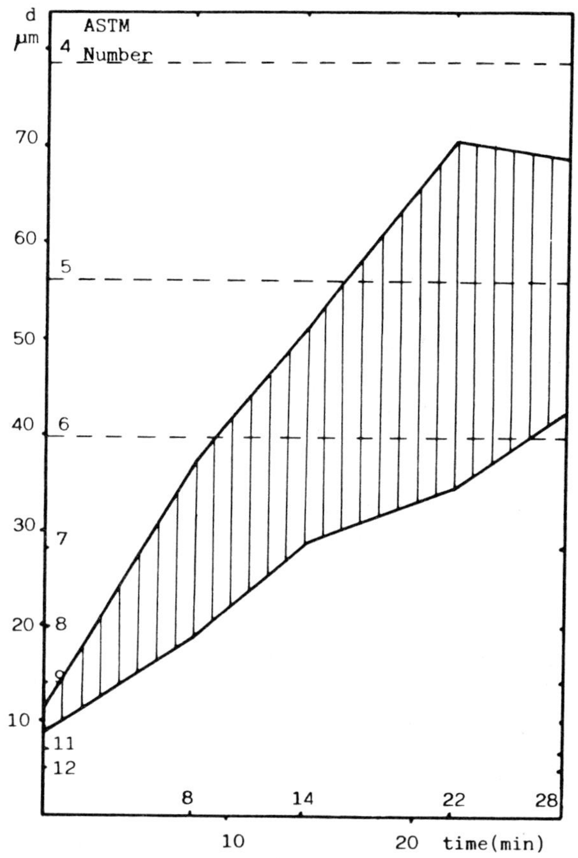

Fig. 7 - Grain size vs. time in 304, 304L and 316 steels heat treated in accordance with the temperature cycles of Fig. 6 (allowed temperature error +0 -15°C, see text)
d : average grain diameter (Snyder & Graff)

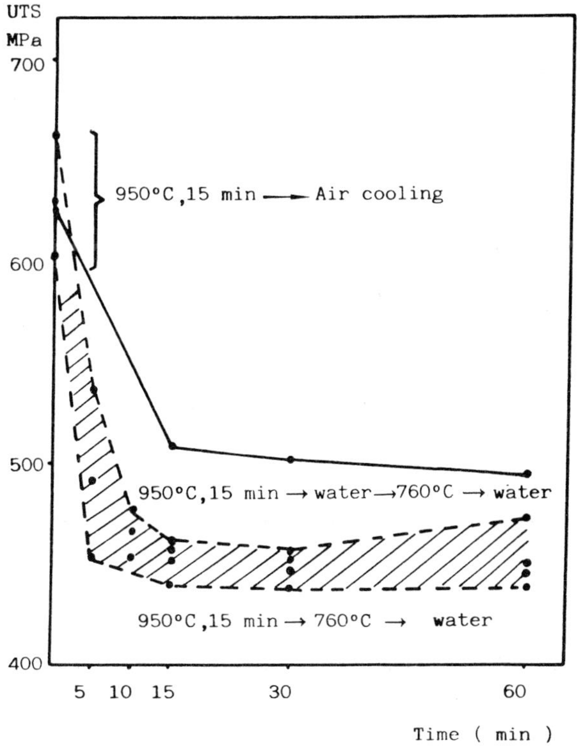

Fig. 8 - Steel 430. Effect of holding time at 760°C on UTS.

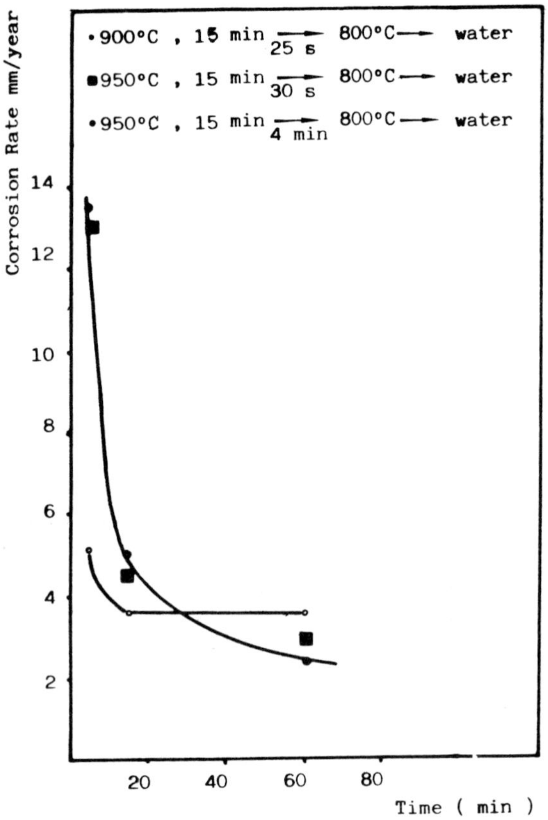

Fig. 9 - Steel 430. Effect of holding time at 800°C on intergranular corrosion rate (Streicher test). Specimens heated to 900°C or 950°C for 15 min and cooled to 800°C at various rates.

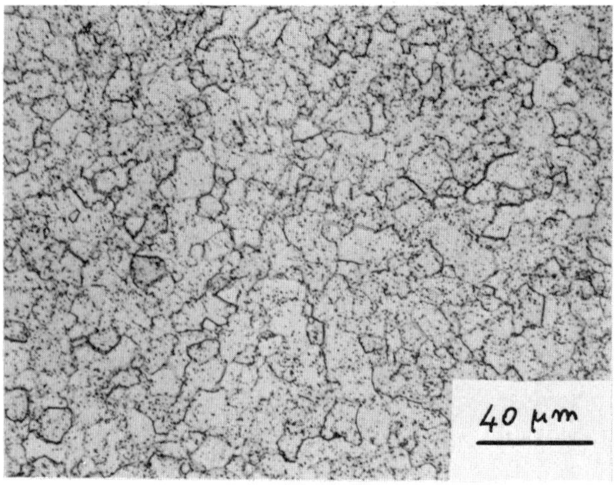

Fig. 11 - Steel 430. Industrially recrystallized wire (800°C for 2 hours). Structure.

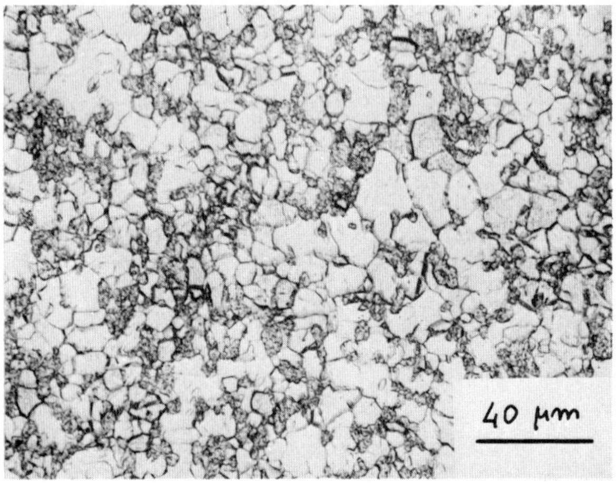

Fig. 12 - Steel 430. Torsion specimen heated to 950°C for 15 min, deformed at 930°C and air cooled. Structure.

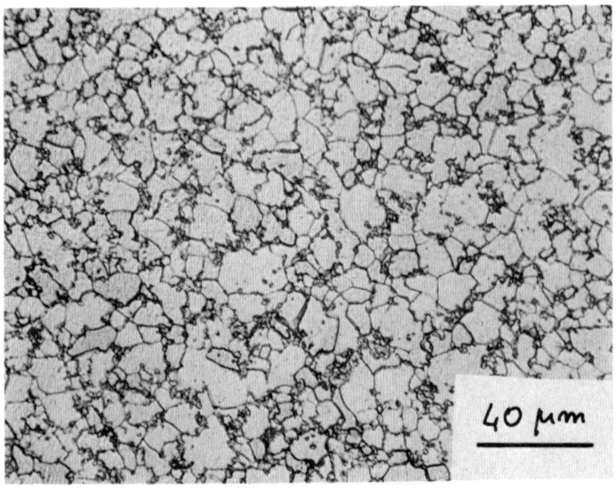

Fig. 10 - Steel 430. As-rolled and air cooled. Structure.

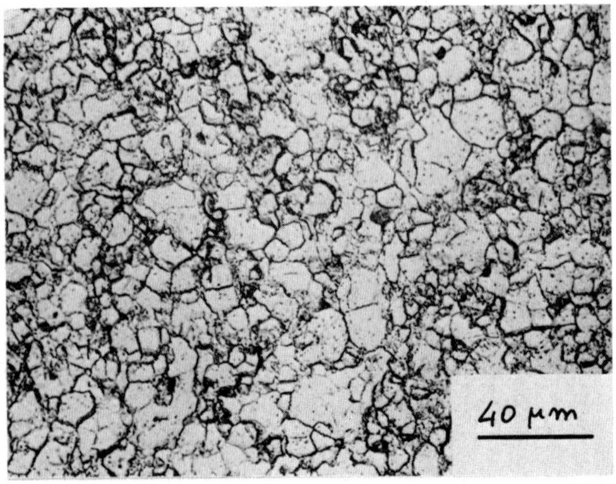

Fig. 13 - Steel 430. Torsion specimen heated to 950°C for 15 min, deformed at 930°C, air cooled and recrystallized at 800°C for 15 min. Structure.

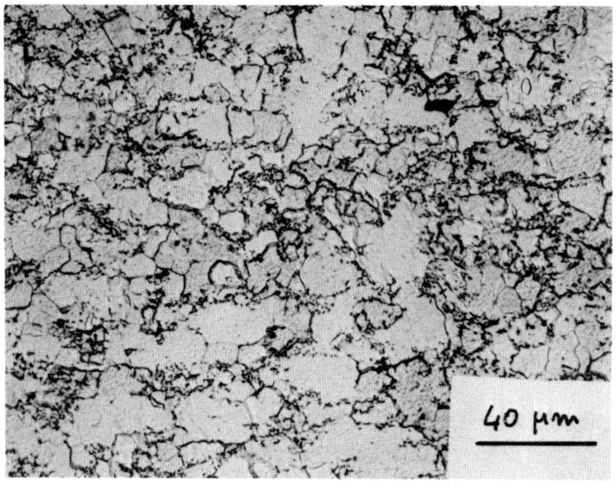

Fig. 14 - Steel 430. Torsion specimen heated to 950°C for 15 min, deformed at 880°C, and taken to 800°C for 15 min. Structure.

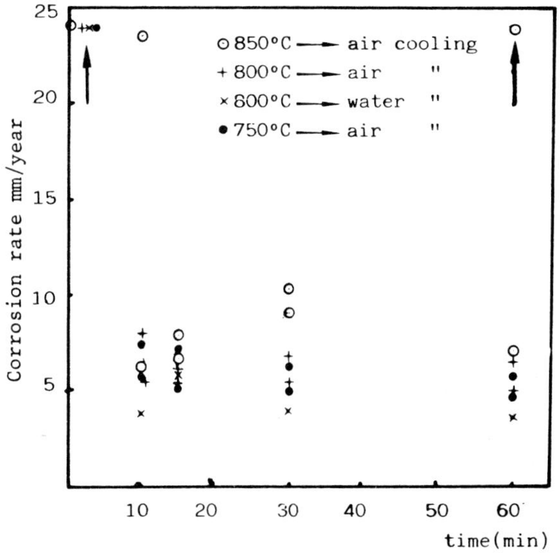

Fig. 15 - Steel 430. Effect of holding time at different temperatures on intergranular corrosion rate (Streicher test).

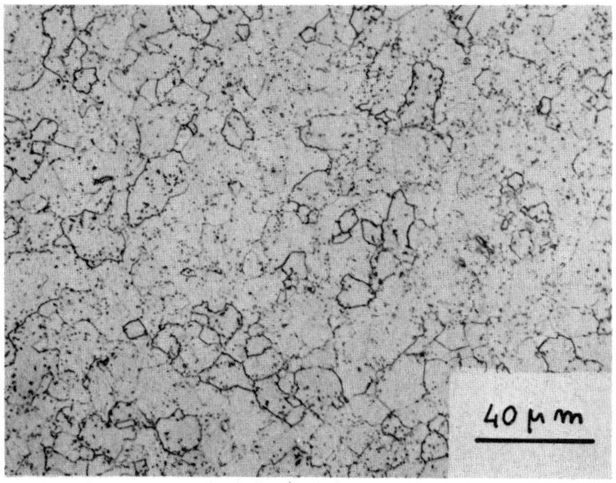

Fig. 16 - Steel 430. Specimen taken from the rolling mill and placed directly in furnace at 800°C for 30 min. Structure.

□

SESSION II
FABRICATION

Progress in welding stainless steels

T G Gooch

Dr. Gooch is Head of Materials Department of The Welding Institute.

SYNOPSIS

The Paper reviews recent developments in welding stainless steels, considering material behaviour, welding procedure and joint service performance. Attention is concentrated on the last few years so that the Paper constitutes an up-date of that presented at the Stainless Steels '84 Conference.

Material developments are presented on the basis of the major microstructural types of stainless steel, namely martensitic, ferritic, austenitic and duplex grades. Particular consideration is given to weld metal corrosion in high alloy fully austenitic materials, and to substantial work carried out on duplex ferritic/austenitic steels.

Conventional arc fusion welding processes remain dominant in industry, although semi-automatic or fully automated systems are being employed increasingly for enhanced productivity and/or reliability. The application of mechanised methods is hindered by the phenomenon of variable penetration, and present understanding of the problem is summarised.

Reference is made to the improved property data generated on welded stainless steels and to activity on nondestructive testing procedures. Finally, attention is drawn to potential welding fume problems and the characteristics of different processes.

INTRODUCTION

The wide range of mechanical and corrosion properties offered by stainless steels has led to their extensive employment in complex process and power generation plant operating under arduous conditions. Moreover, application areas have continued to grow as new materials are introduced and as properties of existing grades are optimised to afford greater economy, and this widespread usage of stainless steels would not be possible without the application of welding.

While the general welding characteristics of the different types of stainless steel marketed are well understood, (1) research has continued on both material and welding process aspects. In the former category, attention has been paid to existing materials or specific welding problem areas for increased understanding and quantification of behaviour. More especially, there has been effort committed to evaluating welding characteristics of recently developed grades, most notably high molybdenum, fully austenitic and duplex, ferritic, austenitic alloys. In terms of the welding process, there is a constant requirement for improved productivity, economy and reliability of fabrication, with consequent study of established processes and trials on the application of newer welding techniques to stainless steels. There has been analogous work on the service usage of stainless steel weldments for better prediction of joint behaviour, to optimise component design and to improve reliability of non-destructive testing (NDT). Consideration has been given also to fume evolution during the welding operation.

The welding of stainless steels was reviewed by the writer in the "Stainless Steels '84" Conference.(2) The present paper considers progress made since that time in the above areas. As previously, the emphasis is on fusion welding by arc processes most commonly employed by industry.

MATERIAL DEVELOPMENTS

Martensitic Grades

There has been little change in welding practice applied to martensitic stainless steels. The major problem to be avoided is that of hydrogen-induced cracking in weld metal or heat-affected zone (HAZ) on cooling to temperatures close to normal ambient. This problem is well known and general guidelines on welding procedures are established.(1) Essentially, these involve minimising the hydrogen potential of the welding process and consumable, with the application of preheat and controlled cooling after welding to encourage hydrogen diffusion away from the joint area while still at a temperature for hydrogen embrittlement to be negligible.

Notwithstanding the good general understanding of the factors promoting cracking when welding martensitic stainless steels, debate still exists regarding optimum control of temperatures during welding, especially for 0.2%C-12%Cr-Mo-V creep-resisting materials. These steels, by virtue of their high total alloy content, display fairly low Ms and Mf temperatures (eg below 200°C), which means that, if a conventional preheat, of say 250°C, is applied, much of the joint area remains austenitic until welding is completed. Hydrogen diffusion from austenite is slow so that cracking can ensue if the joint is then cooled to room temperature with associated transformation of the austenite to martensite. Alternatively, if post-weld heat treatment is carried out before the assembly is cooled to room temperature, a coarse transformation product can develop with consequent reduction in toughness: in fact, it has been reported that the austenite phase may remain stable during the post-weld heat treatment, transforming to martensite on final cooling, again with a risk of hydrogen cracking. The situation regarding preheat has been reviewed by Alberry and Gooch,(3) together with the use of intermediate holds during the welding operation at temperatures below the preheat level to encourage transformation to martensite and hydrogen diffusion out of the joint. For thin material, such as superheater tubing, a preheat of 250°C is normally sufficient to prevent cracking, with a hold on cooling at 100-150°C if necessary. A similar approach is applicable also to somewhat thicker steel, possibly 25-50mm, but the situation is far less clear for heavy section material, such as over 100 mm wall thickness. At least in Germany, a preheat temperature of 400-450°C has been specified with a hold at 100°C prior to post-weld heat treatment, but cracking in trial welds has been found and available evidence strongly suggests that such high preheat temperatures should be avoided.(3)

Ferritic Grades

Ferritic materials remain used almost exclusively in thin section sizes appropriate to sheet or heat exchanger tubing, largely because of the unacceptable grain coarsening when welding is carried out, with reduction in weld area toughness and corrosion resistance.(1,4) These problems, and more especially the latter, have been countered to some degree in recent years by a reduction in steel carbon and nitrogen levels, and by the controlled addition of stabilising elements, titanium and niobium. There is little doubt that materials available on the current market are of advantage relative to those produced in the past, and interest in the wider use of ferritic materials is fairly high (5) because of the potential cost saving which can be obtained by employing simple ferritic rather than austenitic steels. Nonetheless, it remains a fact that the properties of welds in heavy section ferritic alloys made under production conditions tend to be inferior.

As noted previously,(2) the "superferritic" alloys continue to find employment where their particular combination of good corrosion and stress corrosion resistance is paramount. Certainly, fusion welded joints in such steels can have reasonably good properties, but the rigorous control of welding necessary, coupled with the fairly high material cost and significantly lower toughness than austenitic materials, has restricted their more widespread utilisation.

Austenitic Grades

The greater part of stainless steel production continues to be accounted for by austenitic materials, with "traditional" specifications being dominant. From the welding viewpoint, effort has continued on established problem areas, such as intercrystalline attack (6) or weld metal solidification cracking. In the former case, the problem of cracking at butt welds in light water reactor pipework has been prominent with attention to quantifying crack growth rate (7) and to establishing welding procedures giving a compressive residual stress field at the pipe bore.(8) Work on solidification cracking has involved further examination of the roles of residual and impurity elements in promoting low melting point constituents in the solidifying weld pool.(9,10) Consideration is being given also to the allied problem of "microfissuring" or grain boundary liquation cracking in solidified weld metal (or HAZ) during deposition of subsequent runs,(10,12) including a collaborative assessment of the reproducibility of cracking tests under the aegis of the International Institute of Welding, Sub Commission IXH.

Concerning newer materials, there has been an emphasis on evaluating the behaviour of high molybdenum alloys developed especially for sea water service. The presence of about 6% molybdenum greatly improves resistance to pitting and crevice corrosion in chloride media, and these materials are regarded as competitive with copper-based alloys and titanium for sea water systems. However, in chloride media it is clear (13) that the segregation of alloying elements inherent in weld metal significantly reduces corrosion resistance (Fig 1). Following international study, it has become accepted practice for the materials to be welded with non-matching nickel based consumables selected because they offer a sufficiently high molybdenum content (8-10%) that, even in alloy depleted regions of the weld metal, the molybdenum level and associated pitting resistance remain comparable with those of the base steel. Unfortunately, the restriction that non-matching filler must be added complicates the welding operation. In particular, it means that autogenous welds are not permissible (unless perhaps a post-weld homogenising heat treatment is applied), and, moreover, that welding preparations and procedures must be designed so as to ensure adequate filler addition. This is most especially problematic for root runs in pipe: the joint preparation can incorporate a root gap to necessitate filler addition by the welder, but this renders it difficult to ensure maintenance of a completely inert backing gas because of atmospheric entrainment through the gap, with consequent root oxidation resulting from air entrapment and possible reduction in chloride pitting resistance.(14)

Use of commercial nickel-base fillers has enabled satisfactory service to be obtained with the new generation high molybdenum steels. However, this approach does lead to increased consumable costs, and bearing in mind that earlier steels (albeit of only say 4.5%Mo) have been employed successfully with nominally matching composition consumables, consideration could well be given to the general necessity for nickel-base weld metals, for example in chloride-free media. Certainly, manual metal arc (MMA) electrodes are available more closely matching the compositions of current high molybdenum austenitic steels, and further work is desirable to define more closely the environmental, material and welding conditions under which over-alloyed nickel-base fillers are essential.

Multi-phase Steels

Two types of dual phase steels have received attention, namely ferritic/martensitic and ferritic/austenitic alloys. The former variety of steel has been aimed primarily at structural applications, where mild or C-Mn steels will require continued corrosion protection (15) and is being produced in increasing tonnages (Table 1). While the parent steel has a mixed ferrite/martensite structure, the HAZ close to the fusion boundary inevitably transforms to ferrite at high temperature, with associated grain growth. Depending on the cooling rate, intergranular austenite is developed, which transforms to martensite on reaching room temperature (Fig 2), and hence the final HAZ microstructure (and, in principle, that of matching composition weld metal) is considerably coarser than is the case for parent material, and toughness is reduced to a greater or lesser extent. The level of toughness obtained is determined by the mean grain size, proportion of martensite and local yield strengths of the different microstructural constituents, and it is therefore highly dependent on the particular composition and transformation characteristics of the material and on the total weld thermal cycle (Fig 3). However, at present, insufficient data exist for reliable quantitative toughness prediction, and realistic procedural testing remains necessary for any given application to ensure that adequate resistance against unstable fracture is achieved.

Recent years have seen growing usage of ferritic/austenitic stainless steels. In large part, the drive has come from the oil industry, where the materials are seen as of application for line pipe and allied purposes, requiring resistance to acid gas corrosion. This interest has led to extensive research being carried out into the effect of welding on the properties of commercial alloys, most especially those containing about 22 and 25% chromium (UNS S 31803 and 32550 respectively). As with the ferritic/martensitic grades, the high temperature HAZ can reach the delta ferrite field and it is now well established that adequate austenite reformation must take place during cooling to achieve optimum properties.(16) This necessitates correct alloy design in terms of austenite and ferrite forming elements, and avoidance of low heat input welding with concomitant rapid cooling (Fig 4),(17) unless reheating by multipass deposition is employed. For a given composition, austenite contents in weld metal tend to be lower than in the HAZ, probably as a result of the coarser grain size and restricted grain boundary area limiting nucleation and growth of the austenite phase on cooling. As a result, it is normal practice for consumables for ferritic/austenitic stainless steels to be over-alloyed in nickel relative to parent material to increase the final deposit austenite proportion. For most purposes, some latitude is permissible on the weld area ferrite/austenite balance. This is fortunate, since conventional methods for predicting microstructural balance from composition of stainless steel (eg Schaeffler or DeLong diagrams) are not accurate for current duplex alloys; alternative predictive systems are emerging, both empirical (16-18) and theoretical,(19) but more work is required to allow for the effects of single and multipass welding procedural variables on the final ferrite and austenite contents.

For demanding applications, a number of problem areas remain. Weld area toughness is below that of the parent steel, and to some degree this loss must be regarded as inevitable even given high final austenite contents simply because of the coarser structure which is developed. Weld metal inclusion population is also important, properties of gas-shielded deposits tending to be better than those from flux processes,(20) as commonly the case with other alloys.

Moreover, welding reduces the corrosion resistance of ferritic/austenitic steels. As with toughness, it is seldom that the deterioration is of practical consequence, but the effect has led to difficulty in meeting material/welding specifications for offshore service. The reduction in properties stems from two main causes, viz precipitation of carbides, nitrides and intermetallic phases, and segregation of alloying elements between the ferrite and austenite. In the high temperature HAZ and weld metal, precipitation especially of nitrides is most pronounced in areas with a predominantly ferritic structure, and is thus mainly associated with low heat input, rapidly cooled welds.(21) However, loss of corrosion resistance can occur also in the lower temperature HAZ which has not undergone major phase change, whether from carbide/nitride or intermetallic formation; evidently a maximum heat input must be applied, but as yet quantitative limits have not been adequately defined (Fig 5).

Segregation however represents a more vexed problem. The partitioning of elements most strongly endowing pitting resistance (chromium, molybdenum and nitrogen) has been well studied, many investigators having shown concentration of chromium and molybdenum into the ferrite and nitrogen into austenite.(22) In principle, each phase is thereby enriched in one or other element promoting pitting resistance, but, bearing in mind the different diffusion rates and partition coefficients, there is no guarantee that susceptibility to pitting corrosion will be identical for the two microstructural constituents. This is further complicated in multipass welding where segregation occurs not only during the original high temperature cycle but also during both heating and cooling phases of subsequent weld runs with lower peak temperatures. Available information is contradictory and it has been shown that austenite developed during reheating is either more (22) or less (23) resistant to pitting attack than that formed during original cooling. Any effect of segregation, it will be recognised, will be manifest in both weld metal and HAZs. In the former region, it may be possible to achieve corrosion resistance exceeding base metal simply by over-alloying, but this palliative may not be applicable to higher alloy materials in which the precipitation of intermetallic phases would be greater.

Resistance to stress corrosion cracking (SCC) is also structure dependent, whether in chloride or H_2S media.(24) In both environmental regimes, resistance is decreased at high weld area ferrite levels, and again attainment of correct phase balance is essential if joint properties are to equal those of the base metal (Fig 6).

Concern has arisen regarding hydrogen-induced cracking of duplex weldments.(25) With predominantly ferritic welds, cracking can occur at fairly low hydrogen levels (Fig 7), but the risk decreases considerably at higher austenite contents, and, given proper control of phase balance, should not constitute a major practical problem.

WELDING PROCESS DEVELOPMENTS

General Comments

As with material changes there has largely been continued development of existing technology for making welds rather than a major change in process direction. Trends described previously have continued with, for example, increased use of the gas shielded process and of cored wire welding consumables (Table 2). At the same time, data are being generated on the application of the newer processes, with work on high power density fusion methods, laser (26,27) and electron beam welding.(28) The high joint completion rates offered are attractive, and such techniques are finding more widespread use by industry. Investigations have been carried out both to define the particular metallurgical consequences of the resultant rapid thermal cycles and to provide quantitative process information of direct practical utilisation.

Notwithstanding the introduction of advanced processes, the established arc welding methods remain dominant for stainless steels. This has entailed study of diverse aspects of behaviour, for example of flux variables (29) or the use of narrow gap welding for improved productivity.(30) One topic which has received particular attention is the phenomenon of variable penetration in view of the potential drawback this affords to the reliability of welds made by fully automatic welding systems.

Variable Penetration

For some years, it has been recognised that materials to the same specification originating from different heats can show markedly different penetration characteristics in autogenous welding.(31) This is especially the case with austenitic steels for which precision tungsten inert gas (TIG) welding is frequently carried out, and most work on the phenomenon has been aimed at this material/process combination. While it is clear that arc/metal interactions may be contributory, most investigators consider the problem to arise from differences in weld pool fluid flow, the differences stemming at least in part from surface tension gradients on the pool surface. Following Heiple and Roper and others,(32) it is expected that materials of increased sulphur or oxygen levels would give deeper and more consistent penetration, and for sulphur there is general agreement that this is the case (Fig 8).(33) Indeed, fabricators have specified minimum sulphur levels for materials where precision autogenous welding is required in order to avoid the risk of inadequately fused joints. For oxygen, the picture is less clear, but it is likely that the difference in the two elements arises from varying stability of relevant compounds in the molten weld pool. At peak pool temperatures, most sulphides will have dissolved so that the total sulphur content will exercise a direct influence on surface tension. In contrast, while this may also be the case for oxides of manganese and silicon, alumina or calcium oxide particles may persist to higher temperatures so that the net effect of oxygen depends on the deoxidation practice of the parent steel in terms of elements such as calcium, aluminium and possibly titanium.(33, 34)

The phenomenon has stimulated considerable effort on weld pool flow modelling.(35) Because of the complexity of the situation, attention has been paid primarily to static pools rather than a moving pool as would exist in normal joining practice, but understanding of the relative magnitudes of acting forces has increased significantly: certainly, surface tension is as important as buoyancy or Lorentz forces in causing pool flow (Fig 9).

From the practical standpoint, it remains difficult to give specific advice on avoidance of variable penetration. Increasing sulphur is beneficial but the general trend in stainless steel production is towards lower sulphur levels to obtain better corrosion and mechanical characteristics. From investigation of procedural variables, consistent penetration between materials showing extremes of behaviour cannot be achieved by control of individual parameters.(33) Accordingly effort continues on welding systems which monitor some facet of weld penetration with appropriate feedback control of the welding conditions.(36) In principle, greatest reliability can be achieved by back face monitoring but this may be difficult to employ in many practical situations. Thus, at present, if variable penetration is seen as a potential problem, it may be necessary to utilise direct testing on individual heats of steel to assess the penetration characteristics and specify optimum welding procedures.

SERVICE APPLICATION OF WELDED STAINLESS STEEL

Property data continue to be derived on welds in stainless steels with a number of objectives. First, such information is essential if newer materials are to be employed in welded fabrications. Secondly, in many areas, notably service at above normal ambient temperature,(37) greater precision of design is required, involving both long-term tests and studies with complex loading systems to reproduce the practical situation. Third, data are needed to define limits for welded stainless steels, and to optimise specific material or welding variables so that improved performance can be achieved.

High temperature property data are being gathered from a number of programmes, with, inter alia, effort being paid to improving the creep performance of conventional austenitic weld metals.(38–41) This has involved study of the effects of both composition and microstructure, with attention being paid to elucidating the various precipitation processes that occur in these fairly high alloy materials and the significance of such precipitation in terms of creep life. Work in the United States has shown (38,39) that small controlled additions of titanium, phosphorus and boron can markedly improve creep performance and in this regard, it is evident that flux minerals can significantly influence the levels of trace elements in weld metal and resulting properties. (29) Investigations a few years ago indicated that optimum creep ductility with austenitic grades is associated with the presence of a restricted ferrite content, say 4-8%. This work has been continued with evaluation of specific compositional effects on precipitation and decomposition of the ferrite by Farrar and his co-workers.(40,41) Mention must be made also of the international study of dissimilar metal joints between austenitic and low alloy ferritic steels for high temperature service. (42) Stress analyses and metallurgical investigation have been undertaken, nickel-base weld metals currently being preferred to minimise both strain gradients during thermal cycling and the consequences of carbon migration, and thus obtain improved joint life.

Although austenitic grades have been employed for low temperature welded plant for many years, it is not axiomatic that commercial consumables designed for general use will enable common code requirements for weld metal toughness to be attained. Published data have been reviewed by Siewert,(43) the adverse effects of ferrite and carbon, and the beneficial role of nickel being clearly identified, and, if low temperature properties are specified, consumable manufacturers will commonly supply consumables of slightly modified composition relative to normal production.

The coarse solidification structure developed in austenitic stainless steel weld metal poses particular problems for NDT, radiography sometimes being hindered by diffraction mottling, while attenuation causes appreciable difficulties for conventional ultrasonic testing. With improvements to probe design and optimisation of pulse frequency, some advance has been made in reliability of defect detection.(44,45) Further, work involving creep waves close to the material surface has been undertaken, for example in support of the use of duplex stainless steels for pipelines, and a considerable degree of success has been reported in detecting shallow surface flaws.(46) For duplex grades also, the application of acoustic emission for defect detection has been explored.(47)

WELDING FUME

From the occupational hygiene viewpoint, stainless steels pose particular problems because of the potential toxicity of welding fume, especially in terms of the possible formation of hexavalent chromium, so that close attention to ventilation in the working area is needed . The salient characteristics of different processes are now well understood, total fume emission rate (FER) being negligible with TIG and submerged arc welding, but considerably greater with the MMA and metal inert gas (MIG) processes (Fig 10). (48) Fume emission rate for a given MMA consumable tends to increase with increasing welding current: to some degree, this is true also for MIG welding, but Willingham and Hilton found a complex dependence of FER on MIG welding conditions.(49) Occupational exposure limits for CrVI are extremely low, and effort has been paid to examining the extent of CrVI formation for a range of welding processes and consumable types. Although the total amount of fume liberated may be similar for MMA and MIG welding, the fume from the former process contains substantially more hexavalent chromium (although MIG welding can produce significant quantities of ozone). (48) At present, the reason for the difference in fume composition for the two processes is not understood. It is not solely a result of chemical oxidation reactions in the arc, since studies by Carter (50) have indicated addition of conventional deoxidants to flux coatings to have negligible effect. Hexavalent chromium formation is influenced by the binder system employed, sodium and potassium apparently being deleterious, but omission of these elements leads to considerable difficulty in formulating binders for MMA coatings with fully acceptable operating properties. Further study in this area is required, together with more detailed evaluation of the behaviour of cored wire consumables in view of their increasing use by industry.

CONCLUDING REMARKS

Viewed overall, changes have taken place in welding stainless steels even since the 1984 conference. Considering materials application, note must be made of the use of high molybdenum grades for seawater service and of duplex stainless steels for linepipe purposes. In both cases, it will be recognised that achievement of satisfactory service has been critically dependent on the development of reliable welding procedures. In terms of the welding operation per se, the theoretical and practical effort paid to the phenomenon of variable penetration will be noted. More generally, understanding has increased on the usage of established and new materials, and weld processes. In the latter regard, the trend for increased use of continuous wire or fully automated welding systems in other areas of industrial fabrication can be identified also with stainless steels, and must be expected to continue in the future.

ACKNOWLEDGEMENTS

The author thanks his colleagues at The Welding Institute for advice and assistance in producing this paper. Particular thanks are given to the Research Member companies who supplied the data in Tables 1 and 2.

REFERENCES

1. Castro R and de Cadenet J J, "Welding Metallurgy of Stainless and Heat-Resisting Steels", pub by Cambridge University Press, 1974.

2. Gooch T G, Proc Conf "Stainless Steels '84", September 1987, Göteborg, Sweden, pub by Inst of Met, London, 1985, pp 249-261.

3. Alberry P J and Gooch D J, Welding & Met Fab, Nov/Dec 1985, 53, (8), pp 332-338 and Jan/Feb 1986, 54, (1), pp 33-34.

4. Thomas C R and Apps R L, Proc Conf "New Developments in Stainless Steel Technology", Sept 1984, Detroit, USA, pub by ASM, 1985, pp 351-379.

5. Krysiak K F, Weld J, April 1986, 65, (4), pp 37-41.

6. Bruemmer S M, Corrosion, Jan 1986, 42, (1), pp 27-35.

7. Chung P, Yoshitake A, Cragnolino G and McDonald D D, Corrosion, Mar 1985, 41, (3), pp 159-168.

8. Krause G T, Weld J, May 1986, 65, (5), pp 21-29.

9. Katayama S, Fujimoto T and Matsunawa A, Trans JWRI, July 1985, 14, (1), pp 123-138.

10. Kujanpaa V P, Karjalainen C P and Sikanen H A V, Weld J, June 1987, 66, (6), pp 155s-161s.

11. Lundin C D and Chou C P D, Weld J, April 1985, 64, (4), pp 113s-118s.

12. Thomas R D, Weld J, Dec 1984, 63, (12), pp 355s-368s.

13. Lindenmo M, vide ref 2. pp 262-270.

14. Kearns J R, Proc Conf "Corrosion '85", March 1985, Boston, USA, NACE, Paper 50.

15. Hoffman J P, Proc Conf "Inaugural International Conference on 3CR12", March 1984, Johannesburg, South Africa, Middelburg Steel & Alloys (Pty) Ltd, pp 60-81.

16. Gooch T G, Proc Conf, "Duplex Stainless Steels", Oct 1982, St Louis, USA, pub by ASM 1983, pp 573-602.

17. Noble D N and Gooch T G, to be presented at "Fabrication and Welding of Duplex Stainless Steels", ASM Materials Week '87, Oct 1987.

18. Miura M et al, Proc Conf, "Duplex Stainless Steel '86", Oct 1986, Nederlands Instituut voor Lastechniek, The Hague, The Netherlands, pp 319-325.

19. Hertzmann S, Roberts W and Lindenmo M, ibid, pp 257-267.

20. Perteneder E et al, ibid, pp 48-56.

21. Ume K et al, Proc Conf "Corrosion '86", March 1986, Houston, USA, NACE, Paper 155.

22. Ogawa T and Koseki T, Proc Conf "Third International Conference on Welding and

Performance of Pipelines", Nov 1986, London, UK, The Welding Institute, Paper 10.

23. Bower E N, Fielder J W and King K J, ibid, Paper 51.

24. Gooch T G, Welding in the World, 1986, 24, (7/8), pp 148-167.

25. Fekken U, van Nassau L and Verwey M, vide ref 16, pp 268-279.

26. Kimara S, Sugiyama S and Mizutame M, Proc "Fifth International Congress on Applications of Lasers and Electro-optics (ICALEO)", Arlington, USA, Nov 1986, pub by IFS (Publications) Ltd for Laser Institute of America, 1987, pp 89-96.

27. Kujanpaa V P and David S A, ibid, pp 63-69.

28. Tone S et al, Proc "Sixth International Cryogenic Engineering Materials Conference", Aug 1985, Cambridge, USA, pub by Plenum Press 1986, pp 89-96.

29. Marshall A W and Farrar J C M, vide ref 2, pp 271-285.

30. Renelt E, ZIS Mitteilungen, June 1982, 24, (6), pp 664-669.

31. Heiple C R, Burgardt P and Roper J R, Proc Conf "The Effects of Residual Impurity and Microalloying Elements on Weldability and Weld Properties", Nov 1983, London, UK, The Welding Institute, Paper 36.

32. Keene B J et al, ibid, Paper 45.

33. Leinon J, Acta Univ Ouluensis, Series C, Technica No 39, 1987, University of Oulu, Finland.

34. Binard J and Chabenat A, Soud et Techn Connexes, Jan-Feb 1985, 39, (1-2), pp 20-35.

35. Oreper G M and Szekely J, J Fluid Mech, 1984, 147, pp 53-79.

36. Fihey J-L, vide ref 29, Paper 48.

37. Garwood S J, ASTM STP 833, 1984, pp 333-359.

38. Klueh R L and Edmonds D P, Weld J, Jan 1986, 65, (1), pp 1s-7s.

39. Vitek J M and David S A, Weld J, Aug 1984, 63, (8), pp 246s-253s.

40. Farrar R A, vide ref 2, pp 336-342.

41. Farrar R A, J Mat Sci, Nov 1985, 20, (11), pp 4215-4231.

42. Various, Proc Seminar "Dissimilar Metal Welds In Fossil-Fired Boilers", Feb 1984, New Orleans, USA, EPRI CS-3623.

43. Siewert T A, Weld J, March 1986, 65, (3), pp 23-28.

44. Kapranos P A and Whittaker V N, Brit J NDT, Jan 1984, 26, (1), pp 15-19.

45. Erhard A et al, Schweissen und Schneiden, July 1984, 36, (7), pp 315-319.

46. van Nisselroij J J M, vide ref 18, pp 165-175.

47. Vincenzo D R and Bruno B, Proc "Eleventh World Conference on Nondestructive Testing", Nov 1985, Columbus, USA, ICNDT, pp 85-90.

48. Moreton, J, Smärs E A and Spiller K R, Met Con, Dec 1985, 17, (12), pp 794-798.

49. Willingham D C and Hilton D E, Weld & Met Fab, July 1986, 54, (5), pp 226-229.

50. Carter G J, unpublished work, The Welding Institute, 1986.

Table 1 **Sales of 13%Cr ferritic/martensitic stainless steels from a single supplier.**

Year	Tonnes
1981	3400
1982	6800
1983	7400
1984	9600
1985	12700
1986	14400

Table 2 **Change in welding consumable production by a single manufacturer**

Process	Relative consumption, %	
	1981	1986
MMA	61	51
MIG/MAG	30	35
Cored wire	5	10
Submerged arc	4	4

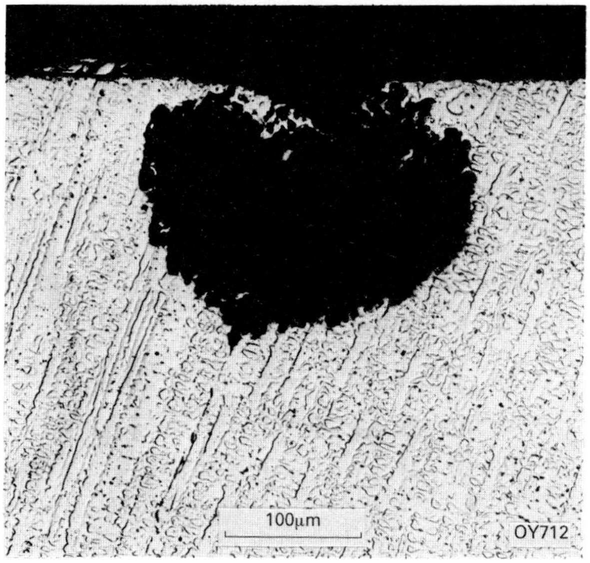

1 *Chloride pitting in high alloy fully austenitic stainless steel weld metal.*

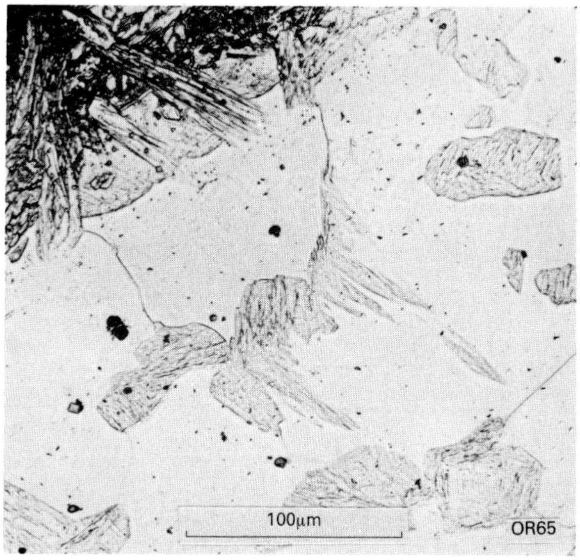

2 *HAZ microstructure of arc weld in ferritic/martensitic stainless steel, showing ferrite grain growth and formation of intergranular martensite on cooling.*

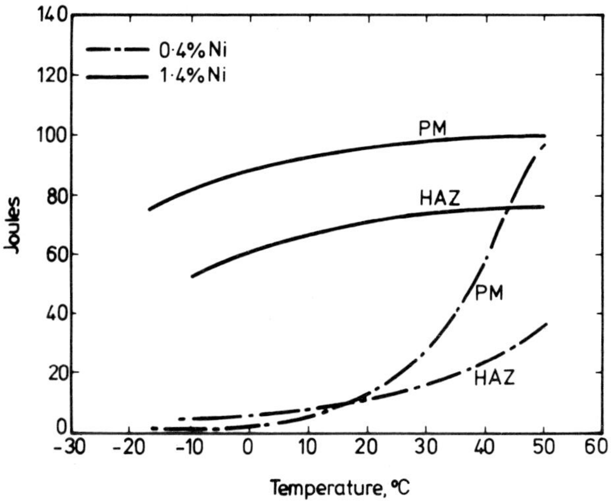

3 *Effects of welding and of varying nickel content on the Charpy impact toughness of low carbon 12%Cr steel: full size samples machined from 25mm plate.*[15]

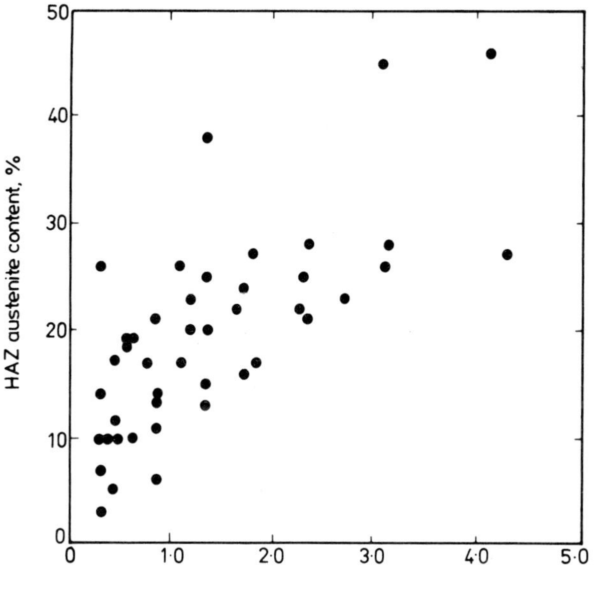

4 *HAZ austenite content versus arc energy for a range of S31803 steels varying in thickness from 13 to 20mm.*[17]

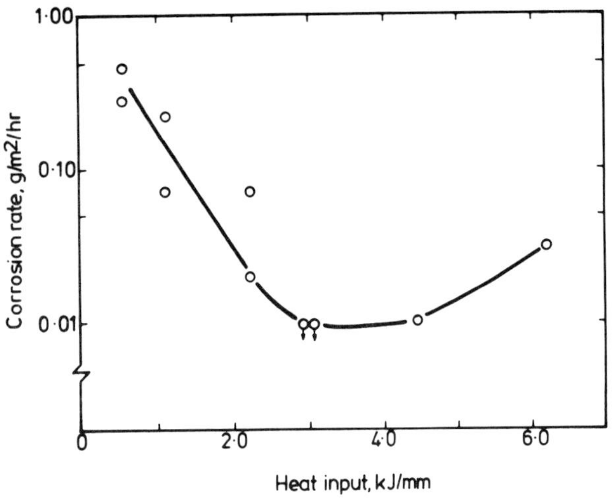

5 *Influence of heat input on pitting corrosion rate of welded 23%Cr-3%Mo-5%Ni-0.12%N duplex stainless steel, tested in 10%$FeCl_3$ at 22°C. At low heat input, attack was at the fusion boundary, but with high heat input was in the HAZ 3mm from the fusion boundary.*[21]

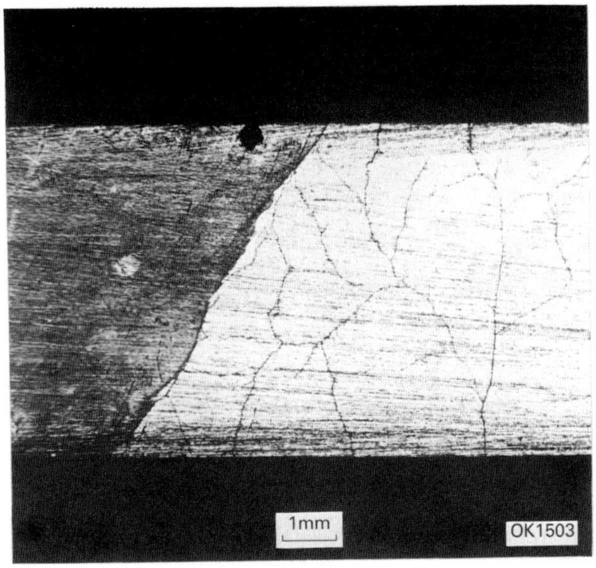

6 *Chloride SCC at TIG weld in ferritic/austenitic stainless steel made with Ni-base filler. HAZ contains c.35% austenite, and SCC has developed generally in parent material, with no preferential HAZ attack.*

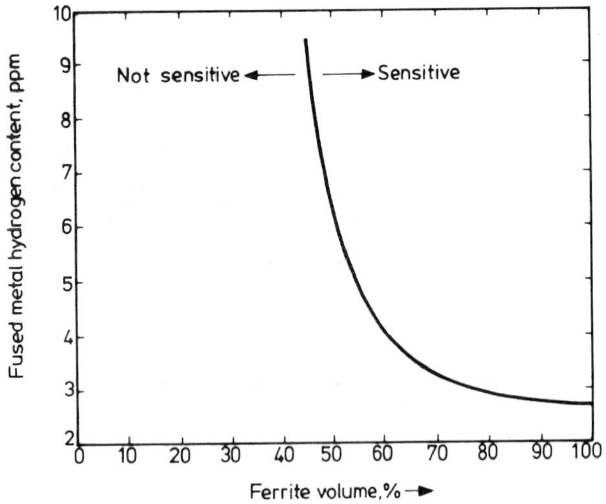

7 *Effect of MMA weld metal ferrite content and hydrogen level on hydrogen cracking behaviour of duplex steels. (After Ref.25.)*

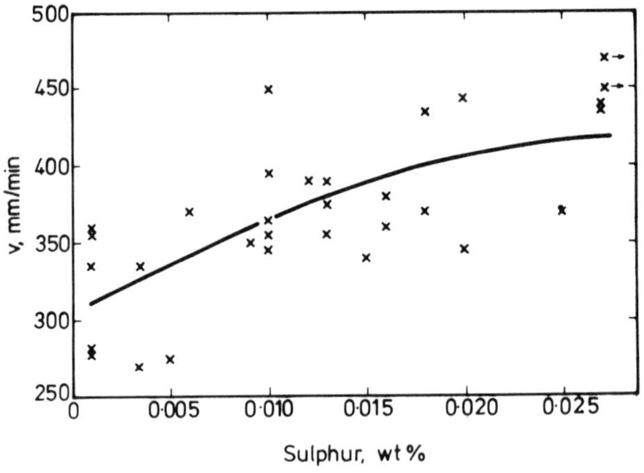

8 *Effect of sulphur content on maximum welding speed giving full penetration in TIG welds produced at 120A for a range of 3mm thick austenitic stainless steels.*[33]

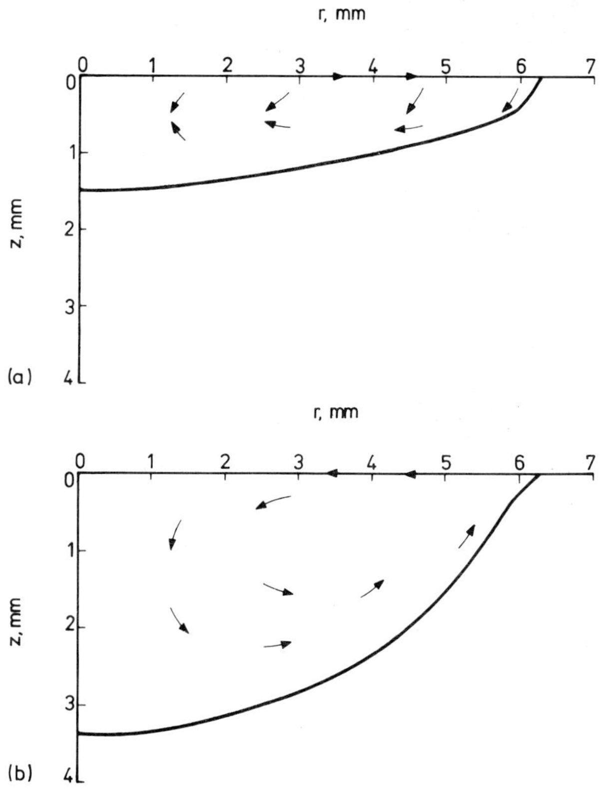

9 *Calculated semi-sections of TIG weld pool profiles, allowing for buoyancy, electromagnetically and surface tension driven flow:*

a) *Negative temperature gradient of surface tension: poor penetration;*
b) *Positive temperature gradient of surface tension: good penetration.*

Note: pool flow directions indicated. (After Ref.35.)

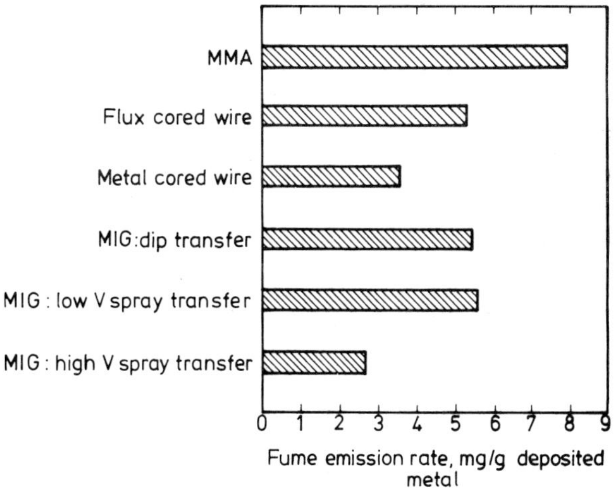

10 *Comparative fume emission rates for MMA and continuous wire welding: 316 consumables. (After Ref.48.)*

Variable penetration in automatic TIG welding of Type 304L austenitic stainless steel

D T Llewellyn, E N Bower and T Gladman

The authors are with the British Steel Corporation, Swinden Laboratories, Moorgate, Rotherham.

SYNOPSIS

A stationary weld test has been developed to assess the current required to achieve penetration of a given material thickness in a given time. The test has been shown to correlate well with penetration in bead on plate welds.

The test was used to evaluate the effects of composition on welding behaviour in both experimental and commercial casts of 304L. However, in spite of the fact that well defined effects for both aluminium and silicon were observed in the experimental steels, the effects are considered to be untypical of those that occur in commercial steels. This is attributed to the fact that the experimental steels contained oxygen contents of the order of 250 ppm, a level very much higher than that encountered in commercial material.

In the commercial steels, the results indicated a very major and beneficial effect of sulphur content. The effect was very similar in both aluminium-deoxidised tube material and silicon-deoxidised plates.

The effects are discussed in terms of Maragoni convection, and suggestions are made for reducing the variability of weld penetration.

1. INTRODUCTION

In recent years, attention has been drawn to a 'cast-to-cast' variability in the welding characteristics of steels and other alloys. The variability manifests itself by marked differences in the depth of penetration of a weld when welding conditions are substantially constant. Alternatively, there is a tendency for the weld to deflect to one side of the joint when steels from different casts of the same nominal specification are welded together. Problems of this kind were reported as early as 1957 when differences were noted in the welding of zirconium, melted in air or in an inert atmosphere.[1] In steels, it has been suggested that differences in penetration characteristics are due to minor variations in the residual or alloying element contents. The variable penetration has been attributed to aluminium,[2,3,4] titanium,[5] silicon,[6] selenium,[7] sulphur,[7,8] oxygen,[8] calcium[9] and manganese[4,6] contents.

The principal mechanisms by which minor changes in composition are thought to influence welding behaviour include arc effects[10] where readily ionised elements such as alkali and alkaline earth metals may change the arc characteristics, and weld pool surface tension effects, where surface segregated species may have pronounced effects.[7] However, in recent years, it has been reported that variable penetration characteristics have been observed in non-arc welding processes such as laser and electron beam welding,[11] adding support to the surface tension mechanism.

The main objective of the work was to evaluate the compositional factors that contribute to variable welding performance in Type 304L stainless steel.

2. EXPERIMENTAL TECHNIQUE

A penetration test was developed, involving a stationary TIG torch, in which a series of spot welds are carried out for a fixed period of time. The heat input of successive welds is increased progressively in order to define the power that will just penetrate through the plate. This procedure was found to be very reproducible and, in general, the critical currents for through penetration were found to be within ±0.5A in a series of tests on a given plate.

2.1 Experimental Heats

Tests were carried out initially on laboratory casts, which were rolled to 6 mm thickness, solution treated at 1050°C, air cooled and finally machined to 3 mm thick plate. The duration of arc in these early tests was 20s and other details of the welding conditions are shown in Table 1. During welding, the test coupons were clamped in a small vice with a self-centred jaw action, the jaws of the vice containing inserts with a 'V' shaped groove in order to minimise thermal contact with the specimen. A small tray, containing an argon feed, was inserted under the specimen to provide a backing gas but this was maintained about 1 mm beneath the specimen to prevent thermal contact. The welds were positioned towards the centre of the test coupon width at a fixed distance from one end. The earth lead of the welding set was clamped directly on to the end of the specimen, in order to produce an efficient, reproducible earth return path.

Single spot welds were carried out on the test coupons, the underside of the plate being examined for evidence of penetration before increasing or decreasing the current for subsequent welds. Having established that the reproducibility of the stationary penetration tests was high, it was also reassuring to observe good correlation with weld depths from bead on plate tests (Fig. 1).

2.2 Commercial Tubing (Al-Deoxidised)

The outside diameters of these tubes varied from 48 to 88 mm, with varying tube wall thicknesses. The tubes were turned on a lathe, both inside and outside, to produce a uniform wall thickness of 2 mm (± 0.05 mm). On selected tubes, the wall thickness was varied over a wider range in order to establish the effect of thickness on current to penetrate.

Whereas an arc duration of 20s had been adopted for the 3 mm laboratory materials, a 5s duration was used for the 2 mm commercial tube materials. The arc was struck on to the outside (convex) surface of the tubes and the welding conditions are detailed in Table 2.

Prior to undertaking the tests, the following burn-in procedures were adopted:

(i) The equipment was switched on to warm up for 1.5h.

(ii) Six spot welds were conducted on scrap material to burn in the electrodes.

(iii) The first spot weld on any test specimen was discarded.

Four tube samples were selected to determine the effects of wall thickness on the current to penetrate and the results obtained are shown in Fig. 2. The relationship obtained in tube No. 6 was 25.6 A/mm and this was used to correct some of the current to penetrate values in materials where the wall thickness deviated from 2.0 mm. However, the maximum correction applied was ± 1.3A corresponding to thickness variations of ± 0.05 mm.

2.3 Commercial Plate (Si-Deoxidised)

Following the work carried out on commercial, Al-deoxidised tubes, attempts were made to obtain samples of Si-deoxidised tubes. However, such materials were not available and therefore attention was turned to Si-deoxidised plate or strip samples.

The materials varied in thickness from 2 to 6.5 mm and some of the samples were in the hot rolled condition. All the samples were therefore solution treated at 1050°C and air cooled, and were ground on both sides to give a test specimen thickness of 2.0 mm. The test specimen thickness was therefore similar to that used for the tube samples and the duration of arc was maintained at 5s. The welding conditions were also the same as those adopted for the tube sample (Table 2) with the exception that the flow rate for the backing gas was reduced from 7 to 4 ft^3/min.

3. MATERIAL FOR INVESTIGATION

3.1 Experimental Heats

Because of the random variations that are likely to occur in commercial materials, it was considered initially that laboratory casts would offer the better opportunity to investigate the effects of specific elements. 50 kg melts were prepared which were split to produce four ingots, each of about 12 kg. Following the pouring of the first ingot, progressive additions of a particular element were made to the residual melt so as to achieve four levels of that element in a common base composition.

Two series of steels were prepared; the first involving variations in sulphur and aluminium contents, whilst the second series was used to examine the effects of silicon and manganese contents.

The chemical compositions of these experimental heats are shown in Table 3. Major problems were experienced in achieving the higher aluminium contents and in fact the highest level obtained was only 0.008% Al. This was attributed to the very high oxygen levels of these small laboratory casts (typically 250 ppm). Only a very limited range of aluminium content was therefore obtained.

3.2 Commercial Tubing (Al-Deoxidised)

A total of 23 commercial tube samples was obtained and the chemical compositions are given in Table 4. The aluminium contents were in the range 0.006 to 0.064%. The sulphur contents ranged from 20-140 ppm and the oxygen contents ranged from 20-70 ppm.

3.3 Commercial Plate (Si-Deoxidised)

Details of the compositions are included in Table 5. Variations in sulphur and oxygen contents were again observed covering similar ranges to those shown in tube material.

4. EXPERIMENTAL RESULTS

4.1 Experimental Heats

The results obtained on the S-Al series are shown in Figs. 3(a) and (b). The addition of aluminium resulted in a significant increase in the power to penetrate in each of the steels, except for those containing 0.004% S. There appeared to be no systematic effect of sulphur but there was a broad correlation between the current to penetrate and the width of the molten pool.

The results obtained on the Si-Mn series are shown in Figs. 4(a) and (b). There was a marked decrease in the current to penetrate in these low aluminium steels as the silicon content was increased up to a level of about 0.6%. The 1.0% Mn steels were the easiest to penetrate but the materials containing 1.2 to 1.8% Mn had similar characteristics. Apart from a delineation with respect to the lowest manganese content, the widths of the bead on the top surface were insensitive to changes in composition.

The variations in deoxidising elements were reflected in the composition and morphology of the non-metallic inclusions. The aluminium-free steels with moderate silicon contents contained elongated inclusions of the $(MnO, MgO)_2 SiO_2$ type together with chromium-galaxite inclusions. An increase in the silicon content in the aluminium-free steels resulted in the formation of more glassy inclusions which were more elongated. A reduction in the silicon content of the aluminium-free steels produced predominantly chromium-galaxite inclusions which were generally not elongated. The aluminium-bearing steels contained mixed Cr-Al galaxite and alumina inclusions which are not elongated.

Although the penetration characteristics observed in these laboratory casts reflected changes in deoxidation practice, it was considered that the data may well be untypical of commercial steels. In commercial steels, the oxygen content is of the order of 50 ppm compared with a range of 200/300 ppm in these laboratory-made steels. For this reason, the work was extended to cover commercial casts.

4.2 Commercial Tubes (Al-Deoxidised)

The results of the weld penetration tests are shown in Table 4. Regression analysis techniques were used to correlate the 'current to penetrate', I, with composition, microstructural features and non-metallic inclusion contents. A simple linear regression equation with sulphur content was found to explain 70% of the total variance:

$$I = 48 - 901 (\%S)$$

This correlation is indicated in Fig. 5. Although a slight improvement in the regression could be obtained by including a positive silicon factor, the statistical significance was marginal and the variance explained was only increased to 78%. There was no evidence of any effect of oxygen or oxide content.

4.3 Commercial Plate (Si-Deoxidised)

The results of the weld penetration tests are included in Table 5. Again, using regression techniques, the major effect on the current to penetrate was the sulphur content. The simple regression equation explained only 50% of the total variance, and was expressed by:

$$I = 47 - 545 (\%S)$$

The effect of sulphur content in the plate materials is also indicated in Fig. 5, along with the results obtained for the commercial tube material. It can be seen that the populations of the tube and plate materials show considerable overlap and indicate a common single effect of sulphur in both materials. Regression analysis of pooled data gave the following expression:

$$I = 47 - 686 (\%S)$$

\- which explained 60% of the total variance. No other factors were statistically significant at the 95% confidence level.

4.4 Variable Welding Conditions

Three plate materials were selected in order to study the effects of variation in plate thickness and arc duration on current to penetrate. The materials used for this study were plates 23, 4 and 24 (Table 5) which would be designated good, intermediate and poor respectively on the basis of the initial 5s duration tests. Tests were carried out on both 2 and 3 mm plate material and the current to penetrate was determined after durations of arc of 2, 5 and 10s. The results obtained are presented in Figs. 6(a) and (b) which illustrate that the relative ranking order is preserved regardless of arc duration or plate thickness. However, in the tests involving 2s duration, the difference between the good and bad casts was increased whereas the variability was reduced in tests of 10s duration. In the tests of 10s duration, there was also no significant difference between casts of good and intermediate penetration characteristics.

4.5 Weld Pool Scum

The three casts used to determine the effects of arc duration and sample thickness were also examined in relation to the composition of the weld pool scum. Scanning electron microscopy was used to provide a qualitative analysis of the surface scum from the energy dispersive analysis spectra. The scum contained a high density of non-metallic particles. The scum and particles within the scum appeared to be concentrated in the centre of the pool and to a smaller extent at the extreme edge of the pool. In all three samples, the vast majority of the particles and areas analysed were found to be essentially Al_2O_3. In some cases, small amount of CaO and TiO_2 were present in conjunction with Al_2O_3. Occasionally, very small amounts of Cr_2O_3, MnO, FeO were also detected. Only one MnS particle was detected in plate 4 and no sulphides were detected in plate 22. In plate 24, a few particles were detected which were rich in iron and sulphur.

4.6 Surface Tension Measurements

In view of the possible effects of surface tension, surface energy measurements were carried out on commercial plate material of good penetration (plate No. 22) and of poor penetration (plate No. 24). The measurements were carried out at the National Physical Laboratory using the levitating drop technique and the results are shown in Fig. 7. Although the results are sparse, they show reasonably good agreement with early studies.[12] Plate 24, which exhibited poor penetration, has a high surface energy at low temperature in the liquid region and a negative temperature dependence. Plate 22, which exhibited good penetration, shows a low surface energy at low temperatures and a positive temperature dependence.

5. DISCUSSION OF RESULTS

5.1 Weld Pool Forces and Marangoni Convection

The forces acting on a liquid weld pool arise from:

(i) Divergence of the electrical currents which will give rise to Lorentz forces. Increased Lorentz forces favour increased penetration.

(ii) Changes in surface energy across the temperature gradients of the weld pool surface which give rise to Marangoni convection.

(iii) Differences in density caused by the thermal gradients will give buoyancy forces.

Oreper, Eagar and Szekely[13] have calculated the effects of these forces on the fluid flow in a weld pool of fixed geometry and concluded that the surface tension effect (Marangoni convection) was of major importance in controlling the direction and velocity of the fluid. This conclusion was also confirmed by Kuo and Sun.[14] The direction of fluid flow can be changed by surface active impurities that cause a change from the normal negative temperature dependence of surface energy for pure materials to a positive temperature dependence when impurity elements segregate to the free surface at low temperatures in the liquid range, as indicated in Fig. 8. Surface flow of the metal occurs from regions of low surface energy to regions of high surface energy with an attendant viscous drag of underlying liquid metal. Thus, in low impurity steels, the surface liquid flow is from the centre of the liquid pool giving an upward axial flow and poor penetration. In steels with a high level of surface active impurity, the surface liquid flow is towards the centre of the liquid pool giving a downward axial flow and good penetration.

5.2 Effects of Sulphur and Oxygen

Sulphur and oxygen are surface active elements that are present in steels. Sulphur in steel is usually present as manganese sulphide, which undergoes substantial dissolution when the steel is melted.

The results obtained for the commercial tube and plate materials (Fig. 5) show a beneficial effect of sulphur content. This is directly in accord with earlier observations and also with the Heiple and Roper[7] mechanism. The surface energy measurements (Fig. 7) confirm that the high sulphur steel, showing good penetration behaviour, has a positive temperature dependence of surface energy, and that the low sulphur steel, showing poor penetration behaviour,has a negative temperature dependence of surface energy, also supporting the Marangoni convection effect.

The absence of any direct effect of sulphur in the experimental steels may be attributed to the dominant effect of the high oxygen contents (in the range 190 - 330 ppm compared with 20 -70 ppm in the commercial steels).

The effect of variations in oxygen content is complex. In the commercial steels, variations in the total oxygen content appeared to have no significant effect on the penetration characteristics. There was also no observable difference in weld penetration between the aluminium deoxidised tube and the silicon deoxidised plate materials at a given sulphur level. There should be significant differences in the free oxygen contents between the two types of steel and this would be expected to be important at low sulphur contents. Mills et al[12] considered the effects of deoxidants in ternary alloys on the levels of free oxygen content that would exist in the liquid metal at 1600°C in relation to the known effects of oxygen on the temperature dependence of the surface energy in pure iron. It was suggested that aluminium additions would reduce the free oxygen to a level that would give a negative temperature dependency (below 20 ppm) and therefore result in poor penetration. On the other hand, elements such as silicon and manganese would permit higher free oxygen contents to remain (without forming discrete oxides) to maintain positive temperature dependency and therefore good penetration characteristics. However, the simultaneous presence of both manganese and silicon is known to produce a synergistic effect on deoxidation, giving lower oxygen levels than can be attained by either element at similar concentration levels.

The oxidic inclusions in the present steels were complex, showing mixtures of chromium galaxite and manganese alumino silicates. Such oxides are undoubtedly stable at the lower temperatures of the liquid phase field, and the free oxygen contents could be significantly lower than the total oxygen levels indicated, particularly in the higher oxygen steels.

Under these circumstances, it is difficult to explain the qualitatively different effects of the deoxidants, aluminium and silicon, in the high oxygen experimental steels unless either the oxide content is high enough to reduce very significantly the deoxidant concentration of the steel matrix, or the kinetics of decomposition of the more stable aluminium bearing spinels are slower than those of the liquid silicates. It is also possible that the presence of high oxide contents may lead to arc instabilities when extensive surface oxides are present on the weld pool surface.

Alternatively, the dissolution of oxides which undoubtedly occurs in the hotter zones of the weld pool may result in high oxygen levels in the weld pool, but the alloying elements such as chromium, manganese and silicon may reduce the activity of the oxygen and therefore reduce the oxygen content at the segregated surface in the cooler regions of the weld pool. Under such circumstances, the absence of an effect of either oxygen content or deoxidants could be understood in the low oxygen commercial steels, but in experimental steels with far higher oxygen contents, the surface oxygen content could be high enough to produce reversals in Marangoni convection effect.

5.3 Practical Implications

In the commercial steels, the weld penetration characteristics were directly related to the sulphur content. In order to reduce variable penetration, the range of sulphur content would have to be restricted, and in order to facilitate penetration, very low levels of sulphur must be avoided. However, other properties e.g. corrosion may set restrictions on the upper level of sulphur that can be tolerated . The steelmaker is faced therefore with the difficult task of controlling the sulphur to a very narrow range.

Alternative procedures have been suggested, including the use of 500 - 1400 ppm SO_2 in the shielding gas.[15] Difficulties associated with this procedure however are the toxicity of SO_2 and discolouration of the weld. Other methods of introducing sulphur into the weld pool include treating the surfaces with a sulphur bearing compound, or inserting strips of sulphur doped metal into the weld preparation or gap region.[11] Both methods have given notable increases in weld penetration. However, the weld surface properties may not be comparable with those of the parent plate.

Control of the welding parameters to minimise the variability in penetration characteristics, as indicated by Fig. 6, can be successful but the conditions which minimise variability are also likely to lead to low productivity.

6. SUMMARY AND CONCLUSIONS

The effects of compositional variables on the weld penetration characteristics of Type 304L stainless steels have been investigated, using a newly developed spot welding test.

The major compositional variable that affects weld penetration in commercial Type 304L steels is the sulphur content. Decreasing the sulphur content changes the temperature dependence of the surface energy causing outward flow of the weld pool surface, and reducing weld penetration. The practical implications of this finding have been discussed in terms of control of sulphur content in the steel, and the possibilities for introducing sulphur into the weld pool. The same basic effect of sulphur was observed in both aluminium deoxidised and silicon deoxidised materials, and neither the deoxidation practice nor the oxygen level of the steel appeared to influence the weld penetration characteristics.

In experimental steels, containing much higher oxygen levels, small additions of aluminium were found to produce small increases in the current to penetrate, whilst silicon additions reduced the current required. The effects of the deoxidants may have been obscured by the high oxygen levels in these experimental steels, although the reasons for the specific effects of aluminium and silicon are not clear. There was no observable effect of sulphur content in these high oxygen steels.

Acknowledgements

The authors would like to thank BNF plc, Babcock Power, the CEGB and BSC Stainless for funding this work. The authors are grateful to Dr. R. Baker. Director of Research, British Steel Corporation, for permission to publish this paper.

References

1. Ludvig H.C. Welding J. 1957, 36, pp 335s - 341s

2. Savage W.F., Nippes E.F., and Goodwin G.M. Welding J.

3. Bennett W.S. and Mills G.S., Welding J. 1974, 53, pp 548s-553s

4. Metcalfe J.C. and Quigley M.B.C., Welding J. 1977, 56, pp 133-139s

5. Patterson R.A., Welding J. 1978, 57, pp 383s - 386s

6. Oyler G.W., R.A. Matuszesk, and Garr C.R., Welding J. 1967, 46, pp 1006-1011

7. Heiple C.R. and Roper J.R., Welding J. 1981, 60, pp 143s-145s

8 Savage W.F., Lundin, C.D. , and Goodwin G., Welding J. 1968, 47 pp 313s, 322s and 336s

9. Savitskii M.M., and Leskov G.I., Automatic Welding, 1980, 33, pp 11-16

10. Dunn, G.J., Allemand and Eagar T.W., Met. Trans A, 1986, 17A, pp 1851-1863

11. Heiple C.R., Roper J.R., Stagner R.T. and Aden R.J., Welding J. 1983, 62, pp 72s-77s

12. Mills K.C., Keene B,J., Brooks R.F. and Husanya, Centenary Conference, Strathclyde University, ed. Baker T.N. 1984

13. Oreper G.M., Eagar T.W., and Szekely J. Welding J. 1983, 62, pp 307s-312s

14. Kuo S. and Sun D.K. Met. Trans.A. 1985, 16A, pp 203-212

15. Heiple C.R. and Bungardt P. Welding J. 1985, 64, pp 159s-162s

Table 1

Welding Conditions for Experimental Steels

Torch shielding gas	= 99.999% Argon - 22 cu ft/min (commence 1.5 s prior to weld)
Backing gas	= 99.999% Argon - 4 cu ft/min (constant flow)
Temperature of vice jaws prior to weld	= <27 °C
Angle of point on electrode	= 20° (2% thoriated tungsten)
Arc gap	= 2.5 mm
Duration of Arc	= 20 s nominal (20.7 s actual)
Welding current (dc)	= various to assess level at which fusion of underside just occurred

Table 2

Welding Conditions for Commercial Tube Material

Torch shielding gas	99.999% Argon flow rate - 22 ft^3/min start 1.5 s prior to weld stop 7.5 s after weld
Backing Gas	99.999% Argon flow rate - 7 ft^3/min start ~60 s prior to weld stop 7.5 s after weld
Electrode	2% thoriated tungsten diameter - 2.38 mm tip included angle - 20° stick out - 6 mm
Arc	gap - 2.5 mm duration - 5 s
Temperature of tube	<27 °C (typically 20 to 25 °C)
Welding current	various (in range 30 to 55A) to assess level at which fusion of underside just occurred

Table 3

Chemical Compositions of Experimental Steels

Sample	Nominal Aim Composition		C	Si	Mn	P	S	Cr	Mo	Ni	Al	N	O
	%S	%Al											
80A	0.003	0.005	0.012	0.32	1.58	0.024	0.004	17.8	0.31	9.32	<0.005	0.024	0.032
B	0.003	0.010	0.012	0.32	1.55	0.024	0.004	17.8	0.32	9.32	0.005	0.024	0.030
C	0.003	0.020	0.012	0.32	1.54	0.024	0.004	17.9	0.32	9.37	0.007	0.024	0.031
61A	0.007	0.005	0.013	0.44	1.53	0.025	0.006	18.0	0.29	9.29	<0.005	0.026	0.020
B	0.007	0.010	0.013	0.45	1.57	0.025	0.006	18.0	0.29	9.30	0.006	0.027	0.024
C	0.007	0.020	0.013	0.45	1.57	0.025	0.006	18.1	0.29	9.36	0.0065	0.027	0.022
62A	0.010	0.005	0.013	0.45	1.57	0.024	0.010	18.2	0.31	9.30	<0.005	0.025	0.020
B	0.010	0.010	0.014	0.45	1.57	0.025	0.010	18.3	0.31	9.36	0.0055	0.025	0.022
C	0.010	0.020	0.013	0.45	1.57	0.024	0.010	18.2	0.31	9.28	0.008	0.025	0.018
63A	0.015	0.005	0.014	0.41	1.54	0.025	0.016	18.1	0.28	9.25	<0.005	0.024	0.019
B	0.015	0.010	0.013	0.42	1.53	0.024	0.016	18.1	0.28	9.23	0.0065	0.024	0.019
C	0.015	0.020	0.014	0.42	1.54	0.026	0.016	18.0	0.28	9.22	0.006	0.025	0.019
	%Mn	%Si											
78A	1.0	0.1	0.018	0.28	1.00	0.022	0.006	17.9	0.31	9.20	<0.005	0.027	0.023
B	1.0	0.3	0.015	0.48	0.98	0.023	0.007	17.9	0.31	9.12	<0.005	0.027	0.023
C	1.0	0.5	0.017	0.67	1.01	0.023	0.006	17.9	0.31	9.19	<0.005	0.026	0.023
D	1.0	0.7	0.016	0.86	0.99	0.023	0.005	17.9	0.31	9.12	<0.005	0.027	0.025
58A	1.3	0.1	0.011	0.25	1.23	0.024	0.006	17.8	0.31	9.32	<0.005	0.021	0.034
B	1.3	0.3	0.013	0.44	1.22	0.025	0.006	17.8	0.31	9.31	<0.005	0.021	0.030
C	1.3	0.5	0.017	0.63	1.22	0.024	0.006	17.8	0.31	9.30	<0.005	0.022	0.031
D	1.3	0.7	0.017	0.77	1.23	0.024	0.007	17.8	0.31	9.30	<0.005	0.022	0.030
59A	1.5	0.1	0.011	0.25	1.51	0.024	0.006	18.1	0.31	9.26	<0.005	0.023	0.031
B	1.5	0.3	0.011	0.45	1.51	0.024	0.005	18.1	0.31	9.24	<0.005	0.024	0.028
C	1.5	0.5	0.013	0.65	1.51	0.023	0.005	18.1	0.31	9.26	<0.005	0.023	0.028
D	1.5	0.7	0.016	0.79	1.50	0.024	0.005	18.0	0.31	9.21	<0.005	0.024	0.032
79A	1.7	0.1	0.011	0.24	1.85	0.025	0.005	18.0	0.29	9.16	<0.005	0.020	0.030
B	1.7	0.3	0.010	0.43	1.84	0.025	0.006	18.0	0.29	9.17	<0.005	0.020	0.028
C	1.7	0.5	0.013	0.63	1.85	0.026	0.006	18.1	0.29	9.24	<0.005	0.019	0.031
D	1.7	0.7	0.016	0.82	1.84	0.026	0.006	18.0	0.29	9.18	<0.005	0.020	0.027

Table 4

Chemical Compositions and Weld Test Results of Commercial Tube

Identity	Tube Dia. mm	Chemical Analysis Wt.%											Current to Penetrate (Amps.)
		C	Si	Mn	P	S	Cr	Mo	Ni	Al	N	O	
1	60	0.020	0.40	1.23	0.010	0.014	18.68	0.02	10.40	0.006	0.029	0.0065	36.7
2	48	0.016	0.38	1.23	0.011	0.013	18.68	0.01	10.32	0.006	0.027	0.0075	36.3
3	60	0.018	0.43	1.66	0.024	0.004	18.46	0.12	10.42	0.012	0.038	0.0035	42.7
4	48	0.019	0.50	1.60	0.023	0.005	18.78	0.15	10.14	0.009	0.063	0.0035	44.0
5	48	0.017	0.64	2.00	0.025	0.008	18.76	0.29	9.72	0.064	0.068	0.0050	42.2
6	48	0.019	0.61	1.82	0.027	0.008	18.57	0.26	9.84	0.060	0.074	0.0060	44.5
7	60	0.021	0.55	1.70	0.030	0.007	18.53	0.28	9.72	0.046	0.058	0.0060	44.5
8	48	0.022	0.54	1.72	0.029	0.008	18.54	0.27	9.68	0.048	0.057	0.0050	41.5
9	60	0.026	0.51	1.23	0.027	0.002	18.52	0.23	9.92	0.023	0.037	0.0030	46.2
10	48	0.026	0.52	1.21	0.027	0.002	18.56	0.23	9.92	0.023	0.032	0.0035	47.7
11	48	0.029	0.35	1.44	0.024	0.012	18.72	0.46	10.20	0.037	0.037	0.0025	36.2
12	48	0.025	0.42	1.38	0.026	0.007	18.62	0.27	10.36	0.028	0.037	0.0025	38.9
13	88	0.026	0.40	1.58	0.023	0.003	18.20	0.24	10.60	0.052	0.032	0.0015	46.7
14	60	0.026	0.40	1.58	0.023	0.003	18.20	0.24	10.60	0.052	0.032	0.0015	45.7
15	88	0.026	0.32	1.68	0.028	0.008	18.36	0.29	10.29	0.028	0.043	0.0020	38.7
16	48	0.016	0.42	1.54	0.023	0.003	18.72	0.26	10.16	0.014	0.045	0.0030	44.5
17	60	0.018	0.42	1.53	0.027	0.003	18.56	0.24	10.32	0.017	0.043	0.0030	47.5
18	60	0.022	0.47	1.45	0.023	0.008	18.48	0.04	10.44	0.040	0.052	0.0035	36.2
19	60	0.019	0.43	1.42	0.025	0.002	18.24	0.36	10.20	0.014	0.036	0.0025	44.7
20	88	0.026	0.50	1.69	0.010	0.007	18.89	0.04	10.20	0.019	0.056	0.0025	43.3
21	48	0.016	0.38	1.52	0.025	0.004	18.52	0.38	10.32	0.035	0.039	0.0045	40.5
22	48	0.030	0.39	1.47	0.021	0.009	18.56	0.22	10.36	0.029	0.050	0.0035	38.5
23	60	0.022	0.44	1.50	0.024	0.005	18.64	0.20	10.24	0.033	0.038	0.0030	45.1

Table 5

Weld Penetration Results and Weld Test Results

of Commercial Plate

Identity	Chemical Analysis, Wt.%											Mean Current to Penetrate (Amps.)
	C	Si	Mn	P	S	Cr	Mo	Ni	Al	N	O	
1	0.029	0.35	1.34	0.026	0.004	18.1	0.28	9.21	<0.005	0.035	0.0054	42.4
2	0.021	0.42	1.49	0.022	0.007	18.3	0.25	9.19	<0.005	0.045	0.0033	43.0
3	0.020	0.30	1.38	0.027	0.013	18.2	0.26	9.22	<0.005	0.051	0.0060	38.9
4	0.017	0.24	1.51	0.027	0.004	18.5	0.28	9.11	<0.005	0.038	0.0036	42.6
5	0.021	0.38	1.49	0.009	0.009	18.9	<0.02	9.79	<0.005	0.023	0.0032	41.4
6	0.012	0.31	1.52	0.010	0.009	18.8	0.02	9.63	<0.005	0.025	0.0052	38.6
7	0.019	0.38	1.45	0.011	0.013	18.6	0.02	9.72	<0.005	0.056	0.0055	38.3
8	0.026	0.32	1.42	0.009	0.011	18.7	<0.02	9.61	<0.005	0.054	0.0054	42.2
9	0.018	0.31	1.51	0.009	0.012	18.6	<0.02	9.62	<0.005	0.057	0.0057	38.4
10	0.016	0.38	1.45	0.009	0.009	18.8	<0.02	9.71	<0.005	0.026	0.0066	42.0
11	0.023	0.47	1.31	0.028	0.008	18.2	0.31	9.16	<0.005	0.045	0.0064	42.6
17	0.024	0.46	1.47	0.026	0.004	18.0	0.33	9.31	<0.005	0.041	0.0035	47.1
18	0.021	0.36	1.45	0.013	0.012	18.9	<0.02	9.85	<0.005	0.033	0.0046	43.7
20	0.020	0.38	1.36	0.025	0.002	18.2	0.36	8.97	<0.005	0.034	0.0026	45.0
22	0.020	0.41	1.45	0.024	0.012	18.2	0.33	9.01	<0.005	0.035	0.0026	39.6
23	0.019	0.41	1.48	0.027	0.012	18.5	0.34	9.10	<0.005	0.035	0.0026	39.5
24*	0.049	0.35	1.49	0.027	0.004	18.2	0.26	8.68	<0.005	0.040	0.0043	47.0

*Type 304

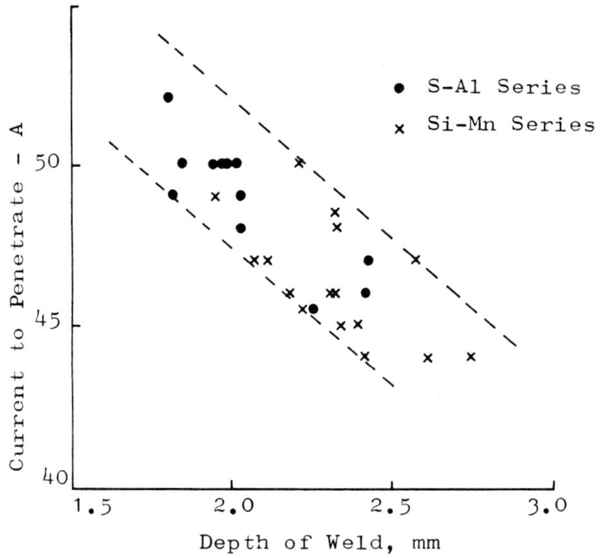

Fig. 1 Correlation between Current to Penetrate and Bead on Plate Tests

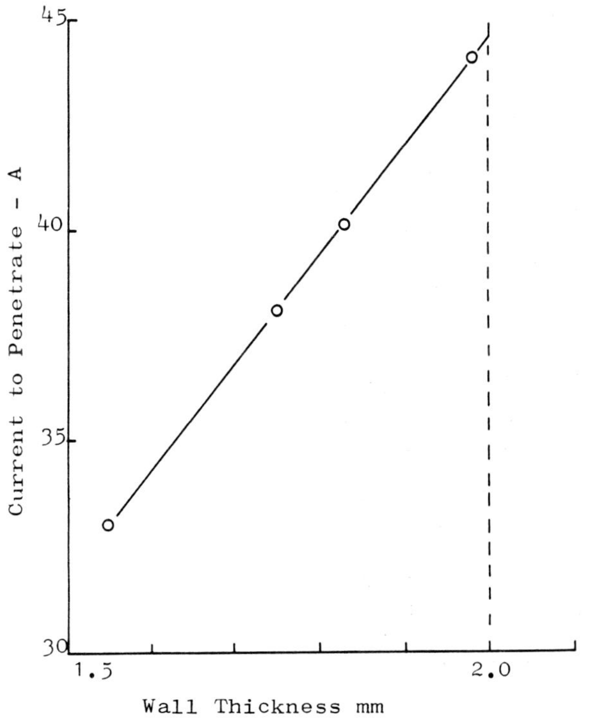

Fig. 2 Effect of Tube Wall Thickness on Current to Penetrate

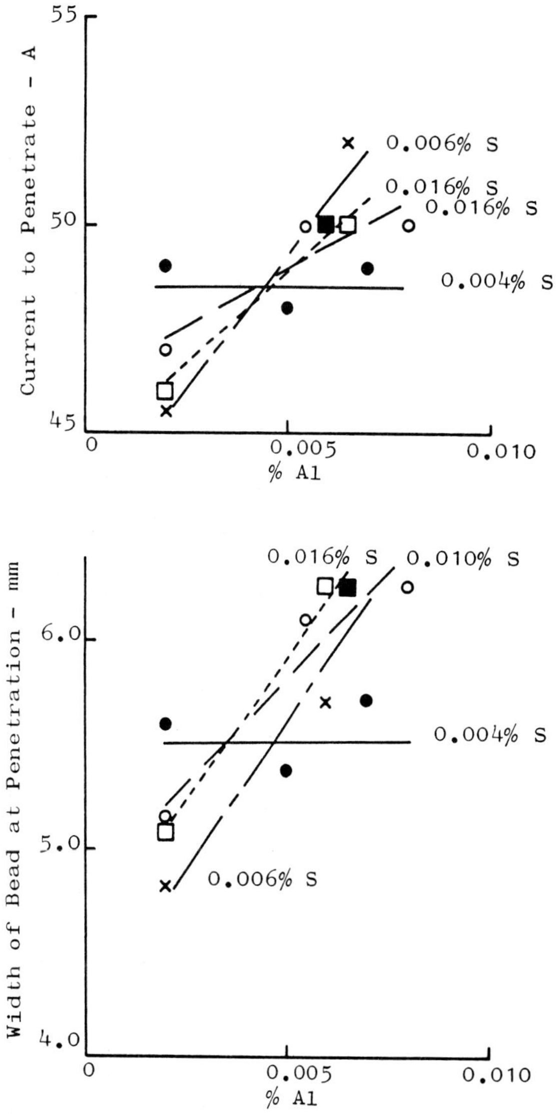

Fig. 3 Effect of Sulphur and Aluminium in Experimental Steels

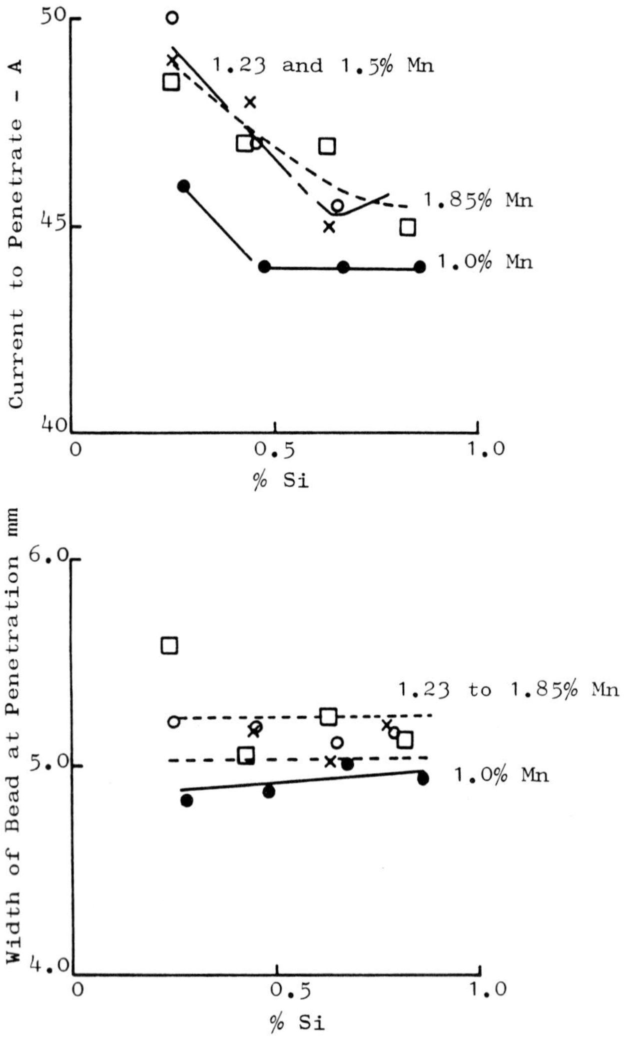

Fig. 4 Effect of Silicon and Manganese in Experimental Steels

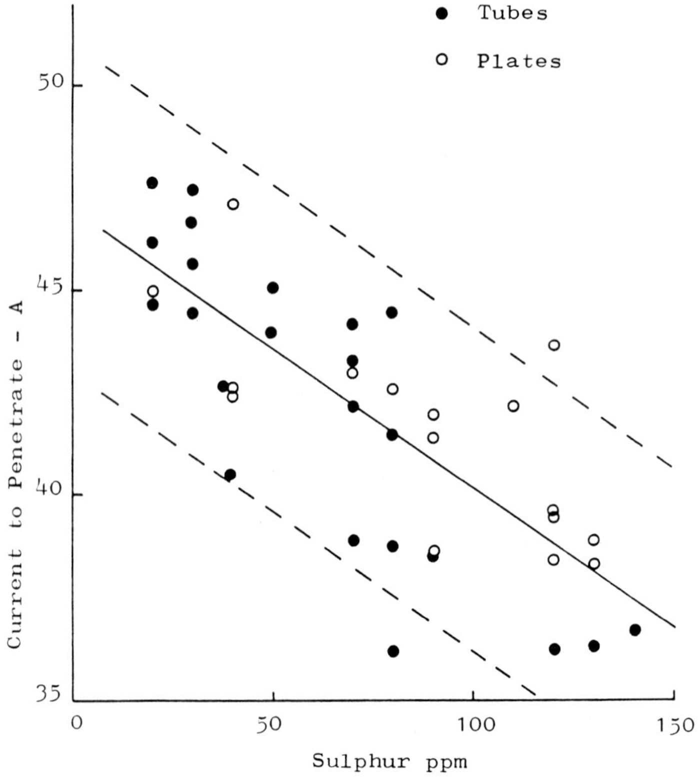

Fig. 5

Effect of Sulphur in Commercial Tube and Plate

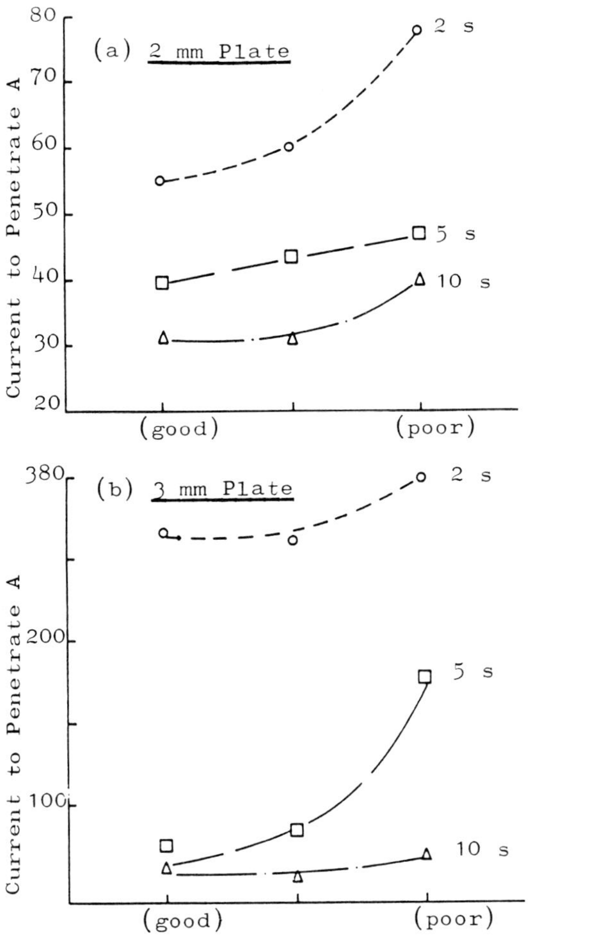

Fig. 6 Variation in Current to Penetrate with Welding Conditions

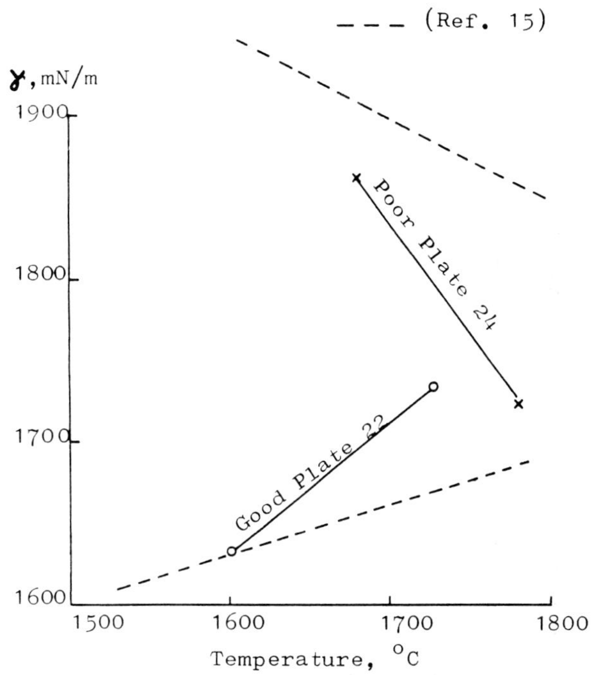

Fig. 7 Surface Energy (γ) v Temperature for Casts Showing Good and Poor Penetration Characteristics

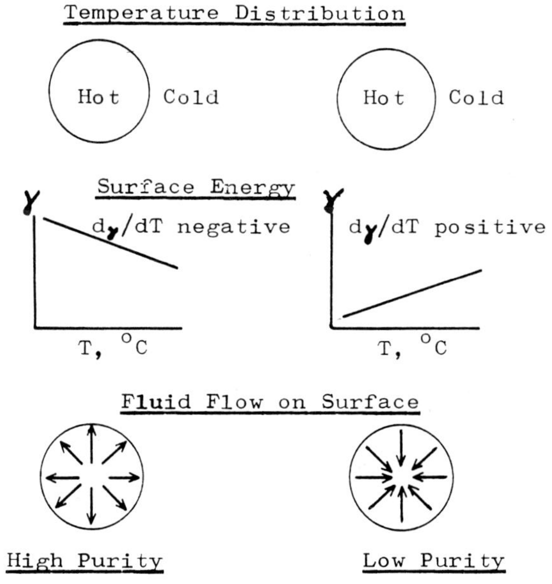

Fig. 8 Marangoni Convection

Effect of shielding gas on variable weld penetration

J I Leinonen

Dr. Leinonen is in the Department of Mechanical Engineering at the University of Oulu, Oulu, Finland.

SYNOPSIS

Considerable cast-to-cast variations in weld penetration characteristics were found to occur in mechanized autogenous TIG welding of three austenitic stainless steel sheets of 2 mm thickness. The maximum welding speed still giving complete penetration was 74 % higher in the best weldable test steel than in the poorest one at a welding current of 120 A using argon as the shielding gas. Hydrogen additions to the argon shielding gas of up to 16.5 % and helium additions of up to 80 % increased maximum welding speeds by as much as 160 % and 90 %, respectively, but could not eliminate cast-to-cast variations.

INTRODUCTION

Cast-to-cast variations in pool penetration during TIG welding of austenitic stainless steels have been reported over the last twenty years.[1-11] The occasional occurrence of variations in penetration is especially harmful in the case of mechanized and automatic welding, because the welding parameters have to be altered accordingly.

Wide, shallow weld penetration means poor weldability, and low welding speeds and high currents have to be used to achieve complete penetration in sheet welding. According to Leinonen and Karjalainen,[5] a high sulphur content in stainless steel sheets improves TIG weldability, as measured in terms of the maximum welding speed still giving complete penetration. Improvements in weldability due to increases in sulphur content have also been reported by other researchers in recent years.[4,6-10] On the other hand, calcium in the form of CaO in non-metallic inclusions is one cause of wide, shallow weld penetration.[5]

The best-known model for controlling the weld fusion zone geometry, based principally on Marangoni convection, is that presented by Heiple and Roper.[6] Surface active impurity elements (Se,S,O_2) are assumed to produce a positive surface tension temperature coefficient on the weld pool surface, resulting in an inward fluid flow pattern with greater heat transport to the bottom of the pool and thus deep penetration. A negative coefficient results in an outward fluid flow pattern and a wide, shallow weld pool.

Many recent investigations support the model based on Marangoni convection.[7-10] On the other hand, Leinonen and Karjalainen [5] suggest that an absolutely clean stainless steel is adequately weldable, which is contrary to the principles of Marangoni convection. This arises from the observation of satisfactory weldability in a steel with very low sulphur content and its oxygen in the form of stable Al_2O_3.

Most investigations regarding variable TIG weld penetration have been performed using argon as the shielding gas, but many researchers have also used $Ar+H_2$ or Ar+He gas mixtures.[1-3,7,9,10] In spite of the improvements in weldability achieved by adding helium or hydrogen to the argon, cast-to-cast variations do not completely disappear under such conditions. On the contrary, 0.3 % CO_2 in the argon shielding gas has been shown to eliminate variable penetration by increasing the depth of penetration more strongly in a poorly weldable steel than in an adequately weldable one.[11]

In this investigation the effects of argon, argon-hydrogen and argon-helium shielding gases on variable weld penetration in three austenitic stainless steel sheets were examined in laboratory welding tests.

EXPERIMENTAL PROCEDURE

The test materials were chosen from austenitic stainless steel sheets on the basis of their variable penetration characteristics. The compositions of the three steel sheets are presented in Table 1.

A mechanized single electrode TIG welding method was used (Table 2). The size of the test piece was nominally 50 by

250 mm, and the weld bead was 200 mm long. The penetration characteristics of a steel sheet were measured in terms of the maximum welding speed still giving complete penetration, and also in terms of the D/W ratio. The trial welding speed was reduced in 10 mm/min steps at speeds below 400 mm/min to determine the maximum welding speed giving complete penetration. The step was 20 mm/min at speeds of 400-800 mm/min and 30 mm/min at higher speeds.

RESULTS

The maximum welding speeds of the test steels varied from 270 to 470 mm/min using argon as the shielding gas (Fig. 1), and the D/W ratio from 0.28 to 0.53 (Table 3).

Additions of hydrogen gave significant improvements in the maximum welding speed, as shown in Fig. 1. An addition of 16.5 % increased the welding speeds of all three test steels by about 160 %, but the relative difference in welding speed between Steels 3 and 1 increased at the same time from 74 to 80 %.

The welding speeds of the test steels also increased markedly upon the addition of helium to the argon shielding gas, as shown in Fig. 2. The relative difference in welding speed between Steels 3 and 1 was reduced to 45 % when using Ar+80 % He as the shielding gas, but it was not appreciably affected by lesser additions of helium.

The D/W ratio increased to some extent when hydrogen or helium was added to the argon (Table 3). Weld pool profiles obtained for the test steels using Ar, Ar+16.5 % H_2 and Ar+80 % He as the shielding gases are presented in Fig. 3.

The arc voltage increased with additions of hydrogen or helium, resulting in a rise in arc power. The voltage was 11.5 V with argon shielding gas and 12.5-16.0 V with the various admixtures of hydrogen or helium.

DISCUSSION

Cast-to-cast variation in weld penetration is one of the few serious problems still associated with the TIG welding of austenitic stainless steels. Variable penetration has most often been studied using argon as the shielding gas, but the addition of hydrogen or helium to the argon is an effective way to improve the productivity of welding. The present systematic study on the effects of shielding gas mixtures on variable weld penetration is therefore justified.

Additions of hydrogen and helium to the argon raise the maximum welding speeds of the test steels markedly, as shown in Figs 1 and 2, the extent of this effect depending largely on the change in welding power due to the higher arc voltage. Another major factor raising the welding speed is the higher D/W ratio, i.e. the narrower weld pool profile (Fig.3).

A decrease in weld defects, i.e. centre cavities and undercuts, can also be achieved by using hydrogen or helium in the shielding gas, since this allows the welding current and cross-sectional area of the weld bead to be reduced.[12]

The tendency for the welding speed to increase seems to continue at least up to a hydrogen content of 16.5 %, and Willgoss [13] advances the view that the beneficial effects of hydrogen on TIG weld penetration are completely realized at a level of 15 %. The exact mechanism by which hydrogen leads to a deeper weld pool is unknown.

The cast-to-cast variations in maximum welding speed between the test steels were not reduced by hydrogen additions of 5 to 16.5 %, but rather the contrary effect was noted. The effects of residual elements (Ca,S), which mainly control variable penetration, remain the same despite hydrogen additions.

A shielding gas containing 80 % helium resulted in a reduction in the variation in welding speed and D/W ratio. The D/W ratio of Steel 1, of poor weldability, increased particularly clearly, because of the narrower weld bead width (Table 3). Similar results have been reported by Ludwig [2] and Bennett and Mills [3], whereas Burgardt and Heiple [10] found the D/W ratio of a poorly weldable steel to decrease with increasing additions of helium to argon. The cause of this discrepancy remains unknown.

Tsai and Eagar [14] show that an Ar+50 % He gas mixture produces a more concentrated heat source with a greater peak intensity and a smaller diameter at an arc length of 2 mm than argon does. It therefore seems plausible that the rather narrow weld bead in Steel 1 with Ar+80 % He shielding gas and the reduction in variation results from a more concentrated heat source on the weld pool.

The use of hydrogen or helium in the argon shielding gas can be recommended for increasing welding speed during TIG welding. The beneficial effects of these gas additions are significant, and no disadvantages, e.g. porosity, were found during the present welding tests. Hydrogen was found to be more effective for increasing welding speed being therefore especially recommended.

Since the variable penetration did not disappear upon modifying the shielding gas, the basic causes of variable penetration evidently have to be removed at the stage of steel production. A relatively high sulphur content (≥0.02 %), the avoidance of CaO in non-metallic inclusions in the parent material and a relatively high oxygen content (≥0.008 %) in the form of improving oxides (such as SiO_2, TiO_2, Cr_2O_3 and MnO) are recommended ways for achieving adequate, homogeneous weldability in austenitic stainless steel casts.

CONCLUSIONS

The effect of the shielding gas on variable TIG weld penetration was investigated here in three austenitic stainless steel sheets of 2 mm thickness.

Bead-on-plate welds were produced autogenously using single electrode TIG welding at a welding current of 120 A.

The maximum welding speed still giving complete penetration was 74 % higher in the best weldable steel than in the poorest steel, and the difference between their D/W ratios was greater than 80 % when argon shielding gas was used.

Hydrogen additions to the argon shielding gas of up to 16.5 % and helium additions of up to 80 % increased maximum welding speeds by as much as 160 and 90 %, respectively. The relative cast-to-cast variations in maximum welding speed and D/W were not eliminated, however, the only appreciable decrease in relative variation being found with the Ar+80 % He gas mixture.

ACKNOWLEDGEMENTS

The author wishes to thank Prof. L. P. Karjalainen for his advice. Support from the Academy of Finland is gratefully acknowledged. The author also thanks Mr. M. Hicks, M.A. for revising the English of the manuscript.

REFERENCES

1 Oyler, G.W., Matuszesk, R.A. & Garr, C.R. (1967) Why some heats of stainless steel may not weld. Weld. J. 46: 1006-1011.

2 Ludwig, H.C. (1968) Current density and anode spot size in the gas tungsten arc. Weld. J. 47: 234s-240s.

3 Bennett, W.S. & Mills, G.S. (1974) GTA weldability studies on high manganese stainless steel. Weld. J. 53: 548s-553s.

4 Fihey, J.L. & Simoneau, R. (1982) Weld penetration variation in GTA welding of some 304 L stainless steels. In: Proceedings of the conference Welding technology for energy applications, Gatlinburg, Tennessee, May 16-19, 1982. American Welding Society, Miami, Florida, pp. 139-153.

5 Leinonen, J. & Karjalainen, L.P. (1983) The effects of composition and inclusions on penetration in the TIG-welding of austenitic stainless steels. In: Proceedings of the conference The effects of residual, impurity and micro-alloying elements on weldability and weld properties, London, U.K., November 15-17, 1983. The Welding Institute, Cambridge, pp. 7.

6 Heiple, C.R. & Roper, J.R. (1982) Mechanism for minor element effect on GTA fusion zone geometry. Weld. J. 61: 97s-102s.

7 Rodgers, K.J. (1983) A study of penetration variability using mechanized TIG-welding. In: Proceedings of the conference The effects of residual, impurity and micro-alloying elements on weldability and weld properties, London, U.K., November 15-17, 1983. The Welding Institute, Cambridge, pp. 8.

8 Tinkler, M.J., Grant, I., Mizuno, G. & Gluck, C. (1983) Welding 304L stainless steel tubing having variable penetration characteristics. In: Proceedings of the conference The effects of residual, impurity and micro-alloying elements on weldability and weld properties, London, U.K., November 15-17, 1983. The Welding Institute, Cambridge, pp. 13.

9 Kemppainen, J. & Kurkela, M. (1985) The effect of chemical composition on the weldability of stainless steels. In: Proceedings of the seminar Steel tubes and their raw-material quality requirements, Helsinki, Finland, May 13-17, 1985. United Nations Economic Commission for Europe, Geneva, Switzerland. pp. 16.

10 Burgardt, P. & Heiple, C.R. (1986) Interaction between impurities and welding variables in determining GTA weld shape. Weld. J. 65: 150s-155s.

11 Leinonen, J.I. (1986) Cast-to-cast variations in GTA weld penetration of austenitic stainless steels. In: Proceedings of the JDC University Research Symposium 1985 International Welding Congress, Toronto, Ontario, Canada, October 15-17, 1985. American Society for Metals, pp. 165-169.

12 Kujanpää, V.P., Karjalainen, L.P. & Sikanen, H.A.V. (1984) Role of shielding gases in discontinuity formation in GTA welding of austenitic stainless steel strips. Weld. J. 63: 150s-155s.

13 Willgoss, R.A. (1980) Weld pool profile and porosity in TIG-welding of BS 1501, Type 316, stainless steel. In: Proceedings of the conference Weld pool chemistry and metallurgy, London, U.K., April 15-17, 1980. The Welding Institute, Cambridge, pp. 133-145.

14 Tsai, N.S. & Eagar, T.W. (1985) Distribution of the heat and current fluxes in gas tungsten arcs. Met. Trans. 16B: 841-846.

Table 1. Compositions (wt%) of the test steels

Steel	C	Si	Mn	P	S	Cr	Ni	Mo
1	.031	.40	1.21	.026	.003	18.4	10.1	.45
2	.034	.43	1.70	.028	.016	18.6	8.9	.05
3	.042	.54	1.59	.030	.130	23.8	13.5	.10

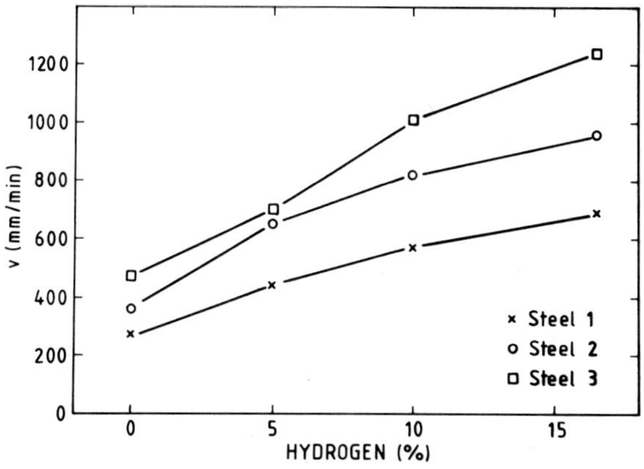

Fig.1. Effect of hydrogen content of the argon shielding gas on maximum welding speed.

Table 2. Welding conditions

Method	Single electrode TIG
Technique	Mechanized stringer bead
Groove	None
Filler metal	None
Welding position	Flat
Polarity	DC, electrode negative
Welding current	120 A
Welding speed	Variable
Arc length	2 mm
Electrode	W+2%ThO_2, diam 2.4 mm and cone angle 60 deg
Arc shielding gas	Ar(99.99 %), Ar+(5, 10, 16.5) % H_2, Ar+(30, 60, 80) % He, flow rate - 8 l/min
Root shielding gas	Ar, flow rate - 8 l/min
Trail shielding gas	Ar, flow rate - 8 l/min

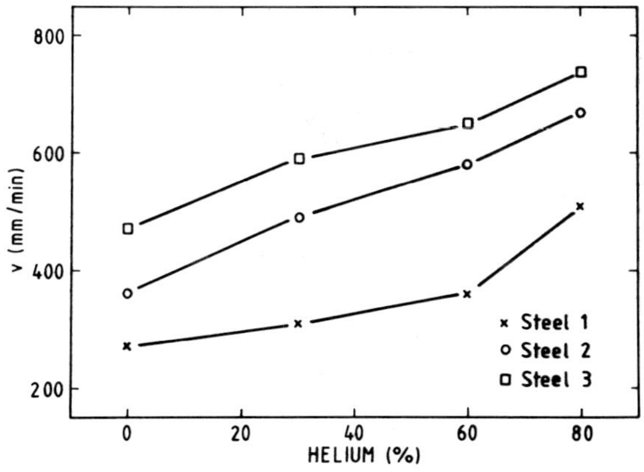

Fig.2. Effect of helium content of the argon shielding gas on maximum welding speed.

Table 3. D/W ratios of the test welds

Shielding gas	Steel 1	Steel 2	Steel 3
Ar	.28	.42	.53
Ar+5 % H_2	.37	.49	.54
Ar+10 % H_2	.35	.55	.61
Ar+16.5 % H_2	.36	.57	.67
Ar+30 % He	.29	.46	.53
Ar+60 % He	.32	.51	.54
Ar+80 % He	.40	.49	.55

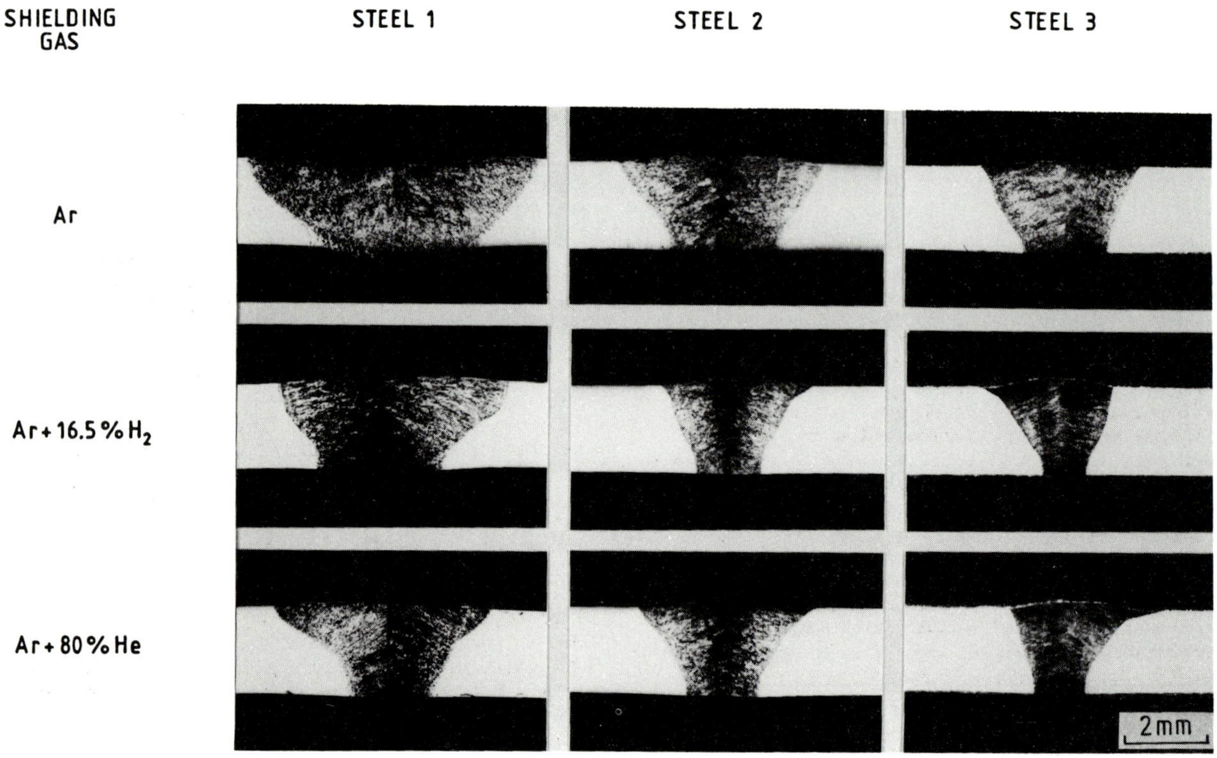

Fig.3. Weld pool profiles of test welds.

Influence of composition and microstructure on properties of a fully austenitic weld metal

S L Andersson and L Karlsson

The authors are with ESAB AB, Göteborg, Sweden.

Abstract

The influence of coating type and weld metal composition on hot cracking resistance, resistance to pitting and crevice corrosion and mechanical properties has been studied for high-Mo, fully austenitic weld metals deposited by MMA electrodes. The weld metals were of the 18Cr 17Ni 4.5Mo 0.13N and 20Cr 25Ni 6.5Mo 1Cu 0.15N types. Transvarestraint testing showed no advantage of the basic electrodes over basic-rutile electrodes with respect to hot cracking resistance. Si was detrimental and Mn beneficial for the hot cracking resistance. However, the effect of Mn was smaller for the more highly alloyed weld metals. Minor amounts of σ, χ, R and Laves phase were detected. Critical pitting temperatures, the measured minimum Mo concentration at dendrite cores and the absence of depleted zones adjacent to intermetallic phases suggest that the composition of the dendrite cores is a most important factor in determining the resistance to pitting. The results are presented in some detail and discussed in connection with microstructural observations.

Introduction

In recent years manual metal arc (MMA) electrodes depositing fully austenitic weld metals with high pitting and crevice corrosion resistance have been developed for welding of austenitic stainless steels with high Mo-content. These high-Mo steels are designed for use in environments with high chloride levels such as seawater or brackish water. Therefore, they have higher Cr (19-25wt.%), and Mo (4-6.5wt.%) content than conventional austenitic stainless steels to get an improved pitting and crevice corrosion resistance and a higher Ni (18-26wt.%) content to ensure a fully austenitic structure. (1,2)

Austenitic stainless steels with a high Mo-content possess a basically good weldability. However, a suitable filler metal has to be used and the well-known rules for welding of fully austenitic stainless steels have to be followed. In particular high heat inputs and high interpass temperatures should be avoided to minimize the risk of hot cracking.

For a given composition, a wrought material is generally less sensitive to pitting and crevice corrosion than as-deposited weld metal. The reason for this is the tendency of Mo (and Cr) to segregate in the weld metal, (e.g. (2-5)), in particular in slowly cooled and in reheated regions. Two approaches may be adopted to overcome this problem. A heat treatment can be used to homogenize the weld metal or an overalloyed filler metal can be used to compensate for composition variations within the weld metal. In practice the latter approach is used in most cases, as post-weld heat treatments are costly and often impracticable.

However, with increasing alloying content the risk of precipitation of intermetallic phases increases. It is well-known that such phases deteriorate the weld metal ductility. Therefore, the composition limits of weld metals for high-Mo austenitic stainless steels are largely determined by the combination of required corrosion resistance and mechanical properties.

Weld metals solidifying as austenite are generally more susceptible to hot cracking than those solidifying as a mixture of ferrite and austenite.(6,7) Therefore, the risk of hot cracking is normally overcome, in the lesser alloyed austenitic weld metals, by formulating the weld deposits to solidify primarily as ferrite. However, this approach is not applicable for the fully austenitic weld metals needed for high-Mo austenitic stainless steels. Consequently, for these weld metals the risk of hot cracking has to be minimized by a detailed balancing of the composition within the limits set by the requirements on corrosion resistance and mechanical properties.

This paper presents results from a study of high-Mo, fully austenitic weld metals deposited by MMA electrodes of the (wt.%) 20Cr 25Ni 6.5Mo 1Cu 0.15N and 18Cr 17Ni 4.5Mo 0.13N types. The study is a part of the development of the two basic-rutile electrodes OK 69.63 (20Cr 25Ni 6.5Mo 1Cu 0.15N) and OK 64.63 (18Cr 17Ni 4.5Mo 0.15N). These electrodes are specially designed to have a high corrosion resistance, mechanical properties matching those of the high-Mo austenitic stainless steels and a high resistance to hot cracking. Therefore, the investigation was concentrated on the influence of coating type and weld metal composition on the hot cracking resistance, corrosion resistance and mechanical properties. Results from the investigation are presented and discussed in connection with microstructural observations.

Experimental

Materials

Two series of MMA electrodes were investigated: 9 variants of the 18Cr 17Ni 4.5Mo 0.13N type, including electrodes with acid-rutile, basic-rutile and basic coatings, and 6 variants of the 20Cr 25Ni 6.5Mo 1Cu 0.15N type. For the latter electrode type only basic-rutile and basic variants were studied. Within each series the Mn- and Si-contents were varied while Cr-, Ni- and Mo-contents were kept constant. The chemical compositions and coating types for the electrodes are given in Table 1.

Hot cracking resistance

The hot cracking susceptibility was tested by the Transvarestraint test.(8)

Three tests were made at a strain of 3.2% for weld deposits from each electrode, except for electrodes A, C, N and O for which two tests were made. After grinding and polishing the test surface the hot cracking susceptibility was evaluated by measuring the number of cracks, maximum crack length and total crack length (Fig. 1). An optical stereomicroscope was used for the measurements.(9)

Only cracks in remelted weld metal were included in the quantitative evaluation. However, a qualitative estimation of cracks in the heat effected zone (HAZ) of the weld metal were obtained from transverse sections of the Transvarestraint specimens (Fig. 2).

For comparison two electrodes (E and H) were also tested by the Varestraint test.(10)
The specimens were evaluated after light etching of the surface and the cracks were divided into two groups: 1) cracks in remelted weld metal and 2) cracks in the HAZ of the weld metal.
Three tests were made for each electrode at each of the three strains: 0.87%, 1.16% and 1.74%.

Corrosion resistance

The critical pitting and critical crevice corrosion temperatures (CPT and CCT) were determined in 10% $FeCl_3 \cdot 6H_2O$ (11) following the method originally proposed by Brigham and Tozer.(12)

Mechanical properties

Tensile testing and impact testing specimens were prepared from all-weld metal deposited according to AWS. Tensile testing was made at room temperature and impact testing was made from -196°C up to room temperature.

Metallography

Weld metals for detailed metallographic investigations were chosen on the basis of the hot cracking test results. The microstructures and cracks resulting from the Transvarestraint test were examined by Optical Microscopy (OM), Scanning Electron Microscopy (SEM) combined with Energy Dispersive X-ray Analysis (EDS) and Transmission Electron Microscopy (TEM). Specimens for OM were prepared by standard grinding and polishing methods. Colour etching according to Beraha (13) was used to render microsegregation and precipitates visible. Segregation was studied quantitatively by elemental mapping and concentration profile determinations using a Link 860-2 EDS-system attached to the SEM.

Precipitates were identified by electron diffraction using a 100 kV transmission electron microscope. Thin foils for TEM were prepared by electropolishing in 15% perchloric acid and methanol at -35°C. SEM-EDS microanalysis was used to determine the composition of precipitates.

Results

Hot cracking resistance

Two types of cracks were found in the remelted weld metal of the Transvarestraint test specimens:

1) solidification cracks at the position of the solid-liquid interface at the moment of straining, and

2) liquation or ductility dip cracks behind the solidification front.

Both types of cracks were found in some of the weld metals (Fig. 1a). However, only solidification cracks occurred in the most crack resistant variants (Fig. 1b).

The solidification cracks were mostly wide and clearly showed the dendritic solidification behaviour of the weld metal (Figs. 2a and b and Fig. 3a). However, the second type of cracks were narrow and followed the grain boundaries (Fig. 2c). The surfaces of these cracks showed smooth regions and regions with drawn-out peaks (Fig. 3b) indicating the presence of a liquid or pasty phase at the grain boundary at the moment of crack opening.

The results of the Transvarestraint testing are shown for the two types of cracks in Fig. 4. A qualitative correlation was found between the degree of liquation or ductility dip cracking in remelted weld metal and the degree of cracking in the HAZ of the weld metal.

The two types of weld metals had comparable resistance to hot cracking. However, differences were noted between different types of coatings. Seen as groups, the acid-rutile electrodes were the least crack resistant, the basic electrodes were more crack resistant and the basic-rutile electrodes were the most crack resistant.

Within each group effects were noted of the composition. In particular a detrimental effect of

Si and a beneficial effect of Mn was found. However, the effect of Mn was less pronounced for the more highly alloyed electrode type.

The Varestraint test ranked the electrodes E and H in the same way as the Transvarestraint test. The basic-rutile electrode E was more crack resistant than the basic electrode H. This was true for all three strains for cracks in remelted weld metal. However, the data for cracks in the HAZ of the weld metal were somewhat more conflicting and no clear difference could be seen between the two electrodes in this respect.

Corrosion resistance

The CPT and CCT values were typically 30°C and 20°C respectively, for the 18Cr 17Ni 4.5Mo 0.13N type weld metals. Typical values for the 20Cr 25Ni 6.5Mo 1Cu 0.15N were 40°C and 17.5°C respectively.

Mechanical properties

The mechanical properties of these two types of weld metals are comparable to those of the austenitic high-Mo steels. Typical values are: 0.2% yield strength ($R_{p0.2}$) 460 N/mm^2, tensile strength (Rm) 620 N/mm^2, elongation (A_4) 35%, impact energy (ISO-V) 90-100 J at 20°C and approximately 40 J at -196°C. The lower impact values were found for Ø 4.0 mm electrodes whereas the higher were found for Ø 3.25 mm electrodes.

Microstructure

Both types of weld metals had for austenitic solidification typical dendritic microstructures (Fig. 3). Significant amounts of intermetallic precipitation were only found in the reheated zones of multipass welds deposited by Ø 4.0 mm electrodes. For the higher alloyed weld metals, typically 1µm sized, isolated precipitates occurred whereas for the lesser alloyed weld metals aggregates of smaller precipitates dominated (Fig. 5). A closer investigation using transmission electron microscopy revealed also the presence of minor amounts of intermetallic precipitates in Transvarestraint specimens, both in the remelted layer and in the reheated regions. However, these were generally much smaller than those found in weld metal deposited using Ø 4.0 mm electrodes.

In the 18Cr 17Ni 4.5Mo 0.13N type weld metals σ, χ and occasionally R-phase was found. σ was the major phase usually found as separate precipitates at low angle boundaries (Fig. 6) and at slag inclusions. χ occurred mainly in aggregates (Fig. 7).

σ, χ, R and Laves phase were identified in the 20Cr 25Ni 6.5Mo 1Cu 0.15N type weld metal. However, in this case χ was the most common phase, normally occurring as separate precipitates within the grains. SEM-EDS analysis of χ precipitates gave a composition of typically (wt.%) 20% Mo, 20% Cr, 20% Ni, 2.5% Mn, 2% Si, 1% Cu rest Fe. The particles were often associated with slag inclusions. Only a few small (<0.1 µm) R- and Laves-phase precipitates were detected. These were found predominantly at high angle boundaries.

Reheated zones of multipass welds, deposited by Ø 4.0 mm electrodes of the 20Cr 25Ni 6.5Mo 1Cu 0.15N type, were chosen as "worst case" specimens to investigate the effects of intermetallic phases and of segregation in the dendrites. Point analysis, elemental mapping and concentration profile determinations using SEM-EDS gave a minimum Mo concentration of approximately 5 wt.% the dendrite cores and a maximum concentration of approximately 9 wt.% in interdendritic regions (Fig. 8). Much smaller variations were found for Cr and Ni. A minimum Cr concentration of 19.5 wt.% at dendrite cores and a maximum of 20.5 wt.% interdendritically was found whereas the opposite segregation behaviour was observed for Ni. A maximum Ni concentration of 25.8 wt.% was found at dendrite cores and a minimum concentration of 24 wt.% was measured in interdendritic regions.

SEM-EDS analysis showed no indications of Cr or Mo depletion adjacent to the precipitates. One example is given for Mo adjacent to a χ-phase precipitate in Fig. 9.

Discussion

Hot cracking resistance

Weld solidification cracking occurs at temperatures approaching the solidus of the weld metal when low melting point constituents segregate to the dendrite boundaries.(14,15) In the final stages of solidification they form liquid films over the interfaces between adjacent dendrites. Cracks then form if the contraction strains developed during cooling impose stresses higher than the instantaneous rupture strength of the weld metal. However, cracking can also occur at grain boundaries in the solid state as a consequence of grain boundary weakening due to segregation.

The Transvarestraint testing was performed at one strain only. A certain scatter was found in the results as is shown for electrode L in Fig. 5. Therefore, slight differences between different electrodes cannot be taken as reliable. However, clear trends in the dependence of the hot cracking resistance on the composition and on the coating type were found. Furthermore, the results of the Varestraint tests are in agreement with the results of the Transvarestraint tests.

The appearance of the fracture surfaces (Fig. 3) suggests that liquid or pasty films have been present in most cases at the moment of crack opening. This is in agreement with segregation of low melting point constituents being the main factor causing hot cracking. Furthermore, the observed detrimental influence of Si and the beneficial influence of Mn on the hot cracking resistance are in agreement with the well known effects of these elements on segregation and formation of low melting point films.(14,16).

P and S are two of the most harmful elements with respect to their effect on hot cracking resistance.(5,6,14,16) Therefore, the content of P and S should be minimized to prevent hot cracking. However, the distribution of elements is often as important as the total concentration. This is illustrated by the ability of Mn to neutralize the detrimental effect of S.

Thus, the differences in hot cracking resistance between weld metals with almost identical compositions (e.g. electrodes L and N or M and O) but deposited by electrodes with different types of coatings are most probably due to differences in the resulting distribution of harmful elements.

Mo is known to be beneficial with respect to the hot cracking resistance of austenitic weld metals.(14,16) Therefore, the beneficial effect of Mn could be expected to be less pronounced for a weld metal with a higher Mo-content. Thus, the smaller effect of increasing Mn-content for the weld metals containing 6.5 wt.% Mo as compared to for the weld metals with 4.5 wt.% Mo is not surprising.

Corrosion resistance

The presence of σ, or other intermetallic phases, can decrease the resistance of stainless steels to pitting and crevice corrosion.(17,18) However, intermetallic phases like σ, χ, R and Laves generally have higher Cr and Mo contents than the matrix. Therefore, it seems unlikely that the lowered corrosion resistance is associated with direct attack on these phases. Attack is more likely caused by a depletion of Cr and Mo adjacent to the precipitates. However, the presence of intermetallic phases in the microstructure is possible without the occurrence of solute-depleted regions.(19)

Intermetallic phases in highly alloyed, fully austenitic weld metals mainly precipitate in the alloy-enriched regions. Thus, even if slight depletion of regions adjacent to intermetallic precipitates would occur, the pitting resistance of the weld is not likely to be affected if the minimum concentration at precipitates is higher than the concentration at the alloy-depleted dendrite cores.

Using the formula PRE = % Cr+3.3% Mo+13% N (5) gives a pitting resistance equivalent of approximately 37.5 for the composition at the dendrite cores in the 20Cr 25Ni 6.5Mo 1Cu 0.15N weld metals. A comparison with published data shows that the expected CPT for a steel with PRE 37.5 is approximately 40°C which is in good agreement with the measured CPT-values of the 6.5Mo weld metal. Thus, microsegregation seems to be a more important factor in determining the resistance to localized corrosion than precipitation of intermetallic phases.

Mechanical properties

For weld metals the impact toughness is greatly affected by precipitation (e.g. (19,20)). The number density of precipitates, the size distribution and the type of precipitates are all important factors in this respect.

The density and type of slag inclusions are mainly determined by the coating type. A basic-rutile coating gives a low density of slag inclusions and a high impact toughness. Even better impact values can be obtained by basic type coatings. However, the welding properties of electrodes with a basic-rutile type coating are much better than for electrodes with a basic type coating.

The risk of precipitation of intermetallic phases is governed by the weld metal composition and the thermal history. Therefore, for these high alloyed weld metals it is not possible to fully eliminate the risk of precipitation of such phases. However, as was shown by the comparison of weld metals deposited by electrodes of different dimensions, even comparatively unfavourable welding conditions causing increased precipitation of intermetallic phases give only a moderate decrease in impact values (here from 100 to 90 J at 20°C). Thus, intermetallic precipitation should not cause any problem in practice if proper welding procedures are used.

Microstructure

It is well known that the tendency of formation of intermetallic precipitates increases with increasing alloying content. σ-, χ- and Laves phases are frequently found in highly alloyed austenitic stainless steels,(18,19) in particular after thermal treatments like multilayer welding or hot working. Therefore, the occurrence of these phases in the reheated regions of the weld metals is not surprising. Nor is the change from σ- to χ-phase as the dominant phase with increasing Mo-content unexpected. However, it is often considered that the comparatively rapid cooling of the last pass in a weld suppresses the formation of intermetallics more or less completely.

It seems, though, that the presence of suitable nucleation sites at low and high angle boundaries and at slag inclusions together with the enrichment of Mo and Cr by microsegregation permits a more rapid precipitation in weld metals as compared with that in wrought steels of similar composition.

R-phase is reported to occur occasionally in austenitic stainless steels.(21) This phase is structurally related to the σ- and χ- phases. The present observation of small amounts of R-phase in both types of weld metal could indicate that precipitation of this phase is favoured by the non-equilibrium conditions and inhomogeneity conditions of a weld metal.

SEM-EDS analysis showed no indications of Cr or Mo depletion at intermetallic precipitates. However, the size of the volume analysed using this technique cannot exclude the possibility of depletion in a very narrow zone. Previous investigations using analytical electron microscopy (19) have shown, though, that the presence of intermetallic phases in the microstructure is possible without the occurrence of solute-depleted regions.

Mo and Cr are the two most important elements in terms of ensuring a high pitting resistance of a steel. These two elements, unfortunately, have a high tendency of segregation in the weld metal. However, as can be seen from the segregation ratios (% max/% min) for Mo (approximately 1.8) and Cr (approximately 1.05) Mo is the most critical element in this respect. The observed variations in concentration between dendrite cores and interdendrite region are consistent with observations made by other workers.(2,4) These results indicate that a weld metal with 6.5% Mo can be used for welding of steels with a Mo-content up to approximately 5% without loss of resistance to pitting corrosion. In terms of the pitting resistance equivalent (PRE) the weld metal should have a PRE-value approximately 25% higher than the base material to ensure the same pitting and crevice corrosion resistance of the weld as the base material.

Conclusions

- During the development of MMA electrodes of the 20Cr 25Ni 6.5Mo 1Cu 0.15N and 18Cr 17Ni 4.5Mo 0.13N types, different coating types and weld metal formulations have been studied with respect to the influence on hot cracking resistance, mechanical properties and corrosion resistance.

- For the coating compositions examined the acid-rutile electrodes were the least crack resistant, the basic electrodes were more crack resistant and the basic-rutile electrodes were the most crack resistant.

- A detrimental effect of Si and a beneficial effect of Mn on the hot cracking resistance was found. However, the effect of Mn was less pronounced for the more highly alloyed weld metal type.

- For the 18Cr 17Ni 4.5Mo 0.13N weld metal typical CCT and CPT values in 10% $FeCl_3 \cdot 6H_2O$ were 20°C and 30°C respectively, whereas for the 20Cr 25Ni 6.5Mo 1Cu 0.15N weld metal typical values were 17.5°C and 40°C respectively.

- Minor amounts of intermetallic phases were found in the weld metals, in particular in reheated regions of multipass welds deposited by Ø 4.0 mm electrodes. σ-, χ- and R-phase, with σ as the major phase, was found in the 18Cr 17Ni 4.5Mo 0.13N type weld metal whereas σ, χ, R and Laves phase, with χ-phase as the dominating phase was found in the more highly alloyed weld metals.

- CPT values, the measured minimum Mo concentration at dendrite cores and the absence of depleted zones adjacent to intermetallic phases suggest that the composition of the dendrite cores is a most important factor in determining the resistance to pitting.

- As a result of this study the basic-rutile electrodes OK 64.63 (18Cr 17Ni 4.5Mo 0.13N) and OK 69.63 (20Cr 25Ni 6.5Mo 1Cu 0.15N) have been designed. OK 64.63 and OK 69.63 deposit weld metals with high resistance to hot cracking, with mechanical properties matching those of the corresponding base material and with a high corrosion resistance.

Acknowledgements

The authors wish to thank Outokumpu OY for performing the corrosion tests.

References

1. A. Garner, Mat. Perf. 21, No. 8 (1982) 9.
2. G. Rabensteiner, IIW-Doc. IXH-164/87.
3. G.E. Hale, Weld. Inst. Rep. No. 250/1984.
4. P.I. Marshall, Weld. Inst. Rep. No. 316/1986.
5. N. Suutala and M. Kurkela, Proc. "Stainless Steel '84", (Göteborg 1984), p. 240, Book No. 320, The Inst. of Metals, London (1985).
6. V.P. Kujanpää, N.J. Suutala, T.K. Takalo and T. J. I. Moisio, Met. Constr. 12, No. 6, (1980) 282.
7. V.P. Kujanpää, ibid 17, No. 1, (1985) 40 R.
8. C.D. Lundin, A.C. Lingenfelter, G.E. Grotke, G.G. Lessmann and S.J. Matthews, WRC Bull. 280 (1982).
9. P. Karjalainen, Research Report, Univ. of Oulu (1986).
10. B. Bonnefois, Research Report, Creusot-Loire Industrie (1987).
11. M. Kurkela, Research Report, Outokumpu Oy (1986).
12. R.J. Brigham and E.W. Tozer, Corrosion 29, No. 1, (1973) 131.
13. E. Weck and E. Leistner, "Metallographic Instructions for colour etching by immersion, Part II", Deutcher Verlag für Schweisstechnik (1983).
14. E. Folkhardt, "Metallurgie der Schweissung nichtrostender Stähle", Springer-Verlag, Wien-New York, (1984).
15. B. Dixon, Austr. Weld. J. (Summer 1983) 16.
16. J.F. Lancaster, "Metallurgy of Welding", George Allen & Unwin, London (1980).
17. A.J. Sedriks, Proc. "Stainless Steel '84", (Göteborg 1984), p.125, Book No. 320, The Inst. of Metals, London (1985).
18. H. Brandis and H. Kiesheyer, ibid, p.217.
19. M. Liljas, B. Holmberg and A. Ulander, ibid, p.323.
20. B. Gretoft, S. Rigdal, L. Karlsson and L.-E. Svensson. Proc. "Stainless Steel '87" (York 1987).
21. D.R. Harries. Proc. "Int. Conf. Mech. Beh. and Nucl. Appl. of Stainl. Steels at Elev. Temp.", (Varese 1981). Met. Soc., London (1981).

Table 1 Chemical composition of weld metals (wt.%)

Electrode	C	Si	Mn	P	S	Cr	Ni	Mo	Cu	N	Coating type
A	0.027	0.62	1.66	0.016	0.010	18.70	16.27	4.27		0.13	Acid-Rutile
B	0.014	0.59	0.81	0.017	0.012	18.03	16.46	4.44	-	0.15	"
C	0.044	0.53	2.50	0.020	0.010	18.93	16.28	4.70	-	0.13	Basic-Rutile
D	0.029	0.58	1.71	0.015	0.011	18.16	17.20	4.51	-	0.13	"
E	0.032	0.58	2.56	0.014	0.011	18.16	17.37	4.56	-	0.13	"
F	0.044	0.52	5.91	0.015	0.009	17.38	16.51	4.45	-	0.15	"
G	0.044	0.80	6.66	0.010	0.007	21.19	16.39	3.11	-	0.28	"
H	0.041	0.52	2.32	0.019	0.015	17.72	17.18	4.79	-	0.12	Basic
I	0.035	0.41	2.91	0.012	0.015	16.06	17.72	4.83	-	0.05	"
J	0.033	0.75	2.53	0.015	0.006	19.63	25.30	6.74	1.03	0.17	Basic-Rutile
K	0.043	0.80	5.35	0.015	0.007	19.70	24.56	6.36	1.25	0.16	"
L	0.040	0.53	5.04	0.015	0.007	19.60	24.85	6.44	1.24	0.15	"
M	0.035	0.57	2.52	0.016	0.007	19.82	25.26	6.70	1.21	0.15	"
N	0.033	0.50	4.84	0.013	0.012	20.10	24.61	6.40	1.13	0.13	Basic
O	0.039	0.35	2.71	0.014	0.008	19.91	25.10	5.98	1.07	0.13	"

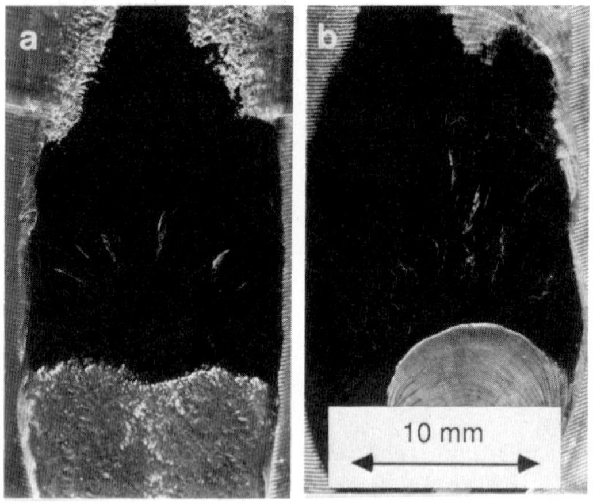

Fig. 1 Polished surface of Transvarestraint test specimen showing: a) solidification cracks (electrode L); b) solidification and liquation cracks (electrode B)

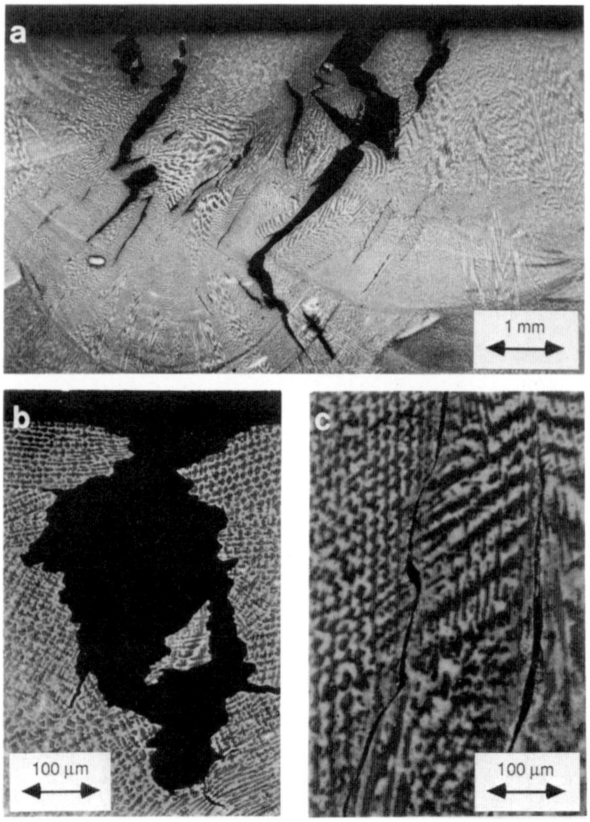

Fig. 2 a) Optical micrograph of transverse section through Transvarestraint test specimen (electrode N); b), c) details from a) showing solidification and liquation cracks respectively (Etchant Beraha II (13))

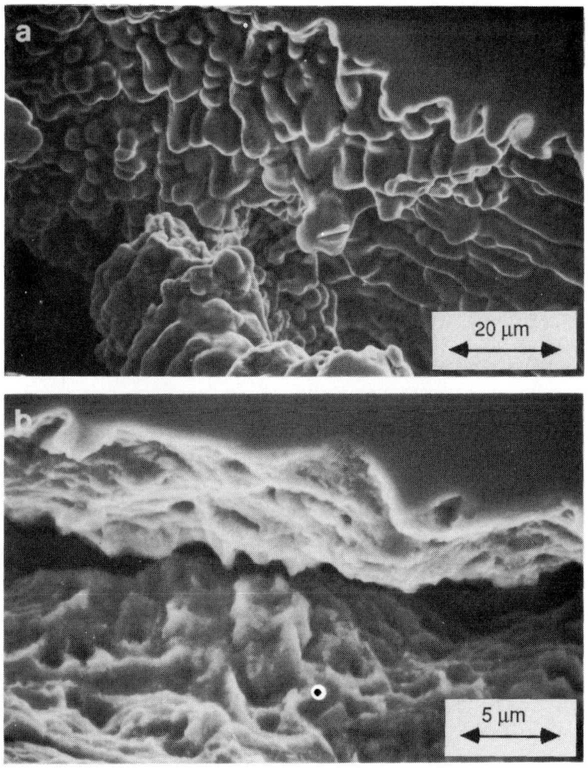

Fig. 3 SEM-micrographs of Transvarestraint test specimen (electrode B): a) solidification crack showing the dendritic solidification; b) liquation crack with drawn-out peaks indicating the presence of a liquid or pasty phase at the moment of crack opening

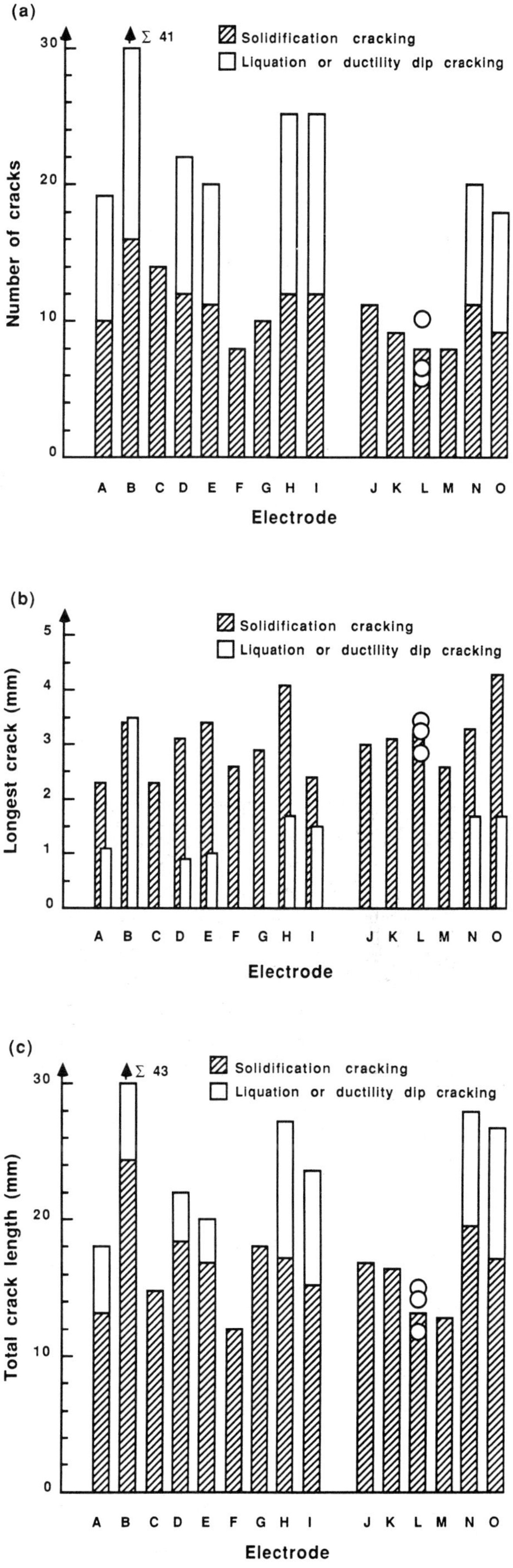

Fig. 4 Transvarestraint test results: a) number of cracks; b) longest crack; c) total crack length. The diagrams show the average results for three tests for each electrode except for electrodes A, C, N, O for which two tests were made. The individual data points are given for electrode L to illustrate the scatter in the results

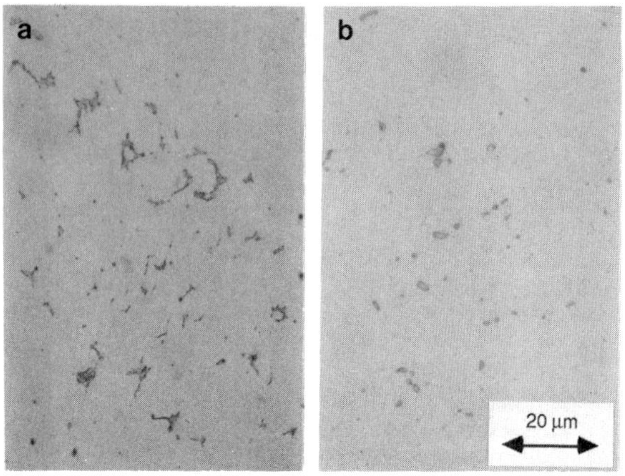

Fig. 5 Optical micrographs showing precipitation in reheated regions of multipass welds deposited by Ø 4 mm electrodes: a) 18Cr–17Ni–4.5Mo–0.13N; b) 20Cr–25Ni–6.5Mo–1Cu–0.15N type weld metals (Murakamis etchant)

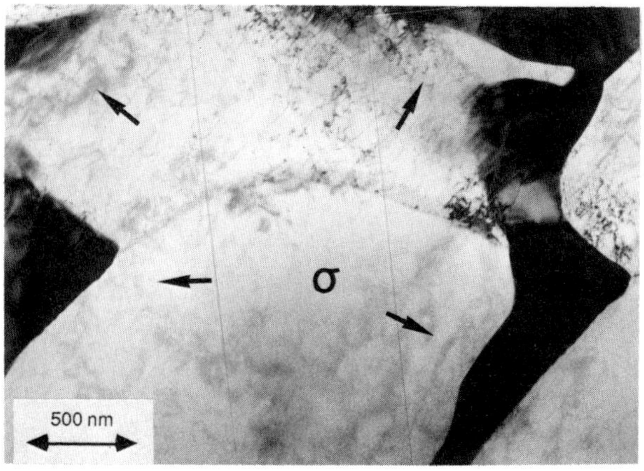

Fig. 6 TEM-micrograph of σ-phase precipitation at a low angle boundary in 18Cr–17Ni–4.5Mo–0.13N type weld metal

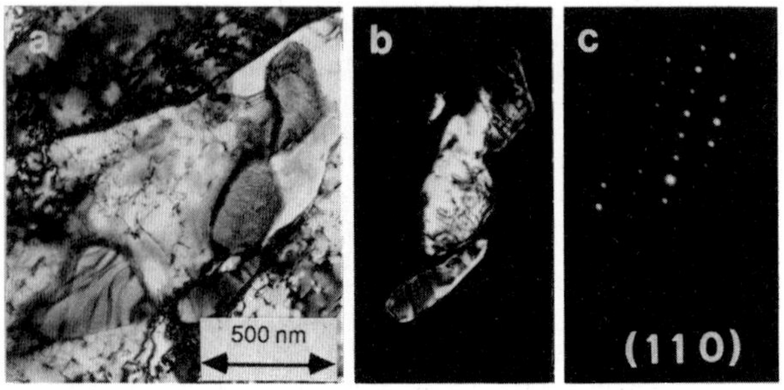

Fig. 7 TEM-micrograph showing aggregate of χ-phase precipitates in 18Cr–17Ni–4.5Mo–0.13N type weld metal: a) bright field; b) dark field; c) diffraction pattern

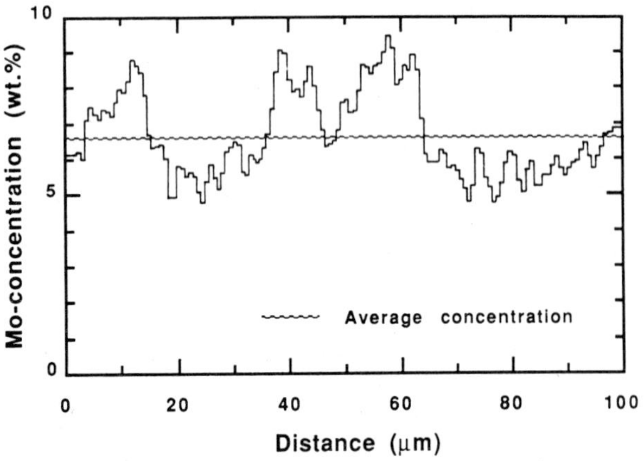

Fig. 8 SEM-EDS concentration profile for Mo in a multipass weld deposited by Ø 4 mm electrodes of the 20Cr–25Ni–6.5Mo–1Cu–0.15N type

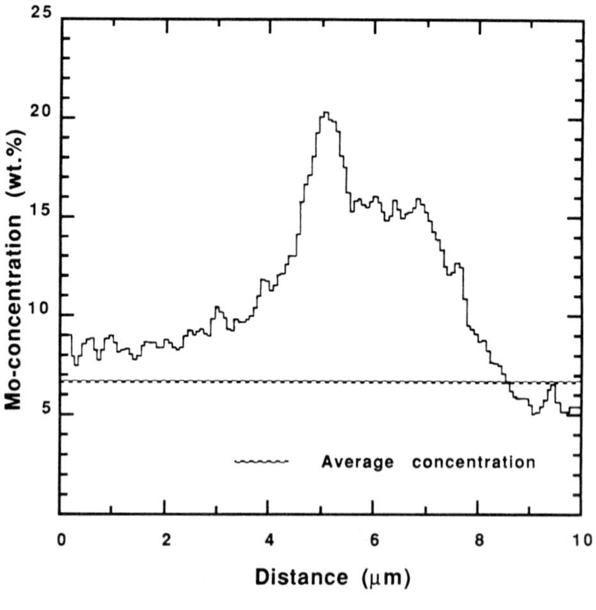

Fig. 9 SEM-EDS concentration profile for Mo at χ-phase precipitate in 20Cr–25Ni–6.5Mo–1Cu–0.15N type weld metal. Multipass weld deposited by Ø 4 mm electrode

□

Duplex stainless steel welding and applications

J Hilkes and K Bekkers

J. H. is with the Technical Staff of Smitweld bv, Nijmegen, The Netherlands and K. B. is the Technical Staff Manager.

INTRODUCTION

Duplex stainless steels are very popular in today's chemical engineering and piping for chloride, hydrogen-sulphide and carbon-dioxide-containing environments e.g. treating and transporting of oil and sour natural gas.

It is applicable for many onshore as well as offshore applications. Duplex stainless steel consists of two phases, austenite and ferrite, both about 50% as illustrated in photograph 1. In this photograph the dark phase, which is the primary structure, is ferrite and the light phase which is the secondary structure, is austenite. The combination of these two phases present about 50%/50% in one structure results in a very high yield and tensile strength as well as an excellent resistance to chloride-induced stress corrosion and pitting. This compares with conventional austenitic stainless steels e.g. AISI 304L and AISI 316L.

The structure of duplex stainless steel strongly depends on its chemical composition and thermal history such as heat-treatment and heat from welding.

In general a duplex stainless steel can have the following chemical composition: < 0.03% C; 19-26% Cr; 5-10% Ni; ~ 0-3.5% Mo; 0-0.35% N; 0-1.5% Cu. Chemical composition of the most used duplex stainless steels are listed in table 1.

METALLURGICAL PRINCIPLES

Due to the chemical composition - e.g.: 0.02% C; 22% Cr; 5% Ni; 3% Mo;0.14% N (DIN W.Nr.1.4462) - of a duplex stainless steel its austenitic/ferritic structure is formed as follows, see figure 1. When the above composition cools down from the liquid phase it starts to solidify as ferrite at 1450°C./1/ Below 1300°C the austenitic phase starts to form by nucleating at the grain boundaries of the ferritic phase.

Depending on the cooling rate a certain amount of austenite will be formed. When the cooling rate is low there is a lot of time hence a lot of austenite will be formed. On the other hand when the cooling rate is high there is very little time, hence only a small amount of austenite will be formed. To obtain the best possible structure the material is eventually homogenized at about 1050°C and water quenched to maintain the optimum structure which is about 50% ferrite and 50% austenite, and to suppress precipitation of any kind.

Duplex stainless steel in fact combines the good properties of ferritic Cr-steels with those of austenitic CrNi-steels which also includes high resistance to hot cracking during welding compared with fully stable austenitic stainless steels. On the other hand, the presence of a lot of ferrite can cause problems when welding.

High ferrite levels, over 80%, show very low toughness and make the structure prone to Hydrogen embrittlement.

For this reason the weld metal ferrite content should be limited within a certain band. The toughness can also decrease severely due to embrittlement caused by grain growth at high temperature and precipitation of intermetallic phases like Sigma(FeCr)- , Chi($Fe_{36}Cr_{12}Mo_{10}$)- or Laves(Fe_2Mo)- phase, Cr_2N, $Cr_{23}C_6$ and 475°C - embrittlement, as illustrated by figure 2 ./2/ Because most embrittling phases originate primarily from the ferritic part of the structure it is obvious that the chances of facing problems increase as the amount of ferrite present increases.

Figure 3 shows the influence of time/temperature on weld metal toughness which illustrates that the maximum working temperature for duplex stainless steel is 280°C. Photographs 2, 3 and 4 show examples of a proper weld metal structure, Cr_2N-precipitation and a weld metal containing Sigma-phase respectively.

To obtain a welded joint with more or less the same corrosion and mechanical properties as the base material, attention must be paid to the welding consumables and the welding itself.

It is obvious that a post-weld heat-

treatment is difficult to carry out from a practical point of view considering field and shop welding of large structures or piping systems. Therefore the weld as well as the heat-affected zone must preferably have all the required chemical and mechanical properties in the as-welded condition. Two major metallurgical principles can be applied to achieve the demanded properties. First of all the chemical composition of the welding consumables should be adapted to give the structure required. Secondly the cooling rate of the weld can be set within certain limits to allow the fused weld metal enough time to form the required amount of austenite.
Adapting the welding consumables' chemical composition is the best and safest way to obtain a weld metal structure which meets the requirements. As already described, a duplex stainless steel starts to solidify ferritically. For weld metal with a matching chemical composition and taking into account the high cooling rates of the weldments, a structure containing over 80% of ferrite is not imaginary.
This high ferrite level will cause embrittlement as previously mentioned.
The formation of austenite should therefore be stimulated. This can be achieved by increasing the amount of elements as Ni, Mn, N and C. The latter element C will clearly not be used for this purpose. The influence of the weld metal chemical composition on the weld metal structure can be illustrated by the Schaeffler diagram, see figure 4.
The austenite- and the ferrite-forming elements are plotted according to two formulae, the Ni- and the Cr-equivalents respectively as shown in figure 4. The Cr-eq.= %Cr + %Mo + 1.5(%Si) + 0.5(%Nb) and the Ni-eq.= %Ni + 0.5(%Mn) + 30(%C) + 30(%N-0.05). Enhanced Ni-content is mostly used. Welding consumables nowadays contain 8-10% Ni to ensure a proper austenite/ferrite ratio.

WELDING
According to the described metallurgical principles, modified welding consumables should be applied when welding duplex stainless steel.
The typical chemical composition and mechanical properties for Smitweld's duplex stainless steel consumables for a number of welding processes are listed in Table 2.
In this chapter the following welding processes will be described regarding specific welding problems and solutions:

- SMAW = Shielded Metal Arc Welding(MMAW)
- GTAW = Gas Tungsten Arc Welding (TIG)
- GMAW = Gas Metal Arc Welding (MIG)
- SAW = Submerged Arc Welding
- FCAW = Flux Cored Arc Welding
- PAW = Plasma Arc Welding

In general, regardless of the type of welding process, a very important process factor is dilution.
Dilution depends on the type of weld preparation, welding process and heat input and can vary theoretically from 0 to 100%.
A dilution of 100% for example could occur when GTA-welding is applied in butt-welds or root runs without any addition of filler metal. It is obvious that in order to ensure a diluted weld metal deposit containing enough nickel to provide enough austenite formation in the weld metal structure, the addition of sufficient filler metal is mandatory for GTA- and PA-welding. For all welding processes the percentage of dilution should be limited to about 30% to ensure a certain nickel content in the fused weld metal.
In order to allow the fused weld metal with enhanced nickel content enough time to form a sufficient amount of austenite, a certain cooling rate is necessary.
The cooling rate is determined by the heat input, pre-heat and interpass temperature and the thickness of the base and weld metal. A proper austenite/ferrite ratio will be achieved when the interpass temperature does not exceed 180°C and the heat input is kept between 0.6 and 2.2 kJ/mm. For heavy sections it is benefical to pre-heat up to about 120°C in order to reduce cooling rate and restraint.

Shielded Metal Arc Welding (SMAW/MMAW)
When welding with stick-electrodes, dilution depends on current which in turn is a function of electrode diameter but is not likely to exceed 30%. Weld preparation therefore has little influence on dilution. Nevertheless it needs attention to assure perfect root runs if only one-side welding is possible e.g. in piping systems. A V-groove is most suitable when it is prepared at a 70° angle, a 1-1.5mm root face and a 2.5-3mm root gap, as seen in figure 5. High-low should be limited. The roundness of the pipe is of great importance to establish a limited high-low.
For welding duplex stainless steel the most weldable electrode is a rutile-basic coated one e.g. Arosta 4462. Due to the rutile-basic coating, slag detachability is very good which is particularly advantageous for the welding of root runs. In general basic coated electrodes do not perform very well when welding root runs in pipe, and slag detachability even without undercut is poor.
Previous to welding, electrodes have to be redried at a temperature of 250-350°C for about 2 hours to prevent start or general porosity during welding.
The only way to overcome the problem of redrying is by using Extra Moisture Resistant (EMR-Sahara) electrodes which have a guaranteed low coating moisture content when used straight out of a vacuum sealed packet. If these electrodes are used up within 9 hours it is not even necessary to keep them in quivers at 80-120°C during the welding job which is still necessary for conventional electrodes.
For root runs it is important that end-craters are ground out completely before starting with the next electrode in order to prevent cracks. Before welding the

second run it is recommended that the root run be smoothed by grinding to create a perfect base for subsequent runs.
In this way slag inclusions and undercut will be prevented. Grinding however should be carried out very carefully because if the root run is ground out too much, it will be too thin and obviously oxidation is too heavy when the second run is being welded. To reduce the heat of the weld pool which will also reduce root run oxidation, the Arosta and Jungo type electrodes should be welded on DC-negative. Other possibilities to reduce heat and prevent oxidation of the root pass and HAZ are to use small diameter electrodes at relatively low amperage and to weld relatively thin weld beads.
A big advantage would be if the inside of the pipe weld was protected by backing gas (Ar or N); this will be discussed in more detail in the chapter GTA-welding.
Due to differences in physical properties such as thermal and electrical conductivity there are some differences in welding practice compared with the welding of unalloyed steels ./3 , 4/ The former is responsible for slow cooling rates of the base material and HAZ, especially for heavy sections which increases the chance of a long period at temperatures at which embrittling phases may occur./5/ During welding, heat is maintained longer in the weld metal and adjacent parent metal causing problems regarding weld pool control, bridging and root-welding. The latter makes it necessary to weld at low amperage to prevent overheating the electrode. As a consequence the chance of getting slag inclusions is increased because the metal/slag separation is not optimal during welding.
To overcome these problems welders have to be properly trained to weld (duplex) stainless steel.

Photographs 5 and 6 show a SMA-welded joint in duplex stainless steel which shows that the SMAW-process is a very good one for welding duplex stainless steels.

GAS TUNGSTEN ARC WELDING (GTAW)
Compared with other welding processes GTA-welding is indeed a very slow process. It is however an ideal process to make perfect root runs in pipe. Due to some practical experiences in the past where corrosion had occurred in SMA-welds in pipe as a result of bad craftsmanship, GTA-welded root runs became popular.
Using GTA-welding for root runs prevents: residual slag, spatter and oxidation of the inside root run when proper backing is applied. However when welding manually, the addition of sufficient filler metal is not assured but determined by the welder. Lack of filler metal during welding will increase the ferrite content tremendously which can obviously result in embrittlement. Therefore when applying the GTA-welding process carried out manually an open V-groove should be used as seen in figure 5 to assure that filler metal has to be used to obtain a fully penetrated rootrun. To increase reproducibility GTA-welding can be automated e.g. orbital GTA-welding. In this way weld settings and the amount of added filler metal are fixed which assures a reproducible weld quality being especially of use for welding pipe lines as shown at photograph 7./6/
A closed weld preparation can be used in this case as seen in figure 5 which also makes lining up the pipe easier.
As shielding gas pure Argon should be used. Despite the fact that addition of H_2 improves the weldability of duplex stainless steel, its use is not recommended. Some years ago a duplex stainless steel pipeline failed during hydrostatic testing by brittle fracture of the longitudinal welds. Investigation showed that fracture had occurred at repair welds. These welds were carried out using the GTA-welding process with Ar + 5% H_2 shielding gas without applying any filler metal. The combination of high ferrite content and a certain H-level provoked cracks, after little deformation, which we simulated in our laboratories,/7/ as illustrated by photograph 8 . Additional tests were carried out using SMA-welding electrodes loaded with extra H_2 and gave similar results as shown in figure 6. Cracks occur when both high ferrite and hydrogen levels are present in the structure and little deformation e.g. by bending is applied./8/ For this reason the use of hydrogen-containing shielding gas is prohibited. Some authorities do not accept any hydrogen in shielding nor backing gas. As a backing gas pure Argon (99.99%) or Nitrogen could be used. Nitrogen is much cheaper than Argon but will obviously influence the ferrite content as shown in the Schaeffler diagram, when it possibly enters the welding arc. Tests have shown that 100% Nitrogen backing decreases the ferrite content by about 4% maximum. When backing is used it should be ensured that the oxygen level is extremely low (< 500 ppm) before welding is started. When welding backing should be applied until about 10 mm of weld thickness has been deposited in order to prevent oxidation of the inside root.

Gas Metal Arc Welding (GMAW)
GMA-welding is a very attractive process because it is relatively clean and gives a fair deposition rate. It can be used for rooting in plate sections but is not very appropriate for rooting in pipe work. If the root and hot pass in pipe are made by GTA-welding, filling up by GMA-welding can reduce the total welding time considerably. To increase weldability and wetting properties of the weld metal, up to 3%Co_2 or 5% O_2 could be added to the Argon shielding gas. The suggested weld preparation for GMA-welding is shown in figure 5. At this moment about 21 miles of offshore duplex stainless steel line pipe up to 9" diameter and 1" wall thickness is under construction. For the reproducability an automatic orbital GMA-

welding system is being used. Photograph 9 shows one of the test welds for this project.
Photograph 10 shows a semi-automatic GMA-weld in pipe with 38 mm wall thickness.

Submerged Arc Welding (SAW)
The biggest problem which may occur when SA-welding is excessive ferrite levels due to too high a dilution. Welding should be carried out very carefully because with high amperage and wrong nozzle position dilution can be up to 70%. As a result the structure will be very coarse. To control dilution the weld preparation should be well chosen, for example as suggested in figure 5. Apart from dilution heat input and interpass temperature should be well controlled. Heavy wall thicknesses for which SA-welding is usually applied will, after a few runs have been deposited, cool down very slowly which increases the possibility of embrittlement in the heat-affected zone and weld metal. Photograph 11 shows a proper SA-weld in duplex stainless steel plate.

Flux-Cored Arc Welding (FCAW)
Due to the high yield strength and work hardening of duplex stainless steel, deformability properties are very poor. Solid duplex stainless steel wires with enhanced Ni are difficult to obtain which obviously results in a high price.
As an alternative flux-cored wire can be introduced. The strip chosen to produce flux-cored wire has very good deformability and the "missing" elements are added to the core. Test welds with Flux Cored Arc Welding using a duplex stainless steel flux-cored wire gave very good results as listed in table 3.
Photograph 12 shows a FCA-weld in duplex stainless steel. For FCA-welding a weld preparation with a 50° angle V-groove is very suitable, see figure 5.

Plasma Arc Welding (PAW)
General advantages of using Plasma Arc Welding are that a close square edge butt weld can be used and little or no filler metal is needed due to the very concentrated heat of the plasma arc. However this is detrimental to the properties when welding duplex stainless steel. Making close square edge butt welds will cause too much dilution and therefore the weld preparation with 50° angle V- or U-groove as for GTA-welding as illustrated in figure 5. Addition of sufficient filler metal should be assured. Because relatively high dilution is inevitable a Nickel-enhanced wire should be used or a post-weld heat-treatment should be carried out.
For the production of pipe the first run of the longitudinal seam is often PA-welded, filled in 2 runs using the SA-welding process and welded on the inside using the GTA-welding process. After completion these welds are then post-weld heat-treated at 1050-1150°C for 5 min, and subsequently water quenched. Photograph 13 shows an example of such a weld.

Acknowledgements
The authors are grateful for the support by their colleagues at Smitweld by Nijmegen, the Netherlands and Welding Rods Ltd, Sheffield, U.K.

References
/1/ Folkhard, Erich; Metallurgie der Schweißung nichtrostender Stähle, 1984 Springer-Verlag/Wien, pp 8.

/2/ Herbsleb, G., Schwaab, R.; Conf. Proc. Duplex Stainless Steels, ASM 1983, pp 15-30.

/3/ Hilkes, J., Bekkers, K.; Praktische Darstellung von Schweißnahtfehlern und ihre Vermeidung beim Schweißen von ferritisch-austenitischen CrNiMo-Stählen (Duplex Stählen), März 1986 Smitweld Veröffentlichung.

/4/ Bekkers, K.; Essential conditions for successful duplex stainless steel welding, Int. DSS conf. The Hague, 26-28 Oct. 1986, paper 27, Dutch Welding Institute.

/5/ Hilkes, J.L.P.,; Influence of residence time and cooling rate on the ferrite content of Duplex stainless steel and corresponding covered electrodes when austenitizing at 1050°C, Smitweld Research Report, Doc.no. 00858, June 1986.

/6/ Hilkes, J.LP., Van Nassau, L.; The welding of heavy wall austenitic-ferritic stainless steel piping, Smitweld Research Report, Doc.no. 00698, June 1985.

/7/ Hilkes, J.; Investigation of Crack-phenomena in GTA-welding DSS (W.Nr. 1.4462) using Argon or Argon + 5%H_2 shielding gas with or without addition of filler metal, Smitweld Res. Report, Doc.No. 00756, June 1985.

/8/ Fekken, U., Van Nassau, L., Verwey, M.; Hydrogen-induced cracking in austenitic/ferritic stainless steel, Int. DSS conf., The Hague, 26-28 Oct. 1986, paper 26, Dutch Welding Inst.

Table 1: Chemical composition of the most applied duplex stainless steels.

Codes/names (examples)	C	N	Cr	Ni	Mo	Others	Average ferrite content %
1.4462	0.02	0.15	22	5	3.0		44
1.4417	0.02		19	5	3.7	Si: 1.7	54
44 LN	0.02	0.15	25	6	1.5		52
Zeron 25	0.03	0.15	25	4	2.5		45
1.4582	0.05		25	7	1.8	Nb: 0.8	54
1.4460 (AISI 329)	0.08	(0.1)	27	5	1.7		75
UR 50	0.06		21	7	2.5	Cu: 1.5	35
Ferr. 255	0.04	0.17	26	5.5	3.3	Cu: 1.7	50
Zeron 100	0.02	0.25	25	6.5	3.5	Cu: 0.7 W: 0.7	50
FMN	0.05	0.2	25	5	2.0		45
UR 47: SM25Cr	0.02	0.16	25	7	3.0	(Cu: 0.5)(W: 0.3)	52
UR 55 M	0.03		26	5.4	2.0	Cu: 0.3	62
UR 52	0.02	0.16	25	7	3.0	Cu: 1.5	50
UR 45 M	0.06	0.16	21	8	2.5		35
UR 50 M	0.06	0.16	21	7.3	2.5	Cu: 1.5	29
UR 46 M	0.04	0.16	24.5	6	1.6	Cu: 1.5	40
UR 52 M	0.04	0.16	25	8	3.0	Cu: 1.5	40
1.4347	0.06		26	6.5	-		60
Ferr. 288	0.06	0.12	27	7	2.5	Cu: 1.3 Si: 1.6	50
Atlas 958	0.02	0.2	25	7	4.5		58
FMS	0.06		19.5	8.7	2.8	Nb: 0.8	40
FMX	0.05		23	9.4	3.4		45
Zeron 26	0.04	0.25	24	6	2.2	Mn: 4.5	30
A 905	0.04	0.37	25.5	3.7	2.3	Mn: 5.8	50
GFCOR 29	0.05	0.15	25.0	6.5	2.3	Cu: 3.0 Si: 1.2	40
Maust. CCM	0.05	0.13	25.5	6.5	2.5	Si: 1.5	48
SAF 2304	0.02	0.1	23	4	-		50

Table 2: Typical chemical composition and mechanical properties of SMIT-WELD´ duplex stainless steel consumables for SMA-,SA-, GTA-, GMA-, FCA-, and PA- welding.

Welding Process	Trade name	C %	Cr %	Ni %	Mo %	N %	FN	Rp0.2 N/mm^2	Rm N/mm^2	A_5 %	A_v-20^{0}C J
SMAW	Arosta 4462	0.02	23.0	10.1	3.1	0.12	35	600	750	24	55
"	Arosta 4463	0.02	24.73	10.7	3.2	0.17	50	710	912	26	41
"	Jungo 4462	0.03	23.6	9.5	3.1	0.15	47	599	807	26	58
"	Jungo 4462 Cu*	0.02	25.1	10.1	3.2	0.14	46	680	850	24	32
GTAW + PAW	SWN-4462-TIG	0.011	22.5	8.3	2.95	0.13	70	668	810	20	120
GMAW	SWN-4462-MIG	0.018	22.46	8.1	2.9	0.17	74	650	803	22	79
SAW	SW4462/P300s	0.025	23.1	8.5	3.0	0.09	54	535	740	36	51
FCAW	Cor-A-Rosta 4462	0.026	22.3	8.4	3.2	0.14	45	554	728	22	52

*:1.23%Cu

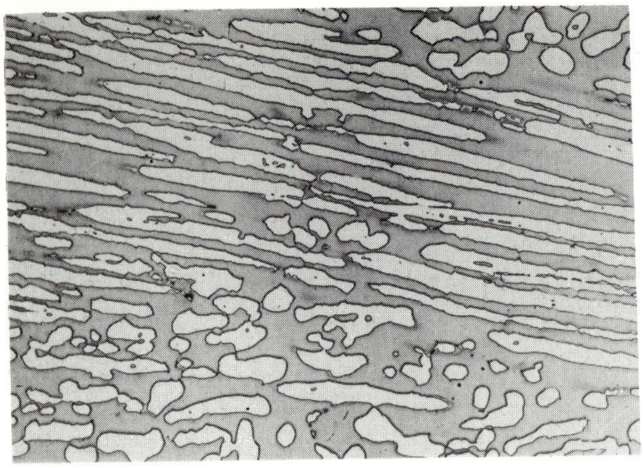

Photograph 1: Duplex stainless steel micro-structure W.Nr.1.4462, M=100x, light phase is austenite and the dark phase is ferrite /6/

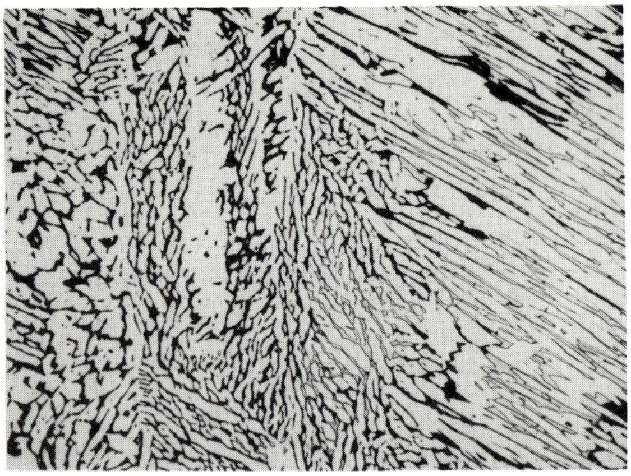

Photograph 4: Sigma-phase in weld metal due to long residence time at 600-800°C, V=200x /5/

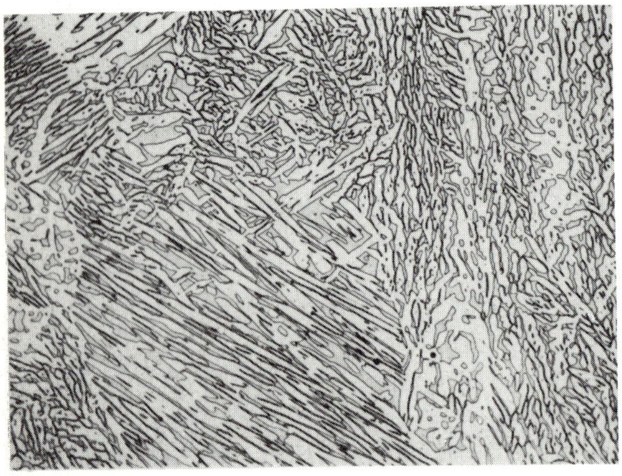

Photograph 2: SMA-weld metal structure containing about 30% ferrite, V=200x /5/.

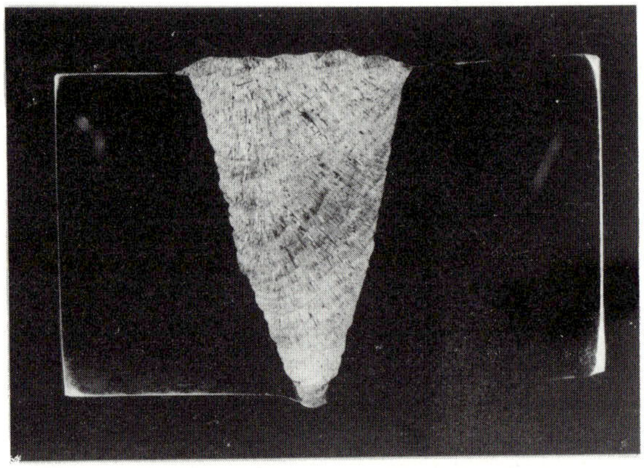

Photograph 5: SMA-weld in duplex stainless steel pipe ø 273*50mm with Arosta 4462

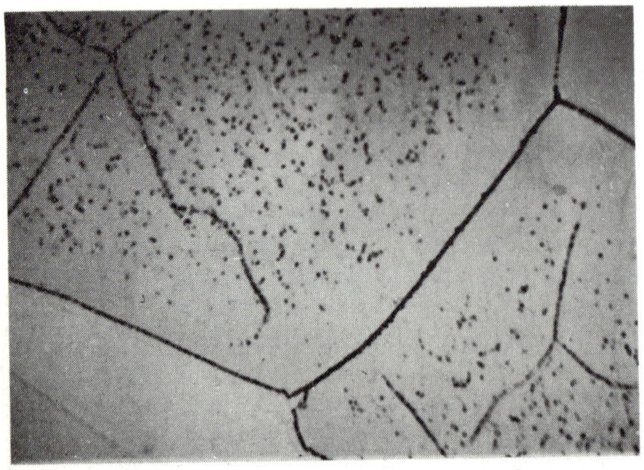

Photograph 3: Cr_2N-precipitation in high ferrite-containing heat affected zone, V=1500x /6/

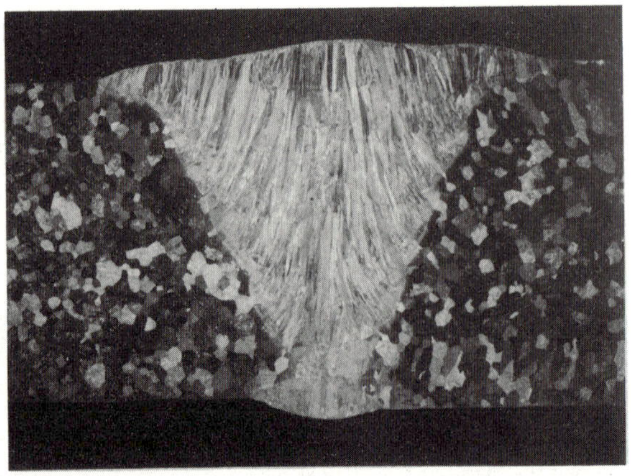

Photograph 6: SMA-weld in duplex stainless steel casting of 20 mm thick with Jungo 4462Cu

Photograph 7: Automatic orbital GTA-welded in duplex stainless steel pipe ø 273*50 mm /6/

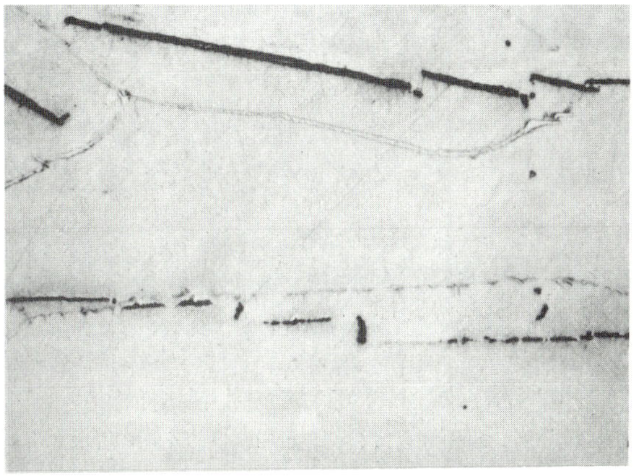

Photograph 8: Hydrogen-initiated crack in duplex stainless steel containing over 90% ferrite

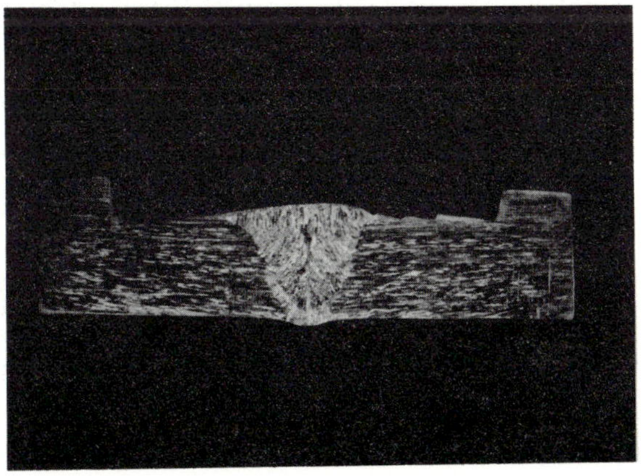

Photograph 9: Automatic GMA-weld in duplex stainless steel pipe ø 323*10mm

Photograph 10: Semi-automatic GMA-weld in duplex stainless steel pipe ø 300*35mm

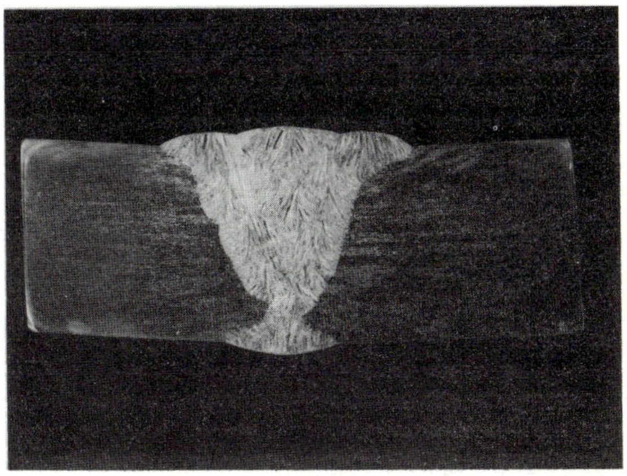

Photograph 11: SA-weld in 25 mm duplex stainless steel plate

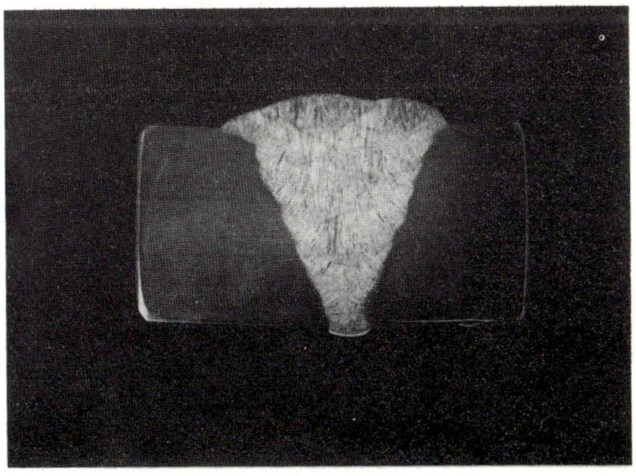

Photograph 12: FCA-weld in 25 mm duplex stainless steel plate

Photograph 13: Longitudinal seam weld in pipe using PA-, SA- and GTA-welding

Figure 1: Pseudobinary diagram for Fe-Cr-Ni steel based on 70% Fe /1/

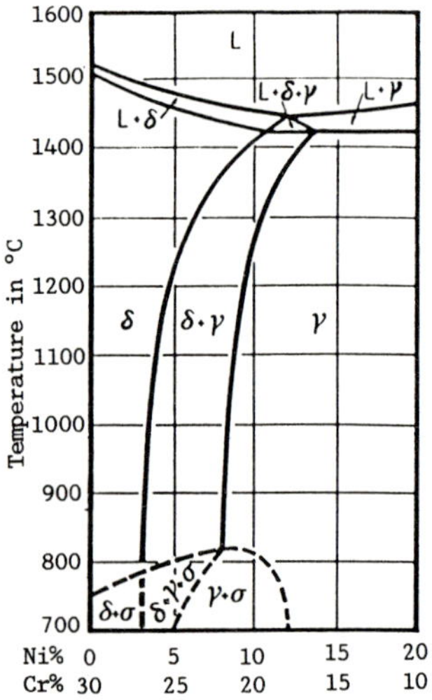

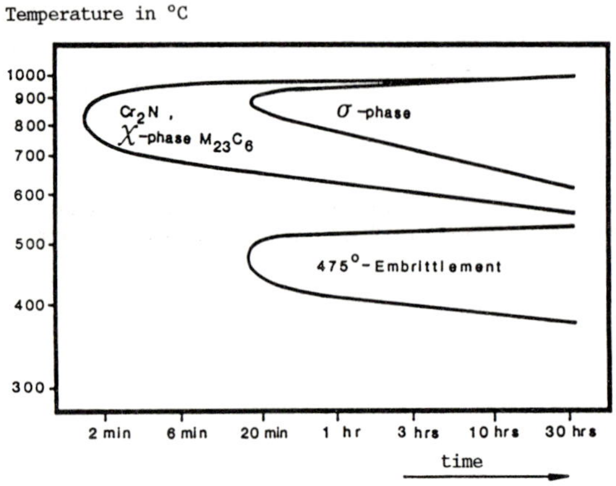

Figure 2: Time/Temperature diagram for the formation of embrittling phases in duplex stainless steel W.Nr.1.4462 /2/

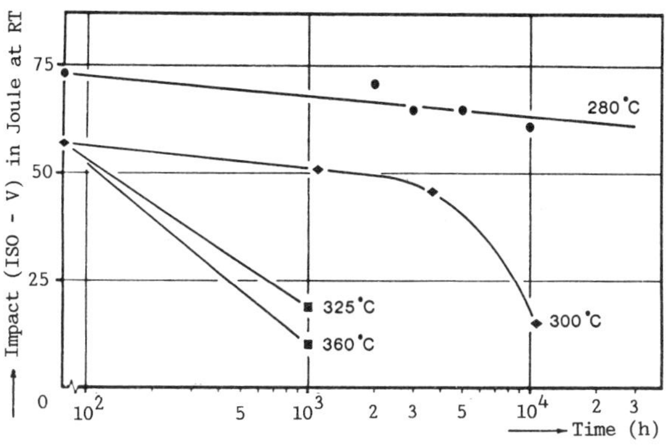

Figure 3: Time/Temperature/Thoughness diagram for Arosta 4462 SMA-weldmetal.

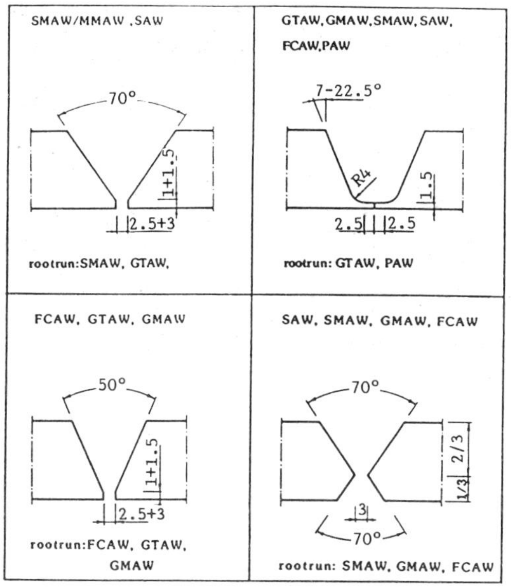

Figure 5: Suggested weld preparation for duplex stainless steel welding regarding the welding process.

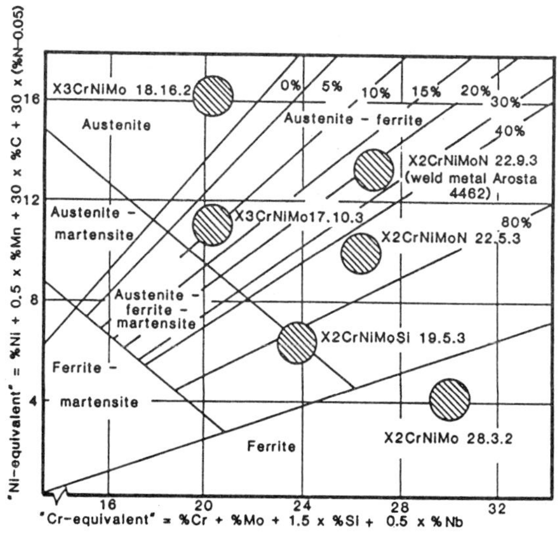

Figure 4: Scheaffler diagram for weld metal in the as welded condition.

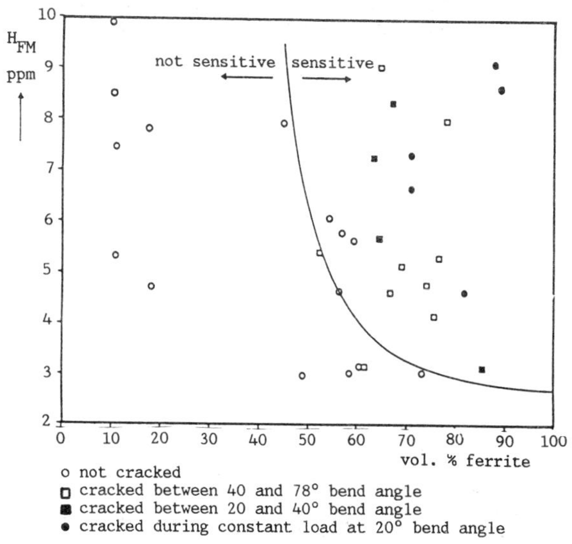

Figure 6: Influence of weldmetal ferrite content and weldmetal hydrogen and crack sensitivity /8/.

Effect of heat input on duplex stainless steel submerged arc welds

M Niset, R Doyen and A Van Bemst

M. N. is R & D Manager, R. D. is R & D Engineer, and A. V. B. is Technical Director at SA SOUDOMETAL, Brussels, Belgium.

ABSTRACT

This paper deals mainly with the presentation of mechanical and corrosion resistance data of submerged arc welds - composition close to UNS S 31803 - made at different practical heat input levels using 2.4 – 3.2 and 4 mm diameter wires and a highly basic agglomerated flux. It describes how, for a given chemical composition, the microstructure, the mechanical properties (such as tensile strength, toughness, bending behaviour, microhardness), the microcracking tendency, the intercrystalline, pitting and stress corrosion resistance of the deposited metal and of the heat-affected zone are influenced by the heat input level.
The results, compared with those measured in the base material, show that submerged arc welding can be utilized in a wide range of heat inputs for UNS S 31803 material. The mechanical properties are better than those obtained with MMA electrodes.

INTRODUCTION

Good mechanical and corrosion resistance properties of austeno-ferritic duplex stainless steels arise from the judicious equilibrium of the two existing phases and from the complex interactions which take place between them.
This equilibrium is secured by the ratio of alphagenous and gammagenous chemical elements on the one hand, and by a suitable heat treatment on the other. In this respect any further heating may lead to a drastic change in the balance of the existing phases, thus impairing the alloy's properties. This is particularly true at the time of welding.

When they were introduced onto the market, austeno-ferritic duplex stainless steels were known to have a rather bad weldability, especially in that it was difficult to monitor the ferrite in the heat-affected zones, making workability quite delicate.
The problem was nevertheless partially solved with a new generation of alloys; also the parallel introduction of welding products with high nickel and nitrogen contents allowed the deposited metals to reach higher standards as regards both mechanical and corrosion resistance properties.
With automatic welding processes, notably the submerged arc, the right choice of welding parameters, ie heat input, was found, as well as the correct definition of the welding product's composition. Steel manufacturers formerly advised users not to exceed 15 KJ/cm as far as welding heat input was concerned, and 150°C for interpasses temperature, thus avoiding the precipitates zones predicted by the TTT diagrams and at the same time keeping the HAZ extension to a reasonable value.

Nowadays it is known that successful welding of duplex stainless steels depends on the time necessary to allow $\alpha \rightarrow \gamma$ transformation to take place in the deposited metal and in the HAZ.
In this regard, the lower the heat input, the higher the cooling rates encountered, thus endangering the reaction. Recent papers (1,2) have been written on heat input effects applied to a wide range of energies within different welding processes. They pointed to a degeneration in corrosion resistance and mechanical properties directly linked with low heat inputs. We have investigated such an influence but are restricted to the SAW process on UNS S 31803 base material, within a range of energies limited to the most frequent values in practical use in stainless steel submerged arc welding.

EXPERIMENTAL PROCEDURE

Six pure weld metal deposits and four joints (referred to in Table 1) identified later by the letters D1 to D6 and J1 to J4 were tested on laboratory-sized samples by the submerged arc welding process, in the flat position.
As shown in Table 1, the groove preparations for all the weldments were those used for the VdTüV approval procedures. We worked with 20 mm thick mild steel plates buffered with 22Cr-8Ni-3Mo-N manual coated electrodes for the pure deposited metal samples, and with 15 mm thick duplex stainless steel plates of UNS S 31803 hue for the joints investigations. The different levels of heat input were obtained by varying the welding wire diameters from 2.4 – 4 mm and also altering the welding parameters; current, tension and welding speed, likewise. They had been selected so that

sound deposits were made and the heat input range was kept between 8 KJ/cm and 29 KJ/cm. The latter deliberately omits the required values for stainless steel submerged arc welding in common practice. The interpass temperature was kept at 128°C. No preheating or post-weld heat treatments were performed.
The wires' compositions were close to UNS S 31803 (DIN W.Nr. 1.4462), with a nickel enrichment (about 8% instead of 5%). Base material and welding wire chemical analysis is referred to in Table 2.
The submerged arc flux is a commercial unalloyed highly basic agglomerated flux also used for other stainless steels, here identified as IND. Its basicity index, according to Boniszewski, is 2.3 and its principal alloying vectors, according to Thier and Adam,(3) are for UNS S 31803-type wire:

C	0	Cr	-1	Mo	-0.2
Mn	-0.4	Ni	0	N	-0.02
Si	+0.15				

A very important feature is that this flux produces deposits with a low oxygen content.

RESULTS AND DISCUSSION

Pure deposited metal

Chemical compositions of the deposits

Chemical analysis of the six pure deposited weld metals is summarized in Table 3. Thanks to the good alloying vector of the highly basic flux used, all our deposits have a composition very close to that met in the welding wires. Also somewhat lower levels of silicon (0.55%) and oxygen (450 ppm) particularly appear when compared with those commonly obtained in submerged arc weldings.

Microstructures of the pure deposited metals

The ferrite measurements are given in Table 3, end of line. They were made with the Magne-Gage instrument. Ascertaining the calibration for the 'Duplex zone' was performed by means of duplex stainless steel Standards kindly supplied by Dr D J Kotecki.
The ferrite contents observed during our experiments vary between 42 EFN and 65 EFN with the higher ferrite increase corresponding to the lowest heat input.
The microstructures of our duplex pure deposited metal samples are typical; that is austenite and ferrite. Micrographs in Fig. 1 show some of the classical duplex structures observed for weldments.

Mechanical properties of pure deposited metals

In dealing with pure deposited metals, this paper focuses mainly on notch toughness, mechanical resistance and hardness; results are presented in Table 4.
The mechanical resistance properties (YS and TS) of the pure deposited metals do not show a clear variation with the heat input; to the contrary, as far as elongation, notch toughness and hardness are concerned, a softening effect increases the ductility of the deposits for the high welding energies. Elongations develop indeed from 23% to 34% - this softening effect is due mainly to an austenite enrichment of the deposited metal.

In so far as this paper deals with notch toughness, it is obvious that from 8 KJ/cm to 29 KJ/cm the absorbed energy increases by about 40 J on average. Hardnesses measured on the deposit surfaces effectively confirm this.
It seems that for a given heat input the notch toughness, hardness and elongation properties are not influenced by the wire diameters.
Also it was noticed that the level of toughness observed in the submerged arc process stays notably higher than the values generally encountered when dealing with commercially available manual coated electrodes.
Fig. 2 gives SAW and MMA average notch toughness values in function of the test temperature.

Joints

Observed microstructures

The micrographs a) and c) illustrated in Fig. 3 show the three main zones for joints J1 and J4. To the right-hand side, the base metal assumes the alternate characters of the ferritic band (in dark) and of the austenitic band (in light).
On the left, the disordered solidification of the deposited metal clearly indicates the cooling speed.
In between the two zones, the HAZ is highly ferritic, particularly when low heat input is applied (micrograph 2a). Up to 75% ferrite content was observed.
With a higher heat input (about 24 KJ/cm), this zone is less marked because of the formation of much more austenite. In the zones neighbouring the base metal an austenitic area with a somewhat jagged morphology can be seen (micrographs 2c, 2d). The width measured for the whole heat-affected zone varied for our test purposes between 100 and 150 μm.
Table 5 summarizes the mean values of the measured ferrite in the joints, but because of the inadequacy of the metallographic average ferrite volume percentage determination and the width of the ferritized area of the heat-affected zones of the joints, these values are not very accurate.

Mechanical properties of the joints

Concerning the joints, we were particularly interested in mechanical resistance, bending states and notch toughness - the results are presented in Table 5.

Mechanical resistance

The tensile strength observed in prismatic samples decreased from 785 Mpa to 740 Mpa with the varying heat input. All ruptures occurred in the base metal.

Bendings

Surface bend tests, carried out on 4 times the thickness, were very valuable and did not show defects of any kind during the present study.

Notch toughness

The notch toughness (KCV) was measured at +20°C and -20°C in the weld metal. At +20°C the values obtained increased from 80 to 114 J with the heat input. At a lower temperature (-40°C) this phenomenon was not observed.
Notch toughness tests were also carried out on plain HAZ in order to show any detrimental effects of the ferritized zone.
The notch toughness was measured every 7/15th of a millimetre, starting at the fusion line, once with the notch perpendicular to the surface

and once with the notch taken parallel to the fusion line. The lowest obtained values KCV at -40°C are more constant than those found at -20°C for the weld metal.
As it is difficult to correlate the position of the notch, and hence the absorbed energy observed, with a specific location of the HAZ due mainly to both beads and joints shapes, we have tried to confirm the good results obtained with those measured on samples submitted to simulated weld heat treatments equivalent to that found at different places by our real joints.
While such simulations do not take into account the mechanical stresses met when dealing with real welding joints, this technique does nevertheless enable us to consider a large enough volume of homogeneous structural metal, thus leading to a better determination of the characteristic microstructure.
Figure 4 clearly demonstrates the agreement between the zones marked on the joint (micrograph a) and the simulated observed structures (micrographs b, c, d).
KCV measurements obtained at -40°C for simulated samples with different heat inputs and thermal cycles are summarized in Table 6. The observed values demonstrate that the toughness levels remained very high, close to those encountered with base metal. Consequently they confirm the values experienced within the HAZ for the true joints.

Microhardness

Microhardness of the base metal, heat-affected zone and weld metal was also measured (Fig. 5). The highest microhardness was found for the lowest heat input - this must be explained by the stronger ferritization occurring in the HAZ when lower heat inputs are involved.

Corrosion resistance of the joints

We made a limited investigation of the corrosion resistance of the joints: pitting, intergranular and stress corrosion cracking tests were performed only to see if all parts of the weldments - weld metal, HAZ and base metal - have an analogous behaviour or not.

Pitting corrosion resistance

Pitting corrosion resistance was estimated according to ASTM G 48 test, but with the following alterations: solution 10% $FeCl_3$, $6H_2O$ at 30°C for 24 hours. The pitting corrosion resistance was investigated particularly for the surface of the joints (results 'S') and in the root of the joints (results 'R'), referred to in Table 7. With regard to the results, it must be pointed out that there is no correlation between the heat input and the pitting corrosion resistance. The values determined for the joints J1, J2 and J4 are roughly the same as those obtained for the base material as far as weight loss, pitting density, size and depth are concerned. For the low heat input tests, pitting occurred mainly within both the heat-affected zone and the overlapping zones between the beads.
For the joints J2 and J4, on the other hand, pitting occurred in the same manner in all parts of the weldments. Photographs of the samples after testing are shown in Fig. 6.

Intergranular corrosion resistance

Resistance to the mentioned corrosion was investigated with the Huey ASTM A262-C test procedure. Measurements were made on the top surface of the joints, here referred to as results S, and within the roots R. It appears that the submerged arc welding joints have a low sensitive reaction to that test; the calculated penetrations have approximately the same value for all the tests and for the base material. These excellent results agree with those mentioned in recent writing and partially account for the elements composition specific to duplex stainless steels.(4)

Stress corrosion cracking resistance

Stress corrosion cracking resistance has been estimated on U-bend ASTM G 30 samples according to the ASTM G 36 test, which has been modified to some extent for a better discrimination between the different hues found in duplex stainless steels. Test conditions were hence $MgCl_2$ 35%, paper grit 1000.
The results measured for samples S and R as for base metal are referred to in Table 7. Failure times obtained for the joints are not good and remain largely under the base metal value. The failures occurred within the weld metal for the joints J1 and J4 and into the heat-affected zone and the base metal in the case of joint J2.
With the aim of reaching a more specific conclusion, we are planning to extend our current study with the NACE TM 01-77 test.

CONCLUSIONS

For duplex stainless steel UNS S 31803, submerged arc welding with a nickel-enriched wire containing 22%Cr-8%Ni-3%Mo + N allied to a highly basic agglomerated flux proves to be an advantageous solution.
It can be successfully used within a range of heat inputs varying from 8 to 25 KJ/cm - mechanical properties, pitting corrosion and intergranular corrosion resistance are satisfactory.

The use of higher energy values definitely leads to better results for weld ductility and notch toughness.

Because of the low weld metal oxygen content obtained with the highly basic flux, the notch toughness values were found to be superior to those commonly achieved with MMA welding, making the SAW process with an appropriate flux very interesting.

The notch toughness levels reached in the HAZ and ascertained by test simulations were found to be averaging in between both weld metal and base metal values.

ACKNOWLEDGEMENT

This work has been carried out with the financial help of IRSIA (Institut pour l'Encouragement de la Recherche Scientifique dans l'Industrie et l'Agriculture), to whom we express our gratitude.

REFERENCES

1 B Lundqvist, P Norberg & K Olsson: 'Influence of different welding conditions on mechanical properties and corrosion resistance of Sandvik SAF 2205 (UNS S 31803)', Duplex stainless steel '86 - Intern. Conf., The Hague, 26-28 October 1986

2 B Bonnefois, D Catelin & P Soulignac: 'Soudage des aciers inoxydables destinés aux milieux corrosifs sévères, Schweissen und Schneiden, pp. 106-112, DVS Berichte 105, 1986

3 H Thier & W Adam: 'Metallurgische und korrosionschemische Eigenschaften von austenitischen UP-Band-plattierungen', pp. 77-83, DVS Berichte 32, Düsseldorf 1974

4 E N Bower, J W Fielder & K J King: 'The characteristics of HyResist 22/5 duplex stainless steel welds produced using commercially available consumables', Welding Institute, November 1986

TABLE 1 - WELDING CONDITIONS FOR SUBMERGED ARC WELDING

	PURE DEPOSITED METAL						JOINTS			
GROOVE PREPARATION	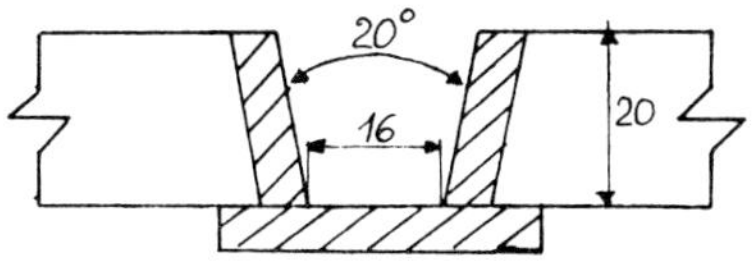						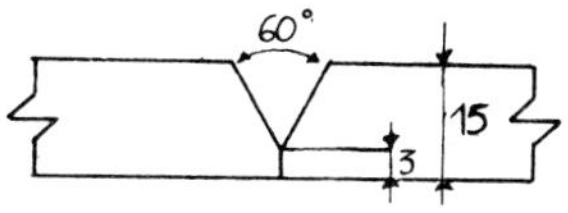			
FLUX	UNALLOYED HIGHLY BASIC AGGLOMERATED FLUX " I N D "									
	D1	D2	D3	D4	D5	D6	J1	J2	J3	J4
WIRE Ø mm	2,4	2,4	3,2	3,2	4	4	2,4	2,4	3,2	3,2
CURRENT =+ A	350	350	350	500	500	600	350	350	400	500
VOLTAGE V	25	28	28	32	32	32	26	28	28	32
SPEED cm/min	66	45	45	40	40	40	66	45	40	40
H I kJ/cm	8,0	13,1	13,1	24,0	24,0	28,8	8,3	13,1	16,8	24,0
NUMBER OF BEADS	19	15	15	8	7	7	8	7	6	3

TABLE 2

BASE METAL AND WELDING WIRES CHEMICAL COMPOSITIONS

	BASE METAL	WELDING WIRES Ø 2,4	Ø 3,2	Ø 4 mm
C *	0,029	0,016	0,020	0,020
Mn	1,69	1,67	1,49	1,64
Si	0,55	0,43	0,45	0,45
Cr	22,0	22,9	22,0	22,4
Ni	5,75	7,8	7,7	7,9
Mo	2,91	2,97	2,82	2,76
N	0,168	0,148	0,150	0,169

* in weight percent

TABLE 3

CHEMICAL ANALYSIS OF THE DIFFERENT DEPOSITS

DEPOSIT	D1	D2	D3	D4	D5	D6
C *	0,016	0,018	0,018	0,024	0,026	0,024
Mn	1,23	1,16	1,15	1,17	1,23	1,28
Si	0,57	0,57	0,58	0,54	0,56	0,55
S	0,002	0,003	0,002	0,004	0,003	0,004
P	0,019	0,020	0,020	0,018	0,019	0,016
Cr	21,4	21,4	21,4	21,2	21,2	21,6
Ni	7,9	7,7	7,8	7,6	7,8	7,7
Mo	2,78	2,74	2,71	2,64	2,68	2,57
N	0,132	0,128	0,128	0,121	0,128	0,127
O2**	n.d.	n.d.	450	414	450	438
EFN	64	65	63	53	42	44

* : in weight percent
** : in ppm

n.d. = not determined

TABLE 4

MECHANICAL PROPERTIES - PURE DEPOSITED METAL

DEPOSIT	D1	D2	D3	D4	D5	D6
H.I. (kJ/cm)	8,0	13,1	13,1	24,0	24,0	28,8
Y.S. 0,2% (MPa)	640	640	649	554	573	678
T.S. (MPa)	780	799	807	739	770	758
E 5d (%)	23	28	29	26	31	34
KCV (J) +20°C	89	80	78	118	112	130
KCV (J) -20°C	72	67	70	98	89	110
KCV (J) -40°C	61	55	62	106	95	107
HARDNESS (HB)	230	234	225	218	212	212

TABLE 5

JOINTS MECHANICAL PROPERTIES AND FERRITE CONTENT

		J1	J2	J3	J4
H.I. (kJ/cm)		8,3	13,1	16,8	24,0
Ferrite %	WELD METAL	43	44	n.d.	43
	H.A.Z.	55	52	n.d.	55
	BASE METAL	48	48	n.d.	48
U.T.S. (MPa)		786	756	769	743
Surf. Bend Tests (4 x th)		GOOD	AT	180°	
Notch Toughness	WELD METAL				
	KCV +20°C (J)	80	98	101	114
	KCV -20°C (J)	71	84	67	75
	H.A.Z. Min.Val.				
	KCV -40°C (J)	76	n.d.	84	82

n.d. : not determined

TABLE 6 - WELD SIMULATION TESTS

HI = 8 KJ/cm, t 12/8 4s, t 8/5 10s		HI = 24 KJ/cm, t 12/8 40s, t 8/5 70s	
Peak Temp. (°C)	KCV -40°C (J)	Peak Temp. (°C)	KCV -40°C (J)
1360	215	1358	218
1335	238	1359	184
1228	145	1209	215
1245	180	1208	192
1144	192	1108	200
1152	226	1107	181
1052	245	1005	262
981	214	1007	193
957	191	907	212
957	229	907	174

Base Metal : KCV -40°C = 200 J.

TABLE 7 - CORROSION RESISTANCE PROPERTIES

		J 1 - 8,3 KJ/cm		J 2 - 13,1 KJ/cm		J 4 - 24,0 KJ/cm		BASE METAL
TEST LOCALISATION (*)		S	R	S	R	S	R	-
PITTING Test G 48 Mod. 30°C	WEIGHT LOSS g/m^2h.	0,070	0,040	0,037	0,037	0,035	0,045	0,039
	P. Density	2	1	1	2	1	2	1
	P. Size	1	1	1	1	1	1	1
	P. Depth	1	1	1	1	1	1	1
HUEY Test A 262-C	Calcul. Penetrat. μ/48 h	1,3	1,7	1,7	1,7	1,7	1,7	2,0
SCC Test G 36 Mod. 35% $MgCl_2$	Time to failure Hr.	166	238	46	167	110	16	> 250
	Failure (**) localisation	W	W	B-H	B-H	W	W	-

(*) S = Surface R = Root

(**) W = Weld metal B-H = Base metal and HAZ

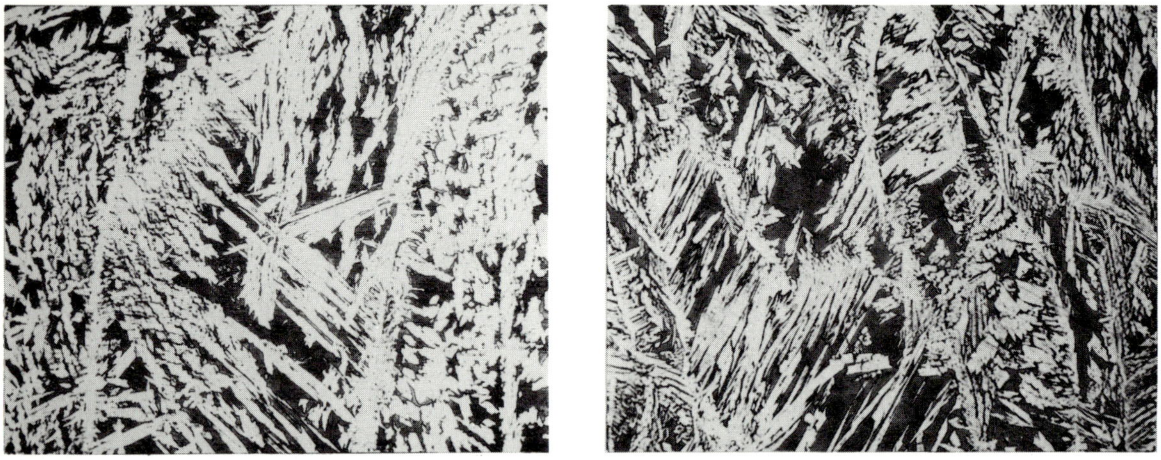

187x D1 : 8,0 KJ/cm

187x D6 : 28,8 KJ/cm

Fig. 1 Pure deposited metal, typical microstructures, Beraha's etchant (ASTM 99). Dark constituent – ferrite, light constituent – austenite

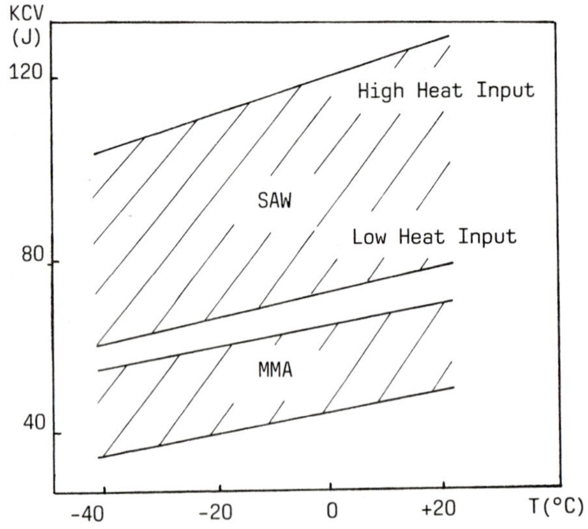

Fig. 2 SAW & MMA notch toughness values in relation to temperature

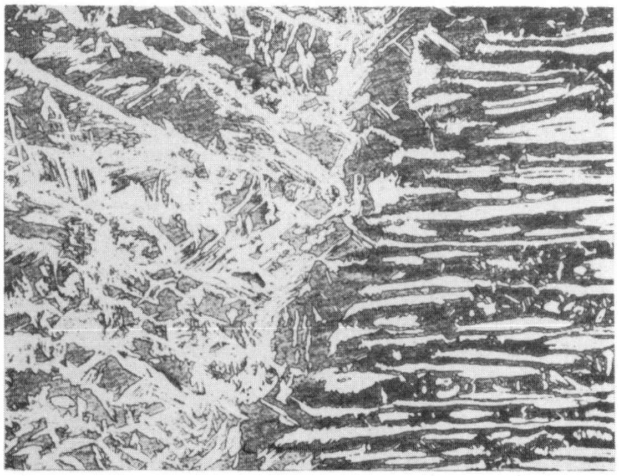

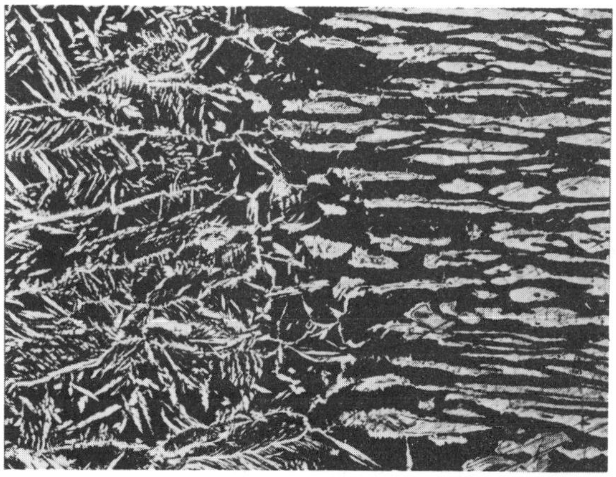

a 75 x c 75 x

a & c : General views : deposited metal on the left side and base metal on the right side.

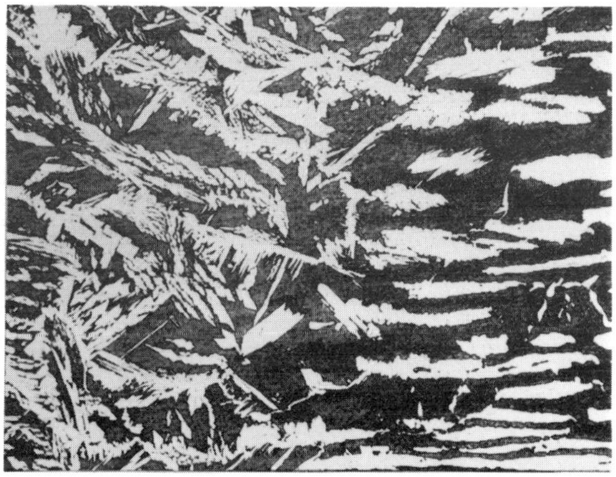

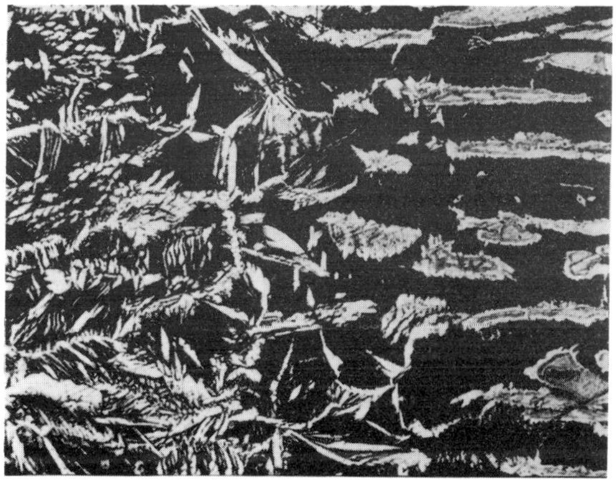

b 187 x d 187 x

b & d : HAZ details : large ferrite areas and transforming austenite.

Fig. 3 Typical microstructures of the joints J1 (8.3 KJ/cm, a & b); J4 (24 KJ/cm, c & d) Beraha's etchant (ASTM 99).
Dark constituent – ferrite, light constituent – austenite

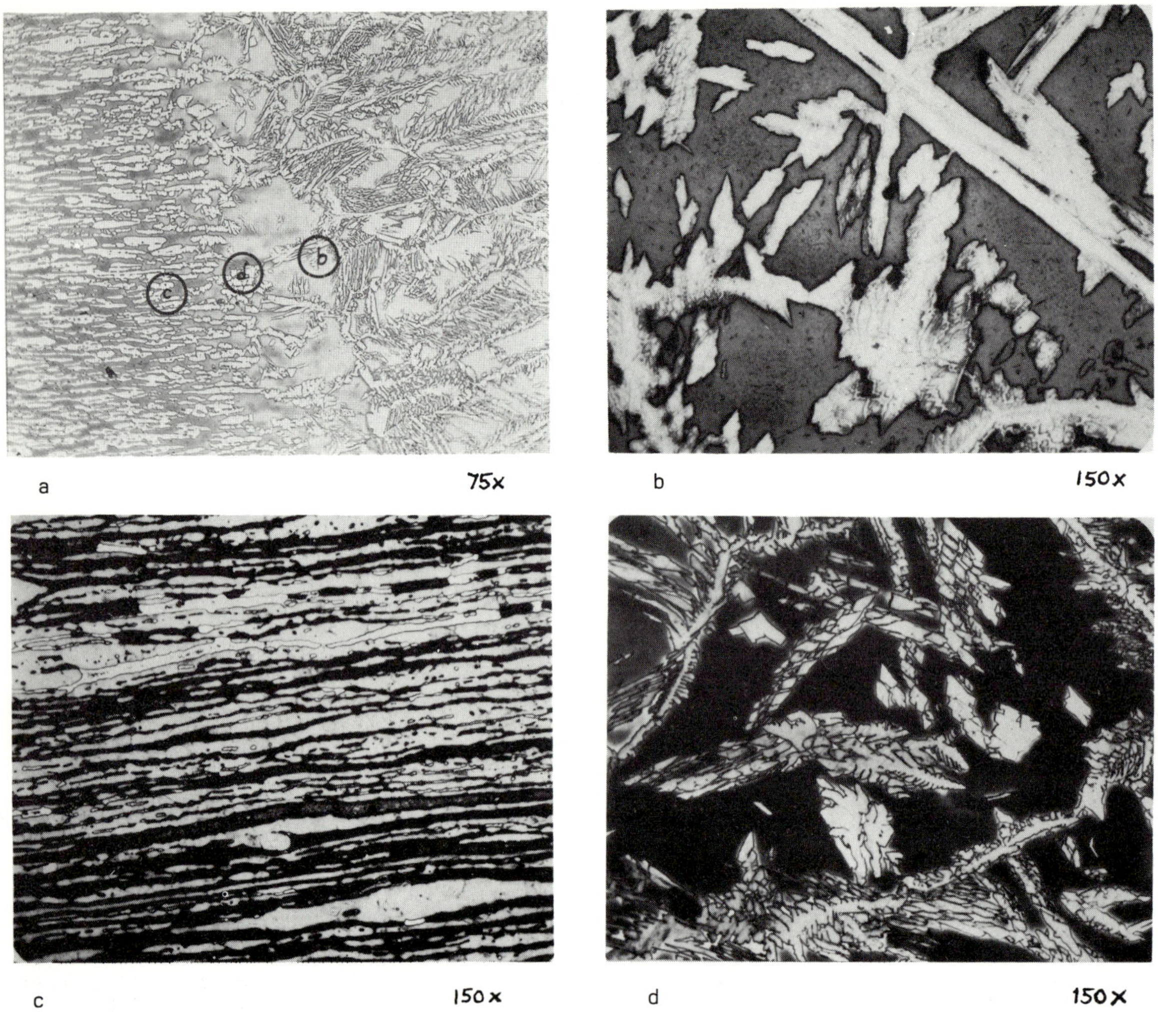

Fig. 4 Weld simulations on base metal, Beraha's etchant (ASTM 99). Dark constituent–ferrite, light constituent – austenite.
a: general view of a welded joint;
b, c, d: simulated HAZ on base metals. Peak T° around 1360°C (b & d) and 1000°C (c)

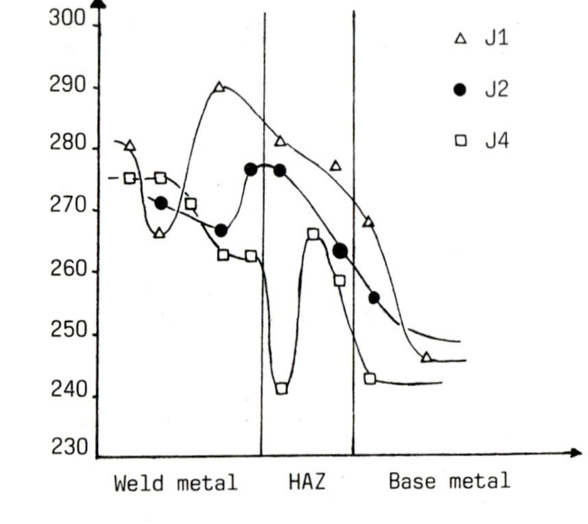

Fig. 5 Microhardness of the joints

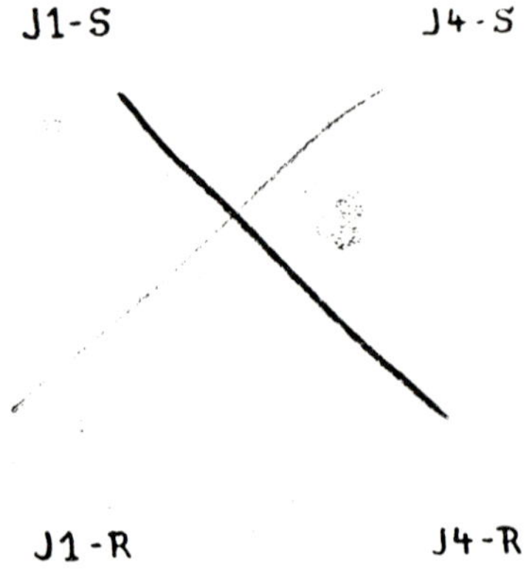

Fig. 6 Appearance of specimens taken on joints J1 (8.3 KJ/cm) and J4 (24 KJ/cm) after ASTM G 48 mod. pitting corrosion test

□

Influence of welding process on mechanical properties of duplex stainless steel weld metals

B Gretoft, S Rigdal, L Karlsson and L-E Svensson

The authors are in the Consumable Product and Marketing Division, ESAB Headquarters, Gothenburg, Sweden.

ABSTRACT

Mechanical properties of four duplex stainless steel weld metals have been measured and the microstructure of the deposits has been examined. The welds were deposited using MIG-welding and submerged arc welding. MIG-welding produced the highest impact toughness, but submerged arc welding with a basic flux gave almost the same properties. After solution heat treatment, the submerged arc weld gave the best properties. The microstructure was in general coarser in the submerged arc welds, both with respect to grain size and second phase particles. The inclusion content seemed to be the most important factor controlling the impact values. Reduction in ferrite content or precipitation of sigma phase did not seem to influence toughness significantly.

1. INTRODUCTION

The advantages of duplex ferritic-austenitic stainless steels in certain applications, such as pipes for gas transports, is now well established. However, some debate still exists regarding the properties of the duplex weld metals and several investigations have been presented during the last years, covering both mechanical and corrosion properties of weld metals.(1-5)

Fabrication of duplex pipes involves both longitudinal seam welding (using submerged arc welding) as well as circumferential welding during actual construction of pipelines. For circumferential welding, it is common practice to use TIG for the root pass.

The filling welds can then be performed with several other methods like manual metal arc, MIG or submerged arc welding. For thin-walled tubes the filling passes can also be made with TIG. In the present paper, the mechanical properties of MIG and submerged arc weld metals are presented and related to their microstructures. This work is a part of a research program, carried out at ESAB´s Laboratories, dealing with mechanical and corrosion properties of high-alloy materials. The mechanical and stress corrosion properties of MMA welds have been presented previously (6) and later papers will discuss pitting corrosion properties, mechanical properties of circumferential welds and also a detailed examination of the microstructures of duplex stainless steel weld metals.

2. EXPERIMENTAL

Four different welds have been examined. They were all of all weld metal type, deposited in a joint according to ISO 2560. Details of the welding procedure are given in Table 1. The first two welds (1 and 2) were made using the MIG-method. The wires were supplied from two different manufacturers, A and B. For the submerged arc welds, (3 and 4) the same wire with two different agglomerated fluxes was used. Weld 3 was made with ESAB OK Flux 10.16, a basic flux giving a deposit with low oxygen content and good mechanical properties. Weld 4 was made, using the neutral flux ESAB OK Flux 10.92, which has very nice operating characteristics, but does not give the same good mechanical properties.

The four welds were tested in two conditions: as-deposited and solution heat treated. The solution heat treatment was carried out at 1050°C for 0.5 hour, followed by rapid quenching. For each weld, one longitudinal tensile specimen and 10 Charpy V impact specimens were tested. The impact specimens were tested at -60°C, -20°C and 20°C, with three specimens at each temperature. In some cases, an additional specimen was tested at -196°C.

The chemical analysis of the welds was determined using an Optical Emission Spectrometer, except for carbon, oxygen and nitrogen, where Leco furnaces were used. The ferrite content was measured using a Magne Gage, calibrated to measure Ferrite Numbers up to FN75.

Specimens were prepared for examination of the microstructures by standard grinding and polishing methods. Etching was made, using a modified Behera II method,(7) producing a difference in colour between austenite and ferrite. The microstructures were examined using a Leitz Metalloplan optical microscope.
Further examination of the microstructures and of fracture surfaces was made, using a JEOL T-200 Scanning Electron Microscope (SEM). Specimens were also examined in a Philips EM 300 Transmission Electron Microscope (TEM). Thin foils

were prepared by electrolytical polishing in a Struers Tenupol, using 15% perchloric acid in methanol at -35°C.

3. RESULTS

The chemical analysis of the weld metals is given in Table 2. (All figures are in weight per cent.) Some clear differences in composition can be noted. The carbon content of the submerged arc welds is higher than in the MIG-welds, while the nitrogen content is lower. The oxygen contents of the two MIG-welds and the submerged arc weld, using the basic flux, are all on the same level, while being considerably higher for the submerged arc weld obtained with the neutral flux.

The result of the mechanical testing is given in Table 3. The 0.2% proof stress and the ultimate tensile strength was approximately the same for all the welds in as-deposited condition. After solution heat treatment, the 0.2% proof stress decreased by approximately 100-150 MPa, while the ultimate tensile strength only decreased by 30-40MPa. Simultaneously, the ductility was increased. The impact toughness of the as-deposited specimens is shown graphically in Fig. 1 and a comparision of impact toughness between as-deposited and solution heat treated specimens is presented in Fig. 2.

The impact toughness showed a similar pattern for all four welds with a fairly constant toughness down to -20°C, followed by a decreasing toughness with further decrease in temperature. The tests made at -196°C were always found to be very brittle.

Examination of fracture surfaces by SEM showed that down to -60°C fracture mainly took place by a fibrous fracture mode, with only very small areas of cleavage fracture. At -196°C, however, the fracture was completely brittle,with cleavage fracture over the whole surface.

Although the same general pattern was found for all welds, the level of impact toughness, from -60°C and upwards, was significantly different between the welds. The MIG-weld from manufacturer A gave the highest impact toughness. The MIG-weld from manufacturer B and the submerged arc weld obtained with the basic flux gave very similar impact toughness, while the submerged arc weld obtained with the neutral flux gave the lowest values. However, it should be noted that the impact toughness for this last mentioned weld did not decrease very much down to -60°C.

For information, the impact toughness from a manual metal arc weld (OK 67.50) (6) has also been incorporated in Fig. 1. This MMA stick electrode has an acid-rutile coating, giving a relatively high amount of oxygen in the weld metal.

The influence of solution heat treatment was somewhat surprising. The MIG-weld from manufacturer A did not improve in impact toughness, while the MIG-weld from manufacturer B improved considerably and in fact gave an impact toughness in excess of the other MIG-weld (see Fig. 2 a and b). Furthermore, the submerged arc weld using the basic flux improved even more and gave the highest impact toughness of all four welds.

The ferrite content of the weld metals is given in Table 4. The ferrite content in as-deposited condition was fairly equal, except for weld 4 (with the neutral flux). After the solution heat treatment, welds 2 (manufacturer B) and 4 (neutral flux) changed in ferrite content considerably.

The microstructure in all four weld metals consisted of primary ferrite grains with austenite mainly precipitated in a Widmannstätten manner. The primary ferrite grain size of the MIG-welds was somewhat smaller than in the submerged arc welds. Representative micrographs of one MIG-weld and one submerged arc weld are shown in Figs. 3 and 4. Using TEM, the difference between the welds was examined in more detail. In the MIG-welds, the small grains also appeared to be more angular in shape (Fig. 5). The density of low-angle grain boundaries was also much higher in the MIG-welds (Fig. 6).

In weld 4 (neutral flux), the slag inclusions were large and rounded, while in weld 3 (basic flux), fewer slag inclusions with a more angular appearance were found. In the MIG-welds the density of particles was much higher, while the average size of the particles was smaller. It seemed that austenite grains could nucleate on these inclusions (Fig. 7). After solution heat treatment, the grains generally had a softer, more rounded appearance. However, this effect was more pronounced in the submerged arc weld metals.

In weld 2 (MIG-weld from manufacture B) some sigma-phase particles were observed in as-welded condition. They appeared mainly on the primary ferrite grain boundaries. After solution heat treatment, the sigma phase precipitation was very extensive (Fig. 8).

4. DISCUSSION

The low nitrogen content of the submerged arc welds arose due to the low nitrogen content of the wires. In fact, a nitrogen content of around 0.06% is similar to what is normally found in stainless steel weld metals without nitrogen alloyed wires.

Despite the low nitrogen content of weld 3, the ferrite numbers of welds 1-3 are very similar in as-deposited condition. Calculation of the Q-factor, introduced by Noble and Gooch (3) shows that all four welds have a Q-factor of approximately 1.6 which should give a ferrite content of approximately 40%. It should, however, be noted that the scatter in the correlation between Q-factor and ferrite content is quite large.

Weld 4 (using the neutral flux), which also had a Q-factor of about 1.6 gave a substantially higher ferrite content. It seems likely that austenite formation has been suppressed in this weld. One obvious difference between weld 4 and the three other welds is the oxygen content but differences in Mn and Si can also be noted. It is possible that the oxide inclusions do play a role in the formation of the austenite and that the large globular inclusions in weld 4 are inefficient in this respect.

The solution heat treatment changed the ferrite content of welds 2 and 4 significantly, while welds 1 and 3 were virtually unchanged. In weld 2, the ferrite transformed to a large extent to sigma phase. The reason for this is not yet understood. It is on the contrary expected that the sigma phase would dissolve during the solution heat treatment.

The ferrite content of weld 4 dropped to a value quite close to the values of welds 1 and 3. If it is assumed that these values are the equilibrium values, predicted by the phase diagram, then once again it seems that the high ferrite content of weld 4 in as-deposited condition is due to inhibited nucleation and not to any other chemical factor.

The impact results follow broadly the trends that would be assumed, looking at the oxygen contents. However, the difference between the two MIG-welds is interesting, since their chemistry is almost identical and yet there is a clear advantage of the wire from manufacturer A for the as-deposited condition. The increase in impact toughness of the MIG-weld with wire from manufacturer B is surprising, since the extensive sigma phase precipitation should lead to a strong decrease in toughness.

The observation that the submerged arc weld, obtained using the basic flux gave the highest impact toughness after solution heat treatment is interesting. This could be due to factors like dispersion parameters of the inclusions, rather than just volume fractions (or total oxygen content). Down to -60°C, fracture was mainly by a fibrous fracture mode, although some examples of cleavage fracture in the ferritic areas could be seen. Anyhow, it seems that it is the toughness of the austenite which is the most important factor at these temperatures. This is also supported by the very slight increase in impact toughness of weld 4 after heat treatment, in spite of the large increase in austenite content.

There are still many factors not understood, concerning the development of microstructure in duplex stainless steel weld metals. Also the interaction between microstructure and properties is not understood in detail. From this investigation, it seems that with a ferrite content of around FN 40 and an oxygen content of around 0.07% all welding processes tested will give approximately the same impact toughness. Submerged arc welding, with its comparatively higher productivity can certainly be used with confidence, if proper choice of consumables is made. Circumferential welding is today mainly made using MMA and MIG. From the point of view of mechanical properties, none of the methods are preferable to the others. Submerged arc welding can also be used for circumferential welding, if roller beds are used for rotating the pipe. Investigations into the effect of different welding processes on the properties of circumferential welded pipes is in progress and will be published in due course.

5. CONCLUSIONS

- A comparison has been made between the mechanical properties that can be obtained in duplex stainless steel weld metals, using manual metal arc, MIG and submerged arc welding.

- The tensile properties were all at the same level, irrespective of welding process.

- The highest impact toughness in the range -20°C — -60°C was obtained with the MIG process. However, MMA and submerged arc welding with a basic flux gave also excellent impact toughness values.

- The impact toughness in this temperature range was mainly influenced by the oxygen content of the welds. Fracture mainly took place by fibrous fracture in the austenite.

- In one of the MIG-welds studied, extensive sigma-phase precipitation took place after solution heat treatment. However, this did not lead to any reduction of the impact toughness.

- Solution heat treatment decreased the tensile strength and improved the impact toughness of the welds.

REFERENCES

1. H. Hoffmeister, R. Mundt and K. D. Buner: IIW-Doc. IX-1339-85

2. J. Honeycombe and T. G. Gooch: Weld Inst. Rep. 286/1985

3. D. N. Noble and T. G. Gooch: Weld Inst. Rep. 321/1986

4. B. Lundqvist, P. Norberg and K. Olsson, Proc. Int. Cont. Duplex Stainless Steel '86, The Hague, The Netherlands, 1986

5. M. Miura, T. Kudo, H. Tsuge, M. Koso and T. Kobayashi: ibid

6. L-E Svensson and B. Gretoft: ibid

7. E. Weck and E. Leistner: Metallographic instructions for colour etching by immersion, Pert II, Deutscher Verlag für Schweisstechnik, 1983

Table 1. Welding conditions of the studied samples

Sample No.	Consumable Wire	Ø mm	Flux/ Shielding gas	Current DC(RP) A	Voltage V	Welding speed m/h	Heat imp. kJ/mm	Heat treatment	°C	h
	MIG *)									
1	A	1.2	Ar + 2% O_2	250	25	17	1.3	AW	-	-
1a	A	"	"	"	"	"	"	PWHT	1050	1/2
2	B	1.2	Ar + 2% O_2	250	25	17	1.3	AW	-	-
2a	B	"	"	"	"	"	"	PWHT	1050	1/2
	SAW									
3	OK Autrod 16.86	3.0	OK Flux 10.16	420	28	25	1.7	AW	-	-
3a	"	"	"	"	"	"	"	PWHT	1050	1/2
4	OK Autrod 16.86	3.0	OK Flux 10.92	420	27	30	1.4	AW	-	-
4a	"	"	"	"	"	"	"	PWHT	1050	1/2

*) Wire A is equal to ESAB OK Autrod 16.86. B is a wire of similar kind, from another manufacturer.

Joint type according to ISO. No preheating. Max 150°C interpass temperature.

Table 2. Chemical analysis of the weld metals. All figures are in weight percent.

Weld No.	C	Si	Mn	P	S	Cr	Ni	Mo	N	O
1. MIG Manufacturer A	0.013	0.45	1.58	0.015	0.004	22.22	7.96	2.86	0.12	0.06
2. MIG Manufacturer B	0.013	0.49	1.68	0.022	0.006	22.63	8.95	3.07	0.12	0.06
3. SAW OK 16.86/OK Flux 10.16	0.021	0.29	1.47	0.026	0.012	21.91	8.60	2.92	0.07	0.07
4. SAW OK 16.86/OK Flux 10.92	0.025	0.66	1.03	0.025	0.013	22.11	8.24	2.92	0.07	0.12

Table 3. 0.2% proof stress, ultimate tensile strength, elongation and area reduction of the welds.

Weld	R_{po2}(MPa)	Rm (MPa)	A (%)	Z (%)
1 (MIG, Manufacturer A)	574	765	32	55
1a (Solution treated 1050°C)	454	724	35	59
2 (MIG, Manufacturer B)	598	772	27	48
2a (Solution treated 1050°C)	470	766	34	55
3 (SAW, basic flux)	559	740	29	52
3a (Solution treated 1050°C)	459	698	32	59
4 (SAW, neutral flux)	606	735	20	36
4a (Solution treated 1050°C)	454	703	30	51

Table 4. Ferrite content of the welds, measured using Magne Gage.

Specimen	Ferrite Content (FN)
1 (MIG, Manufacturer A)	43
1a (Solution treated 1050°C)	36
2 (MIG, Manufacturer B)	47
2a (Solution treated 1050°C)	23
3 (SAW, basic flux)	44
3a (Solution treated 1050°C)	41
4 (SAW, neutral flux)	64
4a (Solution treated 1050°C)	41

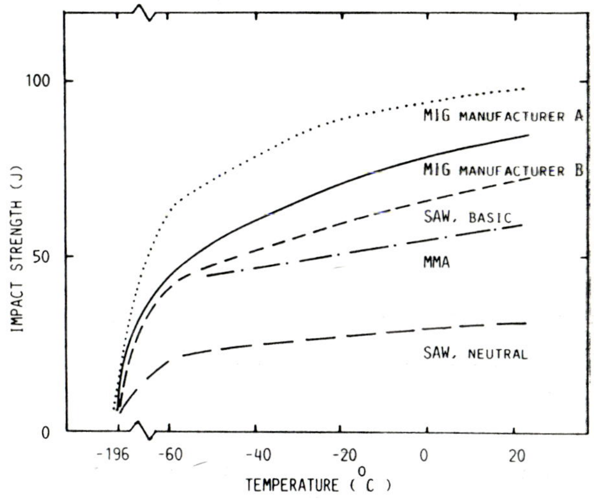

Fig. 1 Impact toughness of as-deposited weld metals, as a function of temperature. For information, impact toughness of an MMA electrode has been incorporated, taken from Ref. (6)

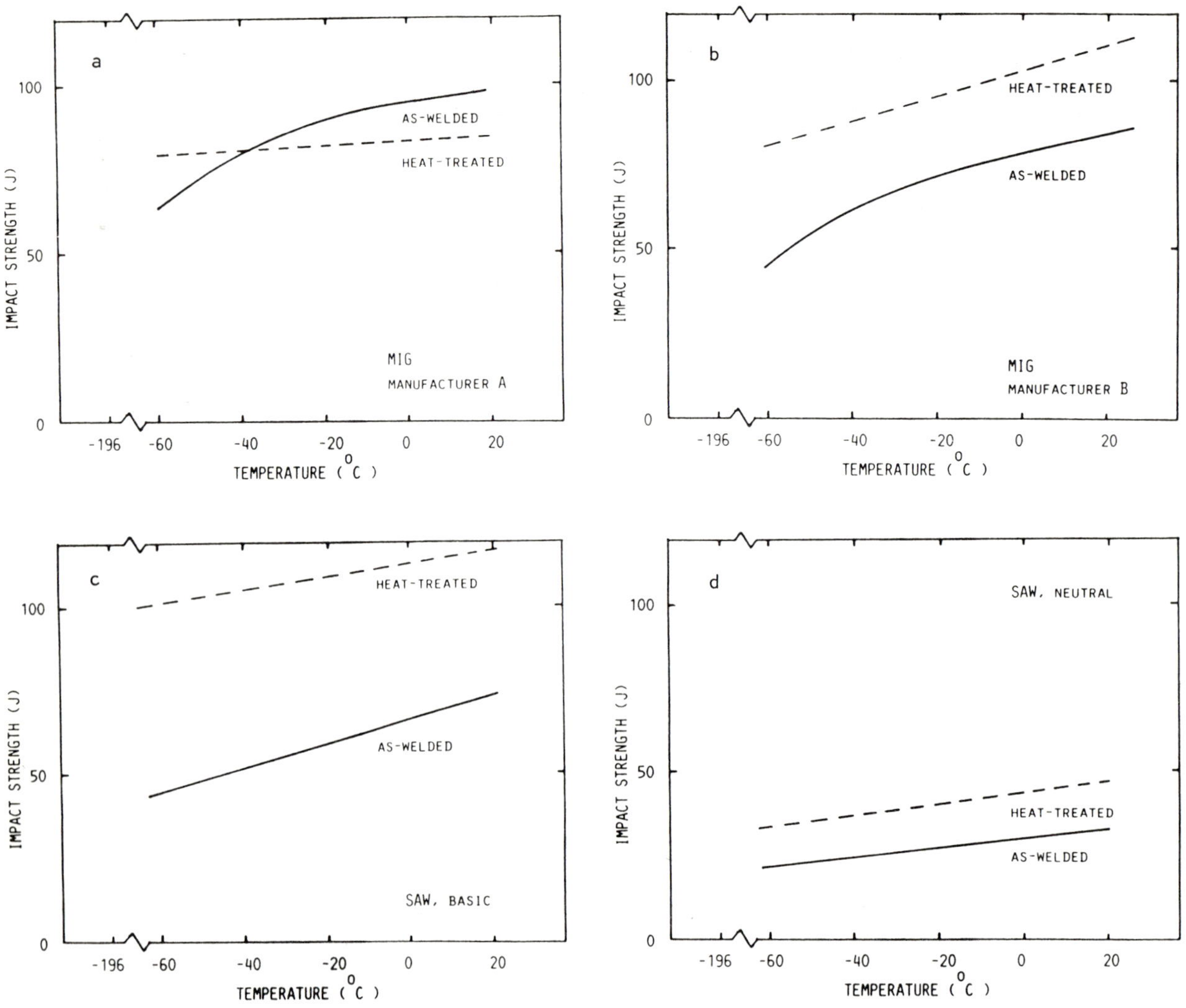

Fig. 2 (a-d)
Comparison of impact toughness between as-deposited and solution heat treated specimens.
a) MIG-weld, manufacturer A
b) MIG-weld, manufacturer B
c) Submerged arc weld, basic flux OK 10.16
d) Submerged arc weld, neutral flux OK 10.92

Fig. 3 Optical micrograph from weld 1 (MIG-weld, manufacturer A), showing the common primary ferrite grains with precipitated austenite

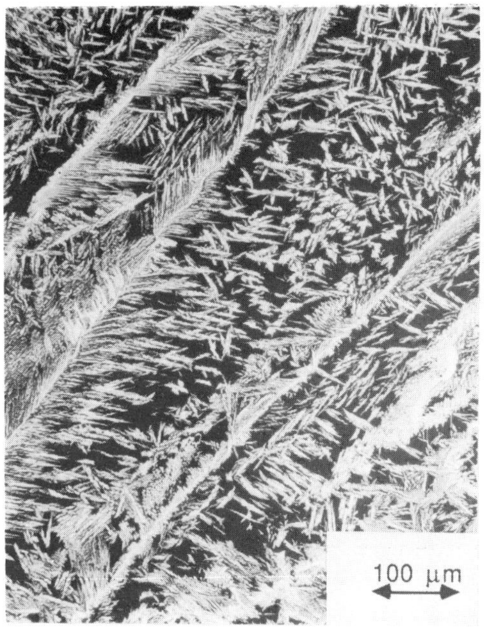

Fig. 4 Optical micrograph from weld 3 (submerged arc weld with basic flux). The ferrite grain size is somewhat coarser than in the MIG-welds

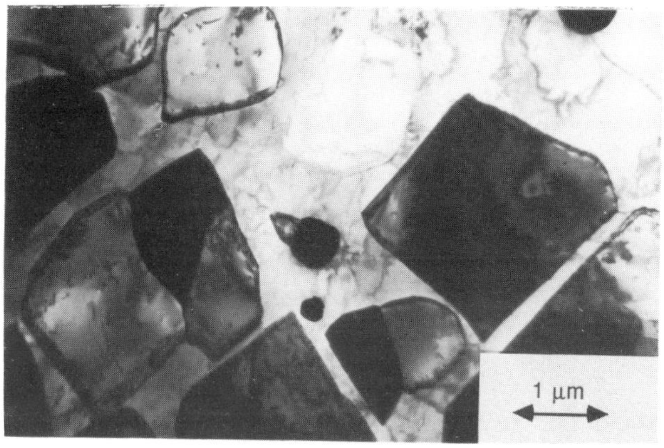

Fig. 5 TEM micrograph of MIG-weld 2 (manufacturer B). Relatively small ferrite and austenite grains are shown having an angular shape

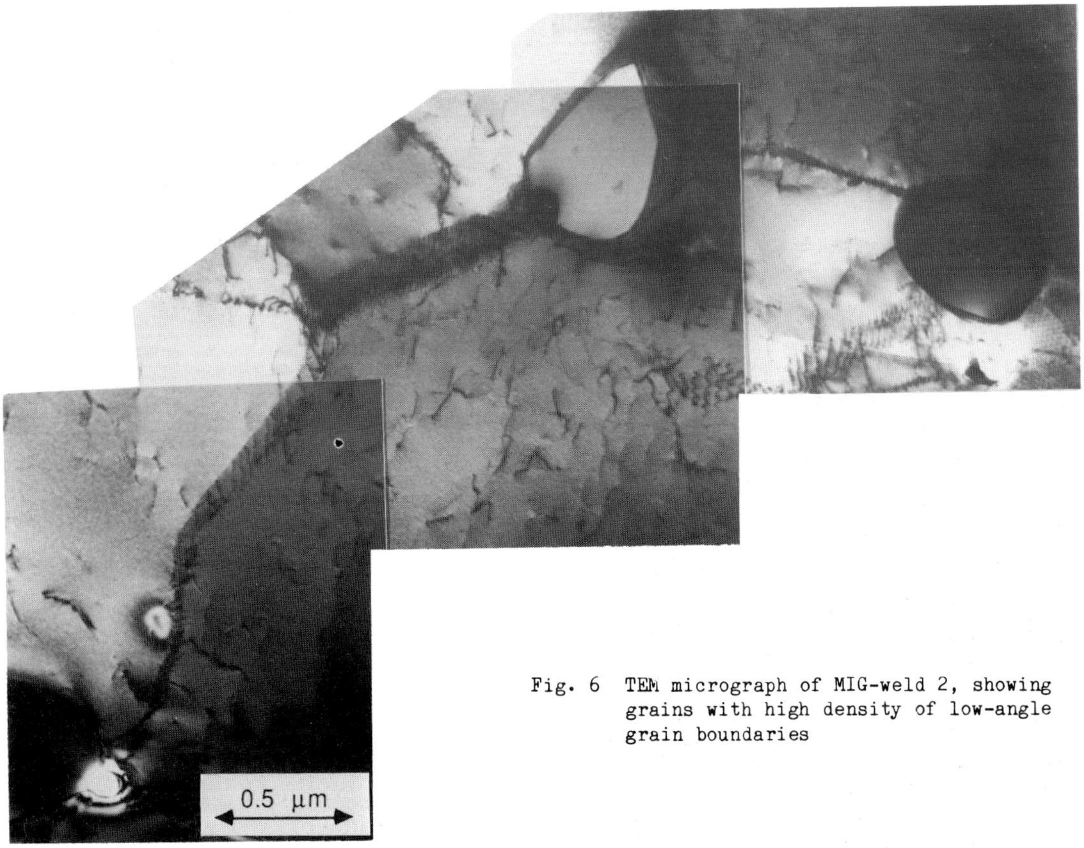

Fig. 6 TEM micrograph of MIG-weld 2, showing grains with high density of low-angle grain boundaries

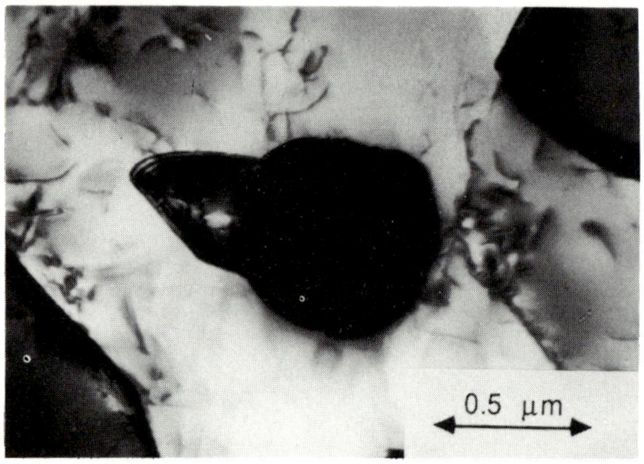

Fig. 7 TEM micrograph of MIG-weld 2. Relatively fine oxide inclusion with angular shape. A small grain of austenite has nucleated on the particle

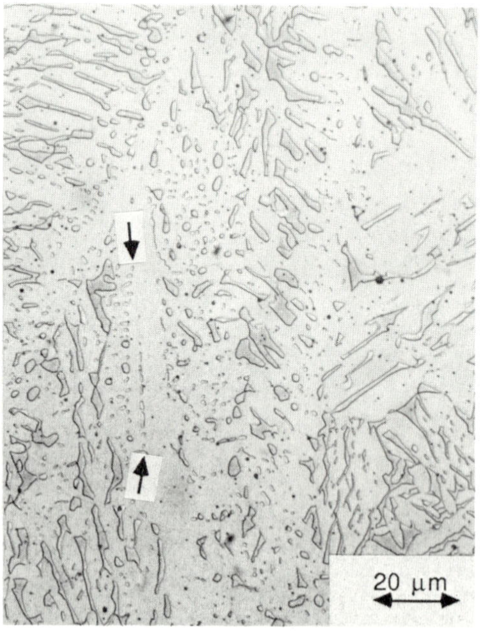

Fig. 8 Optical micrograph of MIG-weld 2, showing heavy precipitation of sigma-phase after solution heat treatment

Phase transformation in duplex 20Cr—9Ni—3Mo stainless steel welds aged at 1050°C

T A Towers and G A Honeyman

Dr. Towers is Head of the Academic Services Centre, Teesside Polytechnic; Dr. Honeyman is with NEI Parsons, Newcastle Upon Tyne.

ABSTRACT

Recent investigations have shown that the heat treatment of duplex austenitic-ferritic stainless steel can result in the readjustment of the compositions of each phase. A series of alloys based on the composition 20Cr 9Ni 3Mo have been manufactured containing various levels of silicon. During post-weld heat treatment at 1050°C, delta ferrite partially transformed to sigma phase and retransformed back to ferrite on further heating, the total time at temperature being four hours. The maximum amount of sigma in any alloy was directly related to the silicon content. Solidification theories and alloy diffusion models are put forward as an explanation for this phenomenon.

1 INTRODUCTION

Duplex stainless steels have developed over a number of years resulting in alloys with improved weldability, increased strength and superior stress corrosion cracking resistance compared to conventional austenitic steels. One important use currently is in the joining of bullet-proof steels.

The problems associated with the welding of these steels are:

(a) Centre weld cracking,
(b) Interface cracking,
(c) Heat Affected Zone (HAZ) cracking.

These difficulties are due to the alloying elements contained in the steels which, after suitable heat treatment, are responsible for its high impact resistance. These elements were primarily chosen for their capacity to produce a very strong steel at moderate cooling rates. If weld material of the same composition were used, then owing to the rapid cooling associated with fusion welding processes, the resultant high hardness would be such as to make cracking of the weld inevitable. Furthermore, severe coarsening of the microstructure can exacerbate these effects. To overcome these problems, a 20Cr 9Ni 3Mo duplex austenitic stainless steel has been developed as the consumable, provided interfusion with the parent metal is the minimum necessary to form a true welded junction and there is no appreciable coarsening of the parent plate due to its low melting point. In addition this weld material undergoes no hardening change, therefore cracking, as a result of hardening stresses, is avoided.

One disadvantage of 20/9/3 austenitic stainless steel is its low strength and an investigation was set up to improve the tensile properties of the alloy by alloying additions. Silicon was chosen as it reduces the stacking fault energy of austenitic steel, increases the delta ferrite content and distorts the lattice network; all three mechanisms resulting in contributions to an overall increase in tensile strength.(2)

In order to study the alloy partitioning effects in the duplex alloys, a post weld heat treatment study was carried out. A temperature of 1050°C was selected to avoid serious second phase dissolution and to, theoretically, avoid the regime of sigma phase formation. However, as is demonstrated in this presentation, alloy diffusion does take place, giving formation of sigma at temperatures above the conventional maximum limit of 950°C and at much shorter times than previously perceived.

2 LITERATURE REVIEW

Rapid solidification of certain compositions of stainless steels gives rise to very fine two-phase structures of austenite and delta ferrite. Castro and Cadenet (3) presented ferrite and austenite mixtures within the dendrite axes and interdendritic regions of a forged stainless steel, the compositions of which were calculated using electron probe microanalysis techniques (**Table 1**). The results clearly indicate the tendency for ferrite to be enriched in chromium and silicon and impoverished in nickel and manganese. Lyman et al (4) using STEM examined the segregation of chromium and nickel in autogenous Type 304 stainless steel welds. Microanalysis along a traverse intersecting regions of retained delta ferrite revealed the extent of partitioning between the two phases (**Fig. 1**). A major contribution to the understanding of partitioning effects has been put forward by Lippold and Savage (5) with their solidification model for duplex stainless steel welds. They studied the microstructures of a number of welds including Type 301 and 312 filler stainless steels and, with the aid of electron probe microanalysis, were able to explain the compositional variances in terms of their model.

The phase stability of duplex steels has been the subject of many investigations. Early work by Weiss and Stickler (6) showed that, for Type 316 austenitic steel, numerous carbides and intermetallic phases appear upon elevated temperature ageing. One of the more significant phases was sigma which is a hard brittle intermetallic compound with a complex tetragonal structure. Since sigma is normally associated with long-term embrittlement, the conditions for its formation are of considerable interest. More recently, Vitek and David (7) have shown that the critical nature of sigma nucleation is not dependent on the diffusion rate of chromium during nucleation, but rather on the structural requirements for establishing sigma phase nuclei. They observed that large-scale movements, such as the advance of a recrystallisation front or deformation can significantly accelerate the sigma phase formation. However, Farrar (8) showed that the rate of transformation of the delta ferrite in Type 316 stainless steel depends on the original segregation of chromium, molybdenum and nickel in the as-welded condition and that an initial transformation to $M_{23}C_6$ at the delta ferrite-austenite boundaries, or within the ferrite laths, is a precursor to sigma formation. Gill et al. (9) found that transformation of delta ferrite was fastest at 700°C and the rate was directly related compositionally to the 'equivalent chromium content' first muted by Hull. (10)

3 EXPERIMENTAL PROCEDURE AND RESULTS

Standard welds were produced from manual metal arc electrodes with a base composition of 0.065 C, 2.0 Mn, 19.58 Cr, 8.35 Ni and 3.01 Mo. A series of increasing silicon content welds were made up by using "doped" coated electrodes and their compositions are indicated in Table 2. Welds were made up from several runs deposited on steel plates, from which 10mm cubes for heat treatment studies and metallographic evaluation were machined. Typical as-welded microstructures are indicated in Fig. 2. Specimens from each alloy were heat treated at 1050°C for various times up to four hours. Examination of the microstructures revealed the presence of both sigma phase and delta ferrite (Fig. 3), the proportions of which are identified in Table 3. SEM studies were carried out on all specimens and the quantitative phase analysis results are shown in Tables 4, 5 and 6. Thin films were also produced by a jetting technique and transmission electron microscopy confirmed the structure of sigma in all the alloys under investigation.

4 DISCUSSION

The microstructural examination revealed that sigma can form at much higher temperatures and at significantly shorter times than previously reported. (**refs 7-11**) Sigma formed rapidly from existing islands of delta ferrite particles, reaching a maximum volume fraction after only two hours in all alloys irrespective of silicon content (**Fig. 4**). However, after further heat treatment, the sigma areas retransformed back to delta ferrite. It was noted that the higher the silicon content the greater the volume of sigma phase present (**Table 3, Fig. 4**) confirming the high sigma-promoting tendency of this element. Using the solidification model of Lippold and Savage, (5) this unusual phenomenon has been previously attributed to alloy diffusion effects as a result of compositional stabilisation during post-weld heat treatment.(12)

(i) Weld Solidification

The composition of the welds investigated indicates that the alloys can be considered to solidify under model 1 conditions where the primary solidification phase is delta ferrite followed by a massive transformation to austenite on further cooling. Retained delta ferrite is positioned at the dendrite cores forming an interconnecting network in the high silicon alloy (**Fig. 2**). The first metal to solidify contains high concentrations of ferrite-forming elements such as chromium and molybdenum, being a maximum at the centre of the dendrites.

Unfortunately, the delta ferrite formed in these alloys has a structure that was too fine to allow definitive analyses at the centre and edges of the second phased particles. However, the partitioning effects between the austenite and ferrite were readily apparent, particularly with the elements silicon, molybdenum, chromium and nickel (**Tables 4, 5 and 6**). The results are in general agreement with those obtained by Castro and de Cadenet (3) and Lyman et al.(4)

(ii) **Formation of Sigma From Delta Ferrite**

It is proposed that due to the coring effects there would be microsized areas of inhomogenous alloy content in the delta ferrite particles with compositions within the gamma-delta - sigma region of the constitutional diagram (**Fig. 5**), but due to the rapid cooling of the welds, the sigma phase has insufficient time to nucleate. However, during heat treatment at 1050°C, the energy for transformation is supplied and a massive transformation to sigma is affected. The compositions of the sigma phases analysed are somewhat complex and their structures difficult to determine. For example, in the 0.59 Si alloy, the sigma phase contained a high molybdenum content, other elements being similar to that found in the delta ferrite (**Table 4**). On the other hand, in the 1.11 Si weld (**Table 5**), three separate sigma phases were determined, one of which contained exceedingly high silicon (6.83%) and manganese (8.84%). Another phase had low silicon, but high molybdenum (3.3%) and chromium (30%). The third form of sigma had a composition with high silicon, molybdenum and chromium. Similar compositions were detected in the sigma phase of the 2% silicon containing alloy (**Table 6**). The presence of high molybdenum intermetallics suggests the inclusion of Chi phase, but confirmation of its transformation has yet to be determined.

Time at temperature had little effect on the composition of either the austenite or ferrite phases (**Tables 4, 5 and 6**). Apart from the chromium, which showed a general tendency to diffuse into the delta ferrite phases at longer soaking periods, the remaining elements showed no such definable trends.

A metallographic study indicated that the formation of sigma phase was not randomly distributed throughout the weld, but restricted to well-defined zones within the material (**Fig.3**). This would tend to suggest some macro-partitioning of the elements which could be attributed to the inhomogenous transfer of some of the alloying elements from the electrode coating to the weld deposit during transfer. This could account for the variation in compositions of all three phases, but particularly sigma, shown in the analysis of the various specimen welds. An analysis of the austenite and ferrite in the untransformed zones showed a tendency for silicon to be lower in content to those values measured in the transformed zones.

It was also established that as well as the presence of sigma phase, particles of $M_{23}C_6$ (**Fig. 6**) had been nucleated at the delta ferrite/austenite boundaries which is known to increase the tendency of the delta ferrite to nucleate sigma.(8) Sigma was seen to be of a 'block' form and not the lamella type observed by Pickering (11) and Hodgkinson.(13)

(iii) **Retransformation of Sigma to Delta Ferrite**

It is proposed that continued heat treatment of the welds at 1050°C permitted further diffusion of the alloying elements resulting in the retransformation of sigma to delta ferrite in the delta + gamma field. Alloy redistribution resulted in volume fractions of ferrite being similar to those measured prior to heat treatment. This is considered to be coincidental and related to the cooling rates during welding. All steels showed an initial reduction in austenite hardness followed by a gradual rise as heat treatment progressed. This phenomenon was probably due to changes in alloy composition, phase proportions and stress relief.

The heat treatment experiments have shown that the microstructure of the welded austenitic 20Cr 9Ni 3Mo stainless steel is formed under non-equilibrium conditions and that short excursions at high temperature can result in the formation of the brittle sigma phase as the material strives to attain equilibrium. The effect of silicon confirms the compositional dependency of sigma formation as observed by many authors including Farrar (8) and Gill et al. (9) However, the rate of formation appeared to be unaffected by any compositional variances as suggested by Vitek and David.(7) A heat treatment of 750°C for four hours resulted in total transformation of all delta ferrite to sigma which could only be retransformed by heat treating at a temperature of 1050°C and above, confirming the temperature dependency of sigma phase transformation.

Although bullet-proof steels will not be subjected to service temperatures within the range described in this paper, it is feasible that in welded joints requiring multi-pass runs, the heat generated from successive layers could result in the nucleation of small amounts of sigma which can ultimately adversely affect the ambient temperature properties of the joint.

5 CONCLUSIONS

5.1 Sigma phase forms at much higher temperatures and at significantly shorter times than previously considered.

5.2 Silicon has a substantial effect on the tendency of austenitic stainless steel to form sigma.

5.3 Doping of electrodes can affect the homogeneity of the weld microstructure resulting in macro-segregation of certain alloying elements. In turn, this may result in the formation of sigma phase.

5.4 Care must be exercised during multi-pass welding using duplex austenitic stainless steels high in ferrite-forming elements, since sigma may form which may impair the ambient temperature impact properties.

REFERENCES

(1) Holt R M et al, JISI 200, 715 (1962)

(2) Honeyman G A and Towers T A, 'Stainless Steel 84' (1984)

(3) Castro R and de Cadenet J J, 'Welding Metallurgy of Heat Resisting Steels', Cambridge University Press (1975)

(4) Lyman C E et al, Scan Elect Mic, 1, 213 (1978)

(5) Lippold J C and Savage W F, Weld Research Supplement, 363 (1979)

(6) Weiss B and Stickler R, Met Trans, 3, 851 (1972)

(7) Vitek J M and David S A, Weld Research Supplement, 106s (1986)

(8) Farrar R A, J Mat Sci, 20, 4215-4231 (1985)

(9) Gill T P S, Vijayalakshmi M, Gnanamoorthy, J B, Padmarabham K A, Weld Research Supplement, 122s (1986)

(10) Hull F C, Weld J, 52, 104s (1973)

(11) Pickering, F B, ISI, Sp Rep, 64, 118 (1959)

(12) Honeyman G A and Towers, T A, Nat Res Soc Symp Proc, 741, 21 (1984)

(13) Hodgkinson, MPhil thesis, CNAA, Teesside Polytechnic (1982)

Table 1

Phase Analysis of 18.6 Cr, 11.8 Ni, 0.12 C, 0.42 Si, 1.72 Mn, 0.49 Ti Duplex Austenitic Stainless Steel (After Castro and Cadenet Ref 3)

PHASE	Composition				
	Fe	Ni	Mn	Cr	Si
Austenitic: dendritic axes	68.6	11.4	1.6	18.0	0.2
Interdendritic regions	64.9	12.8	2.1	19.4	0.5
Ferrite: dendritic axes	66.5	5.5	1.4	26.0	0.45
Interdendritic regions	63.2	6.2	1.7	28.1	0.6

Table 2

Composition of the As-welded Deposits
Produced from the MMA Electrodes (Batch 1)

WELD MATERIAL	COMPOSITION									
	C	Mn	Si	S	P	Mo	Ni	Cr	Co	N
*SE	0.065	2.05	0.50	0.013	0.013	3.05	8.44	20.00	0.00	0.02
1Si	0.060	1.99	0.59	0.013	0.013	3.05	8.88	20.28	0.00	0.03
2Si	0.059	2.20	0.76	0.013	0.015	2.95	8.35	19.92	0.00	0.02
3Si	0.067	2.39	1.11	0.013	0.014	2.76	8.29	19.79	0.00	0.03
4Si	0.065	2.42	1.63	0.017	0.017	3.10	8.26	20.06	0.00	0.03
5Si	0.066	2.53	2.00	0.016	0.018	3.01	8.12	20.00	0.00	0.02

*SE = Standard weld produced from Armoid 1A electrodes

Table 3

Variations in V_f of each phase and hardness of with time

	0.59 Si		0.76 Si		1.11 Si		1.63 Si		2.0 Si	
	%	%	%	%	%	%	%	%	%	%
0	17.7	-	17.0	-	23.3	-	23.2	-	35.0	-
1/2	15.3	-	11.7	4.8	17.7	3.8	20.0	3.2	28.2	1.2
1	10.3	-	6.7	7.3	13.8	4.8	15.8	4.6	16.7	9.7
2	12.6	0.5	8.5	5.0	12.3	7.8	11.2	7.7	1.8	23.8
3	12.2	1.2	12.8	1.2	19.0	6.8	11.6	9.1	13.5	12.0
4	15.0	-	14.8	-	26.0	-	25.8	-	35.2	-

Table 4

Quantitative Phase Analysis of 1 Si Weld

AUSTENITE   1 Si (0.59%)

Exposure time at 1050°c	Composition (wt %)				
	Si	Mo	Cr	Mn	Ni
0	0.56	2.02	22.49	2.62	10.38
1/2 hr	0.69	1.72	21.50	2.40	10.37
1 hr	0.60	1.94	21.11	2.12	8.82
2 hr	0.47	1.38	21.38	2.57	9.96
3 hr	0.59	1.76	21.60	2.19	9.55
4 hr	0.56	1.70	21.37	2.65	9.71

FERRITE

0	0.83	3.38	26.63	2.42	5.29
1/2 hr	0.74	3.34	26.57	2.07	5.93
1 hr	0.64	3.32	25.85	2.14	5.29
2 hr	0.50	2.92	26.26	2.03	5.11
3 hr	0.61	3.27	27.21	1.99	5.17
4 hr	0.64	2.97	27.25	2.12	5.19

SIGMA

0	-	-	-	-	-
1/2 hr	-	-	-	-	-
1 hr	-	-	-	-	-
2 hr	0.78	5.73	27.48	2.36	5.58
3 hr	0.61	5.14	26.85	2.16	6.53
4 hr	-	-	-	-	-

Table 5

Quantitative Phase Analysis of 3 Si Weld

AUSTENITE 3 Si (1.11%)

Exposure time at 1050°c	Composition (wt %)				
	Si	Mo	Cr	Mn	Ni
0	1.08	1.28	22.11	2.13	8.21
1/2 hr	1.00	1.45	21.16	2.71	9.69
1 hr	0.95	1.54	21.60	2.45	9.19
2 hr	1.01	1.42	21.72	2.34	9.81
3 hr	0.90	1.52	22.19	2.19	9.93
4 hr	0.93	1.62	22.03	2.40	9.75

FERRITE

0	1.22	2.89	24.91	3.10	5.45
1/2hr	1.14	2.58	26.40	2.18	5.49
1 hr	1.10	2.81	26.69	2.05	5.06
2 hr	1.16	2.58	26.59	2.00	5.45
3 hr	1.03	2.55	26.71	1.74	5.77
4 hr	1.02	2.84	26.56	2.44	5.97

SIGMA

0	-	-	-	-	-
1/2 hr	6.83	0.92	20.54	8.84	6.85
1 hr	0.68	3.22	28.68	2.37	5.29
2 hr	0.60	3.32	30.20	2.36	4.92
3 hr	1.31	5.84	30.62	2.21	5.21
4 hr	-	-	-	-	-

Table 6

Quantitative Phase Analysis of 5 Si Weld

AUSTENITE 5 Si (2.0%)

Exposure time at 1050°c	Composition (wt %)				
	Si	Mo	Cr	Mn	Ni
0	1.29	1.68	21.92	2.68	9.54
1/2 hr	1.39	1.36	21.45	2.50	9.20
1 hr	1.55	1.49	21.84	2.44	9.70
2 hr	1.92	1.29	20.41	2.99	9.60
3 hr	1.57	1.62	21.31	2.51	9.72
4 hr	1.33	1.44	21.43	2.44	9.38

FERRITE

0	1.56	2.47	25.12	2.71	5.35
1/2 hr	1.59	2.15	25.69	2.02	5.32
1 hr	1.61	2.74	25.80	2.36	5.76
2 hr	2.27	2.02	25.34	2.54	5.26
3 hr	1.91	2.38	26.12	3.05	5.73
4 hr	1.62	2.64	25.91	1.97	5.85

SIGMA

0	-	-	-	-	-
1/2 hr	1.61	5.06	30.06	2.42	4.56
1 hr	0.68	2.14	31.53	2.51	4.65
2 hr	2.19	3.93	28.22	2.79	5.26
3 hr	2.28	5.04	28.55	2.66	5.10
4 hr	-	-	-	-	-

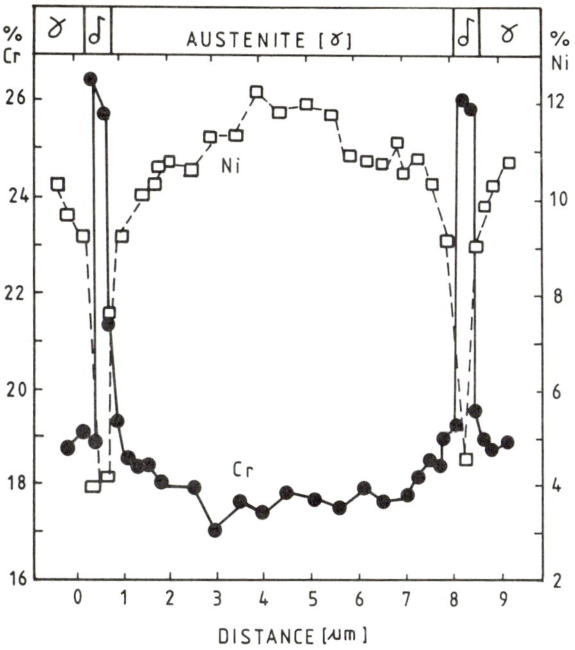

Fig 1. STEM Trace across two retained delta-ferrite regions bounded by austenite. In Type 304L stainless (after Lyman et al Ref 4)

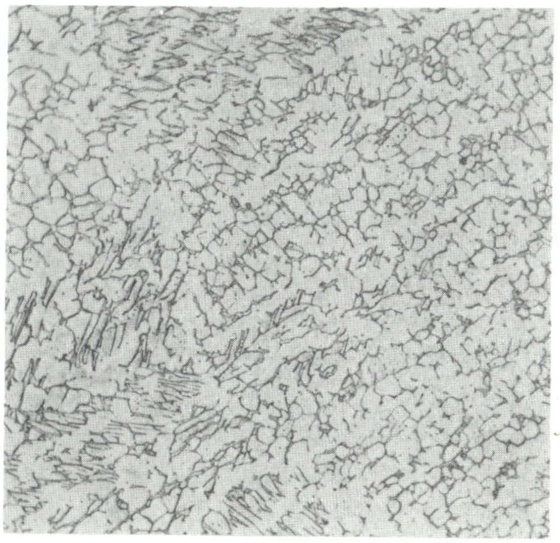

Fig 2. Typical as-welded microstructure. No 5Si showing network of delta ferrite

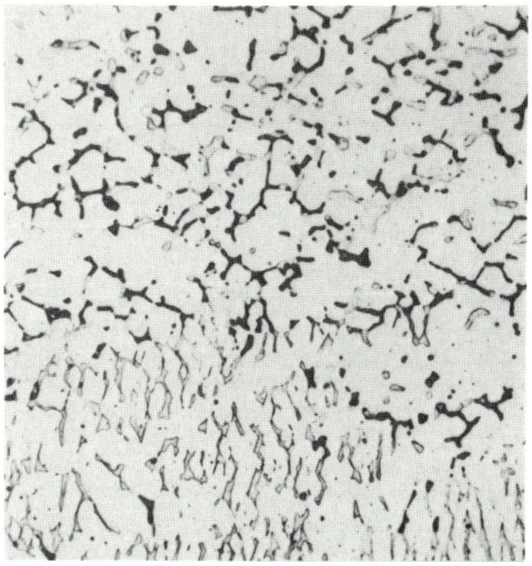

Fig 3. Typical heat treated microstructure showing zoned sigma phase

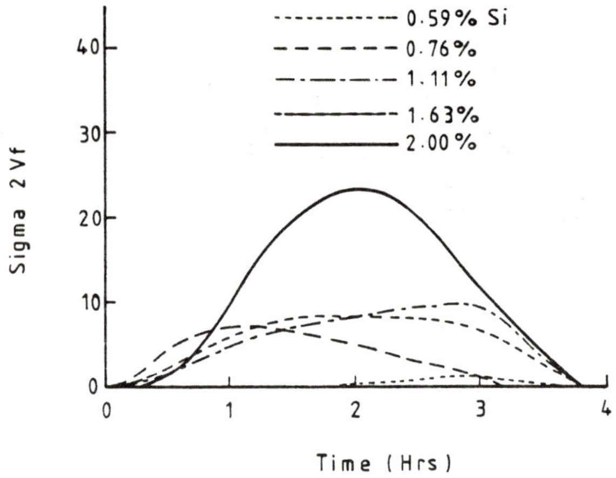

Fig 4. Variation of sigma V_f with time at temperature (Ref 12)

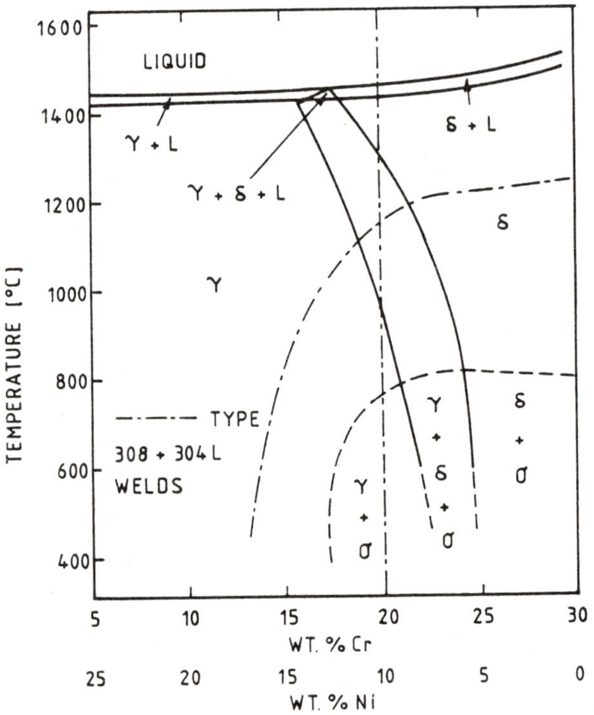

Fig 5. Pseudo binary for solidification of stainless steel weld metals (after Lippold and Savage Ref 3)

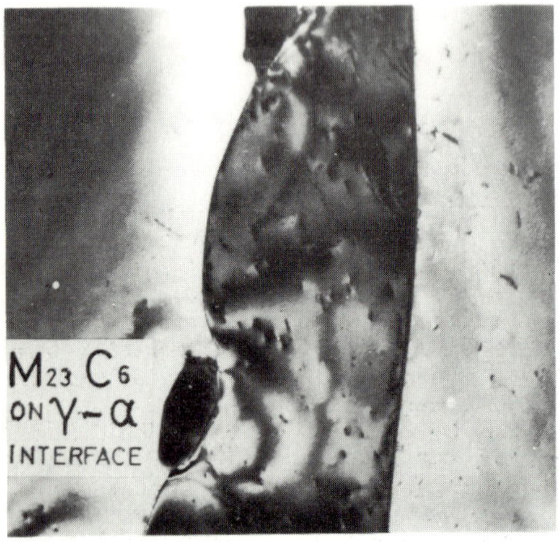

Fig 6. Electron micrograph showing $M_{23}C_6$ at sigma/austenite boundary

Improving properties of weld joints in duplex stainless steel by welding with shielding gas containing nitrogen

K-J Blom

The author is Chief Metallurgist in the Fagersta Tube Mill of Avesta Sandvik Tube AB, Fagersta, Sweden.

SYNOPSIS

The corrosion resistance, ductility and toughness of TIG and plasma welds in duplex stainless can be improved without using over-alloyed filler metal, if the shielding gas is mixed with a small amount of nitrogen (about 5 %). The nitrogen take-up from the gas increases the austenite content of the weld structure so much that precipitation of chromium nitrides is practically eliminated.

INTRODUCTION

Many of the best properties of the ferritic and the austenitic stainless steels are joined in the new duplex i.e. ferritic-austenitic stainless steels. Thus these steels have high resistance both to stress corrosion cracking and to general corrosion, pitting and crevice corrosion. They also combine high strength including corrosion fatigue strength and reasonable formability, toughness and weldability. However the duplex steels are not a new group in the stainless family but a rather old one. In Sweden the first two duplex grades were developed in 1931 (1) and these were standardized in 1947 (2).

But the weldability of these older types was very restricted so they were mainly used in bar form and as forgings and castings. A new era of the duplex stainless steels started about fifteen years ago when nitrogen was introduced as one of the main alloying elements of this steel group (3). The effect of nitrogen is double; it increases the corrosion resistance especially of the austenitic phase, where it is concentrated, and it improves the weldability by accelerating the partial transformation from ferrite to austenite during the cooling after welding. That nitrogen as an interstitial element is an austenite-former and also diffuses much faster than substitutial elements such as Cr, Mo and Ni is of course an effect of its small atomic size. But that it also improves the corrosion resistance is very difficult to explain in passivation terms, which is why a more purely chemical explanation seems to be more probable.

There is also another important difference between the old and the new duplex steels; the austenite to ferrite ratio in the former is about 1:3 while it is between 1:1 and 3:2 in the latter. Also this contributes to the improved weldability of the new grades. In spite of these improvements it is mostly necessary to use filler metal that contains 8 % Ni or more for obtaining a weld with similar toughness and corrosion resistance to the base metal. But using filler metal when TIG and plasma welding thin sheets and making welded thin-walled tubes and pipes is definitely an economic disadvantage. One possibility to increase the austenite content of the weld metal without using filler metal is to TIG and plasma weld with a nitrogen-containing shielding gas. Such TIG welding trials have been made by AST at its tube mill in Fagersta.

RESULTS

All tube welding trials are made in continous welding machines. In these the strip is uncoiled, roll-formed to circular form, melt-welded, bead-worked, calibrated and cut in lengths. Strips of grade AST 2205 (SS 2377, UNS S31803, W.Nr 1.4462) were used for this investigation. In the preliminary trials strips from two heats (A and B) with rather low nitrogen contents (table 1) were welded to 25.4 x 1,65 and 19,05 x 1,65 mm tubes respectively.

Table 1

	Heat A	Heat B
% C	0,020	0,023
% Si	0,57	0,35
% Mn	1,44	1,45
% P	0,027	0,030
% S	0,008	0,005
% Cr	22,7	22,4
% Ni	4,9	5,1
% Mo	2,70	2,64
% N	0,098	0,140

The nitrogen content of the shielding gas was varied up to 7 %, where disturbances in the form of sparks arose. 5 % N in the argon gas increased the nitrogen content of the weld of tubes from heat A from 0,10 to 0,15 % and from heat B from 0,14 to 0,19 %. This resulted in an increase of the austenite content in the weld metal from 5 % to about 30 % (A) and from 13 % to 30 % (B) at the same time as the ductility and pitting resistance was strongly improved. The reason for this is of course that so much austenite content was formed that it eliminated most of the chromium nitride precipitation within the ferrite grains.

With these preliminaries as background more systematic tube welding trials have been realized. As experience (4) has shown that grade AST 2205 must contain at least 0,15 % N, so that the over-heated part of the heat-affected zone of the weld joint shall obtain the same corrosion resistance as the non-affected part of the tube, heats with more than 0,15%N were used (table 2).

Table 2

	Heat C	Heat D	Aim
% C	0,024	0,021	0,020
% Si	0,51	0,45	0,45
% Mn	1,47	1,49	1,50
% P	0,020	0,023	0,020
% S	0,001	0,001	0,002
% Cr	22,0	21,9	22,0
% Ni	5,7	5,8	5,7
% Mo	3,02	3,04	3,00
% N	0,163	0,173	0,175

The composition of these two heats is very close to the aim composition of grade AST 2205. Two tube dimensions were welded 18 x 1,5 mm (heat C) and 42 x 1,0 mm (heat D). Both were TIG welded with Ar + 5%N_2 shielding gas and for comparison with Ar + 10H_2 shielding gas. Higher N_2-content gave as earlier spark formation. Welding with 5 % N_2 in the shielding gas increased the nitrogen content of the weld metal by about 50 % while the nitrogen-free gas gave a slight decrease (table 3).

Table 3

Heat	C	C	D	D
Tube dia. mm	18 x 1,5	18 x 1,5	42 x 1	42 x 1
Gas Ar +	5 % N_2	10 % H_2	5 % N_2	10 % H_2
% N_2 in				
strip	0,16	0,16	0,17	0,17
weld	0,23	0,14	0,27	0,16
% austenite in weld				
Top	46	21	69	19
Centre	46	19	66	21
Bottom	44	14	46	16
HAZ	32	29	26	24

The weld structure of the Ar + 5N_2 shielded welds consists of matrix ferrite and rather thick grain boundary and widmanstätten austenite and there is practically no chromium nitride precipitation in the centre of the ferrite grains (figure 1), while the Ar + 10H_2 shielded welds have a structure of matrix ferrite with fine grain boundary and widmanstätten austenite and heavy chromium nitride precipitation within the ferrite grains (figure 2).

Bend testing showed that the weld of Ar + 5 % N_2 welded tubes were ductile while the weld of Ar + 10 % H_2 welded tubes cracked and thus had a lower ductility.

The pitting resistance was determined as Critical Pitting Temperature in 10 % $FeCl_3 \cdot 6 H_2O$ (ASTM G48). The results demonstrate that welding with Ar + 5 % N_2 shielding gas improves the pitting resistance of the weld so much that it is not far from the resistance of the non-affected part of the tube wall.

Table 4

Heat	C	D
Non-affected tube wall	27,5 (1)	25 (1)
Weld + HAZ Ar + 5 % N_2	22,5 (2)	25 (2)
Weld + HAZ Ar + 10 % H_2	5 (3)	7,5 (3)

(1) = Edge attack
(2) = Outer bead attack
(3) = Inner bead attack

On the other hand welding with Ar + 10% H_2 gives welds with much lower pitting resistance; their CPT-values are no better than those of unannealed grade 316L welds.

DISCUSSION

That the corrosion resistance, ductility and toughness of weld metal of duplex steels are very structure-sensitive properties is well documented (4, 5, 6). Thus it is not astonishing that Ar + 5%N_2 shielded TIG welds have much better properties than Ar + 10H_2 shielded TIG welds since the structure of the former contains more than 40% austenite and practically no chromium nitride (figure 1) while the structure of the latter contains heavy precipitation of chromium nitride and less than 22% austenite (figure 2).
Microprobe analysis of the weld structure of Heat D tubes has shown that independent of heat composition and shielding gas the compositions of the austenite and the ferrite phase have the same chromium and molybdenum content while the nickel content is about 0,5 and nitrogen content about 0,4% higher in the austenite phase. In the ferrite phase the nitrogen content is only 0,01-0,02%, while it varies around 0,45% in the austenite phase and is independent of the shielding gas composition. Simple calculation shows that most of the nitrogen content in the Ar + 5H_2 shielded welds is concentrated in the austenite phase, while less than half of the nitrogen content in the Ar + 10H_2 shielded welds is concentrated in the austenite phase and at least half of the nitrogen content must have formed chromium nitrides. Thus the paradoxical conclusion is that chromium nitride precipitation in welds of duplex steels can be prevented by increasing the nitrogen content of the weld metal.

REFERENCES

1. Private communication from Avesta AB.

2. Swedish standard: Steel SIS 142323 and SIS 142324, December 1947, first edition.

3. Wessling W, Bock HE, Stainless Steel 77 London England, September 26-27, 1977 p. 217 - 229.

4. Liljas M, Qvarfort R, Duplex Stainless Steel '86, The Hague, 26 - 28 October 1986, p. 244 - 256.

5. Lundqvist B, Norberg P, Olsson K, Duplex Stainless Steel '86, The Hague, 26 - 28 October 1986, p. 16 - 29.

6. Marshall AW, Farrar JCM, Duplex Stainless Steel '86, The Hague, 26 - 28 October 1986, p. 26 - 28.

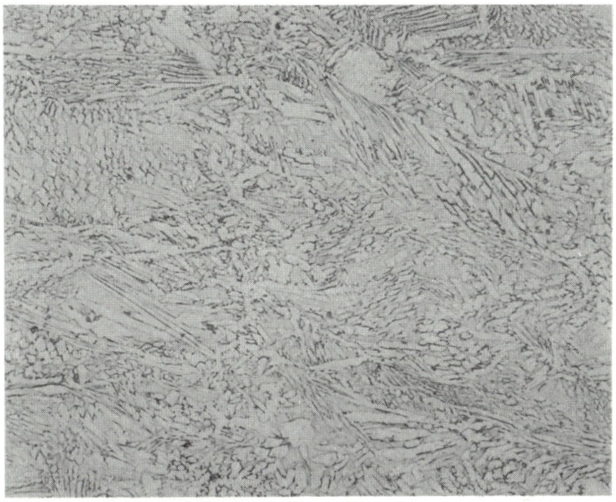

Figure 1. Weld structure of tube from heat D welded with Ar + 5 % N_2. **150 X**

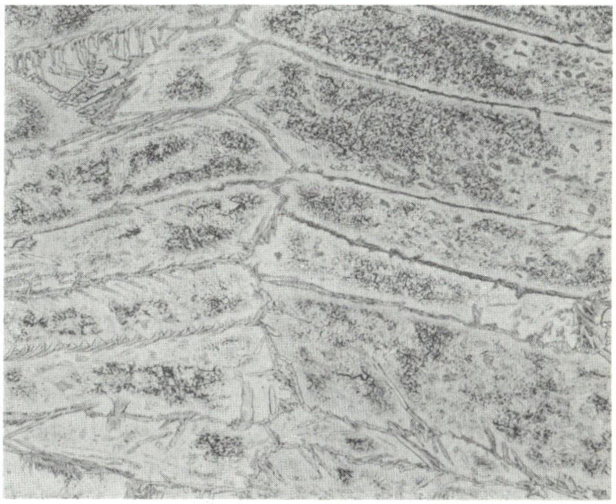

Figure 2. Weld structure of tube from heat D welded with Ar + 10 % H_2O. **150 X**

SESSION III
CORROSION

Metallurgical control of localized corrosion of stainless steels

A J Sedriks

The author is with the Materials Division, Office of Naval Research, Arlington, Virginia, USA.

SYNOPSIS

Recent studies of the effects of microstructure and alloy composition on the pitting and crevice corrosion resistance of austenitic, ferritic/superferritic and duplex stainless steels are reviewed. The joint effect of molybdenum and nitrogen is identified as exceptionally beneficial for the austenitic and duplex grades. The role of inclusions and second phases in determining localized corrosion behavior is also examined. The recognition that manganese sulfide inclusions are the most favored pit initiation sites has led to an exploration of ways aimed at removing them, modifying their composition and changing their morphology, and these approaches are also reviewed. Three new categories of corrosion resistant stainless steels are identified, namely the austenitic nitrogen-containing 6% molybdenum grades, the stabilized superferritics, and the duplex grades containing molybdenum and nitrogen. Exploratory approaches involving laser surface alloying, amorphous alloys, microcrystalline alloys and alloys for which attempts have been made to replace chromium, are noted.

INTRODUCTION

In stainless steels localized corrosion results from the breakdown of the passive film at local sites. In cases where the localized corrosion occurs on an open surface it is known as pitting, whereas in cases where it occurs within tight crevices formed between two contacting surfaces it is known as crevice corrosion. Commercially produced stainless steels contain numerous inclusions, second phases, and regions of compositional heterogeneity, and there has been a growing body of literature reporting results that identify the initiation of localized corrosion with structural heterogeneities at the surface, particularly, but not exclusively, with inclusions (1-16).

Localized corrosion can in fact be a direct consequence of the presence of certain microstructural features, shown schematically in Figure 1. For example, pitting is very often associated with manganese sulfide inclusions, crevice corrosion can initiate at the chromium depleted zones surrounding precipitated nitrides in duplex stainless steels, and intergranular corrosion often propagates along the chromium depleted regions surrounding precipitated grain boundary carbides. Solid solution alloying can also have pronounced effects, with higher chromium levels, for example, being beneficial for both pitting and crevice corrosion resistance.

It is the aim of this paper to review the latest insight gained about the effects of alloy composition and microstructure on localized corrosion resistance. For convenience of presentation the discussion is focussed on three categories of stainless steels, namely the austenitics, the ferritics/superferritics, and the duplexes.

AUSTENITIC STAINLESS STEELS

Inclusions and Second Phases

Inclusions which have been identified in austenitic stainless steels as providing favorable sites for pit initiation are sulfides (13) and delta ferrite (14). Also pitting has been reported to initiate within the chromium depleted zones around precipitated carbides (5) and is believed to initiate within the zones depleted of both chromium and molybdenum around precipitated sigma or chi phases (15).

The role of sulfides, particularly of manganese sulfide, in pit initiation has received by far the most attention and the topic has been the subject of two reviews (13,16).

A recently proposed pitting mechanism suggesting how a manganese sulfide inclusion can initiate a propagating pit in a stainless steel in a chloride solution is shown in Fig.2 (16). The propagation stage is thought to involve the formation of a barrier salt layer at the bottom of the pit (Fig.2f). Salt layers are reported to form at the bottom of pits in stainless steel (17) and are believed to control the kinetics of corrosion via the dissolution of the salt film (17-21).

Thermodynamic calculations also support the idea that manganese sulfide can act as a pit initiation site. The potential-pH diagram calculated for the $MnS-H_2O-Cl^-$ system is shown in Fig.3 (22). The diagram is applicable only to relatively concentrated solutions (e.g. 0.1 M Mn species); it may be somewhat different for more dilute solutions. However, it is useful in defining the dissolution behavior of MnS. For example, it shows that MnS can exist between pH 4.8 and 13.8. At pH values less than 4.8, MnS dissolves to form H_2S and various manganese ions. However, even in the pH range 4.8-13.8, the highest potential at which MnS can exist is about -100 mV (at pH 5). At more noble potentials, MnS dissolves to yield sulfur.

An important conclusion that emerges from the potential-pH diagram shown in Fig. 3 is that manganese sulfide inclusions are active anodic sites at a stainless steel surface in the presence of aqueous chloride solutions. The corrosion potential of passive stainless steels in aqueous chloride solutions is often between 0 and +200 mV SHE. Since manganese sulfide is an electronic conductor it will be polarized to the corrosion potential of the passive stainless steel surface. At potentials between 0 and +200 mV SHE, the sulfide is well outside its domain of thermodynamic stability and will tend to dissolve.

The recognition that manganese sulfide inclusions are the most favored pit initiation sites has led to exploration of ways aimed at removing them, modifying their composition and changing their morphology.

Regarding removal, recent studies of the effect of laser melting of surface layers of type 304 stainless steel suggest that this improves pitting resistance (23). Potentiostatic polarization curves taken from the laser-melted surfaces do not show the increasing current response typical of a stainless steel undergoing pitting. A complex curve suggesting pit nucleation and rapid repassivation was found instead and no pits were found on the surface at the end of the test. This was subsequently attributed to the removal of the manganese sulfide from the laser-melted surface layers. Other removal techniques include the so-called "passivation" treatment often used for resulfurized stainless steels which consists of immersion for 30 minutes in a 50°C solution containing 50% nitric acid and 2% sodium dichromate (24). The improved pitting resistance after this passivation treatment is attributed to the removal of surface sulfides (25,26). It should be noted that the aim of the passivation treatment is to remove only the surface sulfides (and embedded iron) without attacking the stainless steel itself. The surface holes left after the removal of the sulfides may be rinsed with sodium hydroxide to neutralize any entrapped acid (26). If the acid in these surface holes is not neutralized, it could cause further pitting on subsequent exposure to aqueous environments or even to humidity. Electrochemical studies using type 316 stainless steel have shown that the nitric acid passivation treatment raises the pitting potential in de-aerated seawater by some 250mV in the noble direction (27).

Regarding compositional modifications of the sulfide, the main emphasis has been on either reducing the manganese content of the bulk steel or on adding rare earth metals (mainly cerium). A low-manganese stainless steel, in which the resistance to pitting is improved by the presence of higher chromium contents in the sulfides, is now produced commercially in Europe (28). The steel is basically a low-manganese type 316 (Fe-17Cr-12Ni-2.7Mo-0.3 (max.)Mn) and is available as a low-carbon grade with 0.030%C (max.) or as a conventional grade with 0.05%C (max.). The reason for selecting a type 316 base is evident from Fig.4 (29) where it is seen that the effect on the pitting potential of lowering the manganese content to 0.2% is greatest for the type 316 composition. While this effect is also found in type 304 and 904L stainless steels, it is of somewhat lower magnitude for these two compositions.

Rare earth metal (cerium) additions, up to a maximum of 0.07% Ce, to Fe-20Cr-24.5Ni-4.5Mo-1.5Cu stainless steels have been shown to give rise to significantly more noble pitting potentials in 1M NaCl solution at 60°C (30). Since these additions simultaneously alter the volume fraction, the chemistry and the morphology of the sulfide inclusions, it is difficult to ascribe the improved pitting resistance to a single cause. The stringer shaped manganese sulfides are replaced by globular cerium-enriched sulfides, oxysulfides or oxides in the more pitting resistant alloys (30).

The morphology of manganese sulfide inclusions has also been shown to influence their effectiveness as pit initiators (31,32) with transverse sections of the sulfide stringers exhibiting higher reactivity than longitudinal sections, e.g. Fig.5 (31). Morphology effects are also observed in the case of "mixed" inclusions. Commercial stainless steels such as type 304 and type 316 contain a variety of inclusions such as oxides, silicates, aluminates and sulfides. These inclusions are often found associated as "mixed" inclusions, with the sulfides forming shells around the oxides, silicates and aluminates. The manganese sulfide shell represents a particularly favorable location for pit initiation (28,32,34). The dissolving manganese sulfide shells are thought to generate crevices between the insoluble oxide-silicate-aluminate core and the stainless steel matrix, causing conditions favorable for the pit-propagation process, as illustrated schematically in Fig.6 (28). It has also been found that decreasing the size of the manganese sulfide inclusions by producing type 303 stainless steel from rapidly solidified powders improved pitting resistance (35).

Regarding crevice corrosion, manganese sulfide inclusions have also been identified as preferred initiation sites for attack (2). So far, the discussion has centered on manganese sulfide effects, since these have been the subject of the greatest research effort in recent years. Much less effort has been devoted to studying the detrimental effects on pitting of delta ferrite or sigma and chi. Delta ferrite when present in small amounts (e.g. 0.55% by volume) in the austenite has been shown to be detrimental to pitting resistance (14). However, the presence of large amounts of ferrite (e.g. 50%), such as that found in duplex stainless steels, cannot be regarded as detrimental to pitting resistance.

The presence of sigma in type 317L stainless steel has been shown to move the pitting potential in the active direction (13). Sigma and chi generally contain higher chromium and molybdenum contents than the matrix, and it is therefore unlikely that the lowered pitting potential is associated with direct attack on the sigma and chi phases. Attack probably occurs within the matrix alloy immediately surrounding the intermetallic where chromium and molybdenum depletion has taken place (15). In practice the formation of sigma or chi is minimized in high molybdenum stainless steels by adding mischmetal (cerium) or nitrogen to the melt to alter precipitation kinetics, by rapid cooling, or by heat treatment (15).

While inclusions and second phases play a very important role in determining the pitting resistance of austenitic stainless steels, an equally important variable is the chemical composition of the austenitic solid solution.

Alloying Elements in Solid Solution

It is well established that chromium, molybdenum, and to a lesser extent nickel are the main alloying elements required for high pitting resistance in austenitic stainless steels (3). However, a number of other elements can have a significant effect. These are summarized in Fig.7, using data from several sources. The individual alloying elements are described as having a beneficial, detrimental or variable effect on pitting resistance. Much of the data shown in Figure 7 is from a 1967 compilation (36), but also includes data from later studies. These later studies have identified gadolinium as detrimental (37), tungsten as beneficial in the range 2.5 to 12.4% (38) and confirmed silicon as beneficial in the range 1.0-4.4% (39). Uncertainty continues to exist regarding copper additions (which are beneficial for corrosion resistance in sulfuric acid). Earlier studies (36) using copper additions to type 301 stainless steel found a beneficial effect at 0.2 and 0.5% Cu and no effect at 1.0 and 1.9% Cu. More recent studies (40) using an Fe-20Cr-25Ni -4.5Mo alloy found that a 1.5% Cu addition was detrimental and that additions of 0.8% Cu or less produced borderline behavior characterized by the formation of small pits and their re-passivation.

A particularly interesting development has been the identification of strong beneficial effects due to molybdenum plus nitrogen additions, Fig.8 (41), which has now been utilized in the development of several commercial stainless steels (Table 1). It should be noted that while the beneficial effect of nitrogen without molybdenum is observed mainly in solutions containing sulfuric acid (42,43), the beneficial effect of molybdenum plus nitrogen does not require the presence of the sulfate ion. Alloying with nitrogen has also been found to raise the pitting potential in the noble direction for experimental 18% Cr -18% Mn austenitic stainless steels (44).

FERRITIC STAINLESS STEELS

In considering the pitting resistance of ferritic stainless steels a distinction must be made between the AISI standard grades (e.g. types 430,409) and the proprietary superferritic grades (e.g. 29-4C, Sea-Cure, Monit, 29-4-2 etc.). The nominal compositions of several superferritics are shown in Table 2. The superferritics have a much higher pitting resistance in aqueous chloride solutions than the standard ferritic grades.

Alpha prime, which precipitates in the standard ferritic stainless steels in the temperature range 340-516°C, has been shown to be detrimental to pitting resistance (45), as have chromium depleted regions surrounding precipitated carbides in sensitized structures (46). The standard ferritic grades usually contain sufficient manganese (e.g. 1.0% max.) and sulfur (0.03%) to form predominantly manganese sulfide inclusions which, like in the case of austenitic stainless steels, can act as favorable sites for the initiation of pits (47). Studies using a ferritic Fe-18Cr-2.3Mo-0.2S base with very low carbon and nitrogen contents and containing manganese additions in the range 0 to 2.15% and titanium in the range 0 to 0.84% have identified the various sulfide domains shown in Fig.9 (48). This figure is only tentative, with the low-manganese, low-titanium corner remaining to be fully explored. An area may exist there in which only chromium sulfides will be precipitated giving rise to more pit resistant alloys.

Regarding the superferritic stainless steels, there is very little understanding of how pits are initiated in these materials or which sulfide inclusions, if any, are expected to be present. If the data of Fig.9 were applicable also to the higher chromium superferritic stainless steels, whose nominal compositions are shown in Table 2, a number of points would appear evident. Since the solid solubility of sulfur in stainless steels is less than 0.01% at room temperature some sulfide formation is theoretically possible, particularly if the sulfur content is closer to the 0.03% max. level specified for the superferritics. For the titanium stabilized superferritics shown in Table 2, the 0.5%Mn and 0.5%Ti additions would place the sulfide in the titanium sulfide field of Fig.9, and there is evidence that titanium sulfide provides a less effective pitting initiation site than manganese sulfide (2). In the case of the 29-4-2 superferritic stainless steel the absence of titanium and the very low manganese content (Table 2) would ensure that any sulfides formed are predominantly chromium sulfides which are known to be more resistant to attack (29). However, it is by no means clear whether the sulfide domains determined for an 18% Cr steel (Fig.9) are applicable to superferritics (29% Cr) and domains of the type shown in Fig.9 should be determined for the 29% Cr superferritics by further studies.

As noted before, chromium depleted regions around precipitated chromium carbides (and nitrides) can also act as preferred sites for pit initiation in austenitics (5) and the standard ferritic grades (46). Chromium depletion at chromium carbide and chromium nitride precipitates can be minimized or prevented in the superferritics by either keeping the carbon plus nitrogen contents very low (i.e. %C +%N = 0.025 max.) as in the case of the 29-4-2 grade, or by stabilizing with titanium, usually at the minimum level of %Ti = 0.2 + 4 (%C +%N), as in the case of the other superferritics listed in Table 2. The titanium addition is believed to cause the carbon and nitrogen to precipitate as titanium carbide and nitride at high temperatures so that little of these two elements is left to precipitate as chromium carbide and nitride during cooling. These practices must also contribute to the high pitting resistance of the superferritics since they eliminate some initiation sites, namely chromium depleted regions.

It should be noted that sigma and alpha prime, if allowed to form, could degrade the pitting and crevice corrosion resistance of the superferritics. A heat treatment (1 hour at 871°C) selected to produce sigma precipitation has been shown to be detrimental to the crevice corrosion resistance of the 29-4-2 grade (49). However, no problems due to these phases appear to have been encountered in service where the superferritics have been used for thin-walled welded tubing for seawater cooled condensers and heat exchangers. While the absence of pit initiation sites such as manganese sulfide and chromium depleted regions, and procedures to avoid sigma and alpha prime undoubtedly contribute to the high pitting resistance of the superferritics, the chemistry of the ferritic solid solution must also play a major role. The very high chromium contents together with about 4% molybdenum would be expected to give a highly pitting resistant structure, since both of these alloying elements significantly raise the pitting potential in the noble direction. Studies of high purity inclusion-free ferritic stainless steels (50) have shown that increasing the chromium content from 18% to 28% results in thinner passive films and increased chromium enrichment within the films. The exact mechanism by which molybdenum improves pitting resistance is still not understood despite a significant research effort to resolve this question (50,51).

The effects of other alloying elements have been explored for the lower chromium ferritic stainless steels, with tungsten, vanadium and silicon having been identified as beneficial for both 13% Cr and 24% Cr ferritics (52). Other studies (53) have shown that the beneficial chromium plus molybdenum effect is synergistic, Fig.10. Thus, while molybdenum additions are beneficial for both the 13% Cr and 18% Cr ferritics, at the higher chromium content the increase in pitting potential in the noble direction caused by each increment of molybdenum is considerably greater, Fig.10. It has also been shown (53) that for a given carbon plus nitrogen level, the addition of titanium is beneficial, and that the titanium benefit increases with increasing molybdenum content, Fig.11. It is also evident from Fig.11 that the pitting potentials of alloys with low carbon plus nitrogen levels are more noble than those with higher carbon plus nitrogen levels. It remains to be established whether these effects relate to carbide and nitride precipitation and/or possible chromium depletion at the precipitates or whether they are solid solution effects. It should also be pointed out here that in the case of austenitic stainless steels titanium is considered detrimental, nitrogen beneficial and carbon detrimental when precipitated as a carbide and neutral when in solid solution (3). Thus, significant differences appear to exist between the ferritic and austenitic stainless steels in the way in which pitting resistance is affected by titanium, carbon and nitrogen.

DUPLEX STAINLESS STEELS

Duplex stainless steels contain both austenite and ferrite, often in the ratio 50:50. Some of the more recently developed duplex stainless steels contain molybdenum and nitrogen, as shown in Table 3. Corrosion resistance comparisons are often made with type 316 stainless steel. As shown in Figure 12 (54), the pitting potentials of these duplex stainless steels are more noble than those of type 316, which indicates greater pitting resistance. This would be expected because of the higher levels of chromium, molybdenum and nitrogen in these new duplex stainless steels, Table 3. The duplexes are also more resistant to sensitization (3), so the reduction of pitting resistance resulting from chromium depletion at precipitated carbides encountered in austenitic or ferritic stainless steels would not be expected in duplexes. However, the precipitation of chromium nitride, which can lead to chromium depletion in the vicinity of the nitride, has been shown (55) to provide sites for crevice corrosion initiation in cast and annealed duplex stainless steels.

The presence of sigma in duplexes will reduce pitting resistance, as shown in Figure 13 (56), and crevice corrosion resistance (49). Annealing in the alpha prime precipitation region for 4 hours did not have an adverse effect on pitting resistance, possibly because of insufficient precipitation of alpha prime (Figure 13).

Manganese sulfide would be expected to be present in the duplex stainless steels at the manganese levels shown in Table 3. The possibility exists, therefore, that the pitting resistance of the duplex stainless steels could be improved even further by reducing the manganese content to levels at which chromium sulfides are formed. This approach has been explored, for example, in the development of a new duplex stainless steel in England (57). Replacement of manganese sulfide by calcium sulfide has also been reported to be beneficial for crevice corrosion resistance (58), although the reason for this is not clear since calcium sulfide is easily soluble in acids (13).

Regarding effects of alloy composition, it should be noted that some duplex stainless steels contain tungsten additions in the range 0.5-1.0% (57,59). Tungsten additions in this range have been shown to improve the crevice corrosion resistance of duplex stainless steels (60). Nitrogen additions of 0.2% have been shown to improve the crevice corrosion resistance of duplex stainless steels in seawater (61). Copper additions of the order of 1 to 2% are used to extend the application of duplex stainless steels to include sulfuric acid service.

EXPLORATORY APPROACHES

In the stainless steel area the more prominent of the exploratory approaches have involved studies of laser surface alloying, amorphous alloys, microcrystalline alloys and alloys for which attempts have been made to replace chromium. The materials resulting from these studies are still largely in the developmental stage.

Studies of laser surface alloying have explored the concept of surface stainless steels in which chromium is alloyed into a surface layer of steel. In one series of studies (62), specimens of AISI 1018 steel (0.18%C, 0.8% Mn) were coated with electroplated or sputter-deposited chromium layers ranging in thickness from 10 to 20μm. The depth of melting, and hence the alloy composition, was controlled by varying the laser sweep velocity. Surface stainless steel layers $\sim$100μm in thickness, containing average chromium contents in the range of those found in cast and wrought stainless steels, were prepared and evaluated in terms of their passive behavior. The anodic polarization curve of a 20%Cr surface alloy, shown in Figure 14 (62), exhibited an active-passive transition and a transpassive potential in a sulfate solution. The passive current densities of the surface alloys were somewhat higher than those of comparable wrought alloys, which was probably a result of compositional fluctuations. Electron microprobe traces taken across the Fe-20%Cr surface alloy revealed local concentration fluctuations of the order of ±7%Cr. Refinement of the laser alloying technique is likely to reduce these compositional fluctuations and hence reduce the passive current densities. Heat treatments may also be beneficial and should be explored.

The preparation of iron-chromium and iron-chromium-nickel amorphous alloys by various rapid cooling techniques requires the presence in the melt of large quantities of phosphorus (e.g. 13 at.%) and carbon (e.g. 7 at.%) to ensure the retention of the amorphous structure on cooling. The corrosion behavior reflects, therefore, the presence of these metalloids, as well as the absence of crystalline defects such as grain boundaries, dislocations and segregates. Amorphous alloys of the type Fe-10Cr-13P-7C (at.%) and Fe-10 Cr-5Ni-13P-7C (at.%) have been shown to be totally resistant to attack in ferric chloride and in hydrochloric acid solutions shown to cause severe pitting attack in types 304 and 316 stainless steels (63). This high degree of pitting resistance requires the presence in the amorphous alloy of 8 or more atomic percent of chromium (63). Heating of the amorphous alloys causes crystallization, resulting in loss of corrosion resistance (64,65).

Microcrystalline alloys can be produced by rapid solidification processes to yield powders which are compacted and then consolidated by hot extrusion. The final structure consists of crystalline grains with grain sizes of the order of a micrometer. Microcrystalline austenitic and duplex stainless steels produced by compaction and extrusion of rapidly solidified foils have shown greater pitting resistance in chloride solution than their conventional counterparts (66,67). This has been attributed to the higher homogeneity of the microcrystalline alloys.

Studies aimed at conserving chromium resources have been focused on (i) replacement of chromium by manganese and aluminum, and (ii) lowering the chromium content of alloys. It is now recognized that Fe-Mn-Al alloys do not have adequate corrosion resistance in aqueous media and that they do not represent acceptable substitutes for stainless steels for aqueous media (68). Regarding studies of lowered chromium stainless steels, it has been shown that in stainless steels containing nickel and molybdenum, adequate aqueous corrosion resistance is obtained at chromium contents as low as $\sim$9% (69).

ACKNOWLEDGEMENT

Some of the figures are used in this review by permission of the copyright owners. The author acknowledges permissions granted by the National Association of Corrosion Engineers and The Metals Society.

REFERENCES

1. L. L. Shreir, Corrosion, Metal/Environment Reactions, L. L. Shreir, Ed. Vol.1, Newness-Butterworths, Boston, MA., 1976, p.1:182.
2. Z. Szklarska-Smialowska, Pitting Corrosion of Metals, National Association of Corrosion Engineers, Houston, TX., 1986, p.377 and p.309
3. A. J. Sedriks, Corrosion of Stainless Steels, John Wiley & Sons, New York, N.Y., 1979.
4. H. H. Uhlig, Trans. AIMME, Vol. 140, p.411, 1940.
5. M. A. Streicher, J. Electrochem, Soc., Vol. 103, p.375, 1956.
6. K. Lorenz and G. Medwar, Tyssenforschung, Vol.1, p.97, 1969.
7. B. Forchhammer and H. J. Engell, Werkst. Korros., Vol. 20, p.1, 1969.
8. B. E. Wilde and J. S. Armijo, Corrosion, Vol. 23, p.208, 1967.
9. S. Steinemann, Mem. Sci. Rev. Met., Vol. 65, p.615, 1969.
10. M. Smialowski et al., Corros. Sci., Vol. 9, p.123, 1969.
11. Z. Szklarska-Smialowska et al., Br. Corros. J., Vol. 5, p.159, 1970.
12. T. P. Hoar et al., Corros. Sci., Vol. 5, p.279, 1965.
13. A. J. Sedriks, International Metals Reviews, Vol. 28, No.5, p.295, 1983.
14. H. J. Dundas and A. P. Bond, "Effects of Delta Ferrite and Nitrogen Contents on the Resistance of Austenitic Steels to Pitting Corrosion," CORROSION/75, Paper No.159, National Association of Corrosion Engineers, Houston, Texas, 1975.
15. A. J. Sedriks, International Metals Reviews, Vol. 27, No.6, p.321, 1982.
16. Z. Szklarska-Smialowska and E. Lunarska, Werkst. Korros., Vol. 32,p.478, 1981.
17. T. R. Beck, "Fundamental Investigation of Pitting Corrosion in Structural Metals", Report to Office of Naval Research, Contract N00014-76-C-8495, July 1981.
18. T. R. Beck and R. C. Alkire, J. Electrochem. Soc., Vol. 126, p.1662, 1979.
19. K. J. Vetter and H. H. Strehblow, Localized Corrosion, National Association of Corrosion Engineers, Houston, TX., 1974, p.240.
20. I. L. Rosenfeld, I. S. Danilov, and R. N. Oranskaya, J. Electrochem. Soc.Vol. 125, p.1729, 1978.

21. H. J. Engell, Electrochim. Acta, Vol. 22, p.987., 1977.
22. G. S. Eklund, J. Electrochem. Soc., Vol. 121, p.467, 1974.
23. E. McCafferty, P. G. Moore, J. D. Ayers, and G. K. Hubler, Corrosion of Metals Processed by Directed Energy Beams, Metallurgical Society of AIME, Warrendale, PA., 1982, p.1-21.
24. M. Henthorne, Sulfide Inclusions in Steels, American Society for Metals, Metals Park, OH., 1975, p.445.
25. M. Henthorne, Corrosion, Vol. 26, p.511, 1970.
26. M. Henthorne and R. J. Yinger, Cleaning Stainless Steels, STP 538, American Society for Testing and Materials, Philadelphia, PA, 1973, p.90.
27. M. A. Barbosa, Corrosion Science, Vol. 23, No.12, p.1293, 1983.
28. K.-J. Blom, "Corrosion Resistance and Industrial Experience of Low-Manganese Austenitic Stainless Steel Grades", CORROSION/82, Paper No.87, National Association of Corrosion Engineers, Houston, Texas, 1982.
29. J. Degerbeck and E. Wold, Werkst. Korros., Vol. 25, p.172, 1974.
30. L. P. Zhong et al., Stainless Steels '84, The Institute of Metals, London, England, 1985, p.158.
31. P. E. Manning, D. J. Duquette, and W. F. Savage, Corrosion, Vol. 35, p.151, 1979.
32. V. Scotto, G. Ventura and E. Traverso, Corrosion Science, Vol. 19, p.237, 1979.
33. M. Janik-Czachor, A. Szummer and Z. Szklarska-Smialowska, Br. Corros. J. Vol. 7, p.90, 1972.
34. Z. Szklarska-Smialowska, Corrosion, Vol. 28, p.388, 1972.
35. P. C. Searson and R. M. Latanision, Corrosion, Vol. 42, p.161, 1986.
36. A. Moskowitz et al., Effects of Residual Elements on Properties of Stainless Steels, STP-418, American Society for Testing and Materials, Philadelphia, PA., 1967, p.3.
37. D. Warren, Microstructure and Corrosion Resistance of Austenitic Stainless Steels, Sixth Annual Liberty Bell Corrosion Course, NACE, Philadelphia, PA., September 1968.
38. N. Bui et al., Corrosion, Vol. 39, p.491, 1983.
39. B. E. Wilde, Corrosion, Vol. 42, p.147, 1986.
40. H. J. Dundas and A. P. Bond, "Effect of Variation of Cr, Mo and Cu on Corrosion of Austenitic Stainless Steels", CORROSION/81, Paper No.122, National Association of Corrosion Engineers, Houston, Texas, 1981.
41. J. E. Truman, M. J. Coleman and K. R. Pirt, Brit. Corros. J., Vol. 12, p.236, 1977.
42. J.J. Eckenrod and C.W. Kovach, ASTM STP 679, American Society for Testing and Materials, pp 17-41, 1979.
43. C. R. Clayton, "Passivity Mechanisms in Stainless Steels: Mo-N Synergism", Report to the Office of Naval Research, Contract N00014-85-K-0437, August 1986, SUNY, Stony Brook, N.Y.
44. A.G. Hartline, Metallurgical Transactions, Vol.5, p.2271, 1974.
45. E. A. Lizlovs and A. P. Bond, J. Electrochem. Soc., Vol. 122, p.589, 1975.
46. Z. Szklarska-Smialowska and M. Janik-Czachor, Corros. Sci., Vol. 7,p.65, 1967.
47. Z. Szklarska-Smialowska, A. Szummer and M. Janik-Czachor, Br. Corros. J.,Vol. 5, p. 159, 1070.
48. R. Kiessling, Sulfide Inclusions in Steel, American Society for Metals, Metals Park, OH., 1975, p.104.
49. P. B. Lindsay, Materials Performance, Vol. 25, p.23, 1986.
50. W. R. Cieslak and D. J. Duquette, Corrosion, Vol.40, p.545, 1984.
51. A. J. Sedriks, Corrosion, Vol. 42, p.376, 1986.
52. R. Goetz, J. Laurent and D. Landolt, Corr. Sci., Vol. 25, p.1115, 1985.
53. A. P. Bond, E. A. Lizlovs and H. J. Dundas, "Development of Corrosion Resistant Alloys Using Electrochemical Techniques", CORROSION/85, Paper No.68, National Association of Corrosion Engineers, Houston, Texas, 1985.
54. H. Miyuki et al, 25% Cr Containing Duplex Phase Stainless Steel for Hot Sea Water Application, ASM Metals Congress, American Society for Metals, Metals Park, OH, October 1982.
55. H. J. Dundas and A. P. Bond, "Corrosion Resistance of Stainless Steels in Seawater", CORROSION/85, Paper No.206, National Association of Corrosion Engineers, Houston, Texas, 1985.
56. J. E. Truman and K. R. Pirt, Properties of a Duplex (Austenitic-Ferritic) Stainless Steel and Effects of Thermal History, ASM Metals Congress, American Society for Metals, Metals Park, OH, October 1982.
57. C. V. Roscoe, Steels for the Salt Solution, Iron and Steel International, p.9, February, 1985.
58. M. D. Carpenter, R. Francis, L. M. Phillips and J. W. Oldfield, British Corrosion Journal, Vol. 21, No.1, p.45, 1986.
59. M. Kowaka, H. Nagano, T. Kobayashi and M. Harada, Sumitomo Search, No.16, p.64, November, 1976.
60. H. Nagano, T. Kudo, Y. Inaba and M. Harada, "Highly Corrosion Resistant Duplex Stainless Steel", paper presented at 19th Journees des Aciers Speciaux, St. Etienne, France, May, 1980.
61. A.P. Bond and H.J. Dundas, "Resistance of Stainless Steels to Crevice Corrosion in Seawater", CORROSION/84 ,Paper No.26, National Association of Corrosion Engineers, Houston, Texas, 1984.
62. E. McCafferty and P.G. Moore, in Fundamental Aspects of Corrosion Protection by Surface Modification, E. McCafferty, C.R. Clayton, J. Oudar, Eds., The Electrochemical Society, Pennington, New Jersey, p. 112, 1984.
63. M. Naka, K. Hashimoto and T. Masumoto, Corrosion, Vol. 32, p. 146, 1976.
64. R.B. Diegle and J.E. Slater, Corrosion, Vol. 32, p. 155, 1976.
65. M. Naka, K. Hashimoto, and T. Masumoto, Corrosion, Vol. 36, p. 679, 1980.
66. T. Tsuru, S. X. Zhang and R.M. Latanision, Proc. 4th Int. Conf. on Rapidly Quenched Metals, Sendai, Japan, p.1437, 1982.
67. T. Tsuru and R.M. Latanision, J. Electrochemical Society, Vol.129,p. 1402, 1982.

68. C.J. Altstetter, A.P. Bentley, J.W. Fourie, and A.N. Kirkbride, Materials Science and Engineering, Vol. 82, p.13, 1986.
69. S. Floreen, Metallurgical Transactions A, Vol. 13A, p.2003, 1982.

TABLE 1 Typical Compositions of Some New Austenitic 6% Mo Stainless Steels Containing Nitrogen.

Alloy	UNS Number	Cr	Ni	Mo	Mn	Si	C	N	P	S	Other
		Composition (wt%)*									
AL-6XN(1)	N08367	20.75	25.0	6.5	0.50	0.4	0.02	0.2	0.025	0.002	-
254 SMO(2)	S31254	20.0	18.0	6.1	0.50	0.4	0.02	0.2	0.025	0.001	0.7Cu
1925hMo(3)	N08925	20.0	25.0	6.5	1.0	0.4	0.02	0.2	0.025	0.002	0.8Cu

*Balance Fe

(1) Allegheny Ludlum Steel Corporation, Pittsburgh, Pennsylvania.
(2) Avesta Jernverks AB, Avesta, Sweden.
(3) Vereinigte Deutsche Metallwerke AG, Duisburg, West Germany.

TABLE 2 Nominal Compositions of Several Superferritic Stainless Steels.

Alloy	UNS Number	Cr	Mo	Ni	Mn	S	C	N	Si	P	Other
		Composition (wt %)*									
29-4C(1)	S44735	29.00	4.00	0.30	0.50	0.01	0.02	0.02	0.35	0.03	0.50Ti
Sea-Cure(2)	S44660	27.50	3.50	1.20	0.50	0.01	0.02	0.02	0.30	0.02	0.50Ti
Monit(3)	S44635	25.50	4.00	4.00	0.50	0.01	0.02	0.02	0.35	0.03	0.50Ti
29-4-2(1)	S44800	29.00	4.00	2.10	0.05	0.01	0.003	0.015	0.10	0.02	-

*Balance Fe

(1) Allegheny Ludlum Steel Corporation, Pittsburgh, Pennsylvania.
(2) Trent Tube Division, Colt Industries, East Troy, Wisconsin.
(3) Nyby Uddeholm AB, Karlstad, Sweden.

TABLE 3 Typical Compositions of Some New Duplex Stainless Steels and of an AOD[1] Melted Austenitic AISI 316 Stainless Steel.

Alloy	UNS Number	Cr	Ni	Mo	Mn	Si	C	N	P	S	Cu	W
					Composition (wt%)*							
Ferralium[2] 255	S32550	26.0	5.5	3.0	0.8	0.45	0.04	0.17	0.04	0.03	1.70	-
SAF[3] 2205	S31803	22.0	5.5	3.0	1.7	0.80	0.03	0.14	0.03	0.02	-	-
DP 3[4]	-	25.4	6.2	3.3	0.9	0.61	0.02	0.14	0.01	0.01	0.44	0.5
AISI 316	S31600	18.0	11.3	2.1	1.8	0.60	0.04	0.06	0.045	0.01	-	-

*Balance Fe

(1) AOD = argon-oxygen decarburization.
(2) Bonar Langley Alloys, Ltd. Slough, England.
(3) AB Sandvik Steel, Sandviken, Sweden.
(4) Sumitomo Metal Industries, Ltd., Osaka, Japan.

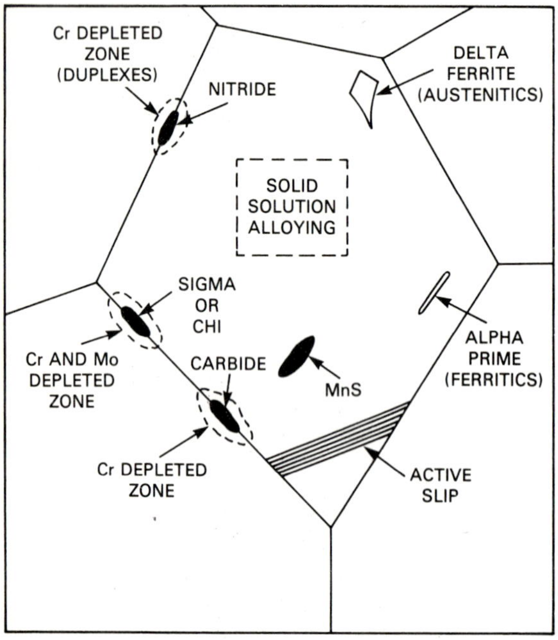

1. Schematic of metallurgical variables affecting the localized corrosion behavior of stainless steels.

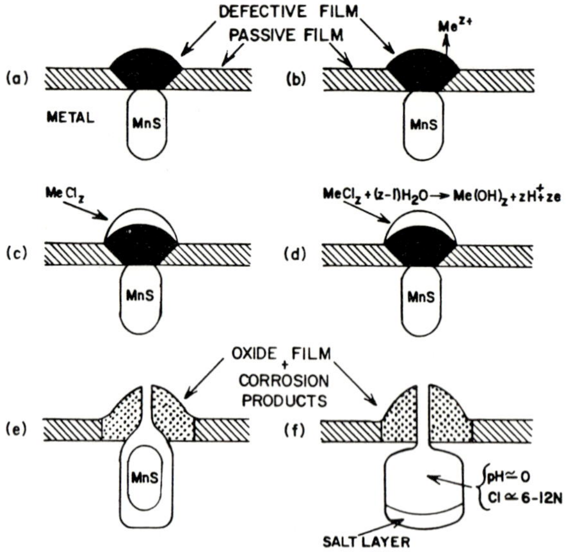

2. Pit nucleation and development at manganese sulfide inclusions in stainless steel (from Szklarska-Smialowska and Lunarska).

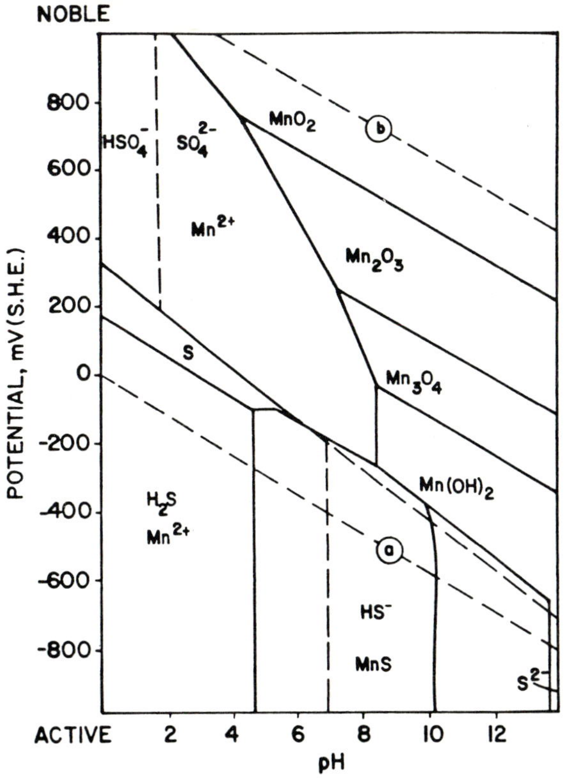

3. Potential-pH diagram for MnS-H_2O-Cl^- system calculated on basis of 0.1M for the ions SO_4^{--}, Cl^- and Mn^{++} (from Eklund).

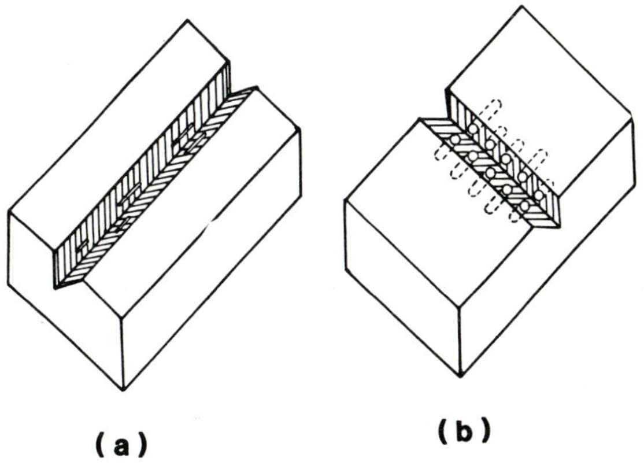

5. Schematic of effect of directionality of grinding with 80 grit paper on sulfide stringers; note that specimens with longitudinal grinding, (a), have pitting potential 60mV more noble than those with transverse grinding, (b), (from Manning, Duquette and Savage).

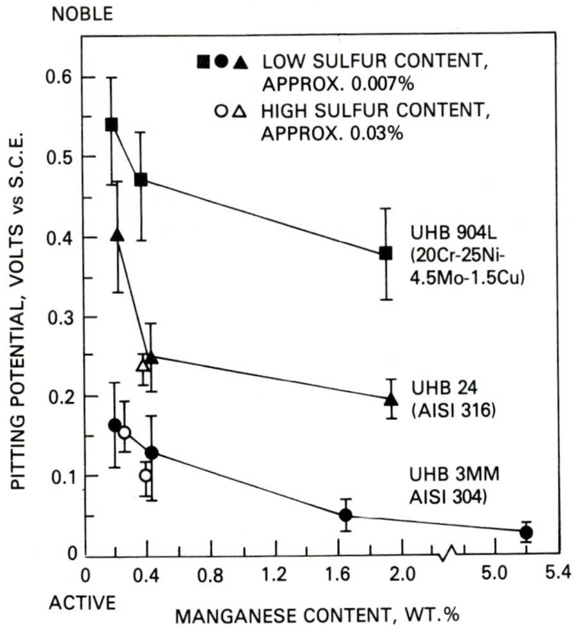

4. Effect of manganese content on pitting potential of various stainless steels in oxygenated 5% NaCl solution (from Degerbeck and Wold).

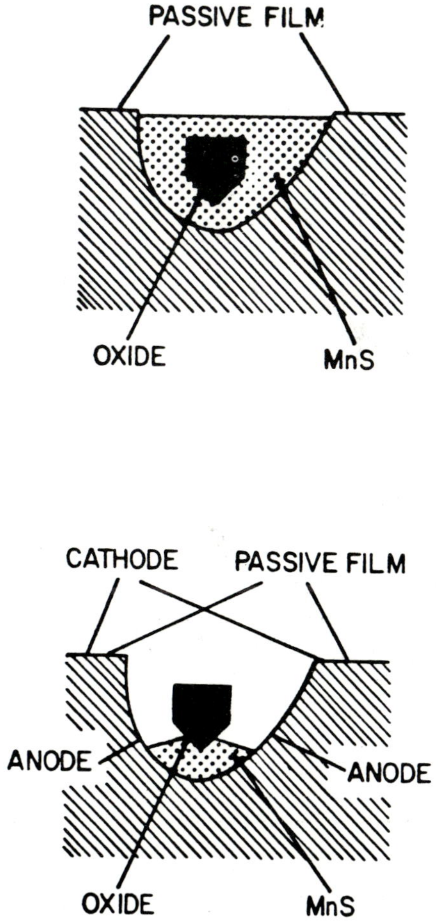

6. Initiation of pitting at a duplex manganese sulfide-oxide inclusion (from Blom).

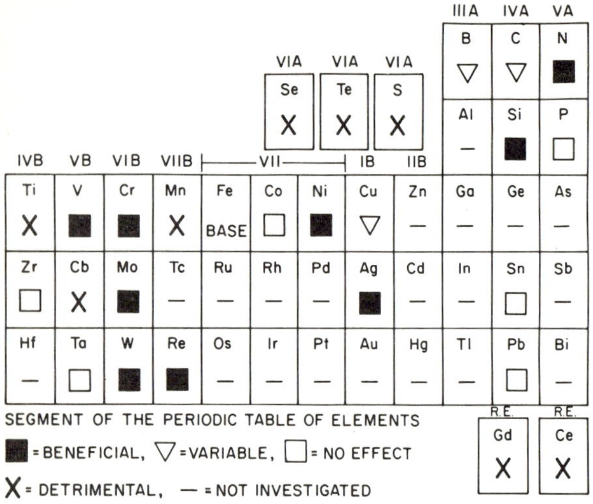

7. Effect of element shown on resistance of austenitic stainless steels to pitting in chloride solutions.

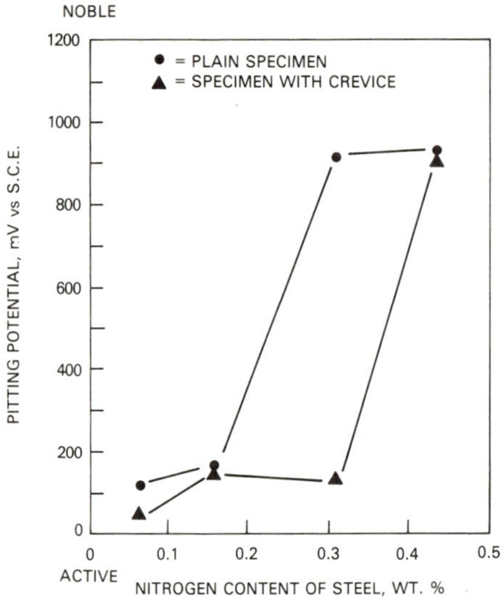

8. Effect of nitrogen content on pitting potential of 22Cr-20Ni-4Mn-2.8Mo-0.03C-0.01S stainless steel in an aerated aqueous solution containing 0.6M NaCl and 0.1M $NaHCO_3$ (from Truman, Coleman and Pirt).

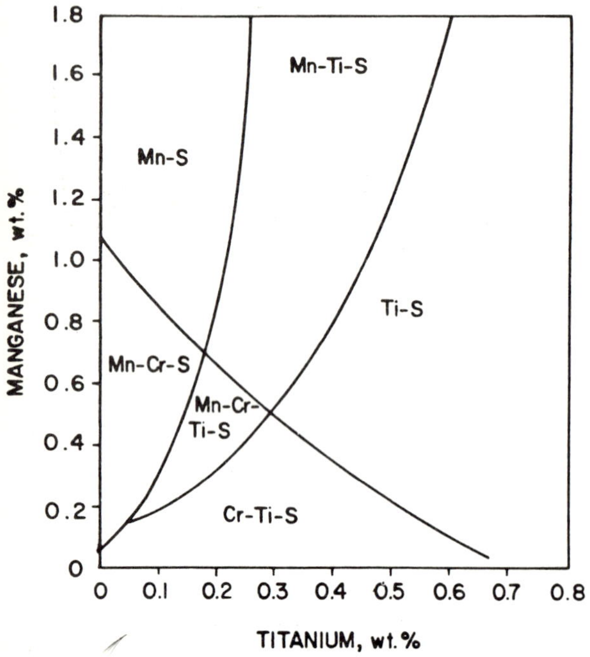

9. Influence of manganese and titanium contents of 18Cr-2.3Mo-0.2S stainless steel on sulfide composition (from Kiessling).

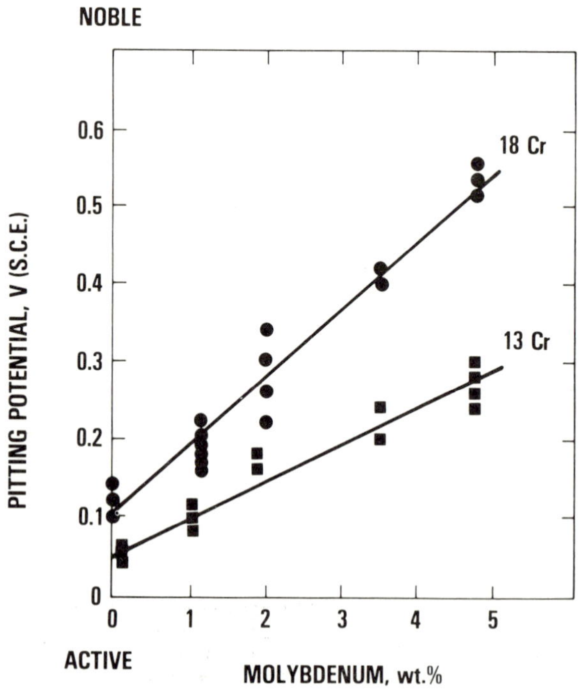

10. Pitting potentials of vacuum melted 13Cr and 18Cr ferritic stainless steels in 1M NaCl at 25°C (from Bond, Lizlovs and Dundas).

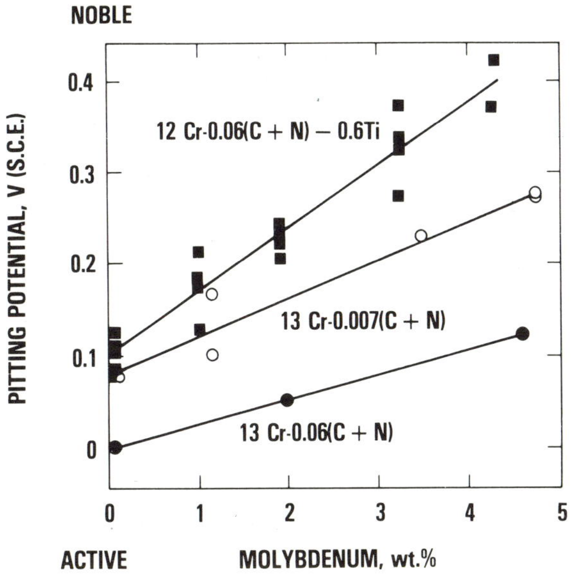

11. Effect of titanium and interstitial content on pitting potentials of ferritic stainless steels in 1M NaCl at 25°C (from Bond, Lizlovs and Dundas).

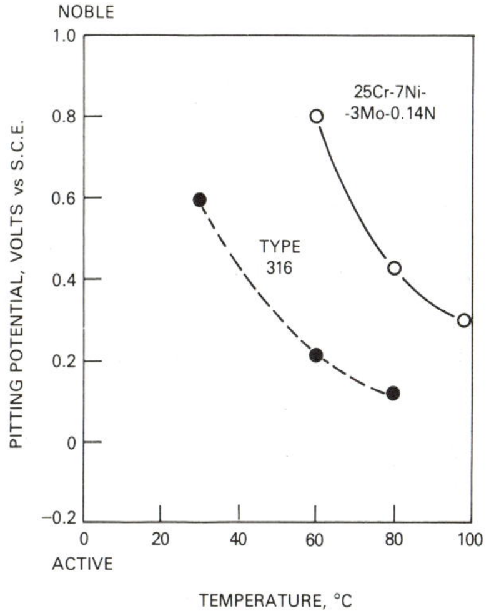

12. Pitting potentials of AISI type 316 and of a duplex stainless steel in de-aerated synthetic seawater as a function of temperature (from Miyuki, et al.).

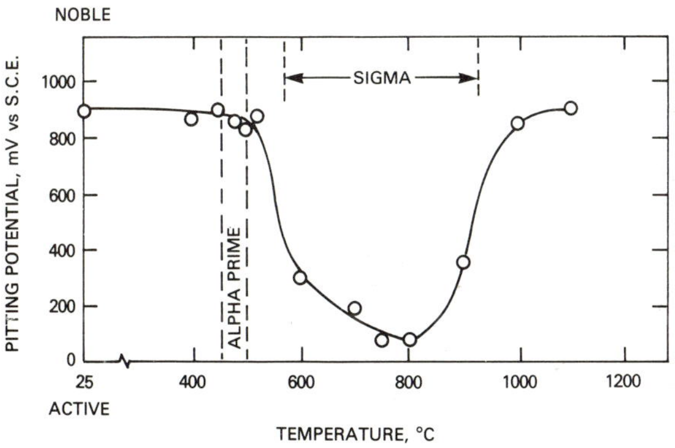

13. Effect of isothermal heat treatment temperature (4 hours) on the pitting potential of a duplex stainless steel (26Cr-5.5Ni-1.5Mo-0.2N) in an aerated 0.6M NaCl + 0.1M $NaHCO_3$ solution at 25°C (from Truman and Pirt).

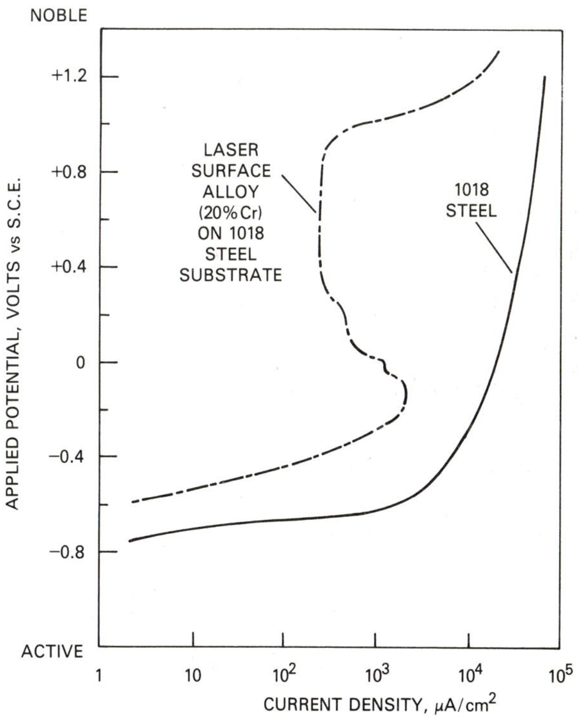

14. Anodic polarization curves in de-aerated 0.1M Na_2SO_4 solution at 25°C (from McCafferty, Moore, Ayers, and Hubler).

Metallurgical aspects of corrosion resistance of stabilized 17%Cr stainless steels

B Baroux, P Pedarre and J Decroix

The authors are with UGINE Research Centre, Ugine, France.

SYNOPSIS

The pitting resistance of Ti-, Nb- and Zr-stabilized 17%Cr stainless steels is investigated with relation to the nature of the non-metallic phases present in the steels in the as-annealed state. A new statistical electrochemical technique is used, from which the potential dependence of the pit generation rate can be deduced. The results confirm the harmful effect of Mn sulphides in the Nb-bearing steels and the beneficial effect of Ti additions. Zr-stabilized steels were found to be as resistant as Ti-bearing steels if the formation of intermetallic eutectics during the steelmaking process was prevented. Such performances are obtained by a precise control of the Zr excess, with respect to the minimum stabilization requirement.

INTRODUCTION

The uses of 17%Cr ferritic stainless steels have increased considerably since stabilizing elements were added in the base 430 grade, providing a good intergranular corrosion resistance of weldments.(1,2) These stabilizing elements are usually Ti, Nb, and more recently Zr. They must be added in a sufficient amount in order to fix both C and N as carbides, nitrides or carbo-nitrides. As a consequence, the microstructure of these steels consists mainly of a fully ferritic phase; however, several metallic phases are also present, the nature and density of which depend on the nature and the concentration of the stabilizing element which is under consideration.

On the other hand, several properties of the steel depend strongly on those non-metallic phases. For instance, it is well-known that pitting corrosion resistance depends heavily on the presence or absence of Mn sulphides.(3) In another field, the creep resistance is considerably improved by Nb additions, (4) due to the beneficial effect of the intergranular precipitation of an intermetallic phase (Fe, Nb), at the temperature of usage.(5) Lastly, the air oxidation resistance has been shown to be strongly dependent on the carbide stability, leading to the choice of Zr as the appropriate stabilizing element when this resistance is needed.(6)

From a practical viewpoint, the composition of the steel must be designed with respect to the expected usage properties. Let us consider, for instance, the steels to be used in the automotive exhaust systems. It is obvious that creep and oxidation resistances are of the highest interest in this case, and Zr + Nb bi-stabilized steels are then expected to be the better choice.(5) However pitting corrosion resistance cannot be neglected, since salts are used more and more to remove the snow from the roads in winter.

It is intended in this paper to present some results obtained on various stabilized steels in the field of localized corrosion resistance and to show how the best composition can be designed, from this particular standpoint.

STUDIED STEELS

Tests were performed on some industrial cold-rolled sheets (thickness 0.8 mm), whose typical composition is shown in Table 1. For each stabilizing element, X = Ti, Nb or Zr, one has to take into consideration the amount ΔX which is in excess in relation to the stoichiometry of the nitrides, carbides or carbo-nitrides. In the first approximation this amount is given by:

$$Ti \simeq Ti - 4\ (C + N)$$
$$Nb \simeq Nb - 7\ (C + N)$$
$$Zr \simeq Zr - 7\ (C + N)$$

Table 2 shows the exact composition of the investigated steels. For Zr-bearing steels, various Zr contents were taken into account, with various ΔZr. For example, the steel 17CrZr 150 is designed to be 0.15%Zr in excess of the carbides' and nitrides' stoichiometry. For Nb-bearing steels, the effect of the S content was investigated; the steel 17CrNb-LS (low sulphur), with S = 20 ppm, was compared to the steel 17CrNb, with S = 80 ppm.

The microstructure of the tested materials was examined using optical and transmission microscopies, scanning electron microscopy and microprobe analysis. The investigated steels are fully ferritic, but also contain some non-metallic phases, whose nature is summarized in Table 3, for the as-annealed state. Attention must be particularly focused on the fact that for Zr-containing steels too high a Zr content produces some eutectic phases containing Fe, Zr and also phosphorus, which can be harmful to various properties.(2) The next figure shows the relation between the density of these globules and the Zr excess ΔZr, as measured at several heats (which are not all shown in Table 2). One can see that for ΔZr larger than 0.2%, this density increases sharply, leading to some poor usage properties (for example those for drawing applications).

EXPERIMENTAL METHOD

The pitting resistance NaCl (0.02M) neutral aqueous media was investigated at room temperature, using a statistical electrochemical technique (presented in other work-7a, b, c), the principle of which (8) is to use a multichannel electrochemical device. One test specimen is connected at each channel and the time needed for pitting (or the pitting potential if the potentiokinetic mode is used) is measured for each single specimen; then the "survival probability" P of a single specimen can be estimated at each time (or at each potential). In previous work, we showed that the "elementary pitting probability" ϖ, per unit area and unit of time, can be deduced from P by the relation:

$$\varpi = -\frac{1}{S} . \text{Ln } P \qquad (1)$$

where S (=0.785 cm^2 in our case) is the area of a single test specimen. The "pit generation rate" (PGR) g can also be assessed:

$$g = \frac{d\varpi}{dt} = -\frac{1}{S} . \frac{dP}{Pdt} = -\frac{v}{S} . \frac{d \text{ Ln } P}{dE} \qquad (2)$$

where E is the electrode potential and v the potentiokinetic scanning rate.

In the present study, samples were first mechanically wet-polished with paper 1200, then aged for 24 h in air, and finally introduced into the electrochemical cell. The potential is then raised to v = 100 mv/mn and the survival probability is measured at each potential; the pit generation rate (PGR) is derived using equation (2).

EFFECT OF Ti AND Nb STABILIZATION

Figure 2 shows the relative behaviour of the reference unstabilized steel and of the Ti- or Nb-stabilized steels. At first approximation, the pitting resistance of the Nb-containing steel is close to that of the reference steel. This result is easy to understand, considering that MnS, which are present in both the reference and in Nb-bearing steels, act as harmful pitting sites, as has been shown by several authors.(3,9) It must be added that the pitting resistance can be significantly increased by lowering the Mn content, which leads to a chromium enrichment of the sulphides, or by lowering the sulphur content. Figure 3 shows that lowering the sulphur content from 80 ppm to 30 ppm results in decreasing the PGR by a factor of 10, or increasing the pitting potentials by 40 mv.

On the contrary, the Ti-containing steel behaviour is much better, whatever the sulphur content, since sulphur is combined with Ti to form insoluble TiS. In this case the pits are believed to initiate at the interface between the metallic matrix and the Ti carbides, but at much higher potential. Moreover, it is worth noting that Ti-containing steel pitting resistance comes close to that of classical 304 steels. Nevertheless, one must keep in mind that the 304 resistance can also be improved by lowering the sulphur and the Mn content. Lastly, we observed that the pitting resistance of all those steels does not depend sharply on the stabilizing element concentration. This is in accordance with the fact that inclusions or precipitates do not significantly change when the stabilizing element concentration changes, as long as it remains larger than the minimum level required for stabilization. In the case of Ti steel, other investigations showed that the improved pitting resistance appears as soon as Ti content becomes larger than 150%, which is much lower than the stabilization requirement.

EFFECT OF THE Zr STABILIZATION

The case of Zr-stabilized steels is more complex. Figure 4 shows the behaviour of some steels for which the Zr excess regarding the carbides' and nitrides' stoichiometry ranges from 0.1% to 0.420%. For low Zr excess, the pitting resistance is equal to that of the Ti-stabilized steel. For high Zr excess, the performances become equal to that of the unstabilized steel, and even lower for very high Zr excess. These findings can be understood by considering the following facts:

- Zr-stabilized steels do not contain MnS, since Zr sulphide is more stable at the melting temperature,
- Zr sulphides are not soluble in the neutral chloride-containing solutions, and the pitting resistance of the Zr steels must be higher than that of the steels containing MnS,
- Some harmful precipitations or inclusions are present for high Zr excess and act as pitting sites. According to the structural observations, these pitting sites are very likely to be the eutectic globules (Fe, Zr, P), the density of which increases with the Zr excess.

Figure 5 shows the relation between the Zr excess of the tested steel and a conventional pitting potential E_1, for which the PGR is equal to 0.1 cm^{-2} s^{-1}. In order to verify both the effect of the Zr excess on the pitting resistance and the correlation between the electrochemical measurements and the actual corrosion resistance, we performed a very simple practical pitting test by dipping the test speciment for 15 s in the corrosive solution, heated to 60°C. This dipping cycle was repeated for 16 h and the test sample was then microscopically examined and the pits' density measured. Figure 6 shows that the pits' density is well correlated to the conventional pitting potential deduced from the electrochemical measurements, and then of course to the Zr excess. In our opinion, this proves both the relevance of the statistical electrochemical test and the detrimental effect of a strong Zr excess.

CONCLUSION

In our opinion, the above results show that designing a particular stabilized grade for a given application is now possible, taking into account the microstructure/usage properties relationships for these kind of steels.

As far as the Zr stabilization is concerned, the Zr excess ΔZr must be kept lower than 0.2%. In other studies we showed that TIG weldments are resistant to the sulphocupric test, providing that $\Delta Zr > 0$. Then, in order to provide good intergranular and pitting corrosion resistances, the Zr content must be adjusted carefully, requiring good control over the steelmaking process. This effort is justified by the very attractive properties of this stabilizing element, either for deep drawability (10) or oxidation resistance.(6) Most of those attractive properties are due to the strong stability of Zr carbides, which are formed at a very high temperature. However, the low solubility of Zr in the metallic matrix implies a high tendency to form non-metallic phases, when the Zr excess is too widely in excess with relation to the minimum requirement for nitrogen and carbon trapping.

For Ti- or Nb-bearing steels, attention was drawn to the marked beneficial effect of Ti additions, and to the necessity of limiting the sulphur content in the Nb-bearing steels.

REFERENCES

1 Y BARBAZANGES, B BAROUX, Ph KRAEMER, Ph MAITREPIERRE: Revue aciers spéciaux, 1980, no. 52, (in French)

2 B BAROUX: Keynote conférence, 25 journee des aciers spéciaux, St Etienne (France), 1986, (English and French)

3 A J SEDRIKS: International Metals Reviews, 1983, vol. 28, no. 5

4 J N JOHNSON: SAE Technical paper series 810035, 1981

5 A new Nb + Zr bearing 17%Cr stainless steel for high-temperature applications, B BAROUX, P GRESSIN, P PEDARRE, J DECROIX: to be published

6 T MOROISHI, H FUJIKAWA, H MAKIURA: Journ. of Electrochem. Soc., Dec. 1979, p. 2173

7a B BAROUX, B SALA, T JOSSIC, J PINARD: Mat. et techn., Avril/Mai 1985, p. 211 (in French)

7b B BAROUX, B SALA: Proceedings of the 7th Euro. Corr. Symposium, 1985, Nice, France, (in English)

7c B BAROUX: In 'Materials Science Forum", Ed. G E Murch, Trans. Tech. Publications, (Switzerland), vol. 8, 1986, p. 91

8 T SHIBATA, T TAKEYAMA: Corrosion NACE, July 1977, vol. 33, no. 7, p. 243

9 G WRANGLEN: Cor. Sci., 1974, vol. 14, p. 331

10 JM HAUSER, B BAROUX, G BLANC: ICOTOM 84, 7th international conference on textures of materials, Sept. 1984

Table 1 : Typical composition(Weight %)

STEEL	Cr	C	N	Ti,Nb,Zr
AISI 430 Reference	16.5	.04	.03	
17 Cr + Ti	»	.02	.01	.30to.600
17 Cr + Nb	»	.03	.03	.50to.800
17 Cr + Zr	»	.02	.01	.25to.650

Table 2 : Composition of the studied steels (Weight %)

STEEL	Cr	C	N	Si	Mn	Ni	Mo	Cu	S	P	Al	Ti	Nb	Zr	Δx
17 Cr	16.3	.045	.045	.30	.62	.27	.03	.11	.003	.03	.05				
17 Cr Ti	16.5	.02	.01	.30	.46	.10	.04	.02	.003	.02	.03	.44			
17 Cr Nb	16.5	.03	.03	.35	.40	.13	.01	.05	.008	.03	.02		.71		
17CrNb LS	16.6	.02	.03	.43	.50	.19	.02	.05	.002	.02	.04		.76		
17CrZr.110	16.1	.037	.007	.33	.50	.15	.04	.05	.003	.02	.04			.410	.110
17CrZr.150	16.3	.025	.010	.38	.44	.16	.01	.04	.003	.02	.03			.390	.150
17CrZr.360	16.1	.025	.006	.35	.48	.12	.03	.05	.002	.02	.03			.580	.360
17CrZr.420	16	.025	.007	.40	.48	.19	.06	.05	.003	.03	.04			.640	.420

Table 3 : Non metallic phases in the as-annealed state

	17Cr-Ti	17Cr-Nb	17Cr-Zr Zr ⩽ .200	17Cr-Zr Zr ⩾ .200
Inclusions	oxydes +(Ti,S)	oxydes + MnS	oxydes +(Zr,S)	- Id -
Coarse precipitation > 2µm	TiN with TiC belt	Nb(C,N)	ZrN with ZrC belts + Fe3Zr belts in some cases ZrC	- Id - + globular eutectics (Fe,Zr,P)
Fine precipitation 1000 to 5000 Å	TiC (Ti,P)	Nb(C,N) Intermetallic phases (Fe,Nb)	intermetallic phase (Fe,Zr) in some cases	- Id -
Ultrafine precipitation ⩽ 500 Å	X	X	Phosphides(Zr,P)	- Id -

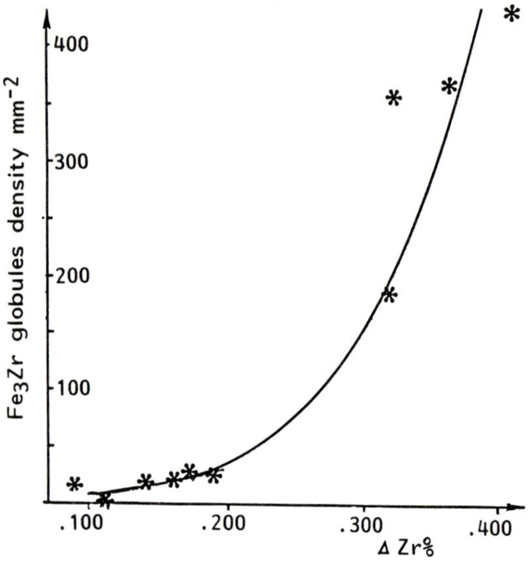

1 Effect of the Zr excess on the eutectic globules' density

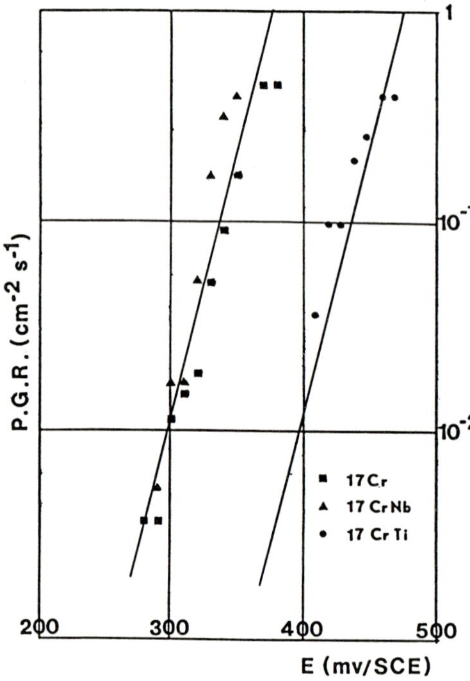

2 Potential dependence of the pit generation rate (PGR) for the unstabilized steels and the Nb- or Ti-bearing stabilized steels

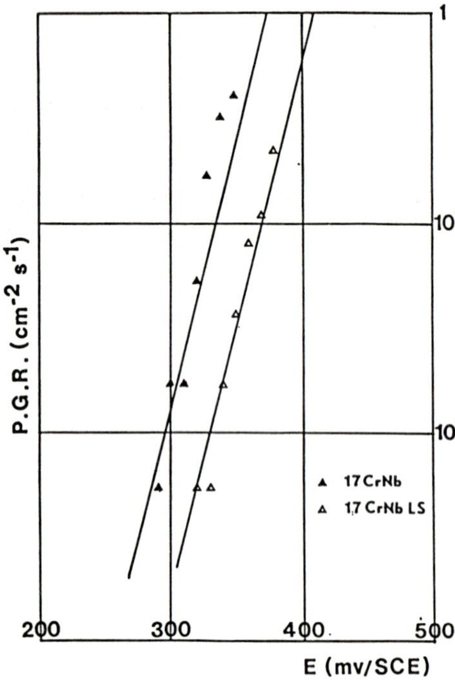

3 Effect of the S content on the PGR of the Nb-stabilized steel: 17CrNb: S = 20 ppm; 17CrNb-LS: S = 80 ppm

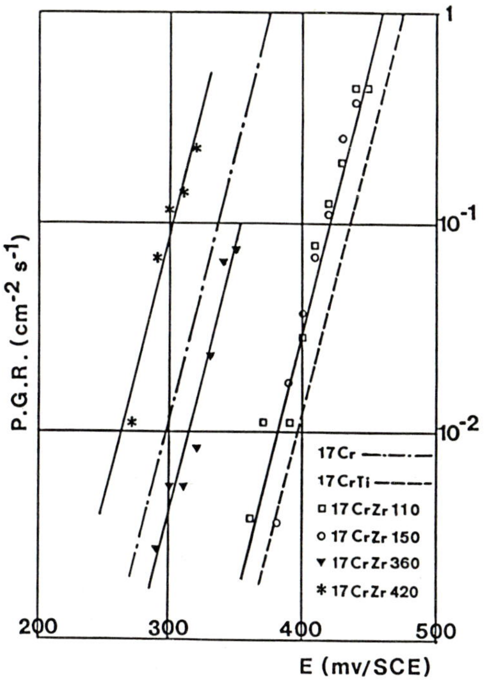

4 Effect of the Zr excess on the PGR of the Zr-stabilized steels; eg for 17CrZr steel, the Zr excess is 0.150%

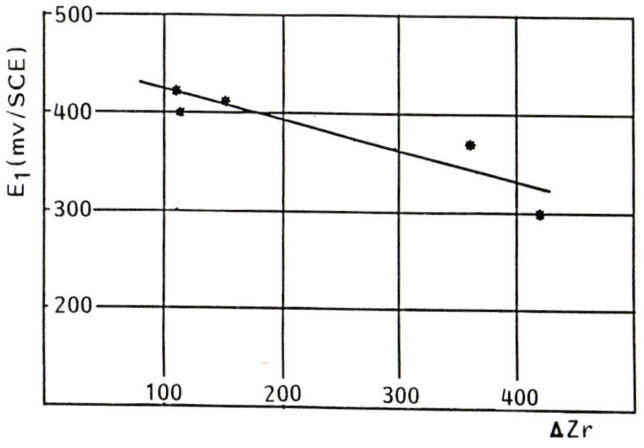

5 Relation between the Zr excess and the potential E_1, for which the PGR is equal to 0.1 cm^{-2}, s^{-1}

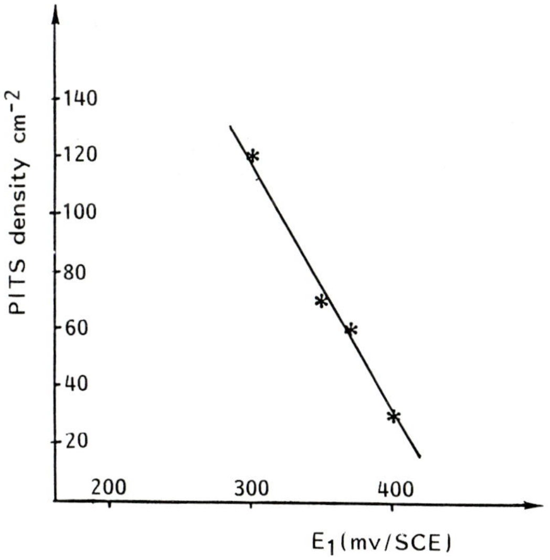

6 Relation between E_1 and the pits' density measured in an exposure test

Development of oxidation-resistant Si-bearing ferritic stainless steels

S Akiyama and H Fujikawa

Mr. Akiyama is in the Naoetsu Research Laboratories of Nippon Stainless Steel Co. Ltd.; Dr. Fujikawa is in the Technical Research Laboratories of Sumitomo Metal Industries Ltd, Japan.

SYNOPSIS

The effect of Si content on high-temperature corrosion of some ferritic stainless steels was investigated in air, kerosine combustion atmosphere and moist atmosphere.

It was found that addition of more than 2%Si not only improved oxidation resistance in air but also prevented formation of a reddish scale which sometimes grew on the commercial ferritic stainless steels in high H_2O atmospheres with low P_{O_2}, such as an oil combustion atmosphere.

On the basis of the results, two types of oxidation-resistant, Si-bearing ferritic stainless steels, low C–11%Cr–2%Si steel and low C–18%Cr–2.5%Si–0.3%Nb steel, have been developed, mainly for use as various high-temperature parts of household heaters.

INTRODUCTION

Si, like Cr and Al, has long been known as an effective element for giving oxidation-resistant qualities to steel.[1-3] Especially austenitic stainless steels containing 2–5%Si have been used in heat-resistant equipment such as exhaust gas purifiers for automobiles, heat exchanger tubes, heaters etc.

On the other hand, several reports refer to the effect of Si addition on the oxidation resistance of ferritic stainless steels, and some of them report that SiO_2 concentration has been detected on the metal surface of the internal oxide layer.[4-6]

Considering the fact of the improved oxidation resistance through Si addition to ferritic stainless steels, it is expected to promote also a high-temperature corrosion resistance in some combustion atmospheres. In the combustion gas atmosphere found in a burning oil fuel, Type 430 stainless steel, the typical material for heating apparatus, easily develops a reddish oxide scale, resulting in a decrease in combustion efficiency and a remarkable deterioration in appearance. T Kawasaki reported that in atmospheres containing water vapour, thick, velvet-like scale formed on Type 430 at the lower temperature of 600, rather than 800°C.[7]

In general, as elements of ferrite-former such as Mo, Si, Al etc are increased in the ferritic stainless steels, so the productivity, formability and ductility of the steels decrease. Due to these difficulties with the steel composition as well as the manufacturing process, types of ferritic steels which contain large amounts of ferrite-former elements have been expensive and have not been so widely used. But recently improvement of the melting process has made it possible to use these steels extensively.

In view of this background, the high-temperature corrosion of Si-bearing ferritic stainless steels has been investigated in air, kerosine combustion atmosphere and moist atmosphere in order to develop the new steels which show improved resistance to reddish scale formation in combustion gas atmospheres.

EXPERIMENTAL PROCEDURE

1 Test materials

The alloys investigated, as shown in Table 1, were ferritic stainless steels containing from 11% to 22%Cr and up to 3%Si. After being melted in air, these steels were cast in 10kg ingots, forged, hot-rolled, cold-rolled to 0.4–1.2 mm width thickness, annealed at 800–870°C and descaled.

Typical commercial stainless steels, such as Type 410L, Type 409, Type 430 and Type 304, were used as relative materials.

2 High-temperature corrosion test

High-temperature oxidation tests were performed in air at up to 1100°C for up to 585 h continuously, followed by weighing of the specimens after tests.

Corrosion tests in a combustion atmosphere were carried out in a few types of kerosine heater for up to 1000 h with coupons of 0.4 mm x 10 mm x 20 mm. Location of coupons in a wick type portable kerosine heater is shown in Fig. 1, and Table 2 shows a typical composition of gas in the flame.

High-temperature corrosion tests in a moist atmosphere at 550–900°C were conducted in environments of Ar–20%H_2O, up to 20%O_2. The schematic diagram of the test apparatus in the moist atmosphere is shown in Fig. 2.

After those corrosion tests the specimens were evaluated by optical microscopy, X-ray diffraction analysis, EPMA and IMMA.

RESULTS

1 Oxidation tests in air

Figure 3 shows the effect of Cr and Si contents on critical temperature for oxidation resistance of Fe–Cr–Si ferritic stainless steels in air for 200 h. As the amount of Cr and Si increases, the oxidation resistance of the steels clearly improves.

Si-bearing steels, as shown in Fig. 4, definitely surpass commercial stainless steels which contain the same amount of Cr as the Si-bearing steels. Figure 5 shows the appearance of specimens oxidized in air at 900°C for 585 h.

2 Corrosion tests in combustion atmosphere

Figure 6 is the appearance of low C–11%Cr steels with various Si contents after corrosion testing in a wick type portable kerosine heater for 8 h. Appearances of low C–11%Cr–2%Si steel, Type 430 and Type 304 after exposure in the same heater for 1000 h are shown in Fig. 7. The steel containing more than 2%Si, even if its Cr is as low as 11%, shows marked prevention of the reddish scale growth as compared with commercial steels such as Type 430 and Type 304.

The effect of Cr and Si contents on the reddish scale formation was investigated, using a burner type kerosine fan-heater in which the reddish scale grew harder than in the wick type heater. As shown in Fig. 8, it is evident that the increase of Cr and Si suppresses the formation of the reddish scale.

Figures 9 and 10 are microstructure and X-ray images of cross-section of the reddish scale formed on Type 430 steel. According to EPMA and X-ray diffraction analysis, the reddish scale consists of α–Fe_2O_3 (Hematite) which has grown on the outside of the protective (Fe, Cr)-oxide layer.

3 Corrosion tests in moist atmosphere

Figure 11 shows the effect of Cr and Si contents on the reddish scale formation in Ar–2%O_2–20%H_2O atmosphere at 600°C for 96 h. It is interesting to note that the reappearance of the reddish scale formation and its prevention by Cr and Si addition in the moist atmosphere are analogous to those in the aforesaid combustion atmosphere.

Figure 12 shows the temperature dependence of weight gain for Type 430 in Ar–2%O_2–20%H_2O for 20 h. It is remarkable that in the moist atmosphere the reddish scale grows more rapidly at around 600°C than at 800°C, in contrast to the monotonous weight change in air.

DISCUSSION

1 Mechanism of reddish scale formation

The reddish scale formed in the combustion gas atmosphere, in respect of its shape and morphology as shown in Fig. 13, is strongly analogous to a scale formed in a superheated steam which ordinarily consists of a high water vapour content with low P_{O_2}. Therefore it is suggested that the environmental factors accelerating the growth of the reddish scale are a moist atmosphere with low O_2 content and a relatively low temperature – around 600°C. Figure 14 shows the relationship of O_2 content and weight change in Type 430 and Si-bearing steels in a moist atmosphere of Ar–20%H_2O at 600°C for 20 h.

From these results, it is estimated that at a relatively low temperature and in a high H_2O atmosphere with low O_2 content as in an oil combustion atmosphere, the protective Cr-rich scale formed at an early stage does not grow so rapidly as in air; consequently break-away of the initial film occurs easily which permits the growth of Fe-rich reddish scale at the Cr-depleted zone under the film. In such a case, the existence of a large amount of water vapour may accelerate preferential oxidation and outward diffusion of Fe, resulting in the growth of the reddish scale which depends on the diffusion of Fe. In addition, there are detrimental ingredients in combustion gas such as SO_2, CO, CH_4 etc, which may also accelerate the break-away of the protective film.

However at the higher temperature of 800°C, in spite of high H_2O and low O_2, the reddish scale hardly grows because sufficient diffusion of Cr, Si and Mn keeps the initial oxide film protective, even if the atmosphere consists of high H_2O and low O_2.

2 Influence of Si addition

The effects of Si addition discussed below have been confirmed in the kerosine combustion atmosphere and the moist atmosphere. Concerning austenitic stainless steels with high Si content, it is known that the formation of the SiO_2 layer in the internal oxide layer improves the high-temperature oxidation resistance.[3]

As may be seen in Fig. 15 about IMMA analysis for 11%Cr–2%Si steel oxidized in a moist atmosphere at 550°C, however, the SiO_2 layer is not formed in the scale, in contrast to austenitic stainless steels, but only a slight concentration of Si is detected in the internal oxide layer. Therefore Si addition to ferritic stainless steels prevents the formation of the reddish scale and promotes the protective oxidation, probably because Si keeps the Cr-rich scale stable by taking oxygen which passes through the scale and moreover, Si promotes healing of the break-down scale.

On the basis of the results of these investigations, two types of low-cost, Si-bearing ferritic stainless steels were developed which show the excellent high-temperature corrosion resistance in an oil combustion atmosphere; low C–11%Cr steel with 2%Si almost completely prevents the formation of the reddish scale, and low C–18% Cr steel with 2.5%Si and 0.3%Nb is more effective, even in extreme conditions.

CONCLUSION

The effect of Si content on high-temperature corrosion of ferritic stainless steels was investigated in air, kerosine combustion atmosphere and moist atmosphere.

The obtained results are summarized as follows:

1 On the typical commercial ferritic stainless steel of Type 430, the reddish scale forms easily in a high-temperature atmosphere with high H_2O and low O_2 contents, such as combustion gas

atmosphere. The scale grows especially at around 600°C, and shows a minimum at about 800°C.

2 This reddish scale consists of α-Fe_2O_3 (Hematite) which has grown on the outside of the protective (Fe, Cr)-oxide layer. It is estimated that the reddish scale forms when the initial protective film is not stable but is apt to turn to break-away in the atmosphere of low temperature and low P_{O_2}, and when Fe in the metal diffuses outward and is oxidized preferentially in the moist atmosphere.

3 Increase of Si addition to ferritic stainless steels is quite effective in preventing the formation of the reddish scale in the same way that Cr addition is. This prevention probably occurs because Si keeps the Cr-rich scale stable by taking oxygen which passes through the scale, and moreover, Si promotes healing of the break-down scale.

4 On the basis of these results, two low-cost types of oxidation-resistant Si-bearing ferritic stainless steels, low C–11%Cr–2%Si steel and low C–18%Cr–2.5%Si–0.3%Nb steel, have been developed, mainly for use as various high-temperature parts of household heaters.

REFERENCES

1 W Hessenbruch: 'Mettale und Legierungen für hohe Temperaturen'; Erster Teil, 1940
2 J F Radavich: Corrosion, vol. 15, 613t, Nov. 1959
3 D L Douglass, J S Armjo: Oxidation of metals, vol. 2, 207, 1970
4 T Mishima, M Sugiyama: Tetsu-to-Haganē, vol. 36, 184, 1950
5 D Caplan, M Cohen: J. Metals, vol. 4, 1057, 1952
6 G C Wood, J A Richardson, M G Hobby, J Boustead: Corrosion Sci., vol. 9, 659, 1969
7 T Kawasaki, S Sato, S Ono: Boshoku-Gijutsu, vol. 31, 164, 1982

Table 1 Chemical composition of alloys tested (wt.%)

C	Si	Mn	P	S	Cr	N	Nb
0.005	0.33	0.52	0.019	0.010	10.77	0.010	-
0.010	0.51	0.45	0.028	0.008	11.24	0.004	-
0.008	1.12	0.70	0.021	0.005	11.31	0.005	-
0.011	1.38	0.56	0.031	0.011	10.78	0.005	-
0.009	1.80	0.56	0.019	0.014	11.01	0.007	-
0.009	1.98	0.66	0.024	0.010	11.23	0.009	-
0.009	2.16	0.68	0.024	0.009	11.41	0.006	-
0.008	2.41	0.61	0.021	0.010	10.98	0.005	-
0.011	2.95	0.52	0.022	0.009	11.03	0.009	-
0.009	2.30	0.34	0.014	0.002	13.60	0.015	0.30
0.011	2.59	0.25	0.018	0.002	16.33	0.011	0.28
0.011	1.56	0.25	0.024	0.002	18.03	0.015	0.30
0.015	2.08	0.25	0.024	0.001	18.04	0.015	0.28
0.011	2.56	0.37	0.024	0.005	18.58	0.014	-
0.015	2.57	0.25	0.020	0.002	18.03	0.013	0.28
0.013	2.60	0.26	0.021	0.002	18.20	0.015	0.54
0.012	3.06	0.25	0.021	0.002	18.24	0.015	0.28
0.010	2.62	0.26	0.022	0.002	20.13	0.012	0.24
0.015	2.50	0.26	0.023	0.002	22.30	0.013	0.23

Table 2 Typical composition of gas in the flame of a wick type portable kerosine heater

CO	CO_2	O_2	H_2	CH_4	SO_2	H_2O
5.38	8.64	6.31	1.12	0.50	tr	11.38

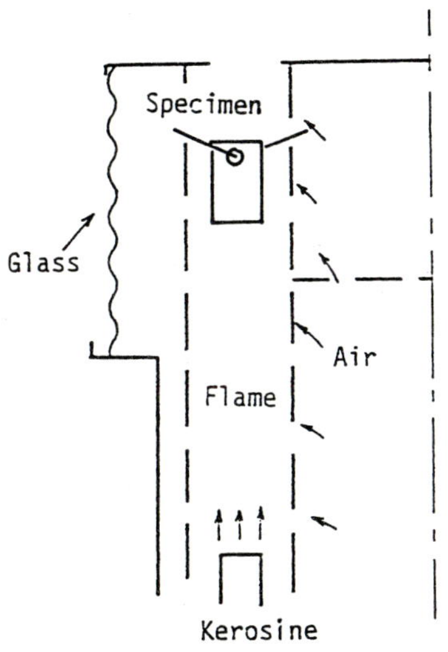

Fig. 1 Location of specimens in a wick type portable kerosine heater

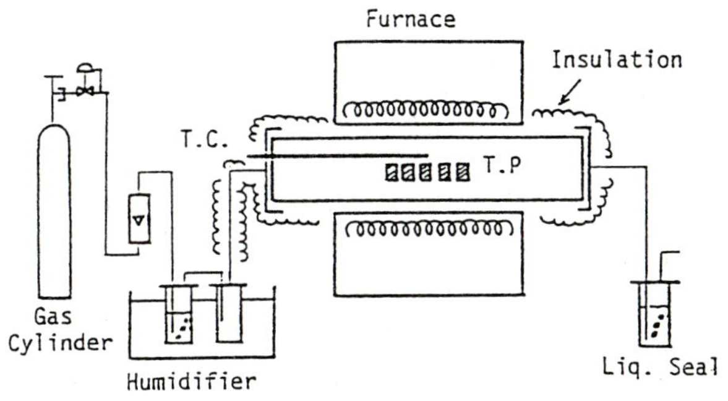

Fig. 2 Schematic diagram of test apparatus in moist atmosphere

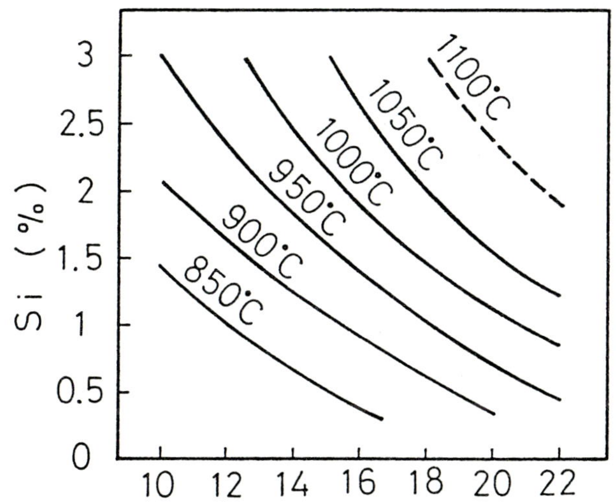

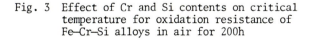

Fig. 3 Effect of Cr and Si contents on critical temperature for oxidation resistance of Fe–Cr–Si alloys in air for 200h

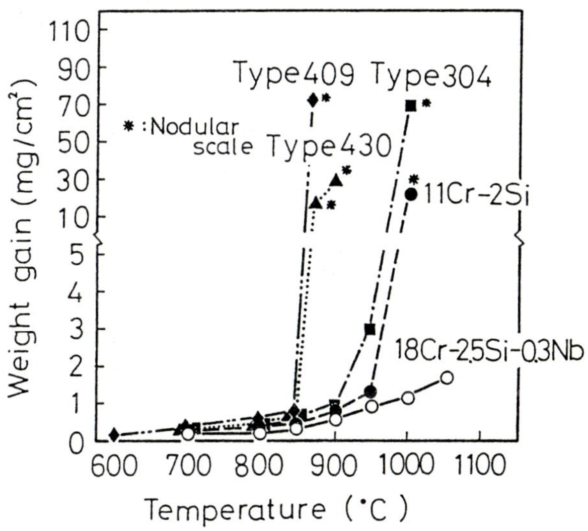

Fig. 4 Result of continuous oxidation testing of commercial alloys and Si-bearing steels in air, at various temperatures for 200 h

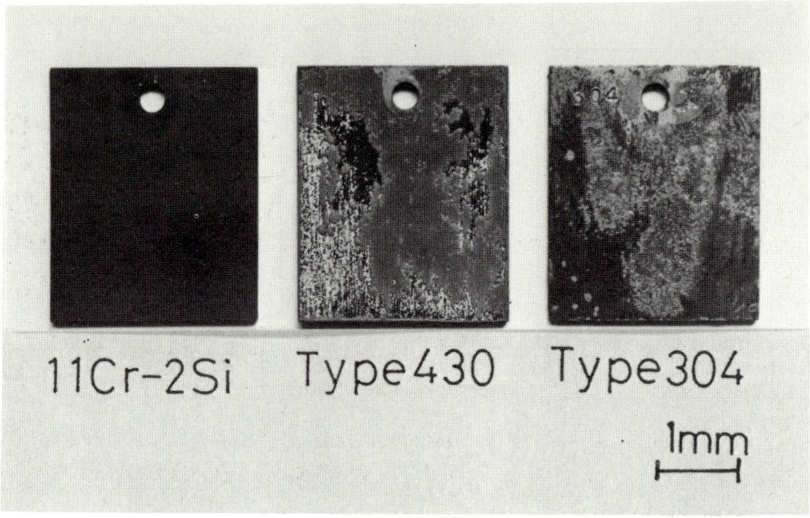

Fig. 5 Appearance of low C–11%Cr–2%Si steel, Type 430 and Type 304 oxidized in air at 900°C for 585 h

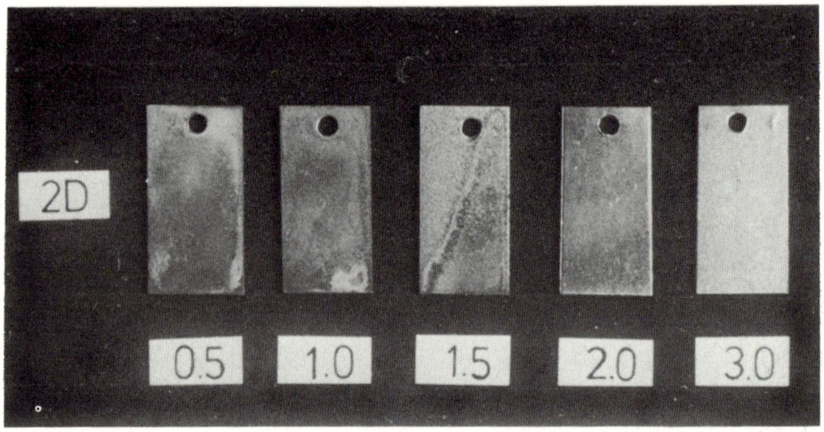

Fig. 6 Appearance of low C–11%Cr steels with various Si contents after corrosion testing in the wick type kerosine heater for 8 h

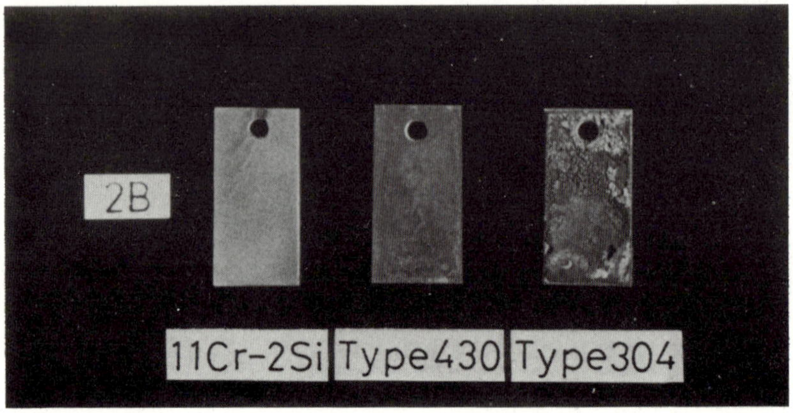

Fig. 7 Appearance of low C–11%Cr–2%Si steel, Type 430 and Type 304 after corrosion testing in the wick type kerosine heater for 1000 h

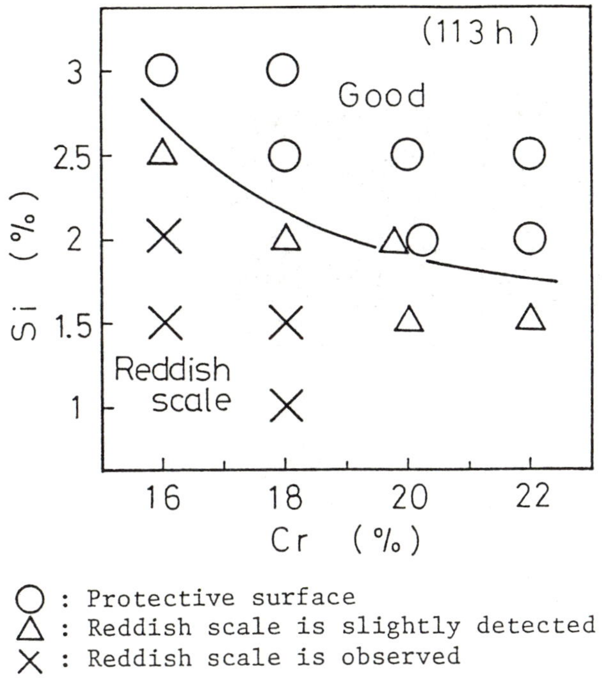

Fig. 8 Effect of Cr and Si contents on the reddish scale formation of Fe–Cr–Si alloys in the burner type kerosine fan-heater for 113 h

Fig. 9 Microstructure of cross-section of the reddish scale formed on Type 430 at corrosion testing in the wick type kerosine heater for 8 h

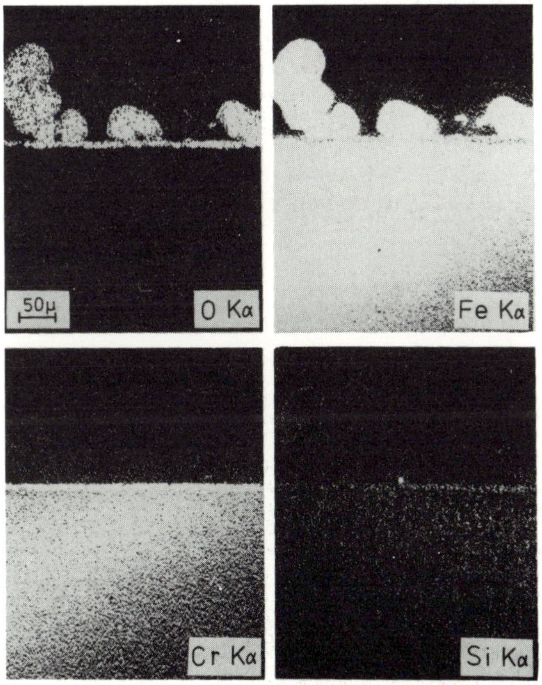

Fig. 10 X-ray images of cross-section of the reddish scale formed on Type 430 at corrosion testing in the wick type kerosine heater for 8 h

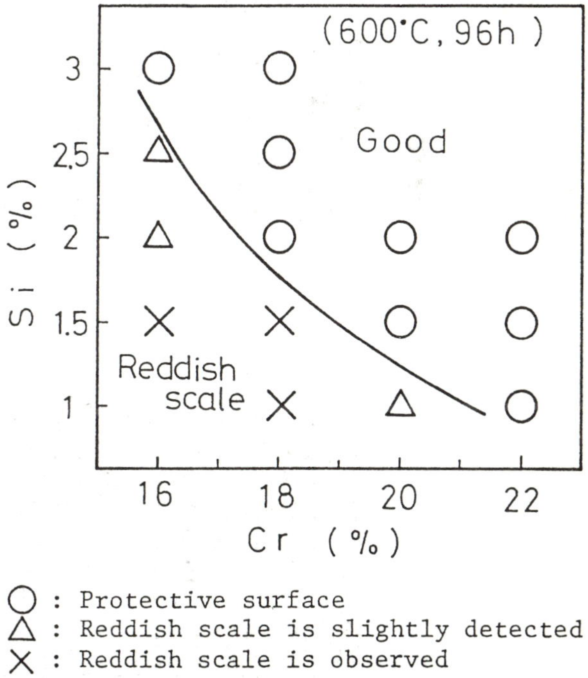

Fig. 11 Effect of Cr and Si contents on the reddish scale formation of Fe–Cr–Si alloys in Ar–2%O_2–20%H_2O atmosphere at 600°C for 96 h

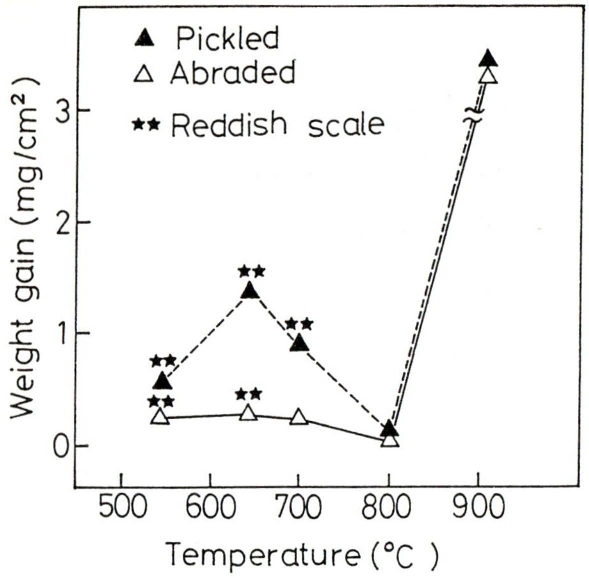

Fig. 12 Temperature dependence of weight gain for Type 430 in Ar–2%O_2–20%H_2O atmosphere for 20 h

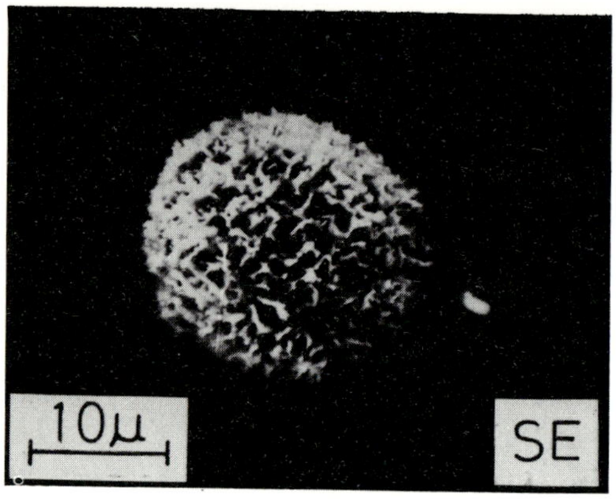

Fig. 13 Scanning electron micrograph of the reddish scale formed on Type 430 in the wick type kerosine heater for 1 h

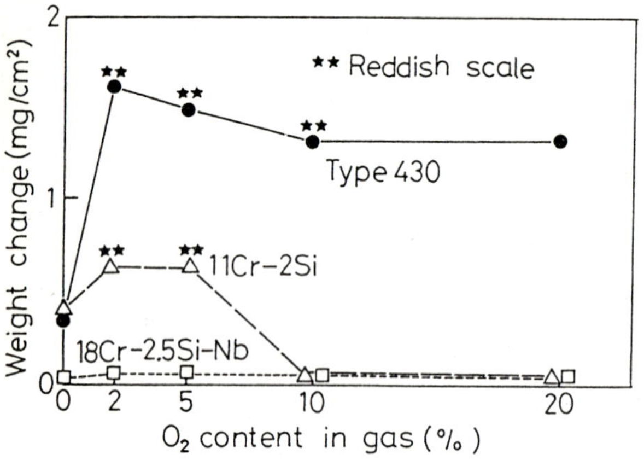

Fig. 14 Effect of O_2 content on the weight change of Type 430 and Si-bearing steels in Ar–20%H_2O atmosphere at 600°C for 20 h

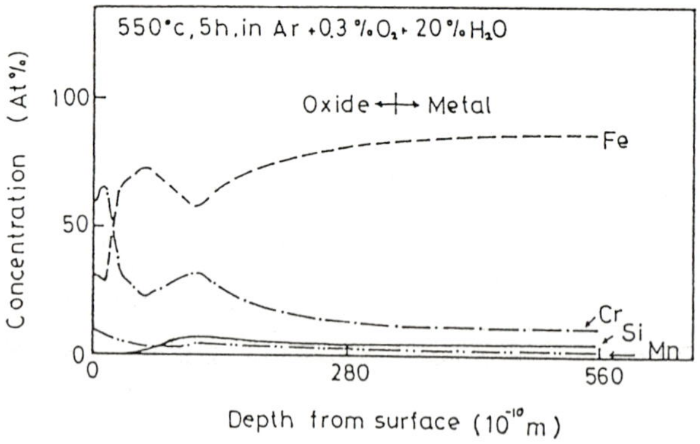

Fig. 15 IMMA depth analysis profiles for the scales formed on low C–11%Cr–2%Si steel in Ar–0.3%O_2–20%H_2O atmosphere at 550°C for 5 h

Microstructural stability and corrosion resistance of 29Cr—4Mo—Ti superferritic stainless steel

J C Bavay, J M Boulet, P Bourgain and P Chemelle

JCB, JMB and PB are in the Research and Development Department of UGINE, Aciers de Chatillon et de Gueugnon, France; PC is in the Department of Physical Metallurgy at IRSID, France.

SYNOPSIS

The alloy, identified by the Unified Numbering System as UNS S44735, is an advanced 29% chromium, 4% molybdenum ferritic stainless steel, stabilized with titanium, which was developed as tubing material for seawater condensers. The effect of heat treatment and welding on the microstructure and the intergranular and pitting corrosion resistance was investigated. Transmission electron microscopy was used to characterize intermetallic phases. The results of corrosion tests were discussed in relation to the presence of titanium carbonitrides and intermetallic phases. In chloride solutions, the corrosion performance of the 29%Cr-4%Mo-Ti alloy was compared with those of some nitrogen-containing duplex and superaustenitic stainless steels.

INTRODUCTION

Stainless steels for seawater applications require high chromium and molybdenum contents. Because they contain little or no nickel, ferritic stainless steels for seawater service are less expensive than more highly alloyed austenitic grades. To-day, there is a variety of high-performance ferritic stainless steels that are candidates as tubing material for seawater-or brackish water-cooled condensers. Commercially available highly alloyed ferritic stainless steels, the so-called "Superferritics", are differentiated by :

- the chromium (25 to 32 %), the molybdenum (1 to 4.5%) and the nickel (0.5 to 4.5%) contents,
- the level of interstitials (carbon and nitrogen) associated with the steelmaking technology (vacuum or AOD process),
- the stabilization by titanium or niobium.

The content of substitutional alloying elements, the amount of total interstitials and the nature of the stabilizer influence toughness and detrimental intermetallic compounds precipitation during hot processing or welding. This could explain a proliferation of high chromium ferritic grades with many combinations of chromium, molybdenum and nickel according to the commercially offered product forms, the areas of application and the desirable properties (1).

The purpose of this investigation was to show the effect of the nominal chemical composition of the 29%Cr-4%Mo-Ti ferritic stainless steel, designated UNS S44735, on :

- toughness and available product range,
- microstructural stability during heat treatment and welding,
- corrosion resistance in chloride solutions and seawater.

The type and the formation rate of intermetallic phases which might form in this alloy were determined . We also examined the influence of the secondary phases precipitation on the risk of embrittlement and susceptibility to intergranular attack.

The AOD processed UNS S44735 alloy is a highly corrosion-resistant candidate as tubing material for seawater- and brackish water-cooled steam condensers and heat exchangers. Its corrosion performance was compared with those of other commercial highly alloyed ferritic, duplex and austenitic stainless steels.

EXPERIMENTAL PROCEDURE

The tested alloys are listed by UNS number or name, structure type and chemical composition in Table 1.
The materials, having a thickness of 0.5-2 mm, were obtained in the mill solution-annealed condition.

TOUGHNESS

The UNS S44735 low interstitial alloy contains an adequate amount of titanium to fix carbon and nitrogen as Ti-carbonitrides and thus to prevent intergranular corrosion due to chromium depletion induced by Cr-rich carbides or nitrides precipitation along grain boundaries.

The effect of titanium on toughness is detrimental. Angular titanium nitrides act as crack initiation sites (2, 3). For this reason, the toughness of the 29%Cr-4%Mo-Ti alloy is not so good as that of the UNS S44800 high purity ferritic stainless steel without voluntary Ti addition produced by a vacuum refining technique (4). Moreover, the UNS S44800 alloy contains nickel whose effect on toughness is beneficial.

The ductile-to-brittle transition temperature (DBTT) of the UNS S44735 alloy increases with the increase in thickness (Fig.1). Consequently, this superferritic stainless steel is used only as thin sheets and strips for which the DBTT is decreased to a point where there is no practical toughness limitation. The significant influence of the specimen thickness on the DBTT level is mechanical rather than metallurgical (5). At thick gauges, the metal at the notch tip is constrained in the direction perpendicular to the strip plane and the stress state is triaxial at the crack tip. A sufficient gauge decrease leads to a plane stress state since constraint does not occur in the perpendicular direction to the strip plane.

INTERMETALLIC COMPOUNDS PRECIPITATION

High chromium-containing ferritic stainless steels are subject to structural instability due to intermetallic phases precipitation that may impair their mechanical and corrosion resistance properties.

The so-called 475°C embrittlement of the UNS S44735 alloy occurs as a result of the precipitation of alpha prime phase. There is a progressive decrease in room-temperature ductility when the UNS S44735 alloy is exposed to the 400-550°C temperature range (1). The alpha prime phase formation constitutes a service limitation with regard to heat-exchange applications but presents few problems in processing. Incubation time of sigma and chi phase precipitation is considerably shortened and loss of ductility is faster when heat treatment of the UNS S44735 alloy is made at about 900°C. The processing difficulties related to instability of the ferritic structure in this temperature range are overcome by using fast cooling.

Fig.2 shows the microstructure of a specimen heated at 900°C for 1 hour. A continuous network of sigma phase is formed at the grain boundaries. Some isolated particles of chi phase are present inside the sigma phase. Chi phase precipitates slower than sigma phase. Fine Ti-Fe-P particles are also identified at the initial ferritic grain boundaries (Fig.3). Electron diffraction results are shown in Table 2. Table 3 gives chemical composition of the different precipitates. Kinetics of intermetallic phases formation, calculated lattice parameters and chemical composition agree well with the literature concerning sigma and chi phases precipitation in ferritic stainless steels (6,7). In contrast to chi phase, sigma phase contains little titanium. Depending on the chemical composition of ferrite, chi phase may form prior to sigma phase (5, 8, 9).

An increase in molybdenum content enhances sigma and chi phase precipitation resulting in loss of bending aptitude (Fig.4). The addition of nickel appears to be an attractive method of improving the toughness of superferritic steels. But nickel greatly promotes sigma and chi phase precipitation as do chromium, molybdenum and titanium (1). Consequently, commercially available titanium-stabilized superferritic stainless steels containing 1-4% nickel have lower chromium and (or) molybdenum contents.

The transmission electron micrograph of the UNS S44735 alloy, held for one hour at 700°C, indicates that two types of precipitates are formed at the grain boundaries (Fig.5). The finer particles are identified as Ti-Fe phosphides. The coarser particles are probably Laves phase whose chemical composition is shown in Table 4.

INTERGRANULAR CORROSION RESISTANCE

In the mill-annealed condition, i.e. with an intermetallic phase-free microstructure, the UNS S44735 stainless steel is found immune to intergranular corrosion in the copper -copper sulfate- sulfuric acid test ASTM A 763 (Practice Y) and in the ferric sulfate - sulfuric acid test ASTM A 763 (Practice X).

In the medium with the weaker oxidizing strength (Practice Y), in-depth attack was very low, not more than five microns, when the UNS S44735 specimens were heated at 900°C for one hour. In the stronger oxidizing environment (Practice X), intergranular corrosion is a little more pronounced, but less than fifteen microns (1). On the other hand, specimens heat treated at 700°C for only five minutes were subjected to extremely severe intergranular corrosion even in the less oxidizing medium where a maximum in depth-attack of about thirty microns was observed. In the more oxidizing conditions, complete dissolution of specimens occured after an exposure time of one hour at 700°C.

From these data, it can be assumed that :

- presence of sigma phase does not result in high intergranular corrosion rate,
- Laves phase formation is responsible for the poor intergranular corrosion resistance.

It appears to be a common characteristic of stabilized superferritic stainless steels to suffer severe intergranular attack in such conditions of heat treatment and environment (10).

WELDING

The TIG process with pure argon is recommended for autogenous welding of superferritic stainless steels. The autogenous weld zone of tubes has demonstrated a significant intergranular attack when exposure was carried out in the ferric sulfate test whereas the base metal has not (Fig.6). Otherwise, the weld zone exposed to the copper sulfate test did not suffer intergranular corrosion.

Intermetallic phases were not found in the weld zone during TEM investigations (Fig.7). Only titanium carbonitrides type MX (M=Ti,Mo,Cr ; X=C,N) with small contents of molybdenum (Mo/Ti : 0.01 - 0.26) and chromium (Cr/Ti $\leq$ 0.15) were detected along the grain boundaries. It can be considered that intergranular attack of the weld zone in the ferric sulfate test is due to dissolution of stabilizing precipitates during welding, resulting in a finer and denser intergranular reprecipitation of titanium carbonitrides, according to a mechanism already suggested (11). It is well-known that titanium-stabilized stainless steels can suffer intergranular corrosion in hot oxidizing acid solutions where the selective dissolution of titanium carbide is taking place (12). Titanium carbide shows high current densities at noble potentials in boiling sulfuric acid solutions and would be expected to present poor corrosion resistance (fig.8).

PITTING CORROSION RESISTANCE

The Pitting Resistance Equivalent :

$$PRE = Cr\ (wt\ \%) + 3.3 \times Mo\ (wt\%)$$

is often used to illustrate the beneficial effect of chromium and molybdenum on the pitting corrosion resistance of ferritic stainless steels. All the materials tested have very high pitting potentials at ambient temperature in NaCl solutions. The use of a high temperature (70°C) and a concentrated (300g/l) NaCl solution was necessary to differentiate their pitting behaviour. The distribution range of pitting potential of each alloy was deduced from the simultaneous test of 12 specimens (13). The exposed surface area (about 1 cm^2) was mechanically polished down to 1 micron finish and aged 24-h in distilled water. The anodic polarization was performed with a scanning rate of 42 mV per hour from the rest potential. The pitting potential was defined as the potential at which the anodic current of a specimen reached 0.0001 A/cm^2. The test solution was deaerated by using nitrogen (1).

Our results conform with the previous statement concerning the beneficial influence of chromium on the pitting corrosion resistance of ferritic stainless steels (Fig. 9). The UNS S44635 Ti-stabilized ferritic alloy with only 25% chromium is less resistant to pitting than the UNS S44735 which has a higher chromium content and a similar molybdenum content. The UNS N08028 and N08904 austenitic stainless steels exhibit poor pitting resistance in this environment. The superaustenitic stainless steels containing a high molybdenum content and an addition of nitrogen have an excellent pitting behaviour like the best superferritic alloys.

The sigma phase precipitated at 900°C impairs the pitting resistance of the UNS S44735 alloy. The formation of Laves phase results in a slight decrease in the pitting potentials and appears less damageable than the sigma phase precipitation. The pitting resistance level of the autogenously welded zone of the UNS S44735 condenser tube is remarkably similar to that of the initial strip.

CREVICE CORROSION RESISTANCE

Crevice corrosion is generally considered the predominant risk limiting the use of stainless steels in seawater and other chloride -containing environments. In the only standardized crevice corrosion test (ASTM G48), the determination of the critical crevice corrosion temperature (CCCT) in 10% $FeCl_3,6H_2O$ is used to rank alloys according to their crevice corrosion resistance. The specimens are placed between two grooved PTFE cylindrical blocks and the entire assembly is held together with two rubber bands stretched at 90 degrees to each other. The decrease of pH (from about 1.6 to 1 by HCl addition) and the increase in exposure time (from 1 to 3 days) greatly increase the severity of the ferric chloride test. Test samples, about 5 x 5 cm, were ground to a 1000 grit finish.

From Figure 10, it is evident that the UNS S44635 ferritic stainless steel has suffered severe crevice corrosion in the ferric chloride test with the following conditions : pH1, 50°C, 3-day exposure. In the same environment, the UNS S44735 ferritic stainless steel appears very resistant. Crevice corrosion resistance of superferritic stainless steels increases with increasing chromium content at the same molybdenum level and with increasing molybdenum content at the same chromium level. A nickel content above 2% can lower the resistance to crevice corrosion when molybdenum is present in a large amount (1).

In these conditions, even the superaustenitic stainless steels containing nitrogen UNS S31254 and AL6X-N exhibit a significant crevice corrosion, their CCCT being under 50°C. Despite a slightly lower molybdenum content, the nitrogen-containing superaustenitic alloy UR SB8 behaves better, perhaps on account of its distinctly higher chromium content. Results of this ferric chloride test correspond with crevice corrosion tests in seawater (1).

DISCUSSION

Tube-to-tubesheet joints between a highly crevice corrosion resistant alloy, such as the UNS S44735 alloy (used for the tube), and another alloy susceptible to crevice corrosion in seawater, such as Type 316L stainless steel (used for the tubesheet), should be avoided (14). In crevices between dissimilar metals, the corrosion of the less resistant alloy results in a decrease in the pH of the solution and in the depassivation of the more alloyed stainless steel. Such an accelerated corrosion can affect a wide range of corrosion resistant alloys in contact with dissimilar materials. In cases where the use of a highly alloyed tubesheet material is not economically viable, cathodic protection is necessary in order to prevent the corrosion of a tubesheet alloy with a poor resistance (15,16).

CONCLUSION

The UNS S44735 Ti-stabilized alloy has superior performance in chloride environments in comparison with many other superferritic and highly alloyed austenitic stainless steels. A combination of 29% chromium and 4% molybdenum yields the optimum alloy composition balance for corrosion performance for use as tubing material in condensers and heat exchangers.

REFERENCES

1. J.C. Bavay, J.M. Boulet, J. Castel and P. Bourgain, NACE corrosion/87 Conference, Paper N°350 San-Francisco, March 9-13(1987)

2. R.F. Steigerwald, H.J. Dundas, J.D. Redmond and R.M. Davison, Stainless Steel '77, p.57, London, September 26-27 (1977)

3. H. Abo, T. Nakazawa, S. Takemura, M. Onoyama, H. Ogawa and H. Okada, Stainless Steel '77 p.35

4. M.A. Streicher, Stainless Steel' 77, p.1

5. R.N. Wright, Toughness of Ferritic Stainless Steels, ASTM STP 706, p.2 (1980), R.A. Lula Editor

6. E.L. Brown, M.E. Burnett, P.T. Purtscher and G. Krauss, Metallurgical Transactions, Volume 14A, P.791 (1983)

7. H. Kiesheyer and H. Brandis, Z. Metallk., 67, p. 258 (1976)

8. K. Mashimo, K. Umeda, M. Haga, T. Sato, A. Kurimoto and T. Sekiguchi, Transactions ISIJ, Vol. 25, N°11, B.311 (1985)

9. H. Brandis, H. Kiesheyer and G. Lennartz, Arch. Eisenhüttenwes., 46, N°12, p. 799 (1975)

10. G. Rondelli, D. Sinigaglia, B. Vicentini and G. Taccani, Proceedings of 9th Intermetallic Congress on Metallic Corrosion, p.57, Toronto, June 3-7 (1984)

11. G.B. Hunter and T.W. Eagar, Metallurgical Transactions, Vol.11A, p.213 (1980)

12. A.J. Sedriks, Stainless Steel'84, p.125

13. B. Baroux, B. Sala, T. Jossic and J. Pinard, Materiaux et Techniques, p. 211 (1985)

14. J.R. Kearns, M.J. Johnson and J.F. Grubb, NACE Corrosion/86 conference, Paper N°228, Houston, March 17-21 (1986)

15. J.F. Grubb and J.R. Maurer, NACE Corrosion/84, Paper N°28, New-Orleans, April 2-6 (1984)

16. R.M. Krishnamurthy and W.H. Hartt, NACE Corrosion/87, Paper n°175, San-Francisco, March 9-13 (1987)

Table 1 : Test alloys and chemical compositions (wt - %)

UNS number or name	MICROSTRUCTURE	C	Si	S	P	Mn	Cr	Ni	Mo	Ti	Cu	Co	N
S 44735	ferritic	0.013	0.37	0.001	0.020	0.28	28.60	0.43	3.76	0.50	0.08	0.02	0.025
S 44660	ferritic	0.021	0.42	0.001	0.017	0.34	27.15	2.04	3.51	0.51	0.11	0.02	0.020
S 44635	ferritic	0.014	0.31	0.007	0.015	0.32	24.92	4.17	3.81	0.50	0.08	0.04	0.018
S 44800	ferritic	0.0039	0.10	0.007	0.004	0.028	28.88	2.31	3.93	<0.01	0.03	0.02	0.015
S 31254	austenitic	0.016	0.48	<0.001	0.018	0.43	19.80	18.19	6.03	<0.01	0.82	0.45	0.210
AL6X-N	austenitic	0.023	0.46	<0.001	0.023	1.58	20.30	24.80	6.14	<0.01	0.19	0.04	0.190
UR SB8	austenitic	0.016	0.26	0.001	0.016	1.01	24.69	25.66	4.80	0.02	1.65	0.04	0.215
N 08904	austenitic	0.010	0.13	0.001	0.028	1.73	19.90	25.71	4.41	<0.01	1.62	0.13	0.052
N 08028	austenitic	0.017	0.39	<0,001	0,013	1,85	27.32	31.49	3.47	<0.01	1.11	0.06	0.062
S 32550	duplex	0.025	0.56	0.001	0.023	1.29	25.84	5.89	3.09	<0.01	1.30	0.14	0.163

Table 2 : Calculated lattice parameters of intermetallic phases formed after heating UNS S44735 specimen 1 hour at 900°C

Lattice Parameter	Sigma (tetragonal)	Chi (Cubic)
a (nm)	0.88	0.89
c (nm)	0.462	-

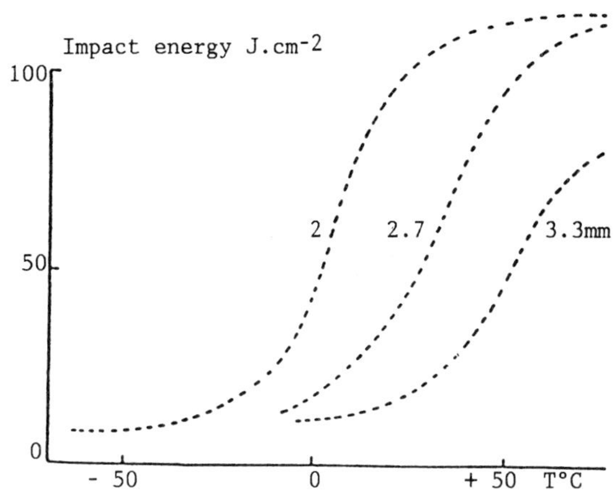

1 - Effect of sample thickness on the impact energy as a function of test temperature (°C) for charpy V- notch UNS S44735 specimens

Table 3 : Composition of precipitates formed after heating UNS S44735 specimen 1 hour at 900°C

Sigma phase

ELEMENTS	Fe	Cr	Mo	Ti	Si
wt-%	59.6	32.6	6.4	0.9	0.5

(on thin foil)

Chi phase

ELEMENTS	Fe	Cr	Mo	Ti	Si
wt-%	59.8	25.2	11.8	2.7	0.5

(on thin foil)

Ti-Fe phosphide

ELEMENTS	Ti	Fe	Cr	P
wt-%	34	36	4	26

(from particles extracted on replica)

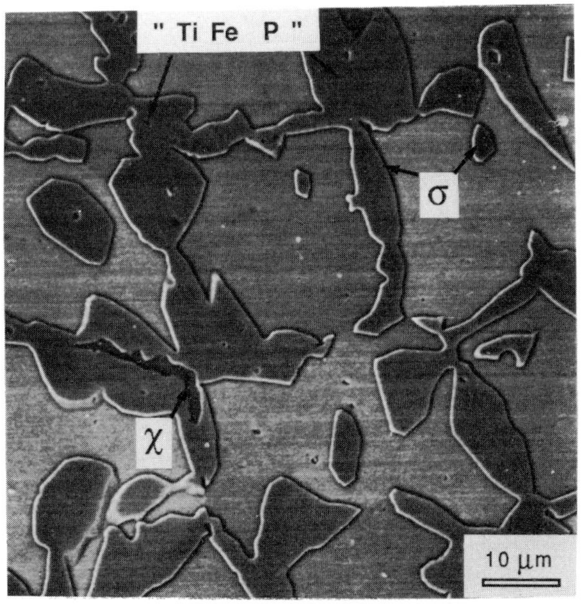

2 - Scanning electron micrograph of UNS S44735 alloy heated 1 hour at 900°C, water-quenched

Table 4 : Composition of Laves phase formed after heating UNS S44735 specimen 1 hour at 700°C

ELEMENTS	Fe	Mo	Cr	Ti	Si
wt-%	45	31	11	10	3

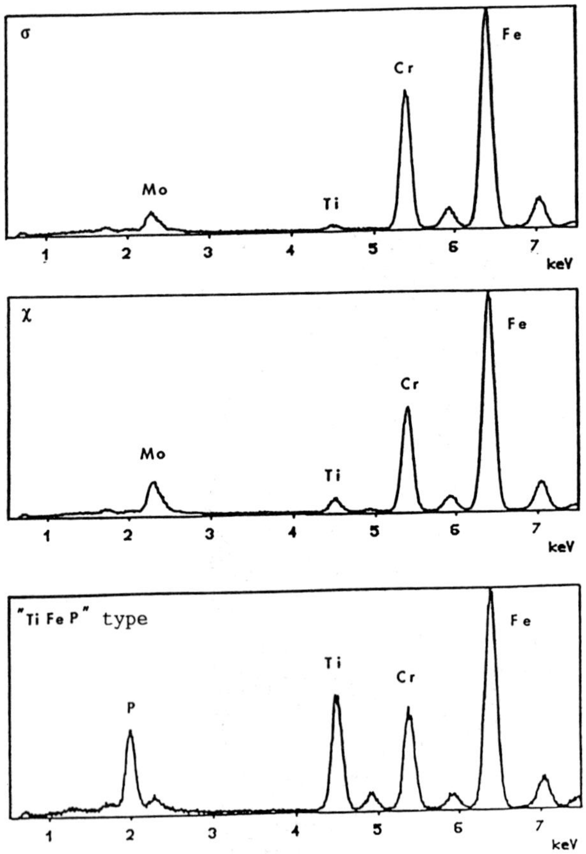

3 - X-microanalysis spectra obtained on thin foil prepared from UNS S44735 specimen heated 1 hour at 900°C, water-quenched

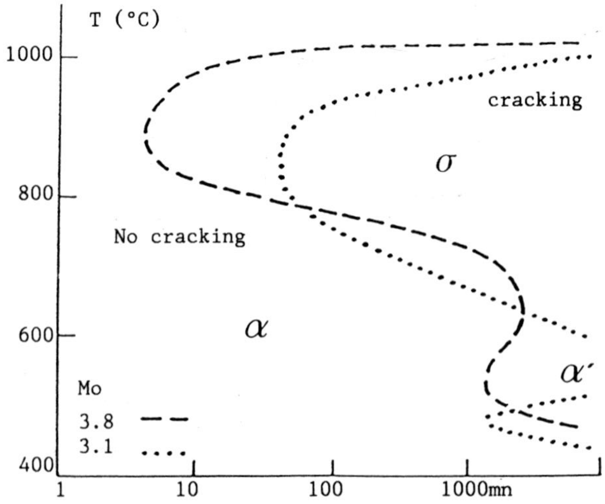

4 - Influence of molybdenum content on the temperature time-precipitation curve determined from room-temperature results of 180-deg bending test carried out after isothermal exposures of 0.8 mm thick UNS S44735 alloy

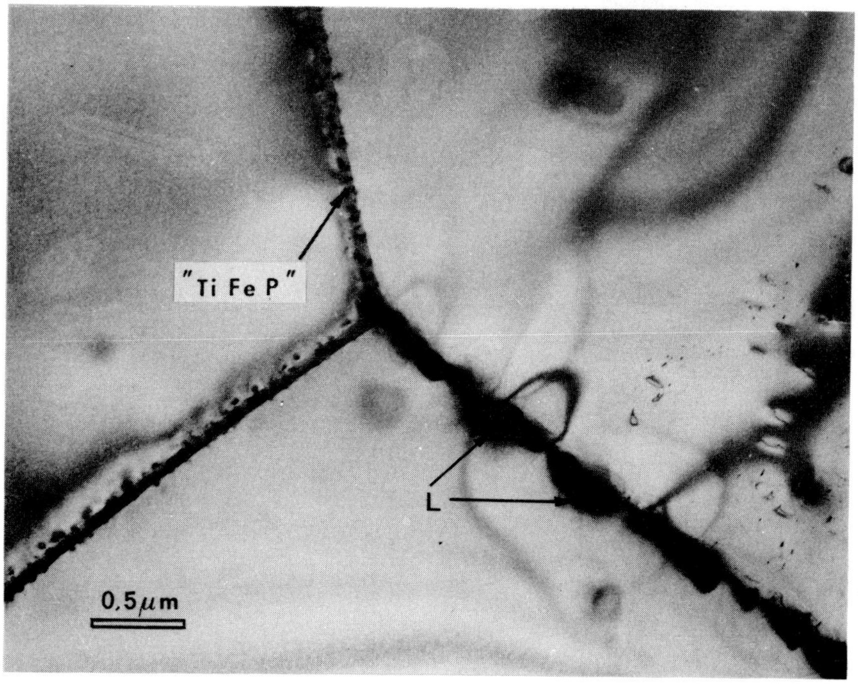

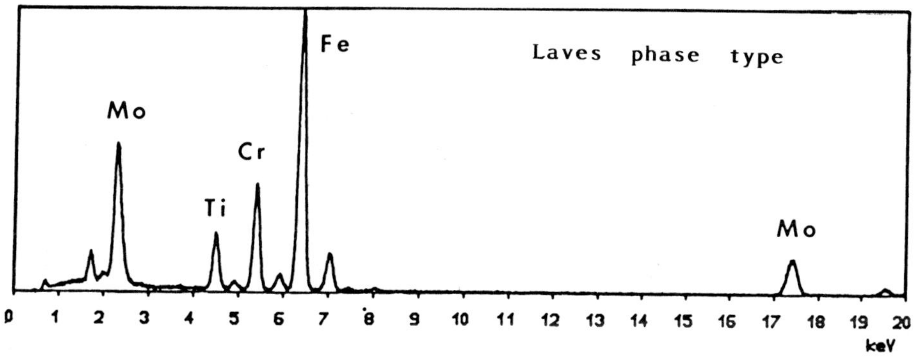

5 - Transmission electron micrograph and X-microanalysis spectra of thin foil prepared from UNS S44735 specimen heated one hour at 700°C, water-quenched

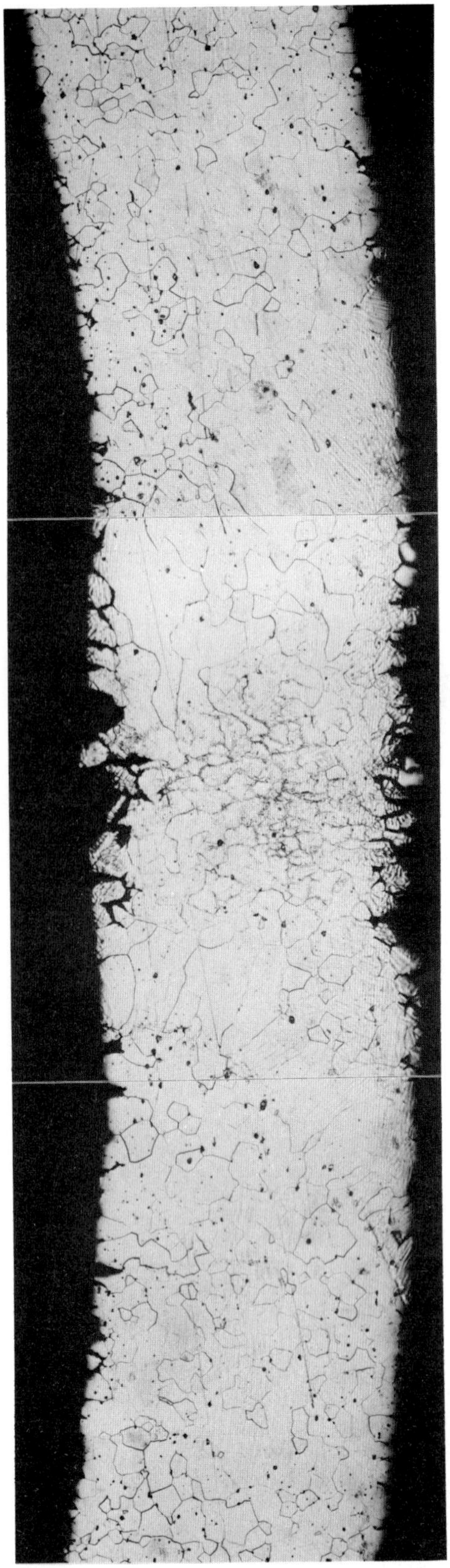

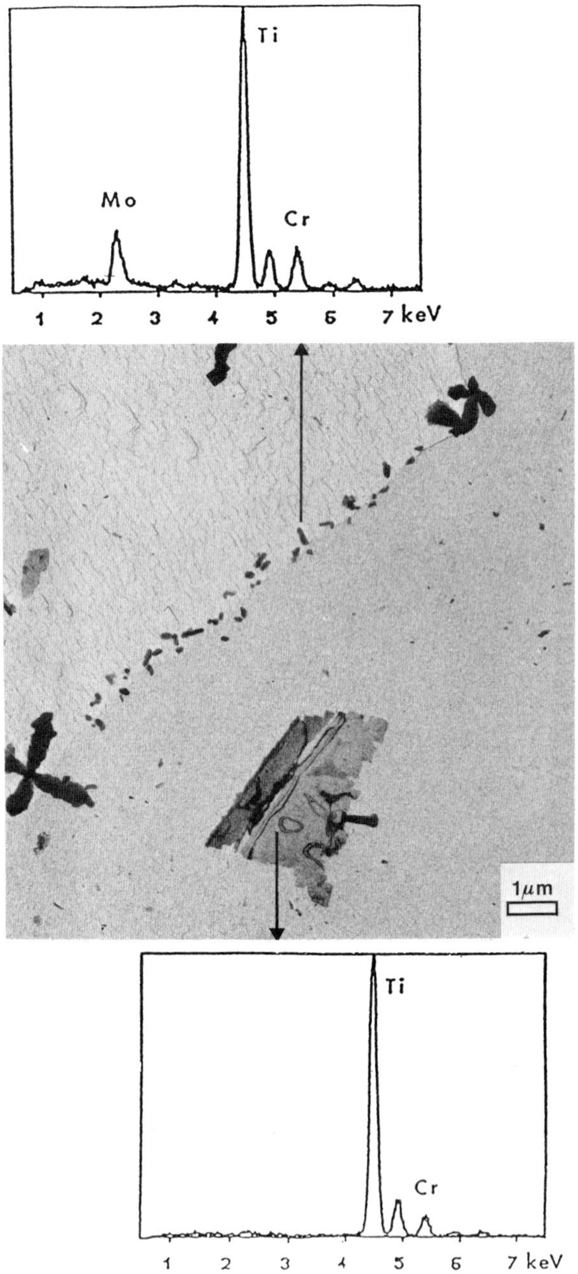

7 - Transmission electron micrograph from replica showing the morphology of precipitates in the weld zone and X-microanalysis spectra

6 - Weld cross-section of as-received UNS S44735 tube (wall thickness : 0.5 mm) after exposure in the ferric sulfate environment ASTM A 763 (Practice Y) and electrolyte etching in oxalic acid

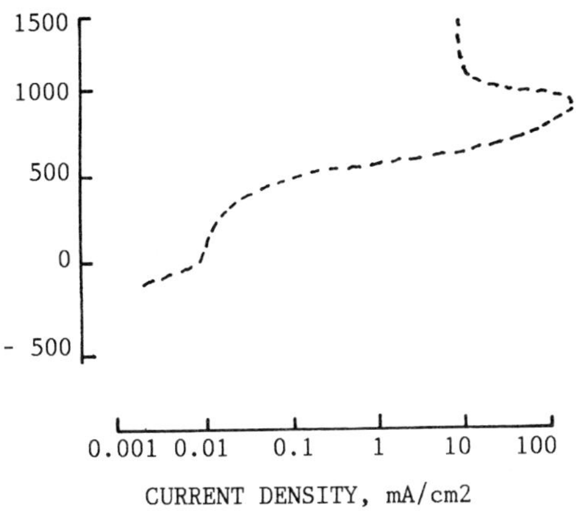

8 - Anodic polarization curve for TiC in boiling 3.4 N H_2SO_4 (12)

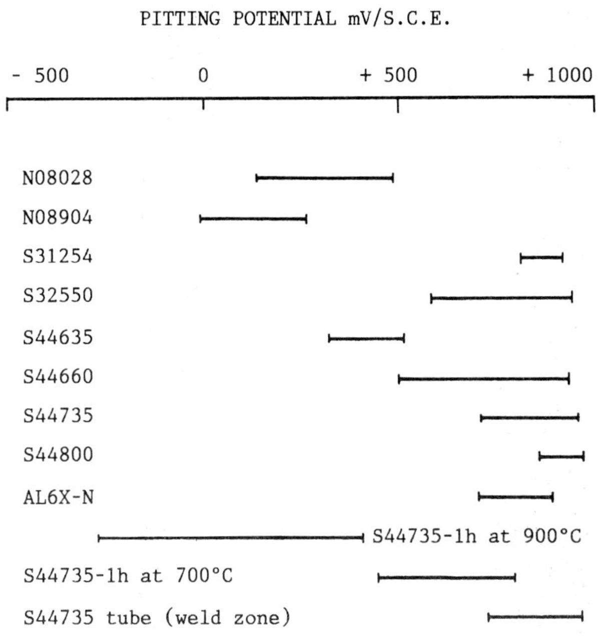

9 - Pitting potential ranges in a deaerated 300g/l NaCl solution at 70°C

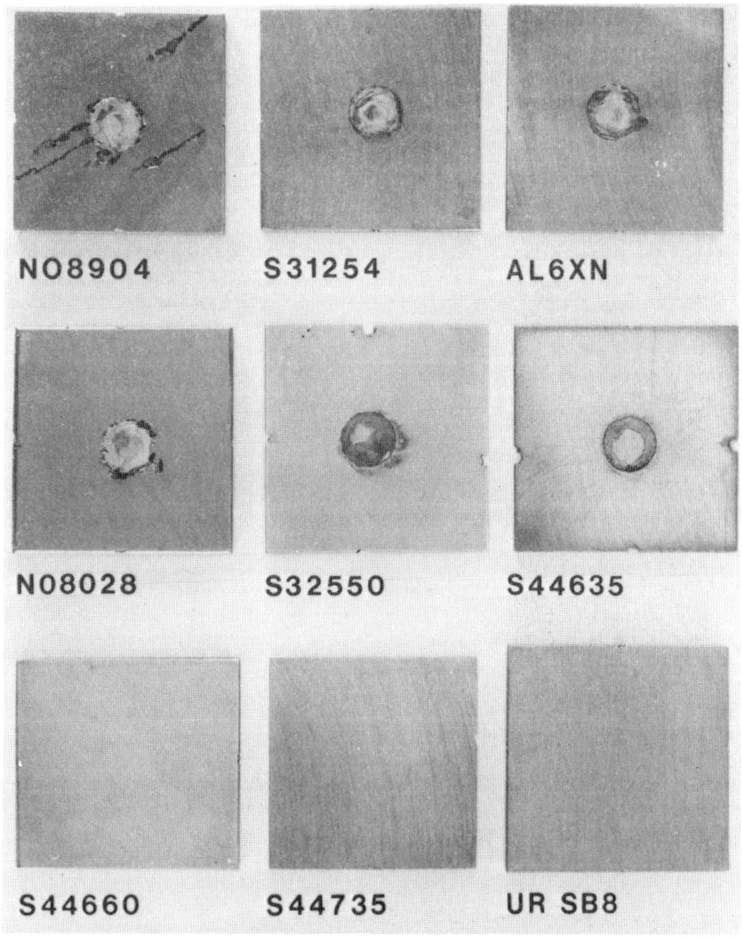

10 - Aspect of ferric chloride test specimens (10%FeCl3,6H20-pH1-50°C-72h)

Processing and localized corrosion properties of 25—29Cr superferritic stainless steels

F Mancia, M Barteri, L Sassetti, A Tamba and A Lannaioli

The work was carried out at the Centro Sviluppo Materiali Spa, Rome, Italy; AL is now with Terni Acciai Speciali Spa, Terni, Italy.

SYNOPSIS

Though Cu, Cu-Ni and Ti alloys have so far been preferred for brackish and sea-water cooled heat exchangers and steam condensers, the new superferritic (Cr25-29, Ni 0-4, Mo 3-4, Ti) and high-alloy austenitic (Cr20, Ni18-25, Mo6) steels have been increasingly used with success since the end of the seventies for both technical and economic reasons.

The ferritic grades can be considered potentially as the best choice on a cost-performance basis, provided certain processing conditions could be applied during their production. The aim of the work reported here has been to single out the metallurgical problems involved in processing 26Cr superferritic grades, as well as to determine their localized corrosion performance by electrochemical tests.

INTRODUCTION

A serious difficulty in the production and transformation of high Cr (⩾ 25%) and Mo ELI steels is their tendency to favour the precipitation of intermetallic phases (mainly sigma and chi phases), the presence of which causes embrittlement, with toughness reduced to practically zero at room temperature/1,2/. It thus ensues that the instability of the high-alloy ferritic structure may create serious problems during transformation from ingot to finished product.

The critical phases of the process are those between heating the ingots in the pit furnace and uncoiling the strip before continuous annealing of the hot-rolled steel.

The aims of the metallurgical study, designed to optimize a works process line are:

a) to determine the temperature and formation times of embrittling phases in the high-temperature and low-temperature ranges (600 to 950°C and 300 to 550°C, respectively) through the measurement of toughness and intergranular corrosion in boiling HNO_3 (Huey Test, ASTM A 262-C);

b) to ascertain the morphology and approximate composition of the phases which affect the possibility of successfully performing the process;

c) to choose - on the basis of a and b - a feasible works line which offers a very good likelihood of success.

Charpy V and Huey tests are used simultaneously because in ferritic stainless steels the toughness and intergranular corrosion (IC) trends are generally markedly parallel. The low solubility of C and N and also of P and S in the high-alloy ferritic matrix at intermediate temperature, and the high rates of diffusion of substitutional atoms in compositions particularly rich in Cr, Mo, Ni and Ti, and with very low C and N activity, result in intergranular segregative enrichments and/or nucleations and coalescence of second phases which occur with extreme rapidity/3,4,5/. These second phases are usually attacked during the Huey test and lead to the onset of brittle fracture in the Charpy-V test.

Resistance to forms of localized corrosion such as pitting and, especially, crevice corrosion, is an essential feature of superferritic stainless steels for seawater-cooled condensers and heat-exchange tubes/5,6,7/.

The aim of predicting localized corrosion risks of ELI steels in industrial and marine cooling waters has been attained by plotting experimental Potential vs NaCl Concentration (E vs NaCl diagrams/8,9) at various temperatures, obtained via potentiodynamic hysteresis electrochemical tests carried out in solutions containing different NaCl concentrations.

EXPERIMENTAL

The metallurgical study was performed on hot-rolled strips of nominal composition 26Cr4Mo3Ni from an experimental heat whose analysis is reported in Table I.

Five-mm thick Charpy-V test-pieces were used to determine Temperature-Time-Precipitation (TTP) curves. The specimens were subjected to:

1) Heat treatments, consisting of solubilization at 1000°C with 30-s hold, followed by rapid cooling in flowing helium gas to treatment temperature, followed by a hold ranging from 1.5 minutes to 20 hours and water quenching (WQ).

2) Determination of toughness at room temperature, utilizing a 15 Kgm pendulum force.

3) Determination of intergranular corrosion rate by the Huey test.

4) Phase extraction by chemical means and X-Ray recognition.

5) SEM and STEM extraction replicas with EDXA microanalysis.

METALLURGICAL STUDY

The maximum toughness (E = 115 J) and minimum IC rate (V_c = 0.009 mm/month) for the steel concerned are attained on test-pieces water-quenched from 1000°C.

The results of the toughness tests are reported in Fig. 1 in terms of isotoughness curves equal to 40% of the maximum energy E = 115 J obtained with the reference test-pieces treated at 1000°C for thirty seconds and water quenched ($E_{40\%}$ = 0.4 x 115 J = 46 J).

Fig. 2 illustrates the isocorrosion curves obtained with the Huey test, for rates V_c of 0.05 mm/month and 0.01 mm/month.

The following points emerge from examination of Figs 1 and 2. The most immediate feature is the nose-like shape of the curves. Reports by other workers/10,11/ and the CSM's own experience indicate that the high-temperature nose (600-950°C) is related to the formation of intergranular intermetallic phases (chi, sigma and Laves) which, of course, have an adverse effect on steel toughness.

Interesting information has been derived from examination of the low-temperature nose (350-500°C) in the Cr-rich (~ 80%) α' formation range which has quite a marked tendency to cause steel embrittlement. The following comments may be made on Fig. 1:

i) At 475°C a holding time of only 1.5 minutes suffices to bring the steel energy to decidedly less than 46 J (40% of the reference value). This result is fairly surprising, because although it has been reported that the maximum rate of formation of the α' phase occurs at 475°C, the times indicated as being necessary for its formation in composition close to that of the steel studied are in the order of an hour/12,13/.

ii) In the 250-300°C range it is possible to observe another secondary nose, different from that attributed to the α' phase.

It is evident from Fig. 2 that the following zones can be distinguished:

a) By adopting a threshold value of Vc = 0.05 mm/month, a single nose is obtained in the 900-475°C temperature range for holding times between 30 minutes and 20 hours.

b) The low Vc values obtained for the specimens treated in the low-temperature range (< 500°C) could be proof that the α' phase is not concentrated at the grain boundaries within the span of holding times investigated (< 20 h).

c) For given holding times, the highest Vc values are in the 600-750°C range.

The nature of the phases was ascertained by X-ray diffraction and inspection under the electron microscope (STEM+EDXA). The phases have been identified which occur in the first embrittlement zone (T = 900-600°C) and in the second one (T = 500-375°C). However, it has not been possible to correlate the adverse effect on toughness in the 375-500°C temperature range with phases that can be detected under the electron microscope or by X-ray examination of extracts.

First embrittlement zone

X-ray diffraction examination of specimens treated both at 900 and 750°C has revealed the presence of the tetragonal sigma phase with the following parameters a = 0.8811 nm, C = 0.4589 nm. These values indicate that the phase examined contains a considerable quantity of Mo. This is confirmed also by the STEM+EDXA microanalysis performed on the 800°C-2h specimen (Fig. 3) which reveals a sigma phase with a Mo content of 6 to 7% by weight. On the 800°C-0.5h specimen the chi phase is by far the most common, generally being found on the grain boundary, while the sigma phase is not very frequent and is small in size (Fig. 3). It must thus be concluded that the sigma phase develops from the chi phase which appears as the precursor.

Specimens treated at 650°C, however, never have any chi or sigma phase. At holding times of 20 h, instead, there is abundant precipitation of the R hexagonal phase (20 Cr, 29 Mo) not just within the ferritic grains but especially on the boundaries, which are completely decorated (Fig. 4).

Hence the first embrittlement zone must be divided into a field of chi phase followed by sigma phase formation (950-750°C) and a field of R phase formation (750-600).

Second embrittlement zone

Specimens treated at 550°C exhibit only the presence of titanium nitride TiN. At 500°C, in addition to TiN, there is abundant orthorhombic M_3C_2 whose parameters are a = 1.154 nm; b = 0.550 nm; c = 0.282 nm. EDXA reveals the presence of Mo (> 8%) and C (> 30% by weight).

A similar situation can be encountered in the specimen treated at 475°C, where the precipitation is not so abundant, however. Carbides of the $M_{23}C_6$ or Cr_3C_2 type are observed. Grain boundaries are generally precipitation-free.

The extremely short times (1-10 min) required for embrittling the steel in the low temperature range (250-500°C) suggest mechanisms involving the high diffusion rate of interstitial atoms C and/or N and their attitude to block dislocations.

The responsibility of precursors of Mo rich carbides (M_3C_2 or $M_{23}C_6$) could be envisaged. Carbide precipitation in fully stabilized ELI steels was previously evident. In particular $M_{23}C_6$ carbides were observed in ELI Ti-stabilized 18Cr2Mo steel after thermal treatments at intermediate temperature/14/.

Ingot/hot-rolled-strip transformation

The industrial process for the production of Ti stabilized 26Cr3Ni4Mo superferritic steel involves Ar refining in the AOD converter.

Transformation from ingot to hot-rolled strip is the most critical part of the process; the formation of embrittling phases in the temperature ranges indicated in Fig. 1 must be avoided.

In particular, the operation of cooling hot-rolled strip to room temperature can result in intense precipitation of embrittling phases. Indeed, the intensity can be so great as to prevent subsequent uncoiling of the strip prior to annealing.

The cycle of operations developed by Terni Acciai Speciali SpA has been worked out on the basis of the laboratory results. The usual process is adopted to transform the ingots into slabs. Care is taken, however, to avoid cooling the slabs to room temperature, because of the intrinsic brittleness of the steel in the presence of precipitates which are inevitable at this stage of the process.

The slabs are subsequently resolubilized and hot rolling starts at around 1250°C. The heat cycle during hot rolling is as indicated in Fig. 5. It will be observed that the finishing-rolling temperature is 940°C to avoid the first zone of embrittlement (Fig. 1).

The hot-rolled strip is coiled at 650°C, at which temperature both the first and second precipitation noses are avoided (Fig. 1).

The last phase of the process consists of rapid cooling of the coil to room temperature to limit embrittlement at 475°C (2nd nose of Fig. 1).

Fig. 6 indicates the temporal variation of temperature on the surface and in the centre of the coil, in the case of air cooling and water cooling. It is evident that only water cooling ensures that in the centre of the coil the holding times in the precipitation range are not excessive (~ 1-h hold). However, even in this case the precipitation of embrittling phases is inevitable. Subsequent uncoiling cannot be easily performed at room temperature, so provision is made for preheating, at least when hot rolled strips of high thickness (> 5-6 mm) are to be treated. Then once the uncoiling has been accomplished the operations of annealing and cold rolling can be performed in the normal manner.

Electrochemical evaluation of corrosion behaviour and plotting of experimental E vs NaCl diagrams

Resistance to localized forms of corrosion such as pitting and especially crevice corrosion is essential for the use of superferritic stainless steels in seawater-cooled condensers.

The cold-rolled and annealed superferritic steel of Table I was used to plot the practical E vs NaCl (Potential vs Chlorides Concentration) diagram for predicting localized corrosion risks at a temperature of T = 64°C.

The practical significance of the E vs NaCl diagrams is illustrated schematically in Fig. 7 which is relevant to AISI 304 stainless steel.

Construction of these diagrams calls for the plotting of anodic hysteresis potentiodynamic curves at neutral pH and at pH equal to or less than the depassivation pH (dpH). Fig. 8 illustrates the E vs NaCl diagram obtained for the steel concerned. From examination of the diagram it can be ascertained that the steel is strongly resistant to the onset of localized corrosion in seawater. Indeed it is seen that the values of the pitting onset potential at 35.000 ppm NaCl and 64°C, both in neutral and acidified conditions, are more noble than the thermodynamic potential of oxygen evolution.

Experimental verification of E vs NaCl diagrams

Numerous exposure tests have been made lasting at least 500 hours in an instrumented circuit using an aerated flowing solution (0.5 m/s) of NaCl at a concentration typical of that of seawater (NaCl = 35 g/l and T = 64°C).

The specimens were polarized at various potentials chosen on the basis of the relevant E vs NaCl diagrams obtained at 64°C or left in free corrosion, following the evolution of the potential.

The campaigns run on non-precorroded specimens show that, in practice, there is no onset of localized corrosion in seawater up to potentials close to those for the evolution of oxygen.

The data obtained on artificially precorroded specimens indicate a protection potential of around -100 mV vs SCE.

Exposure in natural seawater

Various specimens measuring 100 x 100 mm with multicrevice geometry were exposed to natural seawater. The first samples were taken after about six months. They were found to be in perfect condition, thus confirming correctness of the manufacturing process for transforming the superferritic steels in question from ingot to annealed cold rolled sheet.

CONCLUSIONS

1) By means of chemical extraction techniques and electron microscope and X-ray inspection of the residues and thin sections it has been possible to identify the main secondary phases which form in superferritic steels causing embrittlement and/or susceptibility to intergranular corrosion in the Huey test.
At the higher temperatures (875-900°C) the fall off in toughness and the occurrence of intergranular corrosion are due to formation of sigma phase which at a temperature of 750-825°C is preceded by the precipitation of body-centred cubic chi phases. At 625°C, instead, for treatment times of two hours - which have little influence on toughness but which are sufficient to cause heavy corrosion in the Huey test - Mo-rich carbides of the $M_{23}C_6$ type were found to be present. For longer times, when toughness, too, is jeopardized, the dominant phase is an R hexagonal whose composition is 21Cr29Mo3.5TiFe, similar to the chi observed at higher temperatures. For long treatment times at 475-500°C the presence is noted of rare carbides of the M_3C_2 type, probably containing Mo, in a matrix of clean steel. The deleterious effect of toughness in the 400-500°C temperature range cannot be attributed to phases evidenced by inspection under the electron microscope or by X-ray examination of extracts.
Nevertheless, the short embrittling times (1-10 min) in the low temperature range claim for mechanisms involving the blocking of dislocations by the highly diffusible interstitial C and/or N.

2) Determination of the precipitation kinetics of the embrittlement phases has permitted definition of a transformation cycle. The most critical stage is the precipitation of these phases when the hot-rolled strip is cooled to room temperature.
The answer to these problems has been to adopt a procedure involving coiling the hot-rolled strip at 650°C followed by accelerated cooling, then appropriate preheating before uncoiling. The latter is to be considered when the hot rolled strip thickness is rather elevated (> 5-6 mm).

3) Taken as a whole, the localized corrosion (pitting and crevice corrosion) resistance of the ELI superferritic steels in hot (64°C) chloride solutions (NaCl from 20 to 290 g/l) guarantees their reliability for use in seawater-cooled heat exchange equipment.

REFERENCES

/1/ H.E. DEVERELL - ASTM STP 706, R.A. Lula, Ed., (1980), pp 184-201

/2/ T.J. NICHOL, A. DATTA and G. AGGEN, Metallurgical Transactions, 1980, Vol. 11A, no. 4, pp 573-585.

/3/ G. AGGEN, H.E. DEVERELL and T.J. NICHOL - ASTM STP 672, 1979, pp 334-366

/4/ J.R. MAURER and J.R. KEARNS - Proc. Int. Corr. Forum CORROSION/85, Boston, paper no. 172 (1985).

/5/ J.R. MAURER - Proc. Symp. on Advanced Stainless Steels for Seawater Applictions, Piacenza, Italy, Feb. 28, 1980.

/6/ M.A. STREICHER - Stainless Steel 77, Ed. by R.Q. Barr, pp 1-34 (1978).

/7/ M.A. STREICHER - Corrosion, 1974, vol. 30, no. 3, pp 77-91.

/8/ N. AZZERRI, F. MANCIA, A. TAMBA - Corros. Sci. Vol. 22, p 675, 1982.

/9/ F. MANCIA and A. TAMBA, Corrosion, vol. 42, no. 6, p 362, 1986

/10/ G. RONDELLI, B. VICENTINI, M. MALDINI - Metall. It., 1985, vol. 77, p 857-862.

/11/ E.L. BROWN et al. - Metall. Trans. A, 1983, vol. 14A, 791-800.

/12/ R. LAGNEBORG - Trans. ASM, 1967, vol. 60, 67-78.

/13/ P.J. GROBNER - Metall. Trans. A, 1973, vol. 4, 251-260.

/14/ M. BARTERI et al - "The role of Secondary Phase Precipitation on the Intergranular Corrosion of Differently Stabilized ELI Ferritic Stainless Steels" - Stainless Steels 87, Conf., York (1987).

TABLE I Chemical analysis of the 26Cr3Ni4Mo Ti-stabilized superferritic stainless steel

Cr	Mo	Ni	Mn	Si	S	P	C	N_2	Ti	$C+N_2$	$Ti/C+N_2$
25.6	3.80	3.90	.32	.27	.006	.011	.024	.024	.17	.048	3.5

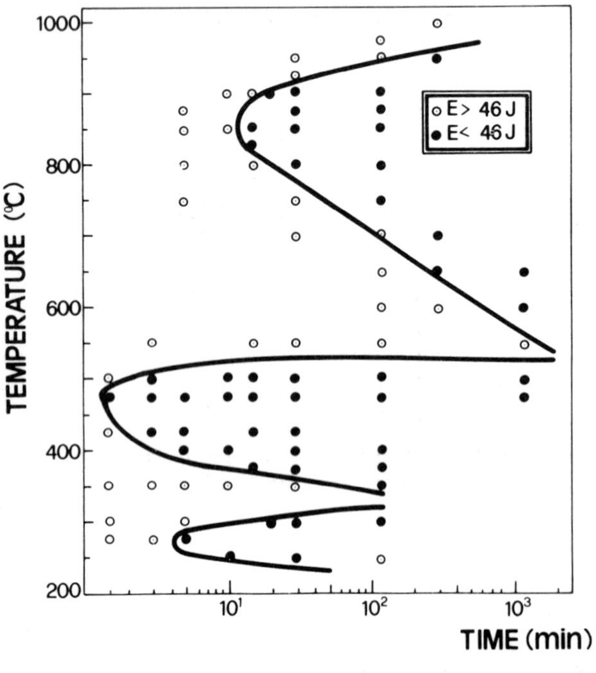

1) Isotoughness curves (E = 46 J) as a function of treatment temperature and time.

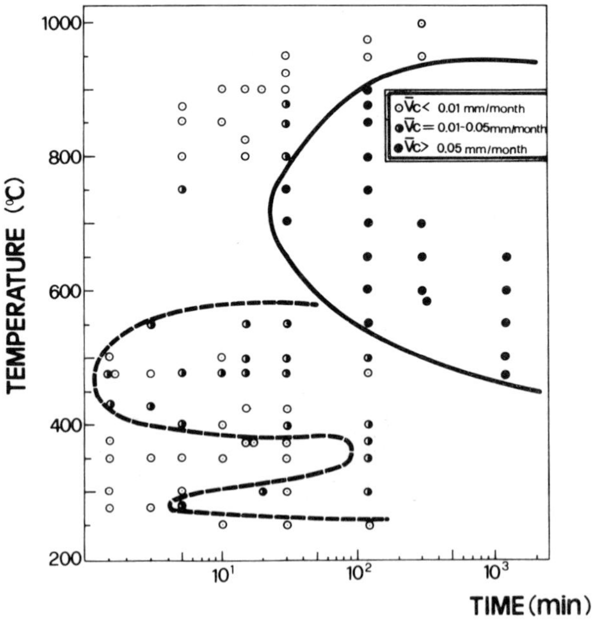

2) Isocorrosion curves (V_C = 0.05 mm/month and V_C = 0.01 mm/month) obtained in the Huey test (ASTM A 262 C) as a function of treatment temperature and time.

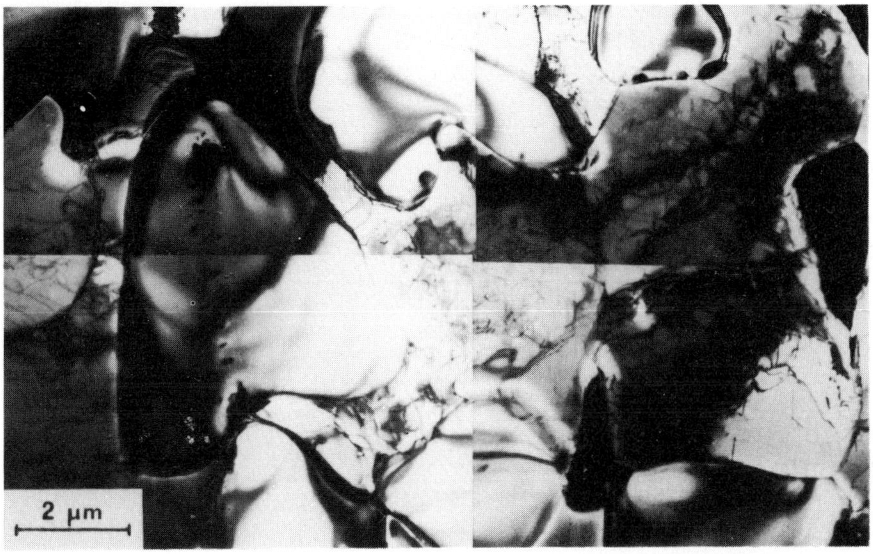

a

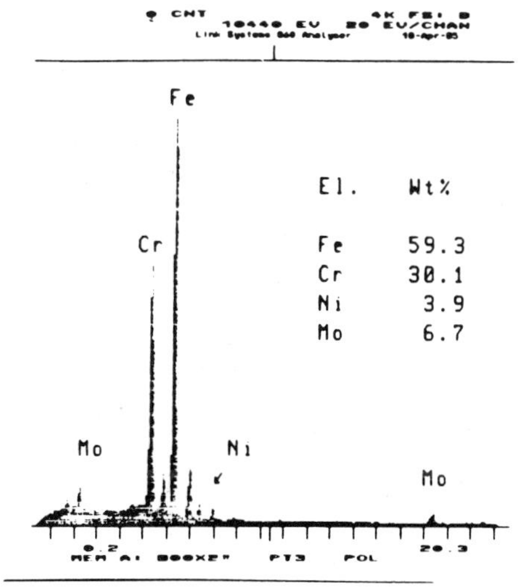

b

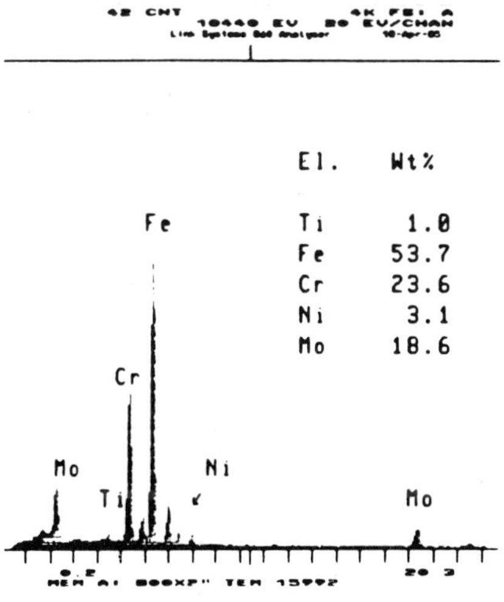

c

3) STEM inspection of specimens treated at 800°C for two hours followed by water quenching.
a) General appearance: large islands of chi phase sometimes with small sigma phase particles around them
b) Sigma phase: EDXA microanalysis
c) Chi phase: EDXA microanalysis

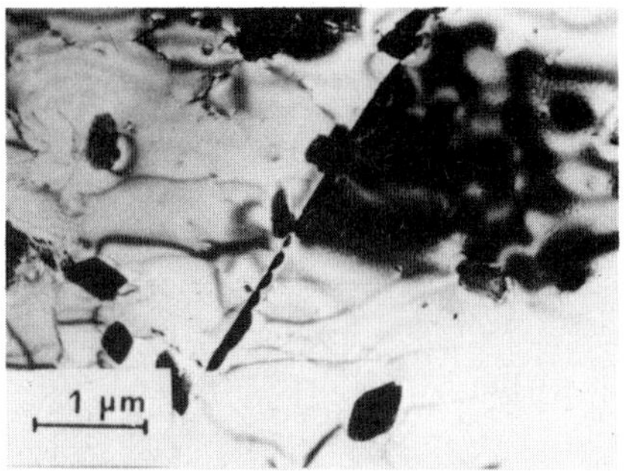

a

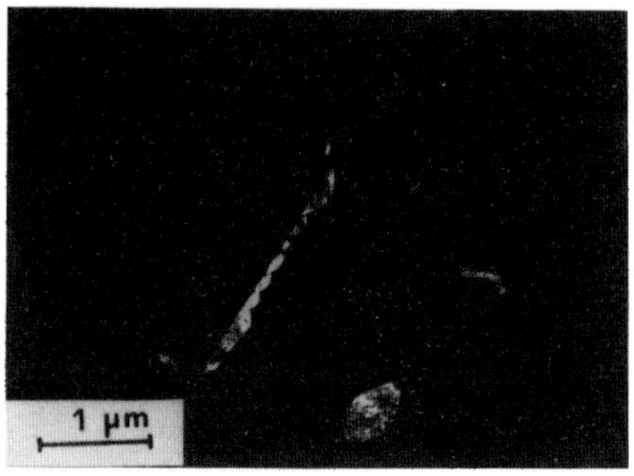

b

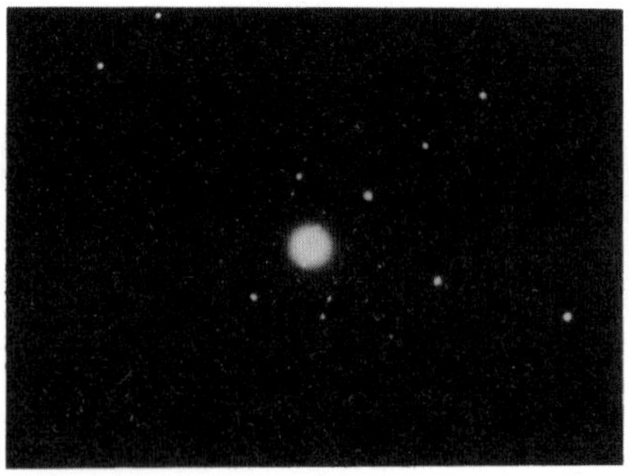

c

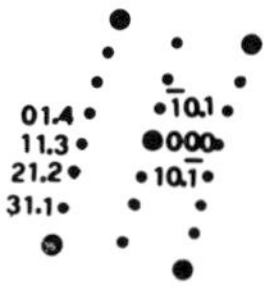

fase R

d

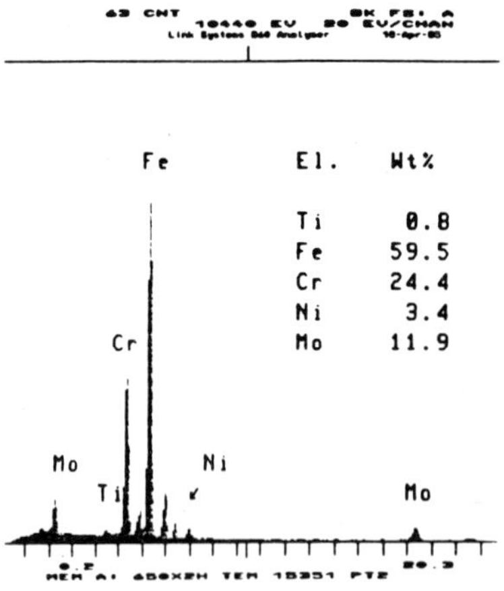

e

4) Treatment: 650°C for 20 hours followed by water quenching
a)Light field
b)Diffraction diagram
d)Interpretation scheme
e)EDXA microanalysis

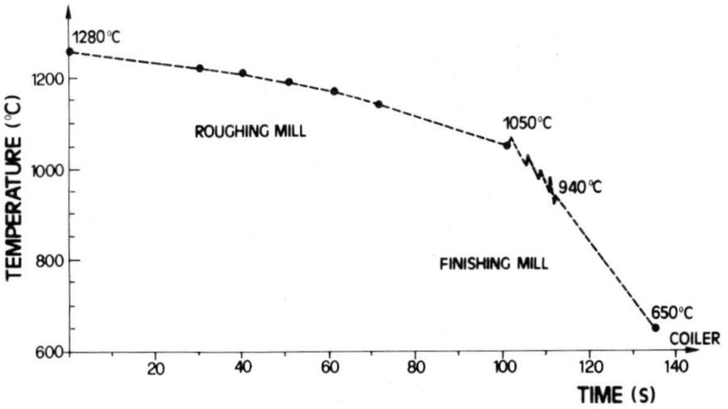

5) Trend of average temperature during rolling of ELI 26Cr3Ni4Mo steel with:
- H slab = 150 mm
- H bar = 22 mm
- H strip = 4 mm

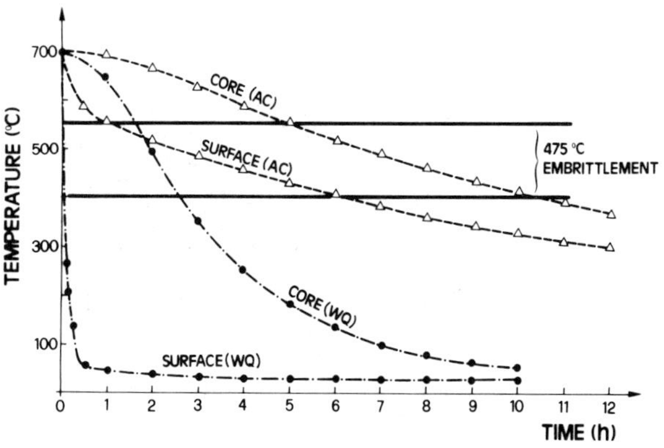

6) Coil cooling in water and still air
- Water: = 800 Kcal/m^{2}h°C
- Air: radiation

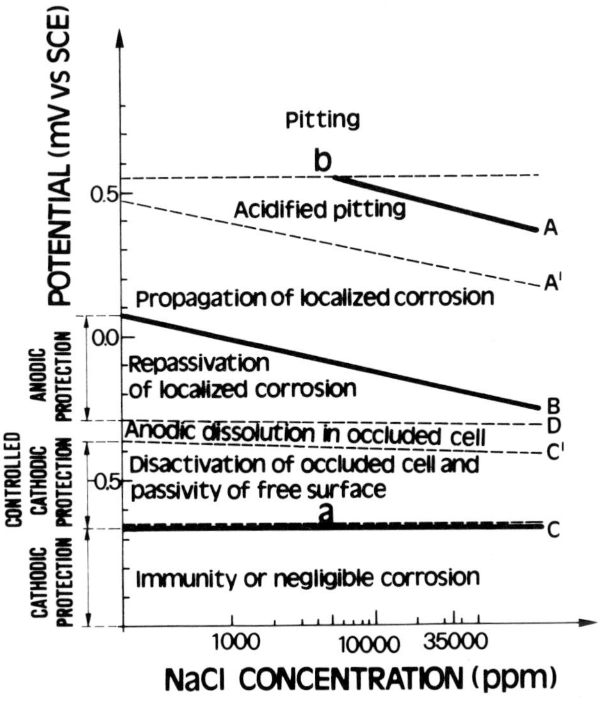

7) E vs NaCl diagram for prediction and control of corrosion risk and definition of corrosion mechanisms (AISI 304, T = 22°C)

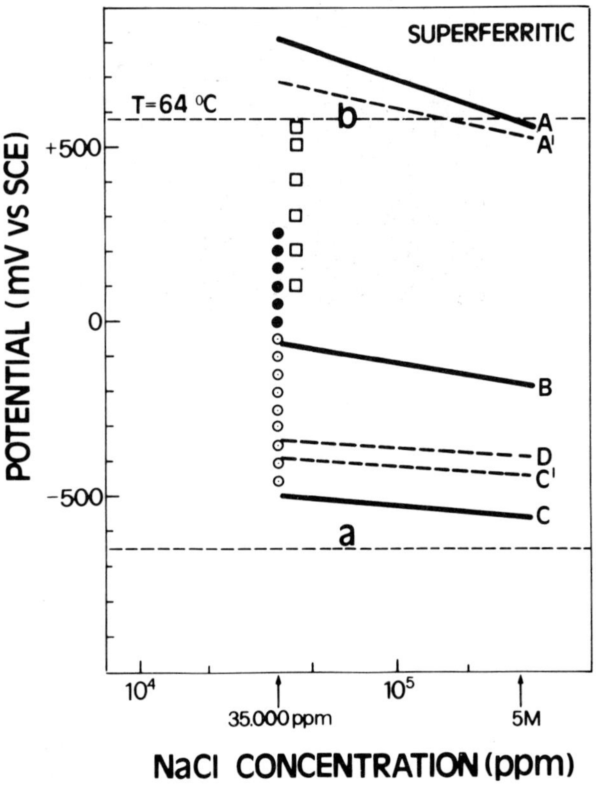

8) E vs NaCl diagram for prediction and control of corrosion risk (Superferritic Stainless Steel, T = 64°C).

□ Specimen exposed in natural sea water non-precorroded, with artificial crevice geometry. No localized corrosion observed.

⊙ ● Specimen exposed in natural sea water precorroded (by an anodic current of 4 mA x 24 hours), with artificial crevice geometry. Open symbols: no localized corrosion advanced. Closed symbols: localized corrosion advanced.

Stress corrosion cracking and hydrogen embrittlement behavior of 25%Cr duplex stainless steel in H_2S-bearing environment

T Kudo, H Tsuge and A Seki

The authors are in the Technical Research Laboratories, Sumitomo Metal Industries Ltd., Amagasakim, Japan.

SYNOPSIS

The resistance to stress corrosion cracking (SCC) and hydrogen embrittlement (HE) of 25% Cr duplex stainless steel was examined in H_2S-Cl^- environment. 25%Cr duplex stainless steel has highest SCC susceptibility around 100°C independent of test method. It is not specific to duplex stainless steel that the maximum susceptibility temperature exists. The materials have an intrinsic temperature at which SCC susceptibility is highest in H_2S-Cl^- environment.

25%Cr duplex stainless steel has high resistance to HE. However it has susceptibility to HE in H_2S containing environments under the condition of both high strength and high applied stress over its yield strength.

INTRODUCTION

The recent research into new sources of hydrocarbon has led to its production from more severe environments. Such wells are deep and are often constructed offshore.

Carbon steel and low alloy steels are not suitable for these wells because of their high corrosion rate in wet CO_2 environment, if the inhibiter system does not work well owing to the high temperature or the economical reason. So highly alloyed materials, for example, duplex stainless steels, high nickel stainless steels and nickel base alloys have been evaluated for the purpose of selecting proper materials depending on the various well conditions [1-6].

Among these highly alloyed materials, duplex stainless steels have been widely used as the production tubing and line pipe because of their high corrosion resistance in wet CO_2 environment. The presence of H_2S, however, deteriorates the corrosion resistance of duplex stainless steels.

The corrosion behavior such as pitting corrosion, intergranular corrosion and SCC of duplex stainless steel have often been investigated [7-10].

The susceptibility to SCC of duplex stainless steels, however, has not been clarified yet. Especially, there is no agreement as to the effect of temperature on the cracking susceptibility of duplex stainless steel in H_2S-Cl^- environment [1,4,11-13].

Moreover HE behavior in H_2S-Cl^- environment of ferrite phase of duplex stainless steel is not understood well.

In this paper, the temperature dependence of the cracking susceptibility of 25%Cr duplex stainless steel was examined by four different testing methods which were different in stressing condition. And the resistance to HE of 25%Cr duplex stainless steel was also investigated.

From these results, the temperature effect of the cracking susceptibility and the possibility of occurrence of hydrogen embrittlement of duplex stainless steel are discussed.

EXPERIMENTAL

1. Material used

The chemical compositions of materials used are shown in Table 1. The 25%Cr duplex stainless steel (Sumitomo DP3, ASTM UNS S31250, $\alpha + \gamma$) was melted in an electric furnace and refined by AOD (Argon Oxygen Decarburization) and made into the slab in the factory. The slab was hot rolled to plates of 15 mm in thickness and then solution heated at 1050°C for 30 minutes followed by water quenching and then 30% cold worked in the laboratory.

The α and γ steels were melted in 50 kg vacuum induction furnaces corresponding to the α phase and γ phase of 25%Cr duplex stainless steel from the chemical analysis of the phases by EPMA method. The 20%Ni and 25%Ni steels were also melted in 50 kg vacuum induction furnaces to investigate the effect of Ni content of γ phase steels on the cracking behavior. The manufacturing process of these steels was the same as the 25%Cr duplex stainless steel.

The mechanical properties of transverse direction to rolling of these steels are shown in Table 2.

2. Test Method

(1) SCC test

Four kinds of autoclave tests were performed, i.e., constant strain test (four point bent beam method), constant load test, slow strain rate test (SSRT) and fracture mechanic test (DCB).

All SCC tests were conducted in 20% NaCl solution with 1 atm H_2S at the temperature range between 30°C and 250°C. The autoclaves were purged with nitrogen gas to deaerate the solution, and then with H_2S gas at the room temperature, and finally heated up to the desired temperature.

a) Constant strain test (four point bent beam method)

The test specimen was 75 mm in length, 10 mm in width and 2 mm in thickness. The stress was applied by four point bent beam method. The tensile stress (σ) was calculated from the strain by the following equation

$$\sigma = Ety(\frac{2}{3}\ell_1^2 + \ell_1\ell_2 + \frac{1}{4}\ell_2^2)^{-1} \quad ---- \quad (1)$$

where y is the displacement at the center point of the specimen, E is the Young's coefficient of the material.

The applied stress range was between 20% of the yield strength (0.2σy) and 100% of the yield strength (1σy) at the room temperature. The specimen was insulated from the jig by the ceramic bars. The test duration was 336 hours. The SCC susceptibility was evaluated by the threshold stress (σth) for cracking.

b) Constant load test

The smooth tensile test specimen which was 20mm in gauge length, 3mm in diameter, was used for the constant load SCC test. The constant load was applied after heating the autoclave up to the desired temperature. The applied stress was 0.8σy at room temperature. The test specimen was insulated from the autoclave by the ceramic jig. The SCC susceptibility was evaluated by the time to failure (σ_{TTF}).

c) SSRT

The test specimen was same as the constant load SCC test specimen. The specimen was extended with the strain rate of 1×10^{-6}/s after heating the autoclave up to the desired temperature. The test specimen was also insulated from the autoclave by the ceramic jig.

The SCC susceptibility was evaluated by the maximum strength (σ_{max}) in the test environment.

d) DCB test

The test specimen with 4 mm fatigue precrack which was 105mm in length, 25mm in width, 7mm in thickness, and 5.9mm in thickness at grooved portion, was used for DCB test. The stress was applied to the desired initial stress intensity factor (k_{10}) with the wedge. The SCC susceptibility was evaluated by K_{ISCC} which was calculated by the following equation.

$$K_{ISCC} = \frac{E}{8} a\delta R^{3/2} \frac{(3.454a + 2.38h)}{(a^4+2.061ha^3+1.416h^2 \cdot a^2)} \left(\frac{B}{Bn}\right)^{1/\sqrt{3}} \quad (2)$$

where E is the Young's coefficient, δ is the displacement of the wedge point, B is the thickness of the specimen, Bn is the thickness of grooved portion, h is the half width of the specimen, a is the final crack propagation length.

For the measurement of the final crack propagation length (a), the first immersion test was conducted for 336 hours, and then 168 hours duration tests were repeated until the crack propagation was stopped. The test solution was refreshed at every immersion test. The stress condition of the DCB test must be in the plane strain region which was evaluated by the following equation.

$$B \geq 2.5 \left(\frac{k_1}{\sigma y}\right)^2 \quad (3)$$

The initial stress intensity factor (K_{10}) was limited by the thickness of the specimen (B). So 7 mm was chosen as thickness and K_{10} was 180 $kg/mm^{3/2}$.

2. Hydrogen embrittlement (HE) test

(1) Constant load test in the cathodically charged condition

The hydrogen embrittlement behavior was examined by the constant load test in the cathodically charged condition. The specimen was 20mm in gauge length, 3mm in diameter and had a 0.3mm notch in the center portion of which the stress concentration factor was 3.1. The test solution was 5% H_2SO_4 with 1.4 g/ℓ$(NH_2)_2CS$ which was added as a poison at the temperature of 35°C. The constant cathodic current density, which was in the range of 0.1 to 50 mA/cm^2, was applied on the test specimen by the galvanostato during the test period. The susceptiblity to HE was evaluated by the time to failure. The maximum test duration was 100 hours.

(2) Hydrogen permeation test

In order to measure the surface hydrogen concentration, the hydrogen permeation test was performed in 5%NaCl with 1 atm H_2S and in 5%H_2SO_4 with 1.4 g/ℓ$(NH_2)_2CS$ under the cathodically charged condition at the temperature range of 35°C to 60°C. The test specimen size was 30 mm x 30 mm x 0.2 mm, which was made by the electric polishing method. The hydrogen which was generated at the surface of the test specimen in the cathode chamber, penetrated through the specimen to the opposite surface. At the anode chamber, the hydrogen was detected as the oxidation current of hydrogen at constant potential. The solution in the anode chamber was 1 N NaOH and the applied potential was -100 mV vs Ag/AgCl.

The surface hydrogen concentration was calculated from the current density by the following diffusion equation

$$\frac{\partial C}{\partial t} = D \frac{\partial^2 C}{\partial x^2} \quad (4)$$

$$C(0,0) = C_o, \quad C(L,T) = 0$$

where D is the diffusion coefficient, L is the thickness of the specimen, C_o is the surface hydrogen concentration. The current density of the hydrogen (Jt) is obtained from the solution of the equation (4).

$$Jt = -D\frac{\partial C}{\partial x} = \frac{DFCo}{L} \frac{2}{\pi^{1/2}} \frac{1}{\tau^{1/2}} \sum_{n=1}^{\infty} (-1)^n e^{-(2n+1)^2/4\tau} \quad (5)$$

where τ = Dt/L and F is the Farady constant.

In this study, τ was very small owing to the small value of D and the short test duration of t. So the current density of the hydrogen (Jt) is obtained as the following expression.

$$Jt = \frac{DFCo}{} \frac{2}{\pi^{1/2}} \frac{1}{\tau^{1/2}} e^{-\frac{1}{4\tau}} \quad (6)$$

The equation (6) indicates that there is the linear relationship between the term of log $(t^{1/2}Jt)$ and the term of 1/t.

$$\log (t^{1/2}Jt) = -\frac{L^2 \log e}{4D} \frac{1}{t} + \log \frac{2D^{1/2}FC}{\pi^{1/2}} \quad (7)$$

The surface hydrogen concentration (Co) and the diffusion coefficient (D) were calculated by the slope and the cross point of the horizontal axis of the equation (7).

Test Results and Discussions

1. Stress Corrosion Cracking (SCC)

SCC test results were shown in Fig. 1 - Fig. 5. The effect of the temperature on the threshold stress for SCC by the four point bent beam test was shown in Fig. 1. The closed marks show the large cracks which propagated mainly through the ferrite phase and avoided the austenite phase of 25%Cr duplex stainless steel. The dotted marks show the shallow selective corrosion of the ferrite phase. The maximum SCC susceptibility existed around 100°C on the base of the evaluation of the large cracks.

The constant load SCC test results were shown in Fig. 2. The time to failure was shortest around 100°C and the test specimen was not broken at 200°C below 200 hours.

The effect of the temperature on the SCC susceptibility by SSRT was shown in Fig. 3. The σ_{max} was smallest around 100°C, and SCC did not occur at 250°C.

The K_{ISCC} values of 25%Cr duplex stainless steel were obtained by DCB test. The crack propagated only at 100°C, and the test duration at 100°C was 672 hours until the crack propagation stopped.

The value of K_{ISCC} at 100°C was 80 $kg/mm^{-3/2}$. The value of K_{ISCC} at 30°C, 60°C, 150°C and 200°C was above 180 $kg/mm^{-3/2}$. These test results show that 25%Cr duplex stainless steel has highest susceptibility to stress corrosion cracking around 100°C in H_2S-Cl^- environment independent of test methods.

In order to confirm that the cracking is the stress corrosion cracking, in other words, the active pass corrosion process, the effect of potential on time to failure of 25%Cr duplex stainless steel was investigated by the constant load test. The results were shown in Fig. 4. The susceptibility to the cracking increased above the corrosion potential the value of which was around -440 mV vs Ag/AgCl, and the susceptibility of the cracking decreased above -350 mV vs Ag/AgCl. At -200 mV, the test specimen was broken owing to the occurrence of pitting. At the cathodic electric potential range between -430 mV and -500 mV, the test specimens were not broken. From this test results, the crack in H_2S-Cl^- environment is thought to be attributed to SCC.

Fig. 3 also shows the SSRT test results of the α and γ steels which correspond to the α and γ phases of 25%Cr duplex stainless steel. This test result indicates that the existence of the maximum SCC susceptibility temperature was not characteristic of duplex stainless steel. The temperature where the γ steel had the highest SCC susceptibility shifted to higher temperature with the α steel. The duplex stainless steel and the α steel had the highest SCC susceptibility at the same temperature. It is thought that SCC of duplex stainless steel is determined by the α phase of the duplex stainless steel. This test results also showed that the α and γ steels had higher SCC susceptibility than the 25%Cr duplex stainless steel which was well known as the keying effect of γ phase of the duplex stainless steel. Fig. 5 shows the effect of Ni content on the maximum SCC susceptibility temperature by SSRT method. The higher Ni content alloy had the higher maximum SCC susceptibility temperature.

The reason why the duplex stainless steels have maximum SCC susceptibility around 100°C was not clear, but it might be explained by the difference in the structure of the surface film. Fig. 6 shows the IMMA analysis of the surface film. The surface film below 100°C was composed of nickel sulfide and chromium oxide, but above 100°C, nickel sulfide film disappeared and the surface film was mainly composed of chromium oxide. This fact indicates that the transition point of the chemical composition of the surface film of duplex stainless steel exists at 100°C; in other words, the surface film is unstable, and depends on the formation and disappearance of the nickel sulfide in the surface film. So at this temperature, SCC susceptibility might be highest.

This might be explained by the fact that the higher the Ni content, the higher the maximum SCC susceptibility temperature.

2. Hydrogen Embrittlement (HE)

(1) Constant load test in cathodically charged condition

Fig. 7 shows the effect of cold work (CW) on HE of 25%Cr duplex stainless steel in 5%H_2SO_4 + 1.4 g/ℓ$(NH_2)_2CS$ solution with notched specimens. The applied stress was 100% of σy, the values of which are given in Table 3. The plastic deformation occurred at the bottom of the notch portion of the test specimen because of the stress concentration factor of 3.1. The test results show that the cold work markedly increased the HE susceptibility of 25%Cr duplex stainless steel. No cold worked material did not show the HE even in severe stress and environmental conditions. So the 30% CW material in which HE was easily caused was chosen for the following investigation.

The HE test results of 25%Cr duplex stainless steel ($\alpha + \gamma$) and the α and γ steels are shown in Fig. 8. The applied stress was also $1\sigma y$. The test result indicated that a critical charging current density exists below which the HE does not occur. The α steel had higher HE susceptibility compared with 25%Cr duplex stainless steel (α + γ). The γ steel showed the good HE resistance and this steel was not broken even under the severe condition of the charging current density of -50 mA/cm^2.

The relation between the applied stress and the critical charging current density of the 25%Cr duplex stainless steel and the α steel is shown in Fig. 9. At the applied stress of $0.8\sigma y$, for example, the critical charging current density of the 25%Cr duplex stainless steel was -50 mA/cm^2 which was much higher than that of the α steel.

Fig. 10 shows the microscopic observation of the specimen of 25% duplex stainless steel which was not broken, but in this test condition the α steel can cause the HE. It can be seen that the micro crack did occur at the α phase of 25%Cr duplex stainless steel and stopped at the boundary of α and γ phases. From this fact, the high resistant of the duplex stainless steel compared with the ferritic stainless steel can be attributed to the keying effect of γ phase as well as in the case of SCC.

(2) Hydrogen permeation test

In order to discuss the HE of the duplex stainless steel in H_2S-Cl^- environment, the measurement of the surface hydrogen concentration (Co) was performed in H_2S-Cl^- environment and in a cathodically charged condition which is shown in Fig. 11. The value of Co in 5% NaCl solution with 1 atm H_2S decreased with increasing the temperature. The Co value in 5%H_2SO_4 + 1.4 g/ℓ$(NH_2)_2CS$ solution in a cathodically charged condition was highest at 40°C, which was different from H_2S-Cl environment. Using these test results, the possibility of HE occurrence of 25% duplex stainless steel in H_2S-Cl^- environment was estimated as shown in Fig. 12.

The surface hydrogen concentration in free corrosion conditions in H_2S-Cl^- environment was 26 ppm at 35°C from the test results shown in Fig. 11, corresponding to the cathodic charging current density of 0.1 mA/cm^2 in $5\%H_2SO_4$ + $1.4g/\ell(NH_2)_2CS$ solution. It can be seen that 25%Cr duplex stainless steel has enough resistance to HE below the applied stress of $1.0\sigma y$ as shown in Fig. 12.

CONCLUSIONS

The stress corrosion cracking and the hydrogen embrittlement behavior of 25%Cr duplex stainless steel were examined in H_2S bearing environments.

The results obtained are as follows:

1) The 25%Cr duplex stainless steel has highest cracking susceptibility around 100°C which is independent of the test method.
   The decrease in the SCC susceptibility at a temperature above 100°C can be explained by the stability of the surface film formed in H_2S-Cl^- environment.
2) The cracking in H_2S-Cl^- environment is attributed to the stress corrosion cracking.
3) It is not specific to duplex stainless steels that the temperature with the highest SCC susceptibility exists. The temperature increases with the Ni content of the alloys, which leads to the fact that the materials have an intrinsic temperature at which SCC susceptibility is highest.
4) The 25%Cr duplex stainless steel has high resistance to hydrogen embrittlement (HE). However it has susceptibility to HE in H_2S bearing environments under the condition of both high strength by cold work and high applied stress over its yield strength. The HE of duplex stainless steel is sensitive to the degree of cold working.
5) High resistance to the HE of duplex stainless steel is attributed to the keying effect of the austenite phase.

REFERENCES

1. J. Ored, S. Bernhardson; CORROSION/82, Paper No. 126, National Association of Corrosion Engineer, Houston, 1982.
2. G. Herbsleb, R.K. Poepperling; CORROSION, Vol. 36, No. 11, p. 611 (1980).
3. A. Desestrel; CORROSION/83, Paper No. 165, National Association of Corrosion Engineers, Anaheim, 1983.
4. P.R. Phodes, G.A. Welch, L. Abrego; Materials for Energy System, Vol. 5, No. 1, p. 3 (1983).
5. S.M. Wilhelm, R.D. Kane; CORROSION/83, Paper No. 154, National Association of Corrosion Engineers, Anaheim, 1983.
6. H. Miyuki, J. Murayama, T. Kudo, T. Moroishi; CORROSION/84, Paper No. 293, National Association of Corrosion Engineers, New Orleans, 1984.
7. G. Herbsleb, P. Schwaadt; Duplex Stainless Steel, American Society for Metals, p. 15, 1983.
8. S. Bernhardsson, J. Ordsson, C. Martension; Duplex Stainless Steel, American Society for Metals, p. 267, 1983.
9. H. Tsuge, Y. Tarutani, T. Kudo; CORROSION/86, Paper No. 156, National Association of Corrosion Engineers, Houston, 1986.
10. H. Tsuge, Y. Tarutani, T. Kudo, K. Fujiwara, T. Moroishi; International Conference on Duplex Stainless Steel, The Hague, Paper 33B, 1986.
11. S. Mukai, H. Okamoto, T. Kudo. A. Ikeda; J. of Materials for Energy Systems, Vol. 5, No. 1, p.59 (1983)
12. J. Sakai, I. Matsushima, Y. Kanemura, M. Tanimura, T. Otsuka; 1982 ASM Metals Congress 8201-010
13. J.C. Prouheze, J.C. Vaillant, G. Guntz, B. Lefebvre; 1982 ASM Metals Congress 8201-012

Table 1 Chemical composition (wt %)

No.	C	Si	Mn	P	S	Ni	Cr	Mo	N	Others	Remarks
1	0.018	0.35	0.91	0.024	0.001	7.45	25.00	3.21	0.13	Cu:0.51, W:0.32	$\alpha + \gamma$
2	0.005	0.48	0.77	0.026	0.003	4.45	27.94	3.62	0.004		α
3	0.039	0.32	1.14	0.014	0.003	10.60	19.78	2.26	0.142		γ
4	0.002	0.30	0.64	0.004	0.001	19.90	20.36	3.48	0.0018		20Ni
5	0.006	0.30	0.64	0.002	0.002	24.67	20.36	3.63	0.0054		25Ni

Table 2 Mechanical properties (30% CW, T)

No.	Y.S. (kg/mm^2)	T.S. (kg/mm^2)	El. (%)	RA (%)
1	114.2	121.6	12.3	52.9
2	93.3	101.7	2.7	16.0
3	98.5	111.3	13.4	46.3
4	91.3	97.4	12.7	55.6
5	85.7	92.6	15.4	71.6

Table 3 Mechanical properties of $\alpha + \gamma$

CW	0%		15%		30%	
	Y.S. (kg/mm^2)	El. (%)	Y.S. (kg/mm^2)	El. (%)	Y.S. (kg/mm^2)	El. (%)
T	59.0	33.0	94.3	13.3	114.2	12.3
L	61.5	36.0	88.5	21.2	101.0	17.3

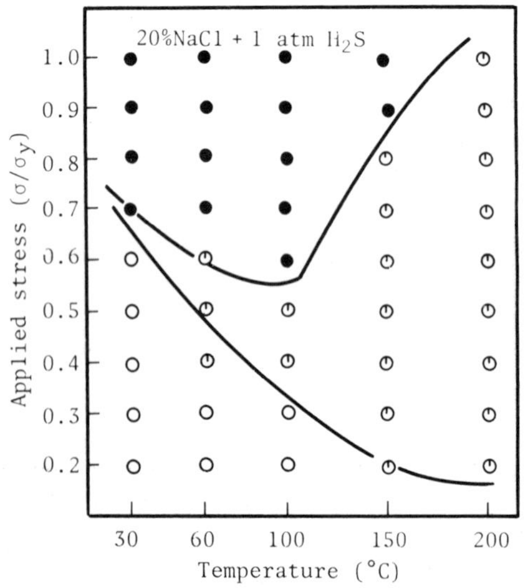

Fig. 1 Stress corrosion cracking test results of 25%Cr duplex stainless steel (30% CW) by four point bent beam method in 20% NaCl solution with 1 atm H_2S for 336 hours. (● : Crack, ⦶ : Selective corrosion ○ : No crack)

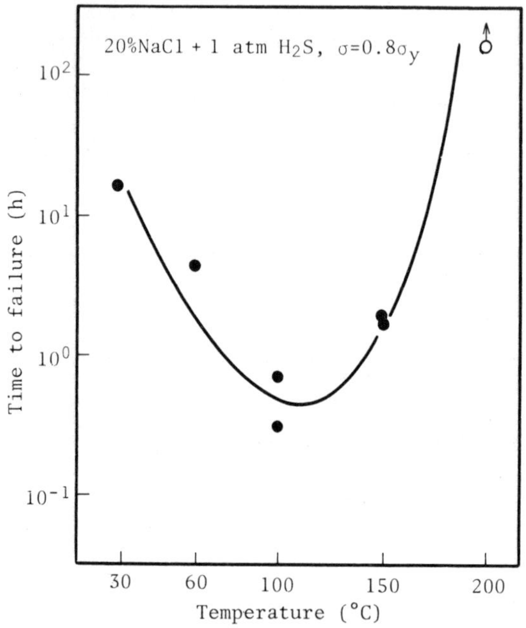

Fig. 2 Stress corrosion cracking test results of 25%Cr duplex stainless steel (30% CW) by constant load method in 20% NaCl solution with 1 atm H_2S. (applied stress : $0.8\sigma y$)

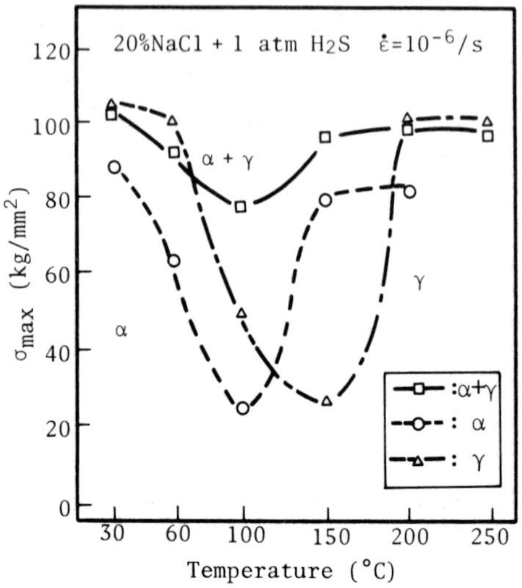

Fig. 3 Temperature dependency of SCC susceptibility of 25%Cr duplex stainless steel and alloys by SSRT method in 20% NaCl solution with 1 atm H_2S (strain rate of $1 \times 10^{-6}/s$).

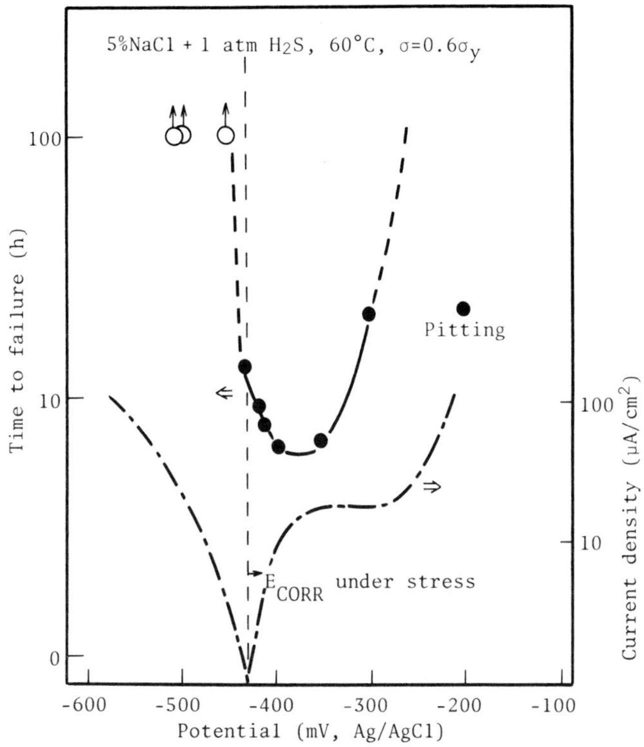

Fig. 4 Effect of potential on time to failure of 25%Cr duplex stainless steel (30% CW) by constant load test in 5% NaCl solution with 1 atm H_2S at 60°C (applied stress of 0.6σy).

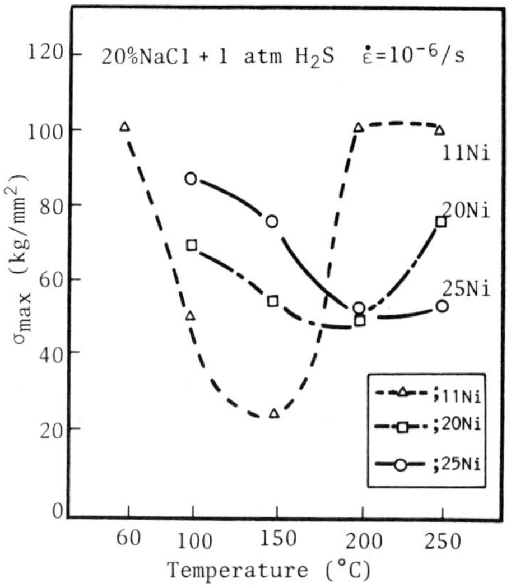

Fig. 5 Effect of Ni content on SCC susceptibility of 20Cr-3Mo alloys by SSRT method in 20% NaCl solution with 1 atm H_2S (strain rate of 1 x 10^{-6}/s).

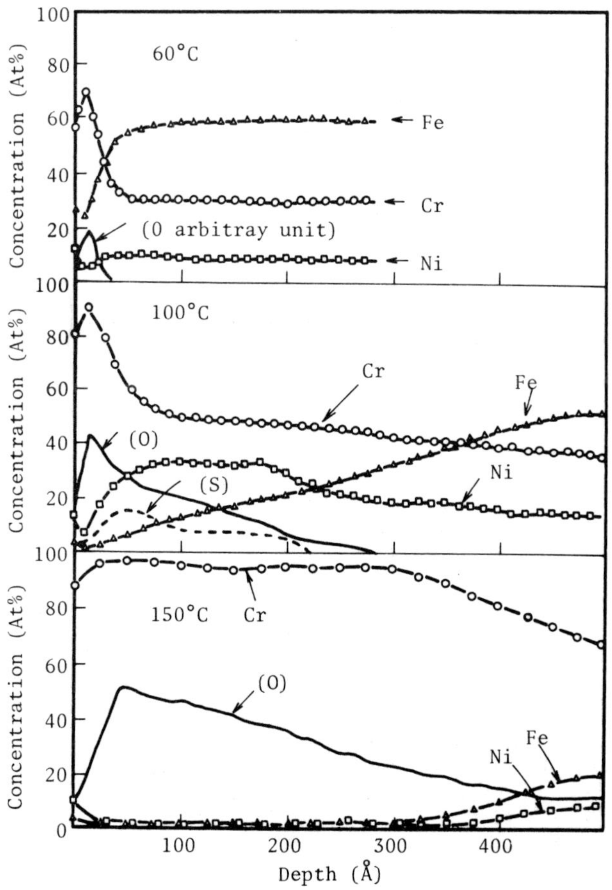

Fig. 6 The surface film analysis of 25%Cr duplex stainless steel in 20% NaCl with 1 atm H_2S for 48 hours.

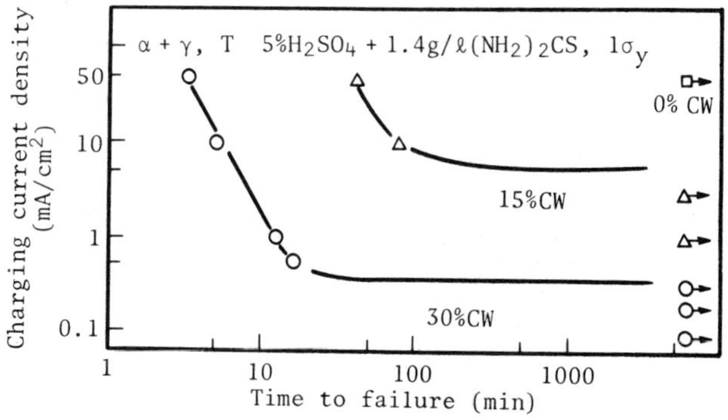

Fig. 7 Effect of cold work on the hydrogen embrittlement of 25%Cr duplex stainless steel by constant load method with notched specimens in cathodically charged condition in $5\%H_2SO_4$ + $1.4g/\ell(NH_2)_2CS$ solution at 35°C.

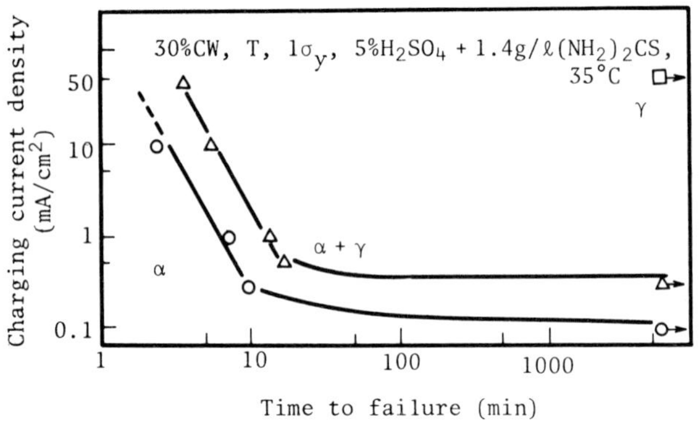

Fig. 8 Effect of charging current density on time to failure of the hydrogen embrittlement of 25%Cr duplex stainless steel (α + γ), α phase alloy and γ phase alloy.

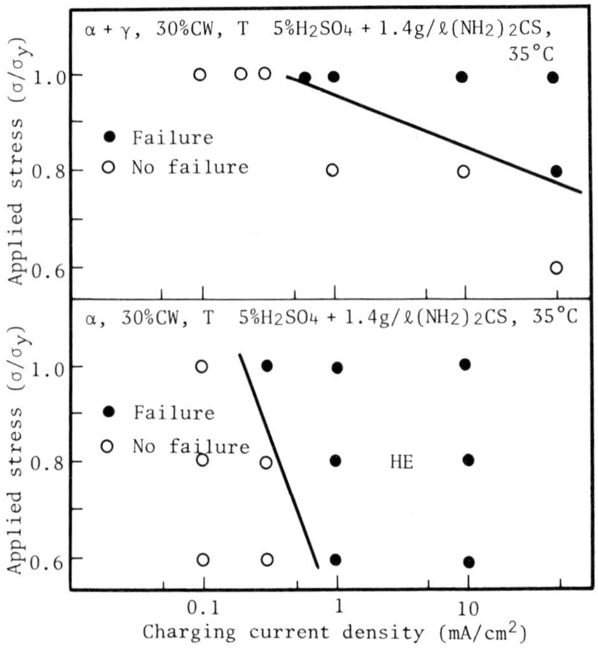

Fig. 9 Effect of applied stress on critical charging current density of the hydrogen embrittlement of 25%Cr duplex stainless steel.

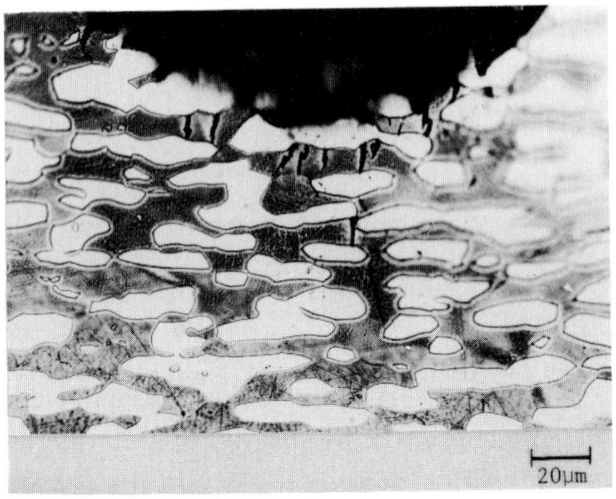

Fig. 10 Microscopic observation of unfailed specimen of 25%Cr duplex stainless steel. (applied stress : 0.8σy, charging current density : 1 mA/cm^2, TTF > 6000 min)

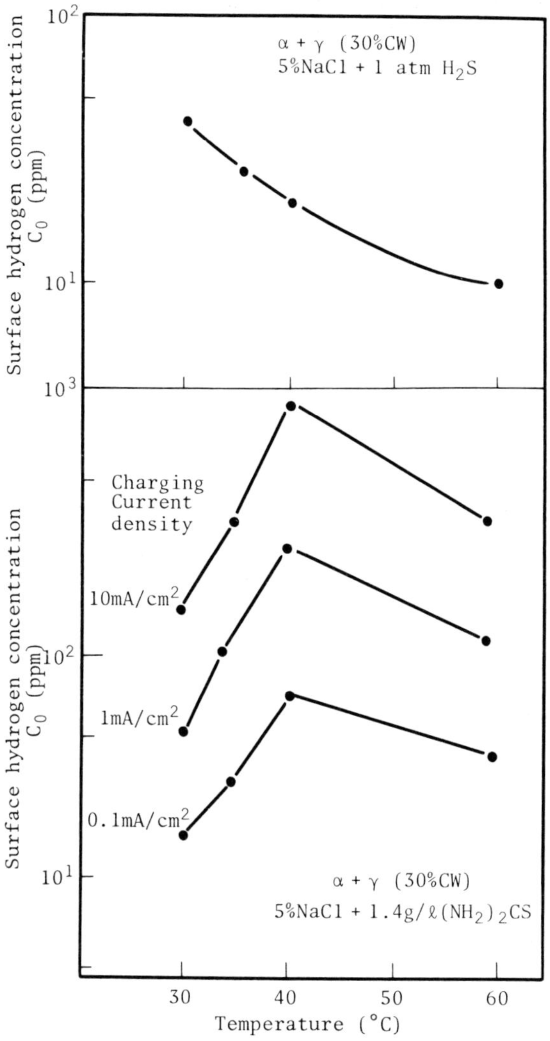

Fig. 11 Surface hydrogen concentration of 25%Cr duplex stainless steel in H_2S-Cl^- environment and in cathodically charged condition.

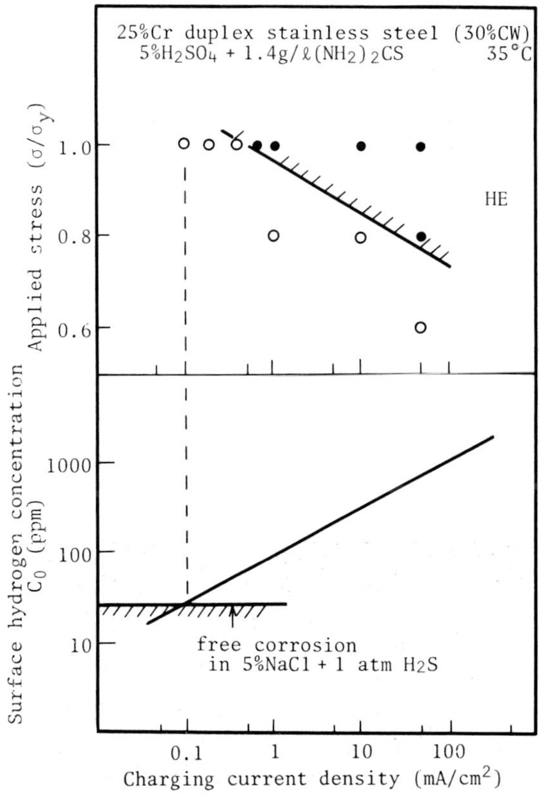

Fig. 12 Relation between HE susceptibility and surface hydrogen concentration of 25%Cr duplex stainless steel.

Selective corrosion of duplex stainless steels

E Symniotis-Barrdahl

The author is with the Swedish Institute for Metals Research, Stockholm, Sweden.

ABSTRACT

An investigation has been carried out concerning selective corrosion in a duplex stainless steel SS 2377 (22Cr 5Ni 3Mo 0.15N) in acid environments. Preliminary tests using a potentiodynamic technique were performed in order to establish the presence of selective corrosion in the different solutions, i.e. solutions containing hydrochloric and sulphuric acids. Comparisons were made with alloys of the same composition corresponding to the individual phases in the duplex stainless steel.

The typical indications expected in a galvanic interaction between the two phases, ferrite and austenite, were observed in a number of environments.

In order to quantify the observed phenomena a potentiostatic weight loss method has also been used. By adapting a model for galvanic corrosion suggested by Mansfeld, good agreement between the experimental and theoretical data has been obtained which supports the assumption that galvanic effects control the corrosion behaviour. In the investigated solutions the ferrite phase has acted as a sacrificial anode in favour of the austenite.

INTRODUCTION

In recent years duplex stainless steels have been used for many applications because of the favourable corrosion properties of these materials. Interest in duplex steels has grown and a number of authors have reported increased corrosion resistance for duplex stainless steels in comparison to standard austenitic steels (1-3). However, the problem of selective corrosion was not always taken into consideration.

Selective corrosion occurs when different components of an alloy, or when different phases of a multiphase material, corrode at different rates. These weight loss measurements were carried out at:

1. A defined potential value which was kept constant by a potentiostatic control, and

2. The free corrosion potential without potentiostatic control.

The first method has been used previously to examine multiphase materials in which every phase dissolves at a different potential (4, 5). This is manifested through the different peaks within the active range on the polarisation curve.

A polarisation curve is the result of all the reactions occurring on the metal surface during the potential sweep. Therefore, the method is not suitable for a quantitative evaluation of the metal dissolution reaction. Another limitation is that the sweep speed can also influence the shape of the polarisation curve leading to incorrect results. However, this method can give an indication that selective corrosion occurs through the appearance of two peaks when the sweep is directed from the passive range towards negative potentials. Although, this is not aways true because for solutions where the rate of the cathodic reaction is high, compared to the anodic reaction, no peaks are present yet the metal dissolution reaction has occurred.

In this work the polarisation curves were used in order to define the behaviour of a duplex material within a certain range of potentials, (active corrosion) and in different solutions with respect to selective corrosion.

The weight loss under the potentiostatic control method allows the anodic curve to be obtained independently of the cathodic reaction. This weight loss at a defined potential during a time period is proportional to the average value of the anodic current during this period. Therefore, in order to define the exact form of attack and to carry out a quantitative comparison between the corrosion properties of the duplex material and the two isolated phases, a potentiostatic weight loss method has been chosen.

Finally, weight loss measurements at the free corrosion potential were carried out for a ferritic and an austenitic steel separately and also in electrical contact. The aim of these measurements was to prove, by a simple and direct method the actual existence of galvanic action between ferritic and austenitic steels.

The phenomenon is well known and occurs in various alloy systems. By using metallography together with electrochemical methods, many workers (4-6) have observed selective corrosion of duplex stainless steels during corrosion in the active region in non-oxidizing acids, such as weak aqueous solutions of sulphuric or hydrochloric acid. Herbsleb (7) observed selective attack of the ferrite phase in the vicinity of the corrosion potential and a selective dissolution of austenite at potentials near the active-/passive transition region. Furthermore Streicher et al (8) have studied selective corrosion and proposed that the selective dissolution of the ferrite is a result of a galvanic effect between ferrite and austenite, where ferrite acts as the anode.
Factors which influence the galvanic corrosion of two dissimilar metals, or alloys, in electrical contact and exposed to a corrosive solution are (9):

a) The corrosion potentials of these two metals in the particular solution

b) The type and rate of anodic and cathodic reactions occurring on the metal surfaces.

c) The solution used and its properties.

Mansfeld (10) has characterised three cases of galvanic corrosion according to the number and type of reactions occurring on the metal surfaces at the galvanic potential. He expressed the galvanic corrosion rate as a function of the parameters of the electrochemical reactions occurring at the galvanic potential as well as the ratio of cathode to anode surface area.

It is intended in this work to study the effect of different solutions upon the selective corrosion of a duplex stainless steel within the active range. Further, a comparison between the corrosion properties of a duplex stainless steel and those of its individual phases has been undertaken in order to investigate the mechanism of selective corrosion.

EXPERIMENTAL TECHNIQUES

The two methods which have been used to study selective corrosion are:

a) Potentiodynamic sweeps to obtain polarisation curves.

b) Weight loss measurements.

Material

The materials used in this investigation were a standard SS2377 grade of duplex stainless steel and two special alloys having compositions corresponding to its two phases. Compositions of these three alloys are given in Table 1. It should be pointed out that the Ni-content of the austenitic alloy was increased by 1% compared to the Ni-content of the austenitic phase in order to eliminate a ferritic rest. This material was heat treated at 1125°C for several hours for the same purpose. However, this heat treatment did not achieve the total elimination of ferrite and the austenitic steel used in this study contained a rest of ~0,6% ferrite, fig 2.
Electrochemical measurements obtained using a purely austenitic steel containing slightly more Ni, i.e. 1% gave practically the same results indicating that the ferrite content is of minor importance in the solutions under investigation.

Potentiodynamic sweeps

The specimens were ground to 600 mesh and then washed with water and acetone. When the samples were dry the surfaces were then covered with a lacquer, except for an area of 1 cm² which would be exposed to the acid solution.

The polarisation curves were determined using a Wenking potentiostate and an electrolytic cell consisting of a glass vessel, containing a platinum net as the counter electrode and a calomel electrode as the reference electrode. This was then connected to the cell through a KCl/Agar bridge. Air was continuously bubbled into the solution through a glass filter tube. All the measurements were obtained at room temperature which was 22°C.

After the specimen had been placed in the cell a potential of -1V was applied for 2-3 minutes in order to clean the surface with evolution of H_2. Then a potentiodynamic sweep towards positive potentials at a speed of 30 mV/min was carried out. This sweep was interrupted at a maximum value of +100 mV and then repeated in the opposite direction at the same speed. The experiment was halted when the active range had been passed. Polarisation curves were determined for the ferritic and duplex steels, but not the austenitic steel, because this was not available at the beginning of this investigation. Compositions of the solutions tested are given in Table 2.

Potentiostatic weight loss measurements

The preparation of these samples was the same as given above except that instead of a lacquer being used a holder was used to provide an electrode area of 1 cm² which would be exposed to the solution, fig 1.

A measurement of the free corrosion potential was carried out before a certain value of electrochemical potential was imposed on the specimen.

The solution used was 4N H_2SO_4+0,1N HCl. Each measurement was obtained over a period of 4 hours. This time period was used because it gave a measurable total weight loss at every potential and a steady state of dissolution velocity was reached. After the end of the test the specimens were washed, dried, and weighed using a balance which was accurate to the nearest 0,05 mg. A typical weight loss was about 1,2 mg/cm².

Standard metallographic techniques were used to prepare the samples for optical microscopy. By etching in Murakami's reagent the austenite and ferrite phases can easily be distinguished because, the ferrite is etched a darker shade of brown.

Weight loss measurements at the free corrosion potential

One ferritic and one austenitic specimen were prepared and weighed before being placed in separate holders and then inserted into the cell containing a solution of 4N H_2SO_4 + 0,1N HCl for a period of 20 hours. In these experiments a 100% austenitic alloy was used having a 1% higher Ni-content compared to the composition of the austenitic steel given in Table 1.

Afterwards, they were cleaned and weighed as before. This same procedure was then repeated except that the two holders were connected by a metallic wire so that the exposed specimens were in electrical contact. Finally a weight loss measurement was obtained for the duplex steel under the same experimental conditions.

RESULTS AND DISCUSSION

1. Polarisation curves

These results are given in Table 2 and show that for some of the solutions tested there is a clear tendency to give rise to selective corrosion. However, it follows from these results that the polarisation curves cannot always give an answer as to whether or not selective corrosion has occurred. The reasons for this have been mentioned earlier but in many cases where pitting and/or crevice corrosion occurred, due to the presence of NaCl in the test solution, the shape of the polarisation curve was distorted such that it was extremely difficult to identify the two peaks. For the test solutions having a composition of the type 4N H_2SO_4+ xHCl where 0,2 N$\leq$ x $\leq$2N two peaks were obtained during a sweep in the negative direction. The peak which occurred at the greater negative value corresponded to the dissolution of the ferritic phase, while the one closer to the passive range corresponded to the dissolution of the austenitic phase which is in agreement with the literature (5,6).

Fig 3 shows the relationship between the surface area A, i.e. area between the anodic curve and the x-axis, and the concentration of HCl. Each point on the diagram is the mean value obtained from 4 or 5 measurements. The surface area A is proportional to the total amount of metal that has been dissolved within the active range. Therefore it would appear that a larger amount of duplex steel compared to ferritic steel will be dissolved within the active range, when all the other parameters are held constant.

Both ferritic and duplex steels were found to have a larger active peak area during a sweep towards positive potentials as opposed to negative potentials,. This is probably due to the passive film which forms prior to the second sweep. Furthermore it was observed that this phenomenon became more pronounced the further the first sweep proceeded into the passive range before returning in the opposite direction. Hence it is probable that the passive film structure has a strong influence upon the heights of the active peaks.

A strong dependency of the metal dissolution rate on the Cl^-concentration is shown in fig 3 and more clearly in fig 4. In this latter figure the two polarisation curves are recorded for two different solutions of the same $H^+{}_3O$ concentration but different Cl^-concentration. The very strong influence of the Cl^-concentration on the dissolution rate of SS 2377 can be seen in fig 5 where the active peak in pure sulphuric acid is ~10 times smaller than in hydrochloric acid even though the $H^+{}_3O$ concentration was 6 times greater in the sulphuric acid.

In fig 6 the maximum current density of the active peaks is shown as a function of HCl concentration at a constant concentration of sulphuric acid. The most interesting features are firstly, the ratios between the height of the active peaks for ferrite acting as a single phase specimen and in the duplex steel and secondly, the ratio between the ferrite and austenite active peak heights. This diagram shows a clear difference between the dissolution rate of the ferritic phase in the duplex steel and the ferritic steel but only at large values of Cl^-concentration. In the case where [HCl]=2 N the anodic reaction proceeded at a much greater rate than the cathodic reaction and therefore the polarisation curves obtained gave more information about the real dissolution rates of the two alloys tested compared to the situation where [HCl]=0,2 N. For this latter case a relatively weak anodic current together with a strong cathodic current flow with the result that the polarisation curve does not give any quantitative information about the dissolution rates. However, the results obtained when [HCl] =2N have shown that there is a relationship between the height of the ferritic and austenitic peaks. Although the standard deviation was found to be large, it does appear that a difference in dissolution rate does occur for the two phases, being higher for the ferritic phase.

2. Potentiostatic weight loss measurements

The results of the potentiostatic weight loss measurements are shown in fig 7 where the anodic current which is proportional to the rate of weight loss, expressed in mg/cm² h is plotted against the electrochemical potential. The measurements were performed on SS 2377 and the single phase ferritic and austenitic steels in 4N H_2SO_4 + 0,1 N HCl solution. The polarisation curves did not show any selective corrosion in this solution.

The results in fig 7 together with optical microscopy clearly show that selective corrosion occurs at the "negative" side of the active curve for the ferritic phase. For the austenitic phase selective corrosion occurs at the "positive" side.

Further, the selective corrosion of the ferritic phase is strongly accelerated compared to the dissolution rate of the single phase ferritic steel. This is more obvious if one takes into account the fact that the phase ratio of the steel is approximately 1:1, fig 8.

The two single phase materials have a potential difference of ~ 45 mV in this solution. Their anodic curves show that in the vicinity of the corrosion potential for duplex steel, the austenitic phase is more noble than the ferritic phase. Hence, this results in an accelerated corrosion rate for the ferritic phase and a protection of the austenitic phase. This situation corresponds to the second case exposed by Mansfeld (10), where two electrically connected metals, A and the more noble C are exposed to an acid solution. At the galvanic potential anodic and cathodic reactions are supposed to occur on A at appreciable rates but only the cathodic reaction occurs on C. In this case the anodic reaction is considered to be the metal dissolution reaction and H_2 evolution as the cathodic reaction. Since all the reactions involved are supposed to follow the exponential law $i = i_o e^{bnF/RT}$, the following equation is valid:

$$\log i_d^A = \log i_{corr}^A + \frac{b_c}{b_c + b_A} \log\left[1 + \frac{i_o^C}{i_o^A} \frac{A^C}{A^A}\right] \quad (1)$$

where:

i_d^A = The dissolution rate (current density) of the anodic metal A when it is connected to metal C (mA/cm²)

i_{corr}^A = The dissolution rate at the corrosion potential for the metal A in the same solution as above but not connected to any other metal (mA/cm²)

b_A, b_c = Tafel slopes for the anodic and cathodic reactions. b_c is supposed to be the same for the cathodic reactions on A and C.

i_o^C = Exchange current density for the cathodic reaction on the cathode [mA/cm²]

i_o^A = Exchange current density for the cathodic reaction on the anode [mA/cm²]

A^C = Area of the cathodic surface [cm²]

A^A = Area of the anodic surface [cm²]

Fig 9 shows the weight loss measurements transformed to current density with the help of the assumption that $i_{an} = i_{cath}$ at E_{corr}. The value for i_{cath} at this potential is evaluated from the extrapolation of the cathodic line which was constructed from measurements of current density at different potentials. These measurements were started at ~ 200mV from the value of E_{corr} in order to avoid any influence of the anodic reaction. Equilibrium potential for the cathodic reaction was calculated using the equation

$$E_{H_3^+O/H_2} = 0,59 \log [C_{H_3^+O}]$$

and is the value of potential where the exchange current densities for the two cathodic reactions are estimated.
It follows from the above that:

b_A= 46 mV/decade $\quad i_o^C$ = 2,18 mA/cm²
b_c= 136 mV/decade $\quad i_o^A$ = 1,65 mA/cm²

From eq. 1 the dissolution rate of the ferritic phase in the duplex steel is 0,58 mg/cm² h.
The experimentally obtained values were,
1,10 mg/cm² h: average value for a period of 4h
0,80 mg/cm² h: average value for a period of 0,50h

Differences between the two values indicates that the phenomenon is time dependent and this is probably due to
(a) changes in the total surface area, and
(b) changes in the surface area ratio between the ferrite and austenite phases produced by selective dissolution of the ferritic phase. The effect of the two dissolution times upon the specimens is shown in fig 10.

3. Weight loss measurements at the free corrosion potential

The results are shown in Table 3 and for the single phase materials they show that a galvanic interaction occurs between the ferritic and austenitic specimens when they are in electrical contact.

However, a comparison between the average weight loss for ferrite and austenite in contact and the weight loss of duplex steel shows that the parallel plates approximation is not adequate and therefore the distance between the two phases and/or the phase distribution must be taken into account.

CONCLUSIONS

1. Selective corrosion of SS 2377 within the active range occurs in solutions of 4N H_2SO_4 + xHCl, where 0,1 N ≤ x ≤ 2N.
2. The position of the active peaks obtained by potentiostatic and potentiodynamic measurements, show that ferrite is selectively dissolved at greater negative potentials than austenite near the corrosion potential while austenite is selectively dissolved near the passivation potential.
3. In some cases, the polarisation curves show a higher dissolution rate for the duplex steel than the ferritic steel and this indicates that galvanic corrosion is occurring.
   The experimental results from the polarisation curves do not permit a quantitative evaluation and also the accuracy of the methods does not allow the detection of selective corrosion at low acid concentration.
4. The potentiostatic weight loss measurements showed that a galvanic effect exists between the two phases and allow for a quantitative evaluation of this effect. Ferrite is anodic to austenite and the ferrite phase in the duplex steel dissolves at a higher rate than the single phase ferrite material. This is valid in 4N H_2SO_4+0,1N HCl solution. The total dissolution rate of the ferritic phase can be distinguished in two parts
   (a) - Self dissolution of the ferrite which is sustained by the hydrogen evolution reaction on its surface.
   (b) - Galvanic dissolution of ferrite which is sustained by the hydrogen evolution reaction on the austenite as suggested in eq. (1).

   The dissolution of duplex steel is time dependent and is probably due to the change in the surface ratio which results from the selective dissolution of one of the phases. The galvanic corrosion is expressed mathematically in the present case and the experimental result is in good agreement with the theory.
5. The galvanic action between the ferrite and austenite is verified by a simple weight loss experiment.

ACKNOWLEDGEMENTS

This work was financially supported by Avesta AB, Fagersta Stainless, Sandvik AB which is gratefully acknowledged. The materials were supplied by Sandvik AB. Staffan Hertzman and Bevis Hutchinson are acknowledged for discussions, Håkan Thoors for assistance with SEM and Stephen Preston for reading manuscript.

REFERENCES

1. SOLOMON, H.D., DEVINE, T.M., Duplex Stainless Steels. Conference Proceedings, Ed., R.A. Lula ASM 1983.

2. GROENENWOUD, K., U.K. National Corrosion Conference 1982.

3. BLANCHARD, GUNTZ, JOLLAIN, MUGGEO, VUILLAUME, Métaux Corrosion Industrie, 60:ème année, vol LIX no 711-12, nov-dec 1984.

4. EDELENAU, C., Journal of the Iron and Steel Institute, April 1957, p. 482.

5. EDELENAU, C., Journal of the Iron and Steel Institute, Feb 1958, p. 122.

6. CIHAL, V., PRAZAK, M., Journal of the Iron and Steel Institute, Dec 1959

7. HERBSLEB, G., PÖPPERLING, R.K., Corrosion 36:11 1980, 611.

8. YUNG-HERNG YAU, STREICHER, M.A., Conference Proceedings, Corrosion '85, p. 228.

9. SHREIR, Corrosion vol. 1, Newnes-Butterworths 1976.

10. MANSFELD, Conference Proceedings Corrosion (Houston) 1972, 27, 436.

Table 1 The compositions of the three alloys used are given in the columns 1, 2 and 4. Column 3 and 5 show the compositions of the corresponding phases in SS 2377

	1	2	3	4	5
	SS 2377	ferrite		austenite	
C	0.013	0.015	(0.002)	0.024	(0.02)
Si	0.43	0.35	(0.40)	0.36	(0.37)
Mn	1.54	1.30	(1.40)	1.63	(1.80)
P	0.009	0.011			
S	<0.003	<0.003			
Cr	21.90	24.52	(24.30)	20.20	(20.50)
Ni	5.54	4.15	(4.10)	8.10	(7.10)
Mo	3.04	3.77	(3.80)	2.2	(2.2)
N	0.14	0.029	(0.046)	0.22	(0.24)

Table 2 The effects of different solutions on the corrosion of SS 2377 at 22°C

H_2SO_4 (N)	HCl (N)	NaCl (N)	Selective corrosion Yes	No	Uncertain	Remarks
0.50	0.50	-			X	No active peak. Crevise corr.
1.00	-	-			X	"
2.00	2.10	-	X			
4.00	-	-			X	No active peak
-	1.00	-			X	Crevice corr.
-	2.00	-			X	Crevice corr.
-	0.50	4.50			X	Crevice corr.
-	0.50	4.10			X	Crevice corr.
-	0.10	4.10			X	Crevice corr.
-	6.00	-			X	No repassivation after the active range
-	1.00	0.10			X	
-	2.00	0.10			X	
-	4.00	0.10			X	
4.00	0.10	-			X	
4.00	0.20	-	X			
4.00	1.00	-	X			
4.00	1.50	-	X			
4.00	2.00	-	X			

Table 3 Weight loss expressed in mg/cm^2 during 20 h in 4N H_2SO_4 + 0.1 N HCl

Specimens isolated		Specimens in electrical contact
Ferrite	0.00	6.20
Austenite	3.55	0.80
Duplex	9.35	

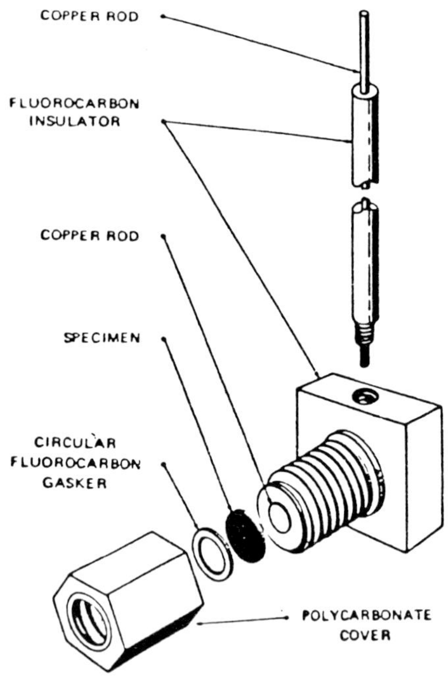

Fig. 1 Holder used for the weight loss measurements.

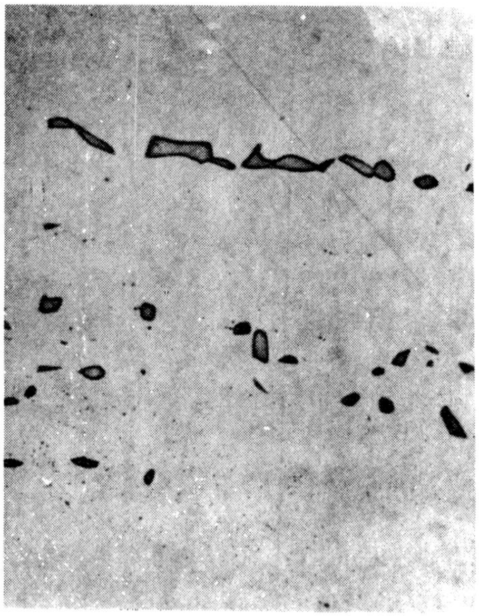

Fig. 2 The amount of ferritic phase in the austenitic steel was low i.e. ~0.6%. Sample etched in Murakami Reagent.

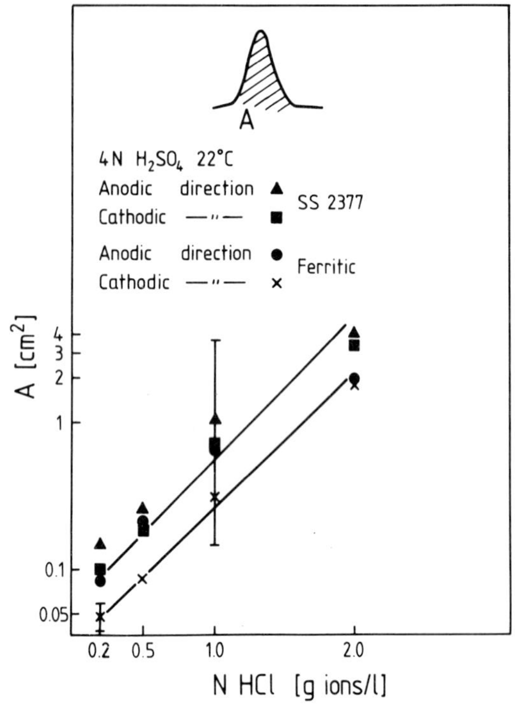

Fig. 3 Area under the polarisation curve in the active range as a function of the HCl concentration. The measurements were carried out in the anodic and cathodic direction.

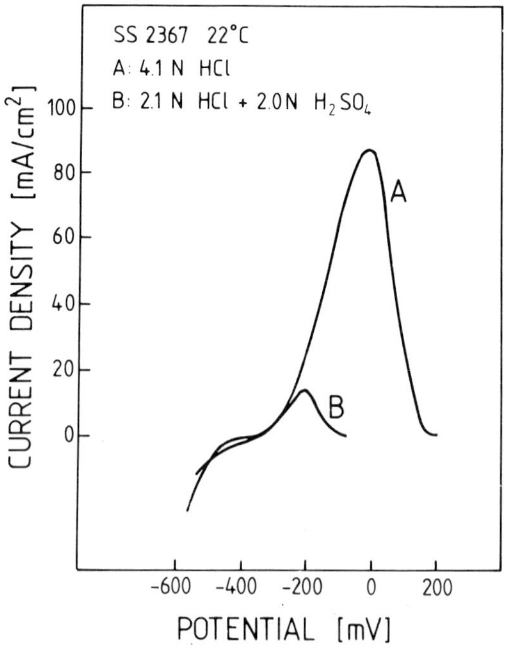

Fig 4. A comparison between two active peaks which correspond to the same $H^+{}_3O$ concentration (4,1N) and different Cl^- concentrations

Cl^-concentrations

A → 4,1 N [Cl-]

B → 2,1 N [Cl^-]

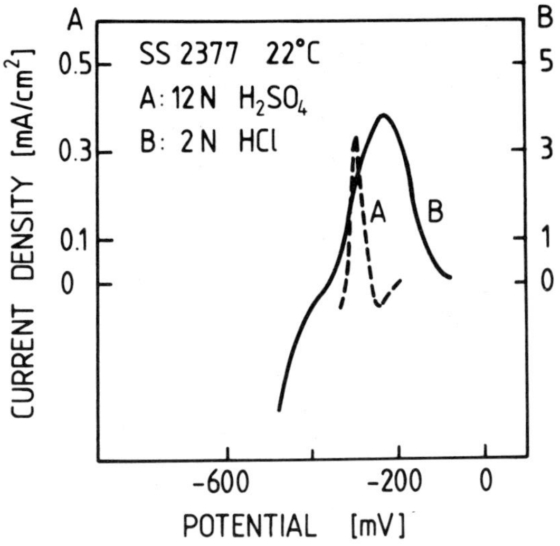

Fig. 5 A comparison between the active peaks taken in A → 12N H_2SO_4 and B → 2N HCl.

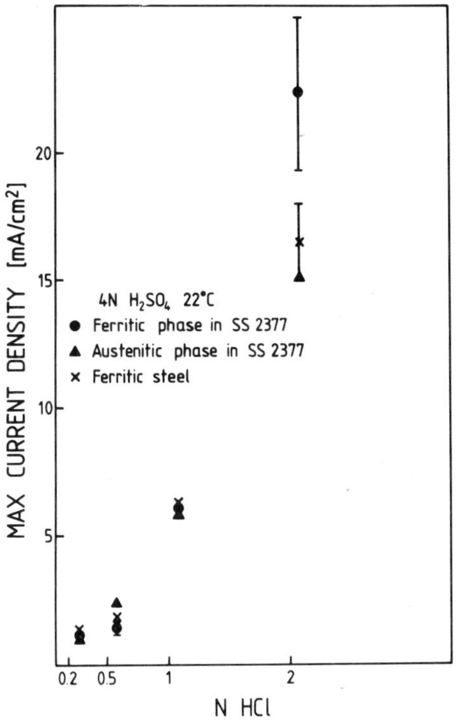

Fig. 6 The height of the active peaks as a function of the concentration. Sweeps were made towards the cathodic direction.

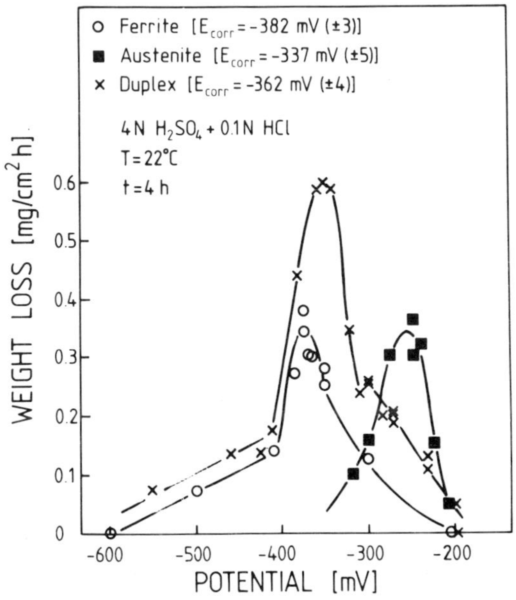

Fig. 7 Potentiostatic weight loss measurements.

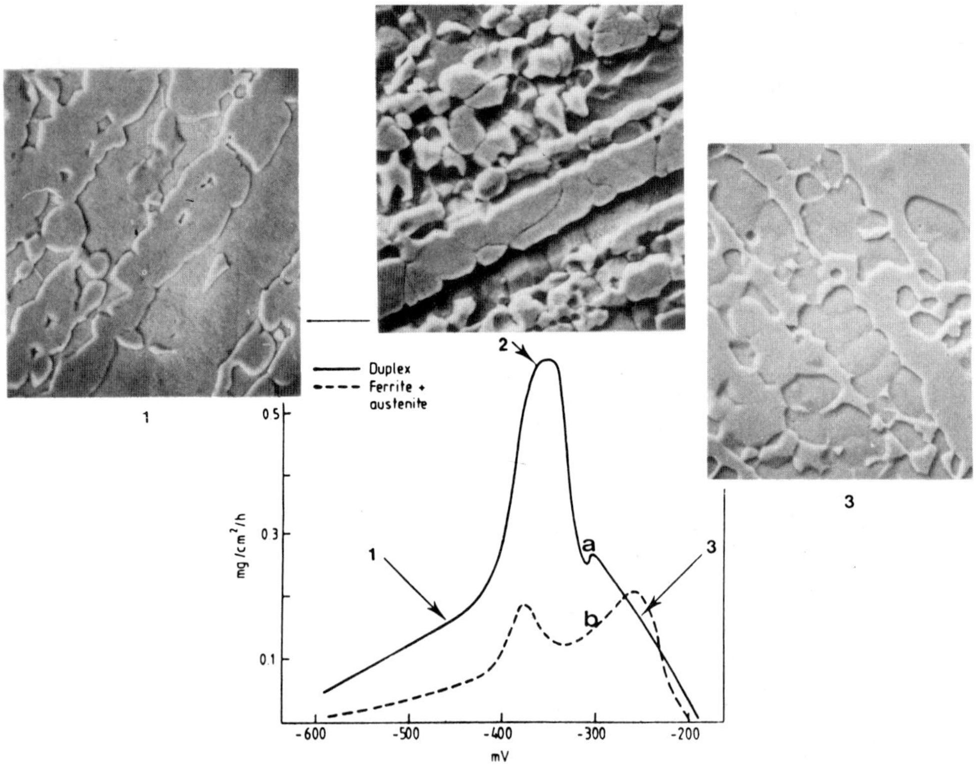

Fig. 8 (a) Total weight loss for both phases in the duplex steel.
(b) Sum of the two weight loss curves for ferrite and austenite according to Fig. 7.

SEM pictures for three typical experimental points (X937.5)

(1) - 460 mV Preferential attack of the ferrite

(2) - 362 mV Corrosion potential. Both phases attacked but mainly the ferrite. The original A^c/A^c is changed

(3) - 250 mV Preferential attack of austenite

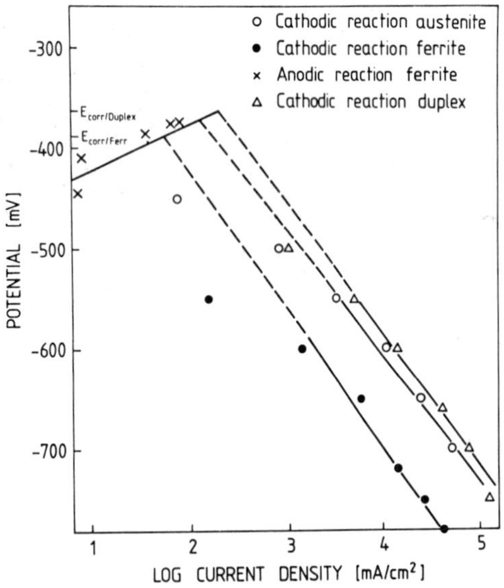

Fig. 9 Anodic and cathodic lines for the evaluation of Tafel slopes and i_o values

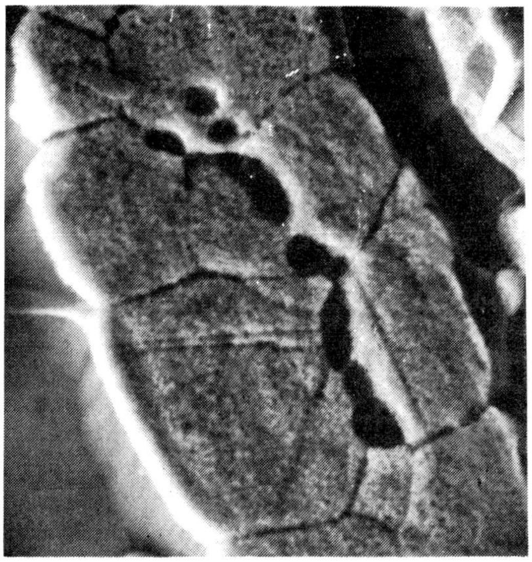

a

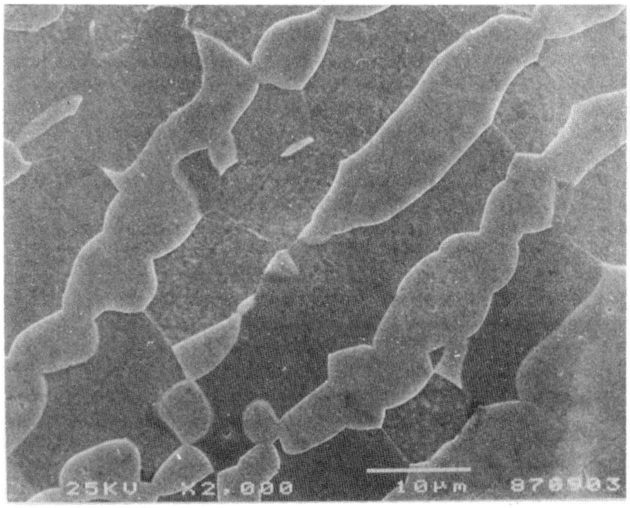

b

Fig. 10 Specimens at -362 mV
A Dissolution time 4h
B Dissolution time 0.5 h

Influence of flow rate on corrosion behaviour of stainless steel

S Lagerberg

The author is in the Research and Development Centre, AB Sandvik Steel, Sandviken, Sweden.

A common way of presenting the corrosion properties of a material is in the form of an isocorrosion diagram, which is a graphic representation of the effect of temperature and concentration on corrosion rate in, for instance an acid. The curve for a corrosion rate of 0.1 mm/year is commonly given and sometimes that for 0.3 mm/year, figure 1.

The solution used when plotting the graph is usually made up of pure chemicals, contrary to what can be expected in real life. Process solutions, as we know, usually contain impurities of varying kinds and quantities. Moreover, in laboratory tests the flow rate is usually very low (only thermal motion and gentle stirring), so the diagram can be said to reflect the corrosion behaviour under stagnant conditions.

All corrosion processes involve an interaction between material and solution. Besides the composition of the solution, the motion between the fluid and the metal surface will influence the corrosion behaviour in a positive or negative direction (1), depending on how the sub-reactions taking part in the process are influenced. For a corrosion process to take place there must be an anodic reaction which emits electrons and a cathodic reaction which consumes electrons.

From a practical point of view it is therefore very important to know how the flow rate influences the corrosion behaviour of a material, as many surfaces in an industrial plant are exposed to process solutions at varying and sometimes very high flow rates. In cases like that the isocorrosion diagram is of limited interest.

In this investigation an account is given of the effect of flow rate on the corrosion behaviour of two stainless steels, AISI 316 and Sanicro 28 (27Cr31Ni3.5Mo) in sulphuric acid at different temperatures and in concentrations ranging from 5 to 60%, with and without oxygen, which can take part in the cathodic process.

At the interface between material and solution laminar or turbulent flow prevails.

At low flow rates the flow will be laminar and run parallel to the material wall. This creates a diffusion layer through which reactive molecules or ions have to diffuse in order to be able to react with the surface of the material. The process will thus be mass transfer controlled (1).

Higher flow rates cause turbulent flows where the turnover of solution at the material surface can be very rapid. This results in a very thin interface layer where the distance to diffuse for reactive components in the solution will be very small. This increases the availability and positive or negative effects will take place more quickly.

The effect of flow rate on the corrosion behaviour can be studied with the aid of several different techniques. In this investigation two were used. One was CTD (Critical Temperature Determination), in stagnant solution and at varying flow rates. The other method was determination of weight loss under the same conditions.

2. ELECTROCHEMICAL MEASURING METHODS

In the following an account will be given of the electrochemical testing method that has been developed and used at AB Sandvik Steel (2), in order to study general corrosion properties of a steel in stagnant solution. In this investigation the method has been extended to cover properties at various flow rates.

<u>Critical Temperature Determination (CTD)</u>

The purpose of this method is to determine the highest temperature at which passivation is possible, for various steels, concentrations and flow rates of a particular solution.

CTD is an electro-chemical method which is suitable for measurement in environments containing reducible species. The species must have an $E_{(redox)} > E_{(corr)}$ for the material in the solution in question.

The method is based on the fact that the potential often provides a definite indication as to whether the material is active or passive, see figure 2.

At an active potential the corrosion rate is high, at a passive potential it is low.

The Critical Activation Temperature (CAT) for stainless steel can be determined by measuring how the potential stabilizes above or below a predetermined potential after activation at a low potential, when testing at constant temperature for a period not longer than 4 h, see figure 3.

In open-circuit measurements there is no applied potential. Instead, the variation of the free potential is studied as a function of time, measured at constant temperature.

$E_{(pp)}$, taken from a polarization diagram, + 100 mV is used as a suitable test criterion for determination of CAT for a metal or alloy, figure 2.

Here, $E_{(pp)}$+100 mV is defined as the potential which the steel has to exceed to be regarded as passivated, at the testing temperature in question.

CAT is defined as, and determined by, the temperature at which the metal or alloy is no longer capable of repassivating. It is known that the passivation temperature is affected by the presence of reducible species, in this case air (oxygen), and by the flow rate.

The measurements carried out at AB Sandvik Steel R&D Centre have shown that the method has a very good reproducibility. The current test method will be described later on (5. Test Procedure).

3. WEIGHT LOSS MEASUREMENTS

In corrosive environments where reducible species are not present it is not possible to use the CTD method. Then the influence of flow rate on corrosion behaviour has to be determined by other means. In this investigation weight loss measurements were used.

The flow rate will affect the corrosion behaviour of the material in two ways, the transportation of species to and corrosion products away from the surface. By measuring the weight loss of the material in activated condition in stagnant solution and comparing with results from tests with rotating specimen it is possible to study the effect of flow rate.

4. EXPERIMENTAL

Material

The tests were carried out on two steel grades, both austenitic. One was AISI 316 and the other a higher alloyed grade, Sanicro 28. Both materials were in the quench-annealed condition, and before the tests the specimen surfaces were ground with 600 grit paper.

Table 1 Chemical composition

Grade	C	Si	Mn	P	S	Cr	Ni	Mo	Cu	N
AISI 316	.043	.63	1.73	.019	.006	17.49	13.36	2.68	.22	.047
Sanicro 28	.012	.13	.16	.012	.003	26.80	30.70	3.75	1.10	.037

Instrument

The electrochemical measurements were performed with the aid of SANTRON EMS, a measurement system developed at AB Sandvik Electronics. This system offers possibilities to control measuring routines and to measure, store the results on floppy discs, process data and plot established relationships with the aid of a plotter. The instrument is equipped with pre-programmed measuring routines as follows

Code	Designation
CTD	Critical Temperature Determination
CPT/CCT	Critical Pitting/Crevice Temperature
CAM	Chronoamperometry
PRT	Polarization Resistance Technique
TST	Tafel Scan Technique
ANP	Anodic Polarization Technique
CAP	Cathodic Polarization Technique
PDS	Potentiodynamic Scan Technique
CYP	Cyclic Polarization Technique
EPR	Electrochemical Potentiodynamic Reactivation (Stainless Steel Sensitization)

Each instrument is equipped with three measuring channels as well as with speed regulation for rotating electrodes, and possibilities for heat control and control of magnet valves.

Test solution

The measurements were performed in an aerated sulphuric acid solution with concentrations in the range 5-60%. Constant bubbling of air was employed for the CAT determinations and nitrogen for the weight loss tests.

Solubility of oxygen in test solution

The solubility of oxygen in the solution varies with temperature and concentration of sulphuric acid, figure 4 (3). By continuous gas bubbling through the solution during the measuring cycle the gas solubility is kept constant at each temperature.

Test cell

The design of the test cell appears in figure 5. Besides reference electrode, counter electrode and test specimen with rotation equipment applied, the test cell is also equipped with immersion heater and temperature sensor for heat control, ceramic distributor for gas bubbling through the test solution, cooling coil and reflux cooler. The volume of the test container is 900 ml.

Flow rates

The flow rates were simulated by rotating a cylindrical specimen in the solution. The rotation speed is variable in the interval 20 to 5000 r/min and can be controlled from the instrument.

The tests in this investigation were performed at 0, 100 and 400 r/min.

The important thing in this investigation is not so much to know the exact flow rate as to determine if there is a laminar or a turbulent flow. *

Specimen

The specimen is cylindrical with 10 mm diameter. This design was chosen in order to get as uniform a flow rate as possible and a simplified surface treatment.

Specimen holder

A problem in electrochemical determination is the risk of crevice corrosion attack which interferes with the measurements. The specimen holder has therefore been given a special design, figure 5. Its lower part is formed as a collar of teflon, which will be filled by the purging gas, thus reducing the risk of crevice corrosion.

In addition to this, to avoid bad electrical contact when testing rotating specimens there are three connectors in the upper part of the specimen holder.

5. TEST PROCEDURE

The measuring procedures for CAT determination are identical in stagnant and flowing conditions.

Surface treatment

Before CAT determination, the specimen surfaces are ground with 600 grit paper and degreased. The grit value used is based on practical experience and gives the best reproducibility.

Determination of CAT

Although simple in themselves, potential measurements normally require a great deal of manual work.

However, with the SANTRON EMS instrument, measurements can be facilitated considerably. The procedure is illustrated in figure 6.

The following parameters are preset

- initial temperature
- activating potential
- activating time
- maximum testing time at each temperature
- potential above which the material is considered passive
- speed control

The test specimen is fixed vertically into the specimen holder in the test cell and exposed to the test solution.

During measurements the temperature is thermostatically controlled by SANTRON EMS. When the solution has reached the preset temperature T_i, the potentiostatic function is switched on and the specimen is activated at the preset potential $E_{(act)}$.

By first activating the material at a low potential, $E_{(act)}$ = -1000 mV(Hg, Hg_2SO_4) for 60 seconds and then measuring the change in potential in an open-circuit connection at constant temperature we find that the potential finally stabilizes at a level above or below a pre-determined potential $E_{(pp)}$ + 100 mV, where $E_{(pp)}$ is the passivation potential of the material, determined by potentiodynamic plotting of current-potential graphs.

The potential is measured for the preset maximum time (4h). If the potential of the specimen exceeds the level for passivation during this period, the temperature is automatically increased by 5°C, and the cycle is repeated.

This measurement sequence will continue at different temperature levels until we finally arrive at a temperature where the material is no longer capable of passivating within the maximum measuring period. This happens when the equilibrium potential of the material does not exceed the defined passivation potential after a certain given period of time at the temperature in question.

The measurement is then terminated and the maximum temperature at which passivation occurred can be read off from the front of the instrument. This temperature is defined as the Critical Activation Temperature (CAT) of the material.

Finally, all measuring results as well as the parameters used are stored on a floppy disc. The data can then be printed out on a suitable plotter.

Once the parameters have been set, the only manual operation required is the one which starts the test procedure. This, in combination with the continuous sensing of the potential has substantially reduced the total time and work needed for the measuring process.

Weight loss measurements

The weight loss measurements were carried out in stagnant solution and at a flow rate equal to 400 r/min. The same type of specimen set up was used as in the CTD test. Upon immersion in the test solution the material was immediately activated by a zinc rod, so as to reproduce, to some extent, the conditions of CTD-tests. The testing temperature chosen was equal to the critical temperature obtained in the corresponding CTD test and the exposure time was 168h.

The test conditions were for AISI 316, 15% H_2SO_4 and +45°C; for Sanicro 28, 30% H_2SO_4 and +80°C, in stagnant solution and at a flow rate of 400 r/min.

The corrosion rate was then calculated by converting the measured weight loss to mm/year.

6. RESULTS

In order to obtain a reliable result, at least three measurements were performed under each test condition. The maximum difference between the highest and lowest measured value is not >5°C.

There are two competing phenomena which to some extent complicate the evaluation. One is the decrease of oxygen content at increasing temperature, the other the increasing concentration of the main species in the solution.

CTD (aerated solution)

Table 2 Results from CAT determination for AISI 316

H_2SO_4	Isocorr	Stagnant		100r/min		400 r/min	
%	°C	°C	ΔT	°C	ΔT	°C	ΔT
5	60	85	30	90	30	95	35
10	45	60	15	60	15	85	40
15	33	45	12	50	17	60	27

ΔT = difference in °C compared with isocorrosion test

Figure 7 shows that the presence of oxygen in stagnant solution increases the CAT 10-30°C above what can be expected from the isocorrosion diagram. At 100 r/min CAT increases 15-35°C and at 400 r/min 25-35°C in a concentration interval of 5-15% H_2SO_4.

Table 3 Result from CAT determination for Sanicro 28

H_2SO_4 %	Isocorr °C	Stagnant °C	ΔT	100r/min °C	ΔT	400 r/min °C	ΔT
10	90	90	0	-	-	-	-
20	60	90	30	95	35	95	35
30	55	80	25	90	35	90	35
40	60	75	15	75	15	70	10
60	65	90	25	90	25	85	20

Figure 8 shows that the presence of oxygen increases the CAT by 0-25°C in stagnant solution and by 20-35°C at flow rates equal to 100 and 400 r/min, compared with what can be expected from the isocorrosion diagram at 10-60% H_2SO_4.

The test results imply that the presence of oxygen (air) in the solution has a greater effect on the practical service temperature for Sanicro 28 than for AISI 316. An increase of up to 35°C can be observed for Sanicro 28.

AISI 316 shows a slightly smaller increase in the same temperature.

Weight loss measurements

Table 4 Weight loss in mm/year

Steel	H_2SO_4	°C	Stagnant	400 r/min
AISI 316	15%	45	0.10 mm/year	0.82 mm/year
Sanicro 28	30%	80	0.25 mm/year	1.30 mm/year

As appears from the test results, there is a considerable difference in corrosion rate as compared with the isocorrosion diagram even at fairly low flow rates. It is also obvious that the relatively low alloyed material AISI 316 is more sensitive to flow rates than the higher alloyed Sanicro 28.

7. DISCUSSION

The equilibrium potential of a material in a solution often provides an indication as to whether the material is in the passive state or not (4). At passive potentials the corrosion rate (log i) is low, and at active potentials it is high, figure 9.

The drop in potential observed at CAT is taken as a sign of what is described in figure 9. As long as the material repassivates it strives to reach a potential equal to the one valid for the actual oxygen content in point "B". When the steel no longer can repassivate, because the system has reached a very low concentration of oxygen, which is dependent on the concentration of the electrolyte and temperature, it is suggested that it drops to point "A" in the figure. The potential at point "A" is the equilibrium potential for the cathodic process = reduction of O_2, and for the anodic dissolution of the material.

The following reactions are suggested

"A" (active conditions)

Cathodic reaction

$$O_2\,(g) + 4\,H^+ + 4e^- \rightarrow 2\,H_2O \quad \text{or}$$
$$[2\,H^+ + 2\,e^- \rightarrow H_2(g)]$$

Anodic reaction

$$2\,Me \rightarrow 2\,Me^{2+} + 4\,e^-$$

$$2\,Me + O_2(g) + 4\,H^+ \rightarrow 2\,Me^{2+} + 2\,H_2O \quad \text{or}$$
$$[Me + 2\,H^+ \rightarrow Me^{2+} + H_2(g)]$$

"B" (passive conditions)

Cathodic reaction

$$O_2\,(g) + 4\,H^+ + 4e^- \rightarrow 2\,H_2O$$

Anodic reaction

$$2\,Me + 2\,H_2O \rightarrow 2\,MeO + 4\,H^+ + 4e^-$$

$$2\,Me + O_2\,(g) \rightarrow 2\,MeO$$

8. CONCLUSION

The investigation shows that the flow rate has a positive influence on the corrosion properties of a material in environments containing reducible species, such as oxygen, which control the cathodic reaction (at anodic dissolution and passivation). An increase in the practical service temperature in the order of 30-40°C has been measured for the alloys AISI 316 and Sanicro 28.

In environments that do not contain reducible species, a negative effect of flow rate on the corrosion behaviour is observed. This is due to an increased transportation of corrosion products away from the material surface. The exact effect on the practical service temperature is not possible to tell, but the results from the weight loss measurements indicate that the temperature will be lower as compared to what can be understood from the isocorrosion diagram.

REFERENCES

1. T. Sydberger, Br. Corrosion J., 1987, Vol. 22, No. 2, p. 83
2. S. Bernhardsson, J. Degerbäck, Extended abstracts, The 8th Scandinavian Corrosion Conference, Helsinki, August 1978, p. 31
3. G. Berglund, Ch. Mårtenson, Corrosion 87, Paper # 21, San Francisco
4. S. Bernhardsson, P. Lau, 9th Conference on Metallic Corrosion, Toronto, Canada, June 1984

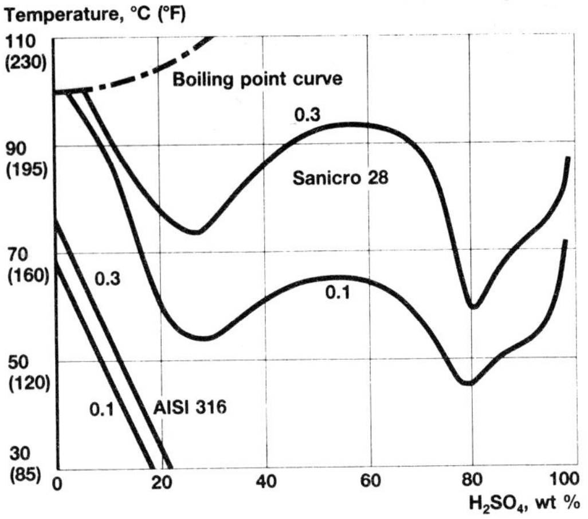

Fig. 1 Iso-corrosion diagram for Sanicro 28 in sulphuric acid (0.1/0.3 mm/year)

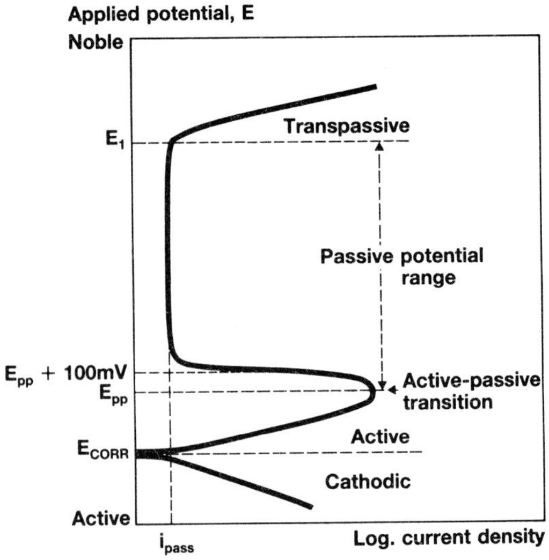

Fig. 2 A schematic polarization curve for a stainless steel in a sulphuric acid solution

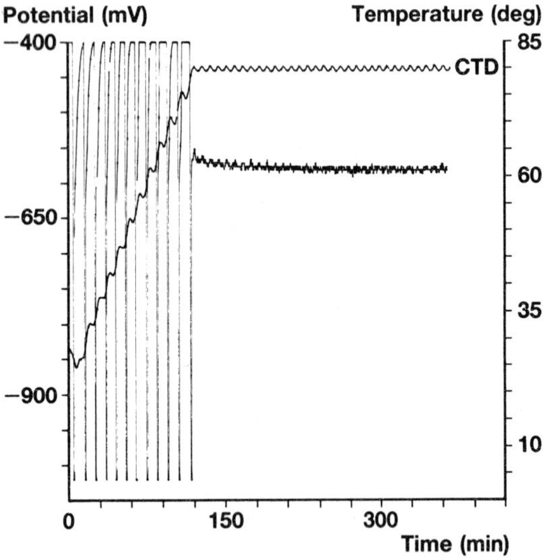

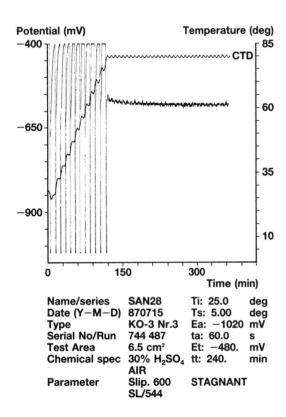

Name/series	SAN28	Ti: 25.0	deg
Date (Y−M−D)	870715	Ts: 5.00	deg
Type	KO-3 Nr.3	Ea: −1020	mV
Serial No/Run	744 487	ta: 60.0	s
Test Area	6.5 cm^2	Et: −480.	mV
Chemical spec	30% H_2SO_4 AIR	tt: 240.	min
Parameter	Slip. 600 SL/544	STAGNANT	

Fig. 3 CTD-diagram

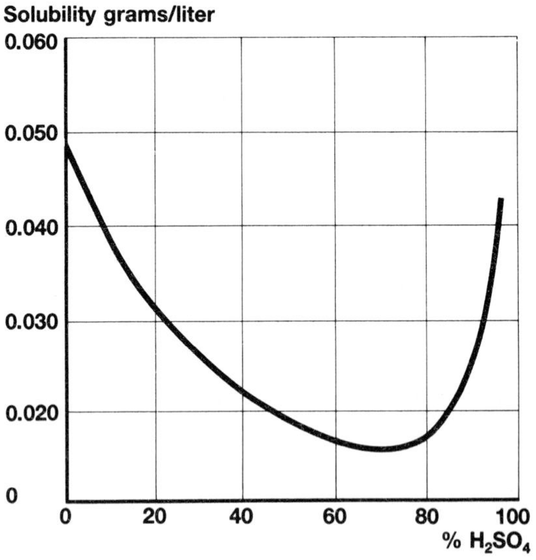

Fig. 4 Solubility of oxygen in sulphuric acid at 15.5°C and atmospheric pressure

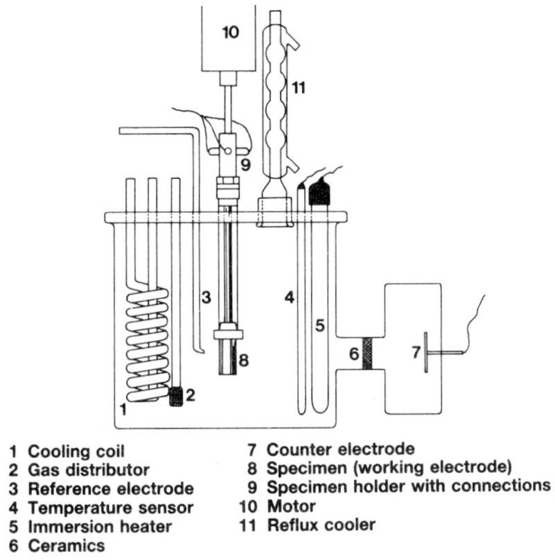

Fig. 5 Test cell for CTD-tests

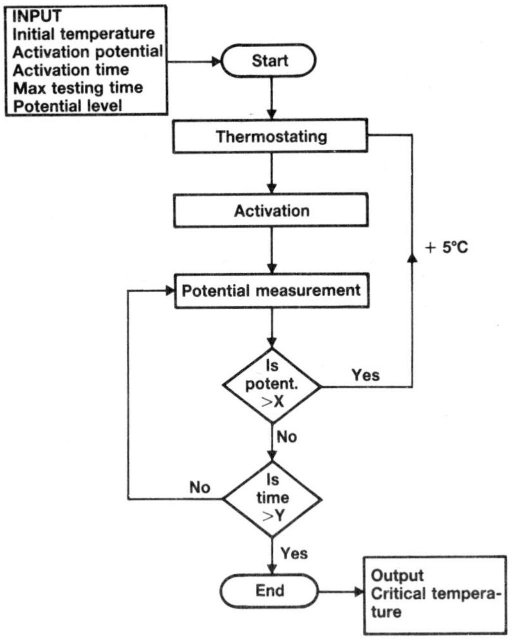

Fig. 6 Flowchart CTD-method

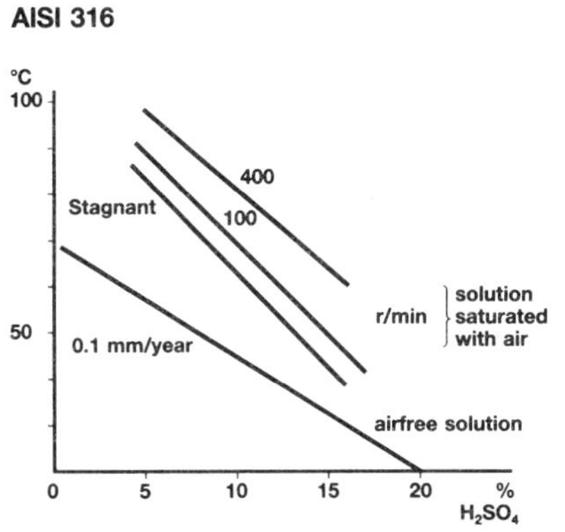

Fig. 7 Effect on CTD for AISI 316 at various flow rates and concentrations of sulphuric acid

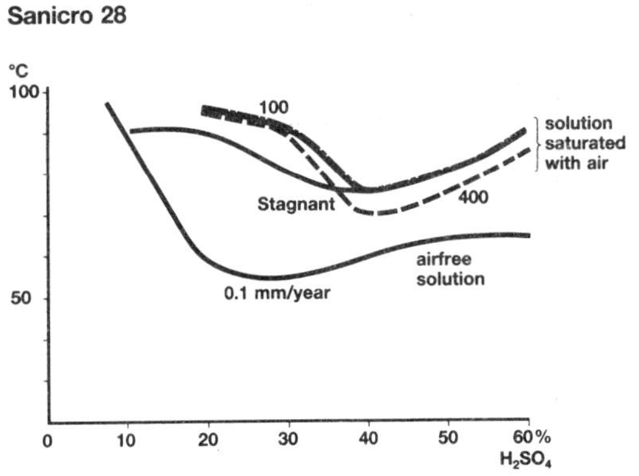

Fig. 8 Effect on CTD for Sanicro 28 at various flow rates and concentrations of sulphuric acid

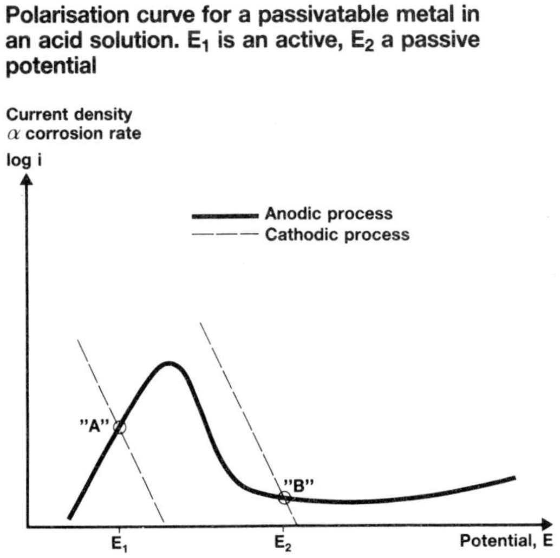

Fig 9 Polarisation curve for a passivatable metal in an acid solution. E_1 is an active, E_2 a passive potential

□

Effects of chlorine on corrosion of high alloy stainless steel in seawater

R Francis

The author is with BNF Metals Technology Centre, Wantage, Oxfordshire, England.

* * * * *

SYNOPSIS

High alloy stainless steels are being considered for sea water piping for the new generation of offshore platforms.

For stainless steels to perform satisfactorily chlorination will be necessary to prevent fouling, and corrosion problems associated with H_2S, produced when the organic material in sea water decays under stagnant conditions. There is little data on the performance of high alloy stainless steels in sea water dosed with chlorine and BNF has been carrying out a preliminary study of some of the alloys being offered for offshore service for a major oil company. BNF has tested two well known duplex alloys and two high molybdenum-containing austentic alloys, in addition to standard alloys for comparison purposes. Panels of each material with INCO-type crevice washers were exposed to once-through sea water with different concentrations of chlorine for up to 3 months. The results showed that the high molybdenum austenitic alloys suffered no crevice corrosion at chlorine concentrations up to 5mg/l. A little crevice corrosion occurred with the duplex alloys but the occurrence of attack was erratic and did not correlate with the chlorine concentration. Further tests investigated the effect of sea water temperature at a chlorine concentration of 1mg/l. At 15°C little or no crevice corrosion was observed, on all the alloys while at 40°C all the alloys suffered some attack .

1. INTRODUCTION

Calculations demonstrate that for every tonne of weight saved on the topside of a long-life offshore platform, £100,000 could be saved on structural steel below the water line. Much of the weight on the topside of a platform derives from the large volumes of sea water and modified sea water in the various piping systems, and this could be reduced by decreasing the pipe diameter. For instance, halving the pipe diameter would reduce the weight of water to 25%. This would also mean that the flow velocity must be increased by a factor of four to maintain the same bulk water flow. Most of the sea water piping systems currently used on offshore platforms are made of 90/10 copper-nickel, which has a design velocity limit of $\sim$3 m/sec, and it is commonly used with flows of 2.0 to 2.5 m/sec.

In recent years a large number of proprietary, high alloy stainless steels have been developed which the manufacturers claim are suitable for sea water service. To prevent fouling and problems due to hydrogen sulphide (produced when the organic material in sea water decays under stagnant conditions), it will be necessary to inject the biocide chlorine into the sea water. There is little or no data on the performance of the new alloys in sea water containing chlorine, particularly at the concentrations likely to be used on offshore platforms. BNF was asked to carry out preliminary tests to identify those conditions under which problems might occur. There are three applications where different kinds of corrosion problems might occur:

1) Pipework.
2) Heat exchangers.
3) Pumps.

This report presents the results of tests to examine the suitability of some high alloy stainless steels for pipework.

2. EXPERIMENTAL

It is well known that stainless steels are highly resistant to flowing sea water at velocities well in excess of 10 m/sec, but concern was expressed that chlorine might increase the susceptibility of these steels to crevice corrosion. The INCO type of crevice jig[1] was chosen as being capable of creating the type of crevice likely to occur at flanged joints in real piping systems. The crevice assemblies were manufactured from polyacetal, and were held in place with insulated stainless steel bolts. Two tests were then set up, one to explore the effect of different chlorine concentrations and the other to examine the effect of sea water temperature.

2.1. Effect of Chlorine Concentration

These tests were set up at BNF's once-through sea water facility at Portland Harbour. A typical sea water composition is shown in Table 1. Three tanks (capacity ~140 l) were constantly fed with fresh sea water at about 30 l/min. Chlorine was generated electrolytically in a by-pass loop and then fed back into two of the tank inlet lines to give the desired chlorine levels. One tank was operated with fresh sea water only as a control. A second tank was dosed to maintain a chlorine residual of 1.0 to 1.5 mg/l which is typical of the concentrations currently used by offshore platforms. The third tank was dosed to maintain a chlorine residual of 4.0 to 5.0 mg/l to represent overdosing, a not uncommon occurrence in real systems. The chlorine level was monitored at intervals using the D.P.D. test[2].

The specimens were degreased prior to testing and the crevice jig was then assembled under water to ensure wetting in the crevice. The assemblies were tightened to a torque of 7Nm and fastened to polypropylene racks which were immersed in each tank.

After one month some specimens were removed from each tank and replaced with fresh samples. The tests were terminated after three months. The specimens were then washed and dried and examined for attack in the crevices, and the depth of any attack was measured.

The sea water temperature was not controlled during the tests, and the average temperature was 16.3°C with absolute variations of +0.4° and -1.3°.

2.2. Effect of Temperature

Although sea water temperatures in the North Sea typically vary from 5° to 15°C, the sea water discharged from some heat exchangers could have a temperature as high as 40°C. To investigate the effects of temperature on crevice corrosion tests were carried out at 15°C and 40°C. Because of the difficulties in controlling the temperature of large volumes of once-through sea water, these tests were carried out in temperature controlled tanks with recirculated seawater obtained from Portland Harbour. The tanks contained ~100 l sea water and half the contents of each tank were changed three times a week. The sea water temperature was controlled at 15 ± 0.1°C or 40 ± 0.2°C. Four separate tests were carried out:-

(1) Natural sea water at 15°C.
(2) Sea water + 1 mg/l chlorine at 15°C.
(3) Sea water + 1 mg/l chlorine at 40°C.
(4) Natural sea water at 40°C.

The chlorine was added as sodium hypochlorite solution by a small peristaltic pump, and the chlorine level was monitored and controlled by a modified Portacel A62 Chlorine Annunciator.

The creviced specimens were attached to acrylic frames in each tank and exposed for 40 days. After exposure the specimens were cleaned and examined in the same manner as the ones from the once-through sea water tests.

3. MATERIALS

Stainless steels suitable for use offshore in sea water handling systems fall into three categories:-

1) Ferritic.
2) Austenitic.
3) Duplex.

Ferritic alloys are not readily available for sea water piping, and so no alloys of this kind were included in the present tests. There are two types of commonly available high alloy austenitic materials; one containing 4.5% molybdenum and the other containing 6% molybdenum. It is already known that pitting can occur with the 4.5% molybdenum alloy under severe crevice conditions, and so only the 6% Mo alloy was considered. BNF tested two alloys of this kind, one containing 18% nickel and the other 25% nickel, as shown in Table 2.

There is a large number of proprietary duplex alloys available, but the majority have one of two common composition types:

1) 22Cr/5Ni/3Mo/0.2N.
2) 25Cr/5Ni/3Mo/0.2N.

Samples of both these types were included in the present tests, as shown in Table 2.

In addition to these four alloys it was felt desirable to test two alloys of known performance for comparison purposes. The ones chosen were 316, an alloy known to suffer crevice corrosion very easily in sea water, and BS3072; NA21, commonly known as alloy 625, a nickel-based alloy known to be highly resistant to crevice corrosion. The composition of these two materials is also shown in Table 2.

All the materials were tested in sheet form in thickness varying from 2 to 6mm. The test panels were 100mm by 100mm and each material was tested in duplicate, or in triplicate where space was available. Because of space limitations only 316, Ferralium and 254 SMO were exposed in the laboratory tests to investigate the effects of sea water temperature.

The crevice created by flanged joints appeared to be the most likely region for localised attack, and so a surface roughness representative of the finish on a flange face was desirable. Under critical operating conditions, such as high pressure service, a good surface finish is required for flange faces. A typical specification is a finish of 0.4 to 1.6 microns. The surface roughness of all the materials was measured by Talysurf. The roughness of the two duplex alloys as supplied was greater than 1.6μm and so the panels of these materials were abraded with 240 grit silicon carbide. The surface roughness of all six materials is shown in Table 3. The value is the average of 24 values; 12 on each side, six with and six across the rolling directions, on a typical panel.

It can be seen that all the alloys except Ferralium and 316 fall within the surface finish limits. Ferralium is only just below the 0.4 micron lower limit, which for the present tests was considered acceptable. The much lower roughness of the 316 means that the crevices would be tighter with this material, and hence crevice corrosion would be more likely.

4. RESULTS AND DISCUSSIONS

4.1. Effect of Chlorine Concentration

There was little difference between the specimens exposed for 1, 2 and 3 months, and the results after 3 months exposure are shown in Table 4. It can be seen that the alloys fall into three groups; deeply pitted, slightly attacked and unattacked.

316 stainless falls into the first group, and all except two of the specimens showed pitting at a number of crevice sites. The number of sites attacked and the depth of attack varied from specimen to specimen, but did not seem to be related to the chlorine concentration. This is not altogether surprising as this alloy is known to suffer crevice corrosion readily in sea water, and the variability is probably a result of variability in the tightness of the crevice. Oldfield[3] has pointed out the variability in crevice gap which is possible with the INCO-type crevice jig.

Both of the duplex alloys showed somewhat erratic behaviour in the present tests. Sometimes no attack at all occurred, while in others some rather shallow attack occurred, and in just a few instances some deeper attack occurred (>0.1mm) with the 22/5 alloy. The occurrence of pitting did not seem to be related to the level of chlorine present, which is a little surprising. While there was no discernible difference between the two alloys after one month, the Ferralium alloy did seem to be slightly superior as the time of exposure increased, as one would expect with an alloy containing more chromium.

The final group, of unattacked materials, included the nickel-based alloy 625, 254 SMO and 1925 hMo. Alloy 625 is well known to be resistant to crevice corrosion under very severe conditions, but the two austenitic alloys containing six percent molybdenum are largely untested. Both of these alloys suffered no attack under the conditions of the present test even with 4 to 5 mg/l chlorine in the sea water. Some very shallow superficial attack was observed on one specimen of alloy 1925 hMo but as it was only an isolated occurrence it may not be significant.

4.2. Electrochemical Measurements

In order to obtain more information on the effect of chlorine on the stainless steels the potentials of some specimens were monitored during the once-through tests. Chlorine had no discernible effect on the potential of 316, and potentials were in the range 0 to -200mV Ag/AgCl throughout the test. Both the duplex and high molybdenum austenitic alloys showed similar behaviour to each other. In natural sea water the potential moved electropositively rather slowly, finally stabilizing at about +200 to +300mV Ag/AgCl after a week or two. In the presence of chlorine the potential moved positively much more rapidly and stabilized at about +500 to +650mV Ag/AgCl. Some typical potential vs. time curves are shown in Figure 1. The potentials of some of the specimens were recorded after two and three months exposure and the results are shown in Table 5. All the alloy potentials show the same trend as the chlorine level was increased i.e. a much more electro-positive potential in sea water with 1mg/l chlorine compared to natural sea water, with a further, smaller increase in potential when exposed to sea water containing 4mg/l chlorine. However, the potentials attained varied from alloy to alloy.

The reason for the marked electro-positive change in potential in natural sea water is generally thought[4] to be due to the formation of a biofilm which catalyses the cathodic reaction i.e. the reduction of dissolved oxygen. The addition of small quantities of chlorine would be expected to remove this biofilm and thus the potential should move electronegatively. This has been observed by Malpas et al[5] and requires no more than 0.25mg/l chlorine. As the chlorine level is further increased an alternative cathodic reaction is provided, the reduction of the chlorine/bromine to the chloride/bromide ion.

i.e. $OX^- + 2e^- + H_2O \longrightarrow X^- + 2OH^-$

where X = Cl or Br.

This will result in an electropositive change in potential. The effect of increasing chlorine concentration on potential is shown schematically in Figure 2. The greatly increased potential in the presence of chlorine suggests that crevice corrosion would be more likely in chlorinated sea water than in natural sea water. The increase in potential in the presence of chlorine varied from alloy to alloy and the more resistant alloys such as 625 and 254 SMO showed smaller potential increases than less resistant alloys such as 22/5. The increase in potential for Ferralium was much less than that of 22/5 and yet the observed difference in corrosion resistance between these two alloys was small. Similarly the potential of 254 SMO was more electropositive than that of Ferralium in sea water containing 4mg/l chlorine, and yet 254 SMO suffered no attack while Ferralium did. The reason for the potential variation from alloy to alloy and its correlation with corrosion resistance is not known at this time.

4.3. The Effect of Temperature

The depth of attack on specimens from the laboratory tests is shown in Table 6. The results at 15°C both with and without chlorine are similar to those from the once-through sea water tests for a similar exposure, demonstrating the validity of the recirculating technique. The results show that attack due to crevice corrosion was more severe at 40°C compared with 15°C, both in natural sea water and in sea water containing 1mg/l chlorine.

The increase in the severity of attack with temperature was most noticeable on 254 SMO which suffered no attack at 15°C but was more severely attacked than Ferrallium at 40°C in sea water containing 1mg/l chlorine.

The presence of 1mg/l chlorine generally increased the severity of attack compared to natural sea water at the same temperature, particularly for Ferralium at 15°C and 254 SMO at 40°C.

This suggests that at any temperature there is a chlorine concentration below which crevice corrosion is unlikely to occur, for a particular crevice tightness and geometry. Thus it should be possible to determine experimentally for each alloy a curve of the type shown in Figure 3 for any particular crevice configuration. The results suggest that the shape of such curves for Ferralium and 254 SMO are different, and the determination of the curves has obvious benefits for engineers engaged in materials specification for particular applications.

5. CONCLUSIONS

The following conclusions were obtained for the crevice geometry and exposure times used in the present tests.

(1) 316 stainless steel pits in crevices in sea water at 16°C and is not noticeably affected by the chlorine concentration in the sea water.

(2) Both duplex alloys tested showed erratic susceptibility to pitting in the crevices at 16°C, which appeared to be independent of the chlorine concentration. The depth of attack was generally much less than on 316 stainless steel.

(3) Alloy 625, 254 SMO and 1925 hMo were immune to pitting in the crevices at 16°C at chlorine concentrations up to ∿ 5 mg/l.

(4) Ferralium, 316 stainless steel and 254 SMO all showed an increased susceptibility to pitting in crevices in sea water containing ∿1 mg/l chlorine when the temperature was increased from 15°C to 40°C.

REFERENCES

[1] Anderson, D.B. Galvanic and Pitting Corrosion ASTM STP 576. Feb. 1976, p.231.

[2] Palin, A.T. Water and Sewage Works 108 (1981) 461.

[3] Oldfield, J.W. 19th Journees des Acier Speciaux, Saint- Etienne, France. May 1980.

[4] Scotto, V., Cinto, R. DI. and Marenaro, G. Corr. Sci. 25 (1985) 185.

[5] Malpas, R.E., Gallagher, P. and Shone, E.B. Chlorination of Sea Water Systems and its Effect on Corrosion. Conference organised by Society of Chemical Industry, Materials Preservation Group, Birmingham. March 1986.

TABLE 1

A Typical Analysis of Portland Harbour Sea Water

(all figures are mg/l unless otherwise stated)

Chloride	19,890
Sulphate	2,400
Bromide	73
Bicarbonate Alkalinity (as $CaCO_3$)	120
Sodium	11,100
Calcium	413
Magnesium	1,210
pH	8.0-8.2
Dissolved Oxygen	90-95% saturated

TABLE 3

Surface Roughness of the Test Alloys
(average of 24 readings)

Alloy	Roughness (μm)		
	Average	Max.	Min.
316	0.032	0.064	0.013
625	0.83	1.11	0.56
22/5	0.57	0.76	0.38
Ferralium	0.26	0.33	0.19
254	0.73	0.91	0.58
1925	0.92	1.09	0.76

TABLE 2

Composition of the Alloys under Test

Alloy	Supplier	Composition (wt %)							
		Fe	Cr	Ni	Mo	Cu	N	Mn	Others
254 SMO	1	Bal	20.1	17.9	6.1	0.60	0.20	0.5	
1925/hMo	2	Bal	20.8	24.8	6.2	0.90	0.19	0.8	
22/5	3	Bal	21.6	5.7	3.0	0.10	0.16	1.3	
Ferralium	4	Bal	25.1	5.9	3.2	1.90	0.19	1.0	
316*	5	Bal	17	12	2.5	-	-	1.0	
625*	5	4	21	Bal	9	-	-	-	Nb; 3.15-4.15 Ti; 0.4 max. Al; 0.4 max.

Bal = Balance.
* Nominal composition only.
1) AVESTA AB, Sweden.
2) V.D.M, West Germany.
3) B.S.C. Special Steels, Sheffield, UK.
4) Langley Alloys, Slough, UK.
5) Standard alloy supplied by a local stockist.

TABLE 4

Depth of attack after 3 months in sea water at 16°C

Alloy	Chlorine Conc^{n} (mg/l)					
	0		1		4	
	No. of Sites	Max. Depth (mm)	No. of Sites	Max. Depth (mm)	No. of Sites	Max. Depth (mm)
316	9 1	0.66 1.01	12 8	0.40 0.26	24 5	0.46 0.40
625	0 0	0 0	0 0	0 0	0 0	0 0
22/5	0 1 0	0 0.22 0	0 0 1	0 0 0.14	0 0 0	0 0 0
Ferralium	0 0 0	0 0 0	1 0 0	0.03 0 0	0 0 0	0 0 0
254	0 0 0	0 0 0	0 0 0	0 0 0	0 0 0	0 0 0
1925	0 0	0 0	1 0	0.01 0	0 0	0 0

TABLE 5

THE EFFECT OF CHLORINE ON THE POTENTIAL OF STAINLESS STEELS AFTER 2 OR 3 MONTHS IN SEA WATER

Alloy	Potential (mV Ag/AgCl)		
	No Cl.	1 mg/l Cl	4 mg/l Cl
254 SMO	150 244 170	510 510 495	650 645
22/5	285 ; 193 300 ; 295 310 ; 274	615 ; 610 610 ; 615 620 ; 620	655 ; 655 645 ; 660 650 ; 655
Ferralium	225 ; -35 165 ; 172 180 ; 162	535 ; 505 505 ; 495 520 ;	595 ; 560 560 ; 550 540 ;
625	235 198	330 340	395 400 400

TABLE 6

THE DEPTH OF ATTACK AT 15° AND 40°C AFTER 40 DAYS IN RECIRCULATING TEST RIGS

Alloy	Temp. = 15°C				Temp. = 40°C			
	No Chlorine		1 mg/l Cl		No Chlorine		1 mg/l Cl	
	No. of Sites	Max. Depth (mm)	No. of Sites	Max. Depth (mm)	No. of Sites	Max. Depth (mm)	No. of Sites	Max. Depth (mm)
316	3 3	0.24 0.12	2 6	0.32 0.68	7 7	0.92 0.60	5 1	1.04 0.47
Ferralium	0 1	0 0.04	2 2	0.10 0.06	1 1	0.12 0.02	1 1	0.11 0.15
254 SMO	0 0	0 0	0 0	0 0	2 1	0.04 0.02	5 6	0.14 0.12

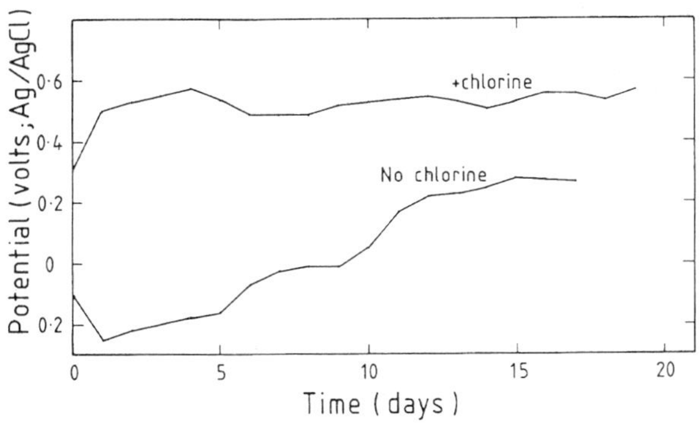

Figure 1 : Potential-time curves for 254 SMO in natural sea water and in sea water dosed with 1mg/l chlorine.

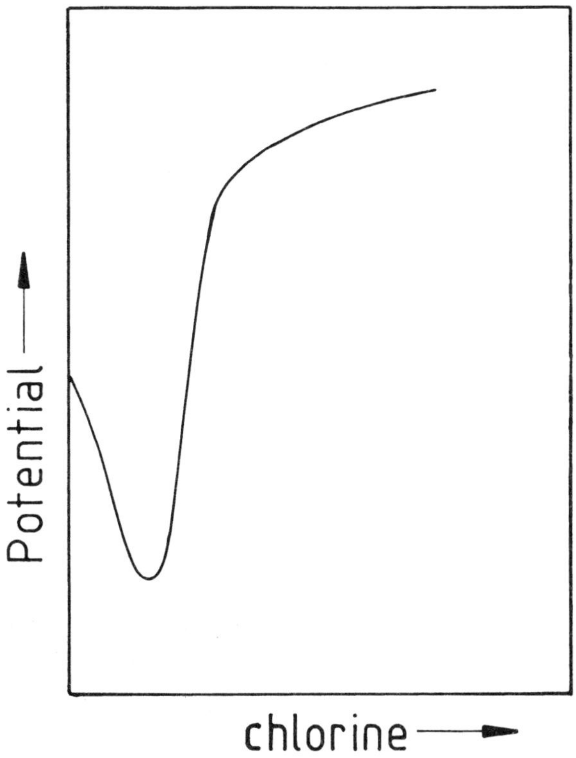

Figure 2 : Schematic diagram of the variation in potential with chlorine concentration.

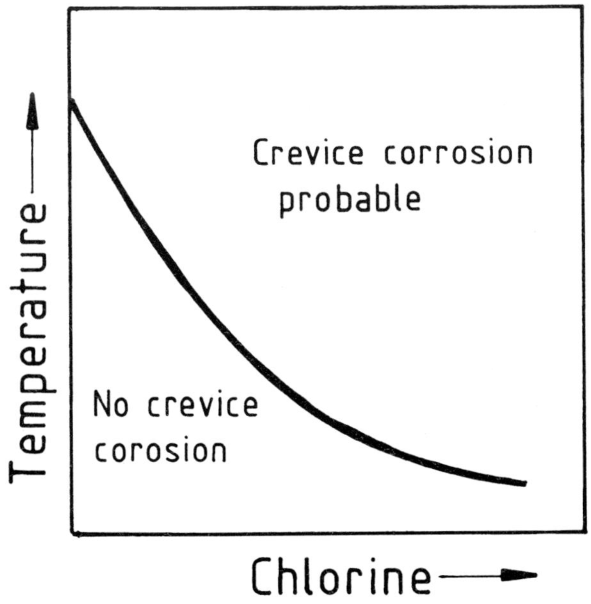

Figure 3 : Schematic diagram of the effect of temperature and chlorine concentration on the susceptibility to crevice corrosion.

□

SESSION IV
CORROSION

In-situ ellipsometric determination of thickness and optical constants of passive films on austenitic stainless steels

K Sugimoto and S Matsuda

The authors are in the Department of Metallurgy, Faculty of Engineering, Tohoku University, Sendai, Japan.

SYNOPSIS

The change in the thickness and optical constants of passive films on austenitic stainless steels as a function of Cr content of the steels has been examined using in-situ ellipsometry in pH 6.0, 1.0 kmol.m^{-3} Na_2SO_4 and pH 0.2, 0.5 kmol.m^{-3} H_2SO_4. The thickness of the films increased with increasing potential and the thickness at a given potential decreased with increasing Cr content of the steels and decreasing solution pH. Distinguishable transitions from passive films to transpassive films were observed on the steels containing > 14% Cr in the neutral solution. Optical constants of the films also changed in dependence on the Cr content of the steels and the solution pH. These results were compared with those obtained on ferritic stainless steels.

INTRODUCTION

The corrosion resistance of stainless steels has been understood to be controlled by the chemical and physical properties of passive films formed on the steels. Therefore, it is important to know the properties of the films under as-formed conditions in environment where the steels are in use. Since aqueous solutions are a usual environment for the steels, extensive attempts have been made to measure in-situ the properties of the films formed in the solutions (1,2). Among these, ellipsometry has been regarded as an effective method for the in-situ analysis of passive films in such an environment and used in the determination of the thickness and optical properties of the films on Fe-Cr (3), Fe-Cr-Ni (4-9), Fe-Cr-Ni-Mo (9-13) alloys and an austeno-ferritic stainless steel (14) in a variety of solutions.

It is a well known fact that the corrosion resistance of passive films increases with increasing Cr content of steels. To get an exact reason why the performance of passive film is improved by an increase in the Cr content of steels is an important subject in the field of the corrosion science of stainless steels. Consequently, changes in the thickness (3) and the Cr content (15,16) of passive films with the change in the Cr content of ferritic stainless steels have been examined.

The purpose of the present investigation is to acquire knowledge about the change in the thickness of passive films on austenitic stainless steels as a function of the Cr content of the steels by ellipsometry. The comparison between the results obtained on austenitic stainless steels and those on ferritic stainless steels is also of great interest.

EXPERIMENTAL

Specimens

Cr-Ni steels containing 0 - 27% Cr at the fixed level of 10% Ni were used as specimens. Their chemical composition is listed in Table 1. Discoid specimens ϕ 20 mm x 2 mm were cut from cold rolled sheets of 2 mm thickness of these materials and heated under vacuum at 1323 K for 3.6 ks, followed by quenching in water at 273 K. All the specimens, except for the 27Cr-10Ni steel which has 50% of α phase, had single γ phase structures. Surfaces of the specimens were ground with emery paper up to No. 1500 and then finished with diamond paste. After polishing, they were degreased in ultrasonic baths of acetone and absolute ethyl alcohol.

Solutions

Deaerated pH 6.0, 1.0 kmol.m^{-3} Na_2SO_4 and pH 0.2, 0.5 kmol.m^{-3} H_2SO_4 were used as electrolytic solutions. Deaeration was achieved by bubbling high purity N_2 through the solutions in a reservoir for more than 14.4 ks. The deaerated solutions were admitted into the electrolytic cell immediately before each experiment. All experiments were carried out in a N_2 atmosphere at 293 ± 0.5 K under stationary conditions.

Electrolytic cell

The electrolytic cell was made of Pyrex glass and had two optically flat windows of synthetic quartz. It was also equipped for electrochemical measurements.

Electrochemical polarization

The passivation of the specimens was accomplished under potentiostatic control. In neutral solutions, the potential was changed from the corrosion potential in the noble direction at intervals of 100 mV and held for 3.6 ks at each setting potential after which the current and ellipsometric parameters were noted. To avoid the surface roughening by active dissolution, however, measurements were omitted at active potentials in acid solutions and started from potentials at the lowest end of passivity region for each steel (7). A saturated calomel electrode was used as the reference electrode and the potentials reported here are referred to this base.

Ellipsometry

A DV-36M Mizojiri-Kogaku ellipsometer, which has a common polarizer - compensator - specimen - analyser configuration, was used. Monochromatic light of wavelength 546.1 nm was used for the measurements. The angle of incidence of the monochromatic light was 74.42°. The compensator was fixed at an angle of -45°. Two optical parameters, the relative phase retardation Δ and the relative amplitude reduction Ψ , were obtained from the angles P and A of the polarizer and the analyser at the extinction positions.

The thickness and the complex refractive indices of passive and transpassive films were determined using the theoretical Δ versus Ψ curve which fits the experimental Δ versus Ψ curve with minimal error. The theoretical Δ versus Ψ curves were obtained by solving Drude's exact equations with the aid of an Acos-600 computer assuming that parallel-sided isotropic homogeneous films with various complex refractive indices $n_2 - k_2 i$ in the ranges $1.5 \leq n_2 \leq 3.0$ and $0.0 \leq k_2 \leq 0.8$ grow from 0 to 10 nm in thickness. These ranges for n_2 and k_2 almost cover the complex refractive indices of oxide and hydroxide films on Fe, Cr, Ni, and their alloys (3,7). For refractive indices of 1.0 kmol.m^{-3} Na_2SO_4 and 0.5 kmol.m^{-3} H_2SO_4, the values of 1.356 and 1.337 were used respectively.

The optical constants of film-free surfaces of the steels used as specimens were determined in this study using the technique of triboellipsometry (14,17) in nonaqueous environment. That is, the surfaces of the steels were polished in dehydrated deaerated methyl alcohol and ellipsometric parameters of polished surface were measured immediately without changing the environment. The ellipsometric parameters thus obtained were used for the calculation of optical constants of the film-free surfaces of the steels.

RESULTS

Optical constants for the film-free surfaces

The optical constants $N_3 = n_3 - k_3 i$ for the film-free surface of the steels, which were obtained by triboellipsometry, are given in Table 2. In the calculation of these values, the errors resulting from surface roughening by polishing in methyl alcohol and from the birefringence of the cell windows were corrected in accordance with the procedure given in a previous paper (7). It is seen from Table 2 that the real part of N_3 for the Cr-Ni steels decreases, while the imaginary part increases, with increasing Cr content of the steels. There are marked changes in the values of both the parts between 0% and 10% Cr.

Optical constants and thickness of passive and transpassive films formed in a neutral solution

Experimental Δ versus Ψ curves were obtained for all the Cr-Ni steel specimens after potentiostatic oxidation in pH 6.0, 1.0 kmol.m^{-3} Na_2SO_4 for 3.6 ks. Theoretical Δ versus Ψ curves were fitted to the experimental results by the computer. Figure 1 illustrates the case of the 18Cr-10Ni steel; the experimental Δ versus Ψ data are compared with two theoretical Δ versus Ψ curves which correspond to the growth of films having optical constants of 2.5 - 0.4i and 2.9 - 0.4i.

In the passivity region extending from -0.1 to 0.8 V the experimental Δ versus Ψ curve give a straight line which coincides with the theoretical curve calculated for a film with an optical constant of 2.5 - 0.4i. It can be seen from this theoretical curve that the thickness of the passive film increases from 1.1 nm at -0.1 V to 2.8 nm at 0.8 V with increasing potential.

Transpassive dissolution started above 0.8 V and was accompanied by characteristic changes in the ellipsometer readings between 0.8 and 0.9 V. These changes correspond to the transformation of the passive film into the transpassive film (3,7).

In the transpassive range 0.9 - 1.3 V, the experimental Δ versus Ψ curve coincides with the theoretical curve for a film with an optical constant of 2.9 - 0.4i and the thickness of the transpassive film thus ranges from 2.6 nm at 0.9 V to 3.2 nm at 1.3 V. Large changes occurred in the leadings of Δ and Ψ above 1.3 V. This should be due to the surface roughening of electrode by the strong transpassive dissolution (7). Some meaningful analysis was difficult in this region of the experimental Δ versus Ψ curve.

The same analyses were performed on the other alloys; the changes in the film thickness for the steels containing 0 - 27% Cr are shown in Fig. 2 as a function of potential and the optical constants, N_2, and the increase rates of thickness with an unit rise in potential, k , for these films are given in Table 3.

From Fig. 2 it is seen that a distinguishable transition from the passive film to the transpassive film takes place on the steels containing > 14% Cr and the thickness of the passive or transpassive film at a given potential reduces as the Cr content of the steel increases. It is shown in Table 3 that the real parts of N_2 of passive films on the steels containing < 10% Cr are larger than those of the films on the steels containing > 14% Cr. The value of k for the passive film decreases with increasing Cr content of the steel. The real parts of N_2 of passive films are smaller than those of transpassive films on the steels containing > 14% Cr.

Optical constants and thickness of passive and transpassive films formed in an acid solution

The same measurements and analyses as described in the former section were performed on the steels in pH 0.2, 0.5 kmol.m^{-3} H_2SO_4. Figure 3 shows the results for the 18Cr-10Ni steel. From this figure it is evident that the experimental Δ versus Ψ curve in the passive region between 0 and 0.8 V coincides with the theoretical curve for a film with an optical constant of 2.0 - 0.2i. The film thickness in the passive region increases from 0.7 nm at 0 V to 1.6 nm at 0.8V. In the transpassive region above 0.9 V, there occurred a large change in the leadings of Δ and Ψ . Since such a change was thought to be due to an increase in the surface roughness of the electrode by the transpassive dissolution, no analysis of experimental curves in this potential region was done.

The results of similar analyses on the other steels are shown in Fig. 4, where variations in the thickness of the passive films are given as a function of potential. The optical constants, N_2, and the increase rates of thickness with an unit rise in potential, k , for these films are shown in Table 4.

In Fig. 4 it can be seen that at a given potential the passive film becomes thinner as the Cr content increases consistent with the observation at higher pH shown in Fig. 2. The tendency of the real part of N_2 to decrease with increasing Cr content is shown in Table 4 on the steels containing 6 - 14% Cr. The values of the real part of N_2 become almost constant on the steels containing > 14% Cr. The value of k looks independent of the Cr content of the steels in the pH 0.2 solution.

DISCUSSION

Effect of pH on the thickness of the passive film

The relationships between the thickness of passive films and the Cr content of the steels obtained in pH 6.0, 1.0 $kmol.m^{-3}$ Na_2SO_4 and pH 0.2, 0.5 $kmol.m^{-3}$ H_2SO_4 are shown in Fig.5. The comparison between thicknesses of the films was made at 0.5 V in both the solutions. It is clear that the thickness of the films formed in the pH 6.0 solution is larger than that of the films formed in the pH 0.2 solution at any Cr content of the steels. As stated above, the thickness of the film decreases with increasing Cr content. This tendency is remarkable in the range of 5 - 10% Cr in the pH 6.0 solution. From Tables 3 and 4, optical constants, N_2, and the increase rates of thickness with a unit rise of potential, k , of the passive films formed in the pH 6.0 solution is larger than those in the pH 0.2 solution. The increased optical constants of the films should mean an increased Fe oxide content in the films (7). These observations correspond totally to those obtained on ferritic stainless steels (3).

Difference in the thickness of the passive films on austenitic and ferritic stainless steels

Figures 6(a) and (b) compare the relationship between the thickness of passive films and the Cr content of the steel obtained on austenitic stainless steels in this study with that obtained on ferritic stainless steels previously by the authors (3). The relationships were taken from the results obtained at 0.5 V in both the pH 6.0, 1.0 $kmol.m^{-3}$ Na_2SO_4 (Fig. 6(a)) and pH 0.2, 0.5 $kmol.m^{-3}$ H_2SO_4 solutions (Fig. 6(b)). From these figures, it is clear that the thickness of the films formed on austenitic stainless steels are smaller than that on ferritic steels at a given Cr content of the steels in both the neutral and acid solutions. This could be explained by the effect of Ni oxide contained in the films on austenitic steels (18).

CONCLUSION

(1) Optical constants N_3 of film-free surface of Cr-Ni steels containing 0 - 27% Cr and 10% Ni have been decided by triboellipsometry in non-aqueous environment. The real part of N_3 decreased sharply with increasing Cr content of the steel in the range between 0 and 10% Cr. It became almost constant above 14% Cr.
(2) The thickness of the passive films on the Cr-Ni steels decreased with increasing Cr content of the steel in both the solutions of pH 6.0, 1.0 $kmol.m^{-3}$ Na_2SO_4 and pH 0.2, 0.5 $kmol.m^{-3}$ H_2SO_4. This tendency is more remarkable on the steel having Cr content in the range 5 - 10% in the neutral solution. The real part of the optical constant N_2 of the passive film decreases with increasing Cr content of the steel.
(3) If compared at a given Cr content ,the thickness of the passive films on austenitic stainless steels is smaller than that on ferritic stainless steels under the same film formation conditions.

REFERENCES

1. K. Sugimoto : Tetsu-to-Hagane, 70, 19(1984).
2. K. Sugimoto : Bull. Jpn. Inst. Metals, 24, 754(1985).
3. K. Sugimoto and S. Matsuda : Mater. Sci. Eng., 42, 181(1980).
4. V. V. Andreeva : Corrosion, 20, 35(1964).
5. G. Okamoto and T. Shibata : Corros. Sci., 10, 37(1970).
6. K. N. Goswami and R. W. Staehle : Electrochim. Acta, 16, 1895(1971).
7. S. Matsuda, K. Sugimoto and Y. Sawada : J. Jpn. Inst. Metals, 39, 848(1975): Trans. Jpn. Inst. Metals, 18, 66(1977).
8. S. Matsuda, K. Sugimoto and Y. Sawada : "Passivity of Metals," R. P. Frankenthal and J.Kruger, Editors, p. 699, The Electrochemical Society, Inc., Princeton, NJ (1978).
9. K. Sugimoto, S. Matsuda, Y. Ogiwara and K. Kitamura : J. Electrochem. Soc.,132, 1791 (1985).
10. G. M. Bulman and A. C. C. Tseung : Corros. Sci., 13, 531(1973).
11. K. Sugimoto and Y. Sawada : Corros. Sci., 17, 425(1977).
12. S. Matsuda, K. Hamano, K. Sugimoto and Y. Sawada : J. Jpn. Inst. Metals, 42, 808(1978).
13. S. Matsuda and K. Sugimoto : Boshoku Gijutsu (Corros. Eng.), 29, 19(1980).
14. K. Sugimoto and S. Matsuda : J. Electrochem. Soc., 130, 2323(1983).
15. N. Hara and K. Sugimoto : J. Electrochem. Soc., 126, 1328(1979).
16. K. Asami, K. Hashimoto and S. Shimodaira : Corros. Sci., 18, 151(1978).
17. J. R. Ambrose and J. Kruger : Corrosion, 28, 30(1972).
18. A. Kasamatsu, S. Matsuda and K. Sugimoto : Proc. Annu. Meet. Jpn. Soc. Corros. Eng., p. 150 (1981).

Table 1 Chemical composition of specimens (mass%).

Specimen	Cr	Ni	Mn	C	Si	S	P	Cu	Mo
10Ni Steel	0.01	10.04	0.0010	0.0020	0.01	0.006	0.003	0.01	0.01
6Cr-10Ni Steel	6.33	9.27	0.0010	0.0050	0.025	0.0104	0.003	0.028	0.0013
10Cr-10Ni Steel	9.56	10.19	0.0011	0.0032	0.025	0.0128	0.003	0.032	0.0016
14Cr-10Ni Steel	14.10	9.73	0.0010	0.0029	0.026	0.0141	0.003	0.034	0.0016
18Cr-10Ni Steel	18.80	8.92	0.0017	0.0014	0.026	0.0083	0.003	0.007	0.0018
27Cr-10Ni Steel	27.40	10.00	0.0010	0.0036	0.029	0.0129	0.003	0.007	0.0013

Table 2 Optical constants ,N_3, for film-free surfaces of stainless steels at wavelength 546.1 nm.

Specimen	N_3
10Ni Steel	3.14-3.50i
6Cr-10Ni Steel	2.85-3.74i
10Cr-10Ni Steel	2.79-3.87i
14Cr-10Ni Steel	2.75-3.89i
18Cr-10Ni Steel	2.73-3.94i
27Cr-10Ni steel	2.74-3.98i

Table 3 Optical constants, N_2, and increase rates of thickness with potential, k, for passive and transpassive films formed in pH 6.0, 1.0 $kmol.m^{-3}$ Na_2SO_4.

Specimen	Passive film		Transpassive film	
	N_2	k /nm·V^{-1}	N_2	k /nm·V^{-1}
10Ni Steel	2.8-0.4i	2.2	-	-
6Cr-10Ni Steel	2.8-0.4i	2.3	-	-
10Cr-10Ni Steel	2.8-0.3i	1.9	-	-
14Cr-10Ni Steel	2.4-0.4i	1.7	2.9-0.4i	1.4
18Cr-10Ni Steel	2.5-0.4i	1.6	2.9-0.4i	1.5
27Cr-10Ni Steel	2.4-0.4i	1.5	2.9-0.4i	1.8

Table 4 Optical constants, N_2, and increase rates of thickness with potential, k, for passive films formed in pH 0.2, 0.5 $kmol.m^{-3}$ H_2SO_4.

Specimen	Passive film	
	N_2	k / nm·V^{-1}
10Ni Steel	-	-
6Cr-10Ni Steel	2.5-0.3i	0.9
10Cr-10Ni Steel	2.3-0.3i	0.9
14Cr-10Ni Steel	2.0-0.3i	1.0
18Cr-10Ni Steel	2.0-0.2i	1.0
27Cr-10Ni Steel	2.0-0.2i	0.9

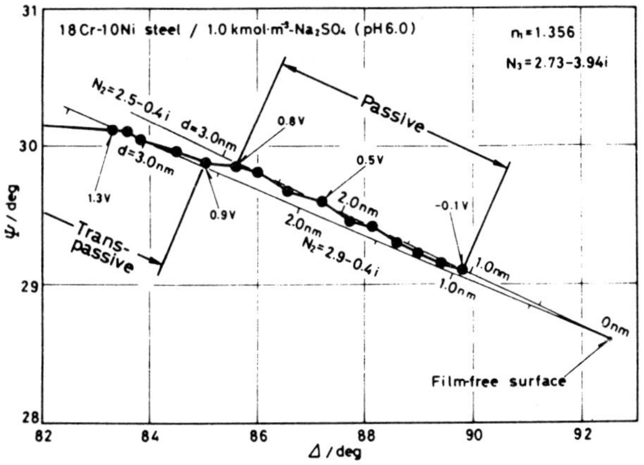

Fig. 1 Experimental Δ - Ψ locus for the potentiostatic oxidation of 18Cr-10Ni steel in pH 6.0, 1.0 kmol.m^{-3} Na_2SO_4 and theoretical Δ - Ψ curves for the growth of films with optical constants N_2 = 2.5 - 0.4i and 2.9 - 0.4i.

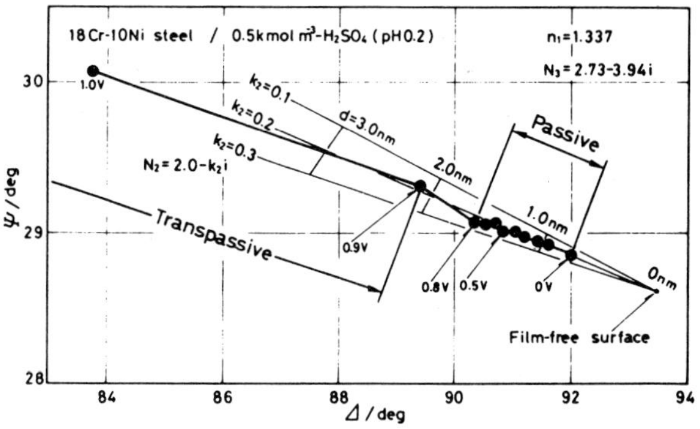

Fig. 3 ExperimentalΔ - Ψ locus for the potentiostatic oxidation of 18Cr-10Ni steel in pH 0.2, 0.5 kmol.m^{-3} H_2SO_4 and theoretical Δ - Ψ curves for the growth of films with optical constants N_2 = 2.0 - k_2i (k_2 = 0.1 - 0.3).

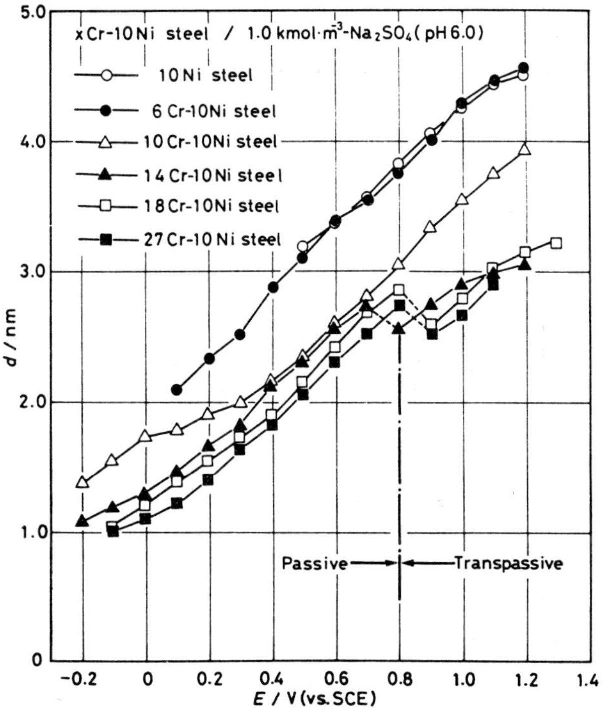

Fig. 2 Thickness, d, vs. potential, E, relations for passive and transpassive films on xCr-10Ni steels in pH 6.0, 1.0 kmol.m^{-3} Na_2SO_4.

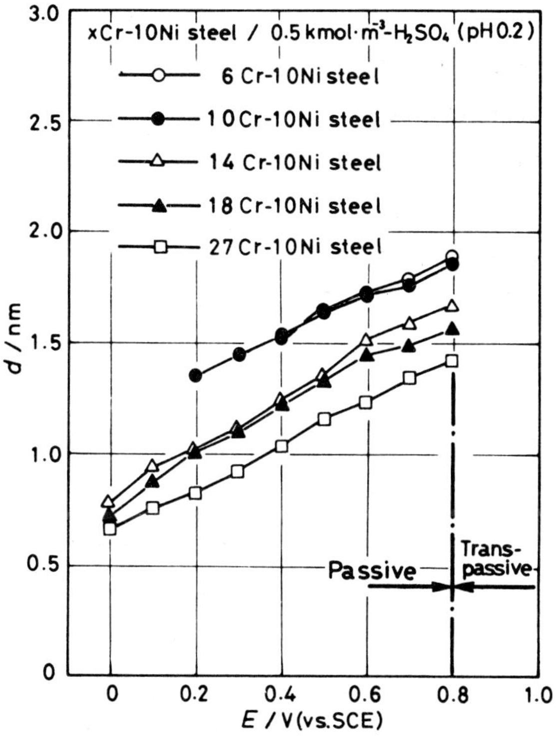

Fig. 4 Passive film thickness, d, vs. potential, E, relations for xCr-10Ni steels in pH 0.2, 0.5 kmol.m^{-3} H_2SO_4.

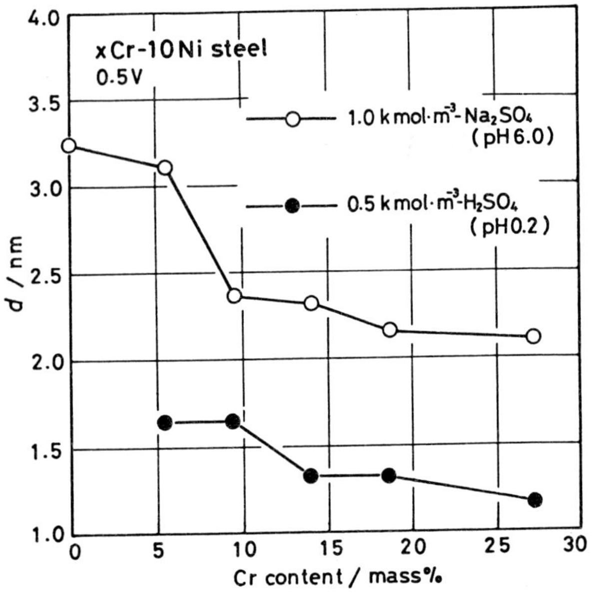

Fig. 5 Passive film thickness, d, vs. Cr content relations for xCr-10Ni steels at 0.5 V in pH 6.0, 1.0 kmol.m^{-3} Na_2SO_4 and pH 0.2, 0.5 kmol.m^{-3} H_2SO_4.

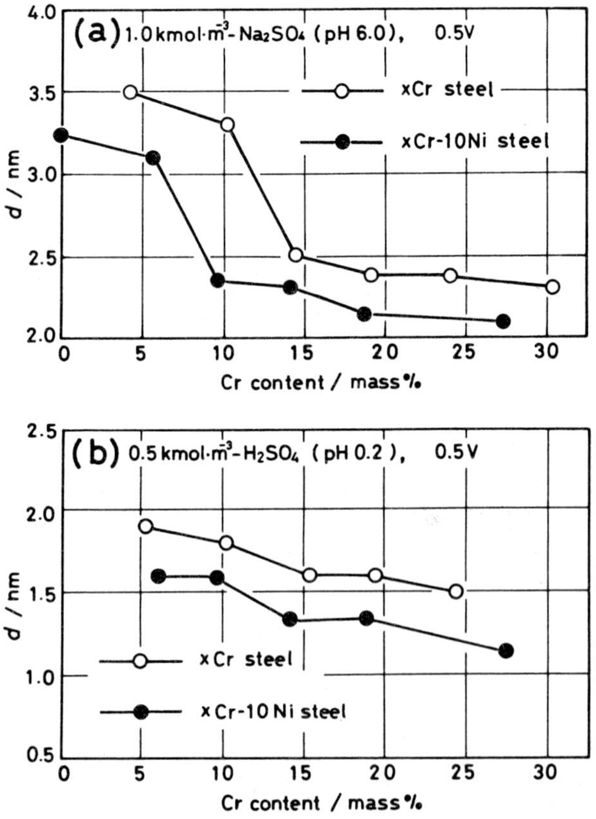

Fig. 6 Passive film thickness, d, vs. Cr content relations for xCr steels and xCr-10Ni steels at 0.5 V in pH 6.0, 1.0 kmol.m^{-3} Na_2SO_4 (a) and pH 0.2, 0.5 kmol.m^{-3} H_2SO_4 (b).

ESCA characterization of passive films on stainless steels

S Jin and A Atrens

The authors are in the Department of Mining and Metallurgical Engineering, University of Queensland, St. Lucia, Queensland, Australia.

SYNOPSIS

The present work is a systematic investigation into the passive film formed on two commercial stainless steels, 18-12 and 29-4-2, using a surface analysis technique ESCA (Electron Spectroscopy for Chemical Analysis) combined with ion etching. The specimens were exposed to 0.1 M NaCl to simulate service environments, at different immersion times (0.5 to 44 h), different temperatures (25 to 90 °C) and different controlled potentials (OCP-1600 mV_{SCE}). A three factor model was developed to describe the passive film under conditions of stable passivity: in contact with the solution there is a hydrated layer, followed by an oxide layer consisting of Fe and Cr oxides having maxima at depths of 3 and 10 Å respectively, and a depleted/enriched metallic layer. There is a smooth transition between the layers. The thickness of the outer two layers is about 15 Å.

INTRODUCTION

Understanding the influence of aqueous environment on the structure and composition of the passive films formed on stainless steels is the first step in understanding passivity breakdown as occurs during pitting, crevice corrosion and stress corrosion cracking. Previous surface studies [1-3] on stainless steels in Cl containing solutions have shown that the enrichment of alloying elements in the outer atomic layers forms the major factor protecting the materials from corrosion. Hydroxides and water occur in the outermost layers and the central region consists mainly of Fe-Cr oxides. A continuous transition occurs from the outer hydroxide layer to the inner oxide and thus the layer can be regarded as an oxy-hydroxide. Earlier investigations [2,4-6] have also shown that at lower applied potentials the sample surface film contains lower oxidized state of the components, such as Cr^{3+}, oxyhydroxide and small concentration of Fe^{2+}; at higher applied potentials the sample surface contains the higher oxidized states, Cr^{6+} and Fe^{3+}, as well as an increased total content of oxide.

The passive film formed on the surfaces of a range of commercial stainless steels exposed to 0.1 M NaCl has been examined by Bruesch et al. [1]. As a logical extension a systematic investigation was carried out to concentrate on the passive film formed on the surface of two stainless steels, 18-12 and 29-4-2. The materials were exposed to 0.1 M NaCl to simulate service environments [7-10], at various immmersion times, temperatures and controlled potentials. The major purpose of this paper is to report on the characterization of the passive films on stainless steels and the changes in the film during the transformation from passivity to passivity breakdown as evidenced by pitting.

EXPERIMENTAL PROCEDURE

The experimental procedures have been reported in the previous papers [1,7-9]. Two stainless steels used are 18-12 and 29-4-2, containing respectively 18%Cr, 12%Ni and 29%Cr; 4%Mo and 2%Ni. The surface preparation consisted of polishing to a mirror-quality surface finish and finally sputtering with argon-ion. Exposure conditions have included different immersion times (0.5 to 44 h), different temperatures (25 to 90 °C) and different controlled potentials (OCP-1600 mV), in 0.1 M NaCl.

Analysis was carried out in a PH1 Model 560 ESCA (PERKIN ELMER Model 560) electron spectrometer using Mg $K_{\alpha_{1,2}}$ (1253.6 eV) as the exciting radiation of the photoelectrons. The structure and composition of the passive films were determined using the depth profiling technique, which consists of periodic sputtering with argon ions and recording the electron spectra emerging from the specimen surface after each etching period. Peak deconvolution has been carried out using a manual and a computer curve fitting procedure, based on the actual individual peak shapes and peak position. A computer program has been developed and used in the deconvolution of the composite peaks into their elemental peaks. This analysis has the advantage that added confidence is given to conclusions drawn from the data. Deconvolution of a composite peak gives the measured intensity for element i in state s after sputtering time t and the measured number of atoms per unit volume is related to the measured concentration. The true concentration, which reflects the depth profile of the passive film on the stainless steels, can also be obtained by means of a correction from the measured concentration.

RESULTS

Fig. 1 shows the true concentrations C_{is}^{T} for the main species of interest for the alloy 18-12 obtained as a function of depth after an immersion time of 44 h. Similarly the true concentrations were also obtained for the alloy 18-12 after immersion times of 1 and 21 h and for 29-4-2 after immersion times of 0.5, 15, 22 and 44 h. These results are to be compared with those measured previously by Bruesch et al. [1], using a different ESCA apparatus and somewhat different experimental and data reduction procedures on the same material exposed for 15 h. To facilitate comparisons critical points from the curves have been tabulated in Table 1.

With increasing depth there is a peak of Fe and a peak of Cr in the oxidized state at approximately 3 Å and 10 Å for both 18-12 and for 29-4-2, respectively. For both alloys the maximum concentration of oxidized Fe decreases and the maximum concentration of oxidized Cr increases with increasing immersion time. In all cases there is a residual of 2~3% of Cr in the oxidized, even after sputtering time of 15 min. This probably is associated with the residual amount of oxygen. The saturation values for the metallic states differ somewhat from the measured bulk values due to the fact that the oxygen concentration at long sputtering times was still significant. The depth at half the saturation value has been chosen to characterize the rapid increase in the concentration of Fe and Cr in the metallic states. In both specimens the oxygen profile decreases sharply with increasing distance from the surface.

The oxidized state of Ni has not been observed in the passive films either before or after sputtering. The results indicate that for both 18-12 and for 29-4-2 a marked accumulation of metallic Ni exists in the region of the passive film where the concentration of Cr in the oxidized state is rapidly decreasing and the concentration of Fe and Cr in the metallic states is rapidly increasing and moreover there is a strong depletion of Ni in the outer part of the passive films.

A film formed on 18-12 by a 3 min exposure only to air was also investigated. This film has a similar structure to those exposed to the solution. Comparison with the film formed by 21 h exposure to 0.1 M NaCl and the film formed by 22 h exposure only to air shows that the air formed film has a structure which is quite similar to the film formed by exposure to the solution. The similarities between the films formed on exposure to solution and those only exposed to air indicate that the total history of exposure of the specimens is important.

The passive films formed by solution exposure, with minimum prior air exposure between the specimen preparation stage and the solution exposure, were also investigated. For a 2 min exposure to the solution the profile shows that this brief exposure is sufficient to establish the basic structure of the passive film. The profile for 14 h exposure to 0.1 M NaCl shows profiles very similar to those produced by 12 to 72 h exposure to air plus 21 h exposure to 0.1 M NaCl. This shows that long exposure to solution produces a passive film which no longer depends strongly on prior air exposure.

Also measured was the profile formed on 18-12 by 15 h exposure to 0.1 M NaCl followed by 3 and 90 min exposure to air respectively. These data show that the structure of the passive film remains largely intact for quite a long time after the specimens have been taken out of contact with aqueous solution, although in some respects the tendency is towards the kind of film formed by exposure only to air.

In both specimens the presence of chlorides within the passive film has not been observed at lower controlled potentials. This is consistent with the prior measurements [11] for film formed in oxygenated water; these films were very similar to those measured herein. Consequently, there is unlikely to be any significant change to the passive film resulting from the brief exposure to distilled water during transfer from the Cl solution to the ESCA apparatus for analysis.

The data show that passive films formed at 60 °C and 90 °C on the surface of 18-12 and 29-4-2 stainless steels are very similar to those formed at 25 °C. The solution temperatures seem to play a less important role than immersion time, due to the less aggressive environment conditions, i.e., no passivity breakdown in the form of pitting corrosion. For 18-12 there are significant differences for Fe and Cr in the films formed at the elevated temperatures compared with the film formed at 25 °C. The data could be interpreted as being consistent with more Fe and less Cr dissolving at the higher temperatures [12-15]. However, the details are different for 29-4-2. For this alloy, at the higher temperatures the Fe parameters are similar to those at 25 °C, the Cr^{ox}_{max} concentration is somewhat higher, the position of Cr^{ox}_{max} is shallower and the position for $Cr^{m}_{0.5sat}$ at 60 °C is much shallower whilst at 90 °C much deeper. The lack of consistency between data trends for the two alloys leads to the conclusion that the data variability masks any clear trends and that all the data are from within the same statistical sample.

Two measures have been chosen for film thickness in Table 2. The first one is the point of the intersection of the profile for the metallic and oxidized states of Cr. At this point the enrichment of Cr in its oxide form and the depletion of Cr in its metallic state are coming to an end and the amount of its metallic state is increasing monotonically to the bulk value. The other point is the one at which oxygen has the same concentration as metallic Fe in their respective profiles: the oxygen content within the film is decreasing rapidly and its influence begins to be limited; and the concentration of metallic Fe in high (about 50% of the bulk concentration) which implies that for greater depths the bulk is approaching. In fact these two points give very similar values for the film thickness, and the average of these two parameters is an ideal value to represent the thickness of the surface films. The data of Table 2 indicate that the film thickness is independent of exposure time. The thickness of the passive film on 29-4-2 is slightly less compared with that on 18-12. This correlates with the higher Cr content of 29-4-2.

For specimens of both materials exposed to more positive potentials, above the pitting potential, there was a very significant change in the nature of the profiles of the surface films. For 18-12, there was a mixture of iron and chromium oxides with quite a constant value for Fe^{ox}, Cr^{ox} and O over the range of 0.5 to 5 min sputtering time. For the film formed at an applied potential of 350 mV_{SCE}, which corresponds to the pitting potential for 18-12, Fig. 2, there was Fe^{ox} about 16%, Cr^{ox} about 11% and O about 70%. In contrast, at the higher potential of 500 mV_{SCE}, the oxidized chromium signal was substantially the same at Cr^{ox} about 11% but there was a significant reduction in the oxidized Fe signal to Fe^{ox} about 3%. For the specimen exposed at the still higher potential of 800 mV_{SCE}, the oxidized chromium was still substantially the same, Cr^{ox} about 11%, and the oxidized iron signal has decreased somewhat further to Fe^{ox} about 2%.

To investigate the existence of chloride ion on the surface, two additional experiments were performed for 18-12 at higher applied potentials 500 and 800 mV_{SCE} under the same experimental conditions. The evolutions of the chloride peaks are shown in Figs. 3a and 3b drawn to the same scale. In Fig. 3a it is clearly shown that there was only a slight amount of chloride in the outer part of the surface at 500 mV_{SCE} and it was completely removed after sputtering. A much greater content of chloride appeared on the sample surface at 800 mV_{SCE} and it still existed even after 15 min sputtering, in Fig. 3b. This corresponded to a much greater accumulation of corrosion products on the surface at the higher applied potential.

Figures 4a-4c show the concentrations of the various states of molybdenum as a function of depth at various controlled potentials for 18-12. At lower controlled potentials OCP and 0 mV_{SCE}, the concentration of total molybdenum appeared only in the metallic state and at relatively low concentrations. On increasing the controlled potential to 350 mV_{SCE} (the pitting potential), the metallic molybdenum was reduced to a negligible level. However, at the still higher potentials, 500 and 800 mV_{SCE}, which were above the pitting potential, the total concentration of molybdenum increased to a considerable level. This can be clearly seen in Figs. 4b and 4c. This oxidized molybdenum was largely in the form of MoO_2 with a significant amount of MoO_3. After a long sputtering time there was only metallic molybdenum left. A similar relationship of the molybdenum components was also found for 29-4-2 at potentials more positive than the pitting potential. It should be kept in mind that there were corrosion products on the surface at the higher controlled potential and they may be the main reason for the presence of these oxides and the enrichment of molybdenum.

DISCUSSION

(1) Formation and Stability of The Passive Film of Stainless Steels

a) The Structure and Composition of The Passive Film

Under conditions of stable passivity the passive film can be described by a three factor model: a hydrated layer in contact with the solution, an oxide layer consisting of Fe and Cr oxides having maxima at depths of 3 and 10 Å respectively, and a metallic layer enriched in Ni. There is a smooth transition between these layers and into the bulk alloy. The thickness of the outer two layers is about 15 Å. There seem to be two possibilities which can explain the observed coexistence of the metallic Fe and Cr within the oxide region: one possibility is the nonuniformity of the thickness in the passive films, and the other is the presence of metallic clusters within the films [1]. A metallic layer containing a Cr concentration slightly less than that of the bulk and enriched in Ni not only affects the dissolution rate and the corrosion potential of the alloy, but also improves the passivation of the alloy [15].

b) Experimental Pretreatment and the Passive Film

The films formed by exposure to air have a structure similar to the passive films formed by exposure to the solution for 18-12 and 29-4-2. The similarities between the films formed on exposure to solution and those only exposed to air, and films formed by short time exposure only to the solution indicate that the total history is important in determining the structure of passive films. An air formed film is rapidly converted (<0.5 h) to a passive film. These results imply the mechanisms and physical and chemical principles underlying film formation are not wholly determined by the environment, but are determined by the metal and the intrinsic properties of the film itself. The data in Table 1 also show that the structure of the passive film remains largely intact for quite a long time. Therefore, the effect of air exposure after the electrochemical treatment and of distilled water in transferring should be limited to a negligible level.

c) Immersion Time and The Passive Film

With increasing immersion time, the stability of passive films on stainless steel changes non-linearly, in terms of the surface accumulation of Cr in the outer part of the films. This is probably related to a selective dissolution of Fe during all times of exposure [12, 13]. The stability of the films seems to be high since they all have a very similar film structure, independent of immersion time. Since the formation of its oxide has a higher free energy, a greater amount of Cr oxide appears within the film. Both the reactions, the selective dissolution of Fe and the selective oxidation of Cr contribute to the result that Cr is enriched in outer parts of films. The enrichment of Cr, however, limits the area of Fe atoms contacting the solution and thereby slows down the further selective dissolution of Fe. Because vacancies are produced and must be reoccupied, the depletion of Cr and the enrichment of metallic Ni were caused. The enrichment of Ni limits the further production of Cr oxide and this may become the second factor in the protective layer when the passivating ability of the Cr content has been reached at the surface.

d) Effect of Controlled Potential and Temperature

The influence of elevated temperature and controlled potential, in the region between the OCP and below the critical pitting potential, on the structure and chemical composition of the passive film is limited in agreement with the prior works [7, 8]. The three-factor model also characterizes satisfactorily the structure of the passive films formed on the material surface. It showed that the structure and composition of passive films formed by short time solution exposure (as was the case of the films studied herein) are also quite dependent on the total exposure history including prior air exposure. Chromium is still the most important alloying element at various temperatures and at lower potentials than the pitting potential and it plays the most important role in the corrosion resistance compared with all the other alloying elements. As the applied potential is lower than the critical pitting potential, the chloride ion is only weakly incorporated in the outermost layer [2, 7, 16]. The film becomes slightly thicker and chromium enrichment increases with increasing aggressivity of corrosive environment although no pits are observed. When passivity breakdown has taken place as evidenced by pitting, then the ESCA studies reveal another significantly different structure.

(2) The Surface Films on Stainless Steels During Pitting Corrosion

A significant amount of the ESCA signals come from the corrosion products between pits with only a small contribution from the passive film itself. The amount of each is uncertain at the various etching depths. This gives a considerably different meaning to the depth profiles of the specimens exposed under high potential conditions, such that there occurs passivity breakdown in the form of pitting compared with those profiles obtained at lower applied potentials. This could be explained as due to a higher dissolution rate at the higher controlled potential [13] of the oxidized elements. It was not possible to determine a relationship between the passive film and the controlled potential because the film was difficult to define. The chromium oxide existed to much greater depths in the specimens exposed at potentials above the pitting potential. It seems that the signal comes mainly from the corrosion products deposited around the pits combined with the relatively lower signal intensity from the passive film between the pits. The trends for the metallic iron signal are also in agreement with the interpretation of relatively thick oxide corrosion products at potentials above the pitting potential for specimens exposed at higher potentials. These two effects of the applied potential, i.e., on both oxidized chromium and metallic iron, formed the major differences in the depth profiles for specimens exposed at potentials more positive than the pitting potential, in comparison with the previous results for the passive films.

With increasing applied potential, most of alloying elements were strongly oxidized and were changed to the oxides which had higher valences. It is certain that molybdenum played a more active role on the surface during the period of pitting, existing in a very complex oxyhydroxide. Its content increased with the controlled potential. However this did not lead to a decreased Cr^{3+} content as observed previously [15] as it was affected by the corrosion products in our experiment. The concentration of oxidized molybdenum was greater on the unetched surface of 18-12 compared with 29-4-2 as higher controlled potential above the critical pitting potential; at greater depth the concentration was much greater for 18-12. From these results it could be seen that the presence of oxidized molybdenum and the enrichment of its metallic state is closely related to pits which occurred at the higher controlled potentials; and the chemical composition of the material also reflects the ability to resist corrosion and pitting corrosion at various controlled potentials, in terms of showing a higher aggressivity on the sample surface with the higher content of alloy elements, such as chromium and molybdenum.

In chloride containing solutions, the breakdown of the passive film is often observed at localized sites, especially at elevated applied potential, and the pitting potential is closely related to chloride concentration and solution temperature [17]. Chromium and nickel enrichment seems to be insensitive to chloride concentration at controlled potentials in the lower region, i.e., in the lower aggressivity of corrosive environment at potentials below the pitting potential. At applied potentials more positive than the critical pitting potential the experimental results correspond to the film incorporation of chloride, thicker films and a decrease in chromium enrichment as the applied potential increases and of course there is passivity breakdown on the sample surface in the form of pitting corrosion. Since the chloride ion is weakly incorporated within the passive film, i.e., the chloride signal disappears after short sputtering times, passivity breakdown cannot be interpreted to indicate chloride penetration through to the metal/oxide interface, i.e. in a process of material transport. Passivity breakdown is more probably due to chloride adsorption [2, 16, 18].

(3) Passivity and Its Breakdown in Stainless Steels

The possibility of the adsorption of chloride ion in the passive film is increased with increasing controlled potential which may cause a higher field strength on the surface. The controlled potential increases the reaction rate of some preferred elements with OH^- ion and the OH^- ion is finally replaced by chloride ion under the assistance from the controlled potential. Consequently the passivity breakdown and pitting are caused by a process of charge transport to form some reaction products like chlorides with preferred alloying elements, in which the mechanism of ion diffusion is important.

At higher controlled potentials above the pitting potential the nickel enrichment disappeared and there was only a steady increase in metallic nickel with depth. It cannot be explained simply as being caused by the covering corrosion products. In addition, the oxidized nickel seemed to, if it existed, have had little effect on pitting and was dissolved into the solution because there was no trace of oxidized nickel on the surfaces. Consequently the influence of nickel on the surface at higher controlled potentials is limited since it has either a relative lower activity or a high dissolution rate in comparison with other alloying elements in such highly aggressive and corrosive environments.

It is well known that either increasing the molybdenum content within the alloy or moving the applied potential in the active direction has a significant and beneficial effect on the resistance to breakdown of passive films, especially for pitting corrosion by altering the distribution and susceptibility of weak points in the passive film, with little change in the macro-characteristics of the film [2, 16, 19]. With increasing aggressivity of the corrosive environment it is possible that oxidized molybdenum replaces iron and chromium ions to be the major diffusion barrier at the oxide-metal interface and thereby slow the corrosion rate. Another possibility is that the oxidized molybdenum reflects corrosion reactions that occur within the pits themselves. However the enrichment of the alloying elements is controlled by the dissolution rate, controlled overpotential and pitting potential of the alloy. When the controlled potential reaches a certain level and produces a more corrosive environment, the effectiveness of the enrichment of the alloying elements, in particular chromium and molybdenum, will be limited and the less active elements will be also oxidized, producing oxidized molybdenum with various valence states. As a result there is passivity breakdown corresponding with pitting at localized sites.

The enrichment of alloying elements is the major factor in protecting the material surface from corrosion and the selective dissolution of alloy elements has occurred at all potentials and times of exposure [2, 5, 7, 13]. At lower controlled potentials the dissolution rates of iron and chromium are higher than the dissolution rate of other alloying elements such as molybdenum and nickel; i.e., iron and chromium are less noble than molybdenum and nickel. This causes the formation of iron and chromium oxides existing within the passive film. The interatomic forces between nickel atoms (probably molybdenum atoms too) slow down the dissolution rate of iron and chromium, and thereby nickel is enriched on the surface inner passive film. A relative steady passive film is formed after the initial enrichment and the dissolution rate is limited for all alloying elements. Earlier studies [20] explained the electrochemical behavior of molybdenum by potential-pH equilibrium diagrams for the Mo-H_2O system and indicated that molybdenum slowed down the propagation of localized corrosion by forming a film of hydrated molybdenum oxide. At high controlled potential the strong appearance of oxidized molybdenum shows that molybdenum may be the more active and effective factor to slow down pitting corrosion, replacing chromium and nickel and tending to protect the surface from corrosion as the aggressivity increases.

CONCLUSIONS

(1) An air formed film is rapidly converted (<0.5 h) to a passive film.

(2) The passive film consists of three layers: a hydrated layer in contact with the solution, an oxide layer consisting of Fe and Cr oxides having maxima at 3 and 10 Å respectively, and a depleted/enriched metallic layer. There is a smooth transition between the layers. The thickness of the outer two layers is about 15 Å.

(3) The maximum of the Fe in the oxidized state decreases, the maximum of Cr in the oxidized state and the maximum of Ni in the metallic state increase with increasing immersion time.

(4) Air formed films have similar structures to films formed by exposure to the solution for 18-12 and 29-4-2. Longer air exposure thickens the air formed film, with Fe^{ox}_{max} and Cr^{ox}_{max} nearly equal in magnitude in comparison with the solution formed films.

(5) Films formed by short time exposure only to the solution are somewhat thinner indicating that the total history is important in determining the structure of passive films. Passive films formed by exposures of 15 h to the solution are no longer strongly dependent on prior air exposure.

(6) The influence of controlled potential in the lower region, between the OCP and the critical pitting potential, on the structure and chemical composition of the passive film is limited in agreement with the prior work.

(7) At various controlled potentials above the pitting potential there are two components, corrosion products and the passive film, on the sample surface.

(8) At the higher controlled potentials, oxidized alloying elements tend to the higher oxidation states on the outermost part of surface together with a possibility of enhanced adsorption of chloride ions.

(9) There is a great amount of oxidized molybdenum, existing in Mo^{6+} and Mo^{4+}, and chloride ions mainly from the corrosion products on the surface at the higher applied potentials. Their contents increase with the controlled potential.

ACKNOWLEDGMENTS

This work was supported by a University of Queensland Special Project Grant and by the Australian Research Grants Scheme. Material was supplied by BBC Brown, Boveri and Company, Switzerland.

REFERENCES

1. P.Bruesch, K.Muller, A.Atrens, and H.Neff, *Appl. Phys.*, **38**, pp. 1-18, (1985).
2. I.Olefjord, B.Brox, and U.Jelvestam, *J. Electrochem. Soc.*, **132**, no. 12, pp. 2854-2861, (1985).
3. J.E.Castle and C.R.Clayton, *Corros. Sci.*, **17**, pp. 7-26, (1977).
4. R.Goetz and D.Landolt, *Electrochimica Acta.*, **29**, no. 5, pp. 667-676, (1984).
5. V.Mitrovic-Scepanovic, B.MacDougall, and M.J.Graham, *Corros. Sci.*, **24**, no. 5, pp. 479-490, (1984).
6. K.Hashimoto and K.Asami, *Corros. Sci.*, **19**, pp. 251-260, (1979).
7. S.Jin and A.Atrens, *Appl. Phys.*, **42**, pp. 149-166, (1987).
8. S.Jin and A.Atrens, *Appl. Phys.*, In press, (1987).
9. S.Jin and A.Atrens, *Appl. Phys.*, To be published, (1987).
10. S.Jin and A.Atrens, *5th Asian-Pacific Corros. Conf.*, Melbourne, Australia, (1987).
11. I.Olefjord and H.Fischmeister, *Corros. Sci.*, **15**, pp. 697-707, (1975).
12. G.Hultquist, C.Leygraf, and D.Brune, *J. Electrochem. Soc.*, **131**, no. 8, pp. 1773-1776, (1984).
13. C.Leygraf, G.Hultquist, I.Olefjord, B.-O.Elfstrom, V.M.Knyazheva, A.V.Plaskeyev, and Ya.M.Kolotyrkin, *Corros. Sci.*, **19**, pp. 343-357, (1979).
14. N.S.McIntyre, Uses of Auger Electron and Photoelectron Stectroscopies in Corrosion Science, in *Practical Surface Analysis*, pp.397-428, ed. M.P.Seah, John Wiley & Sons:Chichester (1983).
15. I.Olefjord and B.Brox, *5th Int. Symposium on Passivity*, pp. 561-570, Bombannes, France, (1983).
16. W.R.Cieslak and D.J.Duquette, The Effects of Alloying on the Resistance of Ferritic Stainless Steels to Localized Corrosion on Cl Solution, in *Passivity of Metals and Semiconductors*, pp.405-412, ed. M.Froment, Elsevier:Tokyo (1983).
17. A.Atrens, Z.*Werkstofftech.*, **15**, pp. 309-314, (1984).
18. B.-O.Elfstrom, *Mater. Sci. Eng.*, **42**, pp. 173-180, (1980).
19. H.J.Mathieu and D.Landolt, *Corros. Sci.*, **26**, no. 7, pp. 547-559, (1986).
20. R.C.Newman, *Corros. Sci.*, **25**, no. 5, pp. 341-350, (1985).

Table 1a. Characteristic values for the passive film formed on 18-12 for various exposure conditions at 25 °C.

Chemical Species		Immersion Time (h)***					solution and air exposure+		air exposure++		air and solution exposure+++		
		0.5	1	15**	21	44	3(min)	1.5(h)	3(min)	22(h)	2(min)	0.5(h)	14(h)
Fe^{ox}_{max}	depth (Å)	5	3	6	4	0	3	4	8	8	5	2	4
	conc. (%)	16	15	13	11	10	13	16	12	16	12	12	13
$Fe^{m}_{0.5sat*}$	(Å)	18	15	15	17	18	17	23	15	21	14	16	18
Cr^{ox}_{max}	depth (Å)	7	9	11	12	12	10	14	8	15	5	10	12
	conc. (%)	19	20	26	26	26	23	17	18	18	16	17	19
$Cr^{m}_{0.5sat}$	(Å)	24	25	28	30	34	31	34	14	38	32	34	32
$Ni^{m}_{0.5sat}$	(Å)	12	16	10	16	16	17	21	16	20	14	15	16
Ni^{m}_{max}	conc. (%)	14	14	17	16	16	14	14	13	15	15	14	15
$O_{(max-min)/2}$	(Å)	22	18	11	17	22	19	20	22	19	14	14	19

* Depth corresponding to a concentration equal to half the saturation concentration.

** Data of Bruesch et al [1].

*** Exposure conditions before ESCA analysis were 12 to 72 h air exposure and stated time in 0.1 M NaCl.

\+ 12 to 72 h air exposure, 15 h immersion in 0.1 M NaCl and stated time air exposure.

++ No prior exposure before stated air exposure.

+++2 min exposure to air and then exposure to 0.1 M NaCl.

Table 1b. Characteristic values for the passive film formed on 29-4-2 for various exposure conditions at 25 °C.

Chemical Species		Immersion Time (h)					air exposure***
		0.5	15**	15	22	44	60(min)
Fe^{ox}_{max}	depth (Å)	4	0	2	2	5	5
	conc. (%)	8	21	8	7	5	11
$Fe^{m}_{0.5sat}$*	(Å)	16	13	16	17	19	20
Cr^{ox}_{max}	depth (Å)	10	10	10	10	11	10
	conc. (%)	21	26	27	28	28	25
$Cr^{m}_{0.5sat}$	(Å)	19	24	23	24	27	28
Ni^{m}_{max}	conc. (%)	2.8	3.7	5.9	-	7.1	-
$O_{(max-min)/2}$	(Å)	18	-	16	16	17	21

* Depth corresponding to a concentration equal to half the saturation concentration.

** Data of Bruesch et al [1].

*** After 15 h exposure to 0.1 M NaCl.

Table 2. Values of the two main factors governing the passive film thickness with immersion times for 18-12 and 29-4-2.

Alloy	Pretreatment	Depth (Å)	
		$C_{Cr^{ox}}=C_{Cr^{m}}$	$C_{O}=C_{Fe^{m}}$
18-12	*immersion time(h)*		
	0.5(25 °C)	22	20
	0.5(60 °C)	20	20
	0.5(90 °C)	24	23
	1.0	20	16
	1.0(E=0 mV)	17	19
	15*	23	13
	15**	(16)	(16)
	21	22	17
	44	22	19
	air exposure		
	3(min)	12	15
	22(h)	19	22
	air effect		
	3 (min)	20	18
	1.5(h)	25	21
	solution exposure		
	2 (min)	18	15
	0.5(h)	18	15
	14(h)	20	19
29-4-2	*immersion time(h)*		
	0.5(25 °C)	17	17
	0.5(60 °C)	15	16
	0.5(90 °C)	20	19
	1.0(E=0 mV)	18	18
	15	17	17
	15*	20	-
	15**	(13)	(13)
	22	16	17
	44	14	17
	air effect		
	60(min)	21	21

* Results from the depth profile of prior study [1].

** Thickness from the prior study [1].

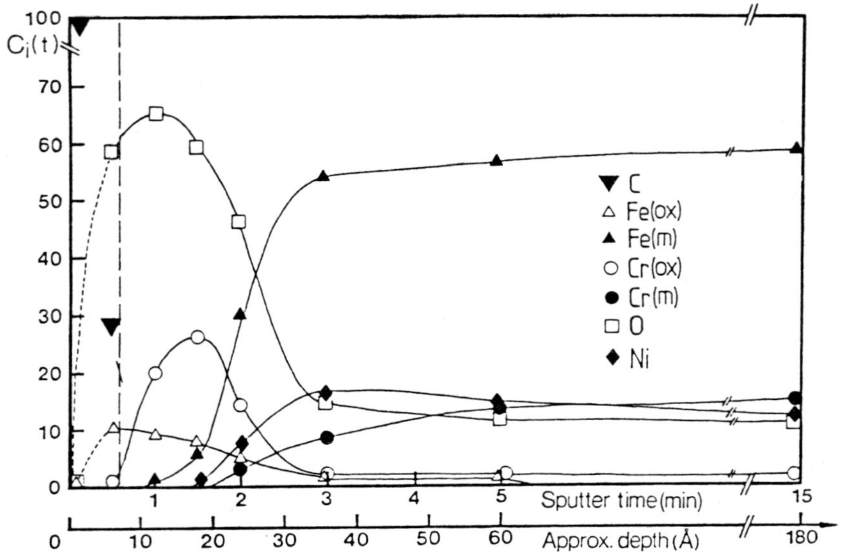

Fig 1. Actual composition profiles for 18-12. Preparation of 21 to 72 h air exposure and in solution for 44 h. Dashed line represents zero point of the passive film.

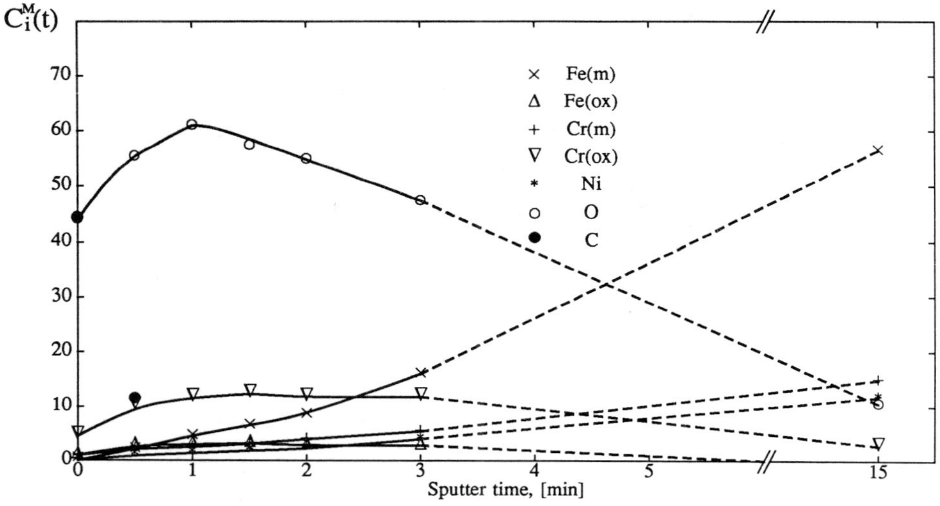

Fig 2. Measured composition profiles for 18-12. Preparation of 21 to 72 h air exposure, followed by exposure at potential E = 350 mV_{SCE} in 0.1 M NaCl at 25 °C for 1 h.

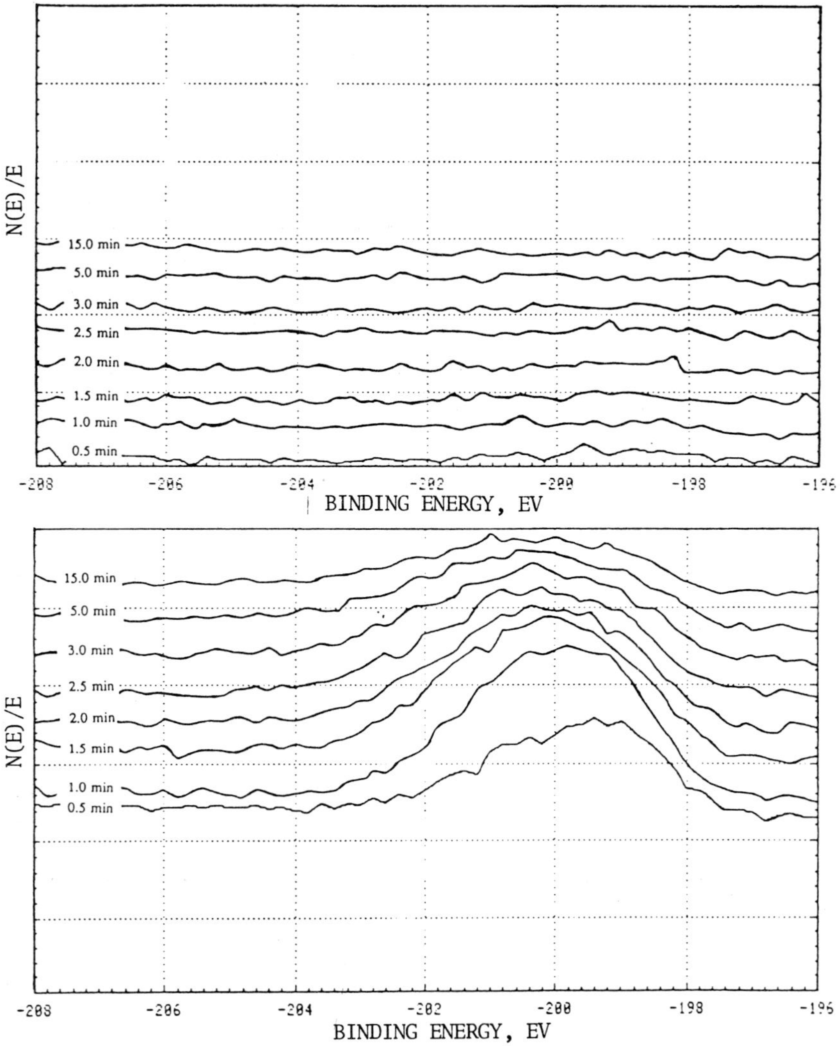

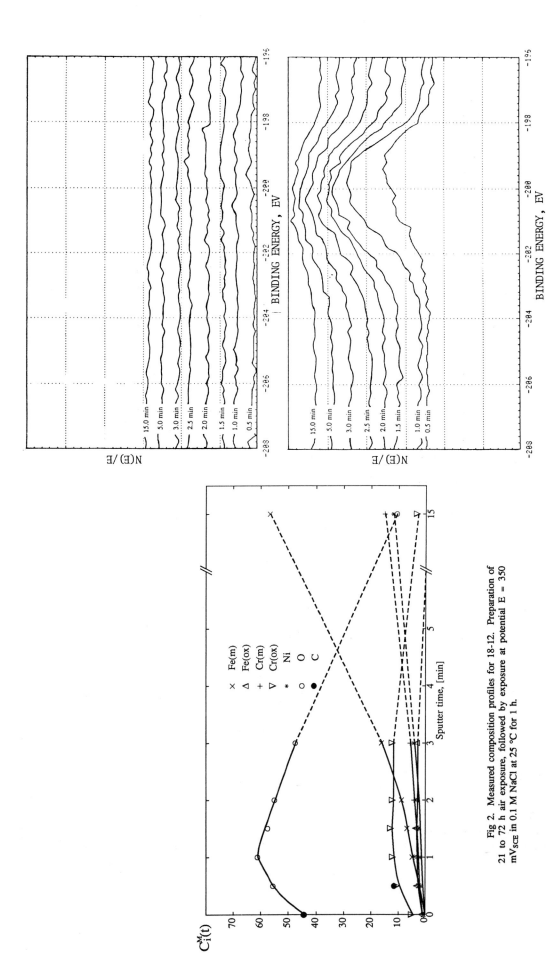

Fig 3. Evolution of the chloride peak consisting of $2p_{1/2}$ and $2p_{3/2}$ states, for 18-12 as a function of sputtering time: (a) E = 500 and (b) E = 800 mV_{SCE}.

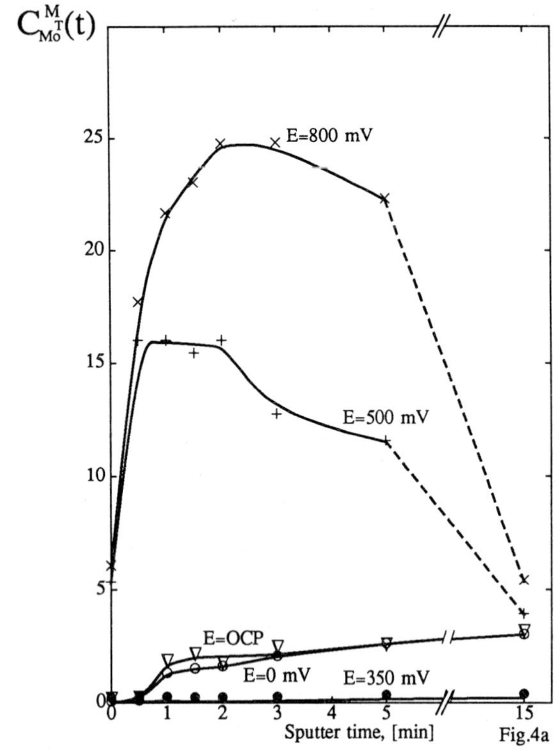

Fig.4a

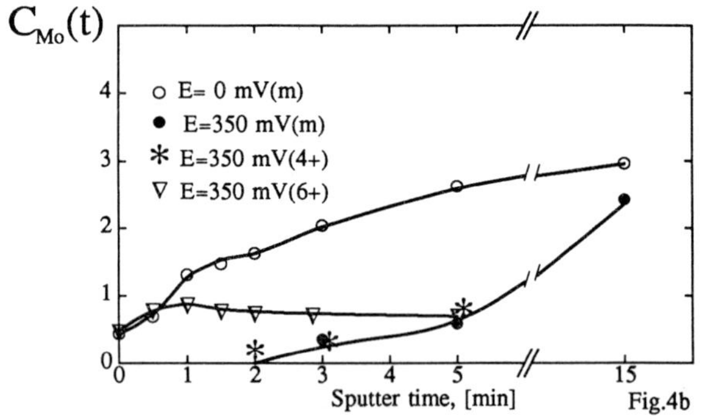

Fig.4b

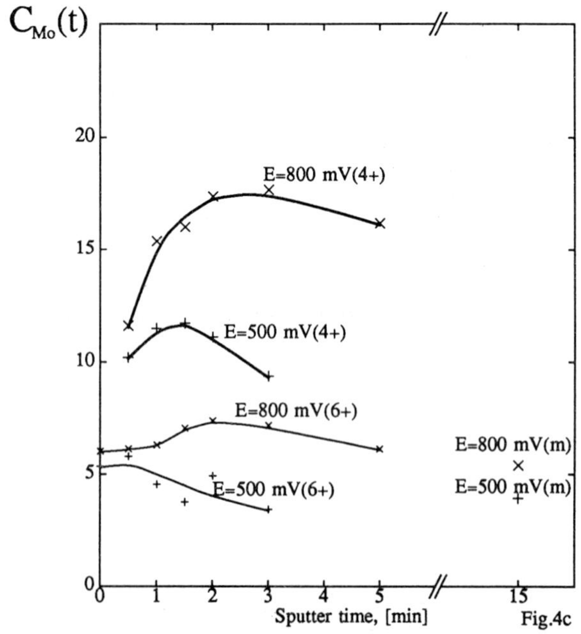

Fig.4c

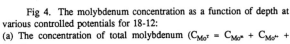

Fig 4. The molybdenum concentration as a function of depth at various controlled potentials for 18-12:
(a) The concentration of total molybdenum ($C_{Mo^T} = C_{Mo^m} + C_{Mo^{4+}} + C_{Mo^{6+}}$);
(b) The concentration of the individual chemical species: C_{Mo^m}, $C_{Mo^{4+}}$ and $C_{Mo^{6+}}$, at controlled potential 0 and 350 mV_{SCE};
(c) The concentration of the individual chemical species: C_{Mo^m}, $C_{Mo^{4+}}$ and $C_{Mo^{6+}}$, at controlled potential 500 and 800 mV_{SCE}

Atmospheric corrosion resistance of stainless steels

N G Needham, P F Freeman, J Wilkinson and J Chapman

The authors are with the British Steel Corporation. NGN and PFF at Swinden Laboratories, Rotherham; JW and JC at BSC Stainless, Sheffield.

SYNOPSIS

The degree of pitting and aesthetic appearance of four austenitic stainless steels have been evaluated after long term exposure at sites in the UK and USA. There is a significant effect of the environment on pitting with the highest degree of pitting being found in a heavy industrial atmosphere. The known effect of molybdenum on the pitting performance has been quantified and this is reflected in lower levels of surface staining in higher molybdenum containing steels. Additionally, the effects of washing and design have been shown to be factors affecting corrosion behaviour.

Surface finish has been shown to have a significant effect on corrosion performance. Wide variations in behaviour have been found for surfaces with an abraded No. 4 finish as the result of surface damage introduced during mechanical abrasion. By eliminating this damage, the corrosion performance can be considerably improved.

The data generated have shown that corrosion by pitting is, in general, superficial. The work highlights the need to optimise the selection of material, surface finish and design of the components.

INTRODUCTION

Stainless steels are used primarily for their resistance to corrosion and for many years, it has been recognised that these materials have excellent long term corrosion properties in naturally occurring environments. This, coupled with their low maintenance costs, means that despite their higher initial material costs, structures with a long design life fabricated from stainless steel have a low life cycle cost compared to their traditional competitors. Thus, stainless steel is now becoming increasingly used for conventional architectural applications for cladding buildings and is starting to be considered for general engineering applications, particularly in aggressive environments. An example of this is the growth in the use of stainless steels in a variety of applications such as module wall cladding and cable ladders, on the topside of North Sea oil platforms. Although many architectural applications require that the surface appearance of the steel must be maintained, this requirement is not so stringent for a number of structural applications. For the latter, the mechanical integrity of the component must be maintained and the loss in net section due to corrosion must be limited. In natural environments, the predominant form of corrosion of stainless steels is localised pitting corrosion. Thus, to predict the behaviour of stainless steels, it is necessary to understand the factors affecting their long term pitting corrosion performance and ensure that the correct grade of material is used for each application.

The factors affecting the atmospheric corrosion resistance of stainless steels have been documented in a number of publications[1-4]. However, these studies have either been short term in nature or have concentrated on the changes in visual appearance of the surfaces of the stainless steel. In natural environments, it has been shown that pitting of stainless steels is controlled by the type of environment, the composition of the steel, the cleaning frequency and the surface finish. This paper examines the long term pitting corrosion performance of a number of austenitic stainless steels and the factors that must be considered when selecting a stainless steel for a particular application.

EXPERIMENTAL TECHNIQUE

Materials

Long term tests have been undertaken in the UK on four grades of austenitic

TABLE 1 - MATERIAL COMPOSITION, %

Type	% C	% Mn	% Si	% Ni	% Cr	% Mn
304	0.047	1.67	0.60	10.30	18.50	0.31
315	0.045	1.53	0.42	9.38	17.20	1.44
316	0.047	1.72	0.36	11.86	17.38	2.70
317	0.044	1.61	0.52	10.97	18.64	3.45

stainless steel, with an abraded (similar to that defined as a No. 4 finish in BS 1449), electropolished, and cold rolled softened, descaled (2B) finish. The composition of the steels is given in Table 1. Additionally, short term cyclic salt spray tests have been undertaken on commercially produced Type 316 S31 samples with a No. 4 abrasive belt polished finish.

Site Testing

Panels 200 mm x 200 mm x 1.17 mm have been exposed for up to 18 years at each of four sites, representing a severe marine, heavy industrial, semi-industrial and rural environment in the UK. Additionally, samples of Type 304 and 316 stainless steel in the 2B condition, tested at a marine site in the USA, were available for comparison.

At the UK sites, the steels were positioned such that they were either directly exposed to the environment or were in a sheltered situation. Half the panels were washed every six months with soap and water to assess the effect of routine maintenance.

Laboratory Testing

Traditionally, continuous salt spray testing has been used to assess rapidly the performance of steels in the laboratory. However, this test was found not to reproduce the performance of stainless steels in natural marine environments. Thus, a cyclic salt test has been developed which allows the performance of stainless steels to be ranked rapidly in the laboratory and simulate their performance in marine environments. This test involves spraying the panels for one hour with a test solution containing 3% sodium chloride at a temperature of 35 °C followed by a period of two hours during which the test chamber is exhaust ventilated and the panels allowed to dry. This three hour cycle is repeated for a total of 21 days.

Pitting Assessment

On the site exposed panels, a small area was cleaned with a non-abrasive cleaner and pit size distributions and pit densities were measured using optical microscopy and 'Talysurf' surface finish equipment, as described previously[5]. Surface staining on laboratory exposed panels was assessed by point counting, using a standard 400 point grid.

TEST RESULTS

Environment

The environment has a significant effect on the level of pitting measured on panels directly exposed to the weather, Fig. 1. Variations in the degree of pitting, assessed from measurements of the pit density and maximum pit depth reflect the level of pollutants in the atmosphere, particularly the sulphur dioxide and chloride ion content. The data show that the pit densities were highest at the heavy industrial site, being double those found at the rural site which had the lowest density of surface pits. However, the pit depth measurements show that pit growth is greatest at the marine and heavy industrial sites which were a factor of four higher than those in the rural and semi-industrial environments.

Composition

The principal element affecting the corrosion performance of the Type 300 series of austenitic stainless steels is molybdenum. Increasing the molybdenum content from a Type 304 steel (0.3% Mo) to a Type 317 steel (3.45% Mo) reduced the pit densities on both 2B and dull polished material by an order of magnitude at all sites. The data for 2B material at the two industrial sites are shown in Fig. 2(a) and the results are similar to those observed at the marine site, Fig. 3(a). Additionally, higher levels of molybdenum also reduce the pit growth rate, the effect being greatest in the heavy industrial and marine environment, Fig. 3(b). The data from the marine site in the USA are similar to that observed at the UK site, Fig. 3(b).

Molybdenum not only affects the degree of pitting but, as would be expected, it has a similar effect on the degree of staining observed on the surface of the panels. Increasing the molybdenum content of the steels significantly reduces the degree of surface staining observed, particularly in the more aggressive environments, Fig. 4(a) and (c).

Surface Finish

The finish applied to the surface of a stainless steel can have a considerable effect on both the degree of staining and the level of pitting incurred on a surface. After 18 years exposure, the Type 304 steel polished to a highly reflective smoother finish, such as that produced after electropolishing, shows little staining compared to the less reflective rougher surfaces on a cold rolled and descaled surface (2B) or an abraded (No.4) finish, Fig. 4. These differences in susceptibility to staining can be duplicated in the cyclic salt spray tests. The modern, highly reflective bright annealed finish has good corrosion properties with low levels of staining, which is similar to that found for electropolished samples. However, increasing the roughness of the surfaces from a 2B to an abraded No. 4 finish results in an increase in the degree of staining, Fig. 5.

These differences in appearance reflect the susceptibility of a material to pitting with the less reflective smoother surfaces having a larger number of sites at which pits can initiate. This variation is clearly illustrated by the pit density data, Fig. 3(a). Irrespective of steel type, the density of pits is higher on material with a dull polished (No.4) finish than a 2B finish.

Abraded Finishes

These finishes are commonly used for architectural applications and are applied by mechanically abrading the surface using an abrasive belt. Depending on the procedure adopted, the surface can have a range of roughness from R_a values of 0.2 um up to and in excess of 1.5 um. These roughness values compare with typical values of 0.1 um and 0.05 um for material with a 2B and bright annealed finish respectively.

Assessment of the corrosion performance of commercially produced material with a No.4 finish has shown that there is a wide range in the corrosion behaviour of material with nominally the same finish. Cyclic salt spray corrosion data on Type 316 stainless steels show that at best these finishes have a corrosion performance similar to the 2B finish to which they are normally applied. However, the degree of staining observed on test panels can vary by a factor in excess of eight.

Examination of the surfaces has shown that pits initiate at surface microdefects. Additionally, highly reflective smoother surfaces have fewer defects than rougher surfaces such as those with an abraded finish.

The best corrosion performance of material with a No. 4 finish was found on samples which had a clean cut abraded finish with little surface damage, Fig.6(a). By comparison, samples with poor corrosion performances have a high degree of surface damage, in the form of gouges, tears and laps which constitute micro-crevices, at which pits can initiate and grow, Fig. 6(b). To optimise the corrosion performance of material with a No.4 finish, this type of surface damage must be prevented during polishing.

In the UK, the abraded No.4 finish is commonly produced using alumina abrasive belts. Surface examination suggests that the cutting action of this type of belt makes it difficult to avoid the deleterious surface damage features which result in pit initiation. The surfaces produced by using alumina belts have R_a values ranging from 0.4 to 1.5 um and the surface topography is typified by that shown in Fig. 6(b). Cleaner cut surfaces with good corrosion resistance can be produced. However, commercially reproducible surfaces with this topography are more commonly produced using silicon carbide abrasive belts. Using this abrasive results in smoother surfaces with R_a values less than 0.4 um. A typical example of such a surface is shown in Fig. 6(a).

Washing

Washing, either by natural or artificial methods, has been found to improve the resistance to pitting, primarily by reducing the rate of pit growth. This is particularly noticeable for the mill finished 2B product, Fig. 7, where the maximum pit depth for the sheltered samples is considerably higher than that for the exposed samples which are washed by the natural action of the rain. The effect of washing on pit density is small for both 2B and No.4 dull polished finish material.

DISCUSSION

Both long term testing in natural environments and short term laboratory tests have shown that there are a number of important factors that influence the corrosion resistance of the 300 series austenitic steels. These include the environment, the composition of the steel, the design of the component, the surface finish applied to the material and the washing/maintenance procedure adopted. The choice of material for a specific application is dependent on all these factors but is primarily governed by the required aesthetic appearance of the surface together with the cost of the product.

The environment has a significant effect on the degree of pitting and, hence, staining observed on the surfaces of the steel. Previous work has shown that the important species dictating the aggressivity of an environment are the sulphur dioxide and chloride ion content in the environment. The increased aggressiveness of marine environments due to airborne chlorides is well documented and has been confirmed in the long term tests undertaken in the present work at the UK marine site. The similarity in pitting performance at this and the USA marine site, despite differences in climatic conditions, emphasises that it is the presence of the chloride ion which is the principal factor controlling pitting in such circumstances. However, the highest levels of pitting were, in general, observed at the heavy industrial site. Although the level of airborne chlorides is low and similar to that at the semi-industrial and rural site, sulphur dioxide in the atmosphere results in higher rates of attack. It has been shown that there is a synergistic effect of the chloride ion and sulphur dioxide on pitting[6] so that care is needed in assessing the aggressivity of environments when these species are present.

For many applications such as building cladding, architects utilise the aesthetic appearance of stainless steels and a prime requirement is that the degree of pitting must be such that staining is a minimum. In this context, the composition of the steel, the applied surface finish and the washing/maintenance frequency affect the corrosion behaviour of the steel. It is well documented that increasing the level of molybdenum in stainless steels decreases the susceptibility to pitting. This has been confirmed by the long term data at both the UK and USA sites, where increasing the molybdenum content from 0.3% to 3.45% reduces the pit density by an order of magnitude and has a similar effect on pit growth rates. For most applications in natural environments, Type 316 stainless steel will have sufficient corrosion resistance to maintain the aesthetic appearance without significant staining. However, the surface applied to the steel should be free of sites at which corrosion can initiate and an adequate maintenance procedure must also be adopted.

Stainless steels are becoming more widely used for structural applications where the aesthetic appearance of the material is of little importance. For these steels, the prime requirement is structural integrity. Many structural and architectural components manufactured from stainless steel have a design life in excess of 50 years. Thus, there is a need to ensure that perforation of such components does not occur or that the loss in section due to corrosion does not affect their load bearing capability. Assuming a linear rate of pit propagation and that the environment does not change, then the estimated time for the largest pits to perforate a 1 mm thick sheet in a marine environment is well in excess of the design life of most structures, even for the Type 304 steel which has the lowest corrosion resistance. Further, examination of the micro pit size distribution shows that,for Type 316 steels, in excess of 95% of the pits are below 5 um in size so that loss of load carrying capacity as the result of corrosion should not present a significant problem. In sheltered conditions, the time to perforate is clearly reduced but even in the more aggressive heavy industrial and marine environments, perforation times for the Type 316 stainless steel are high.

These values should only be used as a guide principally because the pit propagation rate assessed is higher than might be expected. However, as insufficient long term pit propagation data are available, it was decided to take the worst case available and obtain a pessimistic estimate of the rate of pit propagation. Additionally, over the period that the work was undertaken, atmospheric pollution has declined dramatically and the environments, particularly the industrial environments, have changed significantly. Thus, it is considered that higher rates of attack occurred in earlier years and that the rate of corrosion is now much lower.

Atmospheric pollutants, particularly salts and inert matter, collect on any surface exposed to natural environments. Washing, either manually or through the natural action of rain water, will affect the level of pollutants on the surface and, hence, the degree of corrosion. Washing the panels manually every six months or direct exposure to the washing action of the rain can significantly reduce the rate of pit propagation, particularly in the more aggressive environments, Fig. 7(b). Care must be taken to design components to minimise overhangs and sheltered areas and hence maximise the natural washing action of rain. Additionally, manual cleaning of components is beneficial, not only to remove inert matter and maintain surface appearance but also to remove aggressive salts and hence optimise the corrosion resistance.

Although surface finish was known to affect corrosion performance, the present work has highlighted the dramatic reduction in performance that can occur if precautions are not taken when surfaces are mechanically abraded to produce an abraded (No. 4) finish. Smooth, highly reflective surfaces are less susceptible to pitting and staining than less reflective, rougher surfaces because defects at which pits can initiate on the surfaces are minimal. Thus, bright annealed or electro or mechanically polished surfaces are less susceptible to staining and pitting than cold rolled, soften and descaled (2B) or dull polished (No. 4) finishes. This can clearly be seen in the data from the cyclic salt spray tests, Fig. 5, and the long term exposure samples, Fig. 3. Although the surface finishes applied to stainless steels are classified in BS 1449, the description for a No.4 finish is loose and embraces a wide range of roughness and surface topography, Fig. 6. The introduction of defects such as tears, gouges and crevices have been identified as sites at which pitting initiates and results in pitting and staining of the surfaces. The variability in corrosion behaviour is marked, Fig. 5, and considerable care needs to be taken to ensure the optimum surface topography is applied. The use of alumina abrasive belts has been associated with the introduction of defects on surfaces and this is thought to be linked with the nature of the cutting action of the alumina particles. It has been shown that silicon carbide abrasives result in a cleaner cut with less surface defects and the resulting corrosion behaviour of these surfaces is similar to that of 2B material to which it is applied. However, surfaces produced using silicon carbide belts are, in general, much smoother than those with alumina belts and hence work is on-going to identify other methods of producing surfaces with different No.4 finishes.

As the No. 4 finish is widely preferred by architects for building cladding material, it is necessary to pay close attention to surface topography to ensure optimum corrosion resistance, especially in more aggressive natural environments.

CONCLUSIONS

Long term exposure of a series of austenitic steels to sites with different environments, both in the UK and USA, has shown that corrosion by pitting is, in general, superficial. The highest level of pitting was experienced at a heavy industrial site due to the high level of pollutants in the atmosphere.

Estimates of the rate of pit propagation have shown that perforation times are well in excess of the design life of most structures. Coupled with low maintenance costs, this means that stainless steels are cost effective materials for architectural applications.

The work has shown that surface finish can have a significant effect on corrosion behaviour, particularly for the surfaces with an abraded No. 4 finish. This results from surface damage introduced during mechanical abrasion. By optimising the abrasion polishing process route, clean cut surfaces can be produced whose corrosion performance is similar to that of 2B material.

The design and maintenance schedules used on stainless steels have been shown to affect corrosion behaviour.

ACKNOWLEDGMENTS

The authors would like to thank Mr. A.P. Pedder, Director of BSC Stainless and Dr. R. Baker, Director of Research and Development of British Steel Corporation for permission to publish this paper.

REFERENCES

1. Shirley, H.T. and Truman, J.E., JISI 1948, Dec., 367.

2. Truman, J.E., Stainless Steel Industry, 1979, Jan., 11.

3. Evans, T.E., 4th International Congress in Metallic Corrosion, Amsterdam, 1969.

4. Chandler, K.A., ISI Publication 117 1969, p.127.

5. Stone, P.G., Hudson, R.M., and Johns, D.R., Stainless Steels 84 The Metals Society 1985, p.478.

6. Johnson, K.E., Cor. Sci. 1982, 22(3), 175.

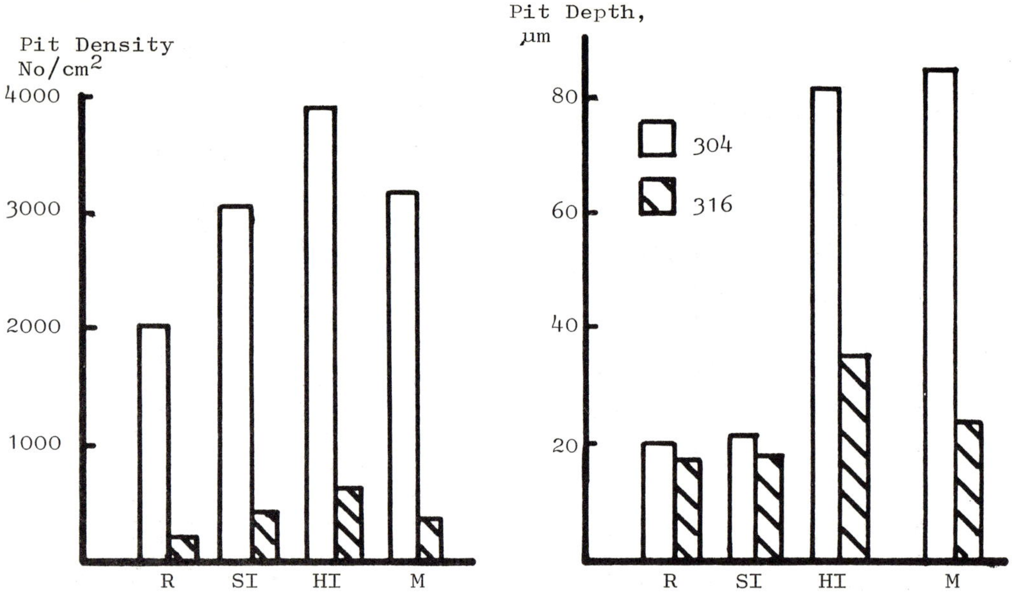

Fig. 1 The pitting behaviour of Type 304 and 316 stainless steels at a rural (R), semi industrial (SI), heavy industrial (HI) and marine (M) site.

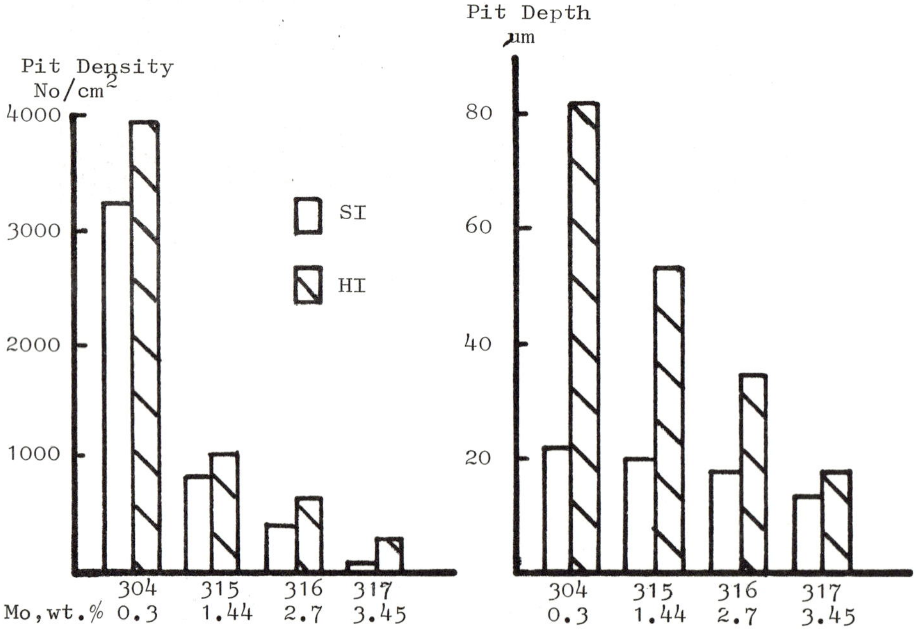

Fig. 2 The effect of composition on the degree of pitting at a semi industrial (SI) and heavy industrial (HI) site.

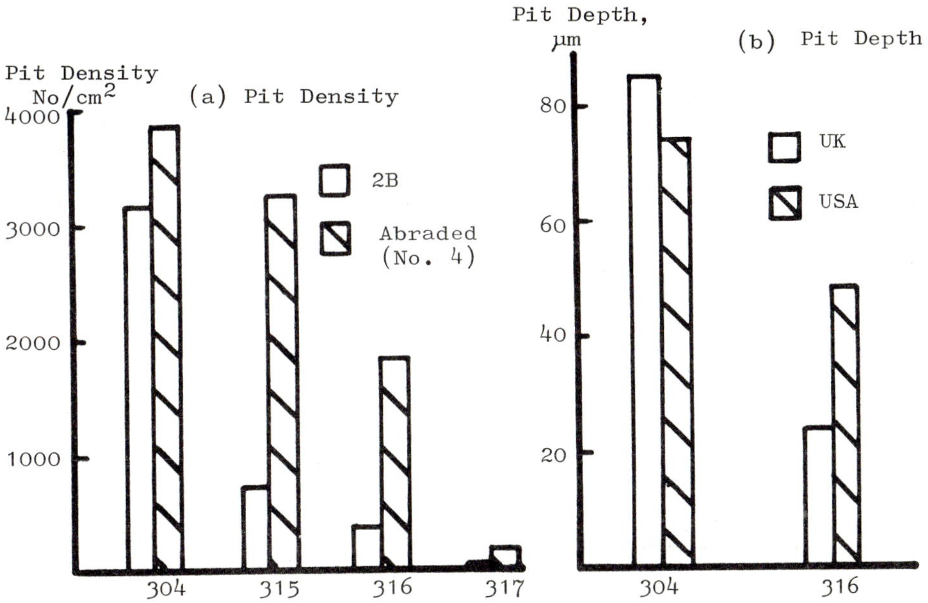

Fig. 3 The effect of composition and surface finish on pitting behaviour at the marine sites.

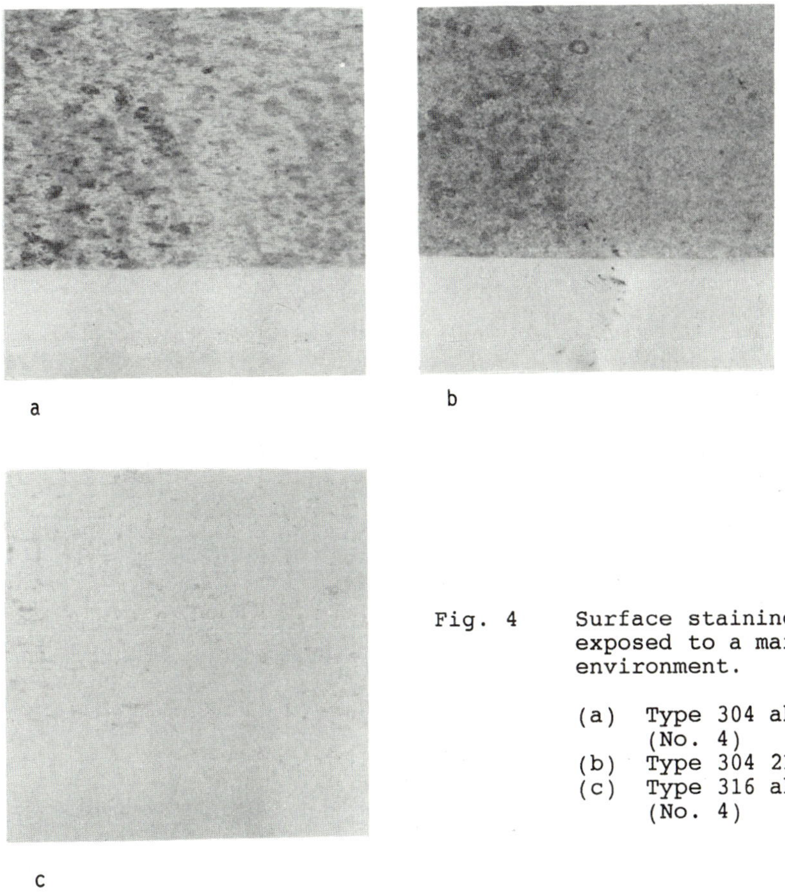

Fig. 4 Surface staining on panels exposed to a marine environment.

(a) Type 304 abraded finish (No. 4)
(b) Type 304 2B finish
(c) Type 316 abraded finish (No. 4)

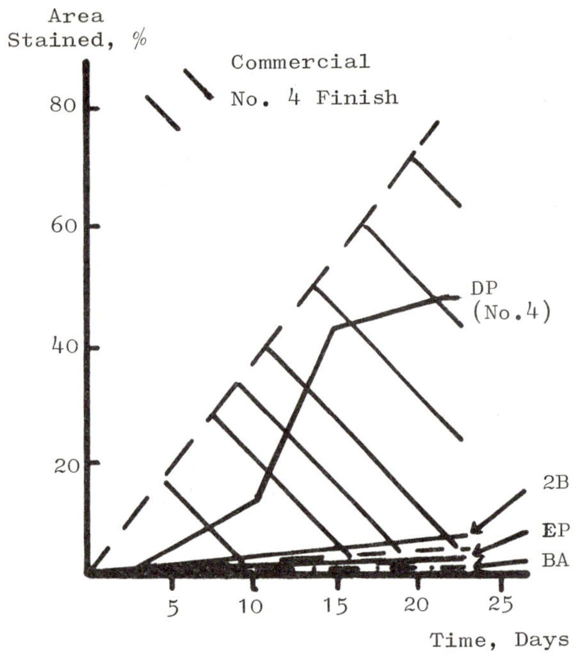

Fig. 5 The corrosion performance of Type 316 samples in the cyclic salt spray test with a dull polished (DP), 2B and electro polished (EP) finish compared to commercially polished material with a No. 4 finish and a bright annealed (BA) finish.

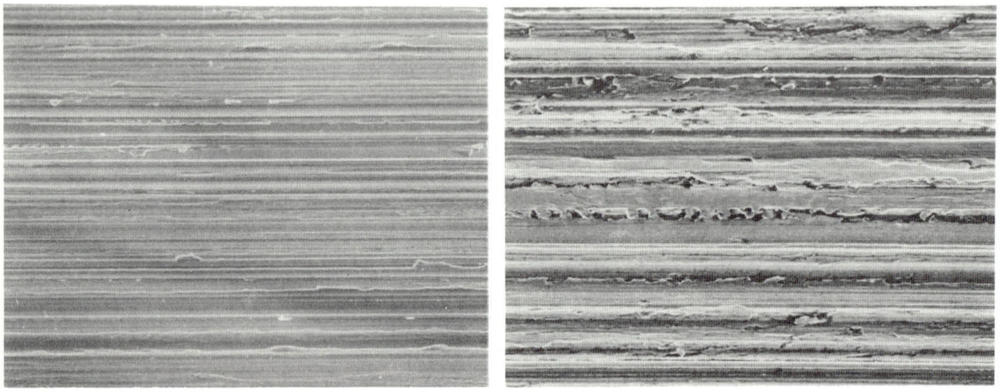

a b

Fig. 6 Surface topography of commercial polished Type 316 stainless steel with a No. 4 finish (x375).

(a) Clean Cut
(b) Surface Damage

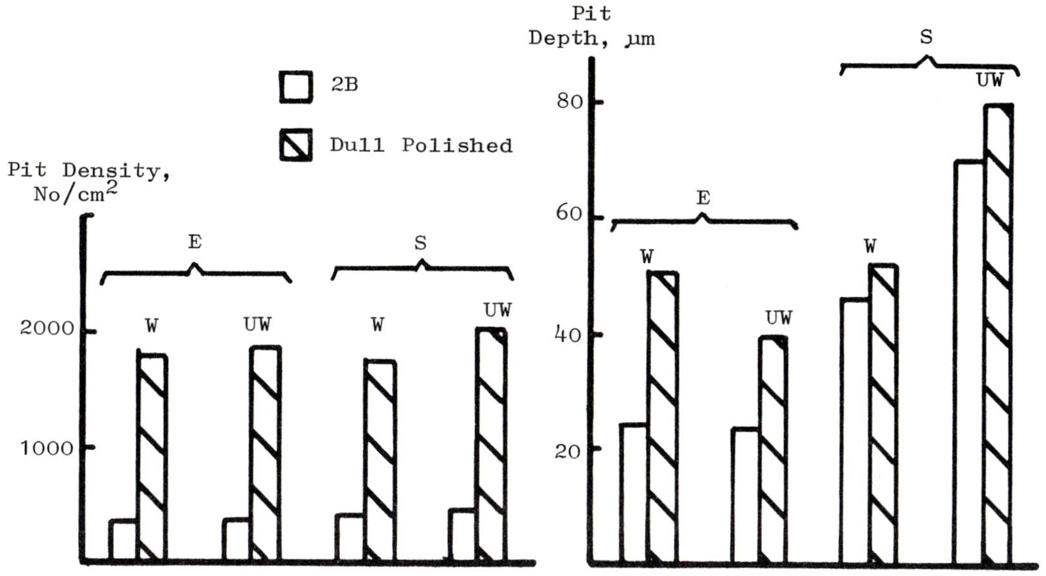

Fig. 7 The effect of sheltering (S) and direct exposure (E) on the pitting behaviour of Type 316 stainless steel on washed (W) and unwashed (UW) panels at the marine site.

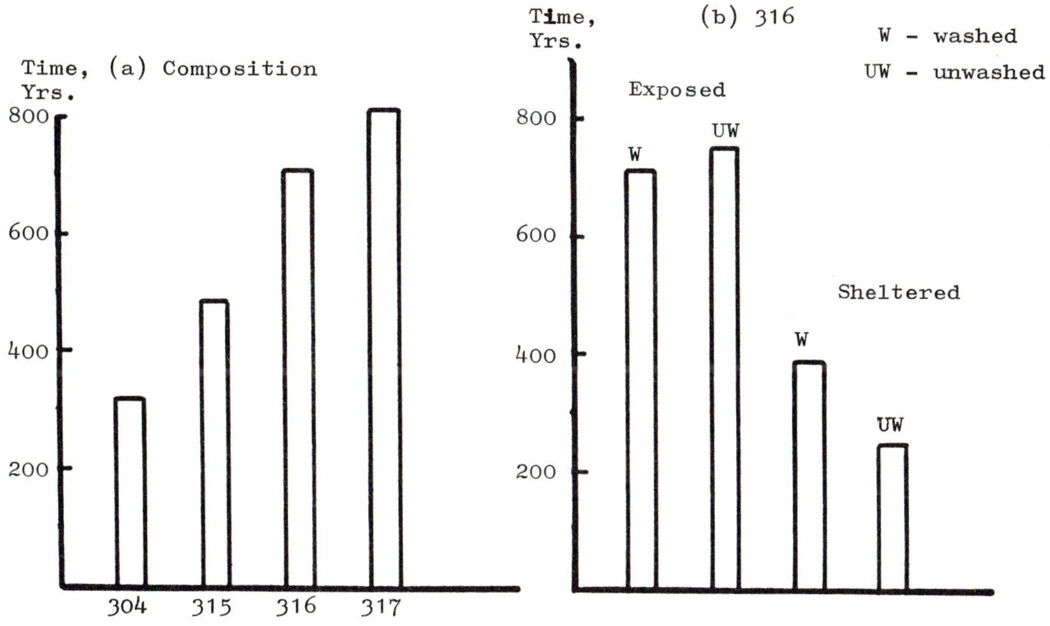

Fig. 8 The effect of composition, design and washing on the time for the largest pits to perforate 1 mm thick strip exposed at the marine site.

Low-molybdenum austenitic stainless steel: improved pitting corrosion resistance

Y Fujiwara, R Nemoto and K Osozawa

The authors are with Nippon Yakin Kogyo Co. Ltd., Kawasaki, Japan. YF at the Technical Research Center, RN and KO at the Development Department.

SYNOPSIS

The effect of type of inclusion and alloying elements on the pitting corrosion of 18-8 austenitic stainless steel was studied on an laboratory scale.
The beneficial effects of the alloying elements such as molybdenum and nitrogen on pitting corrosion resistance are greater in alloys from which those harmful inclusions are removed.
Based on these experimental results,the optimum composition was obtained and an extra low sulfur(<0.0010%)18-8 austenitic stainless steel was produced through commercial process and it was confirmed that 0.3Mo-0.13N-0.4Mn-<0.0010S containing alloy had equal pitting corrosion resistance to that of 316 type stainless steel.

INTRODUCTION

It has been known that pitting corrosion of austenitic stainless steel frequently initiates at sulfide inclusion[1~4] and by controlling the type and amount of sulfide the pitting corrosion resistance can be improved.[5~8] Accordingly there have been many investigations on the effect of sulfide inclusions on the corrosion resistance of stainless steel, but the oxide inclusions also act as the nucleation of pitting corrosion.[9]
In this study the effect of sulfide inclusion on the pitting corrosion resistance of austenitic stainless steel was investigated and then the effect of types of oxide inclusion on pitting corrosion resistance was studied on extra low sulfur and low manganese containing alloys which were treated by various deoxidizers.
Furthermore the effect of alloying elements on the pitting corrosion resistance was investigated on the alloys from which harmful inclusions were removed.
Based on these experimental results, molybdenum saved austenitic stainless steel improved for pitting corrosion resistance was produced by commercial process and its properties such as atmospheric corrosion resistance were evaluated by a new accelerated test method simulated to an outdoor atmospheric corrosion process.

EXPERIMENTAL

Chemical composition ranges of tested materials are shown in Table 1 and Table 2. Induction melted steel ingots weighing approximately 6~10kg were forged,hot-rolled and cold-rolled to approximately 1mm thickness. These steel sheets were heattreated at 1050°C for 10 minutes before test.
In the alloys of group(A) in Table 1,the contents of sulfur and manganese were varied.The alloys of group(B) containing 0.4%Mo,0.15%N and extra low sulfur (<0.0010%) were deoxidized with various deoxidizers such as Ca or Al. Extra low sulfur alloys(S<0.0010%) were obtained by desulfurization refining with Li_2O slag.
In order to investigate the effect of the alloying element,the alloys in Table 2 from which the harmful inclusions to pitting corrosion had been excluded were used.
These samples were examined by optical microscope and non-metallic inclusions of the samples were analyzed by use of X-ray microanalyzer.
Pitting corrosion resistance was evaluated by pitting potential measured potentiodynamically in 3.5% NaCl solution at 30°C.
Atmospheric corrosion resistance was evaluated by a new accelerated test method simulated to an outdoor atmospheric

corrosion process[10] The process of this test is as follows;

Dropping of artificial sea water on the specimen
↓
Dried by infrared lamp
↓
Dewed in constant humid atmosphere
↓
Rust

Details of this test condition are shown in Fig.1.

RESULTS AND DISCUSSION

Effects of S and Mn on Pitting Corrosion Resistance.

The effect of the amount of S and Mn on pitting potential is shown in Fig 2. In proportion to decreasing S and Mn contents the pitting potential becomes noble. It is considered that the dependence of pitting corrosion on S and Mn contents is related to the amount of sulfide.

In Fig.3 the dependence of the amount of sulfide on sulfur content is shown. The amounts of sulfide were measured by microscopical point counting. The critical value of sulfur content at which no sulfide was detected increases with a decrease in the amount of Mn in steels. This critical value was plotted in Fig.4.

As Mn has a strong affinity with sulfur, the lower the Mn content in steel is ,the less the sulfide inclusion precipitates. Effect of sulfide inclusion on pitting corrosion resistance is related to not only the amount of sulfide but also the type of sulfide. As shown in Photo.1 the morphology of sulfide changes with Mn content. In high Mn containing alloys Mn-rich sulfide precipitates and in low Mn alloy Cr-rich sulfide precipitates. Corrosion resistance of Cr-rich sulfide is superior to that of Mn-rich sulfide. Therefore Cr-rich sulfide hardly acts as the initiation of pitting corrosion compared with Mn-rich sulfide.

Effect of Types of Oxide

In order to investigate the effect of types of oxide, the alloys from which the sulfide inclusion harmful to pitting corrosion had been excluded were used. Namely the alloys of Group(B) in Table 1 which were low Mn(0.2%Mn) and extra low sulfur(<0.0010%S) base alloys were deoxidized by various deoxidizers listed in Table 3.

In Table 4 the results of X-ray micro-analysis of oxide inclusion were shown. The type of oxide varies with the kind of deoxidizer used. In Fig.5 the dependence of pitting potential measured on the alloy deoxidized with various deoxidizers on nitrogen is shown. The beneficial effect of nitrogen to pitting corrosion resistance is greater in alloys treated mainly with Al deoxidizer than in alloys treated mainly with Ca deoxidizer.

Photo.2 indicates that $CaO-Al_2O_3$ inclusion in the alloy treated with Al-Ca composite deoxidizer acts as the initiation site of pitting corrosion. The tendency of oxide to become the initiation site of pitting corrosion depends on the affinity of the oxide with water and the shape of oxide inclusion.The stronger the affinity of oxide with water is, the greater the tendency of the oxide to become the initiation site of pitting corrosion. If stringer shaped inclusions dissolve into water a crevice is formed in steel and consequently it becomes easily the initiation site of pitting corrosion.

Fig.6 shows the phase diagram for the system $CaO-Al_2O_3-SiO_2$[11] The oxide composed of a large amount of CaO has the affinity with water and becomes a stringer shaped inclusion during hot rolling because of a low melting point. Al_2O_3 inclusions hardly become the initiation site of pitting corrosion because of high corrosion resistance and high melting point. Therefore in order to exclude the harmful inclusions to pitting corrosion Mn and S content in steel should be lowered and the steel should be treated with Al deoxidizer.

Effect of Alloying Element

The effect of alloying elements on corrosion resistance was investigated on the alloy from which the harmful inclusions of sulfide and oxide were excluded. Fig.7 shows the dependence of pitting potential on molybdenum content. The lower the content of S and Mn in steel is, the greater the beneficial effect of Mo, and particularly multiple effects of N and Mo were found. Fig.8 shows the effect of V, W,Si and Cu. From these investigations controlling S, Mn, Mo and N content is effective for increasing pitting corrosion resistance. By taking the steps indicated in Fig.9 the empirical equation of pitting potential was obtained. By regression analysis V-cal was obtained and the correlation between Vc' measured by potentiodynamic polarization method and V-cal is shown in Fig.10.

Molybdenum Saved Austenitic Stainless Steel

Based on laboratory investigation, the optimum composition of molybdenum saved austenitic stainless steel having pitting corrosion resistance equal to that of type 316 was obtained. Its composition is 0.3Mo-0.13N-0.4Mn-<0.0010S-18Cr-8Ni.

The steel sheet having the above composition was produced in commercial process and corrosion test results are shown in Table 5.

Pitting potential of alloy A is equal to that of type 316. As shown in Fig.11 the frequency of ascending current which is corresponding to the occurrence of repassivation pit is low in the case of alloy A.

Pitting corrosion rate of alloy A by immersion test is lower than that of type 316, because pitting corrosion easily occurs at the cross section where non-metallic inclusion have an effect on pitting corrosion because the shape effect of inclusion is large at the cross section.[4]

In order to evaluate the atmospheric corrosion resistance a new accelerated test was conducted. The correspondence between this accelerated test and an outdoor atmospheric exposure test in a marine environment was good, as shown in Fig.12. Alloy A has the atmospheric corrosion resistance equal to that of type 316 stainless steel.

CONCLUSION

The following conclusions can be drawn from the results obtained in the present investigation on 18-8 austenitic stainless steels.

(1) Although sulfide inclusions become the initiation site of pitting corrosion, their harmfulness disappears in extra low sulfur and low manganese alloys.

(2) Besides sulfide inclusion, oxide inclusion is also harmful to pitting corrosion resistance. The oxide inclusion harmful to pitting corrosion resistance contains mainly CaO which exists as stringer shape. The harmfulness of the inclusion containing mainly Al_2O_3 is small.

(3) Based on these experimental results, an extra low sulphur(<0.0010%) 18-8 austenitic stainless steel was produced by commercial process and it was confirmed that 0.3Mo-0.13N-0.4Mn-<0.0010S containing alloy has pitting and atmospheric corrosion resistance equal to that of 316 type stainless steel.

REFERENCES

1) M.Smialowski,Z.Szklarska-Smialowska,M.Rychcik & A.Szummer ; Corrosion Sci. 9(1969)123
2) Z.Szklarska-Smialowska,A.Szummer & M.Jamik-Czachor ; Br.Corros.J. 5(1970)159
3) Z.Szklarska-Smialowska ; Corrosion 28(1972)338
4) V.Scotto,G.Ventura and E.Traverse ; Corrosion Sci. 19(1979)237
5) B.Rondot,M.C.Belo,J.Montuelle ; C.R.Acad.Sc. Paris 273(1972)1028-SericC
6) J.Degerbeck ; Werkstoff u. Korr. 29(1978)179
7) K.Takizawa,Y.Shimizu,Y.Higuchi & K.Tamura;J. of Iron and Steel Institute of Japan 70(1984) 741
8) T.Adachi & T.Kanamaru ; SAE paper 830586(1986)
9) S.Yano,T.Nakanishi,H.Oi,K.Fujimoto,Y.Mihiri & S.Iwaoka ; J. of Iron and Steel Institute of Japan '73-A109 No.9
10) K.Osozawa,R.Nemoto & Y.Fujiwara ; Pre-Print "Corrosion and Corrosion Control '82 " p.189
11) A.Muan & E.F.Osborn ; Phase equilibria among oxides in steelmaking (1965) p.95 ADDISON-WESLEY PUBLISHING COMPANY,INC.
12) J.F.Bates & E.H.Phelps ; Proceedings Second International Congress on Metallic Corrosion (1963) p.462

Table 1 Chemical Composition of Test Materials (mass %)

	C	Si	Mn	P	S	Ni	Cr	Mo	N
Group(A)	0.05	0.6	0.2 ~ 1.0	<0.01	≤0.0010 ~ 0.005	8.5	18.5	0.04	0.10
Group(B)	0.05	0.6	0.4	<0.01	≤0.0010	8.5	19.0	0.4	0.15

Table 2 Chemical Composition of Test Materials (mass %)

Group	C	Si	Mn	P	S	Ni	Cr	Mo	N	Others
(I)	0.05	0.5	0.1 ~ 1.0	0.03	≤ 0.0010 ~ 0.005	8.5	18.5	0.0 ~ 2.5	0.04 ~ 0.22	
(II)	0.05	0.5	0.5	0.03	≤ 0.0010	8.5	18.5	0.05	0.04 ~ 0.10	Si,V,W or Cu 0 ~ 2.0

Table 3 Deoxidizer (%)

	Al	Ca	Si	Mg
(a)	0.08	-	-	-
(b)	0.08	0.04	-	-
(c)	0.02	0.04	-	-
(d)	-	0.04	-	-
(e)	0.08	0.04	0.55	-
(f)	0.05	0.02	0.33	0.05

Table 4 EPMA Results of Inclusions

Deoxidizer	Counts Al	Ca	Si	Mg	Mn	Ratio Ca/Al	Type of Oxide
Al	●	-	-	-	-	< 0.1	Al_2O_3, Al_2O_3-SiO_2
	●	-	-	-	-		
	●	-	-	-	-		
	●	-	○	-	-		
	●	-	-	-	-		
	●	-	-	-	-		
	●	▽	-	△	-		
Al (Ca)	●	△	-	△	-	0.1 ~ 0.2	Al_2O_3-(CaO), Al-(Ca)-(Mg)-(Si)-O
	●	△	-	△	-		
	●	△	●	▽	△		
	●	△	△	-	-		
	●	△	○	△	△		
	●	△	▽	△	-		
	●	△	-	△	-		
Ca (Al)	○	●	○	▽	△	1 ~ 1.5	Al-Ca-Si-(Mn)-O
	●	●	○	△	△		
	○	●	○	-	△		
	○	●	○	-	△		
	●	●	△	▽	△		
Ca	△	●	△	▽	△	3 ~ 10	(Al)-Ca-Si-O
	△	●	●	△	△		
	△	●	△	-	△		

●>○>△>▽>－

Table 5 Corrosion Test Results

Alloy	Pitting Corrosion Resistance: Pitting potential, 3.5%NaCl 30°C (V vs SCE)	Pitting Corrosion Resistance: Immersion Test, 5%NaCl+2%H_2O 35°C,20hr ($g/m^2 h$)	Atmospheric Corrosion Resistance: Accelerated Test (R.N.)	Atmospheric Corrosion Resistance: Outdoor Exposure, Marine environment 1 year (R.N.)
Alloy A	0.500	0.02	9.0	7.0
Type 304	0.280	0.47	6.5	5.0
Type 316	0.470	0.21	9.0	7.0

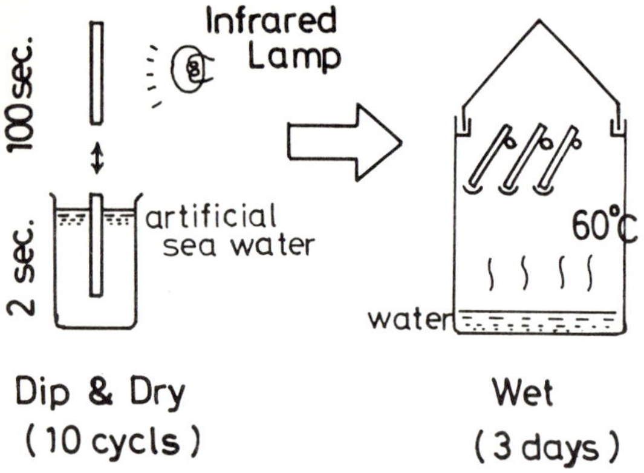

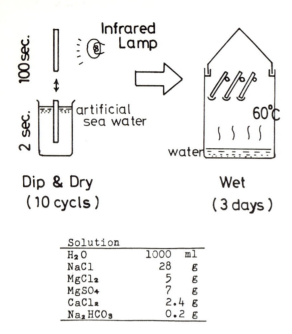

Solution		
H_2O	1000	ml
NaCl	28	g
$MgCl_2$	5	g
$MgSO_4$	7	g
$CaCl_2$	2.4	g
Na_2HCO_3	0.2	g

Fig.1 Details of accelerated atmospheric corrosion test condition.

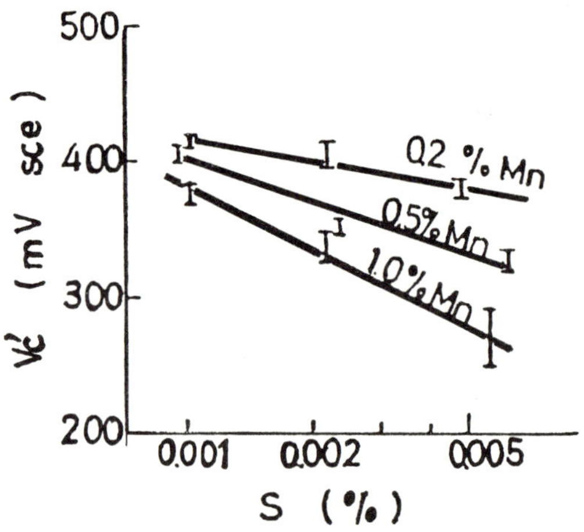

Fig.2 Dependence of pitting potential for various Mn containing 18-8 stainless steels on S content.

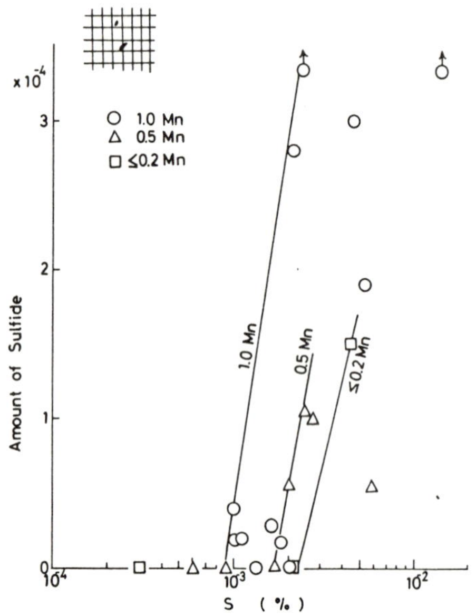

Fig.3 Dependence of the amount of sulfide on S for various Mn containing alloys

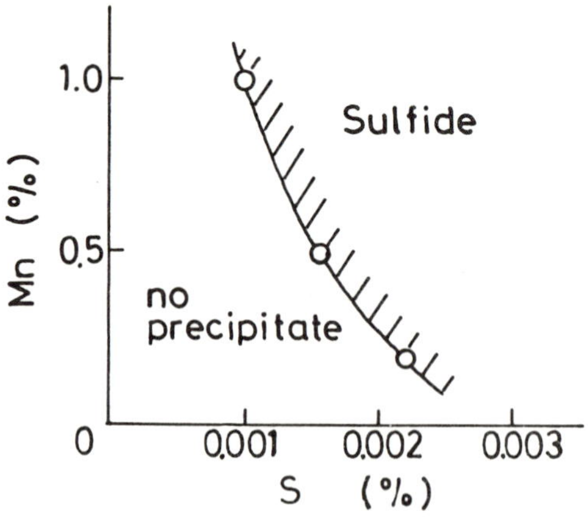

Fig.4 Dependence of the precipitation of sulfide on S and Mn contents.

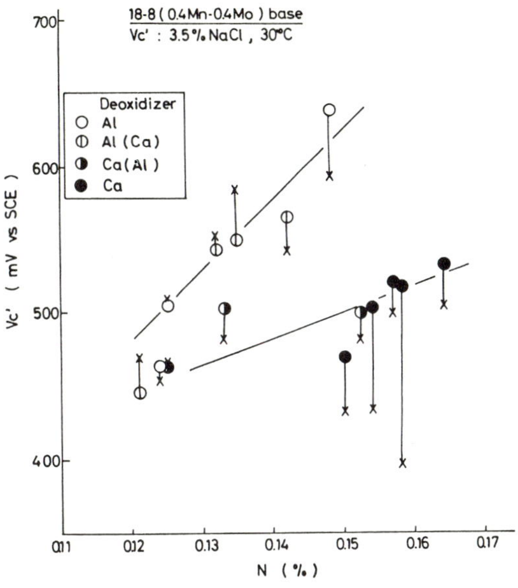

Fig.5 Effect of N on pitting potential for the alloys deoxidized with various deoxidizers.

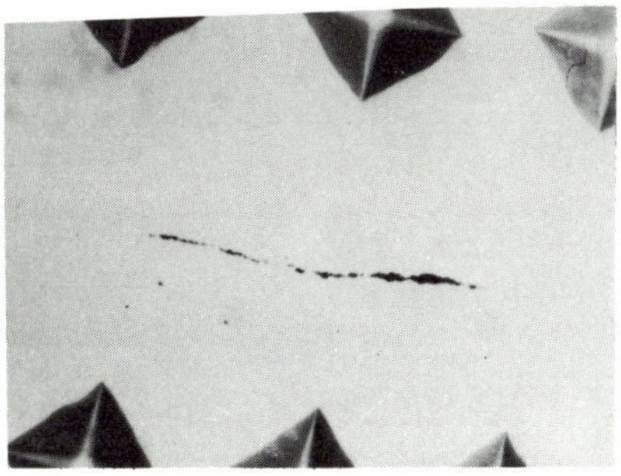

Al-Ca-Si-O

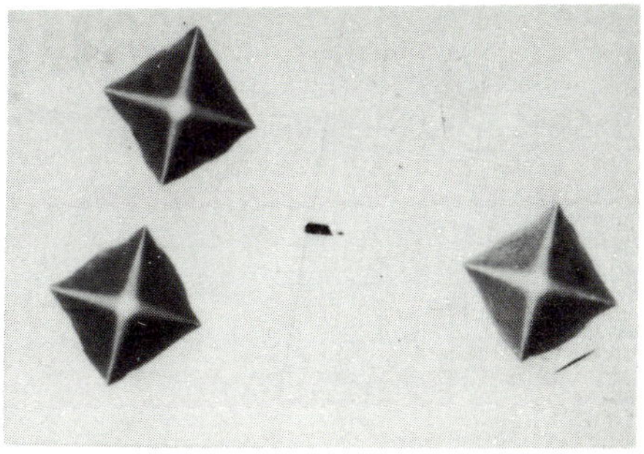

Al-(Ca)-O

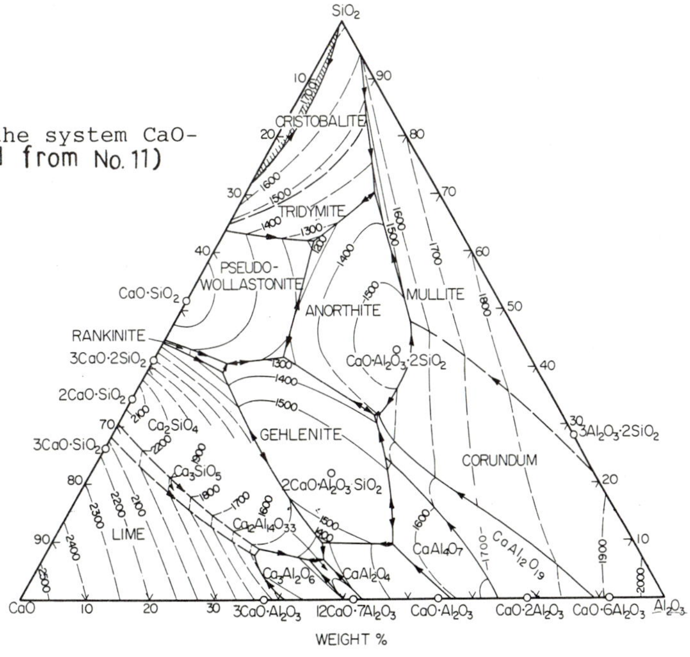

Fig.6

Phase diagram for the system CaO-Al_2O_3-SiO_2. (refered from No.11)

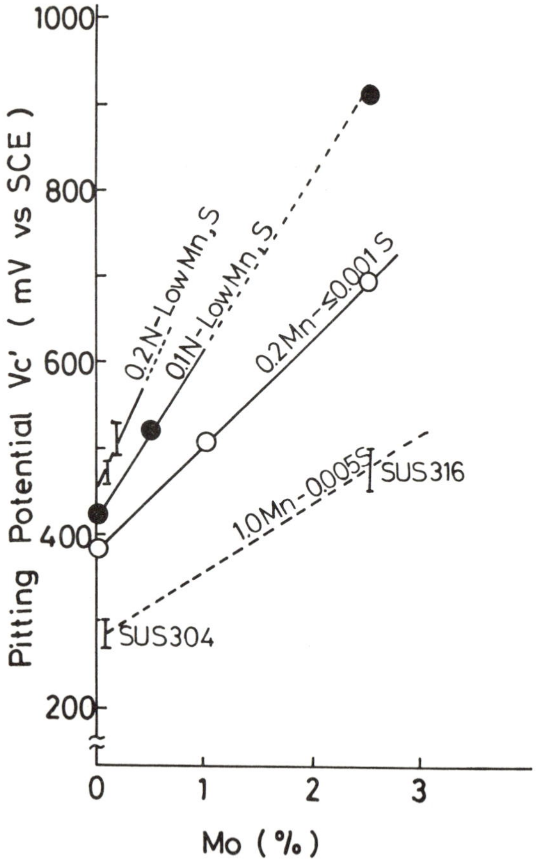

Fig.7 Effect of Mo on pitting potential for various Mn and S containing alloys.

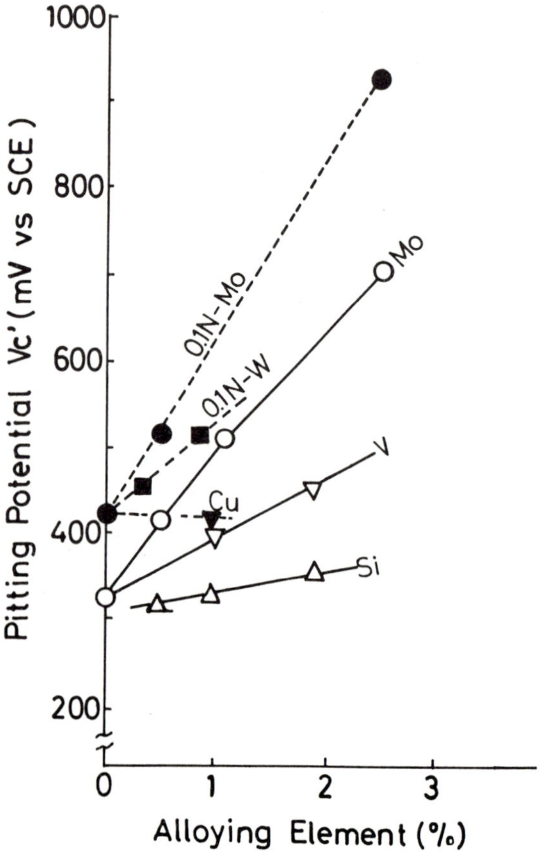

Fig.8 Effect of alloying element on pitting potential.

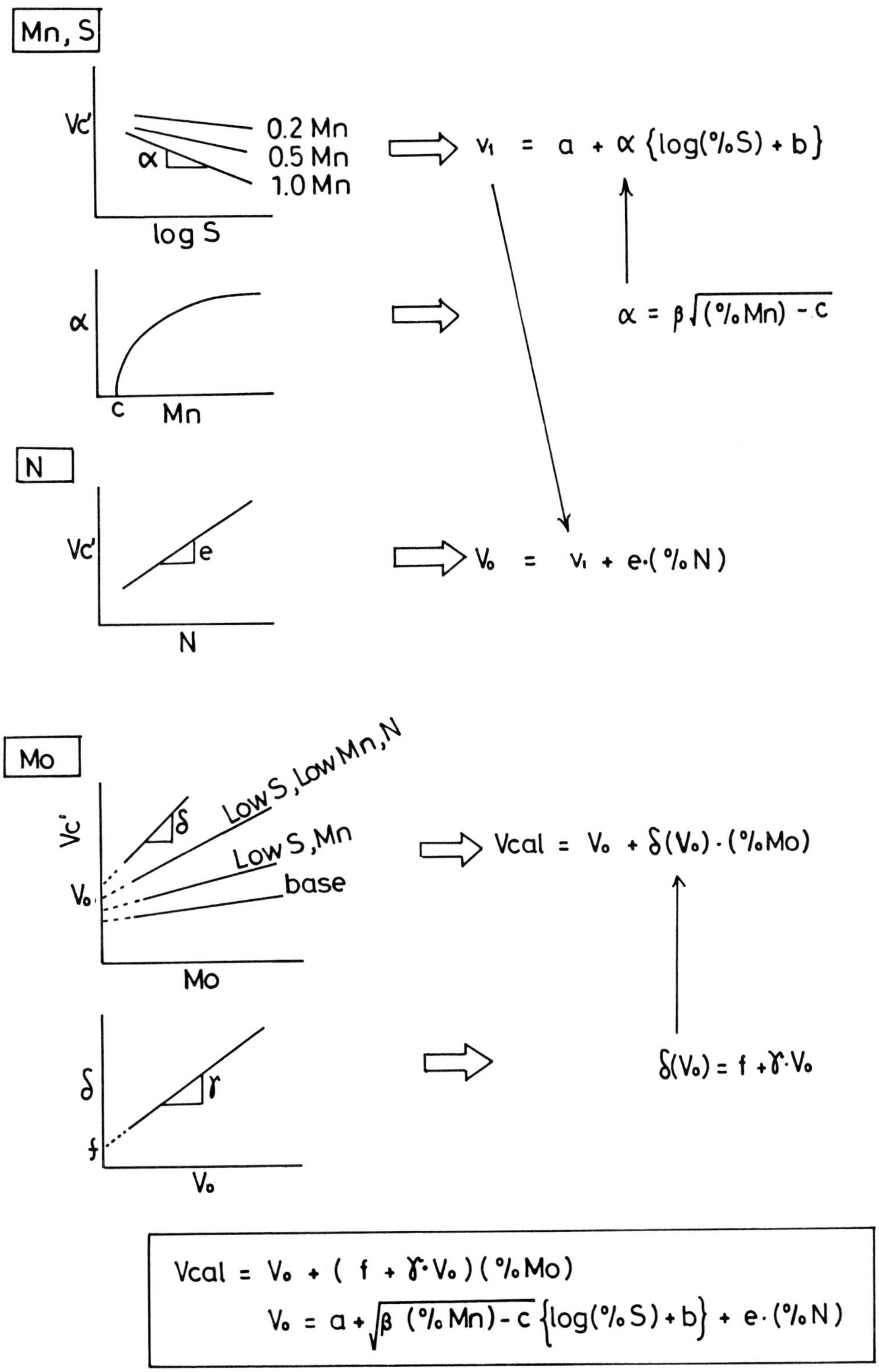

Fig.9 Induction method of the form of empirical equation for pitting potential.

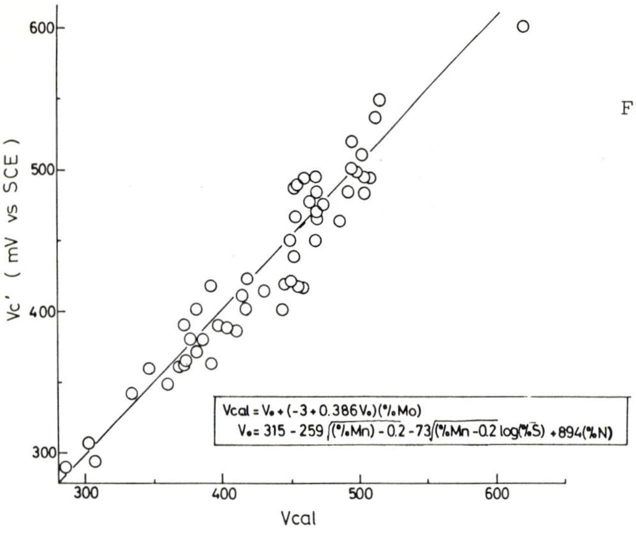

Fig.10 Correlation between V-cal and Vc'.

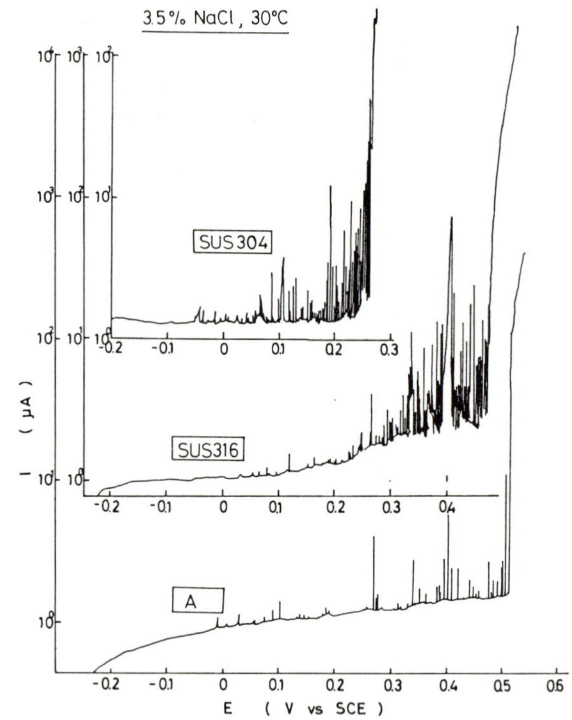

Fig.11 Polarization curves of Alloy A, type 304 and type 316 measured in deareated 3.5% NaCl at 30°C.

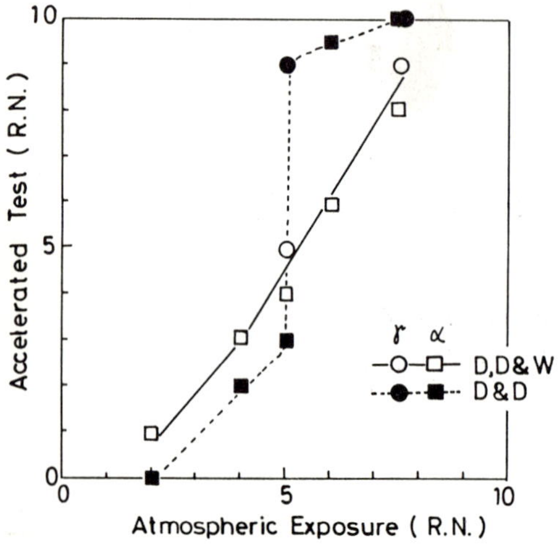

Fig.12[10] Correlation between accelerated test and outdoor exposure test.
D,D & W ; New accelerated test
D & D ; Dip and Dry test[12]
α ; Ferritic stainless steel
γ ; Austenitic stainless steel
R.N. ; Rating Number

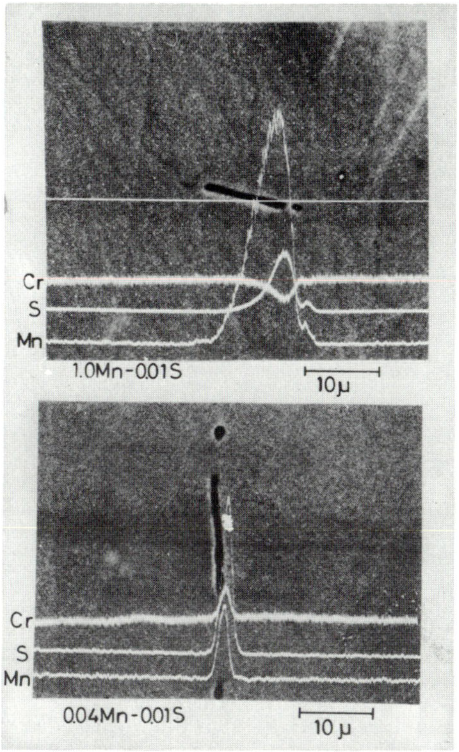

Photo.1 Result of X-ray microanalysis on the sulfide of different Mn containing alloys.

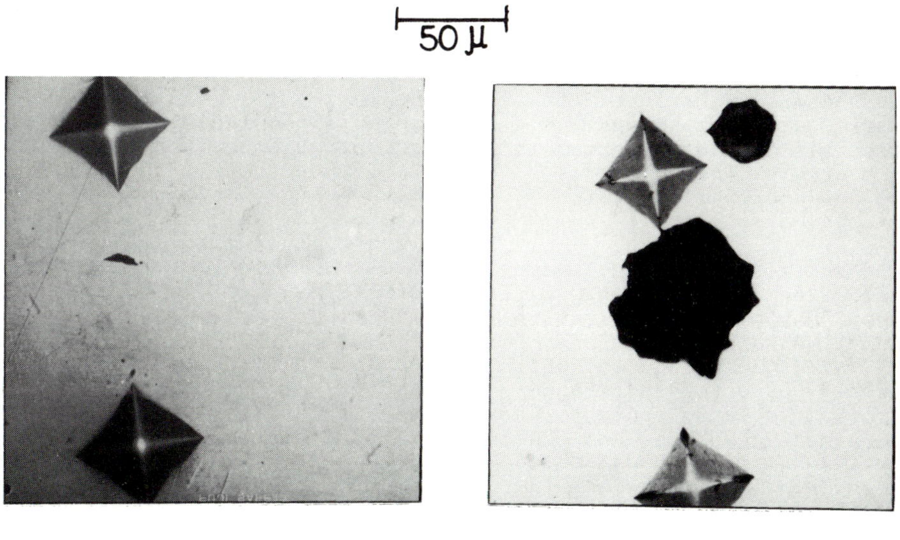

Photo.2 Micrograph of $CaO-Al_2O_3$ inclusion before and after immersion in 10% $FeCl_3 \cdot 6H_2O$ solution at 60°C under ultrasonic vibration.

Effects of residual element contents on corrosion resistance of Type 304L stainless steels in boiling nitric acid

F R Beckitt, B Bastow and T Gladman

BB is with British Nuclear Fuels plc, Sellafield; FRB and TG are with the British Steel Corporation's Swinden Laboratories, Rotherham.

SYNOPSIS

The effects of carbon, silicon, phosphorus and boron contents on the corrosion resistance of Type 304L stainless steels in boiling nitric acid have been investigated. Carbon, silicon and phosphorus contents have no effect on the corrosion rates of material in the solution treated and water quenched condition. Increased carbon contents have a marked adverse effect on the corrosion resistance of sensitised material, and high silicon levels have an additional but much smaller effect. These effects have been attributed to classical sensitisation phenomena. A small effect of silicon was observed in slack quenched steels which has been attributed to non-equilibrium grain boundary segregation.

Studies of the effect of boron have shown that high free boron contents will result in chromium boride precipitation at grain boundaries which cannot be suppressed by water quenching, and which adversely affects corrosion resistance. The small chromium depleted zones formed during rapid quenching can be 'healed' by conventional sensitising treatments, resulting in a dramatic improvement in corrosion resistance.

1. INTRODUCTION

Austenitic stainless steel of the 18% Cr, 13% Ni, 1% Nb type has been used for over thirty years in the nuclear fuel reprocessing industry for plant and waste-storage vessels involving hot concentrated nitric acid liquors.Though successful in these applications, this steel presented welding problems and the material was also particularly prone to end grain attack. In order to obviate such problems, attention was turned to the low carbon, unstabilised, Type 304L steel. To meet the requirements of such applications, BSC have developed a version of the above steel which has been given the name Nitric Acid Grade (NAG) 304L.

It has been known for some years that minor changes in the level of residual elements (within the maximum levels allowed in the specifications of austenitic stainless steels) can have major effects on the susceptibility of these materials to intergranular corrosion. Such effects sometimes have been identified in the solution treated condition (1050-1100 °C and rapidly cooled) as well as in the sensitised condition (typically 1 h at 675 °C and air cooled), particularly after testing in highly oxidising solutions. A preliminary literature survey identified the most likely detrimental residual elements and indicated their critical concentrations. From an initial programme of work, the elements carbon, silicon and phosphorus were selected for further investigation of their joint roles on the corrosion of NAG Type 304L steel as indicated by the Boiling Nitric Acid (Huey) test. The effects of boron have also been examined.

In addition, BSC development work had indicated adverse effects of post solution treatment cooling rates on the BNA corrosion rate, and a limited amount of work has been carried out to investigate these effects in steels with different silicon contents.

2. EXPERIMENTAL PROCEDURE

In a factorially designed experiment to investigate the linked effects of carbon, phosphorus and silicon, four casts of steel were made at each of three levels of carbon, namely <0.010%, 0.015% and 0.022% C. Within these three groups of steel, two casts had silicon contents controlled to as low a level as possible (~0.05%) and the remaining pair had silicon contents of about 0.50%. The two steels at each carbon and silicon level had phosphorus contents of either 0.009% or 0.018%. The base charges of these steels were vacuum melted using high grade electrolytic iron, pure nickel and pure chromium. To produce steels with manganese, sulphur and molybdenum contents comparable with those of commercial steels, the relevant additions of metallic manganese, ferrous sulphide and ferro-molybdenum were made. Adjustments to the phosphorus contents of these steels were made by additions of ferro-phosphorus and the higher levels of silicon and carbon were achieved by addition of ferro-silicon and Warner carburising iron, respectively. A further 25 kg vacuum melt, steel 13, with slightly higher carbon content (0.027%) was used to extend the work on the effects of carbon. The full analyses of steels 1-13 are given in Table 1.

The effects of boron were studied using two 25 kg air-melted research casts and using material from a commercial cast containing boron. The research steels were made as split casts from a 50 kg base melt using similar high purity materials. The minimum level of boron achieved was 0.0007% and the high boron split cast, containing 0.0038% B, was produced by the addition of ferro-boron. The full analyses of these steels again appear in Table 1, the research casts being steels 14 and 15, and the commercial cast was steel 16.

All the research ingots were reduced to strip of 4.5 mm to 5.0 mm thickness by a combination of hot forging and hot and cold rolling, the cold rolling being confined to a final pass of about 10% in the deformation sequence to ensure as flat a product as possible. The commercial material was taken from part of a 15 tonne melt which had been reduced to a 9.5 mm plate product.

The early corrosion testing to investigate the effects of boron was carried out by BSC, using the Boiling Nitric Acid (BNA) corrosion test, ASTM A262-81 (Practice C) and involved single specimen testing with cold-finger condensers. The tests on the vacuum melted steels were carried out using slightly revised testing conditions and were carried out at both BSC Swinden Laboratories and BNF Sellafield. The major modifications in the latter tests were the use of a single batch of nitric acid of tightly-controlled analysis for all tests, the use of reflux condensers, and the use of specimens sectioned from the strip material so that 33% of the surface area was normal to the rolling direction. The tests on all the research melts were carried out on material in one of three heat treated conditions, either solution treated for 1/2 h 1050 °C WQ, or for 10 min at 1300 °C WQ or solution treated 1/2 h 1050 °C WQ followed by 'sensitisation' for 1 h 675 °C AC.

3. RESULTS

The results of the joint BSC, BNF corrosion testing programme carried out on steels 1-13 are presented in Table 2. It can be seen from these results that the widest range of corrosion rates (0.145 to 3.489 mm/y) occurred in specimens in the sensitised condition (1/2 h 1050 °C WQ, 1 h 675 °C AC). The predominant effect of carbon is shown in Fig. 1. It was also found from the data in Table 2, for unsensitised material, solution treated at either 1050 °C or 1300 °C, that although individual penetration rates fell in the range 0.12 mm/y to 0.22 mm/y, none of the variation could be related to compositional variables. The corrosion rates obtained in specimens solution treated at 1300 °C were in general marginally below the values obtained from the equivalent materials solution treated at 1050 °C. These small differences are reflected in the position of the two lines in Fig. 1, representing the average penetration rates for material from each of the two solution treatment temperatures.

An optical metallographic examination showed that the principal form of attack in the sensitised material was by intergranular corrosion. As expected from the results in Fig. 1, the higher the carbon content of the steel the greater was the depth of intergranular penetration. Examples of this effect, as seen on the rolling plane surfaces of the test specimens, are given in Fig. 2 (a)-(d). Grain-dropping can be seen to have taken place in the specimens containing 0.024% C and 0.027% C and such behaviour was found to be consistent with the breakaway corrosion rate indicated in Fig. 1. Intergranular corrosion on the 'end-grain' surfaces of sensitised specimens also increased as the carbon contents increased, but the depths of corrosion on the end grain surface was much reduced. This can be seen by comparing Figs. 2(a) and (e) for low carbon steels and Figs. 2(d) and (f) for the high carbon steels. There was little evidence of end-grain tunnel formation in these two specimens, Figs. 2(e) and (f).

Electron metallographic observations of the microstructure in replicas taken from the corrosion specimens, revealed the presence of $M_{23}C_6$ precipitates on the grain boundaries of the sensitised specimens, as is demonstrated in the case of the 0.027% carbon steel in Figs. 3(a) and (b). These precipitates were identified by electron diffraction, and from Energy Dispersive Analysis of X-rays; a typical composition of the substitutional elements in the precipitates was found to be 80% Cr, 18% Fe, 2% Ni. After solution treatment at 1050 °C, the steels contained between 0.7% and 1.7% delta ferrite, elongated in the rolling direction. After sensitisation, aggregates of $M_{23}C_6$ particles were again identified in these regions, as can be seen for the 0.027% C steel in Figs. 3(a) and 3(e). The density of such particles on both grain boundaries and at the delta-ferrite interface was related to the carbon content of the steels, c.f. Figs. 3(e), 3(f), 3(g) and 3(h).

The effects of carbon content on grain boundary precipitation in the above specimens could also be seen indirectly by the effects observed using the oxalic acid etch test conducted according to ASTM A262-81 (Practice A). These structures were seen to vary from mainly 'step' structures with only a trace of 'dual' structure in the 0.004% C Steel 1 to the completely 'ditched' structures in the 0.027% C Steel 13.

The effects of silicon and phosphorus are illustrated in Fig. 4. Here the steels have been divided into three carbon groupings and because of the factorial design of the corrosion experiments, together with the favourable distribution of carbon variations, any effects of silicon and phosphorus can be indicated. For instance, after combining all high and low phosphorus steel data, the relationship between silicon content and penetration rate, in the sensitised condition can be reasonably derived, Fig. 4(a). This suggests a small adverse effect of silicon in both the medium and higher carbon steels, where an increase in corrosion rate of 0.070 mm/y and 0.058 mm/y respectively is indicated for a 0.5% increase in silicon content. For the low carbon steel group, the variation is well within the limits of experimental error.

For the medium carbon steels, which show the most consistent effect of silicon content on the sensitised corrosion rate, small increases in the depth of intergranular corrosion were observed at the higher silicon contents. In addition, somewhat higher densities of $M_{23}C_6$ precipitates were detected, both on grain boundaries and in delta-ferrite regions of the higher silicon steels. Increased grain boundary 'ditching' was also noted in oxalic-acid etch tests. There were no detectable effects of silicon content on either the corrosion rate or the microstructure (other than inclusions) in samples in the solution treated condition. What is also perhaps quite relevant is that the higher silicon contents (0.5%) had no detectable effects on end-grain corrosion rates although small end-grain tunnels of 160 um and 90 um maximum length (formed from etched-out duplex chrome galaxite plus manganese silicate stringer particles) were observed in the high silicon steels, steel 11 and steel 3 respectively.

A similar treatment of the corrosion data for samples in the sensitised condition to detect any effects of phosphorus is given in Fig. 4(b). There is little evidence of any effect of phosphorus at any of the carbon levels examined.

As can be seen from the corrosion data for steels 14 and 15 in Table 2, the addition of 0.0038% B has only a marginal effect on corrosion in the research casts. The relatively high corrosion rates may reflect either the extremely high oxygen/oxide contents of the air melted steels or the acid grade used in the BNA tests.

In contrast to the above effect, a very detrimental effect of boron has been noted in a commercial 304L cast in the 'works softened' condition (~10 min 1100 °C WQ). In this material, a BNA penetration rate as high as 6.41 mm/y was measured (which could reduce to as low as 0.2 mm/y after sensitisation) and a boron autoradiographic examination indicates the boron to be concentrated at grain boundary sites, Fig. 5. Electron metallography on the above material revealed numerous grain boundary particles which were found by electron diffraction to have a crystal structure consistent with the boride δ-Cr_2B, the chromium to iron ratio in all such precipitates as measured by EDA being approximately 2:1. Additional work on this commercial material, carried out on foil samples at UKAEA Harwell involving an Electron Energy Loss Spectroscopy attachment (EELS) to a transmission electron microscope, confirmed the location of boron to be within the grain boundary precipitates. It also confirmed the diffraction identification to be consistent with δ-Cr_2B and the chromium-rich analysis of the particles.

Metallographic work carried out on the boron research casts in the solution treated condition showed no evidence of boundary precipitation other than isolated examples of $M_{23}C_6$ type carbides in the high boron steel. Boron autoradiographic work on this material showed the small boron concentration in steel 14 to be randomly distributed throughout the structure, Fig.. 5(c) and the higher boron content of steel 15 to be predominantly associated with non-metallic inclusions, Fig. 5(d). After sensitisation of the above steels, the incidence of $M_{23}C_6$ grain boundary precipitates marginally increased.

Further work to investigate the combined effects of post-solution treatment cooling rate and silicon content was carried out on steels 5 and 7. With the exception of their silicon content (0.04% and 0.52% respectively), these steels have compositions typical of the current commercial NAG Type 304L steel. The BNA corrosion test results are presented in Fig. 6. It can be seen that no effect of the silicon content can be detected at either the fastest or the slowest cooling rates investigated. However, at both of the intermediate cooling rates, a small but consistent increase in corrosion rate was detected in the higher silicon cast. Oxalic acid etch testing showed a small amount of 'dual' structure throughout the vermiculite cooled, high silicon steel and also at the centre of the same steel when air cooled. The two faster cooled high silicon specimens and all the low silicon specimens were found to develop only 'step' structures. In the most highly corroded of the high silicon samples, i.e. that which was oil quenched, no grain boundary precipitates could be detected by replica electron metallography.

4. DISCUSSION

A particularly adverse effect of carbon and a smaller effect of silicon on the BNA corrosion rate of 304L stainless steels in the sensitised condition have been demonstrated in the present investigation. These increases in corrosion take place by intergranular corrosion attack, but there was little evidence of end grain attack. The above effects have been duplicated by their equivalent effects on the degree of grain boundary attack as shown by the oxalic acid etch test. In all the above specimens, increases in intergranular corrosion could be related to similar increases in the density of chromium-rich grain boundary $M_{23}C_6$ precipitates. All the above observations and their total absence in solution treated and water quenched material, are consistent with the widely accepted mechanisms for classical intergranular corrosion i.e. preferential corrosion of chromium depleted zones found in the neighbourhood of chromium-rich grain boundary precipitates.[1,2]

Being one of the principal constituents of $M_{23}C_6$, it is easy to appreciate the predominant effect of carbon, but the role of silicon is less obvious. This effect of silicon can probably best be explained by reference to previously published work which has shown that the presence of silicon can increase the activity coefficient for carbon dissolved in iron.[3] Such an explanation,where silicon accelerates the formation of $M_{23}C_6$,is particularly attractive since it is consistent with the increased effect of silicon in the higher carbon steels.

The effects of silicon, observed in the investigation of cooling rate in solution treated material, cannot be explained in the above manner. In this case, the first stages of classical sensitisation (traces of dual structure in the oxalic acid etch test) were observed in the slowly cooled specimens, but were not observed for the oil quenched specimens with the highest corrosion rate. Similarly , no grain boundary precipitates could be detected in the high silicon steel in which the highest corrosion rate was measured. Most of the observations in this part of the work bear strong similarities to instances of 'non-sensitised' intergranular corrosion of stainless steels which have been reported in the literature. Such corrosion has normally been found to occur in highly oxidising nitric acid solutions containing certain metallic ions in their higher valence state. One of the more detrimental species of metallic ion is the hexavalent chromium ion Cr^{6+} and it is relevant to point out that this is one of the principal corrosion products formed during the BNA corrosion test. The work by Chaudron[4] first indicated that high purity 18% Cr-10% Ni alloys were immune to this form of attack. Subsequent studies by Armijo[5] showed that the presence of silicon in proportions between 0.1% and 2% in solution treated 16% Cr-14% Ni alloys made this material particularly sensitive to such attack (as also were similar alloys containing 0.01% phosphorus). The latter work also showed that these effects of silicon were not related to grain boundary precipitation, since the presence of differing levels of carbon between 0.005% and 0.07% produced no variations in corrosion rate. It is relevant to note that silicon is usually regarded to cause the above effects by solute concentration at grain boundaries. If this concentration is achieved by the mechanism of non-equilibrium (vacancy-linked) diffusion, then boundary silicon contents would be expected to be cooling rate sensitive. The mechanism by which silicon affects grain boundary corrosion rates is still open to debate in the literature. However, Desestret[6] has shown that intermediate silicon levels in 16% Cr-14% Ni iron alloys can accelerate corrosion in the transpassive range, which is the type of corrosion which is induced by the presence of Cr^{6+} ions during the BNA corrosion test. This form of corrosion would be expected to occur to the greatest extent within any surface fissures and could lead to the more rapid extension of intergranular corrosion crevices (and particularly those associated with end-grain tunnels).

The work on boron in the present investigation has given rise to widely differing results. In the case of the air-melted research cast, the addition of 0.0038% B produced little effect, since the boron was extensively associated with the oxidic inclusions. It was not available,therefore,to influence grain boundary corrosion properties. However, in a low oxygen commercial steel with a boron content of

0.0025%, the boron was freely available and led to the rapid formation of chromium-rich grain boundary borides. This resulted in very high corrosion rates in the solution treated condition. This behaviour is in agreement with many of the findings of Brandis & Horn[7] who found that the prevention of grain boundary precipitates in water quenched 10 mm diameter test pieces of 16% Cr-13% Ni steel, became impossible at boron contents as low as 0.009%. These grain boundary precipitates were a mixture of $(Cr,Fe)_{23}C_6$ and $(Cr,Fe)_2B$, and because the carbon was significantly higher than in the present steels (0.06% C cf 0.014% C) the carbide precipitates predominated. However, the boride particles detected by Brandis and Horn had the same orthorhombic structures as those found in the present work. It can be seen from the above findings, that the deleterious effects of boron as seen in the commercial steel, can be fully explained in terms of enhanced intergranular corrosion due to grain boundary chromium denudation effects. The only difference between the above behaviour and that of classical chromium denudation effects is related to the kinetics of the chromium-rich boride precipitation process, which is greatly accelerated compared with that of the normal $M_{23}C_6$ precipitates in boron-free alloys. This accelerated precipitation behaviour could also satisfactorily explain the large unexpected improvement in corrosion resistance found after 'sensitisation' treatment of this material, in terms of equally accelerated 'healing' of the narrow chromium-denuded zones.

5. CONCLUSIONS

The effects of carbon, silicon, phosphorus and boron contents on the corrosion resistance of Type 304L stainless steels have been investigated, using the Boiling Nitric Acid Test.

It has been shown that increases in carbon content up to 0.027% have no effect on corrosion rate in the solution treated condition, but have pronounced adverse effects in the sensitised condition. On the basis of these results, it is suggested that Type 304L austenitic stainless steels for use in nitric acid environments should be made to the lowest, commercially-practicable carbon content.

Increasing the silicon content from 0.05% to 0.50% had a small adverse effect on BNA corrosion rates in sensitised materials. These effects were more pronounced at the higher levels of carbon, and have been explained by the effect of silicon increasing the activity of carbon in austenite which accelerates the formation of $M_{23}C_6$ precipitates. A 0.5% silicon addition can also increase intergranular corrosion in the solution treated condition. Such behaviour was found to occur to the greatest extent in slack quenched material, but at cooling rates which are too fast for any degree of classical sensitisation. The most likely explanation of these 'non-sensitised' intergranular corrosion effects is that they occur by the non-equilibrium segregation of silicon to grain boundaries where it affects the Cr^{6+} induced transpassive grain boundary corrosion rate. Because of both the above effects it would be advisable to keep the silicon content of nitric acid grade Type 304L steels at low levels consistent with adequate deoxidation e.g. 0.2%.

Increasing the phosphorus content from 0.010% to 0.018% had no detectable effects in the BNA corrosion test with material in either the solution treated or the sensitised condition.

In a commercial steel, treated with 0.0025% B, high corrosion rates were found in the solution treated condition. Classical sensitisation treatments significantly improved the corrosion resistance. This behaviour was found to be associated with the rapid formation of chromium-rich grain boundary boride particles after solution treatment, and the healing of grain boundary regions of chromium denudation during the sensitisation treatment. In an experimental steel, no similar effects of boron could be reproduced. In this case, the boron was present in oxidic inclusions due to the high oxygen content of the small research casts. The observed effect of boron in clean commercial Type 304L steels clearly indicates that boron levels should be as low as possible.

ACKNOWLEDGEMENTS

The authors are grateful to British Nuclear Fuels and BSC Stainless for funding this work, and would like to thank Dr. R. Baker, Director of Research, British Steel Corporation, for permission to publish this paper.

6. REFERENCES

1. Strawstrom C. et al, JISI, (1969), 207, 77.
2. Tedmon C. et al, J. Electrochem Soc., (1971), 118, 192.
3. Richardson F.D., JISI (1953) 175,33
4. Chaudron G., EURAEC-976, Quarterly Report No. 6, Oct./Nov., (1963).
5. Armijo J.S., Corrosion, (1968), 24, 24.
6. Desestret A., Thesis, University of Paris, (1964).
7. Brandis H.and Horn E., DEW-Tech. Beirchte, (1969), 9, 213.

TABLE 1 CHEMICAL ANALYSES OF STEELS USED (Wt.%)

Steel Code	C	Si	Mn	P	S	Cr	Mo	Ni	Al	B	N	O_{tot}
1	0.004	0.02	1.53	0.009	0.012	18.8	0.20	9.75	0.006	<0.0005	0.0033	0.020
2	0.005	0.03	1.51	0.019	0.015	18.7	0.19	9.71	<0.005	0.0003	0.0065	0.015
3	0.005	0.54	1.54	0.009	0.012	18.8	0.20	9.71	0.009	<0.0005	0.0033	0.018
4	0.006	0.51	1.52	0.018	0.014	18.9	0.19	9.73	0.008	0.0004	0.0067	0.014
5	0.015	0.04	1.53	0.011	0.014	18.8	0.19	9.75	0.007	0.0001	0.0059	0.017
6	0.016	0.05	1.52	0.018	0.014	18.8	0.19	9.75	0.007	0.0001	0.0070	0.016
7	0.017	0.52	1.53	0.009	0.013	18.8	0.19	9.73	0.008	0.0004	0.0061	0.011
8	0.015	0.52	1.50	0.019	0.013	18.8	0.19	9.72	0.007	0.0002	0.0068	0.009
9	0.021	0.02	1.56	0.009	0.014	18.8	0.20	9.75	0.005	<0.0005	0.0037	0.018
10	0.024	0.05	1.53	0.018	0.013	18.7	0.19	9.74	0.007	0.0001	0.0068	0.020
11	0.024	0.48	1.50	0.010	0.012	18.9	0.19	9.83	0.008	0.0002	0.0080	0.023
12	0.021	0.49	1.54	0.018	0.013	18.8	0.20	9.70	0.007	<0.0005	0.0042	0.016
13	0.027	0.47	1.49	0.019	0.012	18.9	0.19	9.74	0.007	0.0001	0.0084	0.020
14	0.013	0.13	1.72	0.009	0.004	18.3	0.10	9.64	<0.005	0.0007	0.023	0.037
15	0.012	0.13	1.75	0.009	0.004	18.2	0.10	9.61	<0.005	0.0038	0.021	0.044
16	0.017	0.26	1.72	0.016	0.002	18.2	0.08	11.7	-	0.0025	0.039	-

TABLE 2 SUMMARY OF BOILING NITRIC ACID CORROSION TEST RESULTS

Spec. Code	Specimen Heat Treatment	BSC Tests			BNFL Tests		
		Initial mm/year	Repeat mm/year	Ave. mm/year	Initial mm/year	Repeat mm/year	Ave. mm/year
1	½h 1050°C WQ	0.1669	0.1563	0.1615	0.131	0.132	0.132
	½h 1050°C WQ, 1h 675°C AC	0.1880	0.2054	0.1967	0.157	0.148	0.152
	10 min 1300°C WQ	0.1792	0.1783	0.1788	0.157	0.163	0.160
2	½h 1050°C WQ	0.1804	0.1586	0.1695	0.133	0.141	0.137
	½h 1050°C WQ, 1h 675°C AC	0.2183	0.1636	0.1910	0.145	0.161	0.153
	10 min 1300°C WQ	0.1591	0.1514	0.1553	0.140	0.142	0.141
3	½h 1050°C WQ	0.1550	0.1687	0.1619	0.135	0.135	0.135
	½h 1050°C WQ, 1h 675°C AC	0.2274	0.2235	0.2255	0.195	0.183	0.189
	10 min 1300°C WQ	0.1545	0.1423	0.1484	0.115	0.117	0.116
4	½h 1050°C WQ	0.1752	0.1401	0.1577	0.134	0.128	0.131
	½h 1050°C WQ, 1h 675°C AC	0.1739	0.1625	0.1682	0.169	0.172	0.171
	10 min 1300°C WQ	0.2241	0.1604	0.1923	0.131	0.172	0.152
5	½h 1050°C WQ	0.1740	0.1721	0.1731	0.142	0.159	0.151
	½h 1050°C WQ, 1h 675°C AC	0.2112	0.1960	0.2036	0.216	0.218	0.217
	10 min 1300°C WQ	0.1694	0.1490	0.1592	0.128	0.145	0.137
6	½h 1050°C WQ	0.1642	0.1684	0.1663	0.145	0.172	0.159
	½h 1050°C WQ, 1h 675°C AC	0.2267	0.2214	0.2241	0.229	0.242	0.236
	10 min 1300°C WQ	0.1756	0.1553	0.1655	0.159	0.148	0.154
7	½h 1050°C WQ	0.2006	0.1604	0.1805	0.163	0.157	0.160
	½h 1050°C WQ, 1h 675°C AC	0.2743	0.3111	0.2927	0.265	0.289	0.277
	10 min 1300°C WQ	0.1401	0.1335	0.1368	0.116	0.127	0.122
8	½h 1050°C WQ	0.1513	0.1588	0.1551	0.200	0.182	0.191
	½h 1050°C WQ, 1h 675°C AC	0.2874	0.2808	0.2841	0.284	0.296	0.290
	10 min 1300°C WQ	0.1405	0.1172	0.1289	0.125	0.119	0.122
9	½h 1050°C WQ	0.2172	0.1762	0.1967	0.166	0.146	0.156
	½h 1050°C WQ, 1h 675°C AC	0.3033	0.2472	0.2753	0.233	0.241	0.237
	10 min 1300°C WQ	0.1982	0.1604	0.1793	0.162	0.153	0.158
10	½h 1050°C WQ	0.1814	0.1854	0.1834	0.152	0.146	0.149
	½h 1050°C WQ, 1h 675°C AC	0.4930	0.4191	0.4561	0.580	0.609	0.595
	10 min 1300°C WQ	0.1602	0.1601	0.1602	0.130	0.131	0.131
11	½h 1050°C WQ	0.1597	0.1639	0.1618	0.136	0.150	0.143
	½h 1050°C WQ, 1h 675°C AC	0.4624	0.4170	0.4397	0.653	0.626	0.640
	10 min 1300°C WQ	0.1615	0.1660	0.1638	0.142	0.142	0.142
12	½h 1050°C WQ	0.1793	0.1740	0.1767	0.147	0.139	0.143
	½h 1050°C WQ, 1h 675°C AC	0.3525	0.3604	0.3565	0.340	0.330	0.335
	10 min 1300°C WQ	0.1607	0.1596	0.1602	0.151	0.167	0.159
13	½h 1050°C WQ	0.1485	0.1624	0.1555	0.147	-	-
	½h 1050°C WQ, 1h 675°C AC	2.6371	3.4887	3.0629	2.889	-	-
	10 min 1300°C WQ	0.1427	0.1395	0.1411	0.125	-	-
14	½h 1050°C WQ	0.3137	0.4238	0.3687	-	-	-
	½h 1050°C WQ, 1h 675°C AC	0.3701	0.5079	0.4390	-	-	-
	10 min 1300°C WQ	0.3104	0.3674	0.3389	-	-	-
15	½h 1050°C WQ	0.3426	0.5456	0.4441	-	-	-
	½h 1050°C WQ, 1h 675°C AC	0.3555	0.5350	0.4453	-	-	-
	10 min 1300°C WQ	0.4030	0.3510	0.3770	-	-	-

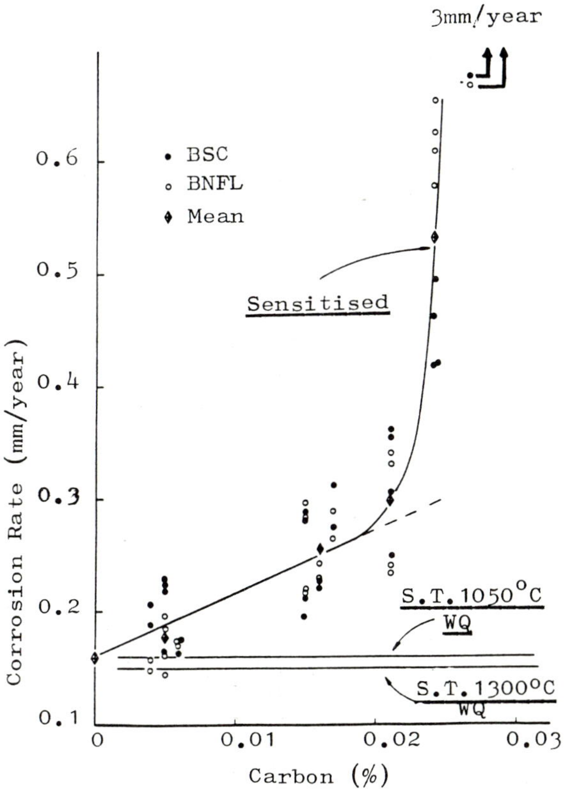

1. Effect of Carbon Content on Corrosion Rate.

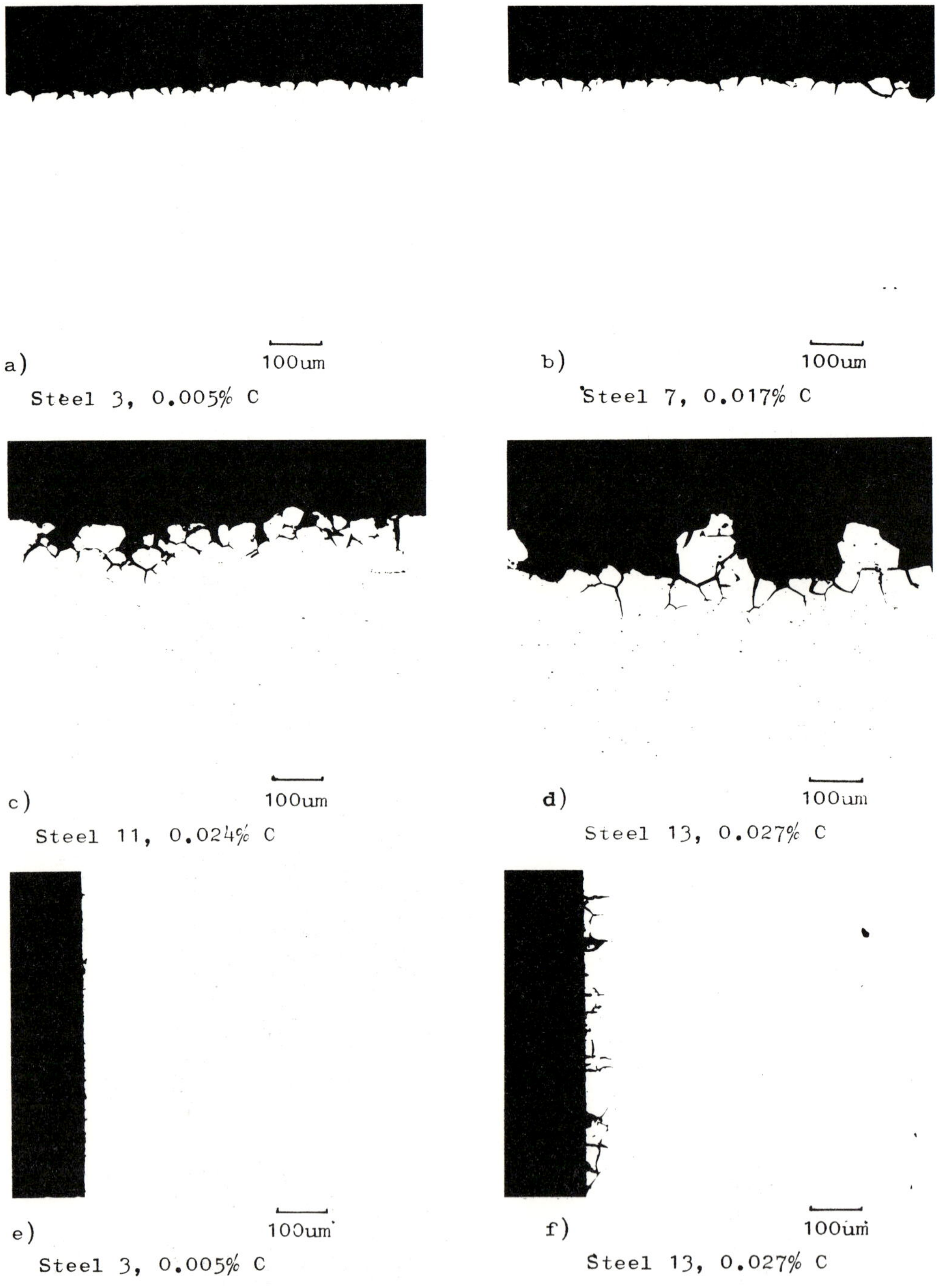

2. Optical Micrographs of Surface Corrosion, Effect of Carbon Content, Sensitised Condition.

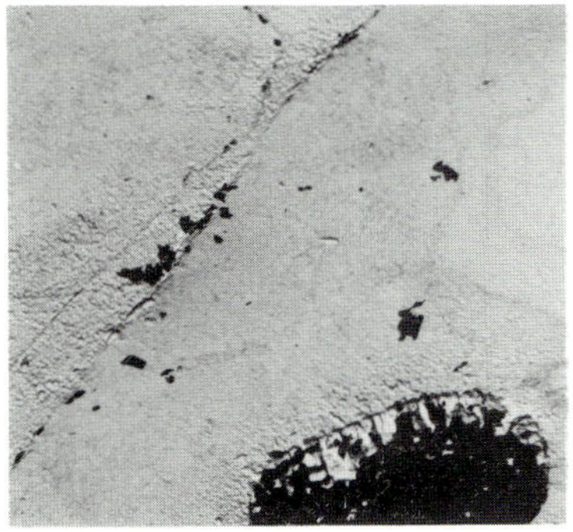

a) 1um

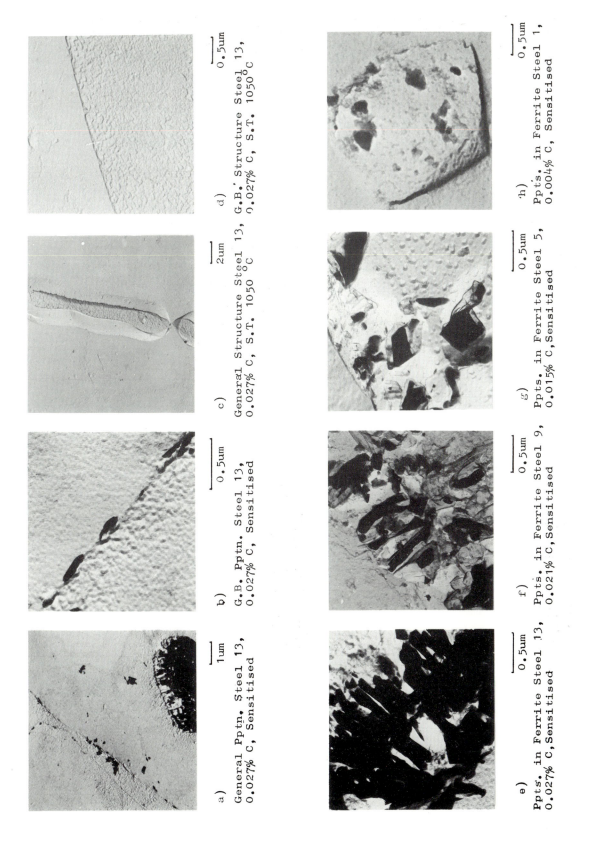

General Pptn. Steel 13, 0.027% C, Sensitised

b) 0.5um

G.B. Pptn. Steel 13, 0.027% C, Sensitised

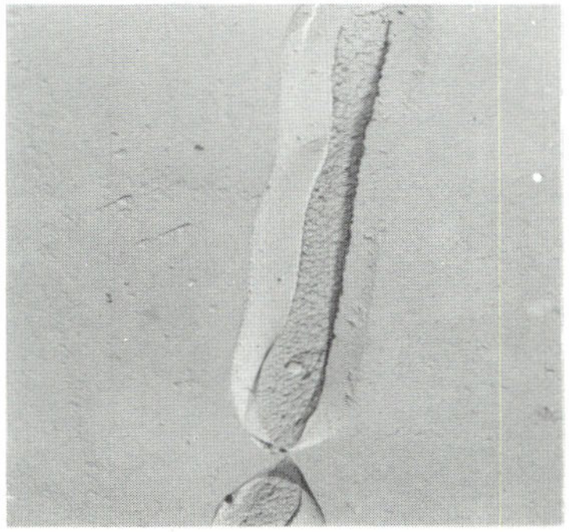

c) 2um

General Structure Steel 13, 0.027% C, S.T. 1050 °C

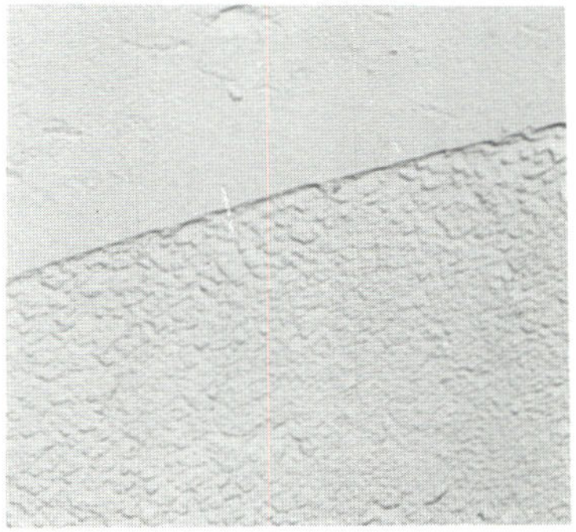

d) 0.5um

G.B. Structure Steel 13, 0.027% C, S.T. 1050°C

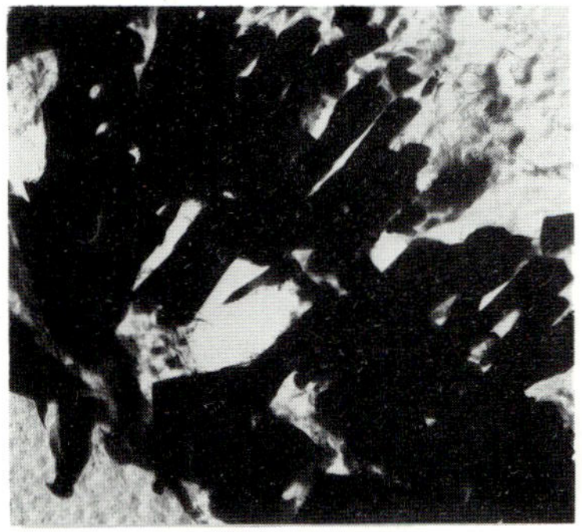

e) 0.5um

Ppts. in Ferrite Steel 13, 0.027% C, Sensitised

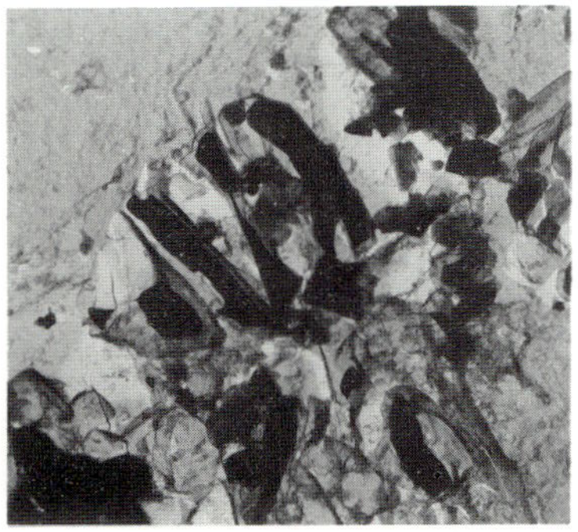

f) 0.5um

Ppts. in Ferrite Steel 9, 0.021% C, Sensitised

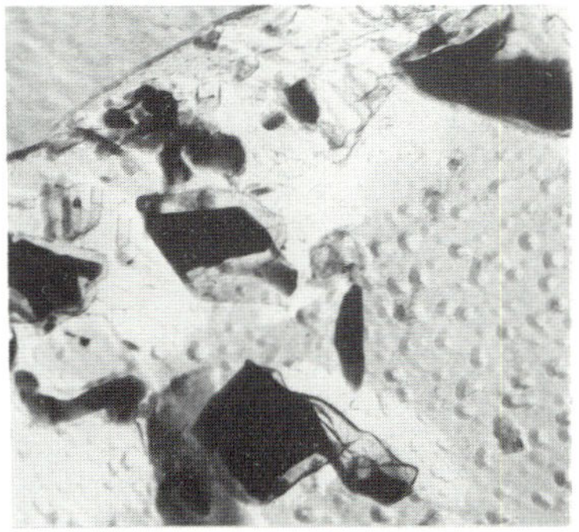

g) 0.5um

Ppts. in Ferrite Steel 5, 0.015% C, Sensitised

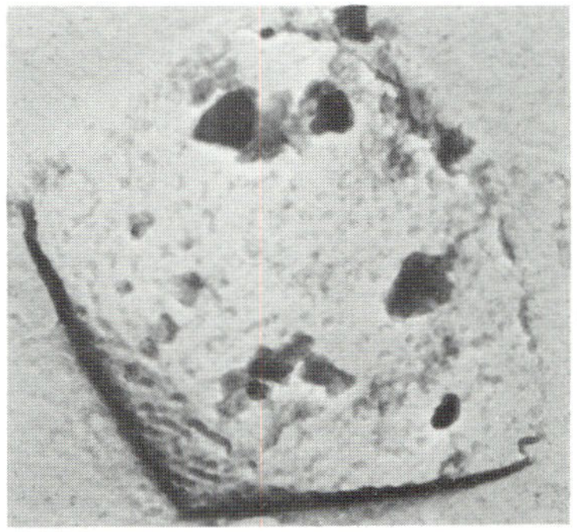

h) 0.5um

Ppts. in Ferrite Steel 1, 0.004% C, Sensitised

3. Electron Micrographs of Replicas Taken from Corrosion Test Specimens.

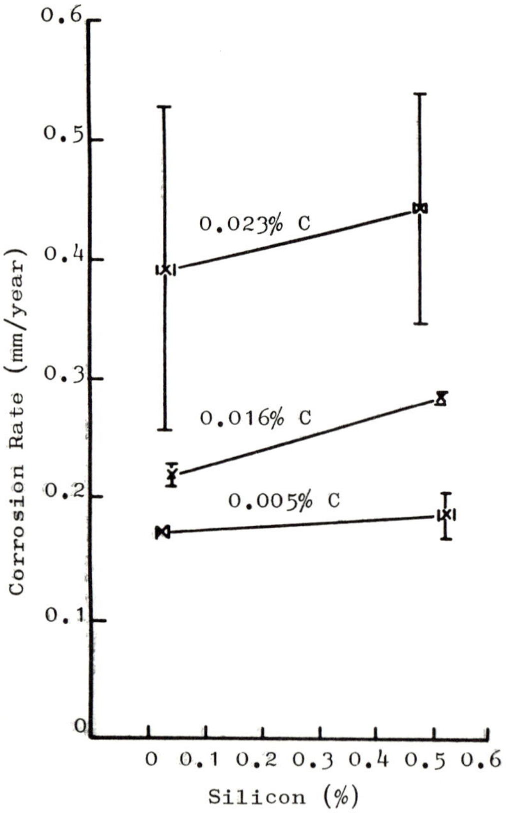

(a) Effect of Silicon

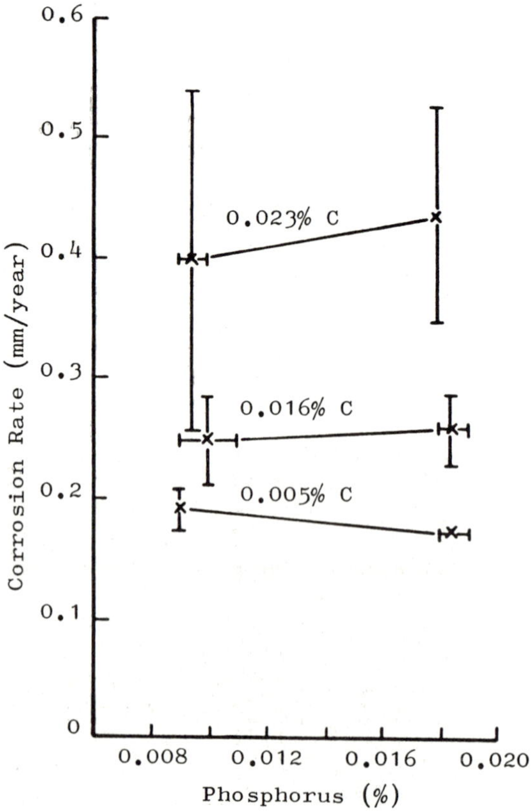

(b) Effect of Phosphorus

4. Effects of Silicon and Phosphorus on Corrosion Rate of Sensitised Material.

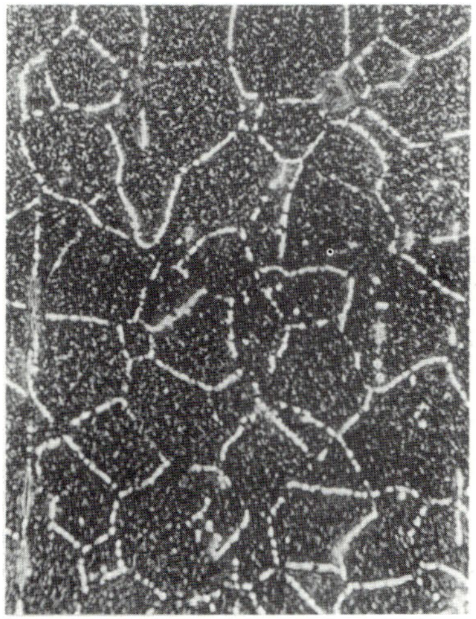

a) 100um

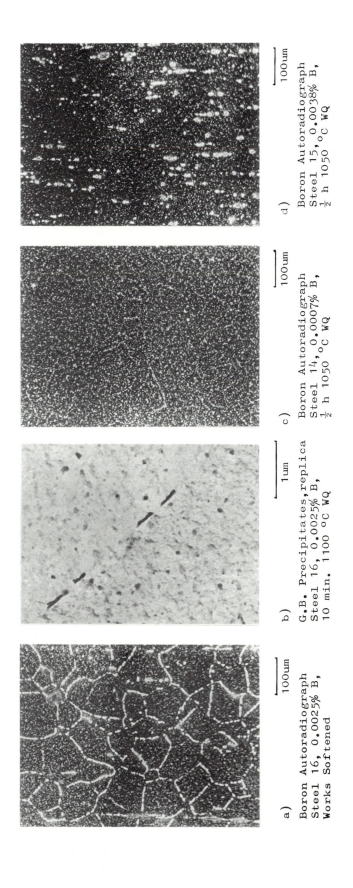

Boron Autoradiograph
Steel 16, 0.0025% B,
Works Softened

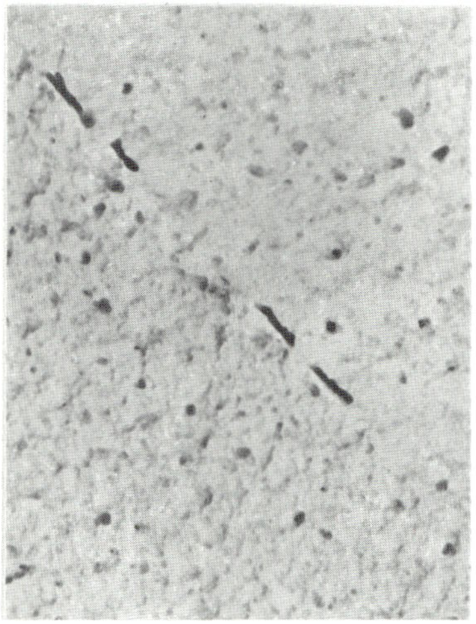

b) 1um

G.B. Precipitates, replica
Steel 16, 0.0025% B,
10 min. 1100 °C WQ

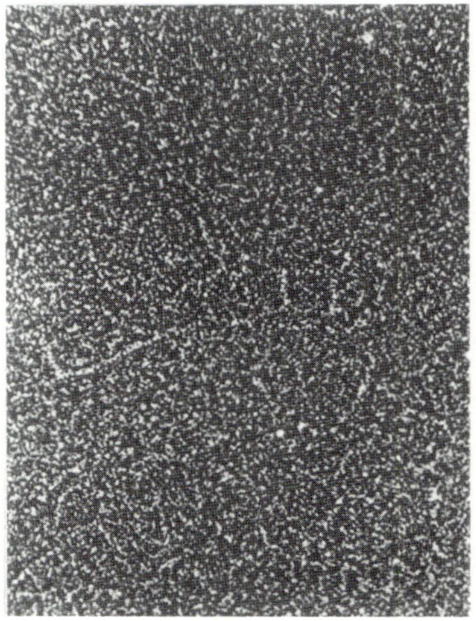

c) 100um

Boron Autoradiograph
Steel 14, 0.0007% B,
½ h 1050 °C WQ

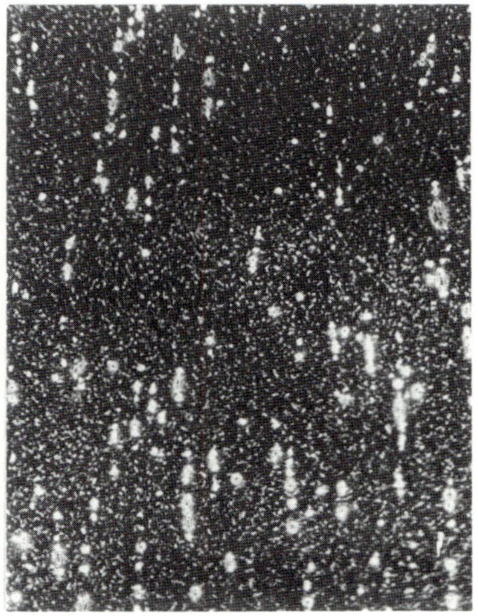

d) 100um

Boron Autoradiograph
Steel 15, 0.0038% B,
½ h 1050 °C WQ

5. Micrographs Showing Location of Boron.

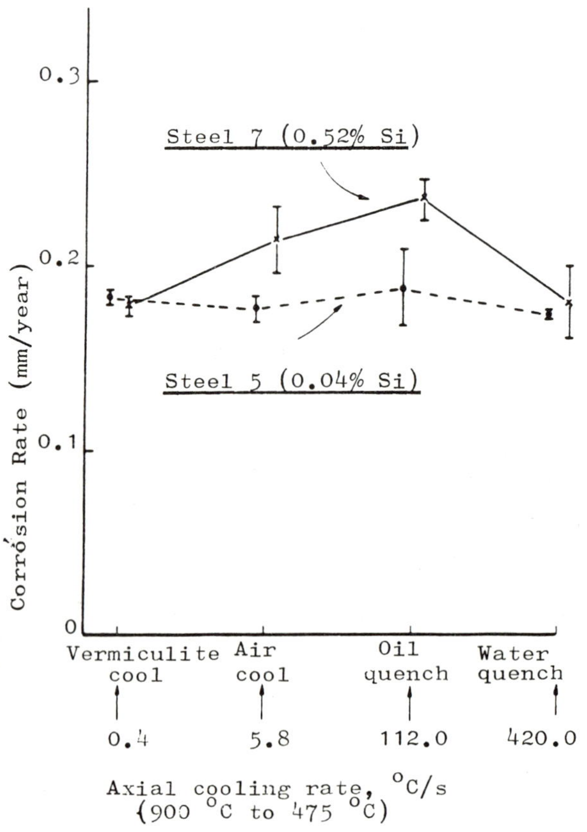

6. The Effect of Post Solution Treatment Cooling Rate and Silicon Content on Corrosion Rate.

High nitrogen stainless steels: austenitic, duplex and martensitic

M O Speidel

The author is with the Institute of Metallurgy, Swiss Federal Institute of Technology, ETH, Zurich, Switzerland.

SUMMARY

In recent years it has become possible to produce in ingots up to 20 tons of stainless steel with more than twice the nitrogen content of customary nitrogen-containing steels. A review of recent achievements in the properties of such high-nitrogen steels is given. New heights have been reached in yield strength, fracture toughness, pitting corrosion resistance, stress corrosion resistance, high-temperature yield strength, and creep rupture strength. Further developments and some limitations are indicated.

INTRODUCTION

Commercial stainless steels of all three kinds discussed here, with austenitic, duplex ferritic-austenitic, or martensitic microstructure, have in the last twenty years been produced with more and more significant additions of nitrogen. However, in nearly all commercial austenitic and duplex stainless steels, the nitrogen content does not exceed 0.4 weight-percent, and in martensitic stainless steels, nitrogen concentrations rarely exceed 0.07 weight-percent. This paper deals with stainless steels in which nitrogen concentrations are twice that or even higher.

Such high-nitrogen steels can now be produced on a commercial scale with ingots weighing up to 20 metric tons, using a pressurized electroslag remelting facility.[1] For laboratory investigations, we have also produced numerous smaller ingots in high-pressure induction furnaces with nitrogen atmospheres up to 200 bar. Perhaps the best-known commercial application of high-nitrogen steels is in generator-rotor retaining rings,[2] but many more applications are presently in various states of development. The present paper summarizes some of those properties of the high-nitrogen stainless steels which make it probable that these types of steel will find increasing commercial application in several fields of modern technology.

HIGH-NITROGEN AUSTENITIC STAINLESS STEELS

High nitrogen contents in solid solution can markedly increase the yield strength of austenitic stainless steels, as shown in Fig. 1. It is remarkable that this significant solid solution strengthening effect of nitrogen is not accompanied by a loss of fracture toughness, at least up to 0.7 weight-percent nitrogen. The solid solution hardening of austenitic stainless steels, and its temperature dependence, are understood from first principles and can be related to the crystal lattice distortion due to nitrogen atoms in interstitial positions.[3] The constantly high fracture toughness level in Fig. 1 is predictable from the fact that nitrogen in solid solution does not influence the coefficient of work hardening, nor does nitrogen in solid solution introduce new void-nucleating sites.[4,5]

Solid solution strengthened high-nitrogen austenitic stainless steels can be further strengthened by cold work.[2,3] This is illustrated in Fig. 2, showing the effect of cold expansion of rings on yield strength and fracture toughness of a steel with 18%Cr, 18%Mn, 0.58%N, and less than 0.05%C. The highest yield strength obtained in massive commercial retaining rings exceeds 1500 MN/m^2. In other product forms, yield strengths of over 2000 MN/m^2 have been achieved, and 2500 MN/m^2 appears feasible as a result of current research programs.

It is evident from Fig. 2 that the fracture toughness is reduced with increasing cold work. However, before cold expansion, clean high-nitrogen austenitic stainless steels, produced by the pressurized electroslag remelting technique, have an enormously high fracture toughness of over 500 $MN\,m^{-3/2}$, as shown in Fig. 2. Thus, even though the fracture toughness is significantly reduced by cold working, it remains at a very acceptable level of about 200 $MN\,m^{-3/2}$ after 40% cold work.

The combinations of yield strength and fracture toughness which can be achieved at room temperature with high-nitrogen austenitic stainless steels are better than any other presently existing steel can provide. An illustration of the high product of fracture toughness times yield strength is given in Fig. 3. It may be noted that the upper technological limit has doubled every decade over the last thirty years, and there is a high probability that if by 1990 a product of 4×10^5 $MN^2.m^{-7/2}$ (= $MPa^2\sqrt{m}$) is reached, this will be achieved with a high-nitrogen, cold-worked austenitic stainless steel.

An example of a recent high-nitrogen steel development where the combination of strength and toughness was important, is given in Figs. 4 and 5. A rotor steel for cryogenic applications with sufficient strength and toughness at both ambient temperature and at 77 K was required. As shown in Fig. 4, a clean high-nitrogen manganese austenite (steel B) can significantly improve both strength and toughness over a more standard chromium-nickel austenite, thus easily fulfilling the required specifications at ambient temperature. Moreover, a steel of Type B has all the strength a turbine designer can possibly require at 77 K, as shown in Fig. 5. The requirement of sufficient fracture toughness at such high strength levels is much more difficult to fulfill, but the achieved value of 150 $MN.m^{-3/2}$ is still acceptable and in many cases exceeds fracture toughness values of rotors for ambient and high-temperature operation.

* The pitting corrosion resistance of high-nitrogen austenitic stainless steels is illustrated in Figs. 6 and 7. Over thirty such steels with different chromium, molybdenum and nitrogen contents have been subjected to pitting corrosion tests in aqueous 6% $FeCl_3$ solutions, according to ASTM G48, and a critical pitting temperature was measured for each steel. These temperatures are plotted in Fig. 6 versus a chromium equivalent ("Wirksumme") without consideration of the nitrogen concentration. The result shows considerable scatter. Such scatter is greatly reduced if nitrogen is incorporated in the chromium equivalent, as indicated in Fig. 7. According to these results it appears that in austenitic stainless steels, one weight-percent nitrogen might confer as much pitting resistance as 30% chromium. That is, the chromium equivalent of nitrogen in the case of pitting is about 30. There are certainly limits to a general applicability of this figure. The nitrogen must be in solid solution, and not even partially present in nitrides, to be effective. Moreover, more than 12% chromium must be present to make nitrogen effective, and the role of molybdenum and the acceptable impurity concentration must be studied further to clarify the limits of the chromium equivalent of nitrogen.

The stress corrosion cracking resistance of a cold-worked, high-strength, high-nitrogen steel is compared in Fig. 8 to the stress corrosion crack growth rate of a high-carbon, austenitic steel which has been cold-worked in the same manner to a similar yield strength. Both steels are used for generator-rotor retaining rings. Note in Fig. 8 that the high-nitrogen, austenitic steel does not exhibit any measurable stress corrosion crack growth at any stress intensity level in the test period of three to six months.[2,6] Thus high-nitrogen austenitic steels are clearly superior to high-carbon austenitic steels in stress corrosion cracking resistance.*

HIGH-NITROGEN DUPLEX STAINLESS STEELS

Ferritic-austenitic duplex stainless steels with about equal amounts of both phases are well known for their combination of high strength and high corrosion and stress corrosion resistance. In order to achieve high yield strength, such duplex stainless steels often contain nitrogen, and commercial duplex stainless steels with up to 0.4 weight-percent nitrogen are available. The superior yield strength of the duplex stainless steels is illustrated in Fig. 9 and is compared to the average yield strength of the austenitic stainless steels, taken from Fig. 1. At a nitrogen concentration of 0.4%, the duplex stainless steels have a yield strength advantage of about 200 MN/m^2. Thus it should be of interest to see whether this strength advantage can be maintained at even higher concentrations of nitrogen. According to Fig. 9, duplex stainless steels with about 0.5 weight-percent nitrogen can be made with yield strengths exceeding 700 MN/m^2. At present, duplex stainless steels with much higher yield strengths at much higher nitrogen contents are not a very promising prospect for several interconnected reasons.

Nitrogen in solid solution, besides being a potent hardener, is also a potent austenite stabilizer with a nickel equivalent of about 15. Therefore, adding nitrogen alone to a duplex stainless steel just increases the austenite phase fraction and finally the yield strength is dominated by the austenite alone as shown in Fig. 9. In order to maintain 50% ferrite phase, the chromium equivalent would have to be increased very substantially. This would lead not only to very expensive stainless steels, but also to the danger of precipitation of nitrides and intermetallic phases. Thus there might be a limit between 0.5 and 0.6% nitrogen for a further development of useful duplex stainless steels.

HIGH-NITROGEN MARTENSITIC STAINLESS STEELS

Much of the recent development of high-nitrogen martensitic stainless steels is based on 9 to 12% chromium steels for high-temperature applications in turbines, boilers and the like. With the pressurized electroslag remelting technology, 0.30 weight-percent of nitrogen can easily be introduced in such steels, thus exceeding the concentration needed by a comfortable margin.

Figures 10 and 11 show what kind of high-temperature strength can be achieved with high-nitrogen martensitic stainless steels. In Fig. 10 the creep rupture strength of a standard 12%Cr-0.21%C steel is compared with the creep rupture strength (extrapolated by the Larson-Miller method) of a high-nitrogen steel with 10%Cr, 0.16%N and 0.06%C. It appears that the following development goal can be exceeded; a time to rupture of 10^5 hours at 600°C at a stress of 100 MN/m^2.

As a precondition for the achievement of a high creep rupture strength, most carbonitrides must be in solid solution at the austenitizing temperature. After the transformation to martensite upon cooling the 10 to 12% chromium steels then precipitate upon tempering very finely distributed, almost spherical carbonitrides which confer the high creep resistance on the steel. For applications which are creep-rupture controlled, the tempering temperature must be sufficiently high (about 720°C) to stabilize the microstructure. If however, an application at intermediate temperatures is envisaged, say up to 450 or 500°C, then a comparatively low tempering temperature near 600°C may be used. This gives the high-nitrogen martensitic stainless steel a combination of strength and fracture toughness which so far has not been possible in the standard carbon-12% chromium steels.

Figure 11 illustrates this fact. Note that

the design requirements for advanced gas and steam turbines include high yield strengths and charpy-V-notch toughness of A_V = 50 Joule. Presently, nickel-base superalloys are used to fulfill such requirements. The standard, carbon-strengthened 12% chromium steel X20CrMoV121 has either the strength or the toughness required but not both at the same time. The high-nitrogen, martensitic stainless steel presently fulfills the strength requirements together with a good toughness which is however not quite as good as the design requirements, which are perhaps exaggerated.

ACKNOWLEDGMENTS

The author acknowledges the contributions of G. Stein, P. Uggowitzer, H. Feichtinger, B. Anthamatten, M. Harzenmoser, and R. Magdowski

REFERENCES

1 P. Pant, P. Dahlmann, W. Schlump: 'A new technology for massive nitriding of steels', Tech. Mitt. Krupp. Forsch. Ber., 1985, 43, (H3), 67

2 Markus O. Speidel, G. Gabrielli, and R. B. Scarlin: 'Alloy development and service reliability of retaining rings'; Proc. Conference 'Advances in material technology for fossil power plants', 1987, Chicago, ASM/EPRI

3 Markus O. Speidel and Peter J. Uggowitzer: 'Strengthening of austenitic steels by nitrogen in solid solution'; Proc. P900 Kolloquium, 1986, SKK Essen

4 P. Uggowitzer, M. O. Speidel: 'The fracture toughness of nitrogen-alloyed austenitic stainless steels'; Proc. 2nd P900 Colloquium, Sept. 1985, Essen

5 P. Uggowitzer: 'Microstructure and fracture of metallic materials', DVM Bruchvorgänge, 1986, Nr. 18, 9

6 Markus O. Speidel: 'Nichtmagnetisierbare stähle für generator-kappenringe, ihr widerstand gegen spannungsrisskorrosion und wasserstoffversprödung', VGB Kraftwerkstechnik, 1981, Vol. 61, 417

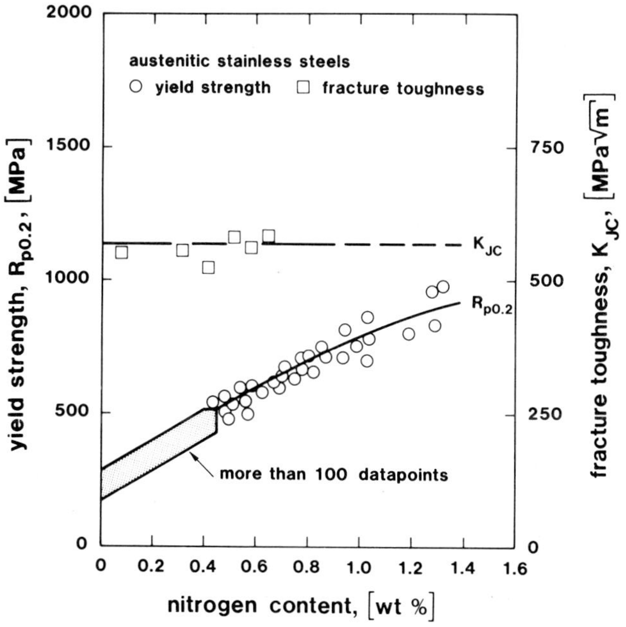

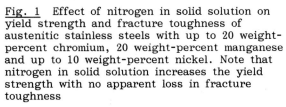

Fig. 1 Effect of nitrogen in solid solution on yield strength and fracture toughness of austenitic stainless steels with up to 20 weight-percent chromium, 20 weight-percent manganese and up to 10 weight-percent nickel. Note that nitrogen in solid solution increases the yield strength with no apparent loss in fracture toughness

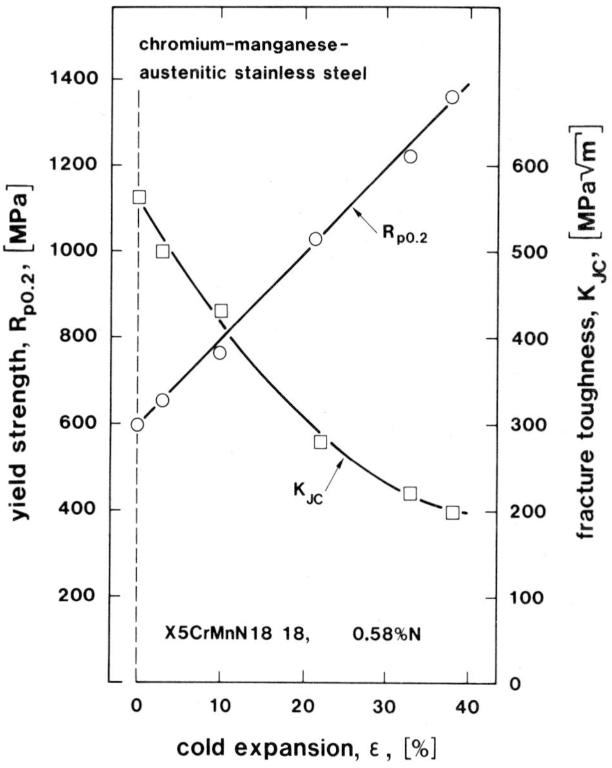

Fig. 2 Effect of cold work on yield strength and fracture toughness of a commercial high-nitrogen austenitic stainless steel. Note that with increasing cold work the yield strength reaches levels which are extremely high for commercial austenitic stainless steels. With increasing cold work and increasing strength, the fracture toughness is reduced but still remains at a very acceptable high level, even after 40% cold deformation.

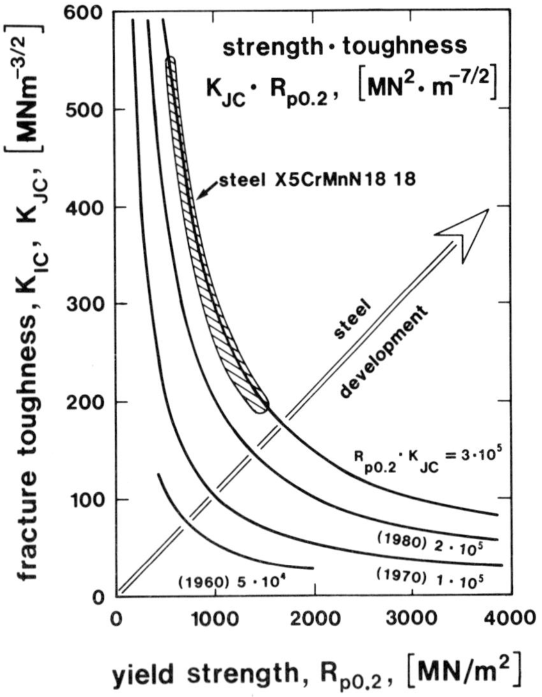

Fig. 3 The upper limit of the product strength times toughness doubles about every decade. Commercial high-nitrogen steel X5CrMnN18 18 (with 18%Cr, 18%Mn, 0.6%N and less than 0.05%C) offers presently the best combination of strength and toughness at ambient temperature.

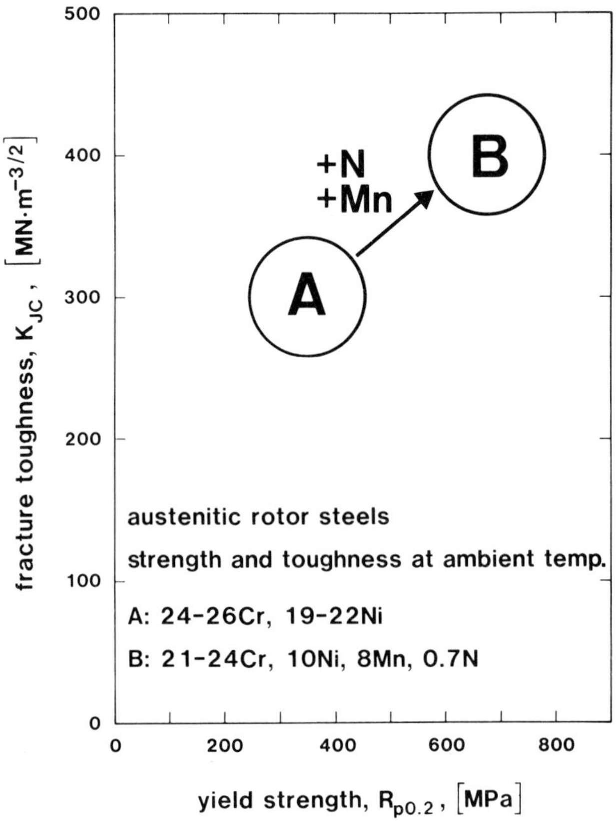

Fig. 4 Rotor steel development: strength and toughness at ambient temperature can both be improved with high-nitrogen austenitic stainless steels

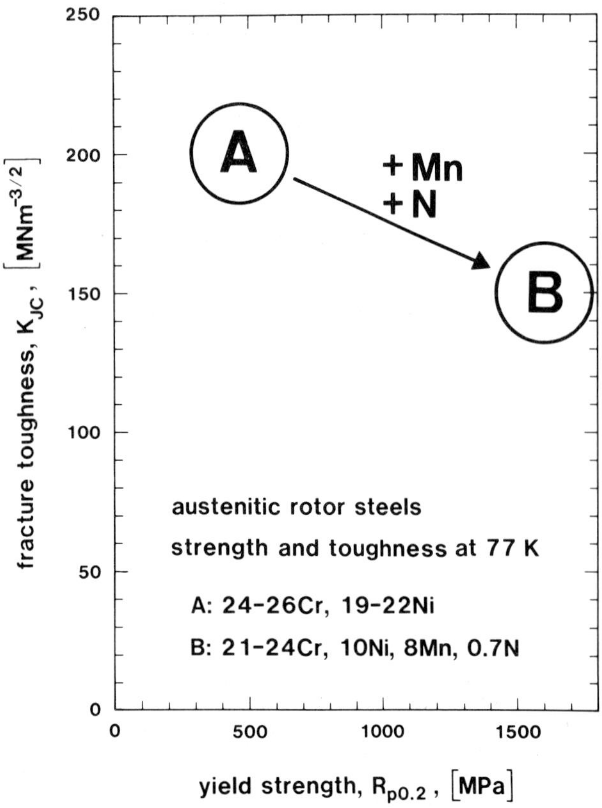

Fig. 5 Rotor steel development: sufficient strength at cryogenic temperatures is no problem at all with high-nitrogen austenitic stainless steels. However, careful selection of alloying additions is necessary in order to maintain acceptable fracture toughness levels at 77 K

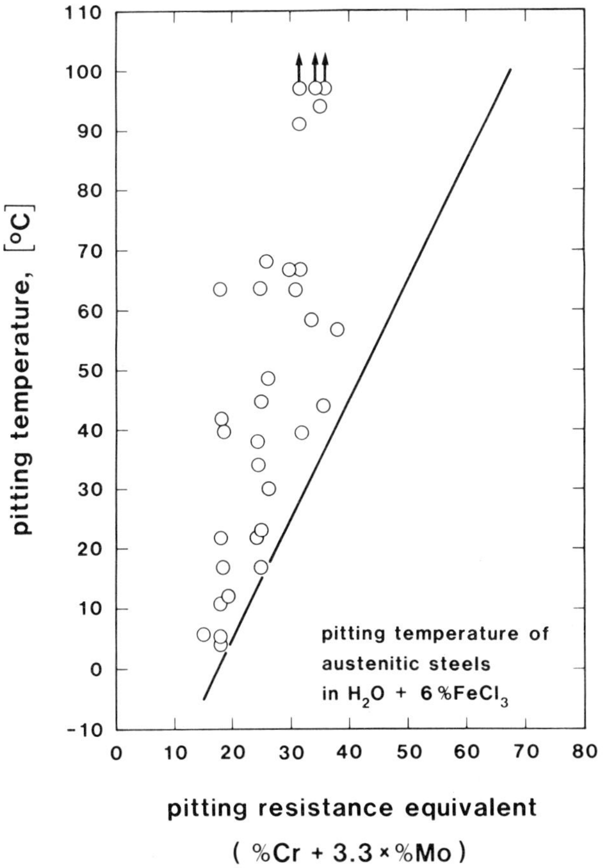

Fig. 6 Critical pitting temperatures of austenitic stainless steels in aqueous ferric chloride solution

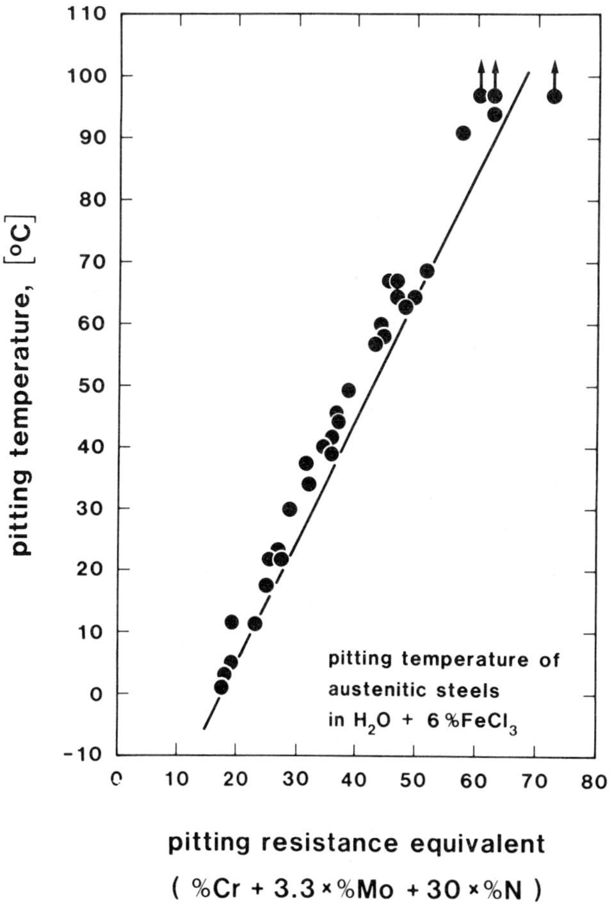

Fig. 7 Critical pitting temperatures of austenitic stainless steels in aqueous ferric chloride solution. Here, nitrogen is included in the pitting resistance equivalent

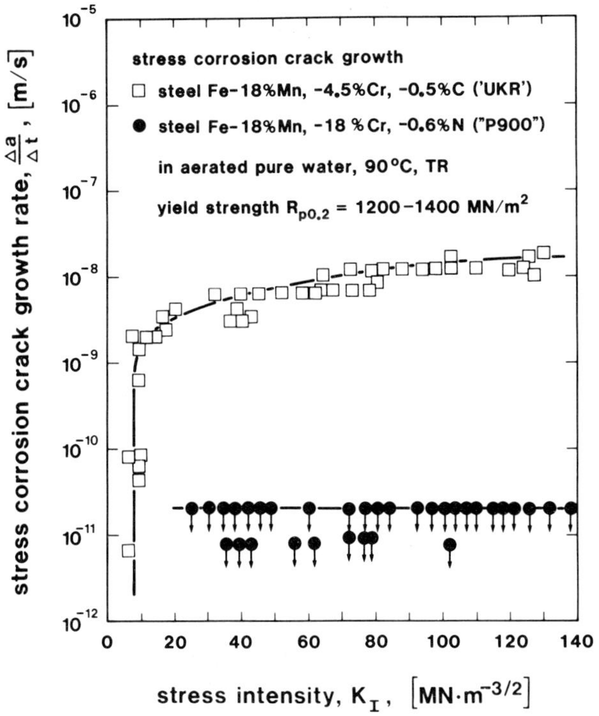

Fig. 8 Effect of stress intensity and steel composition on stress corrosion crack growth rates in cold-worked austenitic steels exposed to 90°C water. Note that the high-nitrogen steel does not exhibit measurable stress corrosion crack growth at any stress intensity

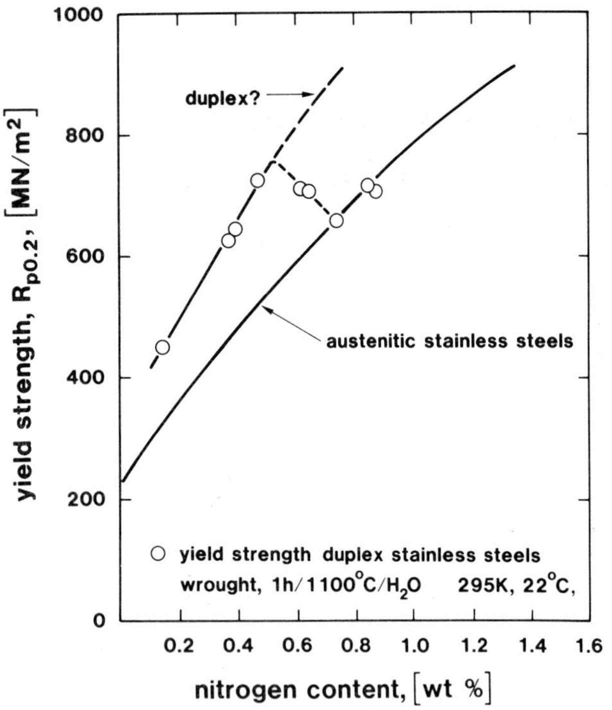

Fig. 9 Effect of nitrogen content on the yield strength of two-phase (duplex) ferritic-austenitic stainless steels. Duplex steels have a yield strength advantage over austenitic stainless steels but at very high nitrogen concentrations, it is difficult to maintain a significant ferrite phase proportion and so the strength is dominated by the austenite

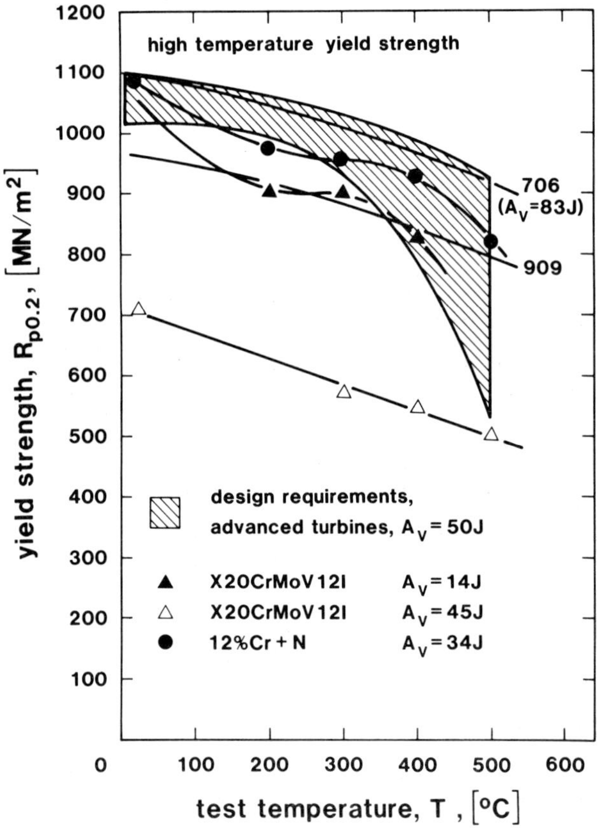

Fig. 11 High-temperature yield strengths of martensitic stainless steels compared with nickel base alloys 706, 909, and design requirements for advanced turbines

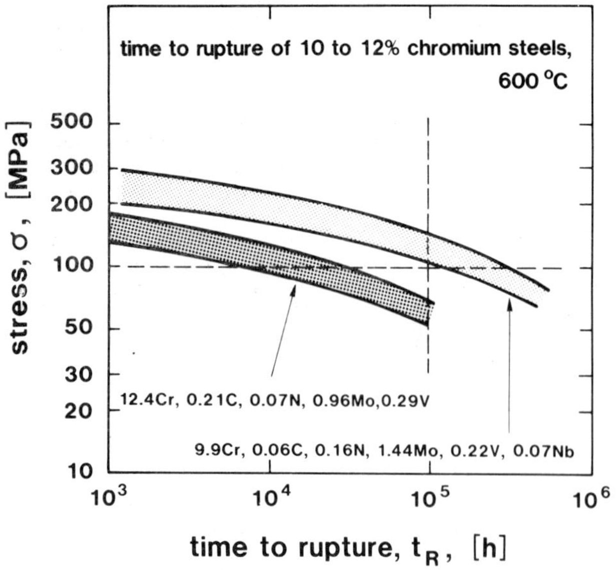

Fig. 10 Creep rupture strength of martensitic stainless steels, indicating an advantage of the high-nitrogen steel

Effect of chlorination on corrosion characteristics of stainless steels used in offshore seawater handling systems

R E Malpas, P Gallagher and E B Shone

The authors are in the Thornton Research Centre, Shell Research Ltd., Chester.

SYNOPSIS

Stainless steels representing those types having low, intermediate and high resistance to corrosion in natural seawater have been exposed to flowing seawater containing various levels of chlorine. This work has shown that low residual levels of chlorine (<0.1 ppm) resulted in lower potentials than normally seen in unchlorinated seawater and exposure tests have shown that this can lead to lower corrosion rates on alloys susceptible to crevice corrosion. Chlorinating to the residual levels usually used in offshore systems (0.5 to 0.8 ppm) gave rise to much higher potentials that were dependent on seawater flow rate. Tests showed that, despite these high potentials, susceptible alloys undergo less crevice corrosion in seawater chlorinated to these levels. Corrosion was not observed on highly resistant alloys of the UNS 31254 type under any test condition.

INTRODUCTION

Most of the seawater handling systems on offshore platforms in the North Sea are constructed from copper alloys. To control fouling in these systems they are continuously chlorinated on a routine basis to residual chlorine levels of between 0.5 and 0.8 ppm. These levels of chlorination do not lead to corrosion problems. However, should the copper alloys be replaced by stainless steel systems which are claimed to be lighter, less expensive, potentially more reliable and easier to handle, then determination of the influence of chlorination on the corrosion of such alloys in seawater becomes an important necessity. With this in mind we have carried out an experimental programme aimed at investigating the relationship between different chlorination levels in seawater and the corrosion properties of stainless steels, taking into account the range of operating conditions expected to be encountered offshore.

The results reported in this paper fall into two sections, both are concerned with corrosion properties in seawater at temperatures typical of the North Sea (between 5 and 9 degrees Celsius) and at chlorination levels currently used in offshore systems.

In the first section we report the results of a study carried out to determine the effect of chlorine on the electrochemical potential attained by a 6% molybdenum containing stainless steel (conforming to UNS 31254) on exposure to seawater at different flow rates. This study is important since previous work has shown that the initiation of corrosion on stainless steel alloys in seawater is strongly dependent on the potential attained by the alloy(1), with more positive potentials resulting in an increased probability that corrosion will occur. The results generated should thus give an indication of changes in the corrosivity of the seawater on introduction of chlorine. UNS31254-type steel is one of the materials that are currently either being used or considered for use in the construction of new offshore seawater systems. It has a high resistance to localised corrosion in natural (unchlorinated) seawater and thus should provide a stable baseline, uncomplicated by corrosion processes, from which to begin the study.

In the second section we report the results of exposure tests on three stainless steel alloys in seawater chlorinated to low residual levels and to the residual levels typical of those used offshore. The results are compared with those obtained from tests on similar alloys exposed to unchlorinated seawater. The alloys chosen for the tests represented those having low (AISI 316L) intermediate (AISI 904L) and high (UNS31254) resistance to corrosion in natural (unchlorinated) seawater.

EXPERIMENTAL

The work described in this paper was carried out in natural flowing seawater at our marine test station in Holyhead. When required, the seawater was chlorinated using an electrochlorinator (Chloropac, Electrocatalytic), and the residual chlorine level measured by a colorimetric method using DPD 1 reagent and a Lovibond comparator.

The 3 stainless steels used in this paper had the following chemical analysis:

	weight %						
alloy	Cr	Ni	Mo	N	Cu	Mn	Si
UNS31254	19	18	6.0	0.21	0.70	0.50	0.056
AISI 316L	17	12	2.0	0.03	-	1.61	0.34
AISI 904L	19	25	4.2	0.03	1.57	1.65	0.24

Determination of the dependence of electrochemical potential on residual chlorine levels was carried out on plate samples of the 6%Mo stainless steel, UNS 31254. These were exposed in tanks to flowing seawater with residual chlorine levels between 0 to 0.8 ppm. The potentials of the samples were measured using a Hewlett Packard 3497A data acquisition unit controlled by a HP85B microcomputer system. Potential measurements were referenced to a saturated calomel electrode and were monitored on a 6 hourly basis throughout the test period which in these experiments extended to 7 months. Results were transmitted back to the Laboratory at Chester by a modem link.

Potential / flowrate / chlorine level relationships were determined in a large-scale pipe test facility. This consisted of a 30 metre length of 50mm bore pipe in alloy UNS 31254, flanged at 6 m intervals, through which a flowrate in excess of 20 m/s of once-through seawater could be maintained at residual chlorine levels up to 1.5 ppm. Flow rates were measured using a magnetic flowmeter (Fischer and Porter), a signal from which modulated the electrochlorinator current to accurately maintain the chlorination level. At each of the chlorination levels studied, the system was allowed to stabilise for a week at a flow rate of 10m/s. Following this stabilising period the potential of the pipe section was continuously monitored using a high impedance voltmeter with the output linked to a chart recorder. The flowrate was adjusted stepwise from 0.5 to 15 m/s during a period of approximately 2 hours, allowing a steady state to be reached at each flowrate. Additionally, this pipeline was exposed for a 7 month period to seawater flowing at 10 m/s and chlorinated to a residual level of 0.4 ppm initially (for 75 days) followed by 0.8 ppm for the remainder of the test.

Small scale exposure tests were carried out on 150 x 100 mm plate samples fitted with a 'multiple crevice washer assembly'. Parallel tests on the alloys were performed in tanks containing both flowing natural seawater and chlorinated seawater in which the residual chlorine level was maintained at either 0.8 ppm or at a level less than 0.1 ppm. During the 100 day (0.8 ppm) or 50 day (<0.1 ppm) exposure period, the potential of each sample was monitored using the computer system described above. At the end of this test period the crevice washers were removed from the samples and the plates examined.

RESULTS

1. Dependence of Corrosion Potential on Residual Chlorine level and Flow velocity

The potential - time relationship for plates of type UNS 31254 stainless steel exposed in tanks of flowing seawater at various residual chlorine levels are shown in Figure 1. Examination of the plates after exposure showed that none of them had corroded; the measured potentials were thus not influenced by corrosion.

In natural (unchlorinated) seawater (Figure 1(a)) the potential increased from about -100 mV (a value dependent on the initial state of the surface of the sample) to a potential of 300 to 350 mV over a period of 15 to 20 days. Other tests have shown that further exposure would produce no further rise in potential. Chlorinating the seawater to residual chlorine levels up to 0.1 ppm prevented this potential rise, giving the potential-time response shown in Figure 1(b). It is well known that corrosion can also lead to lower potentials but it must be stressed that no corrosion whatsoever was evident on the plate after exposure at this chlorine level and the results shown in Figure 1(b) were reproducible over many samples.

Increasing residual chlorine levels above 0.1 ppm produced a rise in both the rate at which the potential increased with time and in the final stable potential attained. The results for a 0.5 ppm residual chlorine level are shown in Figure 1(c). As can be seen, the potential increased rapidly to a value of 500 mV over a period of only 1 day, reaching a stable value of 550 mV after a further 4 days. Increasing the chlorination level above this level to a maximum value of 0.8 ppm produced very little variation in the potential - time response. This is demonstrated in Figure 1(d) where the results for a 0.8 ppm residual chlorine level are shown.

The relationship between the residual chlorine level and the stable corrosion potential attained by the alloy is summarised in Figure 2. This shows there to be a clear minimum in potential at low residual chlorine levels.

The dependence of the potential on flow rate through a 50 mm UNS 31254 pipe at varying residual chlorine levels is shown in Figure 3. It was necessary to allow the pipe to equilibrate for up to 30 minutes at each chlorine level before determination of the potential - flow rate relationship. Carrying out the determination immediately after altering the chlorine level (even to the extreme of turning off the chlorinator) invariably produced little immediate change in the observed relationship.

The results in Figure 3 show that at low residual chlorine levels and without chlorination (Figure 2(a) and (b)), the potential increased by only 100 mV as the flow velocity was increased from 0.5 to 15 m/s. At higher residual chlorine levels, however, the potential increased rapidly with flow rate from values of around 400 mV at 0.5 m/s to 650 mV at 15 m/s, with the actual relationship being dependent on the chlorine level as shown in Figures 2(c) and (d).

The potential of the pipe over a 7 month operating period at a seawater flow rate of 10 m/s is shown in Figure 4. Two chlorine levels were used during this period: 0.4 ppm for the first 70 days and 0.8 ppm for the remainder of the test. The potentials observed at these two levels fluctuated with time but were generally in the region predicted from Figure 3. At the higher residual level, however, the potential did reach values as high as 800 mV for short periods of time. The reason for these fluctuations was not readily apparent.

2. Exposure Tests

In general the crevice washer samples which had been exposed for 100 days to the chlorinated seawater at 0.8 ppm residual levels had less corrosion damage within the crevice area than the corresponding samples exposed to natural seawater. This was most evident on AISI 316L (Figure 5), where the sample exposed to chlorinated seawater showed significantly shallower crevice attack than samples from this and all previous tests in flowing unchlorinated seawater. In contrast, small areas of pitting were visible outside the crevice area on the chlorinated 316L sample that was not seen on the sample exposed to unchlorinated water.

The 6%Mo alloy to UNS31254, which did not show any damage in crevice washer tests in unchlorinated seawater was equally undamaged in the 0.8 ppm chlorinated water. Examples of this type of alloy from several different manufacturers were tested with similar results. The flanged pipeline in a UNS31254 alloy, operated at 10 m/s flowrate and 0.8 ppm residual chlorine level, likewise showed no corrosion damage whatsoever after 7 months of operation.

Samples were also subjected to parallel crevice washer tests in unchlorinated seawater and seawater chlorinated to a very low residual level (less than 0.1 ppm). The results for alloy AISI 904L after the 50 day test period are shown in Figure 6. No corrosion had initiated on the sample exposed to the low residual chlorine level, whereas in unchlorinated seawater considerable corrosion had occurred.

DISCUSSION

The rise in potential of the UNS 31254 sample over the first 15 days of exposure to unchlorinated natural seawater (Figure 1(a)) is a widely observed phenomena on many stainless steel alloys (2-5). The effect has been attributed to catalysis of the cathodic reaction on the steel surface (oxygen reduction) by biofilms of marine organisms and bacteria (3-5), a theory which is supported by the results shown in Figure 1(b), where chlorination of the seawater to a level just sufficient to kill the marine organisms, leaving a barely detectible residual level, prevented the potential rise.

The stable potential attained by the alloy (around 50 mV) in this low residual chlorine seawater was very similar to that seen in artificial seawater, where it has been shown to result in a considerable reduction in corrosion rates (by reducing the chance of initiating corrosion) on materials normally susceptible to crevice corrosion (6). That the low potential in the low residual chlorine seawater also favours a low corrosion rate is seen from the exposure test on the AISI 904L alloy (Figure 6). In this instance the difference between the corrosion on samples exposed to low residual chlorine levels (where no initiation of corrosion occurred) and unchlorinated seawater is quite marked. Low chlorination levels, therefore, appear to have a beneficial effect on the corrosion of stainless steel alloys by maintaining low potentials, thus reducing the probability of initiating localised corrosion.

Increasing the residual chlorine levels above 0.1 ppm, however, leads to a dramatic rise in the potential of exposed samples, to values of around 550 mV at 0.5 ppm. At these higher chlorine levels seawater flow rate also has a strong influence on potential (Figure 3) and values as high as 800 mV have been recorded for short periods during continuous running at high flow rates. The higher potentials are the result of depolarisation of the cathodic reaction by the presence of chlorine; the mechanism for this, however, does not appear to be simple. Increasing the residual chlorine level above about 0.5 ppm appears to produce no further rise in the potential, which appears to indicate activation control of the cathodic process. The flowrate-potential curves in Figure 3 show plateau regions at higher flowrates which could also support this; while the time taken for the potential to reach a steady state suggests that a slow equilibration step in the reduction mechanism is occurring, possibly involving conditioning of the metal surface. It has been suggested that the reduction of chlorine in aqueous chloride solutions at an oxide covered electrode occurs by a two step mechanism involving an adsorbed intermediate (7):

$$Cl_2(aq) + M + e \rightleftharpoons Cl\text{-}M + Cl^- \ (aq)$$

$$Cl\text{-}M + e \rightleftharpoons M + Cl^- \ (aq)$$

where M is the metal surface and Cl-M represents the adsorbed intermediate. Such a mechanism (with one of the steps being rate limiting) could explain the results obtained here, although further work is required to fully investigate this.

The high potentials observed when the residual chlorine level and the flow rate are increased would, by previous argument, be expected to result in a greatly increased probability that localised corrosion would occur, with a tendency to increased pitting as the "pitting potential" for the alloy in seawater is exceeded. The presence of small areas of pitting on samples appears to support this. It is surprising, therefore, that alloys exposed for 100 days to 0.8 ppm residual chlorine levels had less corrosion damage in the crevice area than the corresponding samples exposed to natural seawater. Initiation of crevice corrosion still occurred after approximately the same exposure period on susceptible materials, such as 316L. However, propagation of the corrosion, once initiated, was less than seen in unchlorinated seawater. We can only speculate at the reason for this at present. The anodic reaction could have been retarded by the proposed adsorption mechanism for chlorine reduction; alternatively it has been suggested (6) that the particularly aggressive crevice conditions produced in unchlorinated seawater could be partially caused by the presence of anaerobic bacteria in the crevice. Chlorine, acting in its intended role as a biocide, could conceivably interfere with this mechanism.

The most important result to arise from the exposure tests at the higher chlorination levels, however, is that no corrosion was seen on the 6% Mo alloy, UNS31254, in any of the tests, even in pipe systems over 7 month exposure periods at high flow rates, when high potentials were observed. This gives added confidence for the use of these materials under the operating conditions envisaged in offshore seawater systems.

CONCLUSIONS

Chlorination of seawater to low residual chlorine levels (<0.1 ppm) has been shown to prevent the rise in electrochemical potential normally observed when stainless steel alloys are exposed to seawater. The maintenance of a low potential appeared to have a beneficial effect on the corrosion of alloys that were susceptible to crevice corrosion in natural seawater (such as AISI 904L) by reducing the probability that corrosion would initiate.

The level of residual chlorine used in offshore seawater systems (between 0.5 and 0.8 ppm) produced higher potentials on exposed alloys. The potentials were heavily dependent on seawater flow rate and were observed to reach values as high as 650 to 700 mV at flow rates of 10 m/s and residual chlorine levels of 0.8 ppm. Despite these high potentials, the extent of crevice corrosion on alloys such as AISI 316L was less in chlorinated than in unchlorinated seawater, although the high potentials did lead to the initiation of crevice corrosion and the occurrence of slight pitting outside the crevice area on some materials.

Finally, alloys (such as UNS 31254) that are very resistant to corrosion in unchlorinated seawater, were similarly resistant to corrosion in chlorinated seawater at these higher residual chlorine levels. This provides added confidence for the use of these alloys in the construction of lighter, less expensive and potentially more reliable offshore seawater systems.

REFERENCES

1. Urquidi M. and MacDonald D.D., J. Electrochem. Soc., vol.132 p555 (1985)
2. Bardal E. and Johnsen R., Proc UK Corrosion 86, Birmingham 1986, p287.
3. Scotto V., DiCintio R. and Marcenaro G., Corr. Sci., vol25, 185 (1985).
4. Johnsen R., Bardal E., Corrosion, vol41, 296 (1985).
5. Holthe R., Gartland P.O. and Bardal E., Proc. Eurocorr 87, European Federation of Corrosion, p617.
6. Oldfield J.W., Lee T.S., and Kain R.M., Corrosion chemistry within pits, crevices and cracks, Ed. A. Turnbull. HM Stationary Office 1987, p89.
7. Dickinson T., Greef R. and Wynne-Jones Lord, Electrochim. Acta, vol14, p 467 (1969).

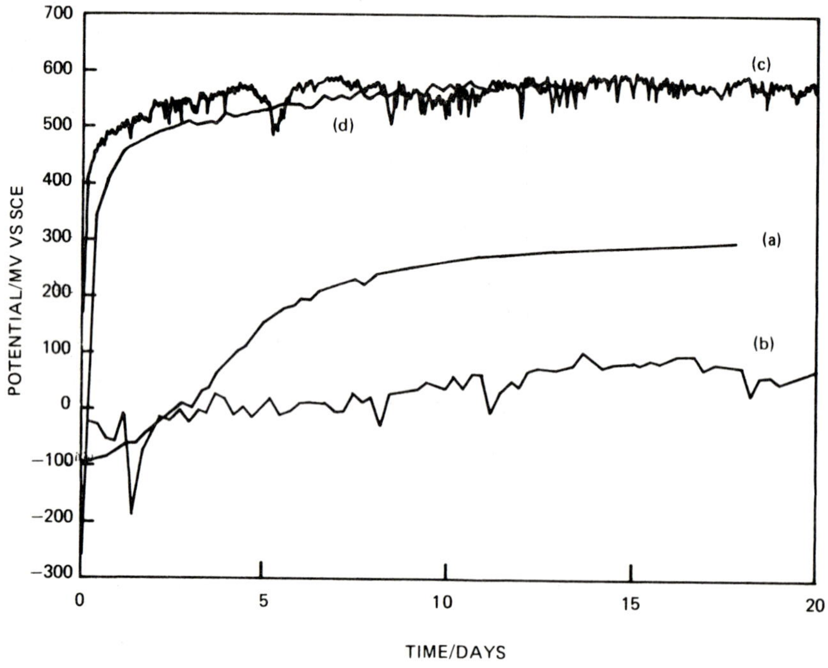

1. Variation of potential with time for alloy UNS 31254 on exposure to flowing seawater containing residual chlorine levels of: (a) 0 ppm; (b) <0.1 ppm; (c) 0.5 ppm; (d) 0.8 ppm.

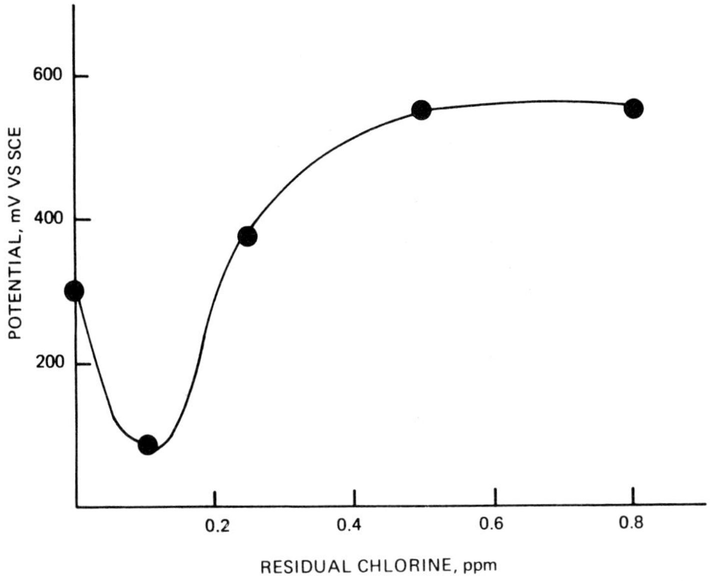

2.

Dependence of the stable potential attained by alloy UNS31254 on the residual chlorine level.

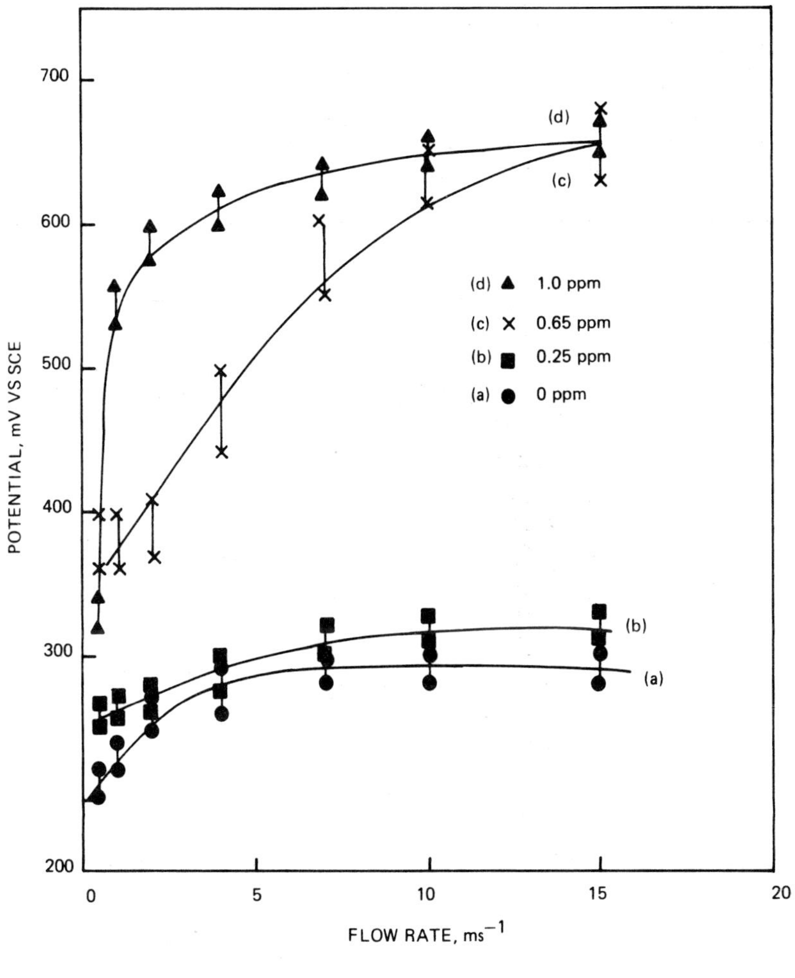

3. Dependence of the potential of alloy UNS31254 on the seawater flow rate at residual chlorine levels of:
(a) 0 ppm; (b) 0.25 ppm; (c) 0.65 ppm; (d) 1.0 ppm.

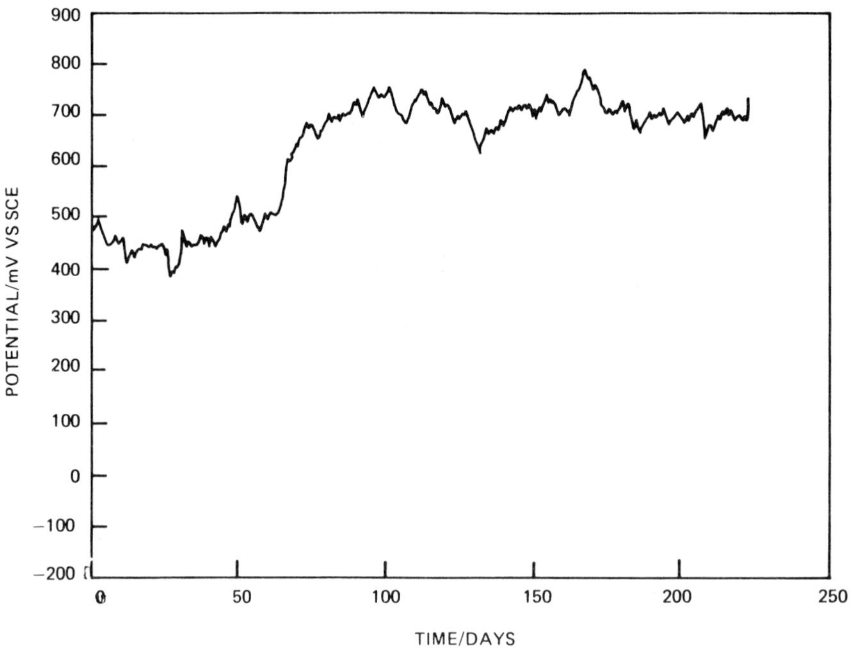

4. Variation of potential with time for a 50 mm diameter pipe in alloy UNS31254 on exposure to flowing seawater at a flow rate of 10 m/s and a residual chlorine content of 0.4 ppm (70 days) and 0.8 ppm (remainder).

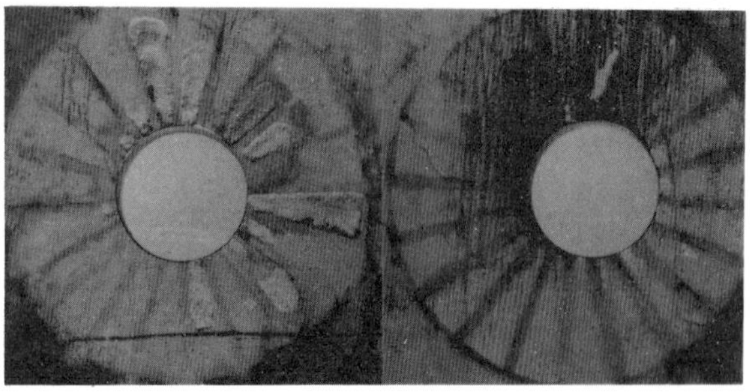

(a) (b)

5. Crevice washer plates of alloy AISI 316L exposed for 100 days to: (a) unchlorinated seawater; (b) seawater chlorinated to 0.8 ppm residual.

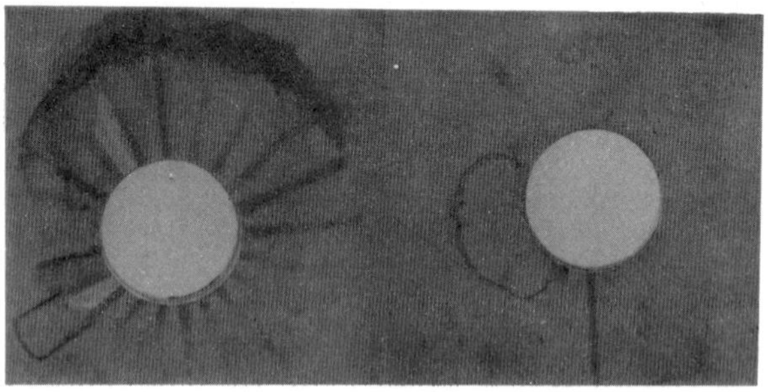

(a) (b)

6. Crevice washer plates of alloy AISI 904L exposed for 50 days to: (a) unchlorinated seawater; (b) seawater chlorinated to <0.1 ppm residual.

Superaustenitic stainless steels for marine applications

J Charles, P Soulignac, J P Audouard and D Catelin

Dr. Charles is Head of the Stainless Steel and Corrosion Department of Le Creusot Research Centre at Creusot-Loire Industrie, Le Creusot, France; P Soulignac is Senior Metallurgist; J P Audouard is Senior Corrosion Engineer at Unirec Research Centre; D Catelin is Product Manager.

SYNOPSIS

The local and general corrosion resistance of stainless steels in chloride solutions can be successfully increased by using higher chromium and molybdenum alloyed steels. These alloys are unfortunately known to be sensitive to the formation of intermetallic compounds.

The purpose of this paper is to present the phase diagram of the 25–25–5–Cu–N superaustenitic stainless steel developed by Creusot-Loire Industrie (URANUS SB8). The kinetics of formation, nature and morphology of intermetallic phases are described, and the influence of the presence of precipitations upon corrosion properties is discussed.

The results are discussed in order to define industrial heat treatment and welding conditions which ensure an austenitic structure free of intermetallic phase precipitations, with particularly improved corrosion resistance properties.

INTRODUCTION

The URANUS SB8 superaustenitic stainless steel alloy has a 25 Cr–25 Ni–5 Mo Cu–N composition. It has been developed to resist acidified and chloride solutions. It is recommended in phosphoric acid and hot marine environments. The pitting index ([Cr + 3.3 Mo]) is particularly high: > 40 (Table 1).

Nitrogen additions stabilize the austenitic phase, improving both yield strength and corrosion properties. Moreover, kinetics of precipitation of intermetallic compounds are generally lowered by nitrogen additions. Molybdenum and chromium additions improve the corrosion resistance of the steel; however, they promote the formation of intermetallic phases.

This paper describes the phase diagram of UR SB8 alloy. Kinetics of precipitation, morphology and nature of precipitations are studied. The influence of heat treatment on grain size and hardness is investigated, and the influence of intermetallic phases on corrosion resistance following ASTM 262 A and G 48 tests is studied. Electrochemical test results are also presented.

All this information is discussed in order to define the conditions necessary to obtain an austenitic steel with improved corrosion resistance properties in base metal as well as in welds.

EXPERIMENTAL CONDITIONS

Material

A 2 mm thick plate taken from a 70-ton heat is used for this study. The plate is water-quenched after a 1180°C treatment.

Phase diagram determination

Samples of 150 x 100 x 2 mm were heated several times (1 minute to 100 hours) at a temperature between 600 and 1200°C. All the heat treatments were followed by water quenching.

The structure was investigated by optical microscopy, electronic microprobe analysis, scanning electron microscopy, X-ray diffraction and transmission electron microscopy. Vickers hardness measurements were also performed.

EXPERIMENTAL RESULTS

Phase diagram

Figure 1 presents the TTT diagram of the 25 Cr–25 Ni–5 Mo superaustenitic stainless steel. Limits of intergranular and transgranular precipitation determined by means of optical microscopy investigations are presented. Optical and SEM micrographs of the structures are presented in Fig. 2. Figure 3 shows the influence of heat treatment on grain size and Vickers hardness.

Nature and morphology of precipitations

Precipitation starts at the grain boundaries and then develops inside the grains. Increase in time exposure results in higher density of precipitations and, particularly at higher temperatures, in coarsening of the precipitates. At 800°C after 100 hours exposure the transgranular precipitations present a plate-like morphology. χ compounds were not detected by means of this technique.

STEM investigations (Fig. 4) show that even for a short time exposure some submicroscopic transgranular precipitations, associated mostly with inclusions or titanium carbonitrides, precipitate. These transgranular precipitations are mostly χ-phases while intergranular precipitations are σ-phases. Figure 5 shows the average composition of the intermetallic precipitations determined by means of STEM investigations. Table II shows the microprobe analysis (CAMEBAX investigations) realized on the thickest precipitations. All the results confirm the presence of big intergranular and transgranular σ-phase precipitations after 100 h exposure. Some chromium carbonitrides are also detected for long time exposure specimens (800–900°C temperature range).

Solution-treated and water-quenched specimens are free of precipitations.

Influence of phase precipitations on corrosion resistance

Figure 6 presents the ASTM A 262-A and G 48 test results. Sensibilization of superaustenitic 25 Cr–25 Ni–5 Mo stainless steel may occur after 5 minutes exposure in the 800–1000°C temperature range. Crevice and pitting corrosion resistance is also affected by intermetallic phase precipitations.

Fracture zones or cross-sections of intermetallic compounds seem to be preferential sites for pitting (Fig. 7).

Electrochemical investigations confirm the detrimental effect of intermetallic precipitations on pitting corrosion potential of the alloy, particularly in acidificated solutions (Fig. 8).

Corrosion resistance of solution-treated specimens

Figure 9 presents the critical pitting and crevice temperatures for several austenitic and austenoferritic steels (Table 1) determined following ASTM G 48 test procedures. Both critical pitting and crevice temperatures of 25 Cr–25 Ni–5 Mo Cu–N superaustenitic steel are particularly high. This is explained by the very high chromium content of the steel (Cr + 3.3 Mo + 20 N $\gtrsim$ 42). These results are confirmed by the curves plotted in Fig. 10, showing the resistance to depassivation of several stainless steels by measuring the maximum dissolution current as a function of acidity in a 30 gr/1 sodium chloride solution at room temperature. Figure 11 shows the minimum chloride content function of the temperature at which several stainless steels can be used in a 54% phosphoric acid solution.

WELDABILITY

SMAW, GTAW, PAW or GMAW welding processes are required for URANUS SB8 alloy since it presents a high susceptibility to the formation of intermetallic phases. No preheating or postheating treatments are recommended. The weld energy must be lower than 15 kJ/cm²; the temperature between weld deposits must be kept under 140°C. The ERNI Cr Mo 3 following AWS 5-14 or ERNI Cr Mo 3 following AWS 5-11 are recommended.

Chemical analyses are presented in Table III. Figure 12 shows some optical micrographs of welded joints obtained with fit metal UR SB8, with and without nitrogen or Inconel 625 - Class 3. This latest choice is the only one which can produce safe heat-affected zones and welded structures. Mechanical properties and corrosion resistance can therefore be guaranteed for welded structures.

CONCLUSIONS

The precipitation phase diagram of a 25 Cr–25 Ni–5 Mo superaustenitic stainless steel is presented. Intermetallic phases (sigma and Chi) are formed when the metal is treated for 5 minutes or more in the 800–1000°C temperature range. These phase precipitations reduce the corrosion resistance of the steel.

After solid-solution and water-quenching treatment the alloy is free of phase precipitations and presents the highest resistance to localised corrosion in chloride solutions of all stainless steels tested. This explains why the 25 Cr–25 Ni–5 Mo austenitic stainless steel is recommended for industrial applications where metals suffer from aggressivity by stagnant cold sea water or hot sea water. In such service conditions even molybdenum 18 Cr–10 Ni austenitic steels are subject to severe corrosion damage due to localized corrosion effects. The superaustenitic 25 Ni–25 Cr–5 Mo (URANUS SB8 CREUSOT-LOIRE Trade Mark) is also to be considered for all chloride-containing, aggressive environments like the pulp and paper industries.

Finally, the high corrosion resistance properties of the 25 Cr–25 Ni–5 Mo stainless steel reported here are also explained by the special melting process used which permits extra-low sulphur content, low phosphor content and close control of the inclusion content.

ACKNOWLEDGMENTS

The authors are grateful to P Chemelle, Senior Metallurgist at IRSID, for the STEM investigations.

REFERENCES

1 J R Maurer and J R Kearns - NACE 85 March 25-29 BOSTON - paper 172
2 J Charles, P Pugeault, P Soulignac and D Catelin - Eurocorr. 87, (Karlsruhe) 6-10 April 1987, pp 601-606
3 J Charles, P Soulignac, J P Audouard, D Catelin - Proceedings of 25e journée des aciers spéciaux 32, 1-11, 1986
4 B Bonnefois, D Catelin, and P Soulignac Lères journées franco-allemande du soudage Karlsruhe 2-3 Oct., 1986, DVS 105 pp 106-112

TABLE I - CHEMICAL COMPOSITION

	Si	Mn	Ni	Cr	Mo	N	Cu	FERRITE	*P.I.
URANUS 35N	0.6	1.5	4.0	23.0	0.2	0.12	-	45	26.9
URANUS 45N	0.4	1.7	5.7	22.0	2.8	0.12	-	45	33.6
URANUS 47N	0.3	1.2	6.5	25.0	3.0	0.17	0.2	45	38.3
URANUS 52N	0.3	1.2	6.5	25.0	3.0	0.17	1.5	45	38.3
ICL 472 BC	0.5	1.5	10	18.5	0.5	0.07	-	2	21.5
UR B6	0.2	1.5	25	20	4.3	0.08	1.5	0	35.8
UR SB8	0.20	1	25	25	4.8	0.2	1.6	0	44.8

* P.I. : Pitting Index
Cr + 3.3 Mo + 20 N

TABLE II - CHEMICAL COMPOSITION DETERMINED BY MICROPROBE INVESTIGATIONS (CAMEBAX)

TREATMENT	Phase	Si	Mn	Ni	Cr	Mo	Cu	Fe
900°C/100 H	γ	0.3	1.1	26.7	23.5	3.3	1.6	43
	σ	0.6	0.8	12.4	35.3	13.9	0.3	36
	?	0.6	0.9	17.7	30	9.7	0.8	38.5
1000°C/100 H	γ	0.4	1	25.2	25.1	4.3	1.4	42.1
	σ	0.6	0.75	12.6	35.3	15.2	0.3	35.4

TABLE III - CHEMICAL COMPOSITION OF THE WELDED JOINTS

Filler Metal	Welds	C	Si	Mn	Ni	Cr	Mo	Cu	N_2	KCV J
NO	GTAW	0.09	0.08	0.9	25.5	25.3	5.1	1.5	0.08	25
UR SB8	GTAW	0.012	0.11	0.9	24.7	24.7	4.6	1.4	0.124	140
ERNI Cr Mo 3	GTAW	0.02	0.10	0.9	56	21	9	0.3	0.04	150

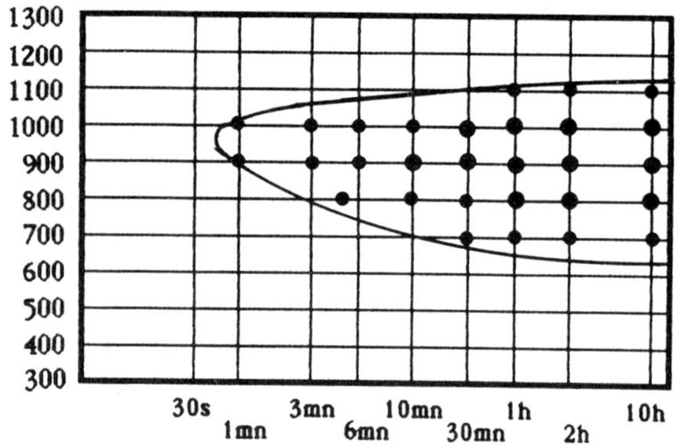

1 TTT diagram of the 25 Ni–25 Cr–5 Mo Cu–N superaustenitic stainless steel (optical microscope investigations)

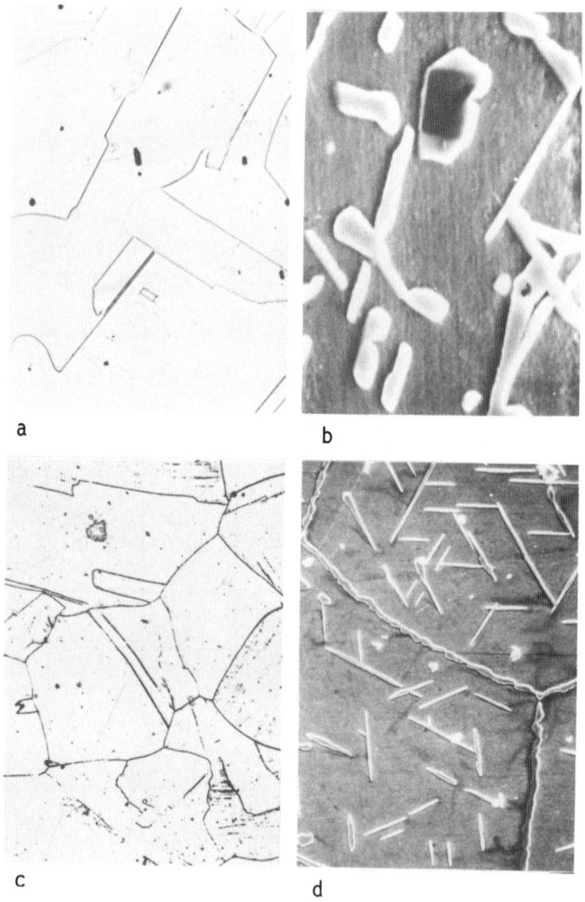

2 Optical and SEM micrographs:
a - water-quenched, solid-solution-treated alloy
b - 900°C/100 H
c - 900°C/5 mn
d - 1000°C/100 H

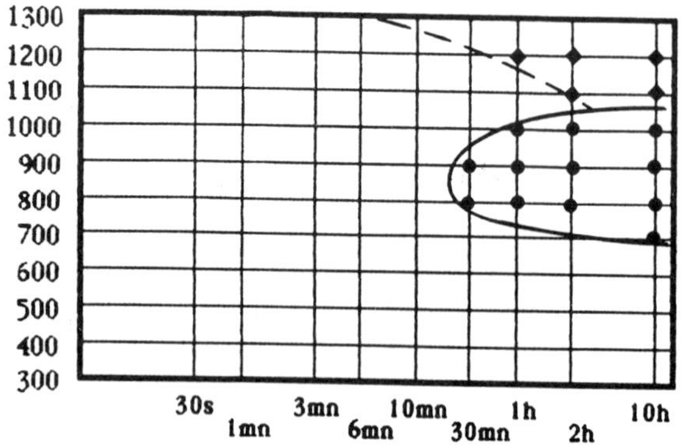

3 Influence of heat treatment on grain size (increase of grain size) and Vickers hardness (HV > 200)

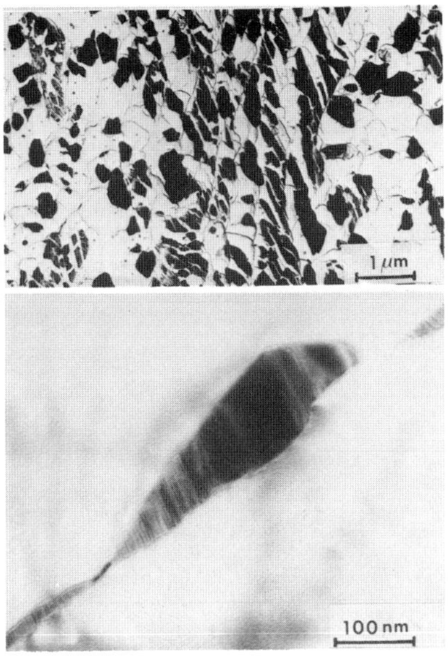

4 STEM micrographs and phase determinations:
800°C/5 mn σ phase

6 25 Ni–25 Cr–5 Mo Cu–N superaustenitic sensibilization diagram:
a - ASTM 262 A test
b - ASTM G48 - crevice - test
(critical temperature lower than 45°C)

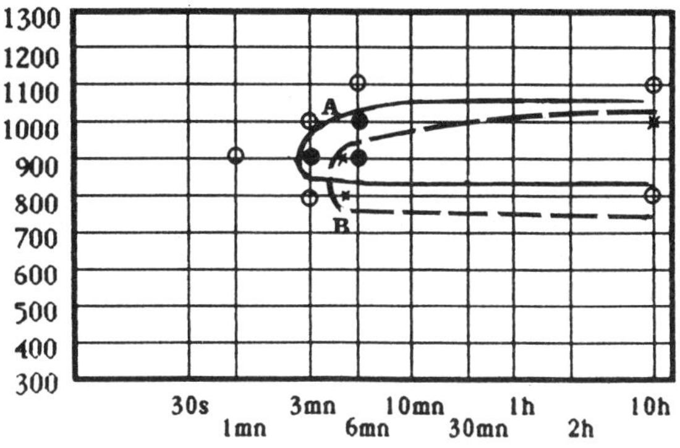

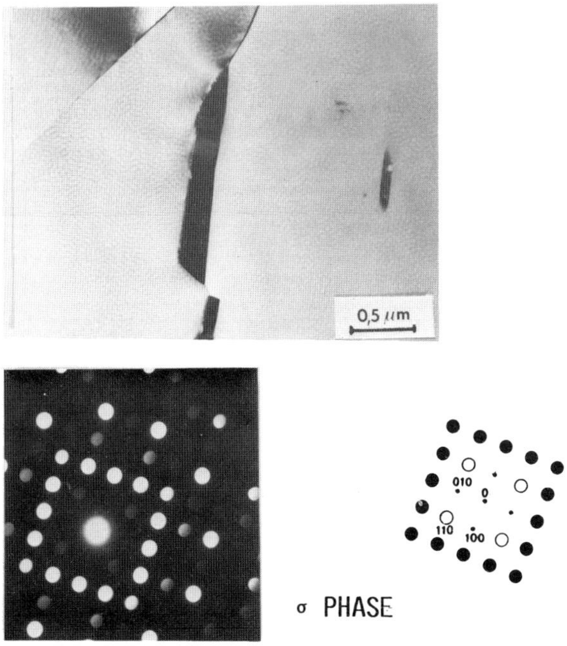

5 STEM micrographs and phase determinations:
900°C/5 mn σ and χ phases

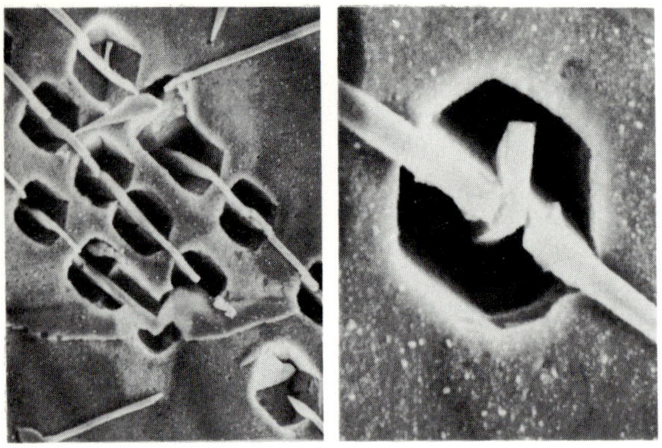

7 SEM micrographs: preferential pitting zones on cross-section of intermetallic phase precipitations

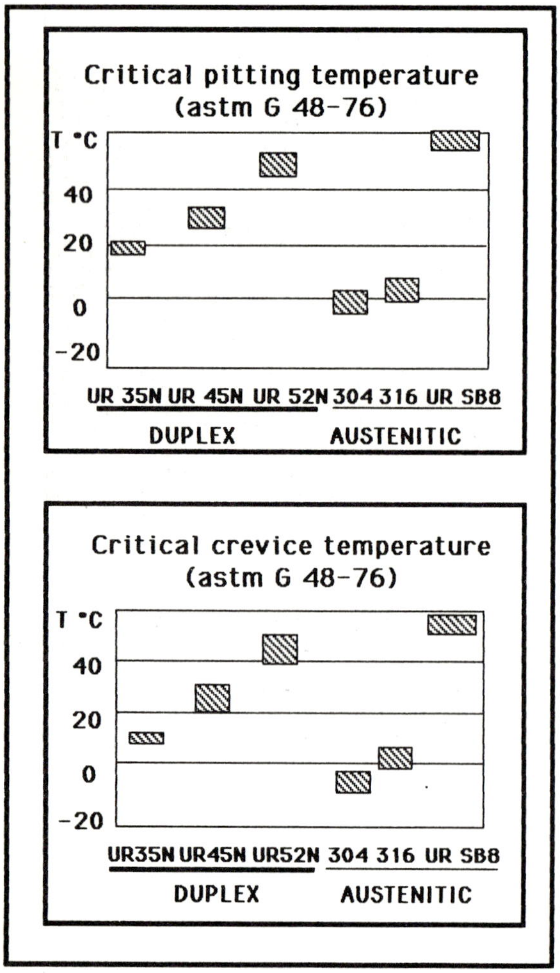

9 Critical pitting and crevice temperatures determined following ASTM G 48 test for several stainless steels

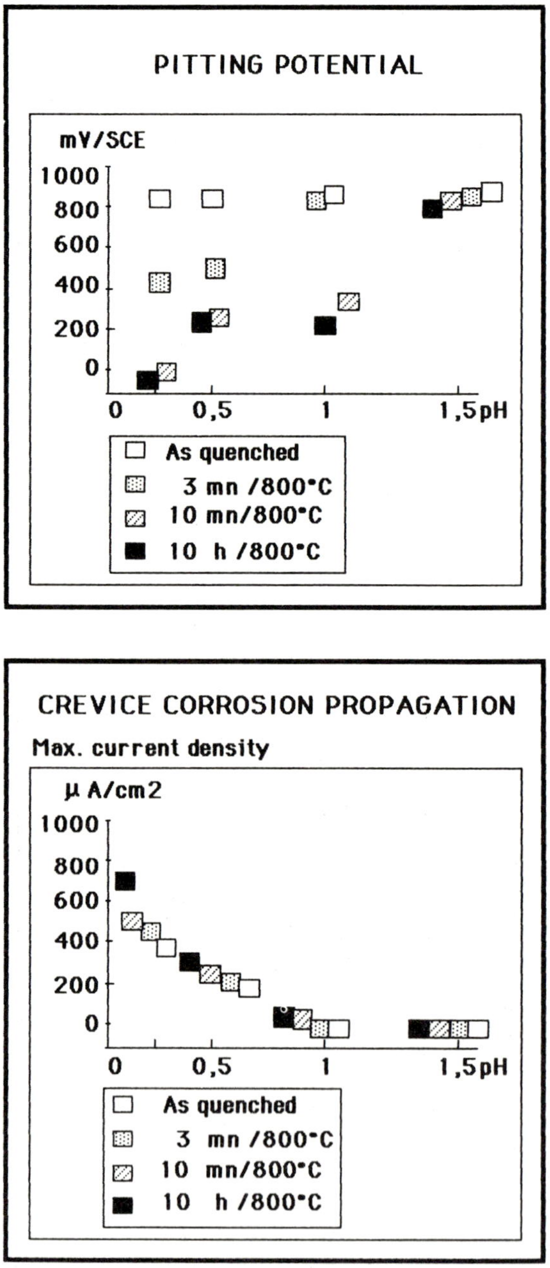

8 Polarization curves results: 300 gr/l NaCl solution, temperature: 60°C, deaerated. HCl acidificated solution.

a - evolution of the pitting potential for 800°C treated samples

b - evolution of the maximum dissolution current for the 800°C treated samples

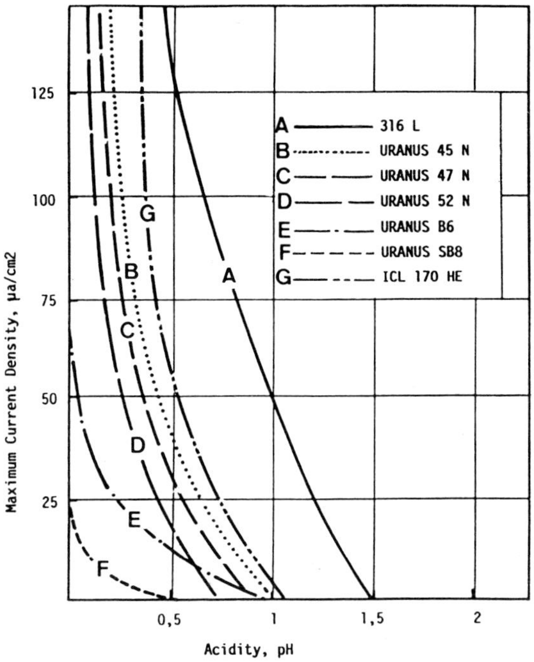

10 Resistance of depassivation of several stainless steels in measuring the maximum dissolution current as function of acidity in a 300 gr/l sodium chloride solution at room temperature

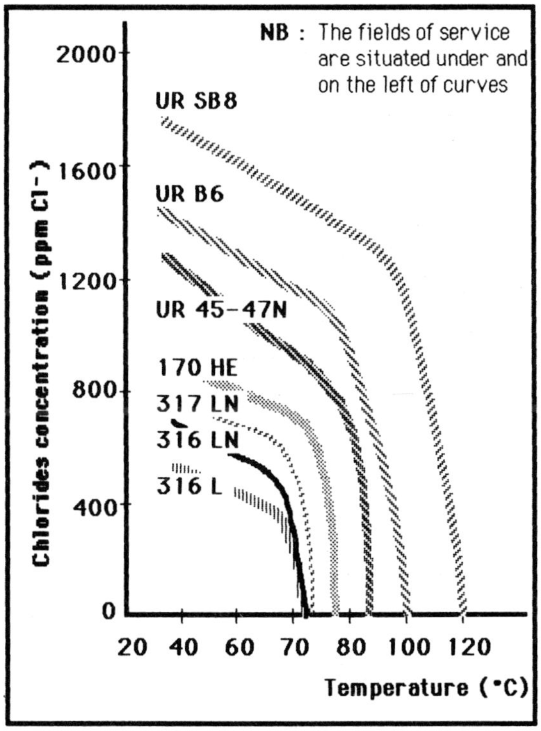

11 Maximum chlorides allowed content according to temperature for an industrial 54% P_2O_5 solution containing H_2SO_4 < 4%, F^- < 1% and HF < 0.2%

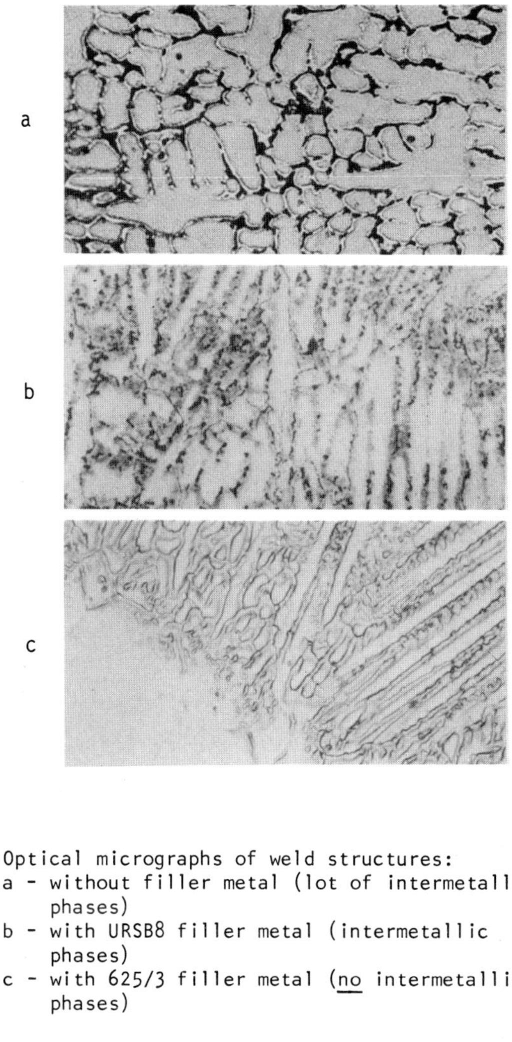

12 Optical micrographs of weld structures:
a - without filler metal (lot of intermetallic phases)
b - with URSB8 filler metal (intermetallic phases)
c - with 625/3 filler metal (no intermetallic phases)

□

Effect of nitrogen alloying on sensitisation behaviour of two highly-alloyed austenitic stainless steels

R F A Jargelius

The author is with the Swedish Institute for Metals Research, Stockholm, Sweden.

SYNOPSIS

The dependence of precipitation kinetics on nitrogen content is investigated for two highly-alloyed low-carbon (<0.02 wt%) stainless steel types: 20Cr 25Ni with 0.064-0.180 wt% N and 20Cr 25Ni 4.5 Mo with 0.015-0.210 wt% N. Oxalic acid etching (ASTM A262 Practice A) and measurements of the critical temperature for pitting corrosion are used to evaluate the effect of ageing between 650°C and 950°C for times of one minute to five hours. Moderate nitrogen contents (0.13%) generally retard the precipitation of intermetallic phases and $M_{23}C_6$ and thus have a beneficial effect on sensitisation resistance. Higher nitrogen contents result in the precipitation of Cr_2N which may be detrimental, particularly after long ageing times.

INTRODUCTION

If stainless steels are subjected to temperatures in the range ~ 600-1000°C, either as a result of welding or heat treatment, there is a risk that both mechanical properties and corrosion resistance deteriorate. This deterioration, or sensitisation, is associated with the precipitation of carbides, nitrides and intermetallic phases primarily along the grain boundaries.

Although nitrogen alloying may be expected to increase sensitisation by facilitating the precipitation of nitrides, it may retard the formation of other precipitating phases and thus have an overall beneficial effect on sensitisation resistance. Such effects have been observed previously, notably in the works of Thier et al (1) on 17Cr 13Ni 5Mo, Brandis et al (2) on 23Cr 17Ni 3MoNb, Blazejak et al (3) on 17Cr 13Ni 2.5Mo, Tuma et al (4) on 16Cr 14Ni 2.5Mo and Mulford et al (5) on types 304 and 316. From these and other studies it is however apparent that nitrogen alloying does not have a consistent effect on precipitation kinetics. While nitrogen generally retards $M_{23}C_6$ precipitation and accelerates Cr_2N and $M_6(C,N)$, its effect on the formation of the intermetallics sigma, chi and Laves phases varies between steel types.

The present investigation is concerned with the effect of nitrogen on the sensitisation behaviour of two highly-alloyed steel types: 20Cr 25Ni and 20Cr 25Ni 4.5Mo. These materials differ, notably in their nickel contents, from the majority of those investigated previously. Attention is also concentrated on relatively short ageing times as these have the greatest practical relevance.

EXPERIMENTAL

The steels investigated comprised six laboratory charges provided by Avesta AB and one commercial charge of 904L as a reference material, compositions are listed in Table 1. The experimental alloys were in the form of band, cold rolled to 3 mm then laboratory annealed at 1100°C for 20 minutes under argon and brine quenched. The reference material, in the form of 4.5mm plate, was annealed at 1150°C/30 minutes before brine quenching in order to remove the intermetallic precipitates initially present. Specimens were then sensitised for times between one minute and five hours at temperatures between 650°C and 950°C. For shorter times (<10 minutes) a salt bath was employed to ensure rapid reheating, while a furnace was used for the longer times. In both cases specimens were brine quenched after heat treatment.

Electrolytic etching in 10% oxalic acid according to ASTM-A262 Practice A (90 seconds at 1 Acm^{-2}) was employed to give an overview of the sensitisation kinetics. In each case the amount of ditched (sensitised) grain boundaries was determined as a percentage of the total grain boundaries using the linear intercept method in the light optical microscope at 250x magnification. When necessary the electrolytic etching was followed by chemical etching in 48% HCl 48%H_2O 4%HNO_3 at 60°C to reveal the unditched grain boundaries. In addition the critical temperature for pitting corrosion in 0.2MNaCl was determined for a number of specimens of each alloy. These specimens had been sensitised for successively long times through the nose of the time-temperature-precipitation (TTP) curve. Measurements of critical pitting temperature (CPT) were made automatically using a Santron Electrochemical Measuring System which held the potential at +600mV SCE and then raised the temperature by 2°C every 15 minutes until a current increase indicated pitting attack.

All such measurements were performed on specimens surface ground to 600 grit SiC paper at least 15 hours prior to the test. In order to demonstrate the relation between grain boundary precipitation and pitting corrosion, pitting initiation studies were performed in 7.98g $AlCl_3$+8.12g $FeCl_3$+50ml ethanol+50ml glycerol (BCMT solution) (6). This produces a large number of small regular pits on the specimen surface so the point of initiation of individual pits may generally be identified.

RESULTS AND DISCUSSION 20Cr 25Ni

The results of oxalic acid etching on the 20Cr 25Ni steels are shown in Fig. 1: the lines indicate the heat treatment conditions giving rise to 25% grain boundary ditching on etching. Increasing the nitrogen content of the steel from 0.064% to 0.130% has a clearly beneficial effect on sensitisation resistance; the nose of the C-curve being displaced from 15 to 25 minutes. The maximum temperature at which precipitation occurs is however raised. Further increasing the nitrogen content to 0.180% has however a marked detrimental effect on the sensitisation resistance over a wide temperature range, this being due primarily to nitride precipitation, see below. This is illustrated by Fig. 2 which compares the grain boundary precipitation and resulting oxalic acid etch structures for the 0.064N and 0.180N steels aged at 800°C for 90 minutes. As can be observed from the figure, the low nitrogen steel exhibits little precipitation or grain boundary attack after ageing under these conditions, but grain boundary nitrides produce severe attack in the high-nitrogen steel.

Similar trends are found when pitting corrosion resistance is examined. Figure 3 shows the critical pitting temperatures of the three 20Cr 25Ni steels after ageing for various times at 700°C. This is close to the temperatures of maximum sensitisation for the 0.064N and 0.130N steels but well below that for 0.180N, and thus shows the latter in a somewhat favourable light. The solid line shows the pitting resistance of quench annealed material which increases monotonically with nitrogen content (7). Ageing for successively longer times reduces the pitting resistance of all materials below this level, although the beneficial effect of even 0.180N is retained at short ageing times (~ 30 minutes). At longer times the detrimental effect of nitride precipitation is however responsible for the rapid fall in pitting resistance of the 0.180N steel and probably also for the poor pitting resistance of 0.130N after 5 hours' sensitisation. Pit initiation studies (BCMT test) indicated that pitting occurred virtually exclusively in association with grain boundary precipitates in all materials. Fig. 4 illustrates the association for two of the steels. It was not however possible to determine whether pit initiation was associated with dissolution of the precipitated particle or attack in the adjacent chromium depleted zone.

Transmission electron microscopy and electron diffraction were used to identify the phases precipitated in the early stages of the sensitisation process. At low nitrogen levels sigma phase and $M_{23}C_6$ carbides were detected. The retardation of sensitisation by 0.130N is in agreement with reported results of the retardation of precipitation of these phases by nitrogen alloying. The increase in the maximum temperature for precipitation is however suggestive of nitride precipitation after long ageing times. In the case of 0.180N only Cr_2N was observed after short ageing times, although from a comparison with the other two steels sigma phase and $M_{23}C_6$ would additionally be expected after longer times below 800°C. Fig. 5 shows the form of grain boundary Cr_2N after ageing at 700 and 950°C.

Thermodynamic calculations made with the Thermocalc system (8-11) indicate that at equilibrium Cr_2N should occur below 1048°C in the 0.180N steel. This is in agreement with the form of the oxalic acid etch curve for this steel in Fig. 1: the curve may reasonably be expected to extrapolate to an upper limit of 1048°C at long ageing times. Use of the same type of calculations indicates that at equilibrium Cr_2N should occur below 985°C in the 0.130N steel and below 870°C in the 0.064N steel. Such precipitation was not observed in these two materials at short ageing times but may be deduced for 0.130% N at long aging times from the form of the curve in Fig. 1. In all three materials the equilibrium presence of $M_{23}C_6$ is predicted below ~670°C.

20Cr 25Ni 4.5Mo

The results of oxalic acid etching on the four 20Cr 25Ni 4.5Mo steels investigated are shown in Fig. 6. These materials are clearly more susceptible to sensitisation than the previous 20Cr 25Ni steels, considerable sensitisation occurring in some cases after only one minute's ageing at 850°C. However, for the molybdenum alloyed steels, nitrogen additions up to 0.210% have a purely positive effect on sensitisation resistance as measured by this test; it being for example necessary to sensitise the steel with 0.210N for 10 minutes at 850°C in order to obtain a comparable amount of grain boundary attack to that produced with 0.015N after one minute. This beneficial effect of nitrogen is illustrated in Fig. 7 which compares the degree of grain boundary precipitation and ditching after oxalic acid etching for the 0.061N and 0.210N steels aged at 850°C for 90 minutes. Here there is clearly no detrimental effect attributable to nitride precipitation although the form of the oxalic acid etch curve in Fig. 6 at high temperatures suggests nitride precipitation occurs after long ageing times.

Measurements of the critical temperature for pitting corrosion in 0.2M NaCl clearly indicate that high nitrogen levels may be associated with detrimental effects after long ageing times (Fig. 8). A nitrogen level of 0.130% had a solely beneficial effect on CPT both before and after ageing to five hours at 850°C. The relatively good pitting resistance of the 0.210N steel in the quench annealed condition and after short ageing times (~30 minutes) was however markedly reduced after ageing for five hours at 850°C, this behaviour may again be related to nitride precipitaion. As in the case of the 20Cr 25Ni steels BCMT testing indicated a clear association between grain boundary precipitates and the sites for pit initiation.

Analytical transmission electron microscopy was used to identify three precipitate types in the 20Cr 25Ni 4.5Mo steels after short ageing times: the intermetallics sigma, chi and Laves phase. The measured compositions of these phases are given in Table 2, their identity was also confirmed using electron diffraction. It may be noted that the compositions in Table 2 as well as the Mo/Cr and Fe/Cr ratios differ in some cases considerably from data reported in the literature for example by Weiss et al (12) and Novak (13). In particular the chromium content of chi phase is higher than that normally observed, the molybdenum contents of both sigma and chi phase are also relatively high. These discrepancies can however be largely explained by differences in alloying level and may also be related to the fact that exceptionally short ageing times have been studied here. Figure 9 shows intermetallic precipitation in the 0.061N steel aged 3 minutes at 875°C and indicates chi and Laves phases to be intermingled along grain boundaries and not exhibiting any distinguishing morphological characteristics. Sigma phase was relatively rare in this case, only two particles being observed, while chi phase dominated. In the steel with 0.210N, Laves phase was the dominant intermetallic precipitate after ageing 10 minutes at 850°C with relatively little chi or sigma phase. A few particles of Cr_2N were also observed in this case and from thermodynamic calculations would be expected to occur below 974°C at 0.21%N.

The delayed onset of sensitisation in the nitrogen-alloyed CrNiMo steels may thus be explained by nitrogen retarding the precipitation of intermetallics, just as it retards sigma phase in the CrNi steels. This is in line with the results of Thermocalc calculations which show the activities of chromium and molybdenum, and thus the driving force for intermetallic precipitation, to be lowered by nitrogen. A further consequence of this is that nitrogen would be expected to have the largest influence on intermetallics such as chi phase with both high Cr and Mo contents. Such an effect is also indicated by the experimental observations in Table 2, although further work is required to clarify the relative effect of nitrogen on sigma, chi and Laves phase. The sharp drop in pitting resistance after ageing, and the increase in the maximum temperature for precipitation (Fig. 6) of the CrNiMo steel with 0.210%N may also be correlated to Cr_2N precipitation. The effect is not however as marked as for the corresponding CrNi alloy. Both kinetic and thermodynamic arguments indicate that chromium depletion would be more severe adjacent to Cr_2N than to intermetallics, so nitride precipitation could be expected to have a more detrimental effect on pitting resistance than intermetallics.

CONCLUSIONS

1. The degree of sensitisation produced in 20Cr 25Ni 0.064-0.180N and 20Cr 25Ni 4.5Mo 0.015-0.210N steels by ageing up to 5 hours between 650°C and 950°C is markedly influenced by their nitrogen content.

2. Sensitisation of 20Cr 25Ni is associated with the precipitation of sigma phase and $M_{23}C_6$ and is retarded by the presence of 0.130% nitrogen. A nitrogen content of 0.18%N causes precipitation of Cr_2N over a wide temperature range and an associated reduction in sensitisation resistance and pitting corrosion resistance.

3. The onset of sensitisation in 20Cr 25Ni 4.5Mo is primarily associated with precipitation of the intermetallics sigma chi and Laves phase and retarded by nitrogen alloying. At longer ageing times Cr_2N precipitation in the steel with 0.21% nitrogen causes a decrease in pitting corrosion resistance although 0.13% nitrogen is still beneficial under such circumstances.

ACKNOWLEDGEMENTS

This work was financed by Avesta AB, Avesta Sandvik Tube, Fagersta Stainless AB, Nyby Uddeholm, AB Sandvik Steel and the Swedish Board for Technical Development (STU). Thanks are also extended to Staffan Hertzman for his invaluable assistance with the Thermocalc system.

REFERENCES

1. THIER, H:, BÄUMEL, A., SCHMIDTMANN, E., Arch. Eisenhüttenwesen 40 (1967) 333-339.

2. BRANDIS, H., HEIMANN, W., SCHMIDTMANN, E., TEW Technische Berichte 2 (1976) 150-166.

3. BLAZEJAK, D., HERBSLEB, G., WESTERFELD, K.-J., Werkstoffe und Korrosion 27 (1976) 398-403.

4. TUMA, H., LANDA, V., LÖBL, K., Mem. Etud. Sci Revue de Metallurgie (1981) 255-259.

5. MULFORD, R. A., HALL, E. L., BRIANT, C. L., Corrosion 39 (1983) 132-143.

6. BIANCHI, G., CERQUETTI, A., MAZZA, F., TORCHIO, S., Corrosion Science 10 (1970) 19-27.

7. JARGELIUS, R. F. A., 10th Scandinavian Corrosion Congress, Stockholm, 1986, 161-164.

8. JANSSON, B., Calphad 9 (1985) 153-190.

9. UHRENIUS, B., Hardenability concepts with applications to steel. Proc. conf. 1977 Chicago, AIME.

10. HERTZMAN, S., Met. Trans. 18A (1987) 1753-66.

11. JARL, M., Scand. J. Metallurgy 7 (1978) 93-101.

12. WEISS, B., HUGHES; W., STICKLER, R., Prakt. Metallog. 8 (1971) 528-542.

13. NOVAK, C. J., Handbook of stainless steels, Ed. Peckner and Bernstein, McGraw Hill 1977.

Table 1. Chemical compositions of the steels investigated, (wt%).

Steel	C	Si	Mn	P	S	Cr	Ni	Mo	Cu	N
9	0.014	0.53	1.31	0.010	<0.003	20.14	24.75	0.08	0.037	0.064
11	0.014	0.55	1.43	0.009	<0.003	20.03	24.87	0.04	0.036	0.130
10	0.013	0.55	1.43	0.010	<0.003	19.96	24.90	0.04	0.036	0.180
5(ref)	0.019	0.67	1.71	0.017	0.005	19.2	24.3	4.14	1.72	0.014
12	0.020	0.54	1.42	0.010	<0.003	19.85	24.86	4.50	0.040	0.061
13	0.015	0.49	1.45	0.009	<0.003	19.87	24.95	4.55	0.039	0.130
14	0.014	0.54	1.44	0.009	<0.003	19.83	24.99	4.59	0.042	0.210

Table 2.
Compositions (wt%) of intermetallic phases in steels 12 and 14 (20Cr 25Ni 4.5Mo with 0.061 and 0.210%N respectively).

Steel 12 875°C 3 minutes						
	Fe	Cr	Mo	Ni	Si	Number analysed
Laves phase	28±2	18±2	41±3	9±1	4±2	17
Chi	31±2	34±2	24±3	9±2	2±1	34
Sigma	34±2	41±6	15±3	9±4	1±2	2
Steel 14 850°C 10 minutes						
	Fe	Cr	Mo	Ni	Si	Number analysed
Laves phase	29±1	17±1	42±2	9±1	4±1	37
Chi	24±3	42±1	28±1	4±2	2±0	2
Sigma	19±0	46±2	18±1	4±2	3±0	2

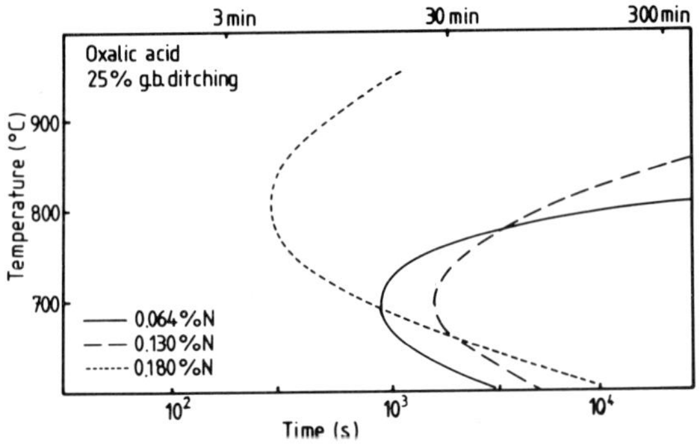

Fig. 1 Effect of nitrogen on the sensitisation of 20Cr 25Ni steels. The lines show the conditions producing 25% grain boundary ditching after etching according to ASTM A262 Practice A.

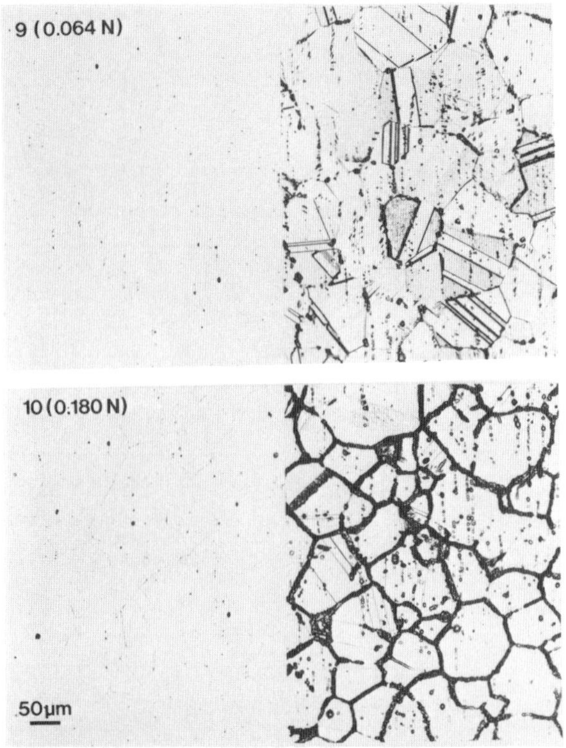

Fig. 2 Microstructures and oxalic acid etch structures resulting from ageing steels 9 and 10 (20Cr 25Ni with 0.064 and 0.180% N respectively) at 800°C for 90 minutes. Severe attack in steel 10 is associated with nitride precipitation.

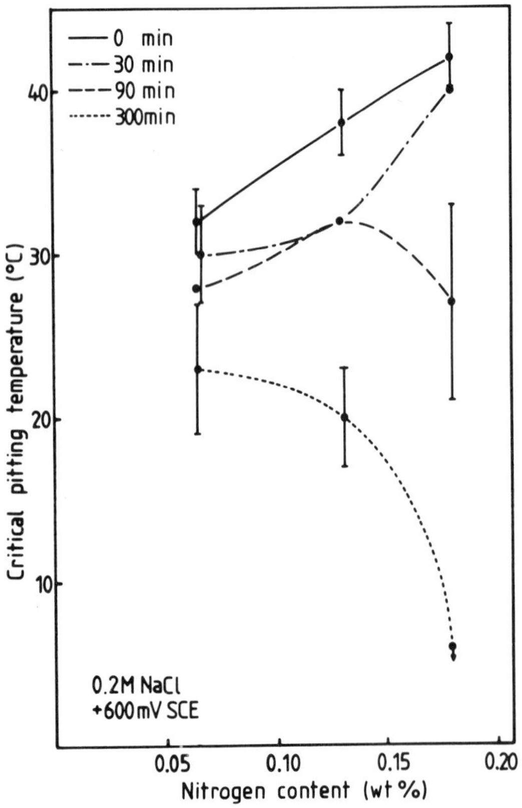

Fig. 3 Critical temperature for the onset of pitting corrosion at +600mV SCE in 0.2M NaCl as a function of nitrogen content for 20Cr 25Ni steels aged for various times at 700°C.

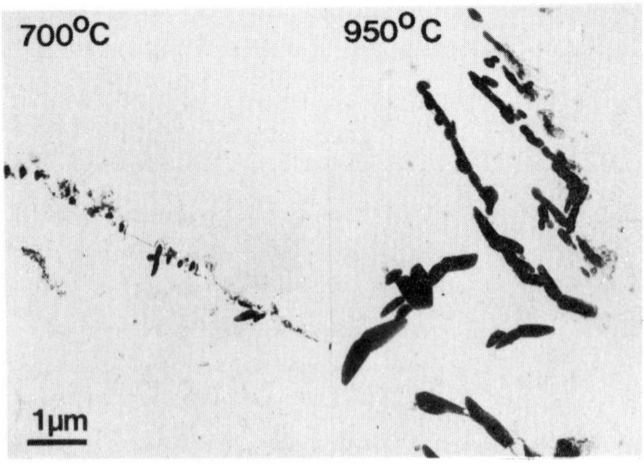

Fig. 4 Results of pit initiation studies in $AlCl_3/FeCl_3$/ethanol/glycerol on steels 9 and 10 aged 5 hours at 700°C. Etching conditions have produced somewhat different distributions of pits but in both cases the association between pitting and grain boundary precipitates is clear.

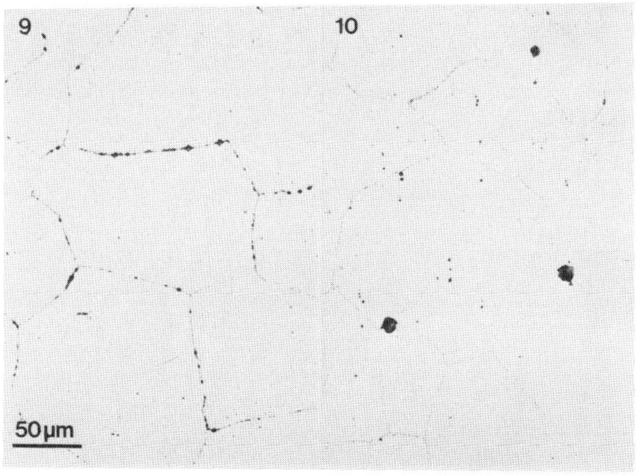

Fig. 5 Grain boundary precipitates of Cr_2N in steel 10 (20Cr 25Ni 0.18N) aged 30 minutes at 700 and 950°C.

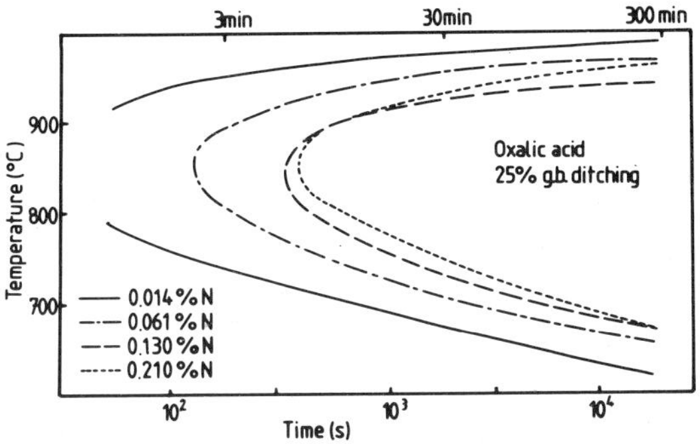

Fig. 6 Effect on nitrogen content on the sensitisation of 20Cr 25Ni 4.5Mo steels. The lines show the conditions producing 25% grain boundary ditching after etching according to ASTM A262 practice A.

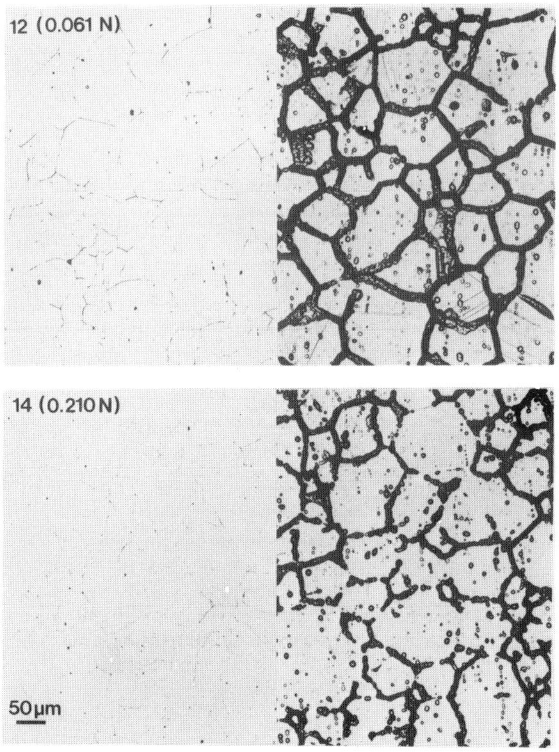

Fig. 7 Microstructures and oxalic acid etch structures resulting from ageing steels 12 and 14 (20Cr 25Ni 4.5Mo with 0.061 and 0.210% N respectively) at 850°C for 90 minutes. Nitrogen clearly reduces the degree of sensitisation.

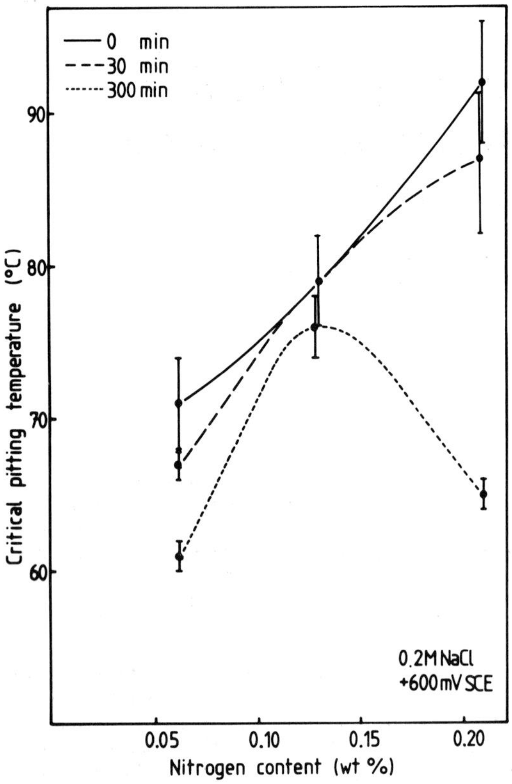

Fig. 8 Critical pitting temperature in 0.2M NaCl at +600mV SCE as a function of nitrogen content for 20Cr 25Ni 4.5Mo steels sensitised for various times at 850°C.

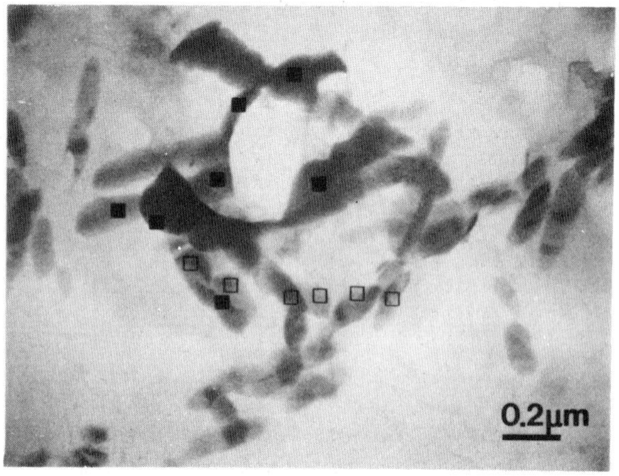

Fig. 9 Grain boundary precipitates in 20Cr 25Ni 4.5Mo 0.061N after sensitising for 3 minutes at 875°C. Open points are chromium-rich precipitates (chi phase) solid points are molybdenum-rich precipitates (Laves phase).

Stress corrosion cracking of steel components used in directional drilling of oil and gas wells

J E Truman and K B Lomax

The authors are with Forgemasters Engineering Ltd., a subsidiary of Sheffield Forgemasters Ltd.

SYNOPSIS

Non-magnetic drill collars are long tubular components run in the lower section of the drill string. They are usually made in austenitic stainless steel. Service can involve prolonged contact with warm, saline solutions and some environmental cracking has occurred. Characteristics and causes are reviewed. Much cracking was due to the use of sensitised steel but there have been cases of transgranular cracking with low carbon steels. Laboratory testing of full section collar lengths has been carried out for comparison with orthodox stress corrosion cracking testing. Results have shown that stress corrosion cracking of the low carbon "Staballoy" material produced by Forgemasters Engineering is unlikely in the normal life of a collar when metal temperatures are less than 100°C. In more aggressive conditions residual stress modification by peening is proving to be a suitable preventative measure.

INTRODUCTION

Heavy tubular components called drill collars are incorporated in the lower part of the drill string assembly when drilling to produce oil or gas wells. The function of these drill collars is to add rigidity to the string, and to maintain pressure on the drill bit. These collars are generally a minimum of 30 feet in length, with outside diameters in the range 4½ inch to 10 inch. They have bore diameters most frequently in the range 2 inch to 3 inch.

The collars are usually made from high strength chromium molybdenum alloy steel of the AISI 4145 type. When deviated bore holes are drilled, magnetic compasses and magnetometers are enclosed in the collars to give information on hole angle direction. In these conditions, it is necessary for the drill collar material to have low magnetic permeability to avoid interference with the instruments. A yield strength of 100ksi minimum (689 MPa), similar to the low alloy steel collars, is still required. These non-magnetic drill collars are known conveniently as NMDCs.

Other non-magnetic items may be used in a drill string such as subpieces, stabilisers and measurement-while-drilling casings. These have different and more complex geometries than the drill collars but because the problems and solutions to these problems are the same, this paper will deal with experiences on collars.

The desired material property combination sets an interesting metallurgical problem which was initially solved by the use of a precipitation hardening copper alloy. Drill collars however, have a limited life expectancy due to mechanical wear and tear and thus less costly alternatives were sought. The austenitic stainless steels were obvious contenders to satisfy the requirements of low magnetic permeability and lower cost, but because the 0.2% proof strength of the austenitic stainless steels is typically 200MPa, it was necessary to improve the strength without increasing the magnetic permeability. Strain hardening was one method of increasing strength, but the degree of strain required with the standard steels to achieve the desired yield strength was excessive. Modified compositions were thus devised with higher initial proof strength. The first generation of NMDC alloys had analyses in the following range:

C	Mn	Cr	Ni
Up to .2%	Up to 19%	12.0/18.5%	Up to 13%

Mo	N	Cu
Up to 3%	.2/.5%	Up to 1%

While cold strain was used for some smaller collars, the power required to cold strain larger items was excessive. Warm working, by forging or rolling at temperatures just below the recrystallisation temperature, was thus widely used. The uniform application of the desired strain within a narrow temperature range demands very tight process control and is difficult to achieve over the long lengths required. In Forgemasters Engineering Limited (FEL), the desired process control is achieved by use of a large fast acting precision forging machine, the GFM-SXP65. Control is such that very similar mechanical properties are demonstrated throughout the bar length. Release tests demonstrate the consistent properties by presenting values measured from each end of each bar sold.

The chemical composition and the mechanical properties of the FEL product "Staballoy" are as below:

C	Si	Mn	S
.05 max	1.00 max	10.0/12.0	.010 max

P	Cr	Ni	N
.030 max	14.0/18.0	5.0/8.0	.4 max

Rp 0.2 100 ksi (689 MPa) min
Rm 125 ksi (861.5 MPa) min
Elongation 30% typical
Izod 65ft.lb min
Magnetic Permeability 1.005 max

Operating Conditions

The whole drill string is tubular and of a smaller diameter than the hole being produced. A drilling fluid, referred to as mud, is pumped down the bore and flows back up the outside of the string. The mud serves a variety of purposes including lubrication, cooling, removal of debris and the establishment of a high hydrostatic pressure to support the hole. Formulation of the mud is complex and as well as having suitable density and viscosity, the possibility of interaction with the strata must be considered. The mud liquid carrying the fine added solids can be water, oil (with water contamination) or an oil/water emulsion. There can also be added or incidental salts in the mud. These are commonly the chlorides of sodium, calcium, potassium or magnesium and can be present in quantities up to saturation point. Strata temperature increases with depth and is usually in the range 40 to 200°C.

Service Behaviour

Two problems of significant magnitude have arisen with NMDCs.

Galling Seizing and galling at the threaded joints has been a problem with the low alloy steel components. It is not surprising therefore that it has been more prevalant with the austenitic stainless steels which are known to have a greater galling tendency. As with other forms of engineering, the industry has learned to cope with pre-torquing, choice of jointing compounds and accurate application of make-up torque. However, operation is often under conditions where the necessary care and attention is difficult to apply and laboratory investigations continue into reducing galling tendency.

Cracking Cracking "downhole" is the second significant problem. Crack growth rate is slow but reports indicate that complete wall penetration has been experienced during operation. FEL have had the opportunity to examine cracking in collars from a wide variety of customers and suppliers. The observed cracking in bores is orientated axially. The cracks initiate at the bore and penetrate radially. If full penetration occurs, drilling mud can flow through the crack causing it to widen rapidly by abrasion and subsequently results in a "washout". This has the effect of suspending the drilling operation with disastrous cost implications. While this extreme case is rare, the fear of it has led to the practice of withdrawing drill collars from service as soon as superficial cracking is detected. A survey of the behaviour of drill collars shows a varied incidence of cracking, depending upon steel composition and geographical location of utilisation.

Characteristics of Cracking

Collaboration with users has allowed FEL to study cracking in samples of drill collars manufactured from the majority of the compositions in use.

Radial sections illustrating the macro-morphology of the two types of deep cracking found are shown in Figure 1. The early stage of cracking involves the development of numerous short, shallow axially orientated cracks. If development of these cracks can occur, only a limited number grow. In 'type A' cracking, there are only two major cracks which develop approximately 180 degrees apart and extend for considerable distances axially. In 'type B' cracking approximately 10 to 18 cracks grow. These appear, regularly spaced around the bore circumference, but axial growth of each individual crack is limited. Parallel, slightly overlapping cracks are formed.

Under metallographic examination, two types of micro cracking also have been observed. Both types are similar in appearance with micro-branching but type 1 is intergranular and type 2, transgranular. Type 1 is always associated with the grain boundary decoration by carbides. Appropriate testing, using the ASTM A262 E method, has shown susceptibility to intercrystalline corrosion due to sensitisation. Some limited grain boundary carbide has also been seen when the cracking has been transgranular. In these cases however, the steel has proved to be resistant to the corrosion test. Only intergranular micro cracking has so far been seen with service failures exhibiting the 'type A' cracking but instances of both intergranular and transgranular microcracking have been observed on samples showing 'type B' cracking.

Susceptibility to Stress Corrosion Cracking (S.C.C.)

The micro characteristics described above are typical of stress accelerated intercrystalline corrosion (type 1) and transgranular stress corrosion cracking (type 2). The former arises from the well-understood metallurgical condition of sensitisation which arises when carbides form in the temperature range 500 to 800°C, and at grain boundaries and with sufficient density to cause a continuous path of adjacent low chromium material. This allows local corrosion attack when the steel is subjected to a suitable corrodent (1). Avoidance of sensitisation involves either rapid cooling through the critical temperature range, or the use of low carbon materials. Some of the early NMDC's were manufactured from steels which had compositions aimed at satisfying the mechanical and physical properties listed above but without knowledge of the corrosive environment that could be encountered during drilling. Thus, high carbon contents, in the range 0.07 to 0.20% were used by some manufacturers. These are certainly high enough to give sensitisation at the slow cooling rate which inevitably exists on cooling the large steel sections used for machining the drill collars. Not surprisingly such higher carbon steels have formed the major proportion of compositions showing evidence of in-service cracking. Carbon contents of less than 0.07% and preferably less than 0.05% are necessary to avoid the sensitised condition and the associated cracking.

Transgranular cracking does not need a special metallurgical condition but does need exposure to critical corrosive conditions in the presence of a suitably applied stress. The corrosive conditions required are quite specific and involve strongly caustic fluids, or solutions of chlorides (1-4). The former fluids are not relevant to the drilling but the latter chlorides are particularly relevant. Even so, more is required for cracking to occur than the simple presence of chlorides.

A substantial amount of corrosion data are available for the standard group of "18/8" types of stainless steels (AISI 302, 304, 321, 347, 316 etc). Parallels and comparisons could therefore be made by assessing the resistance to SCC of the drill collar materials compared to that of the above group of steels.

The majority of collars now in service are either the FEL "Staballoy" composition or are manufactured from the group of alloys based on the following composition:

Mn	Cr	Ni	Mo	Ni
18%	13%	2.5%	.5/1.0%	.25%

Test data for "Staballoy" and one steel from the 18-13-2½-.25 group are compared with some data for AISI 304 steel in Figure 2. The results for one of the high carbon steels are also included in the figure. The tests were on small tensile type test pieces, suitably stressed by weights, and immersed in a boiling, saturated sodium chloride solution. The results of similar tests carried out in boiling 42% magnesium chloride solution are shown in Figure 3. In these figures, the low carbon drill collar materials are shown to be superior to the AISI 304 steel composition. With this knowledge, the available data showing the effects of environmental factors on 304 steel and other similar steels (1-5) can be quoted as relevant to the NMDC alloys, with a high degree of confidence. The relevance of these data to Staballoy has also been assessed at Forgemasters (6,7). The importance of stress is apparent from Figures 2 and 3 and, given a significant chloride content in the corrodent, temperature is also of great significance. Results for "Staballoy" are reproduced in Figure 4. The difference in position of the plots cannot be attributed to the chloride content nor does the type of chloride (cation) have a direct effect (6,7). Solubilities of chlorides do vary however, and there can be much higher chloride content with Mg and Ca than with Na or K. pH is also shown to have a minor effect (3-6) as can also be seen from Figure 2. The boiling point of the solution has no relevance, because boiling is not possible at the high hydrostatic pressures prevailing. The three important factors which control the time to cracking in a given steel are the chloride content, temperature and stress. These three factors interact however and it is not possible to set threshold values.

The features controlling chloride content and temperature are obvious, although complex. The source of the stress however, is less obvious. The nature of the cracking is such that the major stress must be a tensile hoop stress at the bore surface. Such a stress could arise from an applied pressure in the bore. Because the collar is at the bottom of the drill string, the pressure differential between the bore surface and the outside is small. A maximum stress value of a little over 50MPa has been attributed to this source (8). It is therefore unlikely that this is the applied stress causing cracking. Bending or twisting of collars would also give stresses. The patterns developed however would initiate helical or transverse cracks. These have not been observed.

The critical stress, causing the cracking must therefore be residual. Stress analyses of a number of drill collars confirmed the presence of a residual stress. Two methods have been used to measure the stress. The more accurate method has been described elsewhere (8) and involves progressive machining of the outer surface with intermittent measurement of changes in the inside diameter and lengths. The second, less accurate method, is based on changes in diameter when a length of collar is slit axially. Results for bore hoop stress obtained by the two methods have been in good agreement. The hoop stress distribution is shown diagrammatically in Figure 5.

From the data obtained with "Staballoy" collars, although there is substantial scatter, it appears that the magnitude of the bore stress varies with wall thickness. The data are shown in Fig 6. Values due to Kopecki (8) are also shown. These latter values were measured on collars taken from various, unspecified sources and were for Cr-Mn, Cr-Mn-Ni and Cr-Ni steels. In Kopecki's work, the stress values quoted were not identified for each steel type.

To explore the effects of various parameters on cracking susceptibility, test lengths from some drill collars in the 6½inch-8inch OD range, which had been used for stress measurements and which contained residual stress, were subjected to tests in hot chloride solutions without an externally applied stress. The cut ends were protected and the samples were removed periodically for inspection. The solutions used were boiling saturated sodium chloride at 106°C, boiling 40% calcium chloride at 118°C and 60% calcium chloride at 125/130°C. The results of these tests, (Fig 4) are in reasonable agreement with those produced on samples tested under an externally applied stress. They confirm the importance of temperature. Further confirmation is given by the experimental results presented in Figure 7. In these tests, a length of drill collar 7½inch OD x 2.13/16inch ID and with a measured residual stress at the bore of 240 MPa was immersed in a 50% calcium chloride solution. The temperature was increased, step-wise, at intervals with periodic sample inspection as shown in Figure 7. No cracks were detected up to and including assessment at a temperature of 108°C. After a further period at 118°C however, cracking was well advanced.

The cracking seen in the above tests, which were carried out without an applied stress, was allowed to develop. In all cases the cracking resembled that noted after 'downhole' service.

Initiation was at numerous points on the bore surface giving many short axial cracks. In most cases, few cracks developed further, growing as shown in Figure 1B. Only the samples with the highest stress levels, above 300 MPa and at the highest temperature of 130°C developed in the manner shown in Figure 1A. Observation of this cracking mode showed that one of the two cracks grew rapidly after initiation and the second crack, located 180° away grew when the first crack had extended substantially. In the latter stages these cracks developed macro-branching. Because these tests were all carried out on the FEL "Staballoy" composition with its controlled low carbon content all the cracking observed was transgranular.

Improved Collars

The above data demonstrated the importance of environmental conditions. Notably, given a chloride bearing mud, temperature was significant in determining whether cracking will occur in unsensitised collars. This observation explains why service behaviour has been so variable. In fact, discounting the sensitised collars which can crack in mild environments, the incidence of cracking has been small relative to the large number of collars in service. It seems probable from our evidence that the incidence of cracking has been confined to those geographical areas where the use of chloride bearing muds coincides with high 'downhole' temperatures.

Recognising operating constraints and drill collar owner requirements, it would be convenient if NMDCs could be used at any site without reservations. Thus there is a demand for more resistant, if not immune collars.

One approach to satisfy this requirement is to modify the composition of the alloys. This must be done without adversely affecting other properties. Investigations on the effects of variations in common alloying elements and addition of more unusual elements (1, 3, 4, 10) have been carried out but with one exception, no substantial improvement has resulted. The exception is increased additions of nickel. Not only are there laboratory data showing the advantages but additionally high nickel steels are used in some industries for enhanced SCC resistance. The effect on SCC resistance of increasing the nickel content of the "Staballoy" composition is shown in Figure 8. Only when the alloy becomes a nickel based alloy, is near immunity to SCC achieved. The cost penalty for this however is very high. Additionally, galling problems are increased. A lower, more economical, nickel level would be expected to reduce cracking incidence in service but this will not remove it completely and the problem of the 'universal', immune collar would not be solved.

An alternative approach to the problem is to reduce, remove or change the inherent stress pattern contributing to the problem. A convenient and well known method of doing this is by peening, either with a hammer or by bombardment with suitable spheroidal "shot". Surface deformation introduces residual compressive stresses and, although not of a great depth, these are sufficient to prevent crack initiation. Values of the compressive stress at various depths from the surface for two peened collars, one hammer, one shot, are shown in Figure 9. These stress level determinations were made by machining suitable rings from the collars and measuring the progressive dimensional changes caused by removing metal from the peened bores. Metal removal was by electropolishing. The depths at which various stresses are achieved can be altered by varying the peening parameters. The modification of the residual stress pattern in a collar brought about by the peening operation on the bore, is illustrated in Figure 10. It can be seen that the nett result is to leave both the external and internal surfaces in a state of compressive stress. Because SCC initiation is not possible unless there is a substantial applied tensile stress, cracking cannot occur unless a stress is applied which would more than cancel the residual compressive stress value or unless the compressive layer is penetrated in some way. The results of using this treatment have been clearly demonstrated by testing lengths of collar, both peened and unpeened, in a hot, strong chloride solution. Some results are given in Table 1. Furthermore, peened collars have now been supplied for more than 2 years and are giving excellent service with no reports of SCC.

Conclusions

1. Stress corrosion cracking is a service hazard with austenitic non-magnetic drill collars (NMDC's). The extent of the problem however, has been distorted by the use of steels with high carbon contents, which are subject to stress accelerated intercrystalline corrosion.

2. The risk of Stress Corrosion Cracking(SCC)in the low carbon Cr-Mn-Ni-N steels, which now provide the bulk of the in-service items, is much reduced. Some axial cracking has been observed however and has been shown to be transgranular SCC.

3. Laboratory testing has shown that the current low carbon austenitic steels used for NMDCs behave in a similar manner to the AISI 304 type steels in the critical environmental conditions for SCC. Some of the NMDC alloys (eg. "Staballoy") are more resistant than AISI 304.

4. The conditions likely to cause in-service SCC, involve chloride bearing drilling fluids, high metal temperatures and some form of tensile stress. It should therefore be feasible to identify geographical areas where there is high risk of cracking.

5. Study of products from a range of sources, has demonstrated that the stress causing SCC is essentially residual in nature. Since the desired strength of the drill collar is established by strain hardening, a stress relieving treatment, provided by a normal annealing cycle, cannot be utilised because of the reduced strength that would result.

6. Resistance to SCC of NMDC alloys can be improved by increasing nickel content.

Near-immunity to SCC can only be obtained by producing an expensive collar from a nickel-base alloy, however.

7. Use of intermediate nickel content will reduce the probability of SCC in service, but will not produce a collar which can be used anywhere without fear of cracking.

8. The universal, immune collar, can be most closely achieved by modifying the residual stress pattern at the bore surface. Producing a tubular with both outer and inner surfaces in a state of substantial compressive stress by peening, offers one solution.

9. Hammer peened "Staballoy" (XL) has remained crack free for substantial periods in hot chloride solutions which would normally produce very rapid cracking in conventional unpeened NMDCs. This steel and collar treatment offers one solution to the problem. Provided the integrity of the peened layer is maintained, the 'use anywhere' collar is achievable.

References

1. A. J. Sedriks. "Corrosion of Stainless Steel" Pub. J. Wiley & Sons

2. J. E. Truman. Proceedings of Conference on Stress Corrosion Cracking and Hydrogen Embrittlement of Iron Base Alloys (Firminy). NACE, Houston, Texas P111.

3. R. M. Latanisian and R. W. Staehle. Proceedings of Conference on Fundamental aspects of Stress Corrosion Cracking. NACE, Houston, Texas 1969, P214.

4. G. J. Theus and R. W. Staehle, as ref 2, P843.

5. J. E. Truman. Corrosion Science, Vol 17, P737 (1977).

6. P.J. Smith, J. E. Truman, H. T Gisborne and G. Oakes. J. Materials for Energy Systems. Vol 6, No.4 P300 (1985).

7. Unpublished work, Forgemasters Engineering Limited.

8. D. Kopecki. Effects of Residual Stress on Stress Corrosion Cracking of Non-Magnetic Drill collars. Presented at Corrosion 86 NACE, Houston, Texas (1986).

9. J. E. Truman and R. Perry, Brit. Corr. J. Vol 1, P60 (1966).

10. J. E. Truman and P. M. Haigh. Unpublished work.

TABLE 1

Comparative S.C.C. Tests on As Machined and Hammer Peened Staballoy Drill Collars

Concentrated Calcium Chloride Solution 125-130°C

O.D. Inches	7¼	6½	8
I.D. Inches	5.1/16	3¼	2.13/16
Estimated bore residual stress before peening MPa	82	180	275
Time to crack initiation (as machined) hours	72-144	20-90	20-90
Total time of test for peened sample (no cracks) hours (tests discontinued)	2,000	3,300	1,750

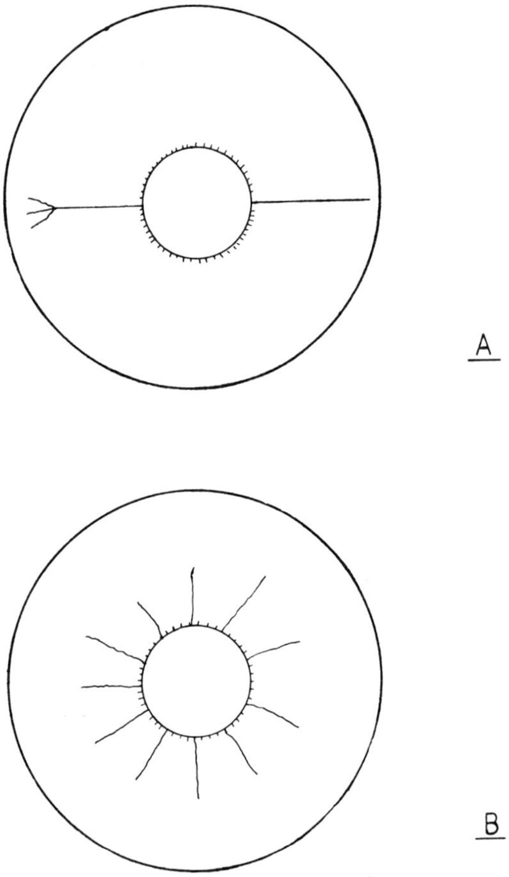

1 Modes of severe cracking

		C	Mn	Cr	Ni	Mo	Cu	N
304	●	.05	1.14	18.74	10.06	.40	-	.03
Staballoy	o	.04	10.44	17.76	6.55	.37	-	.34
	Δ	.03	18.35	12.82	2.66	123	-	.25
	X	.125	17.46	17.79	.46	.97	.94	.46

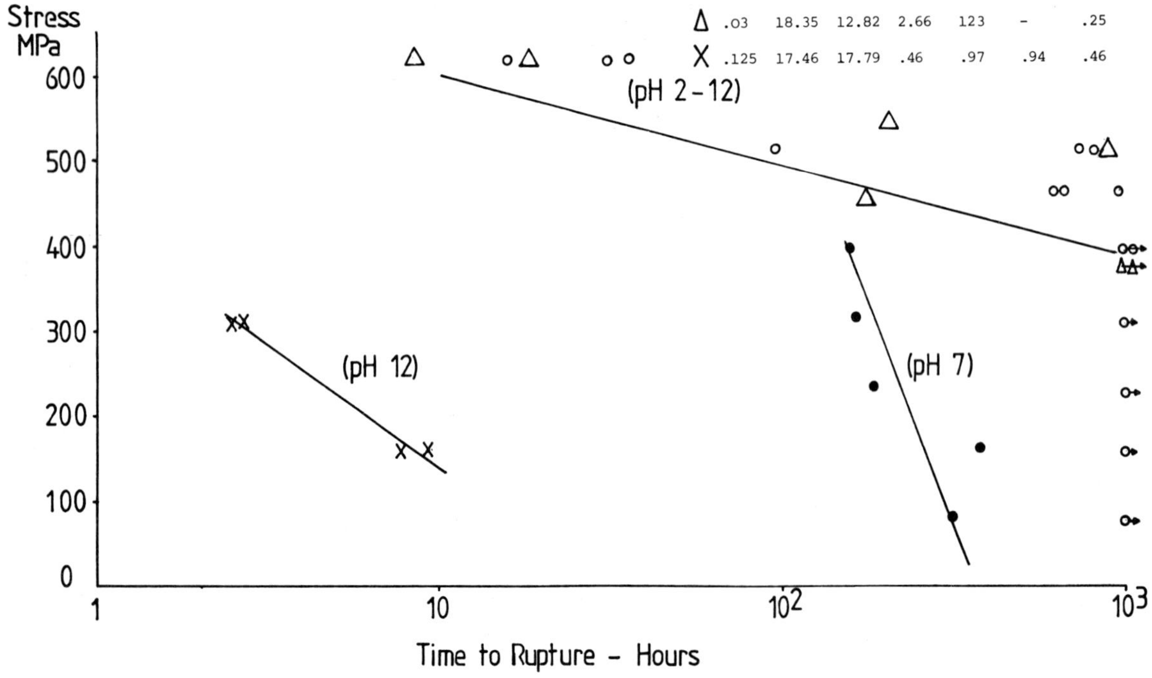

2 SCC tests in boiling saturated NaCl solution

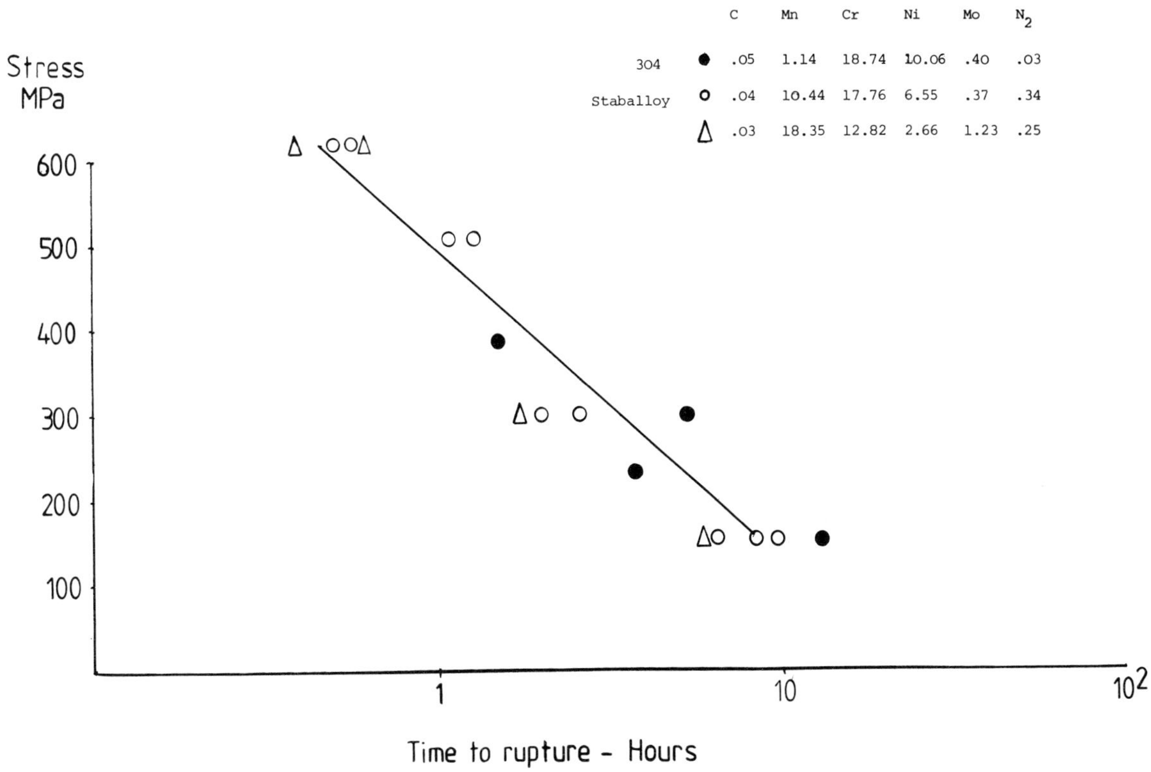

3 SCC tests in boiling 42% $MgCl_2$ solution

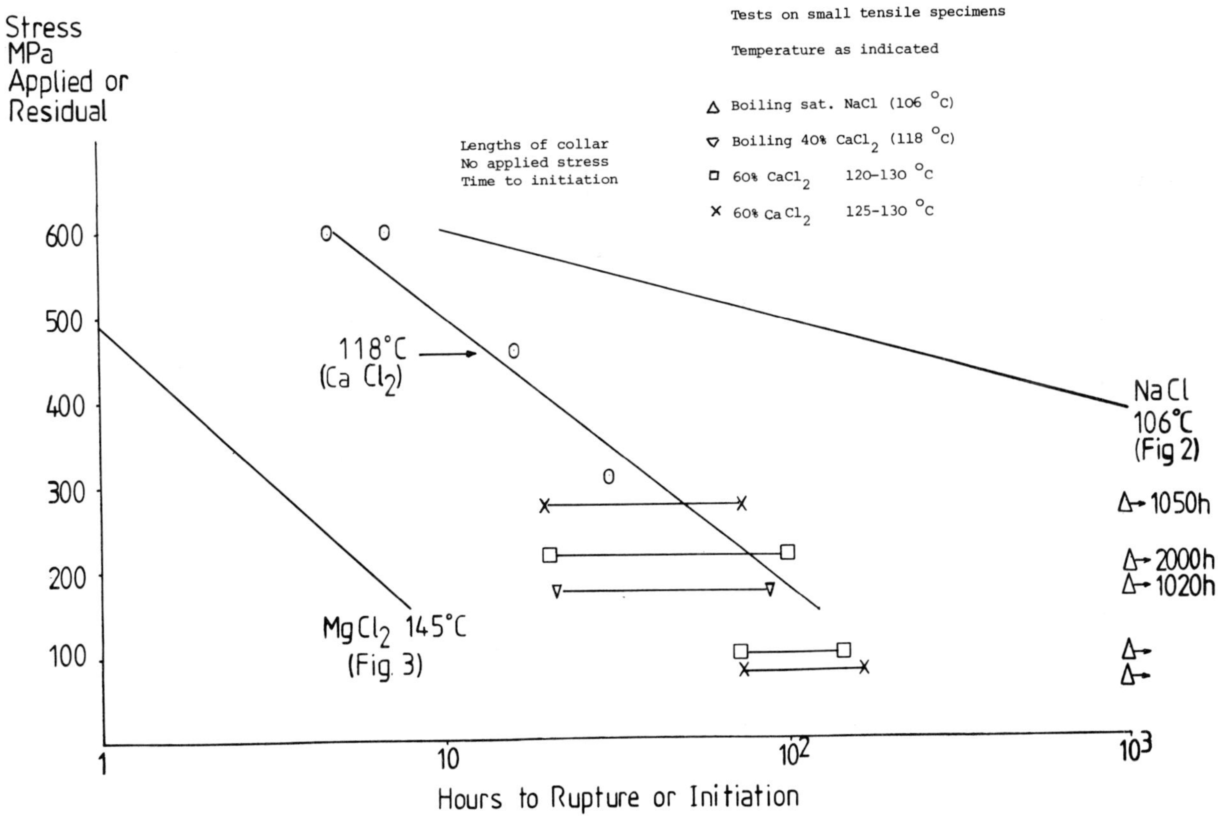

4 SCC tests on Staballoy

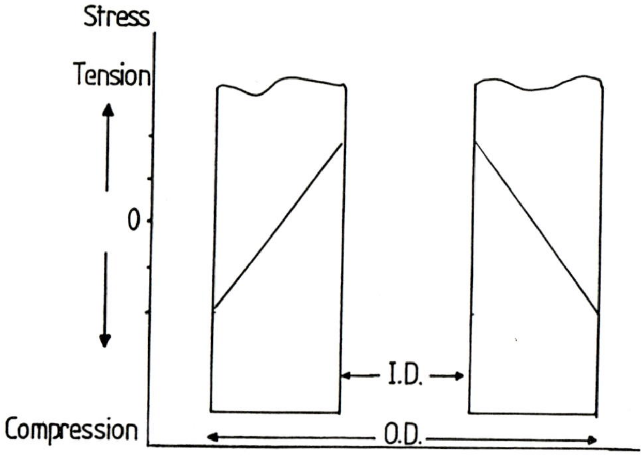

5 Normal stress pattern

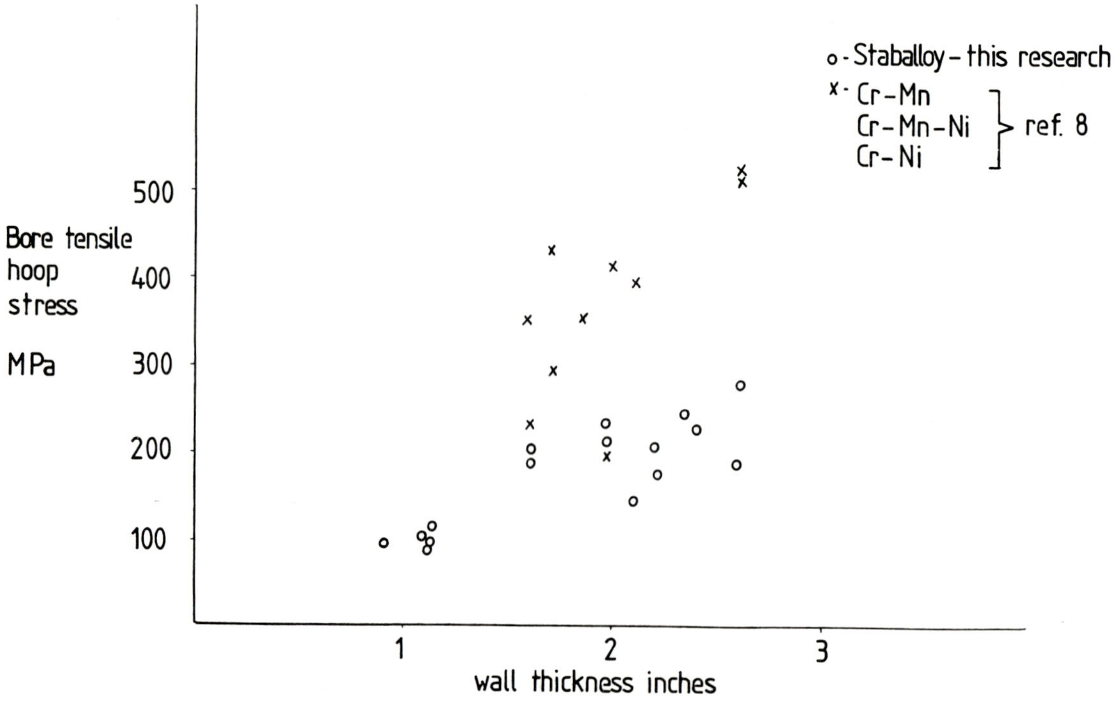

6 Measured residual tensile hoop stress at bore surface

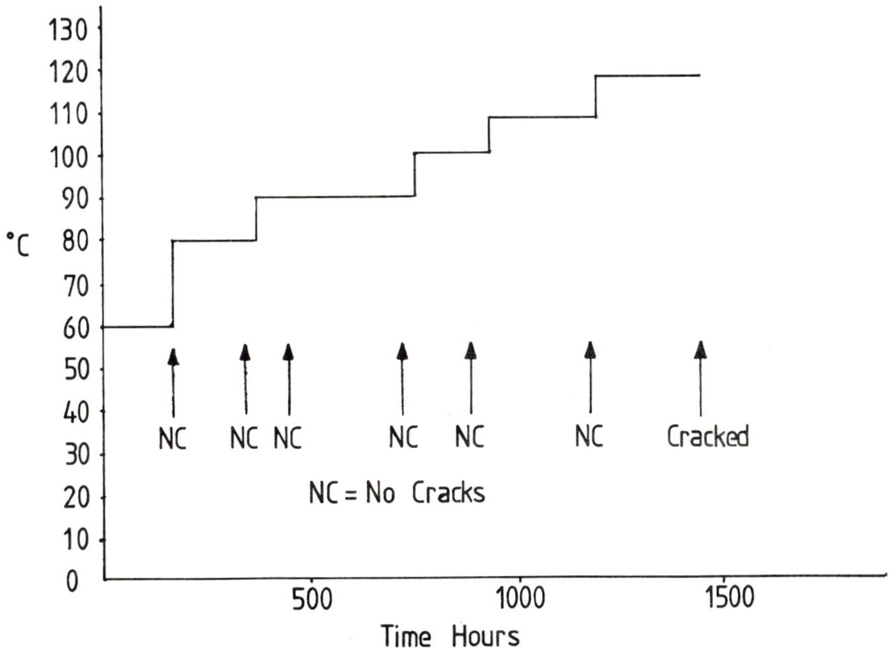

7 7 1/2" x 2 13/16" collar (bore residual stress 240 MPa) immersed in 50% $CaCl_2$

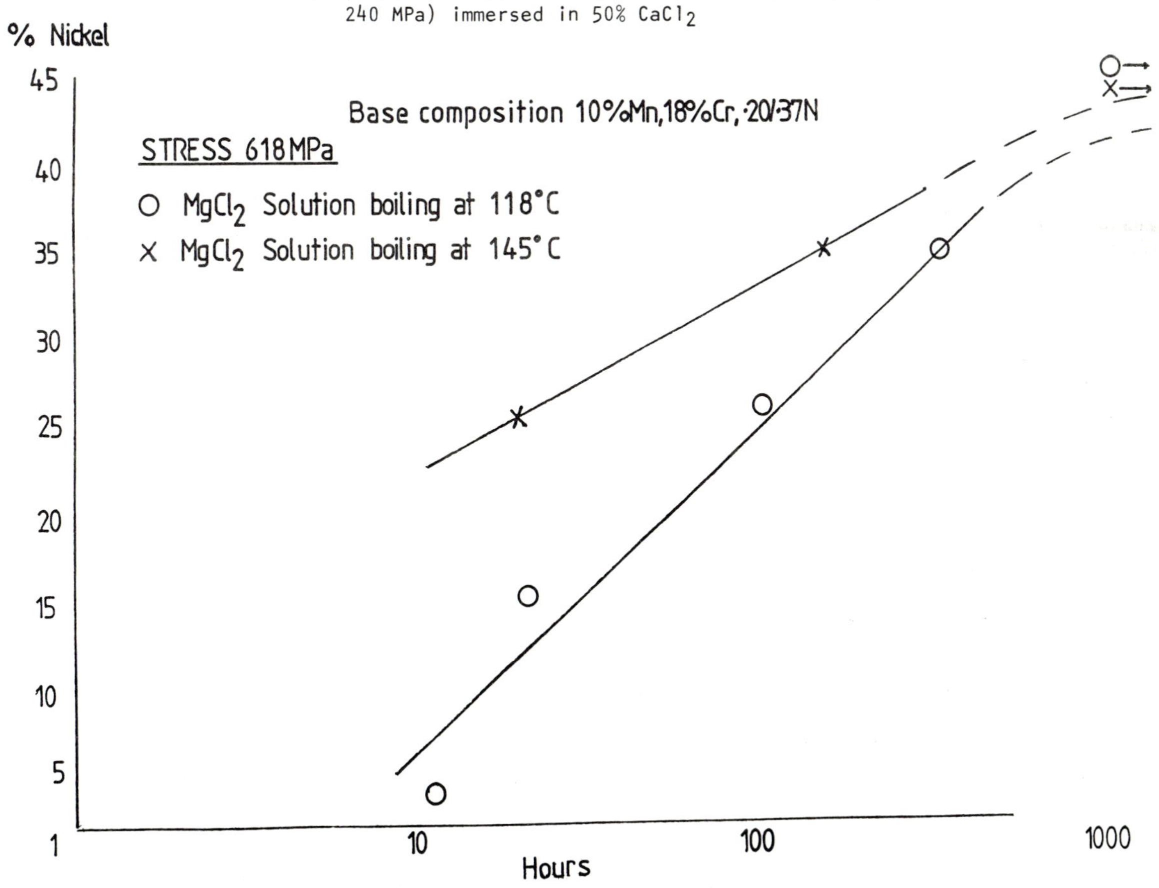

8 Tests on Staballoy variants with various nickel contents

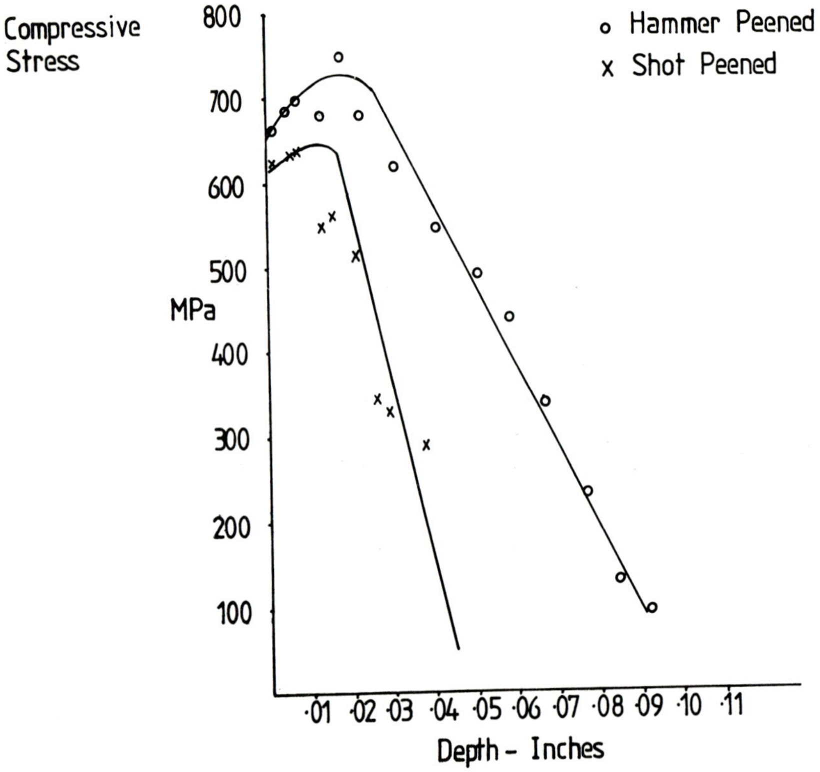

9 Surface stress of peened collars

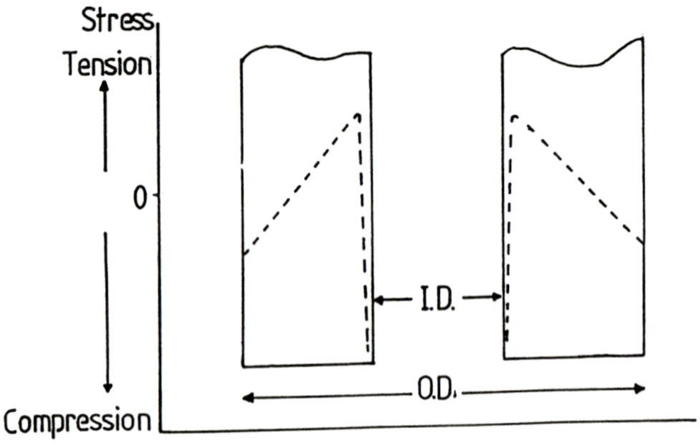

10 Stress pattern after peening

SESSION V
MECHANICAL PROPERTIES

Influence of silicon and nitrogen on properties of some austenitic steels

P Gümpel

The author is a member of the Steel Research Institute of Thyssen Edelstahlwerke AG, Krefeld, West Germany.

SYNOPSIS

The effect of silicon and nitrogen on precipitation and corrosion properties has been investigated in high-alloyed, austenitic stainless steels containing approx. 5% Mo. Increasing the silicon content leads to accelerated precipitation, as has been proved by a shift in toughness values during isothermal heat treatment in a critical temperature range. The corrosion properties of severely oxidizing acids, as determined by Huey testing, are strongly minimized by increasing the silicon content, already in solution heat-treated condition. Increasing the silicon content leads to pronounced sensitization, in the course of isothermal aging at critical temperatures.

The tendency to embrittlement owing to precipitation is limited by an increase of nitrogen. A significant influence of the nitrogen content on corrosion properties determined by the Huey test can neither be observed in solution heat-treated nor aged condition.

Investigation of the pitting corrosion resistance of some steel grades containing molybdenum shows that pitting resistance is improved by increasing the nitrogen content of these steels. Pitting resistance may be estimated by calculating the effective sum $W_{L,N}$ *using the formula:*

$$W_{L,N} = \%Cr + 3.3x\ \%Mo + 30x\ \%N$$

1. INTRODUCTION

In chemical apparatus engineering applications, steels of constantly growing alloying contents are used to meet the constantly growing demands for corrosion resistance. For instance, pitting corrosion resistance is improved by increasing the effective sum W_L[1] as determined by the following formula:

$$W_L = \%Cr + 3.3x\ \%Mo$$

However, increasing the alloying contents in chrome and, above all, in molybdenum also cause disadvantages. Thus, both elements aggravate the tendency for precipitation of phases such as σ, χ, carbides and/or nitrides. The interrelationship described between the alloying content and corrosion resistance is, naturally, applicable to a homogeneous distribution of these alloying elements only. Therefore, in the case of high-alloyed steels, all possibilities in terms of alloying technology should be made use of to avoid precipitation or to move the start of precipitation to prolonged holding times in the critical temperature ranges.

The precipitation behaviour of these steels can be modified by varying their silicon content.[2] The activity of carbon is increased by silicon, resulting in an accelerated precipitation of the mixed carbide $M_{23}C_6$ by a growing silicon content.[3] Additionally, silicon opens the range of sigma phase to lower chromium contents and, apart from that, extends the temperature range of sigma phase development.[4,5]

The sensitivity to intergranular corrosion is aggravated by the alloying element silicon. This has been proven already more than once for austenitic standard steels [6,7,8] and is true both in solution heat-treated and sensitized condition. According to investigations by J.S. Armijo[6], silicon sometimes has effects opposing each other in different concentrations.

In severely oxidizing agents, solution heat-treated specimens with about .1% silicon suffer intergranular corrosion attack that is increased dramatically by increasing silicon contents. As soon as a critical silicon level is passed (approx. 1%), the sensitivity to intergranular corrosion is reduced again. At about 4%, no intergranular attack is visible at all.[6]

In contrast to the precipitation-enhancing effect of silicon, nitrogen reduces the tendency of many phases to precipitation.[2] The precipitation of all phases that are able to dissolve small amounts of nitrogen only, such as $M_{23}C_6$ carbides, σ or χ precipitations, is decelerated by increasing the nitrogen content.[9] On the other hand, the precipitation of carbides with a high solubility to nitrogen is accelerated.[9] There are somewhat contradictory findings that have been made on the effects of nitrogen on the sensitivity to intergranular corrosion, corresponding to the different effect on precipitations, as explained before. According to findings by G. Lennartz[2)] and H. Thier[9)], resistance to intergranular attack is improved, whereas another investigation found an exactly opposite effect.[10)] According to that finding, the effect of the nitrogen content strongly depends on the composition of the steel and/or on the precipitations occurring.

In the past, it has been found that nitrogen, in addition to the elements chrome and molybdenum, may have a considerable effect on the pitting behaviour of high-alloyed austenitic steels. [11-14)] However, it has not yet been clarified in what effective interrelationship the alloying element nitrogen increases the resistance to pitting corrosion. The effective mechanism of nitrogen on the pitting behaviour is not known either.

2. EXPERIMENTAL PROCEDURE

Test heats were produced based on high-molybdenum steel grades with different silicon and nitrogen contents (Table 1). After hot-working to produce the specimens, these were subjected to a uniform solution heat treatment at 1150°C for 2 hrs./water quench, for all steel grade variations.

The so-heat treated specimens were aged afterwards at different temperatures and holding times. Investigation was focussed on the effect of aging on impact toughness and corrosion behaviour, in the Huey test according to DIN 50 921. For more details, if needed, refer to the test procedures and test results enclosed.

3. TEST RESULTS

3.1 Mechanical Properties

By isothermal aging in a temperature range between 700 and 1000°C, impact toughness is reduced considerably, owing to the precipitation occurring at these temperatures. *Fig. 1* shows the limits of 150 Joule notch impact energy on the test specimens A1 to A4.

By increasing silicon contents, i.e. for the steel grades in the order A1, A2, A3, A4, the lines of identical impact toughness are shifted towards shorter incubation times.

The effect of silicon on the embrittlement tendency is clearly shown by comparison of the impact toughness curves at severely reduced toughness levels. Elevating the silicon content clearly reduces the lead time until a specific degree of damage, i.e. a certain impact toughness limit, is reached. *(Fig. 2)*

The effect of isothermal aging on the curve shape of the iso-impact toughness curves is shown in *Fig. 3* , for the specimen grades of group B. Comparing the steel grades B1 and B2 as well as B3 and B4 shows the effect of silicon on the embrittlement tendency also of these grades.

On the contrary, increasing nitrogen from approx. .095% N, as in the grades B1 and B2, to approx. .15 % N, for the grades B3 and B4, reduces the tendency to embrittlement, as becomes evident from a comparison of the corresponding iso-impact toughness curves in aged condition *(Fig. 3)*. The influence of silicon on the tendency to precipitation principally occurs at elevated nitrogen contents. For the elevated-nitrogen heats (grades B3 and B4), a stronger effect of silicon is found than for steels of about .095 %N. *(Fig. 4)*

Microscopic investigations performed with solution heat-treated specimens exhibit a structure free from precipitations. Already after 6 minutes of aging in the critical temperature range, precipitations at the grain boundaries became visible under a light microscope. *(Fig. 5)* The interrelationship between precipitation intensity and embrittlement tendency is evident. Obviously, the tendency to precipitation is reduced by increasing the nitrogen content, whereas increasing silicon contents enhance the precipitation tendency. Investigations of the precipitations occurring, under an electron microscope, show an enhancement of σ-phase precipitation by increasing silicon contents. This shifts the development of χ-phase towards longer lead times.

Elevating the nitrogen content also results in shifting the start of precipitation of the χ-phase. For the steel grade B4, having elevated nitrogen and silicon contents, no χ-phase precipitations could be observed, not even after 20 hours of annealing at 850°C. This grade, however, exhibited M_6C carbide precipitations already after 6 minutes of heat treatment at 850°C.

A pronounced influence both of silicon and nitrogen on the rather high tendency to embrittlement is found with the test specimens of group C, i.e. steels with the highest molybdenum content. *(Fig. 6)* As for the steel grades investigated already, the decrease in toughness after isothermal heat treatment in the critical temperature range is shifted towards longer lead times by elevated silicon and towards shorter lead times by elevated nitrogen contents.

3.2 Corrosion Testing

3.2.1 The Properties of Steels under Huey Testing

A survey of the mean values from the mass loss rates of all steels in relation to silicon content *(Fig. 7)* permits a clear dependence of the corrosion resistance of solution heat-treated specimens on the silicon content to be observed when subjecting them to Huey testing. Up to a silicon content of approx. .15 - .20%, the corrosion properties are not influenced remarkably by silicon. But after passing this limit, the corrosion attack is accelerated. The differences in steel composition between the three steel groups investigated seem to be of a somewhat secondary importance, compared to the effect of silicon.

Similar to the investigation of mechanical properties, the results in terms of corrosion resistance of isothermally aged specimens show that the tendency to precipitation of the steels subjected to investigation is enhanced. *Fig. 8* shows the effect of isothermal aging at 900°C on the mass loss rates of the grades in group A and group B subjected to Huey test. The figures are mean values determined from two boiling periods only, because due to excessively high volume loss rates the tests had to be ceased already after the second boiling period. Particularly, increasing silicon to levels exceeding .2 % leads to a severe aggravation of sensitivity to intergranular corrosion, when aged in the temperature range indicated.

No marked influence of silicon on the corrosion properties of group C materials could be found with specimens aged isothermally in the critical temperature range.

The variation in nitrogen contents existing in steels within the alloying range of the groups B and C does not seem to have any remarkable effect on the attack when subjected to Huey test, neither in solution heat-treated nor in isothermally annealed condition.

3.2.2 *Evaluation of the Influence of Nitrogen on Pitting Corrosion Resistance*

Additional test results from steels produced under real industrial conditions were included in the evaluation of the influence of nitrogen on pitting resistance of steel grades containing molybdenum. Investigation took place in a 3% NaCl solution at a constant potential of 950 mV_H. The so-called "critical pitting temperature" was selected as an evaluation criterion. It is the temperature at which the specimens in the respective testing agent do not yet exhibit any pitting or where a temperature increase by 5°C results in pitting corrosion.

Irrespective of the nitrogen content, the steel grades investigated do not reveal any clearly visible interrelationship between the effective sum calculated from the alloying contents of the steel and the pitting resistance determined by experiment.

However, setting the pitting resistance determined by experiment into relationship with the calculated effective sum

$$W_{LN} = \%Cr + 3.3\% \ Mo + 30 \ \%N$$

with a calculating factor X = 30 for nitrogen, leads to a satisfactory correlation of both characteristics *(Fig. 9)*. It shows a nearly linear interrelationship between the calculated effective sum and the pitting resistance determined by experiment.

4. DISCUSSION

The test results reveal an adverse effect of silicon on the precipitation properties of the high-molybdenum steel grades investigated here. This becomes evident both from testing the impact toughness and from corrosion investigation by Huey testing of specimens isothermally aged in a critical temperature range.

An elevated silicon content has an adverse effect on the resistance during Huey test, already in solution heat-treated condition. The selective attack of grain boundaries gives reason to conclude that, already in solution heat-treated condition, the homogeneity of material in the grain boundary area is disturbed to an increasing extent by elevating the silicon content. This seems to be the initial state of precipitations because the test results from isothermally aged specimens also reveal clearly that the tendency to precipitation is enhanced by elevated silicon content. This will be subject to further investigations, especially to clarify the mechanisms taking place in a early state of precipitation.

Huey testing did not reveal any definite effect of nitrogen on the corrosion properties. According to K.J. Westerfeld [15)], the precipitation of intermetallic phases σ and /or χ does not necessarily lead to an aggravated sensitivity to corrosion attack during Huey test. Although the precipitation rate is reduced by increasing the nitrogen content, as revealed by testing the mechanical properties, the present investigations did not exhibit any marked influence of nitrogen on the corrosion properties when subjected to Huey test.

The relationship between the measured critical pitting temperature and the calculated effective sum, at a differently weighted influence of nitrogen, gives reason to affirm that the factor X = 30 proposed by G. Herbsleb[11)] is rather close to the real relationships, when calculating the effective sum. Such an influence of nitrogen may, however, be linked to certain limits in the levels of other alloying elements. This possibly is an explanation for the differences found in the effect of nitrogen on pitting resistance. In this regard, it becomes evident that the positive effect of elevated nitrogen on pitting resistance occurs for steels where nitrogen, at the same time, reduces the tendency to precipitation, i.e. particularly in molybdenum steels.

Literature:

1) Lorenz, K.; Medawar, G.: Thyssen Forschung 1 (1969) pp. 97/108
2) Lennartz, G.: Mikrochemica Acta 68 (1965) pp. 405/28
3) Horn, E. M.; Kügler, A.: Z. Werkst.Techn. 8 (1977), pp. 362/70
4) Schüller, H. J.: Arch. Eisenhüttenwes. 36 (1965) pp. 513/16
5) Kubaschweski, O. : Iron-Binary Phase Diagrams, Springer-Verlag, Berlin-Heidelberg - New York, Verlag Stahleisen mbH,Düsseldorf 1 1982
6) Armijo, J. S. : Corrosion NACE 24 (1968) S. 24/30
7) Coriou, H. ; Desestret, A. : Grall, L. u. Hochmann, J. : Comptes Rendus des Seances de L`Academie des Sciences 254 (1962) pp. 4467/69
8) Coriou, H. : Desestret, A. Grall, L. u. Hochmann, J. : Revue de Metallurgie, 61 (1964) pp. 177/83
9) Thier, H. : The effect of nitrogen on the precipitation properties of the austenitic chrome/nickel steel X5CrNiMo17 13, Dr.-Ing. dissertation, RWTHAachen 1967
10) Grützner, G. : The intergranular corrosion of nitrogen-alloyed chrome/nickel steels 18/10, Dr. - Ing. dissertation, RWTH Aachen 1971
11) Herbsleb, G. : Werkst. u.K orrosion 33/34
12) Kearns, J. R. : In: New Developments in stainless steel technology American Society for Metals, Metals Park, Ohio, 1985, pp. 117/127
13) Suutala, N. : Kurkela, M. : In: Stainless steels 84, The Institute of Metals, London 1984, S. 240/247
14) Heubner, U. ; Rockel, M. : Werkst. und Korrosion 37 (1986) pp.7/12
15) Westerfeld, K. J. ; The effect of nitrogen on the corrosion properties of solution heat-treated and tempered austenitic chrome/ nickel 18/10 and chrome/nickel-molybdenum steels 18/12, Dr.-Ing. dissertation, RWTH Aachen 1974.

Table 1 Chemical composition of the steels tested

Steel No.	Chemical composition, %						
	C	Si	Cr	Mo	Ni	Cu	N
A 1	0,018	< 0,01	18,57	5,25	14,43	0,07	0,16
A 2	0,018	0,08	18,59	5,25	14,56	0,06	0,16
A 3	0,019	0,23	18,57	5,24	14,45	0,06	0,16
A 4	0,020	0,56	18,41	5,19	14,56	0,06	0,15
B 1	0,018	0,08	18,83	5,45	24,19	1,17	0,096
B 2	0,018	0,41	18,78	5,43	24,13	1,16	0,095
B 3	0,016	0,08	18,78	5,45	24,21	1,18	0,160
B 4	0,018	0,49	18,49	5,35	24,10	1,14	0,150
C 1	0,020	< 0,01	20,70	6,15	17,93	0,76	0,160
C 2	0,019	0,17	20,80	6,12	17,89	0,77	0,160
C 3	0,020	< 0,01	20,56	6,10	17,82	0,75	0,190
C 4	0,020	0,22	20,39	6,07	17,92	0,75	0,190

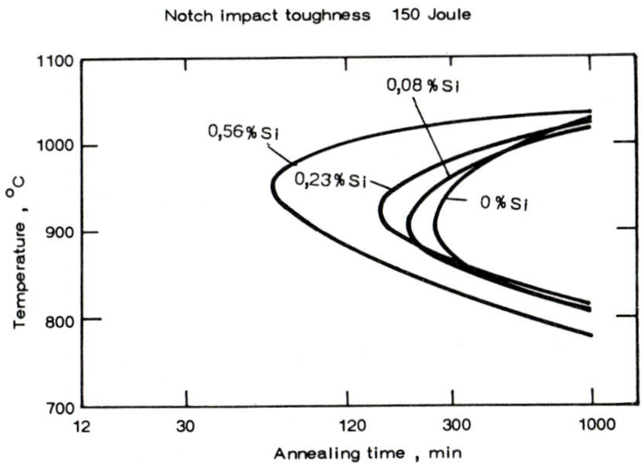

Fig. 1 Influence of an isothermal annealing treatment on the notch impact toughness of the steels A1 to A4

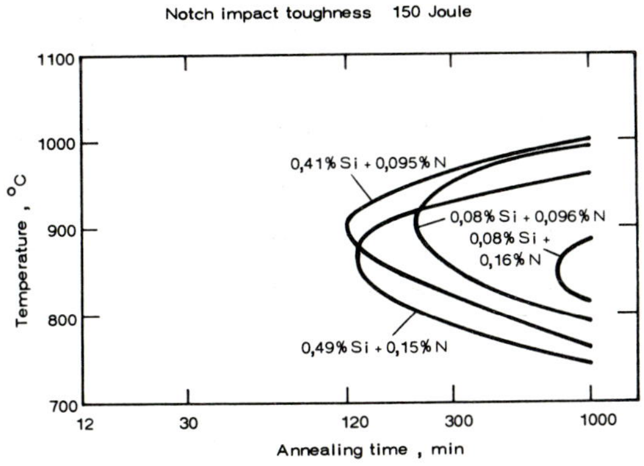

Fig. 3 Influence of an isothermal annealing treatment on the notch impact toughness of the steels B1 to B4

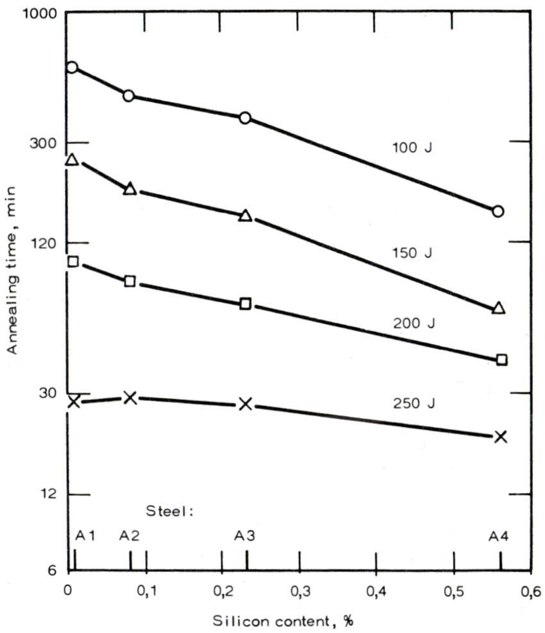

Fig. 2 Influence of the silicon content on the annealing time until a certain notch impact value has been reached (steels A1 to A4)

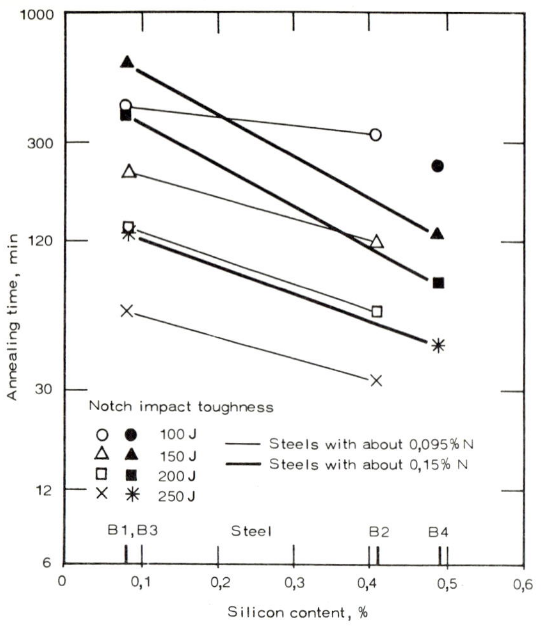

Fig. 4. Influence of the silicon and nitrogen contents on the annealing time until a certain notch impact value has been reached (steels B1 to B4)

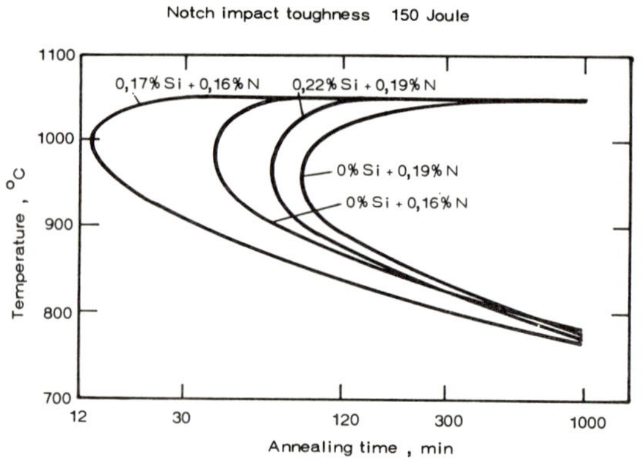

Fig. 6 Influence of an isothermal annealing treatment on the notch impact toughness of the steels C1 to C4

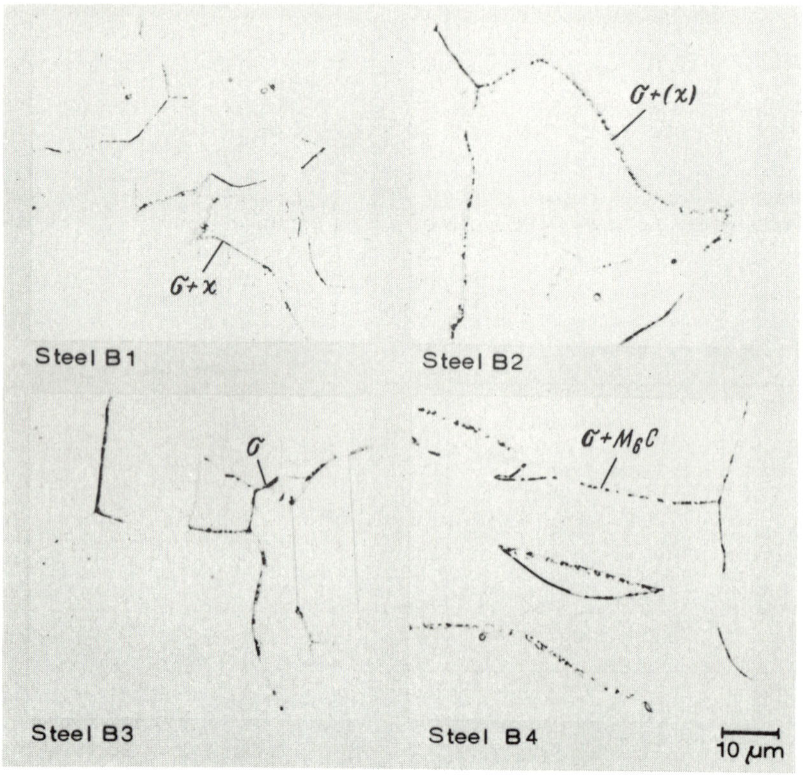

Fig. 5 Microstructure in the steels A1 to A4 after annealing treatment: 1150 °C 2h/W + 850 °C 6 min/W

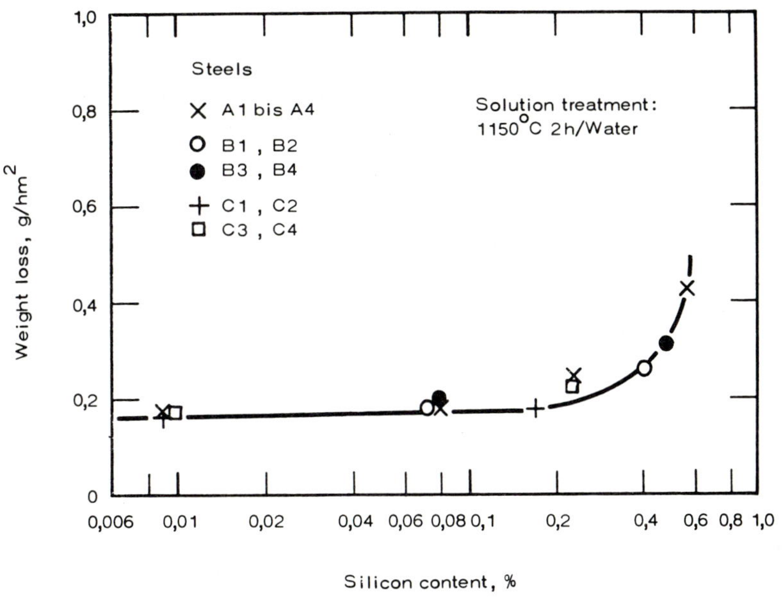

Fig. 7 Influence of the silicon content on the weight loss of the investigated steels in boiling 65 wt.% nitric acid (mean values of five boiling periods)

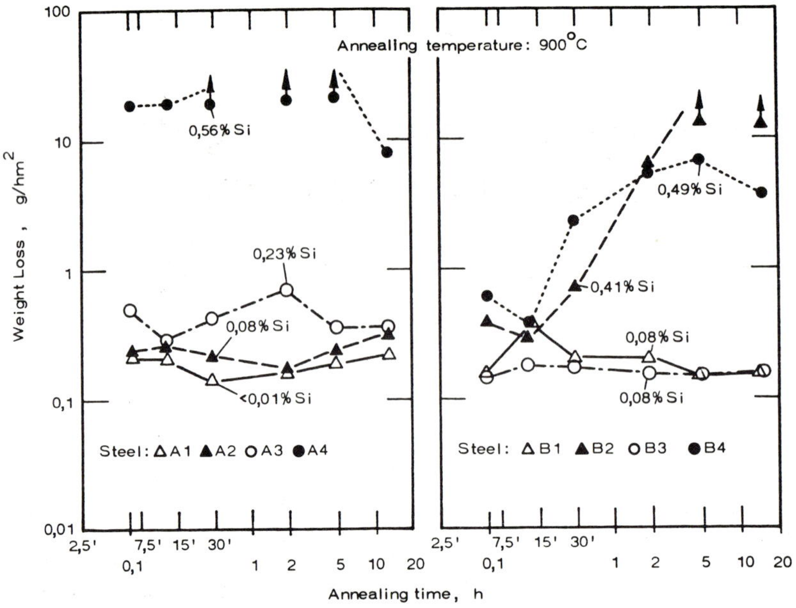

Fig. 8 Influence of an isothermal annealing treatment on the weight loss of the investigated steels in boiling 65 wt.-% nitric acid (mean values of two boiling periods in the Huey-test)

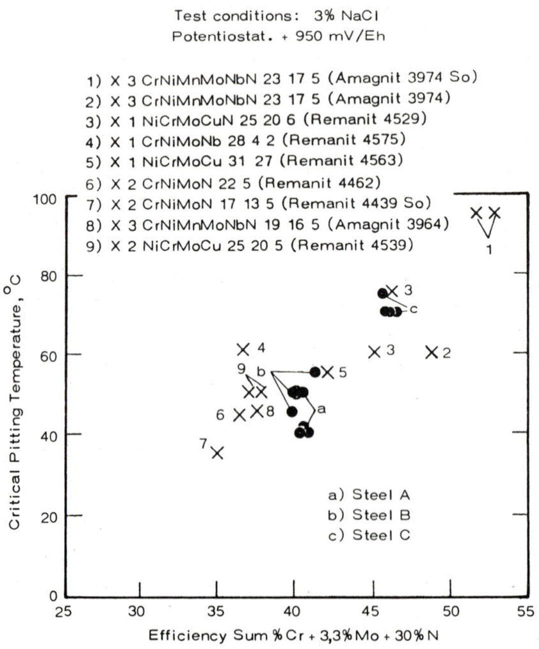

Fig. 9 Influence of the efficiency sum % Cr + 3.3% Mo + 30% N on pitting behaviour of high-alloyed stainless steels with nitrogen

Nitrogen strengthening of duplex stainless steels

G Wahlberg and G L Dunlop

The authors are in the Department of Physics, Chalmers University of Technology, Göteborg, Sweden.

SYNOPSIS

Duplex austenitic-ferritic stainless steels, based on SAF 2205 but containing varying amounts of N, were plastically deformed by either tensile testing or wire drawing. The defect structures in both austenite and ferrite were characterised at different levels of strain using transmission electron microscopy and the results were related to the flow strength and work hardening behaviour. N was shown to provide considerable strengthening to the alloys but the strengthening effect reaches a saturation level at intermediate amounts of N. The strengthening was mainly provided by hardening of the austenite phase. The overall deformation of these duplex stainless steels seems to be determined by the strength of the softer phase and the saturation of strengthening by N is probably caused by strain inhomogeneity. The duplex steels were found to deform by a number of different mechanisms. The most prominent direct effect of N on the deformation structure was to introduce a planar glide mode to the deformation of austenite. The occurrence of deformation twinning in the austenite was shown to be strongly dependent on the crystallographic orientation of the grains relative to the tensile axis.

1. INTRODUCTION

Duplex stainless steels have mechanical and corrosion properties which make them attractive alternatives to conventional stainless steels. As a consequence duplex stainless steels are being increasingly used in applications where high mechanical strength is required in highly corrosive environments which are met in, for example, oil and gas extraction.

The excellent properties of these steels are a consequence of a *complicated interaction between the two phases*, which as yet is not completely understood. Where mechanical properties are concerned (reviews on the subject have been given by Fischmeister and Karlsson (1),and Tomota and Tamura (2)), a variety of factors such as the content, size, shape and distribution of phases and the flow stress difference between the two phases are considered to be of importance. A number of theoretical models, including some which make use of the finite element method, have been used in order to explain the mechanical behaviour of these materials. None of these models have been fully successful and it is clear that a detailed characterisation of deformation microstructures, in combination with mechanical data, is necessary in order to give a conclusive description of the deformation process (3).

Duplex austenitic-ferritic stainless steels are often *alloyed with nitrogen* in order to both strengthen the austenite and to balance the phase content (N can substitute for Ni as an austenite stabiliser). The majority of N concentrates in the austenite phase which is strengthened significantly by even small amounts of nitrogen. In the present work three duplex austenitic-ferritic stainless steels containing different amounts of N were investigated. (The chemical compositions are given in Table 1). Alloying in this way enables the flow stress difference between austenite and ferrite to be varied. The Ni-content was also varied in order to keep the austenite/ferrite volume ratio constant and all other microstructural parameters were kept constant.

The *defect structures* in both austenite and ferrite were characterised at different levels of strain using transmission electron microscopy (TEM) and the results were related to the flow strength and work hardening behaviour. The variation of flow strength of austenite and ferrite, which resulted from the different additions of nitrogen, was used in order to determine the influence of the difference in strengths of these two phases on the overall mechanical properties of duplex stainless steels.

2. EXPERIMENTAL

The duplex stainless steels, which were investigated (see Table 1 for compositions), were manufactured as hot extruded bar and were subsequently heat treated at 1050°C for 20 minutes followed by water quenching. This heat treatment gave equal amounts of austenite and ferrite. The grain size of all three materials was determined to be 5 to 6 μm (linear intercept). The microstructure of the extruded material had the usual characteristic banded fibre-like appearance. Because of partitioning of alloying elements, the compositions of the ferrite and austenite phases in duplex stainless steels are not identical. Table 2 gives the compositions of austenite and ferrite for the three alloys, as analysed by EDX of thin foils. The concentrations of N could not be determined experimentally by EDX and instead it was calculated using the "ThermoCalc" computer programme (4). The results of these calculations are included in Table 2 from which it can be seen that N concentrates strongly in the austenite.

The experimental alloys were plastically deformed by either tensile testing or wire drawing. Homogeneous through-thickness deformation of the specimens was possible by tensile testing while wire drawing enabled higher degrees of deformation to be reached. Tensile specimens were machined from extruded rod of diameter 15 mm. Specimens for wire drawing were delivered as wire rod with a diameter of 10 mm. This rod was drawn to a diameter of 8 mm and then heat treated at 1050°C for 20 minutes and water quenched before being further drawn to different degrees of deformation. Tensile specimens were deformed in the range ε = 0-0.2, while the wire specimens were deformed to true strains of up to $\varepsilon = 1.5$ for alloy LN and HN and $\varepsilon = 2.0$ for alloy SAF 2205.

Microstructural characterisation was carried out using a JEOL 2000FX TEM equipped with a LINK AN10000 fully quantitative energy dispersive X-ray analysis system. Thin foils corresponding to different degrees of deformation were prepared from longitudinal sections using standard techniques. The specimens were machined by spark cutting in order to avoid any unnecessary mechanical deformation.

The textures of the extruded material from which tensile specimens were taken and also the solution treated strain free wire rod prior to wire-drawing were determined by standard x-ray techniques. An <011> fibre texture was found in the ferrite and an <001> fiber texture in the austenite of the rod from which the tensile specimens were made. The corresponding texture in the ferrite for the wire-drawn specimens was <011> while no specific texture was found in the austenite of this material.

3. RESULTS

3.1 Mechanical properties

The yield strengths at room temperature for the three alloys are given in Fig. 1, where the 0.2% proof stress, $\sigma_{0.2\%}$, as well as the ultimate tensile strength, σ_{UTS}, are plotted as a function of N content. The results show that N strengthens these duplex stainless steels considerably, but that the strengthening effect reaches a saturation level at about 0.13% N. Thus the $\sigma_{0.2\%}$ values for SAF2205 and alloy HN are the same (500 MPa), while alloy LN has a $\sigma_{0.2\%}$ value of 440MPa.

The microhardnesses of the austenite and ferrite phases in each of the three alloys in the undeformed state are given in Fig. 2. The hardness of the austenite increased linearly with increasing N content from H_V 237 for alloy LN to H_V 310 for alloy HN. Unlike the yield strength, no saturation level was reached with increasing N content. As shown in Fig. 2 the ferrite phase was also hardened, although to a lesser degree, by increasing the N content. These results indicate how the difference in strengths between the two phases varied for the three alloys. Both phases have the same hardness in SAF 2205, while in alloy LN the ferrite was the harder phase and in alloy HN the austenite was the hardest phase (see Table 3).

The work hardening behaviour of the individual phases was followed by microhardness measurements on deformed materials (Fig. 3). The overall rates of work hardening for austenite in both LN and HN alloys seemed to be somewhat higher than the corresponding rates for ferrite. Initially austenite in alloys LN and HN work hardened at an approximately equal rate, but at the highest strain level the difference in hardness increased considerably. At very low levels of deformation the HN ferrite was harder than both LN austenite and ferrite. At $\varepsilon = 0.2$ the hardness of ferrite in both LN and HN was approximately equal and stayed that way for all higher degrees of deformation. The initial apparent softening of austenite and ferrite in both high and low N alloys should also be noted. This phenomenon might simply be a reflection of the fact that the number of available dislocations for plastic deformation at very low levels of strain is low. This would then result in increased hardness values. Another possible explanation is that internal stresses, originating from the manufacturing of the material, are relaxed during the early stages of deformation.

The work hardening behaviour of the three duplex steels are shown in Fig. 4a, where the tensile yield strengths of wire drawn to various reductions are given. For technical reasons the SAF 2205 alloy was wire-drawn on a different occasion to the other two alloys and because of somewhat different drawing sequences the measured yield stresses are not comparable at an absolute level. On the other hand the characteristic forms of the curves are believed to be representative of their behaviour. In the case of SAF 2205, the flow stress increased rather slowly in the strain range $\varepsilon = 0.5 - 1.5$ after an initial rapid increase and the flow stress again increased at a higher rate after $\varepsilon = 1.5$. The yield stress of alloy HN was higher than the yield stress of alloy LN for all degrees of deformation.

The ultimate tensile strengths of drawn wire for all three alloys are given in Fig. 4b. Once again the curve for SAF 2205 is not strictly comparable with the curves of alloys HN and LN. It is notable however that, for true strains higher than $\varepsilon = \sim 0.5$, the *difference* in the UTS between alloys LN and HN increases at a considerable rate.

3.2 *Deformation martensite*

The amount of deformation martensite in drawn wires of all three alloys was determined by magnetic balance measurements. The results are given in Fig. 5. It can be seen that the volume fraction of martensite, which was formed during deformation, varied significantly between the three alloys. Thus, a considerable amount of deformation martensite was formed in both SAF 2205 and alloy HN, while hardly any deformation martensite formed in alloy LN at all.

3.3 *Characterisation of deformation structures by TEM*

The defect structure of alloy LN developed during tensile testing as follows. At a true strain of $\varepsilon = 0.05$ the *ferrite deformed by dislocation glide* and because of extensive cross slip the structure was quite homogeneous with loose dislocation tangles (Fig. 6a). At $\varepsilon = 0.2$ this structure had changed into a well developed cell structure (Fig. 6b). The cell walls in Fig. 6b are approximately parallel to <111> directions. The deformation mode in *austenite* in alloy LN at $\varepsilon = 0.05$ was dislocation glide with a *tendency to the planar glide* which is common to austenites with low stacking fault energies (Fig. 7a). This structure turned into a loose cell structure at $\varepsilon = 0.2$ (Fig. 7b) and at this strain some austenite grains were also deformed by mechanical twinning (Fig. 8).

The substructure of the ferrite in alloy HN showed similar characteristics to alloy LN. The loose dislocation structure at $\varepsilon = 0.05$ turned into a cell structure at $\varepsilon = 0.2$. By way of contrast the increased N content of alloy HN introduced a pronounced effect in the austenite: at $\varepsilon = 0.05$ the deformation was *strongly planar* (Fig. 9). This planar glide mode was still observable at $\varepsilon = 0.2$ although it was mixed with a tendency for formation of a cell structure. Some grains also deformed by deformation twinning at this level of strain.

Deformation twinning in the austenitic phase of both low and high N alloys showed a dependence on the crystallographic orientation of individual grains, relative to the tensile axis. The relative directions of the tensile axis in a number of grains are shown in Fig. 10, together with an indication as to whether the grain in question had deformed by deformation twinning or not. It is clear that deformation twinning was most predominant for those grains which were oriented so that the tensile axis was far from the <001> direction. Grains, with the <001> direction nearly parallel to the tensile axis and thus also close to parallel to the fibre direction of the banded structure, only deformed by dislocation glide.

As mentioned previously in connection with Fig. 5, *deformation martensite* formed in the austenite of both SAF 2205 and alloy HN at high amounts of strain. A sub-grain/cell structure with very sharp cell boundaries developed in the ferrite at this high degree of strain.

At the lower degrees of deformation the deformation structure of all three alloys, which had been deformed by wire drawing, was found to be markedly different from that of material deformed to similar strains by tensile testing. Deformation twins were already frequent in the austenitic phase of wires drawn to $\varepsilon = 0.05$ and at $\varepsilon = 0.20$ practically all of the austenite grains contained deformation twins. At levels of strain up to $\varepsilon = 0.40$, no cell structure had developed in the ferrite and instead the dislocation distribution remained quite homogeneous.

4. DISCUSSION

4.1 *Mechanical properties*

Tensile tests showed that additions of N provide a considerable increase in strength to duplex stainless steels, but this strengthening reaches a saturation level at about ~0.13% N. As shown by microhardness tests, *hardening of the austenitic phase* is the main contribution to this strengthening. However, unlike the bulk polycrystalline duplex material the hardness of grains of austenite does not saturate at the levels of N investigated here. This is in agreement with previous work by Norström on type AISI 316 stainless steel (5) and also with the work of Nilsson and Thorvaldsson who investigated N-additions of up to 0.35wt.-% in a highly alloyed austenitic stainless steel (Sanicro 28) (6).

The effect of N additions on the hardness of the softer phase in duplex alloys may be seen in Fig. 2 as the lower of the two curves which describe the hardness of the austenite and ferrite phases. The hardness of the softest phase increases rapidly for N additions ranging from 0.04% up to 0.13%, as it is identical to the curve for austenite in this range. Above 0.13% N the softest phase is ferrite whose hardness remains virtually constant. The characteristic features of the curve which describes the hardening of the softest phase are the same as those for the 0.2% proof stress of the duplex alloys as a function of N content (see Fig. 1). Both curves show a rapid increase up to 0.13% N which is then followed by an almost constant part.

It would therefore seem that the most important factor which determines the level of the 0.2% proof stress of the composite duplex alloy, may well be *the strength of the softer phase.*

Under many circumstances it is reasonable that the 0.2% proof stress of materials containing two ductile phases is determined by the flow strength of the softer phase. The yield process of such alloys includes the following sequence of events. First both phases behave elastically. Then the softer phase begins to yield while the harder one remains elastic. Finally, both phases deform plastically. The macroscopic yield point should consequently be identical to the yield point of the softer phase. However, if the duplex alloy deforms under conditions of equal strain in both phases, the 0.2% proof stress should also be influenced by the strength of the harder phase. Under such conditions the stress σ_1 developed in the duplex alloy at a certain strain ε_1 is given by

$$\sigma_1 = \sigma_{1\gamma} \cdot f_\gamma + \sigma_{1\alpha} \cdot f_\alpha$$

where $\sigma_{1\gamma}$ and $\sigma_{1\alpha}$ are the stresses developed in the austenite and ferritic phases respectively at strain ε_1, and f_γ and f_α are the volume fractions of the two phases. This relation is the law of mixtures given by Tamura et.al. (7) which states that the flow stress of the duplex alloy is a linear function of the volume fractions of the phases.

The hardnesses of both austenite and ferrite phases in SAF 2205 (0.13% N) are equal and it seems reasonable that these two phases should also have approximately equal yield stresses and that approximately the same strains should develop in them during the early stages of deformation. The law of mixtures predicts that the 0.2% proof stress should be higher for alloy HN than that for SAF 2205 since its austenite is harder. However, the measured 0.2% proof stresses were the same for both alloys and therefore it may be concluded that the law of mixtures is not obeyed. This indicates that the strain in the austenite and ferrite phases are not the same, but that a *larger part of the strain is concentrated in the softer phase.* It is this strain inhomogeneity that reduces the influence of the harder phase on the 0.2% proof stress of the duplex alloys.

Tamura et.al. have investigated the influence on strength and ductility of two-phase Fe alloys of a parameter C, which is defined as ($\sigma_{0.2\%}$ of the harder phase) /($\sigma_{0.2\%}$ of the softer phase) (7). They found that, when $C \leq 3$, the relation between the 0.2% proof stress and volume fractions of the two phases was described satisfactorily by a linear relationship in accordance with the law of mixtures. However when C was increased above 3 the measured 0.2% proof stress was lower than the value predicted by the law of mixtures. This discrepancy from the law of mixtures was more pronounced for higher values of C and smaller volume fractions of the harder phase. Tamura et.al. explained their observations as being a result of a strain inhomogeneity. This was later attributed by Tomota et.al. to the operation of a relaxation mechanism which reduces internal stresses (8).

In the present investigation the law of mixtures was found to be invalid for alloy HN, where the ratio between the hardness of the austenite and ferrite, (H_V of the harder phase) / (H_V of the softer phase), was only 1.1. This is in contrast with the results of Tamura et. al. who, as already mentioned, found an agreement between experimental data and the law of mixtures for a hardness ratio of 1.8 (corresponding to C = 3.1) in a duplex austenitic-ferritic stainless steel. The reason for this discrepancy is not clear. Factors to consider in this context ought to include the banded two-phase structure of the duplex alloys investigated here, which should act to promote equal strains in both phases and effects of texture which could also be of importance.

4.2 Martensite

As shown by the microhardness measurements, the formation of *deformation martensite in austenite* at the higher strain levels obviously *increases the rate of work hardening* of this phase. This is as expected since it is more difficult to deform martensite than austenite (9). As all of the alloying elements, and N in particular, tend to stabilize austenite against transformation to deformation martensite, it is somewhat suprising that the amount of martensite which forms should be higher in alloy HN than in alloy LN at equal strain. The stress level, however, is lower in the LN austenite and, since the martensite transformation does not start until a critical stress has been reached (9), this might provide a simple explanation to the above observation.

Deformation martensite usually nucleates at the intersection of deformation bands which may consist of twins, plates of ε martensite or ordinary slip planes (10). Considering the high density of deformation bands (slip bands and twins), which are observed in TEM specimens of both alloys LN and HN at high levels of strain, it is unlikely that the number of nucleation sites should be a limiting factor for the formation of deformation martensite.

The increase in hardness of the austenite grains, which follows from the formation of deformation martensite, is also reflected in the increased overall work hardening which occurs in the duplex alloy. This is best seen in the curve for SAF 2205 in Fig. 4a where there is a "knee" in the curve at high strain after which the rate of work hardening increases.

4.3 Planar glide

It has been shown, in this investigation, that N promotes planar glide in the austenitic phase of these duplex stainless steels. This effect has been demonstrated previously for a number of different "single phase" austenitic stainless steels (11-14), but there is no general agreement concerning the mechanism which is responsible for this behaviour. It has been argued that if the stacking fault energy is decreased with increasing N content then this should induce the planar glide mode. The change in stacking fault energy has however been shown to be too small to account for the strong effect of N (11-13). Instead, short-range ordering, induced by N, has been put forward as an explanation for the phenomenon (11,12,). The strong affinity between N and Cr is thought to be the prime reason for the short-range ordering. Experimental evidence, which confirms that there exists *a strong interaction between N and Cr in the austenite phase* of duplex stainless steels in the solution treated condition, has recently been obtained using atom probe field ion microscopy (15). This may be the main reason for planar glide in these alloys.

4.4 *Deformation twinning*

Tensile strain caused the austenitic phase of both low and high N alloys to deform by mechanical twinning at strains of $\varepsilon = 0.2$. As there was no apparent difference in the amount of twinning in low or high N alloys, it would seem that N does not influence the occurrence of deformation twinning. On the other hand there was a *strong crystallographic dependence for the occurrence of twinning* such that only those grains, which had the <001> direction oriented away from the tensile axis, deformed by deformation twinning. Combining this observation with the <001> fibre texture which was found in the austenitic phase of the extruded rod from which the tensile specimens were cut, it is easy to see why the majority of austenite grains in these specimens deformed only by dislocation glide. This is in accordance with results obtained on other fcc materials by Vergnol and Grilhé (16). Their experimental results on Cu-Al alloys showed that twinning was likely to occur if the tensile stress was applied in the <111> direction, but that twinning never occurred if the tensile direction was close to <100>. Vergnol and Grilhé also gave a theoretical explanation for their findings. The occurrence of twinning was said to depend upon the nucleation and growth of extrinsic stacking faults. In the light of this it is possible to explain why practically all of the austenite grains in the wire specimens deformed by deformation twinning, while twinning only occurred for a few austenite grains in the tensile specimens. The absence of any pronounced fibre texture in the wire drawn material obviously created a more favorable situation for deformation by twinning.

CONCLUSIONS

1. N provides considerable strength to duplex austenitic-ferritic stainless steels but the strengthening due to N reaches a saturation level at an intermediate concentration of N. The strengthening is mainly caused by hardening of the austenite phase.

2. The tensile strength of these duplex steels is mainly determined by the strength of the softer phase, which through a strain inhomogeneity, is deformed to a greater degree than the harder phase.

3. A range of different deformation mechanisms can take place in duplex steels of this type. The main effect of N is to introduce a planar glide mode to the deformation of the austenite phase.

4. The occurrence of deformation twinning in individual austenite grains was shown to be strongly dependent on the crystallographic orientation of the grains relative to the tensile axis.

ACKNOWLEDGEMENTS

AB Sandvik Steel and Avesta AB are thanked for financial support and the provision of materials. Discussions with T. Thorvaldsson, J-O. Nilsson and M. Liljas are gratefully acknowledged as is experimental assistance from J. Wasén and A. Wilson.

REFERENCES

1. H. Fischmeister and B. Karlsson, Z. Metallkde., **68**, (1977), 311
2. Y. Tomota and I. Tamura, Trans. ISIJ, **22**, (1982), 665
3. L. Durand, Mat. Sci. & Techn., **3**, (1987), 105
4. B. Sundman, B. Jansson and J-O. Andersson, Calphad, **9**, (1985), 153
5. L-Å. Norström, Met. Sci., **6**, (1977), 208
6. J-O. Nilsson and T. Thorvaldsson, Scand. J. Metallurgy, **15**, (1985), 83
7. I. Tamura, Y. Tomota, A. Akao, Y. Yamaoka, M. Ozawa and S. Kanatani, Trans. ISIJ, **13**, (1973), 283
8. Y. Tomota and K. Kuroki, Mat. Sci. Eng., **24**, (1976), 85
9. T. Angel, J. Iron Steel Inst., **177**, (1954), 165
10. F. Lecroisey and A. Pineau, Met. Trans., **3**, (1972), 387
11. P.R. Swann, Corrosion, **19**, (1963), 102t
12. D.L. Douglass, G. Thomas and W.R Roser, ibid., **20**, (1964), 15t
13. R. Fawley, M.A. Quader and R.A. Dodd, Trans. AIME, **242**, (1968), 771
14. J-O. Nilsson, Scripta Met., **17**, (1983), 593
15. U. Rolander and G. Wahlberg, to be published
16. J.F.M. Vergnol and J.R. Grilhé, J. Physique, **45**, (1984), 1479

Table 1. Chemical compositions of the investigated steels (wt.-%).

	C	Si	Mn	P	S	Cr	Ni	Mo	Cu	N	Fe
Alloy LN	.026	.29	1.60	.009	≤.003	22.05	7.00	3.05	.032	.05	bal
SAF 2205	.025	.32	1.57	.024	.003	21.87	5.54	2.97	.09	.13	bal.
Alloy HN	.025	.29	1.40	.009	.005	21.81	4.50	3.02	.034	.20	bal.

Table 2. Partitioning of alloying elements between the austenite and ferrite phases. The given concentrations result from STEM/EDX analyses of thin foils except for the values for nitrogen which were calculated using the "ThermoCalc" computer programme (wt.-%).

	Alloy LN		SAF 2205		Alloy HN	
	α	γ	α	γ	α	γ
N	.013	.063	.039	.225	.056	.376
Si	.2	.3	.1	0.0	.2	0.0
Mn	1.4	1.7	1.3	1.4	1.3	1.6
Cr	26.0	20.8	25.5	21.9	24.8	22.6
Ni	4.6	8.2	3.4	6.4	3.2	4.9
Mo	3.1	2.0	2.6	1.8	3.1	2.5
Fe	64.7	67.1	67.2	68.6	67.4	68.4

Table 3. N content and hardnesses of the austenite and ferrite phases in the investigated alloys.

		N content (wt.-%) total	N content (wt.-%) phase	Hardness (HVN)	Ratio ($H_{V\gamma}/H_{V\alpha}$)
LN	α	0.05	0.013	261	0.9
	γ		0.063	237	
SAF 2205	α	0.13	0.039	277	1.0
	γ		0.23	273	
HN	α	0.20	0.056	282	1.1
	γ		0.38	310	

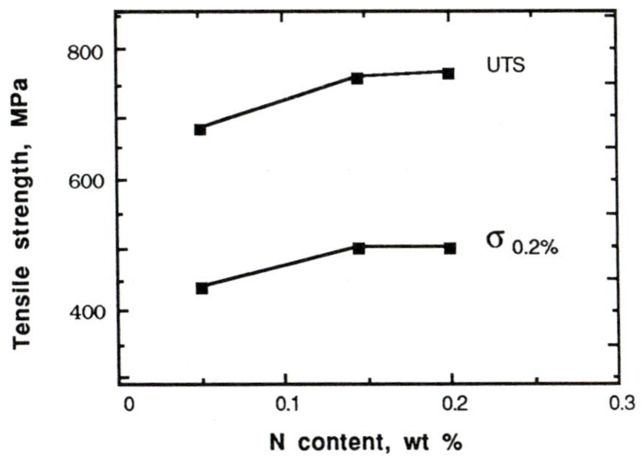

Fig. 1. 0.2% proof stress and ultimate tensile strength as a function of N content.

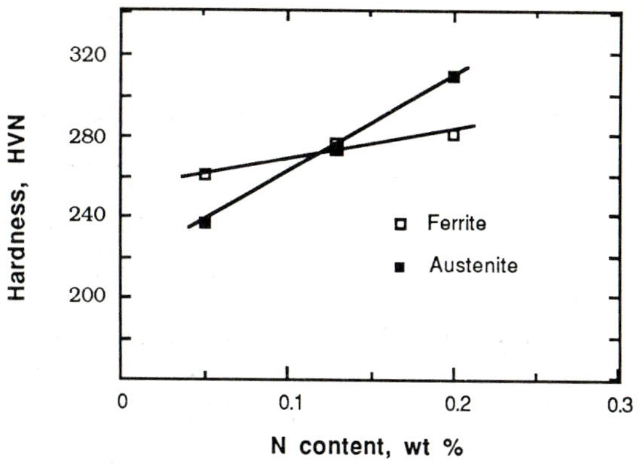

Fig. 2. Microhardness of austenite and ferrite phases as a function of total nitrogen content. The microhardness measurements were carried out using a load of 5g.

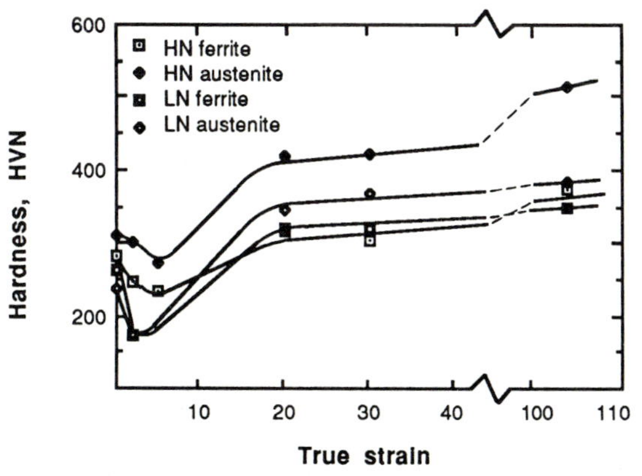

Fig. 3. Microhardness of the austenite and ferrite phases in alloys LN and HN as a function of the degree of deformation. A load of 5g was used in the microhardness measurements.

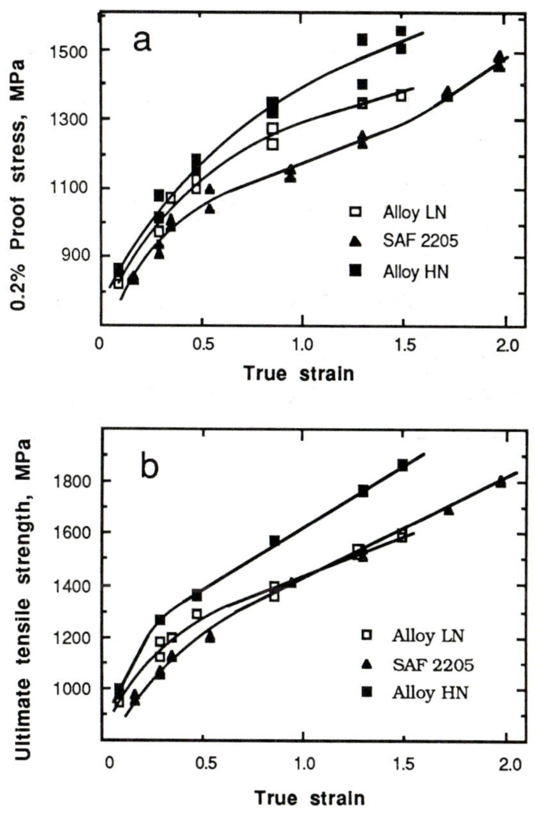

Fig. 4. Tensile strength as a function of the degree of deformation by wire drawing. (a) 0.2% proof stress. (b) Ultimate tensile strength.

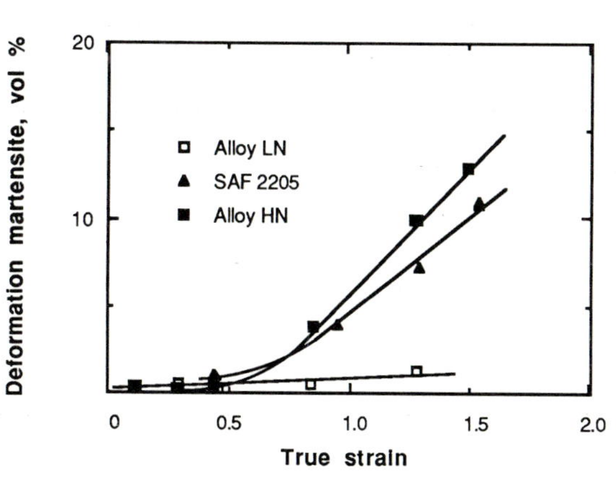

Fig. 5. Amount of deformation martensite formed in wire specimens as a function of the degree of deformation.

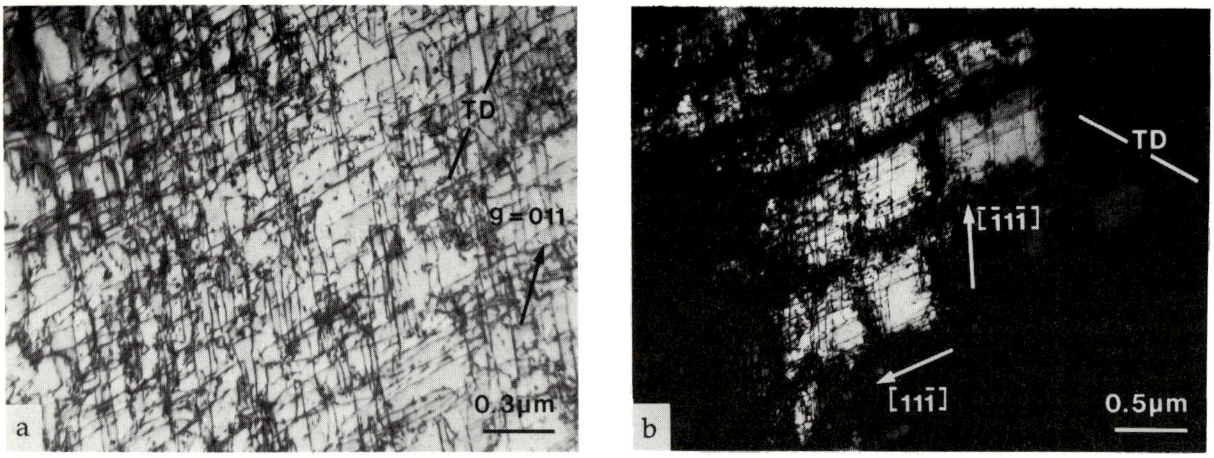

Fig. 6. Deformation structures in the ferrite phase of alloy LN. Tensile test specimen. (TD = tensile direction). (a) ε = 0.05. Homogeneously distributed dislocations resulting from extensive cross-slip.
(b) ε = 0.2. Well developed cell structure. The segments of cell walls are approximately parallel to <111> directions.

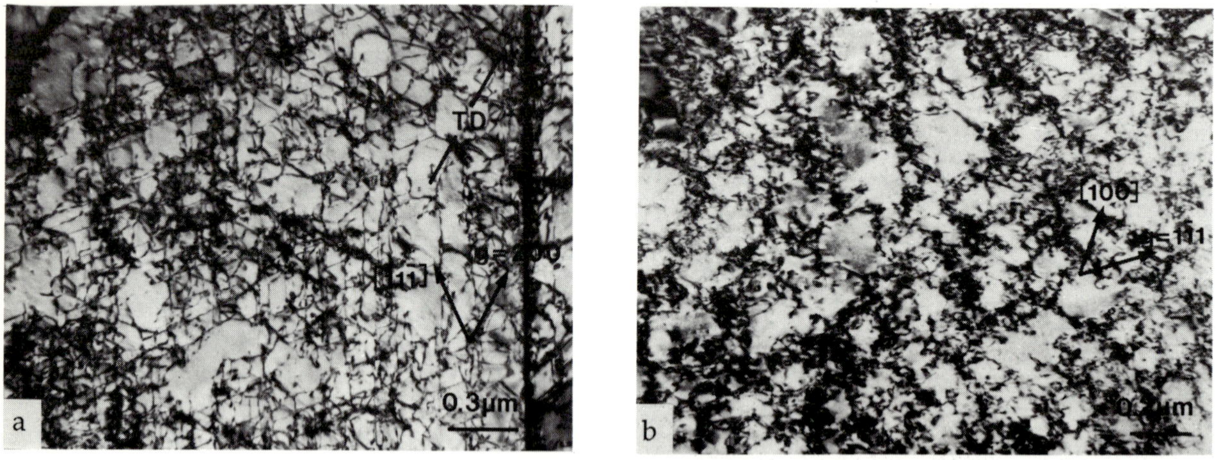

Fig. 7. Deformation structures in the austenite phase of alloy LN. Tensile test specimen. (TD = tensile direction). (a) ε = 0.05. Homogeneous distribution of dislocations with some tendency for planar glide.
(b) ε = 0.2. Loose dislocation cell structure.

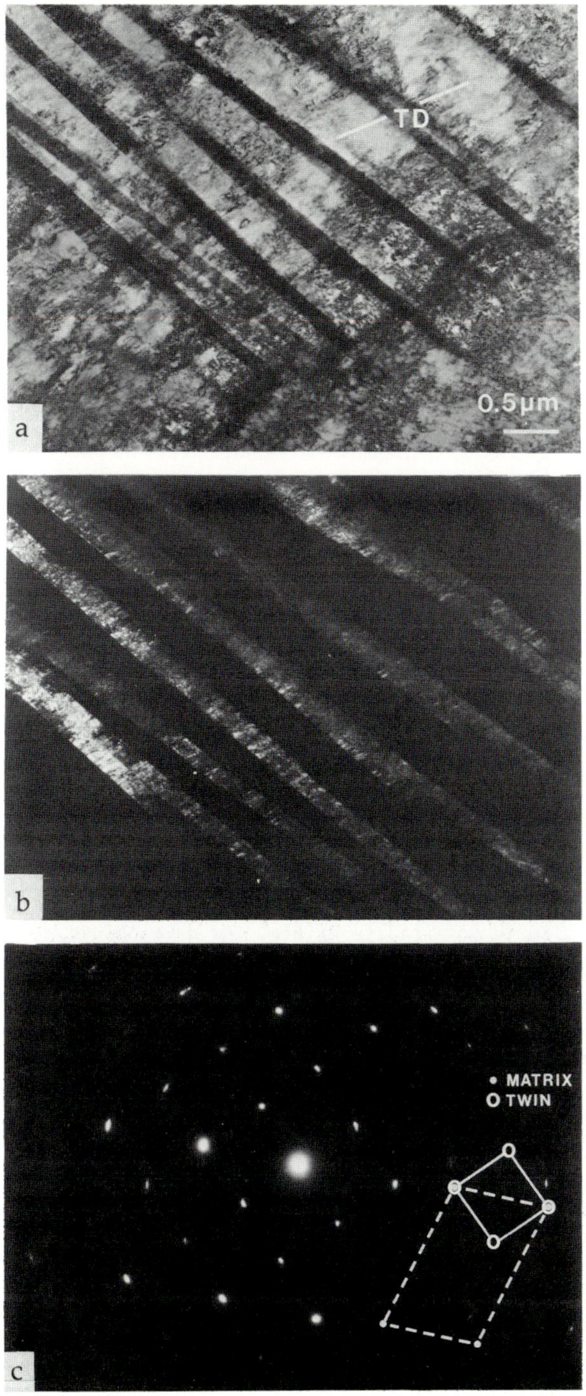

Fig. 8. Deformation twinning in the austenitic phase of alloy LN. Tensile test specimen. ε = 0.2. (TD = tensile direction). (a) Bright field image. (b) Centered dark field image formed using a twin reflex.
(c) Selected area diffraction pattern showing <122>-type matrix and <001>-type twin patterns.

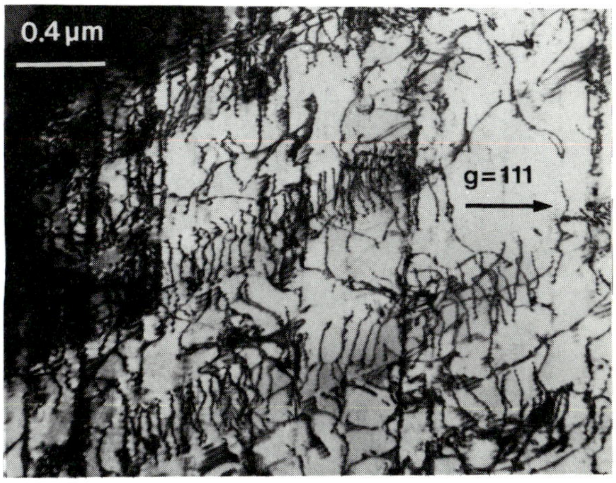

Fig. 9. Deformation structure in the austenitic phase of alloy HN. Tensile test specimen. ε = 0.05. Deformation is dominated by planar glide.

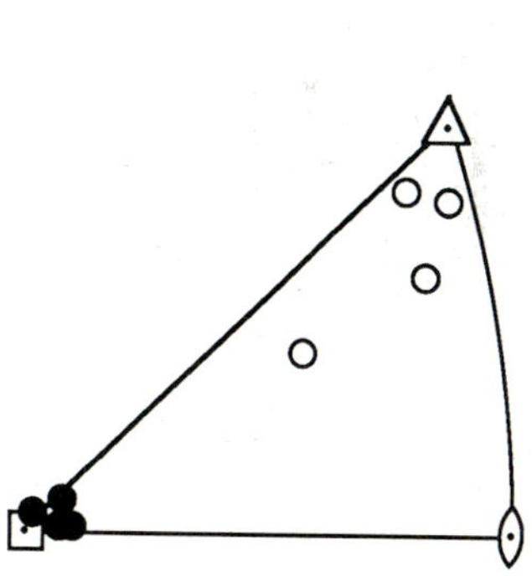

Fig. 10. Crystallographic dependence for the occurrence of deformation twinning in the austenite phase of tensile test specimens. The relative directions of the tensile axis in a number of austenite grains in alloys LN and HN are shown in the stereographic triangle. Open symbols (○) indicate the occurrence of twinning, and filled symbols (●) the absence of twins.

Development of corrosion-abrasion resistant steels for South African gold mining industry

B Metcalfe, W. M. Whitaker and U R Lenel

BM and WMW are with the Research Organization, Chamber of Mines of South Africa; URL is with Fulmer Research Laboratories Ltd.

SYNOPSIS

The need was identified to develop new steels for equipment operating at the working faces of South African gold mines where extremely abrasive and corrosive conditions prevail. From an extensive laboratory research programme four experimental alloy compositions were selected as suitable candidates for industrialization. After further qualifying tests at the intermediate 250 kg level, industrial melts of 15 t were commissioned and processed successfully into a variety of stock forms.

1. INTRODUCTION

The environmental conditions at the working faces of South African gold mines are among the harshest encountered in any mining situation. The rock in which the narrow gold-bearing reefs are embedded is harder than nearly all materials used in the construction of machines and, in deep stopes, is in a heavily fractured condition because of the high rock pressure. When the rock is broken from the face, fragments with very sharp edges are produced, causing severe abrasion when brought into contact with moving machine surfaces.

The mining operation entails the use of large quantities of water to allay dust and to cool the workings. This water becomes contaminated with acids and salts originating from the bacteriological oxidation of pyrite in the reefs, blasting fumes, and minerals dissolved in the fissure water. This, together with the elevated air temperatures of typically 30 °C, produces a humid and highly corrosive environment which is far more severe than most other industrial environments.

The combined effects of abrasion and corrosion cause rapid deterioration of the normal engineering steels used in the construction of mechanical equipment at the faces. In setting out to develop new or improved mechanized approaches to gold mining, it was realized that serious attention had to be given to finding cost-effective materials capable of surviving under these conditions.

Another major consideration in the design of mechanized mining equipment was the choice of power medium. The need for generating high forces within the confines of the working areas necessitated the use of hydraulic power in practically all situations. Under such arduous conditions, however, it is extremely difficult to maintain hydraulic circuits in a leak-free condition and, with conventional oil-hydraulic systems, the loss of oil has proved to be economically unacceptable. Effort was directed at developing a new technology based on emulsion-type hydraulic fluids containing up to 98 per cent water. This technology is now well-advanced and commercially viable, but suffers from the disadvantage that underground electro-hydraulic pump stations are required, needing regular attention to maintain fluid quality standards and to make up fluid losses.

In the course of developing emulsion hydraulic technology, it became apparent that a realistic possibility existed for developing hydraulic equipment which could operate on plain water. One of the advantages of this approach was that the hydraulic pressure could be generated very simply by using the hydrostatic head of chilled water brought into the mine for cooling purposes. This concept, known as hydro-power, minimizes the requirements for skilled labour and eliminates the need for pump stations underground. The main consideration in using water as a hydraulic medium is that the materials of construction need to be corrosion resistant, as well as mechanically strong and tough to withstand high hydraulic stresses, impact loads, and the effects of cavitation and flow erosion that tend to be associated with the use of high pressure water. This requirement is very similar to that for external machine components, in that the material must simultaneously withstand the damaging effects of corrosion and extreme mechanical stresses.

One obvious approach to the problem was to investigate the use of surface coatings, but it was soon found that these had little to offer because they could not provide an effective barrier to corrosion or were easily damaged mechanically[1]. Various classes of commercially available steels were investigated. It was found that abrasion resistant steels performed little better than mild steel; despite the good hardness and toughness of the base materials, the corrosion products generated on their exposed surfaces were soft and easily removed by abrasion. The performance of corrosion resistant steels was better, but was limited by their moderate strength and poor abrasive wear resistance. Moreover, the high cost of stainless steels precluded their economical use for all but the smallest components. It became clear that, for most applications, no ideal steel existed which could cope with the combined effects of corrosion and abrasion. The need was identified, therefore, for the development of a new range of economical, corrosion-abrasion resistant steels for general use in equipment for South Africa's gold mines, and a collaborative programme of research was initiated with steelmakers, universities and other institutions around the world. This paper describes how this work has progressed from the initial stages of laboratory experimentation through to production of steel on an industrial scale.

2. BASIC REQUIREMENTS

In terms of the requirements for corrosion resistance, it was recognized that the developed steels must be economic to produce, and that this would impose constraints on the degree of corrosion resistance that could be incorporated. Experience had shown that steels with no more than 12 per cent chromium, such as 3CR12, exhibited reasonably good corrosion resistance in most situations underground, and it was decided, therefore, to work within a chromium content range of 8 to 12 per cent. In terms of mechanical properties, the specifications of typical abrasion resistant steels were taken as a guideline, and the following targets were set:

Hardness	: 400 to 600 H_V
Charpy impact toughness	: 35 Joules
Yield strength	: 1 200 MPa
UTS	: 1 500 MPa
Elongation	: 10 %

With regard to resistance to the synergistic effects of corrosion and abrasion, the objective was to achieve a wear resistance that was as high as possible, and certainly higher than the best attainable with currently available steels. In addition, it was essential that the steels should be amenable to economic processing and manufacturing techniques, and in particular should be hot workable, castable, and preferably capable of being cut and welded with oxy-acetylene equipment.

3. LABORATORY INVESTIGATIONS

Previous research had shown that careful microstructural control was a key factor in achieving the desired mechanical properties such as resistance to abrasion[2] and erosion[3,4,5]. Two basic approaches were adopted in establishing suitable microstructures. The first involved the production of dual phase Fe/Cr/C alloys with a microstructure consisting of tough laths of low carbon martensite surrounded by plastic films of retained austenite, thereby combining heat treatable mechanical properties with adequate corrosion resistance[6,7,8]. This microstructure is illustrated in Figure 1, and was obtained by close compositional control, and by low temperature tempering (around 200 °C) of heavily hot worked material.

The second approach involved the production of nitrogen strengthened austenitic material in a relatively soft condition which transforms to hard martensite on mechanical deformation[9]. The potential for high wear resistance is derived from extreme work hardening through the wear-induced surface transformation[10], as illustrated in Figure 2. Toughness and ductility levels are high, typical of austenitic steels, and the absence of high levels of carbon contributes to ease of welding. The nitrogen, which increases the bulk hardness by its inclusion as an interstitial solid solution hardener, was used together with manganese to replace nickel for austenite stabilization. These two elements could be employed at lower levels than nickel to induce mechanical transformation. In this way, it was possible to arrange that the normal martensite transformation temperature (M_S) was below room temperature, but that the martensite transformation temperature under mechanical deformation (M_D) was above room temperature.

Arising from the laboratory work, four candidate steels were selected for production on a larger scale. These comprised three steels of the dual phase type (825, 102A and 122) covering between them the full range of chromium content from 8 to 12 per cent, and one mechanically transformable steel (1210). Details of the compositions of these four steels are given in Table 1; the compositions are not complex, and are indicative of a low intrinsic cost.

The results of mechanical testing are shown in Table 2. The mechanical properties of the three dual phase steels comfortably fulfil the specified requirements. The mechanical properties of the mechanically transformable steel in the normalized condition show that the toughness and elongation are more than adequate; also it has been confirmed experimentally that hardness levels exceeding 600 H_V are generated at the surface under mechanical deformation. The extreme work hardening of this steel is illustrated in Figure 3, which includes curves for two commercially available steels for comparison. The experimental steel exhibits much higher work hardening than a commercial unstable austenitic stainless steel (304), and a Hadfields manganese steel which is itself known as a highly work hardening material.

Having established that these four experimental steels had the required mechanical properties, a corrosion-abrasion test, simulating typical underground service conditions, was conducted. In this test, the steels were subjected to predetermined cycles of corrosion and abrasion[11]. It has been found from previous work that results from the test correlate well with actual wear measurements taken on underground shaker conveyors[12]. Results comparing the 'relative wear resistance' of the four development steels with commercially available steels are given in Table 3, where

$$\text{Rel. wear resistance} = \frac{\text{Volume loss of mild steel}}{\text{Volume loss of test steel}}$$

The relative wear resistance of 3.7 for AISI 431 represents the best result obtained for a proprietary alloy in this test. It was very encouraging, therefore, to find that the results for all four of the developed steels comfortably exceeded this value.

4. PILOT SCALE MELTS

As a first step towards commercial scale production, pilot scale melts of 250 kg of each of the four steels were produced. These melts provided material for a further series of tests which included hot workability, weldability, and castability.

4.1 Hot Workability Testing

Both hot tensile and hot torsional techniques were employed. Tensile testing covered the range 800 to 1 300 °C using both direct and indirect heating methods. Results were plotted as ductility (percentage reduction of area) against temperature. All four steels were found to have ductility curves similar to that of AISI 316.

Hot strengths were taken from the maximum torque observed during hot torsion testing. The hot strength of the 1210 steel was again deduced to be comparable with that of AISI 316, with the other three steels being similar to proprietary martensitic steels. A temperature of 1 250 °C was identified as a suitable soaking temperature for all four steels and the satisfactory results obtained at 900 °C suggested the viability of controlled rolling.

4.2 Weldability Testing

Weldability testing was performed on the 825 and 1210 steels, the purpose being to establish general welding procedures for the dual phase and mechanically transformable steels in plate form.

The test coupons were heat treated as follows:

1210 steel: as rolled, 1 h at 1 050 °C,
air cool
825 steel: as rolled, 12 h at 200 °C

All welds were of the constrained butt type. No pre-heat was used on the 1210 steel and interpass temperatures were kept below 100 °C. The 825 steel was preheated to between 100 and 150 °C. Soundness was assessed by X-radiography, magnetic particle and dye penetrant techniques. Weld quality was investigated using side, root and bend tests together with transverse tensile tests. Hardness profiles across the HAZ were plotted by testing at 2 mm intervals. Susceptibility to intergranular corrosion was estimated using 15 day immersion tests in mine waters according to ASTM A262 81.

General conclusions were that sound welds could be made readily in both alloys in thicknesses from 5 to 25 mm. The weld properties in the 1210 steel were similar to those obtained with commercial austenitic steels whereas those in alloy 825 were similar to those in proprietary martensitic steels.

Use of AISI 309 electrodes yielded satisfactory mechanical and corrosion properties in both steels.

4.3 Castability Testing

To determine the feasibility of producing castings with properties conforming to the basic requirements established for the development steels, a limited castability programme was initiated assessing both the suitability of the steels and the influence of section on mechanical properties.

The programme included evaluating the relative fluidity, hot tear resistance, shrinkage, unrestricted contraction, microstructure and mechanical properties.

The fluidities for the four alloys were found to be similar to that of Hadfields manganese steel, which itself is considered to be good. All the steels, with the exception of the 825, exhibited higher unrestrained contractions than those of medium carbon steels, leading one to expect an increased tendency for hot tearing. However, trials showed that the hot tear resistances of all four steels exceeded that of Hadfield's steel, which, in turn, is considered to be excellent. It was noted that gas porosity in the 1210 steel due to gas evolution could occur in sections greater than 75 mm.

The toughnesses of the four steels in cast form were found to be insensitive to ruling section up to 75 mm. The toughnesses of the three dual phase steels were, surprisingly, the same in cast form as in wrought form. The 1210 steel was not as tough in cast form as in the wrought state, but was still some three times tougher than the dual phase steels.

5. INDUSTRIAL MELTS

Following the satisfactory outcome of the tests on 250 kg melts the decision was taken to proceed to industrial melts in all four steels at the 15 t scale. The product range was planned to include plate of 5, 15 and 25 mm thickness, bar of 16 to 164 mm diameter and 3 mm diameter coil for wire drawing.

Each melt was cast into one 7.7 t slab and two 4 t square ingots, both types being bottom poured. Melting and refining was conventional EAF and AOD practice. For the 1210 steel, manganese was introduced as low carbon ferromanganese during the oxidation stage of AOD refining and as manganese metal during the reduction stage. Nitrogen was introduced by blowing after the oxidation stage.

The slab ingots were rolled to slab stock for subsequent secondary rolling to plate. One of the square ingots was rolled to billet for secondary rolling to bar while the other was directly rolled to the large rounds. Generally the rolling was successful and confirmed that the three dual phase steels behaved as standard martensitic steels, and that the mechanically transformable steel behaved as standard austenitic steels.

Rolling to the large rounds, secondary rolling to bar and wire drawing were again achieved using standard procedures and without serious problems.

In general, the experience in processing the 15 t casts was very encouraging and produced only two cautionary notes. The first applied to the 102A steel; some of the ingots, despite having undergone the normal treatment used for martensitic stainless steels after casting (that is, hot stripping from the mould, pit cooling, slow reheating and soaking prior to rolling), were found to have developed mild clink cracking. This problem was overcome by adopting an alternative procedure in which the ingots, after hot stripping, were then hot charged to the reheating furnace. The second point of caution related to the machining of the 1210 steel; as was to be expected, this steel was found to exhibit high rates of work hardening during the machining process. However, subsequent machinability testing has indicated that the material may be machined successfully by appropriate selection of machining conditions.

An extensive programme of characterization testing of the industrially produced material has now begun, and initial results have shown that the guideline specifications for hardness and toughness have been met. In fact, for the three dual phase steels, the mechanical properties have been found to be superior to those achieved with the pilot scale melts.

Corrosion tests, involving the measurement of free corrosion potential, corrosion current, and pitting potential, have been performed on samples of the four production materials, and the results are shown in Table 4. These data were obtained at a scan rate of 1 mV/s in fully aerated unstirred synthetic minewater at 30 ± 1 °C. The specimens were heat treated (1 050 °C for one hour, air cooled) and ground to P1200 on silicon carbide paper. The synthetic water was chosen as being representative of a particularly aggressive minewater, and conformed to the following specification:

pH	: 5.9
Total dissolved solids	: 7 370 mg/ℓ
Cl^-	: 1 940 mg/ℓ
$SO_4{}^{2-}$	: 1 700 mg/ℓ
$NO_3{}^-$	: 1 400 mg/ℓ

Examination of the corrosion currents presented in the Table confirms that the steels corrode at rates significantly lower than those for carbon steels. The free corrosion potentials of the steels indicate that they become corrosion resistant by passivation in this aggressive water. Consequently, pitting corrosion can occur, and the pitting potentials recorded are typical for steels of these compositions. Therefore, in consideration of applications in which these steels may be fully immersed, further work is being carried out to determine their pitting behaviour.

It is anticipated that the three dual phase steels, with their varying chromium contents, will find application over the full spectrum of conditions calling for moderate to good corrosion resistance. It is envisaged that the 825 steel, containing 8 per cent chromium, would be utilized in structural and wear applications requiring predominantly plate material, such as in the construction of rockhandling equipment. At the other end of the scale, the 122 steel would tend to be used where a higher degree of corrosion resistance is required, for example in wet conditions involving moderate levels of abrasive stress, and could be considered for the internal components of water powered hydraulic equipment. The 1210 steel is intended for corrosive applications where particular advantage can be taken of the special properties of this material. It could prove very attractive for making corrosion resistant high pressure piping for hydro-power or backfill slurry transportation systems. The high work-hardening rates experienced during the forming process could generate very high values of yield stress, thus allowing valuable reductions in wall thicknesses to be made.

The steel industry in South Africa has shown considerable interest in producing these steels commercially for mining applications. Since the steelmaking processes involved have been shown to be similar to those for conventional martensitic and austenitic stainless steels, any differences in steelmaking costs will result only from differences in the costs of the charge materials. On this basis, a preliminary cost analysis has shown that the production costs of the four development steels should be, in principle, no more than about 10 per cent above those of existing corrosion-resistant steels of similar chromium content, such as 3CR12 and AISI 420. It is likely, therefore, that these new steels will be highly cost-effective in corrosive-abrasive applications.

6. CONCLUSIONS

The two basic microstructural approaches adopted in the development of corrosion-abrasion resistant steels for gold mining equipment have proved to be successful. Four new steels have been developed, all of which have been shown to comply with important basic requirements. In particular, all four steels exhibit acceptable levels of corrosion resistance; the three dual phase steels have high hardness and toughness (471 to 583 H_V, 35 to 58 J CVN); the mechanically transformable steel has particularly high toughness (195 J CVN) and has demonstrated an extremely high work hardening rate.

Industrial scale production of these steels has been shown to be entirely feasible, without any special processing procedures being required. The four steels appear to be eminently suitable for a variety of applications in the corrosive and abrasive conditions in South African gold mines, and are expected to be relatively cheap to produce.

ACKNOWLEDGEMENT

The work described in this paper formed part of the research programme of the Chamber of Mines of South Africa.

REFERENCES

1. Cruise, W.J. The selection of materials for piping systems in gold mines. Journal of the South African Institute of Mining and Metallurgy, Vol. 85, No. 10, 1985, pp.361-365.
2. Noël, R.E.J. The abrasive-corrosive wear behaviour of metals, M.Sc. Thesis, University of Cape Town, 1981.
3. Forse, C.T. Solid particle erosion in hydraulic machinery, Ph.D. Thesis, University of Cape Town, 1984.
4. Heathcock, C.J. Cavitation erosion of materials, Ph.D. Thesis, University of Cape Town, 1980.
5. Barletta, A. An assessment of polymeric materials and surface treated steels as cavitation erosion resistant materials. M.Sc. Thesis, University of Cape Town, 1983.
6. Sarikaya, M., Steinberg, B.G. and Thomas, G. Optimization of Fe/Cr/C base structural steels for improved strength and toughness, Metallurgical Transactions, Vol. 13A, 1982, pp.2227-2237.
7. Peters, J.A. A microstructural approach to alloy design for superior corrosive-abrasive wear resistance. M.Sc. Thesis, University of the Witwatersrand, 1983.

8. Peters, J.A. Structure-property relationships of chromium-containing martensitic steels and their influence on abrasion and corrosion resistance. Ph.D. Thesis, University of the Witwatersrand, 1986.
9. Lenel, U.R. and Knott, B.R. Microstructure-composition relationships and M_s temperatures in Fe-Cr-Mn-N alloys. Metallurgical Transactions, Vol. 18A, 1987, pp.767-775.
10. Lenel, U.R. and Knott, B.R. Wear resistance of steels designed for use in severe abrasion-corrosion conditions. Wear of Metals, Houston, 1987.
11. Ball, A. and Böhm, H. The design and performance of steels in an abrasive-corrosive mining environment. Int. Conf. on Tribology: Friction, Lubrication and Wear - 50 Years On, Instn. Mech. Engrs., London, July 1987.
12. Ball, A. and Ward, J.J. An approach to materials selection for corrosive-abrasive wear by systematic in situ and laboratory testing procedures, Tribology International, Vol. 18, No. 6, December 1985, pp.347-352.

TABLE 1 COMPOSITIONS OF THE DEVELOPMENT STEELS

Name	DIN Designation	Composition					
		C	Cr	Ni	Mn	Al	N
825	X 25 CrNi 8.3	0.25	8	3			
102A	X 20 CrMnAl 10.1	0.20	10		1	0.5	
122	X 20 CrMn 12.1	0.20	12		1		
1210	X 5 CrMnN 12.10	<0.05	12		10		0.17

TABLE 2 MECHANICAL PROPERTIES OF THE DEVELOPMENT STEELS

Name	Heat Treatment	H_V	CVN (J)	YS (MPa)	UTS (MPa)	El (%)
825	Austenitized at 1 100 °C,	583	35	1 346	1 580	14
102A	oil quenched, tempered at	471	64	1 256	1 576	14
122	200 °C	513	58	1 387	1 565	13
1210	Austenitized at 1 050 °C, air cooled	229	195	361	1 178	44

TABLE 3 RESULTS OF CORROSION-ABRASION TESTS

Steel	Classification	Relative Wear Resistance
Mild steel	-	1.0
Bennox	Abrasion resistant steels	1.1
Wearalloy 400		1.1
A R COL 360		1.2
AISI 420	Corrosion resistant steels	1.9
AISI 410		2.0
AISI 316L		2.2
AISI 304L		2.3
3CR12		2.4
AISI 430		2.5
AISI 431		3.7
122	The newly-developed corrosion-abrasion resistant steels	4.0
102A		4.4
825		4.7
1210		5.0

TABLE 4 CORROSION PROPERTIES OF THE DEVELOPMENT STEELS

Name	Free Corrosion Potential V(SCE)	Corrosion Current 10^{-7} A/cm²	Pitting Potential V(SCE)
825	-0.22	5.6	0.15
102A	-0.22	5.7	0.28
122	-0.20	5.1	0.41
1210	-0.20	5.7	0.09
Mild Steel	-0.64	297.0	

a

b

1 DUAL PHASE MARTENSITE/AUSTENITE MICROSTRUCTURE

(a) Bright field micrograph of the general lath structure

(b) Dark field image of the inter-lath austenite

a

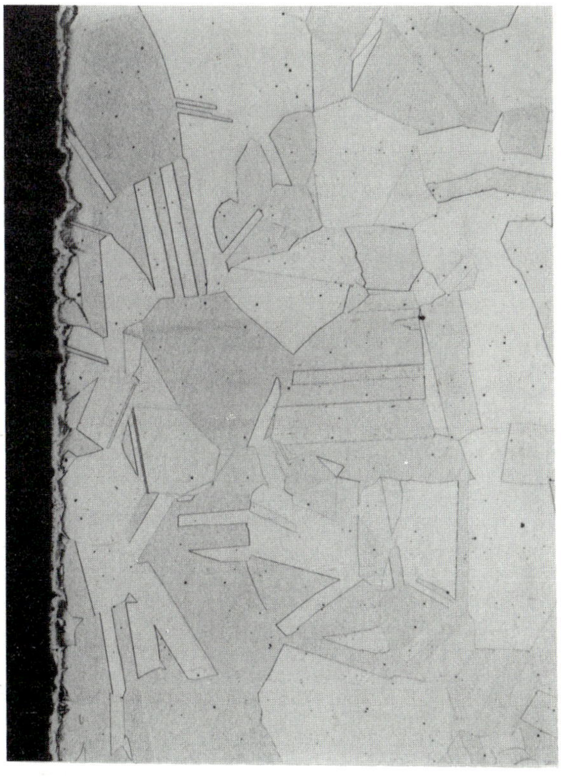

b

2 MECHANICALLY TRANSFORMABLE MICROSTRUCTURE

(a) Austenitic structure of an undeformed electropolished specimen

(b) Structure at the surface of a mechanically polished specimen, showing evidence of the transformation to martensite

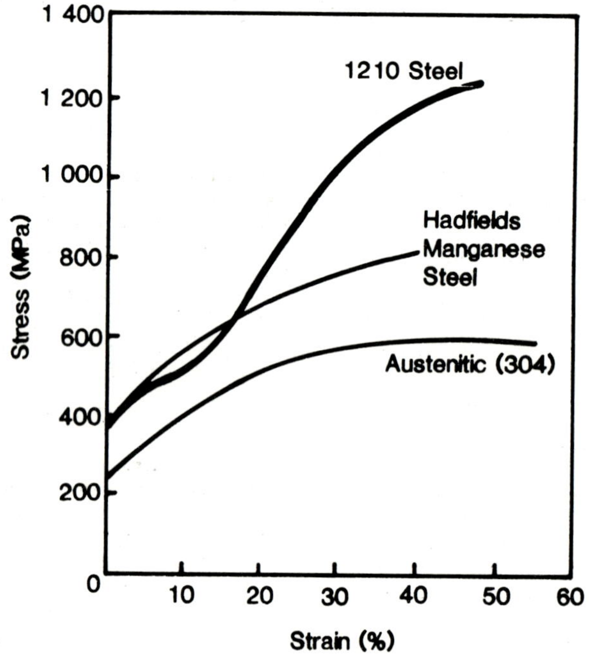

3 STRESS-STRAIN CURVE OF 1210 STEEL COMPARED WITH SCHEMATIC STRESS-STRAIN CURVES OF OTHER STEELS

Application of stainless steels for improved wear resistance and flow characteristics in bulk solids storage and handling plant

W T Cook, J Peace and J R Fletcher

WTC: BSC Swinden Laboratories
JP : BSC Teesside Laboratories
JRF: BSC Stainless

SYNOPSIS

Linings used in bulk solids handling plant are required to resist wear and promote easy flow of material over their surface. In commercial plant bulk solids often contain water, producing corrosive conditions, which can have an adverse effect on both wear and flow characteristics.

The benefits to be gained from stainless steels have been examined using both laboratory tests and evaluation in production plant handling coal, coke and ores. In commercial plant stainless steels can provide a greater life improvement over low alloy steels than may be indicated by laboratory tests. Additionally their ability to maintain a low friction, corrosion-free surface provides for continuous smooth flow, which is a significant advantage in many bunker and chute applications.

Austenitic, ferritic and martensitic stainless steels have each proved satisfactory in a variety of commercial installations and their performances are described. A new martensitic stainless steel, Hyflow 420R, has been developed to combine high hardness, corrosion resistance and good ductility at moderate cost. The properties of this steel and the experience gained in trial installations are discussed.

1. INTRODUCTION

Large quantities of wear resisting linings are used in plant handling bulk solids in the steel, coal, quarrying, chemical and power generation industries. In the UK alone, the steel industry handles well over 100 million tonnes per annum of materials including ore, sinter, coal and coke, whilst the coal industry produces a similar quantity of saleable product along with associated waste. Wear from the handling of bulk solids costs the UK millions of pounds per annum in replacement plant linings alone. Also, restrictions to free flow are often caused by the degradation of the surfaces of lining materials. These restrictions may result in additional high costs from down time and poor control of subsequent processes which depend upon accurate metering and flow.

The mechanisms causing wear include abrasion,erosion, impact and corrosion, either singly or in combination.

Surface wear by dry particles is commonly by abrasion where, for a wrought steel, the hardness level is generally the important factor governing performance. However, bulk solids are often handled in wet or damp conditions, where corrosion can have a very adverse effect on wear and also lead to poor flow characteristics. It is important, therefore, to examine the role of corrosion resistant materials in bulk solids handling.

This paper describes the benefits which have been obtained from the use of stainless steel linings and compares the results of laboratory and plant experience. The need for a durable, low cost, corrosion resistant material has been identified, resulting in the development of a new martensitic stainless steel variant, Hyflow 420R. This grade has been developed to combine corrosion resistance with the hardness level typical of many low alloy wear resisting steels.

2. COMPARATIVE WEAR AND CORROSION PERFORMANCE

Laboratory wear tests offer potential for material selection, but care must be taken in extrapolating the results to the performance under site conditions.

For example, a number of common wear resisting materials have been tested at Swinden Laboratories using a variety of wet and dry abrasives. A summary of the rank order of results obtained from selected tests is shown in Table 1. The ferrous materials tended to show a general effect of hardness on abrasion resistance. Tests with coke and sinter showed a life advantage from the use of a high hardness martensitic stainless steel over a standard, heat treated alloy steel grade. This was almost independent of whether the test was carried out wet or dry and probably relates directly to the hardness levels of the test material. The softer, austenitic stainless steel was at or near the bottom of the rank order, so that the results indicated no conclusive benefit for using stainless materials.

The relative performance of three major classes of steel, under laboratory and site conditions, is shown in Table 2. Although the nature of the abrasive was different in the two test conditions, the austenitic stainless steel showed a life improvement over carbon steel of about 50%, compared with its poorer performance in the laboratory test. The martensitic stainless steel showed about twice the life of the carbon steel during plant testing. The conditions in this plant involved impact and abrasion from waste products consisting mainly of -30 mm shale carried in water. The change in rank order, particularly in relation to the austenitic stainless steel, is considered to be due to the greater significance of corrosion to the wear process in operational plant.

In the laboratory tests the abrasive was introduced continuously, under a constant pressure, either dry or with the simultaneous introduction of water. Abrasion appeared to be the only significant degradation process operating and clearly, the test did not assess the contribution to performance from time dependent corrosion mechanisms.

The benefits from using stainless steel under operational conditions are further illustrated by data from the work of Allen et al[1,2] comparing both laboratory and plant test results. Under dry conditions, their laboratory tests showed little difference in relative life between low alloy and stainless steels, Table 3. Introducing separate periods of corrosion into their laboratory tests showed an advantage for stainless steel over a higher hardness low alloy steel. However, when tested in operational plant, the stainless steels showed an even greater advantage.

3. MAINTENANCE OF FLOW CHARACTERISTICS

In bunkers designed for mass flow, or in low angle chutes, it is necessary to ensure easy movement of a bulk solid over the liner surface from initial filling. This promotes uniform flow and eliminates segregation, arching and rat-holing, thus ensuring that the full storage volume is utilised with good metering at the outlet. This reduces down time and the costs associated with freeing blockages. Good metering also improves the control of subsequent processes. The flow characteristics of a charge over a lining surface depends in part on the initial surface finish, but also upon its ability to maintain a satisfactory surface with time.

Flow is promoted by a low wall/burden friction characteristic which must be maintained with time. Two low friction lining materials have performed well in British Steel Corporation plant. These are Ultra-high Molecular Weight Polyethylene (UHMWP) and stainless steel and both have been successful in overcoming sticking problems. Unfortunately, UHMWP cannot withstand abrasion from coke and similar abrasive media, whilst stainless steel has been found to be substantially more durable when handling similar materials.

Laboratory tests have been carried out to determine the friction angles associated with steels and a number of other common lining materials, when handling a range of wet coals. The results are summarised in Table 4 and show:-

(a) a friction angle comparable to or lower than that of UHMWP can be obtained on stainless steel.

(b) the friction angle on a stainless steel depends on the surface finish

(c) surface finishes produced by rolling resulted in similar friction angles in all directions. Mechanical abraded finishes are directional, giving improved flow in the direction of polishing.

(d) low friction angles, similar to those produced by cold rolling, can be produced on thicker plate by mechanical polishing.

The friction angle shown for the carbon steel in Table 4 is related to a rusty surface at the start of the test.However, corrosion would be expected to increase the friction angle of a polished surface and with a static charge, this could result in sticking.Tests have been carried out to assess the effect of a 24 h dwell period on the friction angle associated with fine wet coal in contact with polished surfaces of carbon and stainless steels. The results are shown in Table 5. All the friction angles were determined in the direction of polishing. The friction angle for the mild steel increased significantly after 24 h contact time, whilst that for the stainless steel remained constant and low. As illustrated in Fig. 1, superficial pitting had taken place on the carbon steel, whilst the appearance of the stainless steel remained unchanged. Clearly, longer dwell times would allow for increased corrosion and adhesion with a non-stainless steel.

The effect of corrosion is significant in bunkers, where conditions may be static for long periods. The use of stainless steel liners minimises corrosion and permits immediate operation when required.

4. STAINLESS STEEL FOR LINING

The choice of a lining material depends on the properties of the bulk solid, the design of the unit, its method of operation and the cost of the lining. Consideration needs to be given to whether free flow is important and whether wear resistance is an additional requirement. Other factors to take into account include corrosion, method of fixing, temperature and the life required.

Unlike polymeric linings, stainless steels can operate satisfactorily at elevated temperatures. They can be installed by mechanical fixings and, in some locations, stainless steels have been fixed using stud welding onto the back of the lining followed by bolting. This preferred method of installation provides a smooth surface and allows the full thickness of the lining to be used.

4.1 Composite Stainless Steel/Rubber Linings

Stainless steel linings have replaced other materials in a number of BSC installations and in two instances, this was due to high wear rates with the use of UHMWP3. Stainless steels have been found to offer flow advantages when handling wet cohesive coal, whilst at the same time being durable enough to withstand abrasion by coke. A further development has been to use a composite of stainless steel bonded onto a thick resilient rubber backing. This allows a reduction in the thickness of the stainless steel used and an ability to reduce operating noise. The rubber also compensates for slight irregularities in the structural wall of a bunker permitting simplified installation. The composite may be fixed by bolting in a similar manner to UHMWP or by stud welding and bolting. A typical installation is shown in Fig. 2.

4.2 A New Martensitic Stainless Steel Variant, Hyflow 420R

From the laboratory work and site evaluations, the potential for combining corrosion resistance with a high hardness in a martensitic stainless steel, appeared to offer attractive benefits, especially if a low cost process route could be developed.

Accordingly, BSC Stainless and research personnel have developed a martensitic stainless steel - HYFLOW 420R, optimised in composition to achieve high hardness combined with low cost processing. Containing approximately 0.14% C and 12% Cr, the steel has sufficient hardenability to give typical hardness levels of 400/450 HB in sections up to 40 mm or greater in the as-rolled condition.

Whilst producing a hardness typical of, or better than, that of many low alloy steel wear plate grades, the carbon content is sufficiently low to ensure good ductility in the as-rolled condition. The typical mechanical properties of Hyflow 420R are shown in Table 6, where they are compared with those of other stainless steels.

5. PLANT INSTALLATIONS

Stainless steels have been installed and their performance monitored in a number of commercial installations handling ores, coal, coke, sinter and coal waste. In many cases, the objective was to achieve free flow and examples are summarised in Table 7. Selected installations are described in more detail below.

5.1 Feed Chute to a Coal Crusher (Installation 5, Table 7)

When lined with abrasion resisting tiles, poor flow characteristics led to fines sticking to the liner surface and an inconsistent feed into the crusher. After lining with Type 316 austenitic stainless steel, the flow improved to a satisfactory level, but the lining life was short due to heavy impact damage producing local deformation and eventual holing. Replacement with a Type 420 (0.22% C.13% Cr) in the softened condition (∼200 HV) continued to give good flow with a life 2 to 3 times that of Type 316, although failure again resulted from heavy impact. The lining has been replaced with hardened Type 420 (∼450 HV) and continues to operate, with an estimated life at least 30% greater than that of softened Type 420.

5.2 Feeder Pans for Iron Ore Pellets (Installation 6, Table 7)

When lined with Nihard tiles, the pans gave severe flow problems caused by adhesion of broken pellets and dust. The build up can be seen on Fig. 3(a).

This produced an inconsistent output and problems further down stream. After lining with Type 304 with a dull polished finish. flow problems were eliminated and the output was uniform, Fig. 3(b).

5.3 Wharf Discharge Hoppers (Installation 7, Table 7)

These hoppers at a large marine ore terminal receive the discharge from bulk cargo carriers, including coal, coke, iron ore and various other solids used in ironmaking. The cargoes are removed from the ship, by grabs holding up to 20 tonnes, with loads dropped from a height of about 10 m into the hoppers which in turn meter the delivery of materials onto conveyor belts. The original rubber linings gave major flow problems when handling wet coal resulting in unloading delays and excessive demurrage charges.

Satisfactory flow was obtained by lining the bunkers with UHMWP, but this gave very high wear rates especially when handling coke.

The bunkers were relined with Type 304 12 mm plate, compared with 48 mm of UHMWP. This continued to give good flow and handled a slightly greater total quantity of material than the UHMWP, including a much greater quantity of coke, before a number of plates near the bottom edge had to be replaced.

The bunkers have been relined recently with 12 mm thick Hyflow 420R and the as-installed lining is shown in Fig. 4. These linings continue to operate and monitoring of thickness loss indicates a life expectancy of about 1.5 times that of Type 304.

The Type 304 lining was installed with a dull polished finish to ensure free flow and the original surface is shown in Fig. 5(a). After removal of the worn plate, the surface was re-examined on a section about 1 mm thick, Fig. 5(b). The change in surface appearance was small and the abrasion from the charge had maintained a directionality pattern similar to that on the as-polished surface.

The Hyflow 420R was installed as plasma cut shapes with no preparation of the as-rolled surface. This was quickly polished by early abrasive charges and showed good flow. Both the Type 304 and Hyflow 420R plates were fixed by stud welding to the back face and bolting to the support structure.

5.4 Waste Discharge Chute in a Coal Preparation Plant

Various steels were introduced consecutively into the same position on a line handling shale waste4. Caneloid shale, falling through a height of about 2 m, impacted a panel at an angle of 45° changing the direction of flow through 90° onto a conveyor. The shale was carried in copious water and the panel was subject to impact, abrasion and corrosion. The particles were generally less than 30 mm ∦, although much larger pieces were occasionally entrained.

The relative lives obtained from the various steels are shown in Table 8. The stainless steels all showed an improved performance when compared with the carbon steels with the martensitic stainless steel showing the best result. The surface appearance of the Hyflow 420R, after removal from the test site, is shown in Fig. 6.

Based on the relative life data, Table 8 also includes a relative cost of the various liner materials. Where Hyflow 420R can be installed in the form of as-rolled plate, the relative cost of this steel compares favourably with the other wrought ferrous materials. These figures do not include the savings on the costs of installation and maintenance, resulting from use of a more durable grade.

6. DISCUSSION

Under dry conditions, the abrasion resistance of a steel is closely related to hardness and stainless steels offer no advantage over carbon steel grades. In damp conditions, the benefits from stainless steel for bulk solids handling increase, due to the greater contribution of corrosion to the wear process. Stainless steels offer special benefits when the maintenance of free flow is important.

The advantages relate to the increased corrosion resistance, the durability of the surface layer and the interaction of the surface with the abrasive particles. The protective oxide layer formed on a stainless steel has a similar volume to that of the parent metal, is very thin, strongly bonded to the surface and reforms very quickly if removed by abrasion. In contrast, the rust formed on a low alloy steel can grow to a significant thickness and is only weakly attached to the parent metal.

In the presence of abrasive particles, the corrosion product on a low alloy steel will be removed readily, exposing fresh surface and allowing further corrosion to proceed rapidly. Even where particulate matter may be effectively separated from the surface by a lubricative film such as water, the nature of the corrosion product on a low alloy steel may disrupt the interface more than that of the smooth, continuous film formed on stainless steels. Furthermore, under conditions of intermittent disruption of abrasive flow, a low alloy steel will continue to degrade by corrosion alone whilst there would be no effect on a stainless steel.

When free flow is important, a small amount of corrosion can increase surface friction and prevent easy flow. Should a charge stand on the surface of a non-stainless steel, the corrosion product can grow into the charge causing sticking in addition to surface roughening. This has not been observed in the site tests with stainless steel liners.

Whilst benefits can be obtained from a stainless steel, the choice of grade requires consideration of the operating conditions, life/performance requirements and cost relative to other candidate materials. Also, choice may be dictated by availability, forming or fixing considerations. For many applications, where the environment involves water, the improvement in corrosion resistance offered by a 12% Cr martensitic steel over a low alloy steel has been found to be adequate for wear resistance and flow promotion. The higher corrosion resistance offered by ferritic or austenitic steels may be unnecessary and the higher hardness and lower installed cost of a martensitic stainless steel may make it the optimum choice.

7. CONCLUSIONS

1. It has been shown that there are benefits to be gained from the use of stainless steels in a number of bulk solids handling applications and that consideration should be given to their use for both wear resistance and flow promotion.
2. Corrosion can result in significant material loss in low alloy steels and the benefits of stainless steels increase as the nature of the charge becomes more corrosive.
3. Where corrosion is part of the surface loss mechanism, the benefits obtained from using stainless steel in operational plant can be greater than indicated by accelerated laboratory tests.
4. Austenitic and ferritic stainless steels have been shown to offer advantages in bunkers and chutes, improving life and flow. However, the use of a martensitic grade of stainless steel, combining hardness and corrosion resistance may offer new opportunities, with a reduction in long term costs.

8. ACKNOWLEDGEMENTS

The authors wish to acknowledge the support of colleagues in BSC's Swinden Laboratories and Grangetown Laboratories and in various works for their contributions to this work. Thanks are due to BSC Stainless and to Dr. R. Baker, Director of Research BSC, for permission to publish this paper.

9. REFERENCES

1. Allen, C., Ball, A., Protheroe, B.I., The Selection of Abrasion/Corrosion Resistant Materials for Gold Mining Equipment.J. South African Inst. of Mining and Met. Oct. 1981, p 289.
2. Allen, C., Ball., Protheroe, B.I., The Abrasive/Corrosive Wear of Stainless Steels International Symposium on Wear of Materials San Francisco 1981, March 30-April 1.
3. Peace, J., Material Flow Improvements in Hoppers, Bunkers and Silos Using Low Friction Lining Materials - a British Steel View. Concrete Society Conference 'Design' Construction and Maintenance of Concrete Storage Structures' Newcastle upon Tyne.
4. Moreton, G., Yeardley, D.E., The Evaluation of Wear Resisting Metallic Materials for Coal Mining Equipment I.Mech.E.1984 C348/84 p 243.

TABLE 1
RELATIVE WEAR RESISTANCE OF VARIOUS MATERIALS TESTED UNDER CONTINUOUS ABRASION - BOTH WET AND DRY
(Results in each column show life relative to 0.4% C, Steel - CS1)

Relative Life Bracket	ABRASIVE							
	Coke		Sinter		Sand		Coal	
	Wet	Dry	Wet	Dry	Wet	Dry	Wet	Dry
>100		AL HCC1 HCC2	-	-	-	-	-	-
50-100	AL			HCC2				
10-50	HCC1 HCC2		HCC1 HCC2 AL	AL HCC1				
5-10		MSS1				HCC2		
3-5	MSS1 MSS2	MSS2 AS			HCC2 MSS2	AL	MSS1 MSS2	
2-3	AS		MSS1	MSS1	HCC1	HCC1		MSS1
1.5-2	CB	CB	MSS2 AS	AS MSS2		MSS2 MSS1 MSS3		MSS2
1- 1.5			MSS3		MSS1 MSS3 AL	AS	MSS3 ASS	
1	CS1 MSS3	CS1	CS1 ASS	CS1 CS2 MSS3	CS1 ASS	CS1 ASS CS2	CS1	CS1
0.5 - 1	CS2 ASS	MSS3 CS2	CB	ASS			CB	MSS
<0.5		ASS		CB	CB	CB	G UHMWP	

Material	Code	Material	Code
Alumina - 1500 HV	AL	Alloy steel	AS
High Cr Cast Iron Tile - 640 HV	HCC1	Carbon steel)200 HV)140 HV	CS1 CS2
High Cr Cast Weld Deposit - 580 HV	HCC2	Austenitic stainless steel - 180 HV	ASS
Martensitic) 600 HV Stainless) 450 HV Steel) 200 HV	MSS1 MSS2 MSS3	Cast Basalt	CB
		Glass	G
		Ultra high molecular weight polyethylene	UHMWP

TABLE 2
COMPARISON OF LABORATORY AND PLANT DATA FOR THREE STEELS

Steel	Relative Life (Carbon Steel = 1)	
	Laboratory Test*	Plant Test**
Carbon Steel (350 HV)	1	1
Austenitic Stainless Steel (Type 304)	<1	1.5
Martensitic Stainless Steel (450 HV)	1.6	2

* Abrasive medium, 80% coal, 20% sand (wet)

** Abrasive medium, discard from coal preparation plant (rock/shale/copious water)

TABLE 3
EFFECT OF CORROSION AND ABRASION ON THE RELATIVE PERFORMANCE OF STAINLESS AND LOW ALLOY STEELS (ALLEN et al)

Steel	Hardness	Relative Life (Mild Steel = 1)			
		Laboratory Tests*			Plant Performance**
		Abrasive Only	24 h *** Corrosion + Abrasion	48 h *** Corrosion + Abrasion	
316	190	1.5	2.5	3.1	3.9
304	180	1.9	3.3	3.7	3.5
430	160	1.3	2.0	2.0	4.2
3Cr12 (12% Cr-Ti)	165	1.2	2.1	2.3	3.1
0.2% C 0.6% Mo	400	1.5	1.3	1.2	1.2
0.3% C 1% Cr 0.2% Mo	400	1.6	1.5	1.4	1.0
Mild steel	130	1.0	1.0	1.0	1.0

* Abrasive media - alumina
Corrosive media - synthetic mine water, ph 6.5, temperature 30°C

** Abrasive media - quartzitic rocks and mine water

*** Separate periods of corrosion followed by abrasion

TABLE 4
FRICTION ANGLES FOR VARIOUS LINER MATERIALS (INSTANTANEOUS)

Material	Test Direction	Surface Roughness CLA, μm	Friction Angle (Degrees)
Concrete	All	24	26
Rubber	All	-	27
Carbon Steel (Rusty)	All	16	26
UHMWPE	All	3.5	23
Glass	All	0.005	13
Stainless Steel (Type 304)			
HRS & D	All	5	24
2B	All	0.2/0.8	18/20
2A	All	0.03	15
Dull Polish	Transverse	1.8	23
	Longitudinal	0.9	16

HRS &D - Hot Rolled Softened and Descaled
2B - Cold Rolled Finish (2 mm to 4 mm thick)
2A - Bright Annealed
Dull Polish - Mechanically Abraded
(longitudinal surface finish difficult to measure, the recorded value of 0.9 μm CLA may be regarded as a maximum)

Friction angle determined using Jenike Friction Dist Test.

Results are the average of tests on various fine coals containing an average of 12% moisture.

TABLE 5
EFFECT OF RESIDENCE TIME (CORROSION) ON THE SURFACE CHARACTERISTICS OF STAINLESS STEEL AND MILD STEEL
(Fine Wet Coal Against Steel Surface)

Material	Surface Finish	Surface Friction Angle (Degrees)		Comment
		Instantaneous	After 24 h in Contact	
Mild Steel	180 Grit	17	22	Pitted after 24 h in contact
Mild Steel	Hot Rolled and Descaled	16	21	Pitted after 24 h in contact
Stainless Steel	180 Grit	17	15	No change to surface appearance
Stainless Steel	2B	18	18	No change to surface appearance

TABLE 6
TYPICAL MECHANICAL PROPERTIES OF THE STAINLESS STEELS DISCUSSED (AMBIENT)

Steel	TS N/mm2	0.2% PS N/mm2	% El.	2 mm V-Notch Impact Strength J		HV
				L	T	
Type 304	600	270	57	230	170	160
Type 409 (Sheet)	440	270	33	-	-	150
Hyflow 420R (Hot Rolled Plate)	1520	1050	16	55	16	450

TABLE 7
CASE HISTORIES (SUMMARY)

Installation	Previous Experience			Experience After Stainless Steel Installed		Comments
	Material	Problem	Life	Material	Benefit	
1) 4 x 300 tonne coal Blending Bunkers (Coal size - 25 mm ∦)	Mild steel	Arching, Rat-holing Blockage Poor Blending	Satisfactory (Core flow)	304-2B (1) 409-2B (1)	Free Flow Achieved Good Blending	409 showed very light staining, 304 remained bright. First bunker lined showed no evidence of wear after handling 150 Kt. Now handled 440 Kt and still satisfactory. Other bunkers have each handled 85 Kt.
2) Road Reception Hopper - Coal Preparation Plant				304-2B (2)	Free Flow	No measurable wear after handling 205 Kt of -20 mm ∦ coal. Surface polished to mirror finish.
3) Rail Reception Hopper - Coal Preparation Plant	Glass Tiles	Impact damage 10 mm below Entry	Low	6 mm 304 to Impact Point 3 mm 304 above	No Impact Damage & Free Flow	Handled 2.25 million t - no measurable wear.
4) Coke Oven Charge Car Hoppers (12 off) (Coal size 85% - 3 mm ∦)				6 mm 304 Dull Polish	Consistent Free Flow	Total throughput to date - 4 million t - less than 0.5 mm of wear.
5) Inlet Chutes to Coal Crusher (2 off)	Abrasion Resisting Tiles	Poor flow		3.4 mm 316 4.2 mm 420 (200 HV) (3) 2.9 mm 420 (450 HV) (3)	Free Flow	Replaced after handling 150 Kt) Replaced after handling 440 Kt) Two chutes showed) similar results Still running after 390 Kt) (projected life 600 K))
6) Feeder Pans from Iron Pellet Bunker	Ni-Hard Tiles	Poor Flow Fines Build Up Production Stoppages		6.3 mm 302 Dull Polished	Free Flow	Continuous even feed since introduction of stainless steel - now extended to 50 pans.
7) Wharf Discharge Hoppers, 2 off, 50T	UHMWP		48 mm UHMWP $\approx$ 12 mm of Type 304	12 mm 304 Dull Polish 12 mm Hyflow 420R (450 HV) (4)	Free Flow & Wear Resistance Free Flow	No. 1 Bunker - partial replacement after 2.1 million t No. 2 Bunker - partial replacement after 2.75 million t Wear mainly on bottom edge of converging walls. To date bunkers have handled 1.1 and 1.5 million t, projected life 1.5 x Type 304

1) - 2 mm thick S/SS vulcanised to 8 mm of rubber
2) - 3.2 mm thick S/SS vulcanised to 8 mm of rubber
3) -Installed with heat treated surface) Polished by charge
4) - Installed with as-rolled surface)

TABLE 8
RELATIVE LIFE DATA FOR VARIOUS MATERIALS SUBJECT TO IMPACT, ABRASION AND CORROSION IN OPERATIONAL PLANT
(HANDLING WASTE FROM COAL PREPARATION PLANT)

Material	Relative Life	Relative Price/Tonne of Steel	Relative Cost over Service Life
Mild steel	1	1	1
Hardened Low Alloy Steel (350 HV)	2	1.7	0.8
Ferritic Stainless Steel (Type 409)	2.5	2.3	0.9
Austenitic Stainless Steel (Type 304)	3	3	1
Martensitic Stainless Steel (Type 420, 0.15% C hardness 450 HV)	4	2.3	0.6

Cost figures based on relative price and life only (Mild steel = 1)

(a) Stainless steel after 24 h dwell time

No change from as prepared surface

Carbon steel as polished showed similar surface

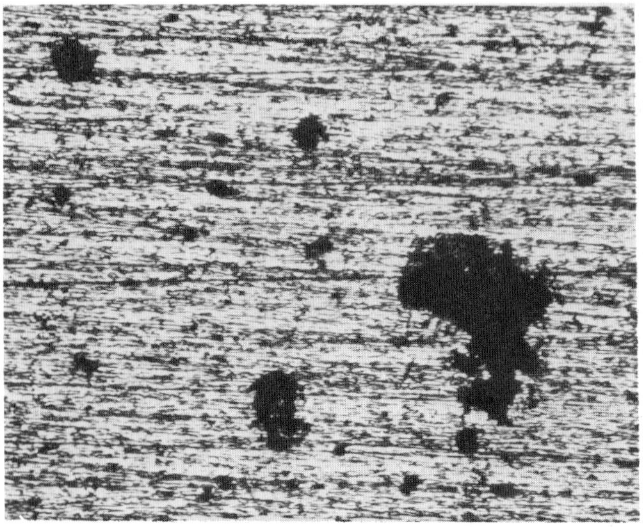

(b) Carbon steel after 24 h dwell time, showing corrosion

FIG. 1

EFFECT OF DWELL TIME IN CONTACT WITH WET COAL (12% WATER) ON THE SURFACE OF CARBON STEEL AND TYPE 420 STAINLESS STEEL

FIG. 2

COAL BUNKER WITH COMPOSITE STAINLESS STEEL (TYPE 304)/RUBBER LINING

(a) Lined with Ni-hard tiles.
Note sticking of fines

(b) Lined with Type 304 Stainless steel
Note: Homogeneous flow

FIG. 3

FEED PANS FOR IRON PELLETS

FIG. 4

WHARF DISCHARGE BUNKER - NEW HYFLOW 420R LINING

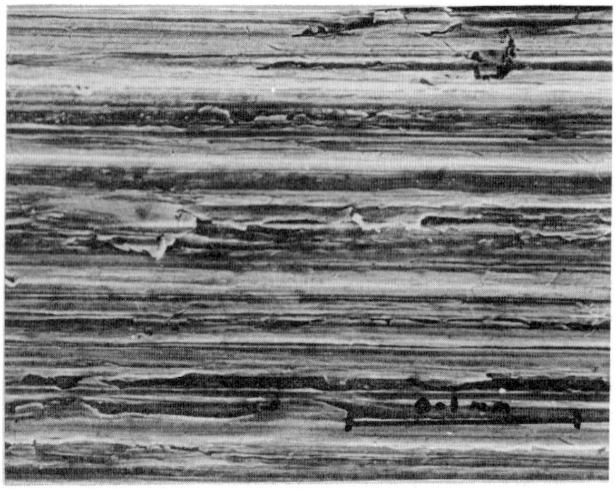

(a) As polished surface - 12 mm plate No. 4 finish

FIG. 6 ▲

SURFACE OF HYFLOW 420R AFTER SERVICE IN A WASTE DISCARD CHUTE FROM A COAL PREPARATION PLANT
(AREA WORN TO 6 mm THICK, INSTALLED AS 12 mm PLATE HARDNESS 450 HV)

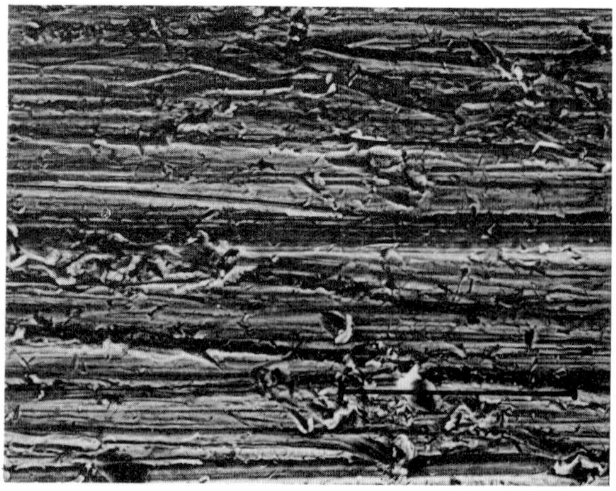

(b) Region of plate worn to 1 mm in thickness

◄

FIG. 5

SURFACE OF TYPE 304 PLATE IN WHARF DISCHARGE BUNKER

□

Application of as-quenched low-carbon martensitic stainless steels to wear-resisting use

K Yoshioka, K Miura and S Suzuki

The authors are in the Technical Research Division of the Kawasaki Steel Corporation.

SYNOPSIS

Medium-carbon martensitic stainless steels widely used for the manufacture of wear-resisting parts have certain shortcomings; the indispensability of quenching and tempering to get suitable hardness, and the deterioration of corrosion resistance due to tempering.

With a view to obtaining a new stainless steel which can derive specified hardness readily from quenching, a development guideline was set up to fully enlarge an austenitic temperature range by the addition of Mn, and to control the hardness of martensite by the amount of (C + N).

Thus an examination was made of the effects of Mn and (C + N) contents on the properties of a 13Cr quenched steel.

Having high Mn and controlled (C + N) contents, the steel can readily obtain the hardness necessary for the wear-resisting parts by merely quenching, without strict control of the heat-treatment conditions, and it is also superior in toughness and corrosion resistance to conventional steels. Wear resistance is simply in proportion to hardness.

The steel is suitable for wide use in a field where existing medium-carbon martensitic stainless steels are used after quenching and tempering processes.

INTRODUCTION

Medium-carbon hot-rolled martensitic stainless steel plates, represented by types 420 and 429, have been used as a material for wear-resisting use that requires good corrosion resistance and wear resistance.

In these steels, however, an extremely high dependency of hardness on quenching temperature necessitates tempering, causing not only cumbersome processing and added costs, but also the deterioration of corrosion resistance as a chromium-depleted zone is formed around chromium carbonitride as a side-effect of the tempering.[1-4]

In martensitic stainless steels generally, the microstructure is very sensitive to heat-treatment conditions and chemical composition; the amount of austenitic phase at high temperature becomes larger with an increase in temperature up to 1100°C.[5] The austenite at high temperature is transformed into martensite during cooling, and so the quenched-in hardness is determined by the hardness and the amount of martensite formed.

With a view to obtaining specified hardness at a stable level readily from quenching in a wide temperature range, a development guideline was set up to fully enlarge an austenitic temperature range by the addition of Mn,[5,6] a low-cost austenite-former, and to control the hardness of martensite formed by subsequent quenching by the amount of (C + N). Thus an examination was made of the effect of (C + N) amount and Mn addition on the properties of a 13Cr quenched steel.

EXPERIMENTAL PROCEDURE

The specimens used were 30kg ingots produced by vacuum-induction melting; their chemical compositions are shown in Table 1. They were basically 12.5%Cr–(0.01–0.015)%N steels containing two levels of Mn content, 1% and 1.5%, and C content ranges from 0.045–0.081%. These ingots were hot-rolled into 6 mm-thick plates and heat-treated at 725°C for 30 min. The plates were heat-treated in an electric furnace at 900–1050°C for 10 min followed by quenching at a cooling rate of 30°C/s, or heat-treated by induction heating at 900–1050°C for 15 s followed by quenching at a cooling rate of 5–30°C/s. These heat-treated specimens were subjected to hardness measurement, microstructure observation, Charpy impact test, salt-spray test and Ogoshi-type abrasion test.[7]

The conventional steels in 6 mm thickness; types 410, 420 and 429, were also examined for comparison.

TEST RESULTS AND DISCUSSION

Effects of (C + N) content and quenching temperature on quenched-in hardness

The effects of (C + N) content and quenching temperature on hardness are shown in Fig. 1. In 10 min holding at high temperature, the quenched-in hardness increases with higher quenching temperature, and becomes a constant value which increases with (C + N) content, within a quenching temperature range of 900–1050°C regardless of Mn content. In a short heating such as a 15 s holding, on the other hand, the increase in Mn content lowers the critical quenching temperature at which quenched-in hardness begins to become constant, enlarging the quenching temperature range to obtain a constant quenched-in hardness.

Figure 2 shows the relation between (C + N) content and hardness obtained by quenching in a temperature range of 925–1050°C. In the steel containing 1%Mn, the quenched-in hardness suddenly decreases when the (C + N) content is as low as 600 ppm, while the quenched-in hardness of 1.5%Mn-containing steels is linearly proportional to (C + N) content. The microstructures as quenched from 1000°C are shown in Photo 1. The microstructure of the 12.5%Cr–1%Mn steels containing a low (C + N) content exhibits a duplex structure composed of ferrite and martensite. Thus the deviation from a linearly proportional relationship between the (C + N) content and the quenched-in hardness of 12.5%Cr–1%Mn steels is considered to be caused by the forming of ferrite. In 12.5%Cr–1.5%Mn steels, on the other hand, a fully martensitic structure can be obtained even in a low (C + N) content.

The dependency of the hardness of 12.5%Cr–1.5%Mn–0.07%(C + N) steel on quenching temperature is shown in Fig. 3, in comparison with those of the conventional steels. In the medium-carbon martensitic stainless steels such as types 429 and 420, the dependency is extremely great so that the required hardness is difficult to obtain merely by quenching, while the hardness of type 410 is too low. In a 12.5%Cr–1.5%Mn–0.07%(C + N) steel, the quenched-in hardness remains constant within the requirement (for example 35 ± 3HRC) by quenching in a wide temperature range of 900–1100°C.

Figure 4 shows the relation between the quenched-in hardness of 12.5%Cr–1.5%Mn–0.06% (C + N) steel and the heat-treating condition using induction heating followed by cooling at 30°C/s. The heat treatment below 900°C or short-time heat treatment such as within 5 s at 925°C results in a lower quenched-in hardness than required because of the remaining coarse ferrite. The longer and higher temperature heating, such as 925°C or over for 10 s, leads to a constant hardness within the required range, and to a fully martensitic structure.

Effect of cooling rate on quenched-in hardness

The relation between hardness and the cooling rate after heat treatment at 950°C is shown in Fig. 5.

In general, hardness has a tendency to decrease with a decrease in cooling rate, which is especially noticeable in 12.5%Cr–1%Mn steels. In 12.5%Cr–1.5%Mn steels, the tendency is minimal above a cooling rate of 10°C/s. Their optical microstructure, however, revealed no structural changes due to the differences in Mn content and cooling rate. So, the effectiveness of Mn addition for hardenability is thought to relate to the retardation for the precipitation of sub-microscopic chromium carbonitride at a temperature range of 600 to 700°C during cooling.

Thermal stability of quenched-in hardness

Figure 6 shows the isochronal hardness curves of 12.5%Cr–1.5%Mn steels quenched at a cooling rate of 30°C/s after heat treatment at 950°C for 10 min. The hardness remains nearly constant up to 500°C, although a small decrease is caused by the precipitation of cementite at 200–300°C. In tempering above 500°C, a marked decrease in hardness occurs due to the precipitation of M_7C_3.

Wear resistance

Figure 7 shows the relation between the hardness of plates and wear resistance determined by the Ogoshi-type abrasion test, in comparison with those of types 420 and 429. Wear resistance is in proportion to hardness regardless of the type of steel. Therefore, the wear resistance of martensitic stainless steels is determined only by the hardness, with no relation to (C + N) content.

Mechanical properties

Figure 8 shows the relation between the Charpy impact value at 20°C and the quenching temperature in comparison with that of type 420 whose hardness is controlled to 33 HRC by tempering after quenching. The low (C + N) steels are superior in toughness to type 420, although toughness decreases with an increase in quenching temperature and (C + N) content. The degradation in toughness with an increase in quenching temperature is thought to be caused by the coarsening of the martensitic structure. A short induction heating even at high temperature, which has often been used in the actual manufacturing process of wear-resisting parts, does not seem to cause the degradation in toughness.

Table 2 shows the mechanical properties of a 12.5%Cr–1.5%Mn–0.07%(C + N) steel plate quenched at a cooling rate of 30°C/s after heat treatment at 950°C for 10 min, in comparison with those of type 429 steel plate with hardness of 35 HRC controlled by tempering after quenching. The ductility of 12.5%Cr–1.5%Mn–0.07%(C + N) steel is superior to that of type 429.

Rust resistance

A salt-spray test was carried out for 12.5%Cr–1%Mn and 12.5%Cr–1.5%Mn steels plates quenched at a cooling rate of 30°C/s after heat treatment at 900–1000°C for 10 min. Rusting was not observed in either steel regardless of (C + N) contents and quenching temperature. The appearance of salt-spray test specimens of 12.5%Cr–1.5%Mn steel plates quenched from 950°C is shown in Photo 2.

APPLICATION TO BRAKE DISK OF MOTORCYCLE

Application of the low-carbon martensitic stainless steel to wear-resisting use is introduced using an example of a motorcycle brake disk, shown in Photo 3.[8]

Existing medium-carbon martensitic stainless steels have been used as a material for the disk which requires good corrosion resistance and wear resistance. In the manufacturing process of the disk, the plate is subjected to press-quenching with a water-cooled die after a short electro-magnetic induction heating at high temperatures and tempering after quenching. The higher the hardness, the higher the wear resistance of the disk, but when it is too hard a disk tends to screech on operation. Therefore, the hardness of the brake disk is controlled at between 32 and 38 HRC, though with some variation between motorcycle manufacturers.

As shown in Fig. 4, the low-carbon martensitic stainless steel with high manganese –

12.5%Cr–1.5%Mn–0.07%(C + N) steel – can readily obtain the necessary hardness only by quenching over a wide range of high temperature, illustrated as the hatched region in the figure, although the induction heating in the actual manufacturing process is a monotonous heating.

Photo 4 shows the specimens after the salt-spray test, taken from the brake disks with hardness of 35 HRC. The disk of the low-carbon martensitic stainless steel produced only by quenching exhibits good rust-resistance, while that of type 420 is extremely inferior due to tempering after quenching.

CONCLUSIONS

Medium-carbon martensitic stainless steels have been used as a material for wear-resisting use. These steels, however, have some shortcomings in the heat treatment of the disk, such as the indispensability of quenching and tempering, and deterioration of corrosion resistance due to tempering. The addition of Mn to low-carbon martensitic stainless steels and the control of their (C + N) content can solve the problem as described above; addition of Mn enlarges the temperature range at which fully austenitic structure exists at quenching temperature, and the control of (C + N) content realizes a suitable hardness of martensite formed on quenching. The steels can readily obtain the hardness needed for wear-resisting parts by only quenching, without strict control of the conditions for heat treatment, and it is superior in toughness and corrosion resistance to conventional steels.

The steel is suitable for wide use in a field where existing medium-carbon martensitic stainless steels are used after the quenching and tempering processes, such as a motorcycle brake disk and cutlery, etc.

REFERENCES

1 F. K. Bloom: Corrosion 9 (1953), p 56
2 M. Marchinkowski and A. Szirmae: Trans. Met. Soc. AIME 230 (1964), p 676
3 A. Baumel: Arch. Eisenhüttenwes., 34 (1963), p 135
4 G. Herbsleb: Werkst. Korros, 19 (1968), p 204
5 R. N. Wright and J. R. Wood: Met. Trans., 8A (1977), p 2007
6 R. Castro and R. Tricot: Mém, Sci. Rev., Meta, 63 (1966), p 657
7 J. Ogoshi, T. Sada and M. Mizuno: Trans. of the Japan Society of Mechanical Engineers, 21 (1955), 107, p 555
8 K. Yoshioka, S. Suzuki, B. Ishida, M. Horiuchi and M. Kobayashi: Kawasaki Steel Technical Report, 9 (1984), p 47

Table 1 Chemical compositions of specimens

(wt %)

	C	N	Si	Mn	P	S	Cr	C+N
12.5Cr-1Mn	0.045	0.014	0.16	1.03	0.023	0.005	12.55	0.059
	0.058	0.011	0.15	1.03	0.024	0.005	12.46	0.069
	0.081	0.010	0.15	1.02	0.023	0.005	12.43	0.091
12.5Cr-1.5Mn	0.045	0.015	0.16	1.53	0.023	0.005	12.46	0.060
	0.058	0.012	0.15	1.53	0.024	0.005	12.40	0.070
	0.081	0.011	0.15	1.53	0.022	0.005	12.41	0.092

Table 2 Mechanical properties of as-quenched plates

	0.2% proof strength* (kgf/mm²)	Tensile strength* (kgf/mm²)	Elongation* (%)	Hardness (HRC)
12.5%Cr-1.5%Mn-0.07%(C+N)steel	92.8	112.1	19.1	35.1
Type 429 **	73.3	113.2	10.0	32.0

* JIS No.5 specimen, transverse to rolling direction

** 12.5%Cr-1.5%Mn-0.07%(C+N) steel and type 429 were quenched at the cooling rate of 30°C/s after heat treatment for 10 mm at 950°C and 910°C, respectively

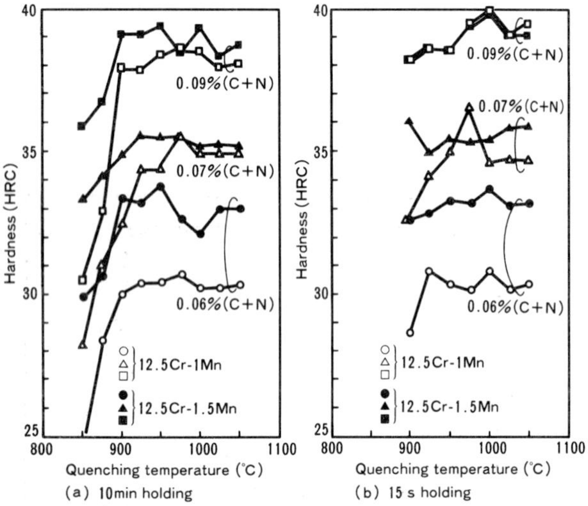

Fig. 1 Effect of (C + N) content and quenching temperature on hardness (cooling rate 30°C/s)

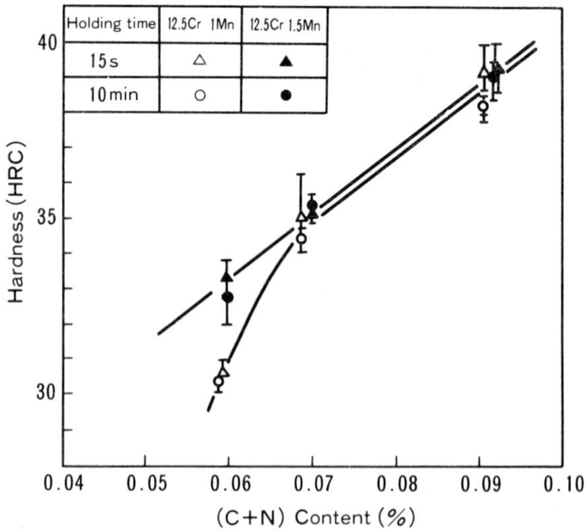

Fig. 2 Relation between (C + N) content and hardness obtained by quenching from 925–1050°C (cooling rate 30°C/s)

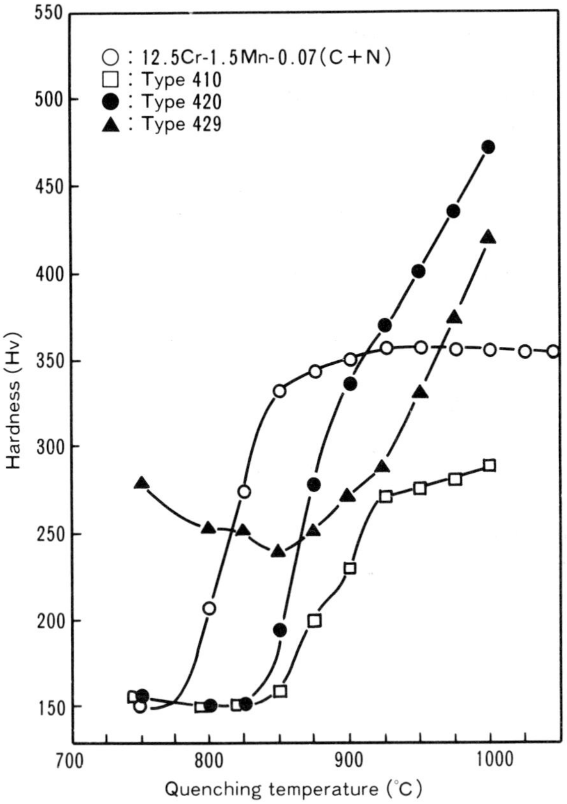

Fig. 3 Relation between hardness and quenching temperature (holding time 10 min; cooling rate 30°C/s)

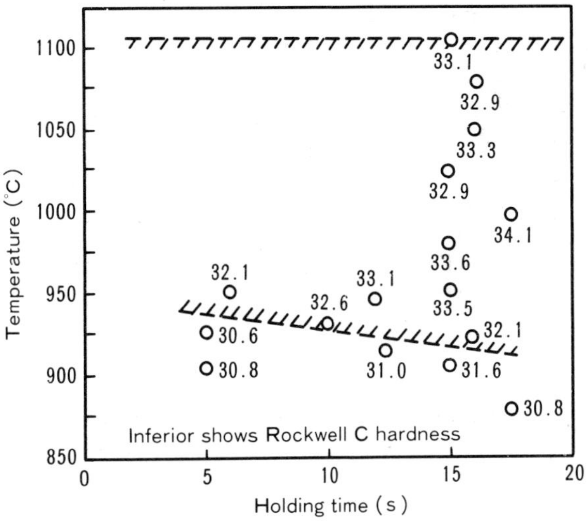

Fig. 4 Relation between quenching conditions and hardness of 12.5%Cr–1.5%Mn–0.06%(C + N) steel (cooling rate 30°C/s); the hatched region shows the constant quenched-in hardness of 33 HRC)

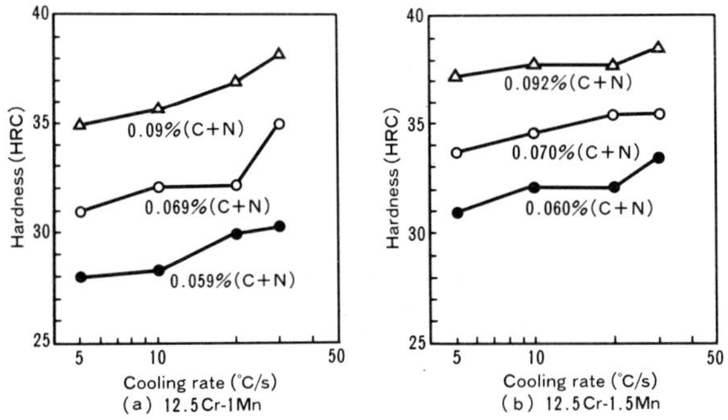

Fig. 5 Effect of cooling rate after heat treatment at 950°C for 10 min on quenching hardness

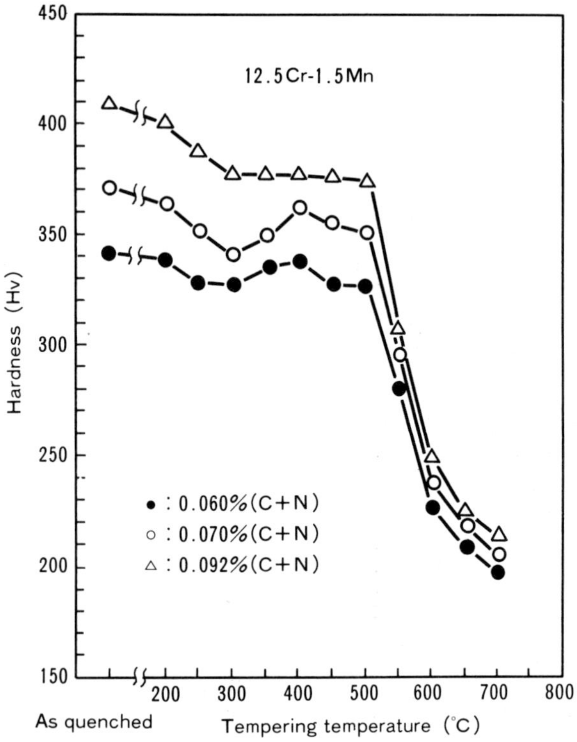

Fig. 6 Change in hardness with tempering for 30 min (quenching condition: cooling rate of 30°C/s after heat treatment at 950°C for 10 min)

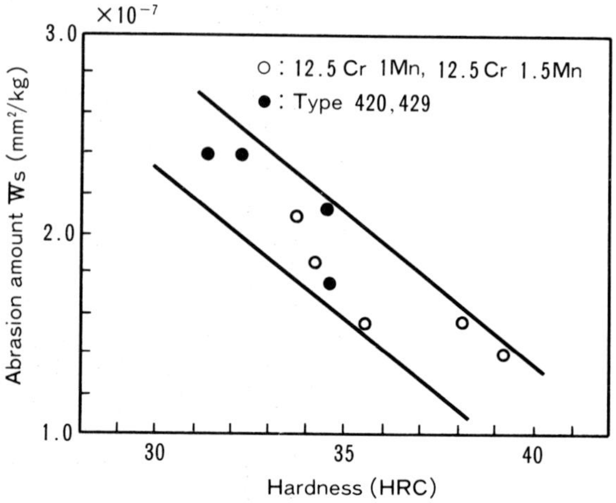

Fig. 7 Relation between wear resistance, measured by Ogoshi-type abrasion test, and hardness

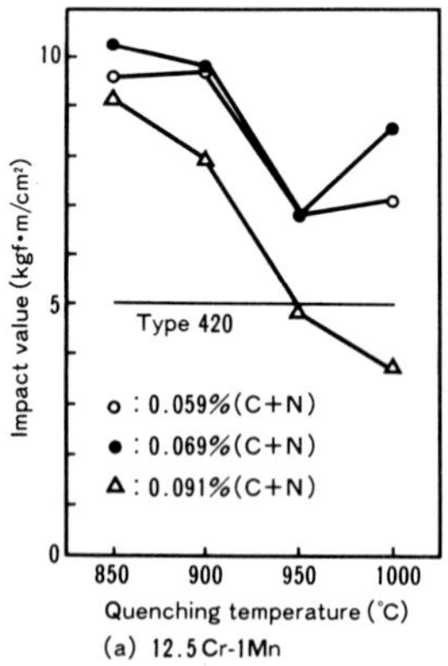

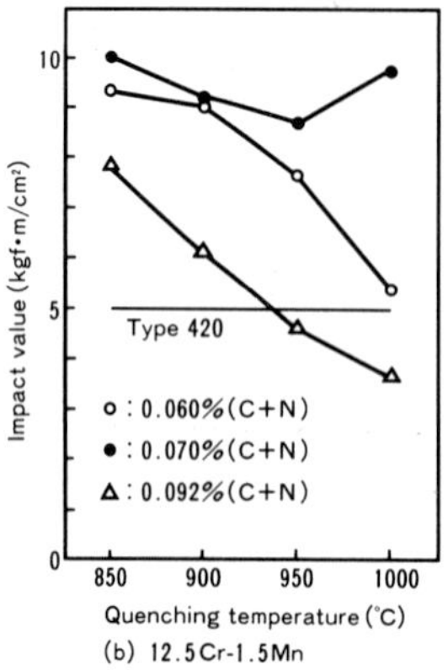

Fig. 8 Charpy impact value at 20°C for plates quenched at the cooling rate of 30°C/s after heat treatment at 850–1000°C for 10 min (JIS No. 4 half-sized specimens were used. The value of type 420 was also shown for comparison, where 420 was quenched and tempered with the aim of obtaining hardness of HRC 33)

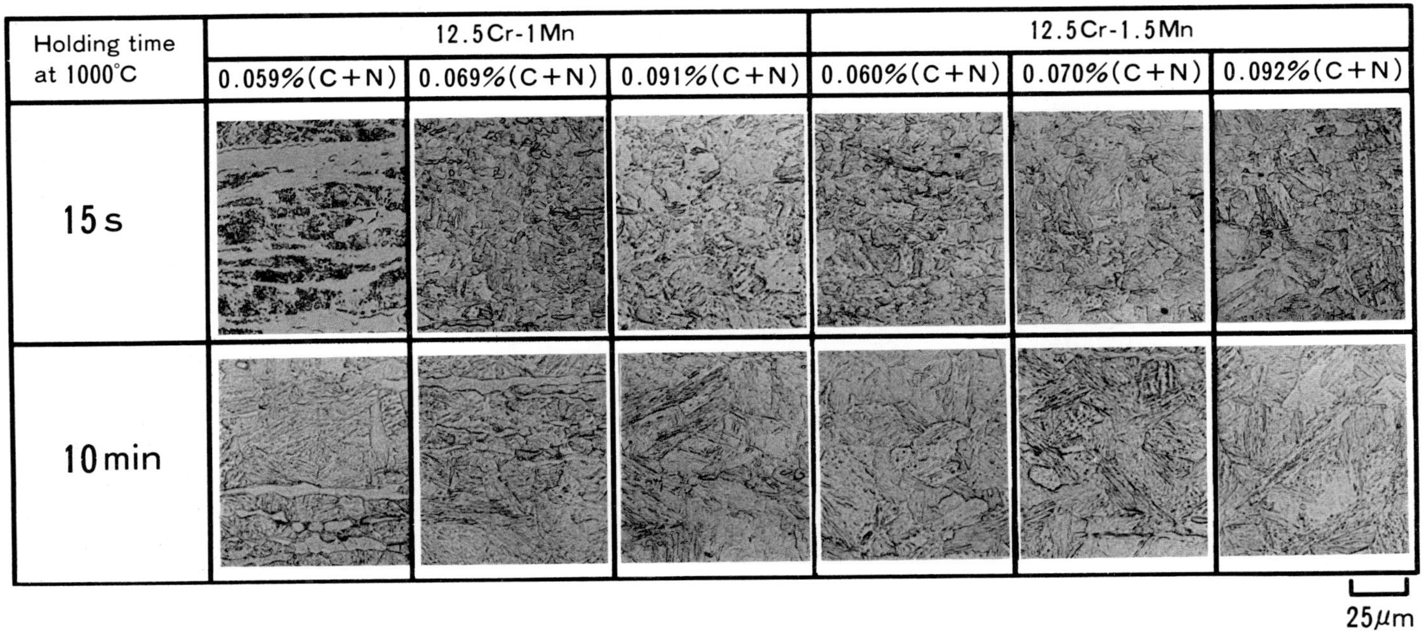

Photo 1 Microstructure of plates quenched at the cooling rate of 30°C/s after heat treatment at 1000°C for 10 min

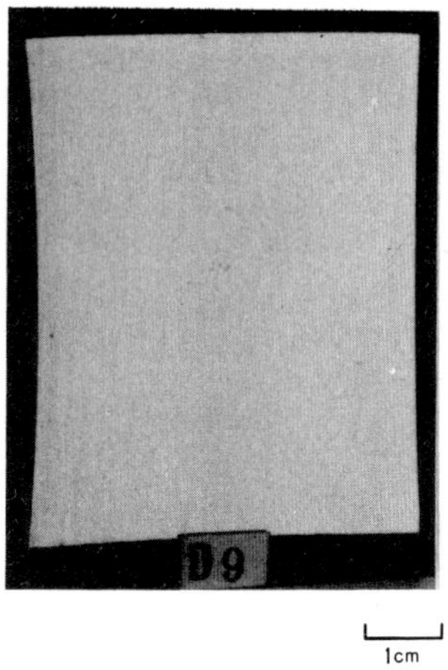

Photo 2 Appearance of a salt-spray test specimen quenched at a cooling rate of 30°C/s after heat treatment at 950°C for 10 min (testing time 24 h, #500 polishing)

Photo 3 Example of a brake disk

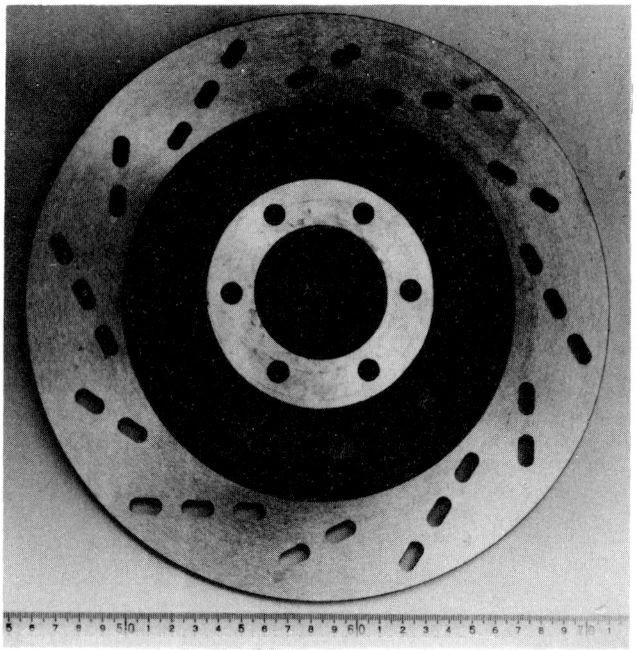

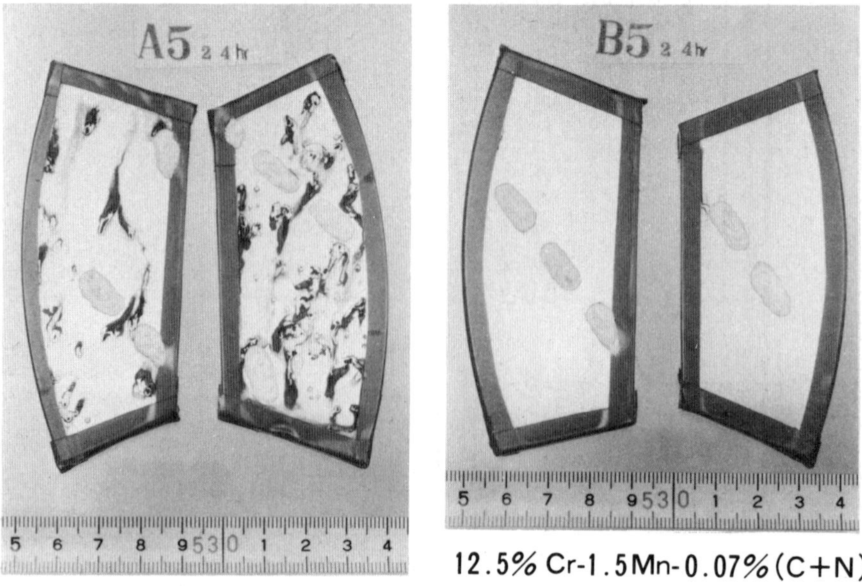

12.5% Cr-1.5Mn-0.07%(C+N)

Type 420

Photo 4 Appearance of salt-spray test specimens taken from brake disks of motorcycles (testing time 24 h, #500 polishing)

Developments in use of stainless steels for offshore pipework systems

C V Roscoe, K J Gradwell, M Watts and W J Nisbet

CVR, KJG and MW are with Mather and Platt Machinery Ltd., Manchester; WJN is in the Department of Metallurgy and Materials Science, University of Manchester/UMIST.

SYNOPSIS

Conventional duplex stainless steels have been used extensively for process pipework systems in view of their excellent combination of mechanical properties and resistance to stress corrosion cracking. However field service experience has shown that these low alloy duplex stainless steels have only a limited resistance to pitting and crevice corrosion in sea water. In contrast the 6Mo super-austenitic grades of stainless steel have excellent resistance to localised corrosion in sea water but their use in process pipework systems is restricted by their relatively poor mechanical properties and high cost.

The new generation of super-duplex stainless steels combine in one alloy the sea water corrosion resistance of the 6Mo super-austenitic stainless steels and mechanical properties which are superior to conventional duplex stainless steels. Super-duplex stainless steels are distinguishable from duplex stainless steels primarily by the addition of higher levels of nitrogen, chromium, molybdenum, tungsten and nickel. These and other alloying elements must be closely controlled to achieve the correct ferritic/austenitic microstructural phase balance in order to optimise the mechanical and corrosion resistance properties. Furthermore, super-duplex stainless steels incorporate minimum Pitting Resistance Equivalent values of 40 as part of their specifications to guarantee consistent localised corrosion performance.

Potentiodynamic studies have been undertaken to compare the pitting and crevice corrosion resistance of the various categories of stainless steels in sea water. A number of accelerated corrosion tests have been carried out on weldments in $FeCl_3$ solutions to assess the localised corrosion performance of the heat affected zones and the weld metal.

In this paper the performance of specific categories of stainless steels are compared and their suitability for use in topside and subsea pipework systems assessed. The relative mechanical properties and corrosion resistance of these stainless steels are discussed and examples given to show the advantages of designing systems in super-duplex stainless steels to reduce wall thickness and overall system weight.

INTRODUCTION

In offshore pipework systems there is an increasing demand for higher strength and more corrosion resistant materials. Until recently the more conventional duplex stainless steels had been considered to have adequate localised corrosion and stress corrosion cracking resistance when used in sub-sea process pipelines. However the more sour nature of the newer oil and gas fields in the North Sea has led to the introduction of the more corrosion resistant super-duplex stainless steels. In topside sea water pipework systems and fire water systems it is now also economic to use high alloy stainless steels in preference to 90/10 cupro-nickel alloys. Some commercial examples of high alloy stainless steels used for these applications are shown in Table 1.

The development of the highly alloyed and super stainless steels required an understanding of the effect of alloying elements on the corrosion performance and mechanical properties of the steels. A summary of the effect of certain alloying elements on the anodic polarisation curve of stainless steels is shown in figure 1[1]. The elements that have the most beneficial effect on the resistance to localised corrosion are chromium, molybdenum, nitrogen and tungsten.

The addition of chromium (>13wt%) to iron results in the formation of a protective chromium oxide film which isolates the steel from the environment. In chloride solutions the formation of the protective film expands the passive potential range by increasing the pitting potential and reducing the passive current density (figure 1). In duplex stainless steels the ferrite stabilising influence of chromium must be balanced by additions of nickel to preserve the phase balance between austenite and ferrite.

For a given chromium content, molybdenum has a strong beneficial effect on a steel's passivity, primarily by increasing the pitting potential and lowering i_{max} (figure 1). The mechanism by which molybdenum exerts its influence is not fully understood[2,3], but may be associated with its inhibition of the active dissolution rate in incipient pits[4]. From a practical point of view it is necessary to have a high chromium and molybdenum content in the stainless steel to prevent crevice corrosion in hot sea water. This has led to additions of 3-4wt%Mo in super-duplex and 6wt%Mo in super austenitic stainless steels.

Nitrogen additions improve the localised corrosion resistance of stainless steels by increasing the pitting potential[5]. The beneficial effect of nitrogen appears to be enhanced by the presence of molybdenum[6]. Nitrogen may improve corrosion resistance by being concentrated at the interface between metal and film[7]. More recent work suggests the beneficial effect of nitrogen on pitting corrosion is associated with the blocking effect of nitrogen on anodic dissolution in the local chemistry of a pit[8]. In duplex stainless steels nitrogen may also improve corrosion resistance by reducing partitioning of the chromium[9].

Tungsten (like Mo) has been shown to extend the passive potential range and increase pitting potential when added to duplex stainless steels[10]. Its effect may enhance corrosion properties by being adsorbed into the passive film as WO_3[11]. WO_3 then interacts with the oxides, resulting in enhanced stability and improved bonding of the film to the base metal.

PITTING RESISTANCE EQUIVALENT

The beneficial effects of alloying elements can be combined to give an indication of a stainless steel's corrosion resistance. Such a compositionally derived empirical relationship for pitting resistance of stainless steels is known as the steels Pitting Resistance Equivalent (PRE_N). For nitrogen containing stainless steels the PRE_N has been derived as

$$PRE_N = \%(Cr) + 3.3\%(Mo) + 16\%(N) \quad (12)$$

Other workers suggest a multiplication factor of 30 should be used for nitrogen in stainless steels[13]. There are reservations about these formula as they do not take into account the beneficial effects of other alloying elements, particularly tungsten. Indeed a more relevant PRE_N formula for tungsten containing stainless steels may be

$$PRE_N = \%(Cr) + 3.3\%(Mo+W) + 16\%(N)$$

Figure 2 shows that the introduction of tungsten into this formula decreases the amount of scatter when PRE_N is plotted against critical pitting temperature for specific stainless steels.

It is generally considered that PRE_N values of greater than 40 are necessary to guarantee the localised corrosion resistance of a stainless steel in oxygenated sea water and both the super-duplex and super austenitic (6Mo) stainless steels satisfy this criterion (Table 1). However further work is required to refine the PRE_N formula so that the beneficial/detrimental effects of all alloying elements and the microstructural factors associated with the breakdown of passivity are considered. In duplex stainless steels it may be more relevant to use two PRE_N numbers, one for ferrite phase and one for austenite. Table 2[14] shows the compositions and the PRE_N numbers of the bulk metal and individual phases of two super-duplex stainless steels. Depending on the element partitioning one phase may be more susceptible to pitting (lower PRE_N) than the other. It is known that ferrite undergoes preferential dissolution in reducing environments whereas austenite undergoes preferential dissolution in more oxidising environments [15]. Thus depending on which phase is likely to be attacked the PRE_N value of the individual phases may be optimised.

PITTING AND CREVICE CORROSION OF WELDMENTS

It is more relevant to consider the combined influence of the PRE_N value and the microstructure to determine the localised corrosion performance of stainless steels. To demonstrate this fact a series of weldments have been produced in conventional duplex, super-duplex and 6Mo austenitic stainless steels.

In respect of the super-duplex stainless steel (Zeron 100) a number of production weldments have been studied to examine the effect of heat input, welding position and the addition or deletion of filler metal for both manual TIG and automatic welding techniques. These investigations are discussed in an internal publication[16].

Accelerated localised corrosion assessments have been carried out on all the weldments in 10%$FeCl_3$ solution in accordance with ASTM G48. The results of these investigations are summarised in Table 3. From the results obtained, it can be deduced that the weldments in both the as-welded and solution annealed conditions can be ranked in terms of their critical pitting and critical crevice temperature in the ferric chloride solution as follows:

Zeron 100 (most resistant) > 6Mo austenitic stainless steels ≫ 2205.

The solution heat treated weldments generally give critical pitting and critical crevice temperatures which are similar to those observed for the parent materials. The beneficial influence of higher PRE_N values can readily be seen by comparing the CPT values obtained on the Zeron 100 weldments (65-75°C) and the 2205 weldments (20-25°C) in the solution heat treated condition. However some of the as-welded plates on the 6Mo weldments displayed very severe pitting in the weld metal on the unmachined test specimens when tested at 40°C. Also the machined 6Mo weldments showed extensive crevice corrosion at 35°C in the region of the unmixed zone, heat affected zones and weld metal. These results are generally in agreement with the findings of other workers[17].

The reduction in the corrosion performance of the weld metal is attributed to the precipitation of phases such as Laves, niobium and chromium rich nitrides and M_6C. These phases can contain up to 50wt% of molybdenum (Laves) and up to 70wt% of chromium (nitride) and therefore denude the matrix of these elements in the surrounding areas. This in turn lowers the pitting and crevice corrosion resistance of the weld metal. Similarly both Chi and Laves phases have been reported to precipitate in the heat affected zones of 6Mo weldments[17]. Consequently it is not surprising that the HAZ regions of the 6Mo weldments show severe crevice corrosion attack in the machined specimens.

In respect of the as-welded super-duplex stainless steel weldments these also exhibited a lowering of the localised corrosion resistance. However it should be emphasised that the heat input levels were purposely varied on these weldments in an effort to determine the permissible variation from the optimum welding parameter. Notwithstanding this fact, only isolated pitting is observed in the unmachined as-welded specimens when tested at 45°C and the crevice corrosion is only just discernable on the machined specimens at 45°C. When welded in the optimum heat input range with metal cored wire having a PRE_N value of 41.6 (Table 3) the Zeron 100 weldments in the as-welded condition gave CPT values of 55-65°C and CCT values of 50-60°C.

A typical microstructure of the weld metal, heat affected zone and parent metal of the Zeron 100 weldment which has been welded in the optimum heat input range of 1.01-2.75kJ/mm is shown in figure 3a. The proportion of austenite varies from 36-44% across the weldment and there is no evidence of deleterious second phase particles.

On these specimens only slight evidence of pitting and crevice corrosion has been observed in the weld metal even though the PRE_N value of the weld metal (40.1) in these instances is less than that recorded for the parent plate (41.2).

In the specimens welded with heat inputs levels outside the optimum range there is some evidence of M_2X precipitation in isolated areas of the ferrite matrix (figure 3b) in the heat affected zone.

Lower heat inputs will tend to give insufficient time at temperature for austenite to re-precipitate in the ferrite during cooling matrix which in turn leads to supersaturation of nitrogen in the ferrite resulting in the precipitation of M_2X particles. The M_2X nitride precipitates typically contain high levels of chromium molybdenum and nitrogen i.e. $(CrMo)_2N$ and are finely dispersed in regions of the microstructure where the proportion austenite is locally less than the average value across the weldment. (i.e. in particular areas of the HAZ). Consequently the matrix is locally deficient in these elements which lowers the pitting resistance of the weldment.

Heat input levels higher than 2.75kJ/mm can also result in the precipitation of M_2X since the HAZ is effectively being aged in the temperature region where M_2X precipitation will occur virtually independently of area fraction of austenite present in the microstructure. Higher heat inputs can also lead to elemental loss of Cr and N from the weld metal which also leads to a lowering in the pitting resistance of the weldment.

Irrespective of the precipitation of these M_2X particles the localised corrosion performance of all the Zeron 100 weldments is better than that observed for the 6Mo austenitic stainless steel weldments. It should also be noted that the solid wire utilised on the Zeron 100 weldments had a PRE_N value of 40.1. Therefore the corrosion results obtained probably reflect the lowest ever values likely to be achieved in the TIG welding of Zeron 100 since the PRE_N minimum in the material specification is 40. Alternatively the P12 filler utilised on the 6Mo weldments had a PRE_N value of 50.0. However the higher molybdenum-containing phases and microsegregation present in the weld metal markedly reduces the corrosion performance of the 6Mo weldments. Despite the obvious limitations of the minimum PRE_N value it does allow a more consistent measure of the corrosion performance to be predicted particularly when second phase particles are avoided.

CASE STUDIES

The beneficial effects of using super-duplex stainless steels in preference to other materials can be shown by considering typical offshore pipework systems, and these are shown in table 4[18].

Firewater Systems

Deluge and Sprinkler System

In order to evaluate the potential benefits of using pipework in stainless steels in preference to 90/10 cupro-nickel, a typical deluge system has been designed using a computer hydraulic programme. The system had to meet the minimum nozzle pressure and flow required, and had 10 bar inlet pressure available from the ringmain.

The design programme ultilised 6m/s velocity limit in cupro-nickel and 10m/s in super-duplex stainless steel. In fact the super-duplex stainless steel is capable of much greater velocities, but the Renolds number of water changes at these higher velocities and a revised calculation procedure is required. The results are shown in Table 5a.

From this study it is possible to deduce the following:

1. Due to velocity limitations with cupro-nickel, the system is unable to utilise the available inlet pressure. As a result it would be necessary to install an orifice plate to create a 4.72 bar pressure drop at the deluge valve set.
2. The super-duplex stainless steel system has smaller pipes and as a consequence a smaller deluge value set would be required.
3. As a result of the smaller bore stainless steel system, the dry weight is reduced by 15% compared to cupro-nickel and the wet weight by 33%.
4. The increased strength of stainless steel enables 38% few pipe supports to be used and eliminates the need for the comprehensive insulation kits.

Typical costs of a stainless steel pipe and fittings package relative to those in a cupro-nickel deluge system are presented in Table 5b. The overall cost of the system is greatly reduced using super-duplex stainless steel in place of 90/10 cupro-nickel.

Support and prefabrication costs are much lower for the super-duplex stainless steel. Other savings would also arise from the use of smaller deluge valve sets and skids. The super-duplex stainless steel is also easier to handle whereas the cupro-nickel is prone to damage.

Extending this analysis to include all the fire water deluge and sprinkler systems (excluding the ring main) for a medium sized platform reveals the information presented in Table 5c. The reduced pipe sizes gave a 10 tonne (20%) reduction in the wet (operating) weight for which the platform must be designed.

Table 6 show a similar case study for a typical firewater ringmain. The study has compared pipe call-off quantities using velocity limitations of 3.5m/sec for cupro-nickel and 7m/sec for super-duplex stainless steel. A velocity of 10-12m/sec could have been used for the super-duplex stainless steel system to parallel the assumptions used in the deluge system example yielding still greater savings in cost and weight. The study reveals the following information:

1. Super-duplex stainless steel pipe is lighter than 90/10 cupro-nickel pipe at the larger diameters due to the reduced wall thickness.
2. Reduced diameter fire water mains can be utilised with super-duplex stainless steel due to their tolerance to higher water velocities. In view of the necessity to supply the helideck on the top of the platform with sea water, there is excess pressure in the ring main for the other duties. Therefore, it is not necessary to specify larger fire pumps when using the additional velocities allowed by stainless steels.
3. Both the dry weight and the wet weight is significantly less when using super-duplex stainless steel (Table 6b).

Table 6c gives the cost comparisons between 90/10 cupro-nickel and super-duplex stainless steel. These figures show cost savings when using super-duplex stainless steel for:

1. Pipe materials
2. Supports
3. Prefabrication

The benefits and savings from the easier handling of the stronger super-duplex stainless steel have not been reflected; nor have the benefits from smaller pump delivery and ring main valves.

Process pipework

There is an increasing tendency, particularly in Norway, to specify 6%Mo austenitic stainless steels in preference to standard 22%Cr duplex stainless steels. The rationale supporting this departure from past practice has little to do with Cl^- or H_2S stress corrosion cracking but is concerned principally with localised corrosion performance.

Process fluids in the latest generation of oil and gas fields have higher water cuts than previously encountered. The aqueous phase is frequently at elevated temperatures and often contains high proportions of chlorides and H_2S. Consequently the 22Cr duplex stainless steels do not give adequate resistance to pitting and crevice corrosion particularly in the root and heat affected zone areas of weldments.

Conversely the super-duplex stainless steels give a localised corrosion performance equivalent to 6%Mo austenitic stainless steels but have strength properties even greater than conventional 22%Cr duplex stainless steels.

Therefore process pipework systems in super-duplex stainless steels give considerable cost savings in comparison to systems in 6%Mo austenitic stainless steels.

Table 7 shows comparative costs for a simple model of a process manifold system constructed in a 6%Mo austenitic and a super-duplex stainless steel (Zeron 100). Assumptions for the model are:

i. Temperature 100°C
ii. Pressure 150 bar
iii. Simple cylinder - no branches
iv. Norwegian general rules for piping systems (TBK6 1983)
v. No corrosion allowance

As can be seen from Table 7 the significant increase in allowable stress afforded by the super-duplex alloy combined with equivalent or superior localised corrosion performance will produce significantly lower system build costs when these alloys are specified in preference to 6%Mo alloys for process pipework.

A natural extension of the proposed move towards alloys with superior localised corrosion performance in process applications will be to specify the alloys for submarine pipelines. A very significant cost in this context is lay barge hire costs. These hire costs are dictated by girth joint welding speeds. Clearly the reduction in wall thickness allowed by the super-duplex alloys will reduce the welding costs substantially.

Since the barges typically lay 0.5-1 kilometre of 6" pipe a day, at a cost of approximately £250,000 per day the potential savings are enormous.

CONCLUSIONS

1. The use of a Pitting Resistance Equivalence (PRE_N) formula incorporating tungsten, provides a simple method of predicting the localised corrosion performance of stainless steels. However microstructural factors that affect corrosion performance must also be considered.
2. The critical pitting and critical crevice resistance in ferric chloride of the weldments tested can be ranked as follows:

   Zeron 100 (most resistant) > 6Mo austenitic stainless steels>> SAF 2205.
3. The use of super-duplex stainless steels affords considerable scope for cost reduction when specified for offshore pipework systems. In respect of firewater systems displacement of cupro-nickel alloys leads to significant direct and indirect cost savings through reductions in material usage (dry weight) and an impressive associated savings in wet weight. In respect of process pipework the savings that accrue from the specification of super-duplex steels in preference to the 6%Mo grades cannot be overstated.

ACKNOWLEDGEMENTS

The authors are grateful to the directors and the Weir Group plc for the kind permission to publish this paper. In addition they are indebted to all those members of staff who have assisted with the development work and preparation of this paper.

REFERENCES

1. C.V. Roscoe and K.J. Gradwell, Proceedings of Conference on Duplex stainless steels 26-28 Oct. p.126. 1986.
2. A.J. Sedriks, Corrosion of stainless Steels - John Wiley and Sons, New York N.Y. 1979.
3. R.C. Newman and E. Franz, J. Electrochem. Soc. p.223, 1984.
4. H.S. Isaacs and R.C. Newman, in Corrosion Chemistry within Pits, Crevices and Cracks, ed. A. Turnbull, HMSO publications, 1987, p.45.
5. J. Chance, K.J. Gradwell, W. Coop and C.V. Roscoe, Duplex Stainless Steels Conference Proceeding, ASM Metals/ Material Technology Series Paper 8201-019) p.371, 1982.

6. J.E. Trueman, M.J. Coleman and K.R. Pirt, Br. Corros. J. Vol.12, p.236, 1977.
7. Y.C. Lu, R. Bandy, C.R. Clayton and R.C. Newman, J. Electrochem. Society, Vol.130, p.1774, 1983.
8. R.C. Newman and T. Shahrab, Corr. Sci. Vol. 27, No.8, p.827, 1987.
9. E.A. Lizlors, Climax Report RP-33-80-08 June 1981.
10. C.V. Roscoe, The Development of Zeron 100, Mather & Platt Ltd., Internal Publication 1984.
11. N. Bui, A. Irhzo, F. Dabosi and Y. Limouzin-Marie, Corrosion NACE, Vol. 39, p.491, 1983.
12. A.J. Sedriks, Stainless Steels 84 Proceedings of Goteborg, Conference Book No. 320, The Institute of Metals, 1 Carlton House Terrace, London SW1Y 5DB, p.125, 1985.
13. P. Gümpel, "The influence of silicon and nitrogen on the properties of some austenitic stainless steels" in this conference.
14. R. Longbottom, private communication, Manchester Materials Centre, UMIST.
15. V.S. Agarwal and N.D. Greene, "Selective Corrosion of Cast 187Cr-8Ni Stainless Steel", Paper No. 57, Corrosion 77, San Francisco, California.
16. C.V. Roscoe, Mather & Platt, Internal Report 21st Sept. 1987 "Investigation to Determine the Optimum Parameters for Welding Zeron 100 Super-Duplex Stainless Steel.
17. M. Liljas, B. Holmberg and A. Viander, Stainless Steels 84, Conference Proceedings, p.323.
18. Roscoe C.V., Gradwell K.J., Watts M.R., Whiteley, Developments in the Use of High Alloy Stainless Steels for Offshore Firewater Systems. proceedings of International Conference on Fire Engineering in Petrochemical and Offshore Applications, 23/24 June 1987, p.103.

Table 1

Nominal Compositions of some Commercially Available High Alloyed Stainless Steels

Steel	Cr	Ni	Mo	Cu	W	N_2	PRE_N
Zeron 100*	25	8.0	3.8	0.7	0.7	0.25	41.5 43.8w
SAF 2205ˣ	22	5.0	3.0			0.14	34.1
Sanicro 28⁺	27	31	3.5	1.0			38.9
AL6XN#	20.8	25	6.5			0.20	45.4
254 SMO#	20	18	6.1	0.7		0.20	43.3

* Super duplex w - Tungsten adjusted PRE_N
ˣ Duplex
⁺ High alloyed austenitic
\# 6Mo super austenitic

Table 2

PRE_N values for two duplex stainless steels. Partitioning coefficients determined from Termocalc Version C, iron data base at 1100°C[14].

wt%	Cr	Ni	Mo	W	N	PRE_N
Bulk	24.2	7.37	4.0	0.63	0.25	43.5
K_γ/K_α	0.93	1.4	0.83	0.5	8.8	
α(56%)	24.97	6.27	4.32	0.81	0.06	42.8
γ(44%)	23.22	8.77	3.59	0.40	0.49	44.2
Bulk	24.2	8.0	4.0	0.63	0.25	43.5
K_γ/K_α	0.91	1.41	0.79	0.49	8.35	
α(50%)	25.34	6.64	4.46	0.85	0.05	43.7
γ(50%)	23.06	9.36	3.53	0.41	0.43	42.9
PRE_N = %Cr + 3.3 (%Mo+%W) + 16%N						

Table 3

Summary of Welding Trials Conducted on Zeron 100, 6%Mo Austenitics and SAF 2205 Stainless Steels

Material	PRE_N (base metal)	PRE_N (filler metal)	Condition of material	Heat Input (kJ/mm)	CPT (°C)	CCT (°C)
Zeron 100 base metal	41.7 (43.7)	-	Mill finished plate	-	65-75	60-65
Zeron 100 GTAW with Zeron 100 SW	41.2 (43.1)	40.1 (42.0)	As welded plate	0.97-3.12	45-53*,**	45-50*,**
	41.3 (43.2)	41.6 (44.0)	As welded plate	1.95	55-65**(1)	50-60**
Zeron 100 GTAW with Zeron 100 MMCW	41.5 (43.6)	42.5 (44.3)	Solution annealed	0.9-2.1	65-75	55-65
6 Mo Austenitics base metal	43.3/45.4	-	Mill finished plate	-	55-65	45-55
6 Mo Austenitics GTAW with P12 filler	43.3/45.4	49.0/49.8	As welded plate	0.9-1.1	35-45*(1)	35-45*,**
6 Mo Austenitics GTAW with P12 filler	43.3/45.4	49.8	Solution annealed	0.9-1.1	50-55	Not assessed
2205 base metal	33.4/34.2	-	Mill finished plate	-	25-35	15-25
2205 GTAW with 22.8.3 filler	33.4/34.2	34.3	Solution annealed	2.0	20-25	Not assessed

Note: PRE_N figures in brackets are adjusted to include tungsten.
* Localised corrosion in the weld metal.
** Localised corrosion in the HAZ.
(1) Optimum Heat Input.

Table 4

Advantages and Disadvantages of Materials in Ring Main, Deluge and Sprinkler Systems.

Material Options	Ring Main	Deluge System	Sprinkler
Carbon Steel	Unacceptable	Unacceptable	Unacceptable
Cement lined	Severe wt. penalities high installation costs not now normally specified.		
Carbon steel lined/galvanised		Corrosion due to evaporation/concentration in seawater droplets after testing. Corrosion products leads to nozzle blockage.	Risk of nozzle failure. Examples of complete failure have been reported.
Cupro-Nickel	Corrosion performance generally good. Occasional damage behind welds and at bends due to eddying. Velocity limitations/ low strength requires larger pipe sections.	Corrosion performance O.K. Dry systems have high potential for heat damage, in the early stages of a hydrocarbon fire. Velocity limitations requires larger pipe sizes.	Generally satisfactory.
Standard stainless (316)	Unacceptable	High salt concentrations combined with evaporation in small pools and droplets leads to severe pitting and perforation & risk of nozzle blockage by corrosion products.	Risk of pitting and perforation.
Super duplex super austenitic	Optimum solution.	Ideal solution.	Ideal solution.
Note	Perhaps surprisingly more problems and failures are reported in notionally dry deluge systems than in ring main or sprinkler systems illustrating the unexpectedly harsh environment that exists.		

Table 5
Case study of a typical 40 nozzle deluge system

5a Alternative designs

Pipe dia. (inches)	lengths of pipe required (metres)	
	90/10 Cupro-Nickel	Super Duplex
1	36.65	52.4
1.5	23.85	33.5
2	25.4	5.9
3	11.5	28.5
4	22.85	25.0
6	25.03	-
Valve Set (mm)	150	100
No of Supports	55	34
Inlet Press Reqd.(bar)	5.28	10
Flow (litres/min)	4368	4880
Pipe Volume (litres)	809	472
Dry Wt. (Kgs)	634	542
Wet Wt. (Kgs)	1442	960

5b Costs

	90/10 Cupro-Nickel (£)	Zeron 100 (£)	Change (%)
Piping Material	6500	6400	-2
Supports	1100	680	-38
Prefab Labour	7000	4000	-43
Total	14600	11080	-24

5c Weights

	90/10 Cupro-Nickel	Super Duplex
Dry Wt. (tonnes)	28.5	24.3
Wet Wt. (tonnes)	46.3	36.8

Table 6
Case Study of a Typical Firewater Ring Main

6a Lengths of Materials Reqired (metres)

Pipe Dia. (inches)	18	12	8	6	4	3	2
Cupro-Nickel (3.5m/sec)	170	-	5	30	40	-	-
Zeron 100 (7m/sec)		170	-	5	30	40	-
Zeron 100 (12m/sec)			170	-	5	30	40

6b Weights

	90/10 Cu-Ni 3.5m/sec	Zeron 100 (7m/sec)
Dry Wt. (tonnes)	17.4	6.7
Wet Wt. (tonnes	46	19
Pipe Supports (No.)	123	82

6c Costs

	90/10 Cu-Ni (£)	Super Duplex (£)	Change (%)
Piping Material	96000	60000	-38
Supports	3700	1650	-55
Prefab Labour	32010	9500	-70
Total	131700	71150	-46

Table 7
Comparative Costs for 6Mo Austenitic and Super Duplex Stainless Steels for a High Pressure Process System

	6%Mo Austenitic	Super Duplex (Zeron 100)
Design Stress 100°C MPa	174	291
20" O.D. Pipe Wall Thickness (mm)	21.5	13
Cost £/ratio	2.4	1

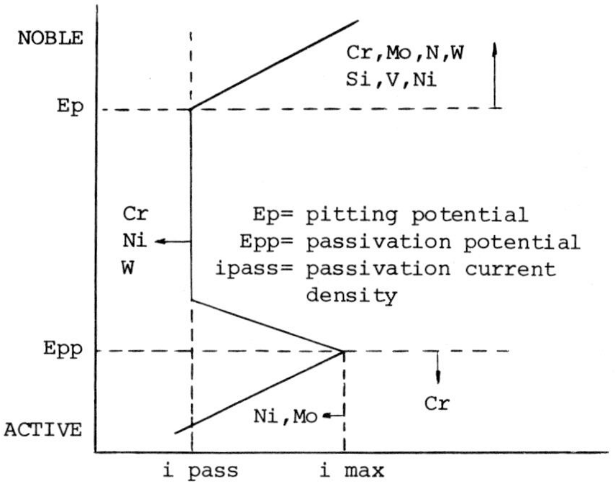

Figure 1 Schematic representation of the effects of certain alloying elements on the polarisation curve of stainless steels(1).

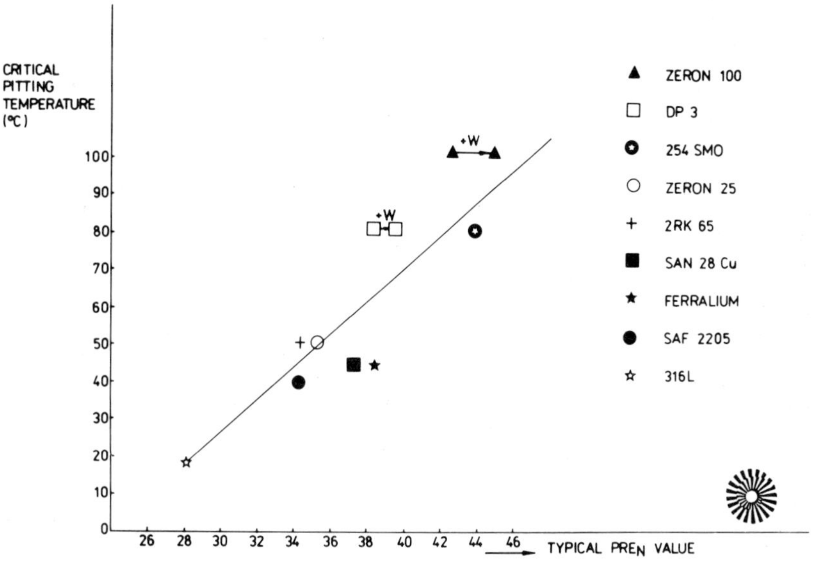

Figure 2 PRE_N (%Cr + 33%Mo + 16%N) and tungsten adjusted PRE_N (%Cr + 3.3(%Mo + %W) + 16%N) versus critical pitting temperature for a selection of stainless steels.

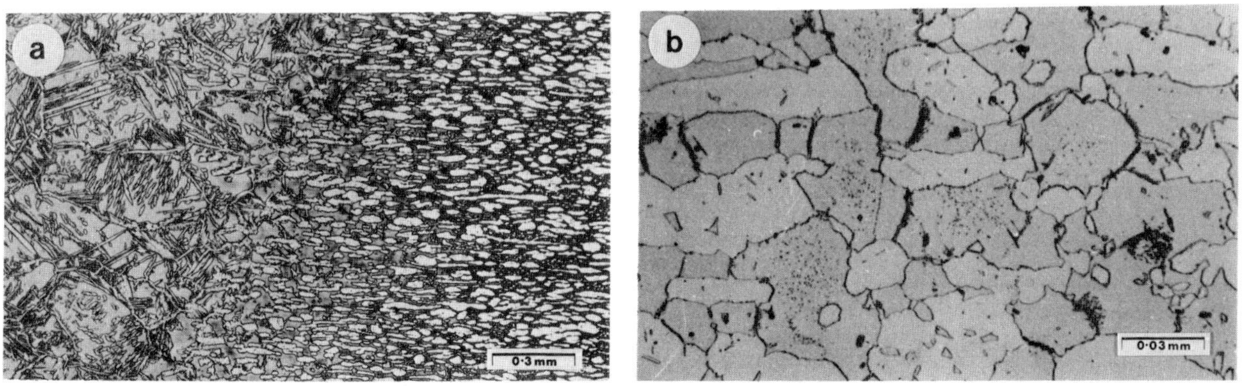

Figure 3 a) Typical microstructure of weld metal + HAZ + parent metal of Zeron 100 welded in the heat input range of 1.25-2.75kJ/mol b) M_2X precipitates in the ferrite of the HAZ of Zeron 100 welded outside the heat input range of 1.25-2.75kJ/mol.

Anisotropy of elastic and plastic properties of 17Cr—8Ni—2Mo and 19Cr—12Ni—3Mo weld metals

S T Kimmins and C A P Horton

The authors are with the Materials Branch, CEGB, Leatherhead, Surrey.

SYNOPSIS

The anisotropy of elastic and plastic properties of Type 316 weld filler materials has been examined. A strong solidification texture has been found and its effect on various mechanical properties rationalized. The properties examined include ductile fracture toughness, tensile and proof strengths and high stress creep deformation. The influence of post weld heat treatments at 800°C and 1050°C on texture and properties were also examined.

INTRODUCTION

Strong directional solidification preferences existing in austenitic steels lead to columnar grains with preferred orientations in their weld metals (Savage, Lundin and Aronson, 1965; Savage and Aronson, 1966). Elastic moduli vary with crystallographic direction and, consequently, the pronounced textures exhibited by austenitic weld metals may lead to significant anisotropy of elastic and plastic properties in the weld metal (Dewey, Alder, King and Cook, 1977). One well known consequence of the elastic anisotropy is the variation in ultrasonic velocity with direction in austenitic weldments (Kapranos, Al-Helaly and Whittaker, 1981).

This Paper presents the results of an investigation of the extent of anisotropy in the elastic and plastic properties of weld metals for Type 316 stainless steel. The welds were in the form of weld pads and a butt weld. Tensile and ultrasonic techniques were used to measure elastic moduli at room temperature, whilst plastic properties were measured at room temperature and 538°C. The tests also included the effects of post-weld heat treatments at 800°C and 1050°C. The elastic and plastic properties were compared with X-ray measurements of texture.

MATERIALS

The weld metal compositions used in this study are listed in Table 1. The chemical analyses were performed after welding. Magnetic measurements of delta ferrite levels are also listed in Table 1, together with the Cr and Ni equivalents calculated from:

$$Cr_{eq} = \%Cr + \%Mo + 1.5\%Si$$

and $Ni_{eq} = \%Ni + 30\%C + 0.5\%Mn$ (Delong, 1974).

The solidification structure of all of the weld materials prepared for specimen machining was revealed by macro-etching with Tuckers Reagent. Such macro-etched sections of a butt weld and a weld pad show the characteristic columnar grain structure associated with solidification, Fig. 1.

EXPERIMENTAL METHODS

The elastic modulus of the weld metal was measured using two techniques. The 'dynamic' elastic modulus was obtained ultrasonically at room temperature by the resonant frequency technique using small rod-shaped specimens ~3.5 mm in diameter and 18 mm in length. These specimens were cut from weld pad A and the butt weld D (Table 1), in the orientations shown in Fig. 2.

Tensile specimens cut from welds pads A, B, C and E, in similar orientations to those of Fig. 2, were tested at room temperature and 538°C using a strain rate of 0.002/min. Values for 'static' elastic modulus, 0.2% and 1.0% offset proof strengths were obtained.

Some constant load creep rupture testing was also performed at 538°C in air on tensile type specimens from weld pad A.

The orientation of a specimen cut from a weld was defined by the angle between the columnar grain direction (equivalent to the solidification direction) and the major axis of the specimen, i.e. either the long axis of the ultrasonic rod specimens or the tensile axis of the tensile specimens. Thus the 0° orientation defines specimens with the columnar grains parallel to the long axis of the specimens, the 90° orientation specimens with the grain structure perpendicular to the same axis and the 45° orientation defines a grain structure at 45° to the same axis.

The preferred orientation, or texture, of the weld materials was characterized by X-ray computer-aided orientation studies (Cross and Kelly, 1979). A pole figure of a polished weld metal specimen is produced by a surface scan of a ~5 mm raster at the appropriate Bragg angle of a selected crystallographic plane, i.e. in this study the austenite (111) or (200) reflections.

The grain structure, particularly its development resulting from heat treatments, was also examined metallographically.

Toughness measurements were carried out at room temperature on specimens cut from weld pad E, using techniques and analysis advocated in the ASTM standard E813-81 (ASTM, 1981).

The elastic-plastic toughness parameter for a given specimen orientation was obtained from a five specimen crack growth resistance curve. The specimens were 12.5 mm thick compact tension specimens. The orientation of these specimens was defined by the angle between the fatigue pre-cracked ($\Delta K \simeq 20$-25 MN $m^{3/2}$) notch and the columnar grain direction. The final notch depth was such that $(a/w) = 0.6$ and the 0°, 45° and 90° orientations were examined. Each specimen was stopped at an arbitrary point on the load v.s. displacement curve such that a range of ductile crack lengths between ~0.1 mm and 0.6 mm were obtained. The data were then reduced to give a J value for each specimen, and this was plotted against the crack growth.

RESULTS

Texture Analyses

A number of pole figures, defining the solidification textures, are given in Fig. 3. For each pole figure details of specimen source and orientation are given. For each of the Figures 3(a), (b) and (c) the specimen surface was in the plane of the major axis of the columnar grains, and the North-South lines of the pole figures represents the columnar grain direction. The reflection selected was $(111)_\gamma$. Fig. 3(d) shows the pole figure obtained from the same material conditions as used for Fig. 3(b). However, the specimen surface was, in this case, perpendicular to the columnar grain direction and the $(200)_\gamma$ reflection had been selected.

A number of experimental factors should be borne in mind before an interpretation is ascribed to these figures. Absorption effects during the surface scan lead to a lack of features in the outer regions of the pole figures. The discreetness of the intensity distribution is associated with the coarse grain size of the weld in relation to specimen size. The askew symmetry of some of the pole figures about the diagram axis is caused by inaccuracies in the cutting of these specimens with reference to the columnar grain direction.

When the above factors are taken into account the $(111)_\gamma$ reflection pole figures (Fig. 3(a)-(c)) indicate a fairly uniform pole distribution along the 35° latitude lines and are symmetrical about the solidification/columnar grain axis. A consideration of stereographic projections shows these figures to be consistent with a <100> fibre-type texture. In fact, as fibre textures are associated with wire-drawing processes, a more precise description is that there is a <100> orthotropic texture, with the major axis of the texture being parallel to the solidification direction. The low intensity features in Fig. 3(a)-(c), lying along the East-West line may be due to a co-existing weak <110> orthotropic texture, also with the major axis parallel to the columnar grain direction.

The above analysis of pole figures 3(a)-3(c) is confirmed by Fig. 3(d) which would be expected if the structure is rotated through 90°. A schematic representation of a <100> fibre texture is given in Fig. 4, indicating the major crystallographic directions. Appropriate stereographic projections from it may be compared to the pole figures. The figure is equivalent to the rotation of a cubic crystal about a <100> direction, parallel to the solidification direction.

The influence of 'solution treating' at 1050°C on the weld solidification texture was studied, as a function of time, using weld pad A. Fig. 5 shows (111) pole figures produced from specimens subjected to various times at 1050°C. As time progressed the regular pole representation disintegrated to a random 'spotty' distribution.

Metallography

The optical micrograph in Fig. 6(a) shows the as-deposited state of weld pad A. A post weld heat treatment of 10 h at 800°C, which completely transformed the δ-ferrite according to magnetic measurements, produced no change in optical appearance, the delta ferrite transformation products remaining in the same regions in which the delta ferrite existed.

Figs. 6(b) to (d), show the microstructural changes associated with the 1050°C solution treatment. Note the disappearance of the characteristic cellular delta ferrite structure with some delta ferrite apparently remaining in a spheroidized form at later times. The long meandering grain boundaries are characteristic of the coarse grained austenitic welds. These gradually re-crystallized to give more regular grains with accompanying annealing twins eventually resulting in a microstructure optically similar to that of wrought annealed '316' plate.

Elastic Properties

The initial characterization of elastic anisotropy is shown in Fig. 7(a), where ultrasonic determinations for specimens cut both from weld pad A and butt weld D are shown. The room temperature wrought, annealed '316' average is also indicated. Fig. 7(b) shows, for comparison, Young's modulus, as obtained from room temperature tensile tests, on weld pads A, B, C and E in an as-welded state. Data resulting from specimens heat treated at 800°C, and from as-welded specimens tested at 538°C, are also shown. In all cases the elastic modulus was maximum for the 45° orientation, i.e. $E(45°) > E(90°) > E(0°)$. Note that the anisotropy is not altered by post-weld heat treatment at 800°C. The two orthogonal 90° orientations ($w + \ell$ in Fig. 7a) exhibited very close values in elastic modulus.

The ultrasonically determined elastic modulus was also obtained for pad A material from specimens orientated at 0°, 45° and 90° after solution treatments at 1050°C for times ranging from 30 minutes to 8 h. These results are presented in Fig. 8. The differences between the three orientations have disappeared after 2 to 4 hours at 1050°C.

Tensile Properties

The proof stress and UTS properties showed the same trend, with regard to orientation, as the elastic moduli. The 0.2% PS values, Fig. 9(a), were in the order 45° > 90° > 0°, although the magnitude of the variation was smaller than that of the elastic properties. The 800°C heat treatment on pad A reduced the proof strengths and the differences between the orientations, although the 0° orientation was still just weakest, Fig. 9(a).

The values for elongation to fracture, Fig. 9(b), exhibited more scatter and showed no trends with specimen orientation. However, the values did indicate that the as-welded material was slightly more ductile than the 800°C post-weld heat treated material, and that testing at 538°C reduced the ductility of the as-welded state.

Clear evidence of anisotropic plastic deformation processes came from an examination of the broken tensile testpieces. For specimens with tensile axes orientated at 90° and 45° to the columnar grain direction, the initially circular gauge length cross section had become elliptical. This effect was not noticeable for specimens tested at the 0° orientation.

A measure of strain hardening was obtained from the log (true-stress, σ) vs. log (true-strain, ε) curves derived from the room temperature tensile tests on pad E. Assuming the simple power law relationship, $\sigma = (\text{const})\varepsilon^n$, a clear difference in work-hardening capacity, as measured by n, was obtained, such that n(45°) = 0.51, n(90°) = 0.42 and n(0°) = 0.31. This comparison again shows a maximum for the 45° orientation.

High Stress Creep Rupture

High stress creep rupture tests were performed to induce short term failures where the failure mode would be primarily in a high temperature ductile manner minimizing the interphase and grain boundary processes which can dominate creep rupture in longer term tests. Table 2 summarizes the results.

The values obtained for elongation to rupture are relatively independent of orientation and material condition. However the rupture times showed a marked influence of orientation with the 45° orientation exhibiting a failure time much longer than those of the 0° and 90° specimens, Table 2.

Examination of the creep specimens after failure showed differences in macroscopic deformation behaviour. The 0° orientated specimens had generally deformed uniformly to retain circular gauge length cross sections. The 45° and 90° specimens exhibited pronounced elliptical cross sections after deformation in the same manner as found for the tensile specimens. All failures occurred in a ductile, trans-granular mode.

Ductile Fracture Toughness

The values derived for J vs crack extension are plotted in Fig. 10. Linear regression analyses were performed to obtain J_{IC} values from the points of intersection with the blunting line. The values for J_{IC} together with errors derived from the regression analyses were: $J_{IC}(45°) = 114 \pm 23$ kJ/m^2, $J_{IC}(90°) = 112 \pm 32$ kJ/m^2 and $J_{IC}(0°) = 86 \pm 25$ kJ/m^2. Thus, while there are indications that the 0° orientation gives the lowest initiation toughness, Fig. 10, and the 45° orientation has the steepest crack growth resistance curve, the test scatter means that no statistically significant anisotropy can be considered to be present, certainly not to the same degree as found for the elastic moduli or high stress creep rupture.

The fractured pieces of the compact tension specimens exhibited some signs of dominant sets of slip planes, whilst scanning electron microscopy revealed that voids had formed on the fracture surfaces and were related to a homogeneous dispersion of manganese and silicon-rich welding inclusions, all ~1 μm in diameter.

DISCUSSION

The Solidification Induced Texture

The X-ray measurements of Fig. 3 indicate a strong <100> fibre texture coexisting with a weak <110> fibre texture. The principal directions of both textures lie parallel to the columnar grain direction and therefore parallel to the solidification direction. This is a consequence of the solidification characteristics of the weld metal. It is well known that cubic metals generally solidify with a <100> direction parallel to the solidification direction (Savage et al., 1965; Savage and Aronson, 1966). The rapidity of growth in the <100> direction of solidifying cubic metals is such that competitive grain growth, even from nuclei such as randomly orientated base metal grains, quickly leads to the dominance of a <100> direction. Thus, a strong texture throughout the thickness of a weld may be expected despite the reheating effect of a large number of weld passes.

The solidification texture consists of a collection of columnar grains, each with a <100> direction parallel to the solidification direction and each rotated randomly about this <100> direction. The weak <110> texture, also observed parallel to the solidification direction, may be ascribed to the transformation of ferrite to austenite during cooling.

The Influence of Heat Treatment on Texture and Microstructure

The 800°C heat treatment reduces the high dislocation density, found in the as-deposited weld-metal, to regular, two dimensional networks. There is no recrystallization and so the crystallographic texture remains unaltered.

The solution treatment at 1050°C, on the other hand, induced the rapid 'dissolution' of the delta-ferrite and this was accompanied by grain growth, together with the nucleation of new grains, and twin formation, Fig. 6. In parallel with this the X-ray measurement showed a randomizing of the crystallographic texture, Fig. 5. There was no indication of a secondary recrystallization texture.

Elastic Properties

The pronounced anisotropy of elastic modulus observed in the weld metal can be related to the observed texture, as schematically represented in Fig. 4. Work on single crystals of Type 305 stainless steel (Kikuchi, 1971) has shown the <100> direction to have an elastic modulus $E_{<100>}$ = 104 GPa at 20°C. The <111> direction has a modulus $E_{<111>}$ = 320 GPa whilst the elastic modulus $E_{<110>}$ = 210 GPa. Given the compositional insensitivity of E for steels, these results help to provide a qualitative explanation of the observed elastic anisotropy.

A dominant <100> orthotropic texture (Fig. 4) in the weld metal will lead to a reduction in elastic modulus along the <100>

direction/solidification axis from that of a random polycrystalline structure ($E \simeq 200$ GPa) towards the single crystal $E_{<100>} = 104$ GPa. The elastic modulus at 45° to the solidification direction will contain major contributions from $E_{<111>}$ and $E_{<110>}$ resulting in a high modulus of elasticity. The elastic modulus perpendicular to the solidification direction will have major contributions from $E_{<100>}$ and $E_{<110>}$ and so will assume an intermediate level. Thus, the order $E(45°) > F(90°) > E(0°)$ is to be expected and is in agreement with the results, static and dynamic, obtained for the elastic modulus (Fig. 7).

The results are also in accord with other results obtained by Dewey et al. (1977) on a Type 308 weld metal at 20°C. These gave E(0°) = 104 GPa, E(90°) = 142 GPa and E(polycrystal) = 195 GPa. The effect of the heat treatments at 800°C and 1050°C on the elastic moduli is clear. The 800°C heat treatment does not affect the texture significantly and consequently does not affect the values for the elastic moduli, Fig. 7. Increasing time at 1050°C re-crystallizes and randomizes the as-deposited texture and consequently appears to induce an isotropic elastic modulus with a value close to the wrought, annealed value, Fig. 8.

The elastic anistropy is unlikely to be of major structural significance. However, a significant effect may become apparent when stresses are calculated from thermally induced strains using the elastic modulus. In the case of a welded component this presupposes sophisticated local elastic analyses, but in the case of austenitic castings the application may be more straightforward. Another application involving use of elastic moduli is found in fracture mechanics. In particular, conversions of J integral toughness to an equivalent stress intensity, K, will be affected (e.g. $K = \sqrt{EJ}$). Care should be taken when performing single specimen determinations of ductile fracture toughness via the unloading compliance technique. E should be measured, not assumed.

Plastic Properties

From the observations of preferred macroscopic slip and its dependence on specimen orientation it appears that the orientation dependence of uniaxial plastic properties can also be related to the weld metal texture. The predominant slip system in f.c.c. crystals is (111) <110>.

Fig. 11 shows three stereographic projections, where the central axis represents the tensile axis of the 0°, 45° and 90° orientated specimens respectively, and the (111) pole and <110> directions are indicated. In Fig. 11(a) (the 0° orientation) all of the (111) poles and most of the <110> directions are close to 45° to the tensile axis, thus giving a maximum Schmid factor and hence the greatest possible maximum resolved shear stress. Since they are symetrically distributed about the tensile axis, the consequence is symmetrical plastic deformation. In Fig. 11(b), the 90° orientation is represented. A lower level of symmetry is apparent. Some regions exist where the Schmid factor will be at a maximum, and deformation will preferentially take place along these slip systems. Thus preferred slip parallel to the columnar grains is to be expected. Fig. 11(c) shows the 45° orientation. The slip systems are completely assymetrical with respect to the tensile axis.

It is unlikely that specimen orientation significantly influenced the yield strength because at each orientation there were (111) <110> slip systems with almost maximum Schmid factors. However, the stress-strain curves exhibited a high rate of work hardening and this is likely to have dominated the proof stress values.

The measurements of work hardening rate showed it to be greatest for the 45° orientation and least for the 0° orientation. Thus, a lower availability of slip planes, with a greater anisotropy of deformation, was related to a higher rate of work hardening and a higher proof stress.

It should be emphasized that the strength of '316' weld metal is also dependent on its hot-worked microstructural condition. The high density of tangled dislocations clearly explains why the weld is stronger than annealed '316' plate. An 800°C heat treatment will induce recovery and produce dislocation networks, but this weaker condition is still stronger than the annealed condition. Therefore, the assymetrical effects must be understood to be superimposed upon the main, independent strengthening mechanism.

At high temperatures, it may be expected that recovery processes will be increased (Foulds et al., 1983) such that the strain hardening capacity, as defined by $n(\sigma = B\,\varepsilon^n)$, is decreased overall, and that differences in n, as a consequence of the assymetrical distribution, are reduced. Even so, at 538°C the true stress/true strain data indicated the assymetrical 45° orientation to be stronger than the other two examined orientations.

The time-dependent plastic property, creep rupture, also showed a clear dependence upon specimen orientation, Table 2. Since elongation to failure, for all orientations, was very similar, it follows that the 45° orientation was also the most creep resistant, the 90° orientations being almost as weak as the 0° orientations. The macroscopic deformations were identical to those seen for the tensile tested specimens, so one may assume the same slip system behaviour as for the tensile tests. Since creep resistance is a function of work hardening rate, it follows that the influence of specimen orientation on the rate of work hardening also explains the creep behaviours.

The J integral toughness data does not show a statistically significant anisotropy of behaviour. Clearly, the shape of the elastic-plastic stress-strain field about the notch will be influenced by the elastic and plastic anisotropy. Related to this will be a certain anisotropy in work-hardening behaviour, which should influence the plastic deformation and any associated tearing. Indeed, anisotropic macroscopic deformation was visible. However, crack growth in ductile materials is strongly influenced by the nucleation and growth of voids around inclusions. The weld metals contain abundant, homogenously distributed slag inclusions which will negate, to some degree, the manner in which the strain is accumulated. Furthermore, the triaxiality of the applied stress in notched testing will increase the number of available slip planes over the case of simple tensile loading, thus further reducing the effect of the underlying anisotropy. It should be noted that where a solidification texture exists under cleaner conditions (e.g. in austenitic castings) an effect on toughness properties may be observed.

CONCLUSIONS

1. The elastic modulus of Type 17Cr-8Ni-2Mo and 19Cr-12Ni-3Mo weld metals shows a strong orientation dependence with respect to the columnar grain direction in the weld metal. This is equally true in weld pad and butt weld material, in the as-deposited and post-weld heat treated (800°C) conditions. The orientation dependence is $E(45°) > E(90°) > E(0°)$.

2. Plastic deformation, produced in tensile tests, exhibited some macroscopic signs of non-uniform deformation. The orientation dependence of tensile strength (S) also tended to be $S(45°) > S(90°) > S(0°)$ but was much less extreme than the variation in elastic modulus. Furthermore, the magnitude of the yield strength variation as a function of orientation is low compared to heat to heat scatter.

3. Plastic deformation induced under creep conditions also shows macroscopic non-uniformity. The 45° orientation is strongest and the 90° and 0° orientations are almost equally as weak.

4. Any apparent anisotropy in toughness behaviour was statistically insignificant. This is due to the dominating influence in the nucleation and growth of voids of the welding slag inclusions. It should be noted that conversions of J integral toughness values to equivalent stress intensities, K, (e.g. $K = \sqrt{EJ}$) will be dependent on the choice of elastic modulus, E.

5. The anisotropic properties can be correlated with the existence of a strong solidification texture. All weld metals examined exhibited a strong <100> orthotropic texture, with a weaker co-existing <110> fibre texture, the principal directions being parallel to the columnar grain directions.

6. A post weld heat treatment of 10 h @ 800°C did not change the crystallographic texture or anisotropic elastic/plastic properties.

7. Solution treatment at 1050°C of the weld metals resulted, at times ~4 h and greater, in a re-crystallization of the weld microstructure and a consequent elimination of the texture and elastic anisotropy.

ACKNOWLEDGEMENTS

The authors would like to thank Dr D.S. Wood (UKAEA, Risley) and Dr K. Gilchrist (UKAEA, Springfields) for their assistance in the ultrasonic determinations of elastic modulus, and Mr E.H. Kelly (ECRC, Capenhurst) for performing the texture measurements. The mechanical testing and metallography were carried out at the Central Electricity Research Laboratories and the paper is published by permission of the Central Electricity Generating Board.

REFERENCES

ASTM, 1981 - Standard Test Method for J_{IC}, a Measure of Fracture Toughness, ASTM Standard E813-81

Cross, A.D. and Kelly, E.H., 1979 - Computer-Aided Preparation of Preferred Orientation Pole Figures, Electricity Council Research Report ECRC/M1223

Delong, W.T., 1974 - Ferrite in Austenitic Stainless Steel Weld Metal, Welding J., 53, 273-5. July

Dewey, B.R., Adler, L., King, R.T. and Cook, K.V., 1977 - Measurement of anisotropic elastic constants of Type 308 stainless steel electroslag welds, Proc. Soc. for Exptl. Stress Analysis, 34, 2, 420

Foulds, J.R., Moteff, J., Sikka, V.K. and McEnerney, J.W., 1983 - Deformation behaviour of a 16-8-2 GTA weld as influenced by its solidification substructure, Met. Trans. A., 14A, 1357, July 1983

Kapranos, P.A., Al-Helaly, M.M.H. and Whittaker, V.N., 1981 - Ultrasonic velocity measurements in 316 austenitic weldments, Brit. Jn. of NDT, 23, 6, 288

Kikuchi, M., 1971 - Elastic anisotropy and its temperature dependence of single crystal and polycrystal Type 18-12 stainless steel, Trans. J.I.M., 12, 117

Savage, W.F. and Aronson, A.H., 1966 - Preferred orientation in the weld fusion zone, Weld. J., 45, 2, 85-s

Savage, W.F., Lundin, C.D. and Aronson, A.H., 1965 - Weld metal solidification mechanisms, Weld. J., 44, 175-s

Table 1: Chemical Analyses of Weld Metals

(wt.%)	Pad A	Pad B	Pad C	Weld D	Pad E
Boron	0.0005	0.0004	0.0002	0.0004	0.0003
Carbon	0.08	0.07	0.04	0.07	0.061
Chromium	17.6	18.2	17.7	18.3	17.7
Copper	0.07	0.05	<0.05	<0.1	0.06
Manganese	2.06	2.40	0.83	2.0	2.08
Molybdenum	2.23	1.70	2.65	1.37	1.59
Nickel	8.4	8.6	11.2	9.1	9.12
Phosphorus	0.027	0.021	0.021	0.017	0.024
Silicon	0.76	0.3	0.34	0.34	0.32
Sulphur	<0.01	<0.002	<0.008	<0.01	0.01
Ferrite No.	8.5-10	5-6	3.5-6	6.3-8.3	6.5-7.5
Cr_{eq}	20.97	20.35	20.86	20.01	19.77
Ni_{eq}	11.83	11.90	12.82	12.20	11.99

Cobalt, Niobium, Titanium and Vanadium <0.1,

Table 2: Rupture Times and Final Elongations Obtained from High Stress Creep Rupture Tests at 538°C (all weld pad A)

Stress	Material Condition		Orientation 0°	45°	90°
350 MPa	As-deposited	Rupture time (h)	37	214	41
		Elongation (%)	20	27	26
"	Post-weld heat treated (800°C)	Rupture time (h)	0.5	105	3
		Elongation (%)	23.5	25	23
300 MPa	As-deposited	Rupture time (h)	223	1669	323
		Elongation (%)	24	24	26

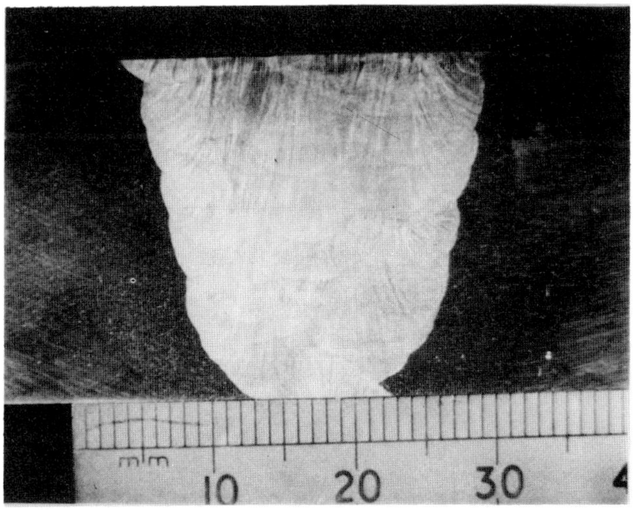

a

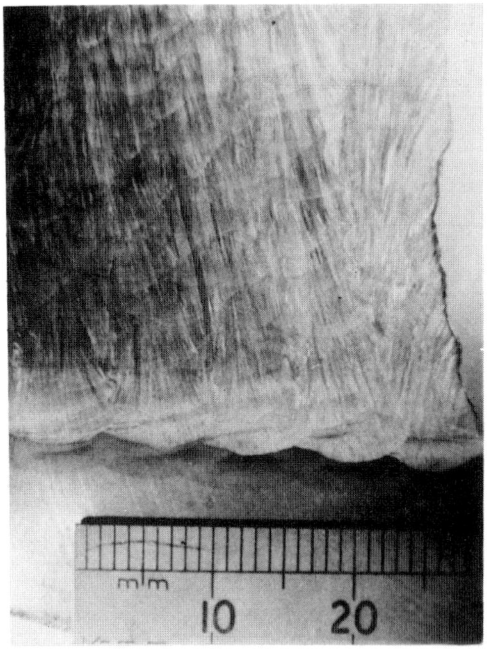

b

1 Macro-etches of 17Cr8Ni2Mo austenitic stainless steel welds (a) Butt weld, (b) Weld pad (as-deposited on base plate)

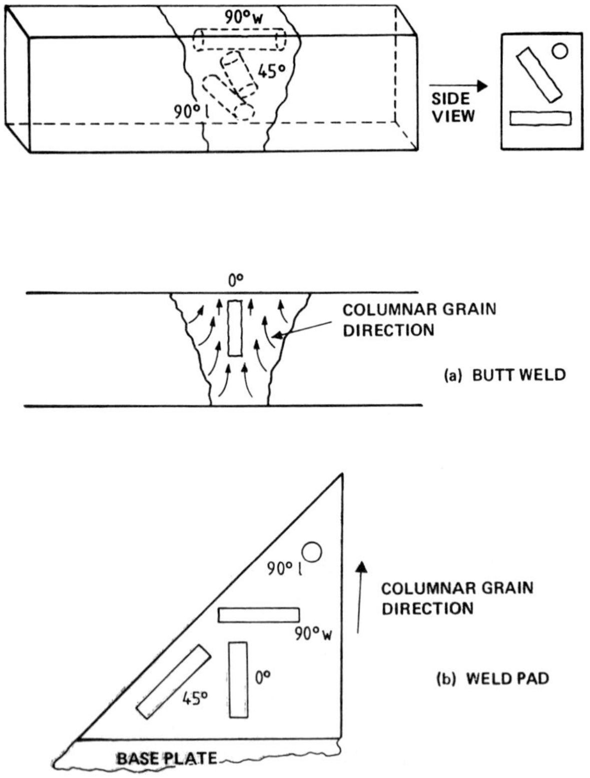

2 Orientations of ultrasonic specimens cut from the weld metal

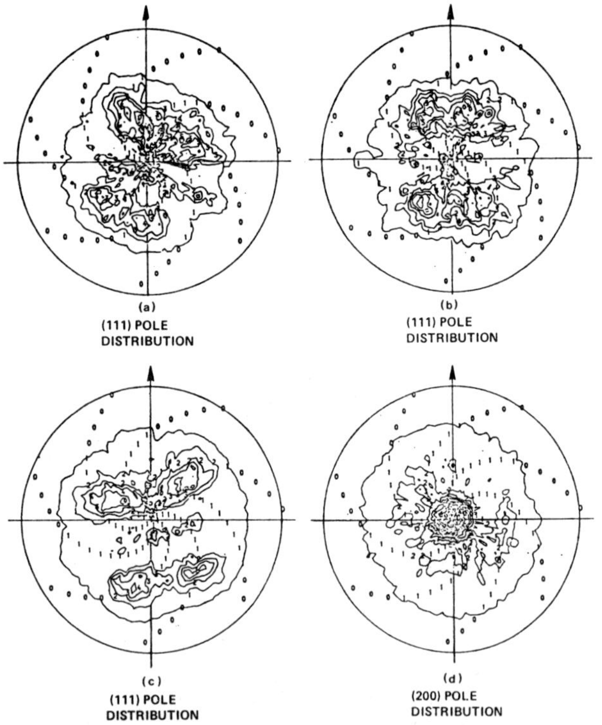

(a) AS-WELDED WELD PAD A
(b) AS-WELDED BUTT WELD D
(c) POST-WELD HEAT TREATED (800°C) PAD A
— SPECIMEN SURFACE PARALLEL TO GRAIN DIRECTION. VERTICAL AXIS MARKS GRAIN DIRECTION

(d) AS-WELDED BUTT WELD D — SPECIMEN SURFACE PERPENDICULAR TO GRAIN DIRECTION, GRAIN DIRECTION PERPENDICULAR TO PLANE OF FIGURE

3 Pole figures from samples of the welds

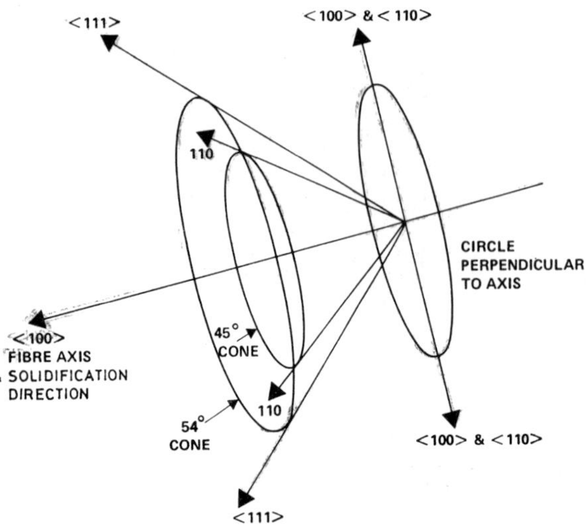

4 Schematic representation of the <100> texture

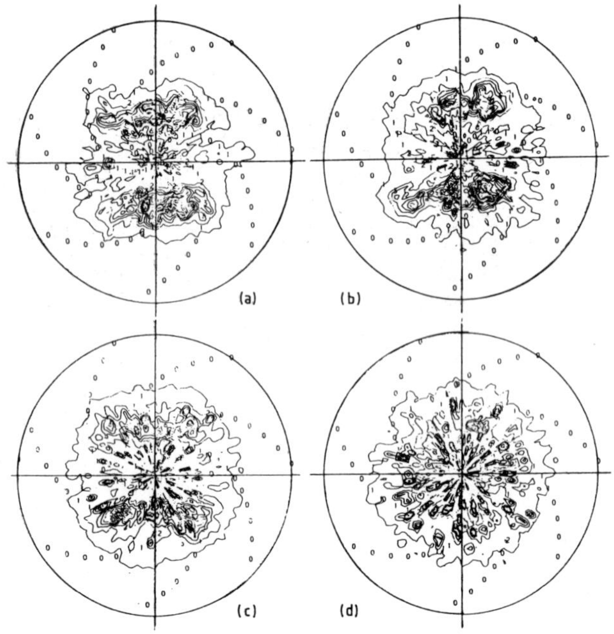

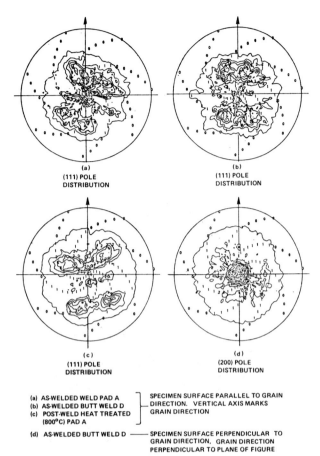

5 Pole figures of weld metal samples after various times at 1050°C. (a) As-deposited, (b) 2 h, (c) 4 h, (d) 8 h

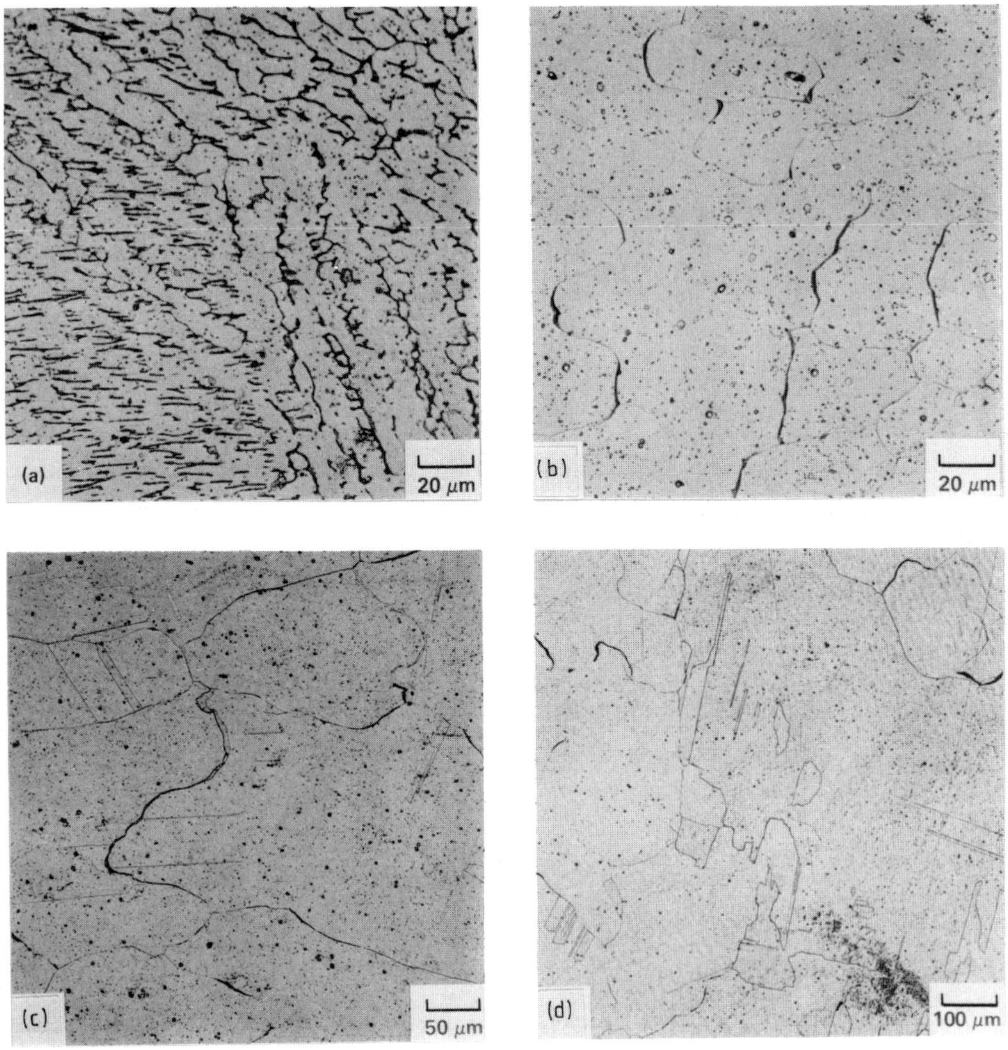

6 Microstructural changes induced by heat treatment at 1050°C, weld pad A. (a) As-deposited, (b) 2 h, (c) 4 h, (d) 8 h

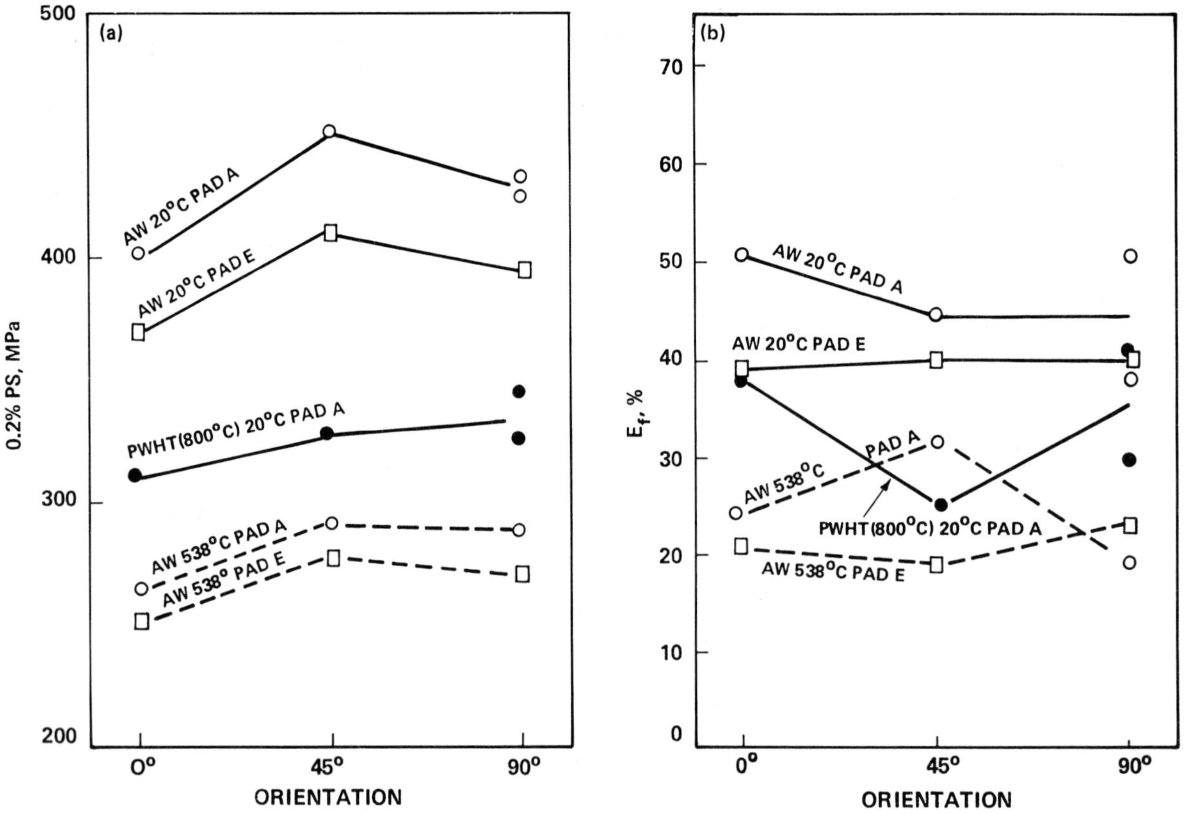

7 Measurements of elastic moduli;
(a) measured ultrasonically at 20°C,
(b) from tensile testing at 20°C and 538°C

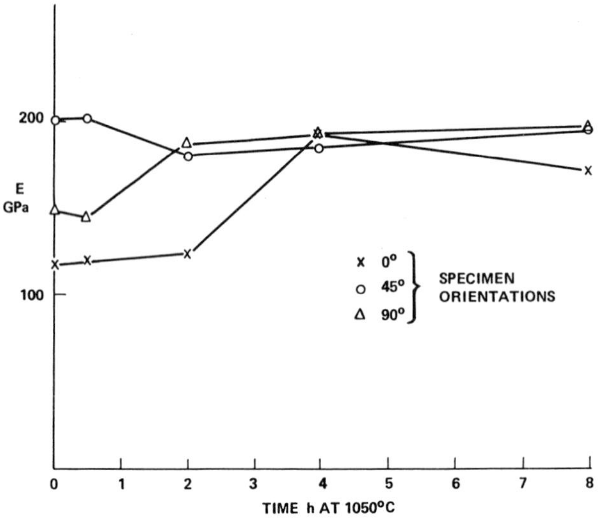

8 Effect of heat treatments on the elastic moduli in weld pad A

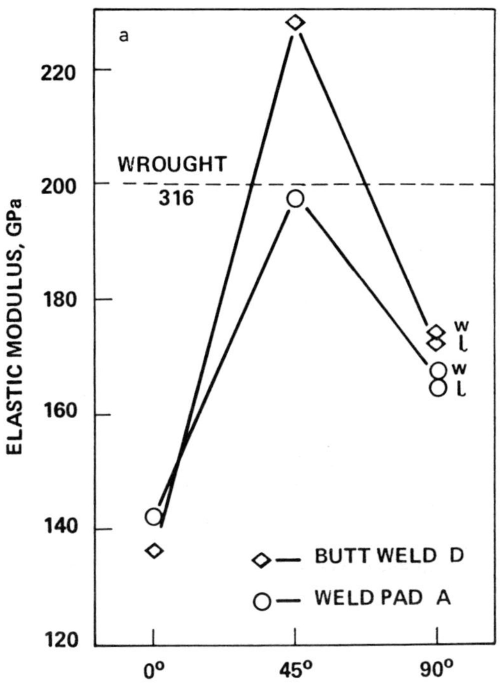

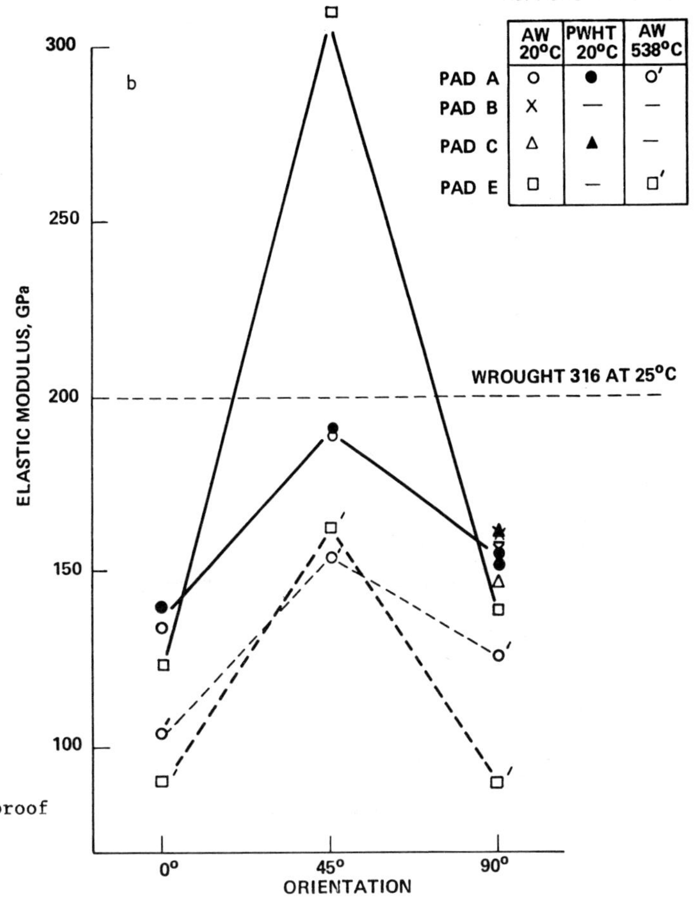

9 Tensile properties as a function of orientation for weld pads A and E; (a) proof stress, (b) elongation at rupture

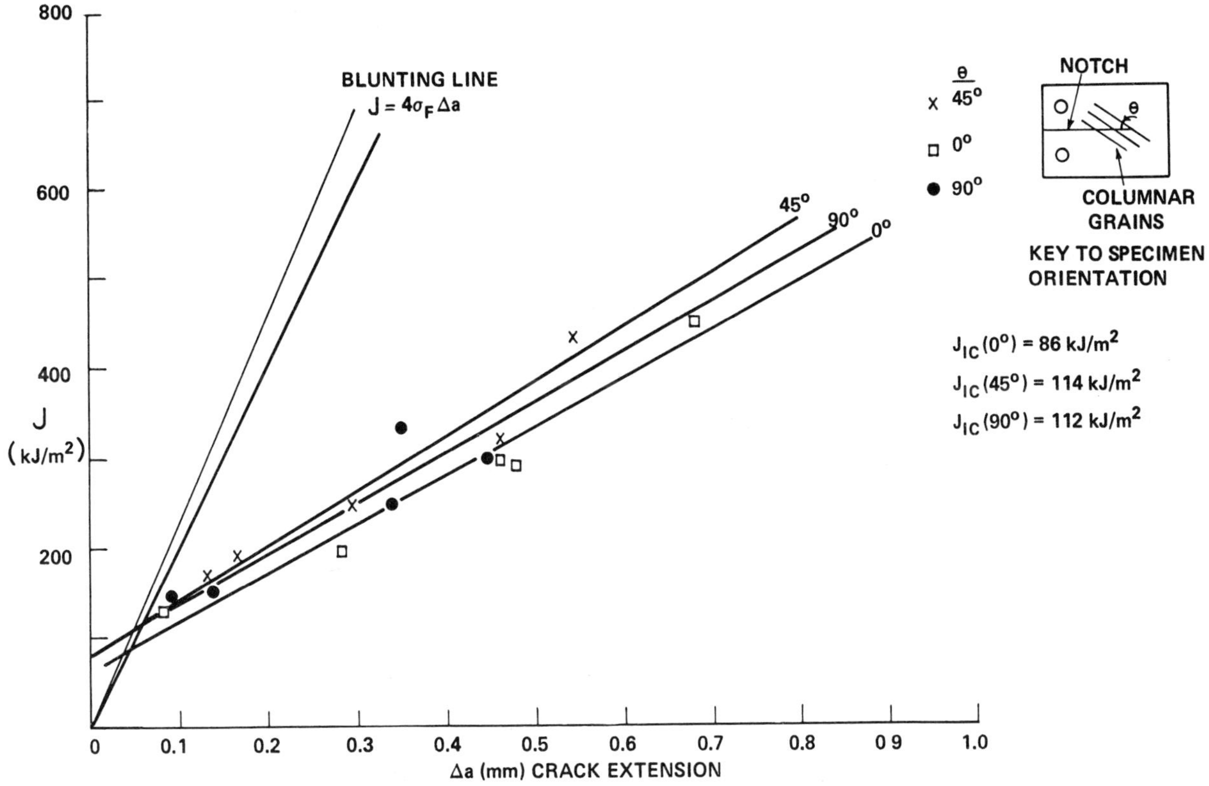

10 Orientation dependence of toughness properties of weld pad E tested at room temperature

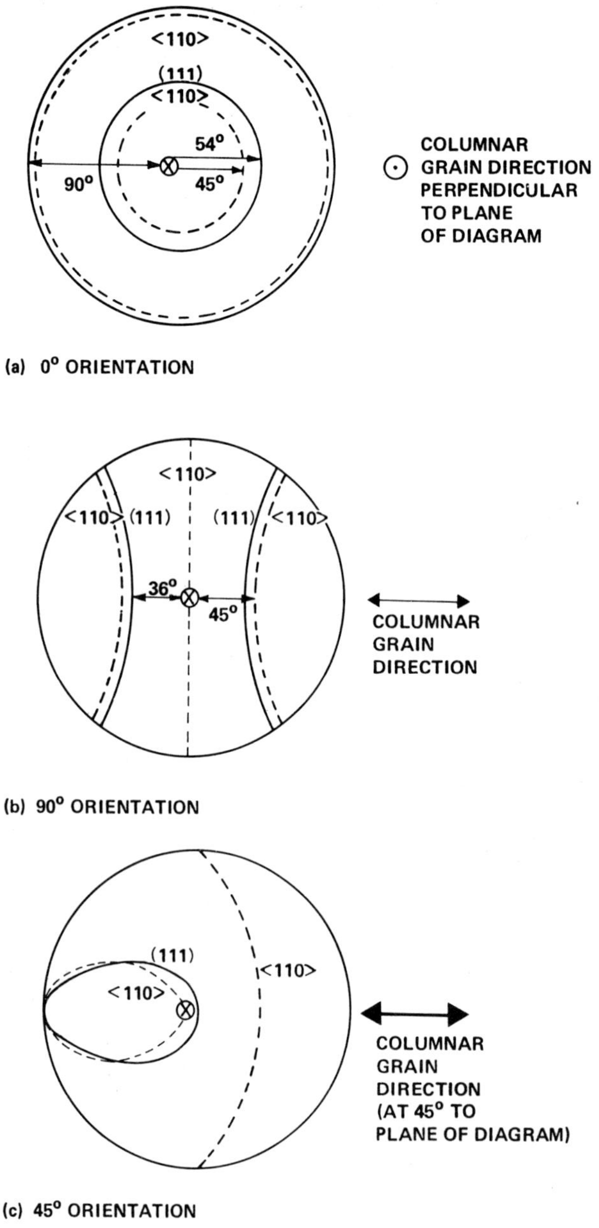

11 Stereographic projection of (111) poles and <110> directions with relation to tensile axes, at diagram centres

Physical metallurgy of 'Fecralloy' steels

S R Keown

The author is with Robert Keown and Associates, Sheffield and a Consultant to Harwell Laboratory.

SYNOPSIS

The development of Fecralloy* steels which contain 15 to 23%Cr, 4 to 5.5%Al and 0.05 to 0.5%Y was initiated at the UKAEA's Harwell Laboratory in 1965 and continuous experimentation and development has progressed to the present time. Initially the work concentrated on the optimisation of steel compositions but during the 1970's applications became a priority and more recently product form has received most attention. The first commercial cast of Fecralloy steel was produced in 1974 and the steels have achieved major industrial success as resistance heating element wire. Currently the steels are candidate materials for a number of high technology applications where the product form ranges from wire and strip to tube and cladding.

This paper concentrates on the physical metallurgy of Fecralloy steels, attempting to review the large number of previously unpublished internal reports produced by Harwell alongside several published papers. Information on constitution, microstructure, oxidation resistance, grain control and mechanical properties is reviewed, but details of applications, fabrication aspects and the substitution for yttrium which have been the subject of many other investigations by Harwell and by the steel industry are not included in this review.

Fecralloy steels emerge as a unique family of alloy steels with the best oxidation resistance of any ferrous material. This is achieved by the combination of 5%Al which gives an excellent oxidation resistant α alumina layer and yttrium which provides adhesion of the alumina to the steel. Yttrium-rich phases prevent grain growth at very high temperatures and Fecralloy steels remain ductile even after extended ageing at 1200 °C.

* Registered Trade Mark of the United Kingdom Atomic Energy Authority.

INTRODUCTION

Fecralloy steels are ferritic stainless steels containing 15 to 23%Cr, 4 to 5.5%Al and 0.05 to 0.5%Y. The Fe-Cr-Al-Y alloys were originally developed in the USA by General Electric for nuclear applications. In an extensive alloy development programme, General Electric established 2 basic compositions-15%Cr, 4%Al, 1%Y; and 25%Cr, 4%Al, 1%Y for steels which had good combinations of fabricability, mechanical properties and oxidation resistance. A collaboration between General Electric and AERE Harwell resulted in the further development of Fe-Cr-Al-Y compositions by Harwell (2)(3) and the registration of a family of steels under the Fecralloy trade mark in December 1973.(3)

The 3 alloying elements chromium, aluminium and yttrium all contribute to the oxidation resistance of the Fecralloy steels to varying degrees. For resistance above 800 °C, aluminium is the most important constituent and its minimum concentration should be 4%. With aluminium-containing steels the chromium concentration is not so critical for oxidation in air but a minimum level of 15% is required for resistance to steam. This level of chromium also ensures stainlessness which is useful for some applications and also for storage considerations. Yttrium additions improve the adherence of the protective scale, extending the useful range of aluminium-containing steels to very high temperatures. Yttrium also reduces nitrogen pick-up by aluminium steels in air and other atmospheres.

Although the Harwell Laboratory was initially primarily interested in the nuclear properties and applications of the Fecralloy steels, they realised the non-nuclear industrial potential of the alloys as early as 1967 when attempts were made to manufacture Fecralloy wire for use in furnace windings.(3) Due to the excellent oxidation resistance of Fe-Cr-Al-Y alloys Fecralloy steels have achieved major industrial success as electric furnace elements in wire and strip form in oxidising atmospheres up to 1375 °C.(4) In 1973 the potential for using Fecralloy steel for car catalyst substrates was identified and subsequent effort was placed on the commercial production of Fecralloy in the form of thin metallic foil. The industrial production of Fecralloy steels in tonnage quantities was achieved by the close interaction of Harwell with the Sheffield steel industry and 2 licensees, Resistalloy Limited and Firth Brown-Fox, started to produce Fecralloy in commercial quantities about 10 years ago. Fecralloy steels are currently being produced by Resistalloy Limited, and Stocksbridge Precision Strip Limited (part of the British Steel Corporation) and the steels are being assessed for a wide range of high technology applications, including fluidised bed components, gas purification plant, printed circuit board substrates and energy conservation systems, where the excellent oxidation and sulphidation resistance of the steels is of major importance. Typical Fecralloy specifications are shown in Table 1.

CONSTITUTION AND MICROSTRUCTURE

The composition range of Fecralloy steels is shown in Fig.1 superimposed on a schematic ternary section of the Fe-Cr-Al phase diagram. It can be seen that Fecralloy compositions are remote from the intermetallic σ Fe-Cr and Fe-Al phases under equilibrium conditions but the σ phase has occasionally been observed in aged specimens.(2) Fig.1 also indicates that the α' chromium-rich ferrite phase might occur in the higher chromium grades of Fecralloy. Cook and Roberts(5) have in fact observed the α' phase in a range of Fecralloy steels using transmission electron microscopy techniques, Fig.2. The influence of 4% Al on the Fe-Cr system, (6) Fig.3, shows the anticipated ferrite stabilisation and it has been suggested that yttrium has a similar effect.(5) Poole has noted that the basic constitution of the Fe-Cr-Al system seems to be little affected by the presence of small amounts of yttrium(7) but the precipitation processes are markedly affected.

A number of yttrium-containing phases have been reported in Fecralloy steels but the major phases identified are YFe_9 and Y_2O_3. Large volume fractions of YFe_9 precipitates have been confirmed by X-ray and electron diffraction analysis.(2)(8) More recently Feest(9) has proposed the Y_2Fe_{17} phase (containing small amounts of Cr and Al) on the basis of microanalysis of 'as cast' specimens from controlled solidification and thermal analysis investigations but this identification was not confirmed by crystallographic techniques. Feest has deduced the schematic phase diagram shown in Fig.4 for the quaternary system at elevated temperatures. This shows the formation of the $Y_2(FeCrAl)_{17}$ phase below 1400°C and the phase was observed metallographically as filamentary interdendritic regions. Feest also reported a globular complex eutectic phase, possibly the 3-component α-Fe_2Y-Y_2O_3 eutectic. Another minor phase identified was YFe_2 in association with YFe_9 and unidentified smaller grain boundary precipitates have been assumed to be carbides, (2) presumably $M_{23}C_6$.(8) The yttria Y_2O_3 phase is commonly encountered in Fecralloy steels aged above 900°C(10) as a discontinuous network at grain boundaries. The Y_2O_3 often forms at the expense of YFe_9, as observed by depleted zones of the intermetallic phase. Most commercial wire material contains significant amounts of Y_2O_3 formed by internal oxidation during the pre-oxidation heat treatment process but experimental specimens of Fecralloy that have not been subject to internal oxidation may not contain Y_2O_3.

Both the YFe_9 and the Y_2O_3 phases are extremely stable up to very high temperatures and they are thought to be responsible for the remarkable grain growth resistance of Fecralloy steel up to at least 1200°C.

The solubility of yttrium in ferrite in the presence of YFe_9 in a 0.32%Y Fecralloy steel has been determined by Feest(9) as 0.03%Y so that large volume fractions of yttrium-containing phases will be present in steels containing relatively small amounts of yttrium.

OXIDATION RESISTANCE

Fecralloy steels have the best oxidation resistance of any alloy steel and in the pre-oxidised condition can be used up to 1375°C as resistance heating elements.(4)

When the steels are heated above 1000°C in an oxidising atmosphere, a protective oxide scale which is predominantly α Al_2O_3 is formed on the surface of the steel.(11) This oxide replaces the normal air-formed chromium oxide film which is usually present on the steels which are of course stainless because of their high chromium content. The alpha alumina has a low native defect concentration which reduces material transport and slows down the oxidation process.(11) However, Antill and Peakall(10) showed that a 15%Cr, 4%Al Y-free steel suffered catastrophic oxidation above 1000°C due to spallation of the oxide film. The addition of the reactive element yttrium was shown to prevent oxide spalling and in combination with the alumina layer to promote oxidation resistance of the steel to very high temperatures. The mechanism by which yttrium imparts adherence to the oxide layer is thought to be associated with grain boundary macro-pegging of the layer by yttrium-rich phases, Fig.5, and/or by the presence of internally oxidised Y_2O_3 precipitates which act as vacancy sinks and thereby prevent void formation at the scale interface.(10) These mechanisms and observations are consistent with those proposed by Whittle and Stringer(12) for the improvement of high temperature oxidation resistance.

The weight gain oxidation data obtained by Antill and Peakall(10) for Fe-Cr-Al and Fe-Cr-Al-Y alloys on exposure to carbon dioxide atmospheres at 800° to 1200°C is shown in Fig.6. The major benefit of adding 0.43% and 0.86%Y to the steel is clearly shown. Fig.7 shows the effect of increasing the oxidation temperature from 1100° to 1430°C on the oxidation resistance of a 25%Cr, 4%Al, 1%Y alloy.(1) A further series of tests, Fig.8, established that as little as 0.04%Y was effective in improving the corrosion behaviour of the steel.(2)(7) It is interesting to note that this amount is similar to the measured solubility of yttrium in the steel. Grain growth considerations can justify larger additions of yttrium as will be shown later.

When Fecralloy is oxidised at temperatures lower than 1000°C, theta alumina with a whisker morphology forms on the steels.(11) It has been noted that this morphology can be advantageous in catalysis applications where the whiskers form an excellent key for the entrapment of washcoatings of noble metal catalysts.(11)(13)

GRAIN SIZE AND MECHANICAL PROPERTIES

The effect of yttrium content on the grain size of 15%Cr, 4%Al Fecralloy steels is shown in Fig.9.(14)(15)Whereas the amount of yttrium necessary to contribute to oxidation is less than 0.1%(2)(15),it has been suggested that about 0.35%Y is required for optimum grain refinement.(2) Fig.9 shows that a significant refining effect occurs between 0.1 and 0.2%Y with the smallest grain sizes at about 0.4%Y. This Figure gives the relationship between grain size and yttrium content after various levels of cold work and recrystallisation at 1000°C for 1 hour. Similar results were obtained after recrystallisation at temperatures between 700° and 900°C. The same data with grain size plotted as a function of cold reduction is given in Fig.10 for alloys containing 0.08 and 0.6%Y.(14) It can be seen that the very significant grain refinement obtained by the addition of 0.6%Y is maintained after cold work and recrystallisation. Fig.10 also shows the abnormal grain growth effects that can occur with 10 to 15% cold work and annealing at about 1000°C (consistent with Figure 9.) Hardness data on the same series of specimens indicates the work hardening response of the 0.6%Y alloy, Fig.11, and also that recrystallisation is complete in the specimens annealed at 800° to 1000°C but not at 700°C.(14)

All the information illustrated in Figs. 9-11 and other unpublished data provides guidelines for the composition and fabrication of the steel where grain refinement is important for subsequent rolling and drawing processes. Steels containing between 0.2 and 0.4%Y should be cold worked less than 5% or greater than 20% followed by annealing at 800°C for about 1 hour to give optimum grain refinement.

The effect of extended ageing of a 15%Cr, 5%Al, 0.3%Y Fecralloy steel and of a 20%Cr, 5%Al, Y-free steel showed that the grain control due to yttrium continues up to 1200°C, Fig.12,(16) and that the Fecralloy steel has superior ductility at 20°C after ageing for up to 1000 hours between 800° and 1200°C, Fig.13 (data replotted from reference 16). This latter Figure contains information on tensile ductility obtained using fast and slow strain rates of 4 and 4 x 10^{-4} sec^{-1} respectively and shows that Fecralloy still exhibits 18% elongation on slow strain rate testing following ageing for 1000 hours at 1200°C.

Taken together with the grain size data in Fig.12, it can be implied that the superior ductility of Fecralloy steel after extended ageing at high temperatures is related to the smaller grain size of the alloy compared with an Y-free steel. In this work Cook and Roberts(16) claimed that the grain refinement in the Fecralloy steel was due to a random distribution of YFe_4 precipitate particles but Roberts has since conceded that the Y_2O_3,produced by internal oxidation during extended high temperature ageing, could also contribute to the grain refinement.(17)

The tensile properties of a 15%Cr, 4%Al, 0.86%Y Fecralloy steel and a 15%Cr, 4%Al, Y-free alloy have been determined at testing temperatures between 20°C and 950°C by Roberts et al(8), Fig.14. This Figure also contains information on irradiation effects but the 'unirradiated' properties are clearly indicated. The steels were annealed at 1000°C for 1 hour and the 0.86%Y steel had a grain size of 17µ compared with 100µ for the Y-free steel. The strength of the Fecralloy steel was marginally greater than that of the 15%Cr, 4%Al alloy but the ductility was little affected with a testing strain rate of 2 x 10^{-4} s^{-1}. The increased strength was presumably due to grain refinement and to the distribution of intermetallic YFe_4 precipitates. These properties are typical of a ferritic stainless steel with relatively poor high temperature tensile strength but good ductility.

Impact properties have not been studied systematically with respect to grain size but data which is available from a wide range of experimental alloys(7) indicates fairly high impact transition temperatures and low shelf energies. Typical impact data on a 16%Cr, 5%Al, 0.4%Y alloy hot rolled from 1000°C in a multi-stand mill is shown in Fig.15 with transition curves for 'as rolled steel' and for material reheated to 1000°C.(18) The much lower impact transition temperature of the as rolled steel was related to a fine-grained recovered structure whereas reheating produced coarse recrystallised grains with a subsequent increase in impact transition temperature to a very high level. Obviously, as with all ferritic stainless steels, the impact properties of Fecralloy can be controlled by grain refinement to give acceptable toughness.

As noted earlier, Fecralloy steels are susceptible to 475°C embrittlement by the precipitation of α' chromium-rich ferrite, Fig.2. Systematic work by Cook and Roberts(5) has related age hardening curves and tensile data to the microstructural detection of α' clusters lying on $\{100\}_\alpha$ planes. Ageing curves for temperatures between 200°C and 550°C for ageing times up to 500 hours are shown in Fig.16 where it can be seen that maximum age hardening occurs at 475°C. Stress-strain data obtained at a strain rate of 4 x 10^{-4} s^{-1} also shows age hardening with reduced ductility, Fig.16. Ductile-brittle transition curves of tensile ductility at 3 different strain rates, for solution treated and aged conditions showed an increase in transition temperature and a lowering of the ductility above the transition temperature, associated with 475°C ageing, Fig.17.

Fecralloy steels have very poor hot strength compared with creep-resisting steels and, understandably, little information is available on the creep properties of Fecralloy. However, one significant paper by Lobb and Jones(19) has provided useful data. Fig.19 shows the stress rupture properties of 14%Cr, 4.5%Al, 0.82%Y Fecralloy between 650°C and 800°C(19) and Lobb and Jones went on to find a transition in the creep behaviour of Fecralloy when plotting the temperature dependence of the secondary creep rate. This transition was close to the Curie temperature (710°C) for an Fe-15%Cr alloy and the transition was attributed to changes in the lattice diffusion coefficient of the steel. Fig.20 gives a comparison between the creep strength of Fecralloy and that of commercial creep-resisting steels.(19) It was shown that Fecralloy is 2 to 3 orders of magnitude weaker than the Nb steel and 5 to 6 orders of magnitude weaker than the nitrided Ti steel.

CONCLUSIONS

Fecralloy steels are emerging after 25 years of development as a very special family of alloys(20) for an expanding range of applications. Fecralloy steels are unique in that they can withstand oxidation up to 1375°C. This oxidation resistance is achieved by the combination of approx 20%Cr for stainlessness, 5%Al to provide an alumina layer on the steel surface and less than 0.1%Y to promote adhesion of the alumina layer. A larger yttrium addition also imparts grain refinement to the steel due to the precipitation of YFe_4 and Y_2O_3, the latter phase occuring due to internal oxidation. Fecralloy steels retain their small grain size on extended ageing above 1000°C and this ensures a high degree of ductility even after extensive service at these very high temperatures.

Apart from their remarkable oxidation resistance, Fecralloy steels exhibit typical mechanical properties for a ferritic stainless steel with difficult fabricability, poor hot strength, low impact resistance and a susceptibility to 475°C embrittlement. However, these problems can either be eliminated or modified by careful attention to processing and heat treatment parameters or they may be tolerated by careful component design. Fecralloy steels can now be readily hot and cold rolled down to 0.05mm strip or drawn to 0.025mm dia. wire. The steels can be fabricated by welding, hot extruded as composite clad tubing, deposited by plasma spraying and used in powder form for hot isostatic pressing. Recent developments include the dispersion of Fecralloy fibres or rapidly solidified fragments in composite materials. Most of these product developments are the result of close collaboration between the Harwell Laboratory and industry which started about 13 years ago when the Sheffield steel industry became involved in the up-scaling of Harwell's experimental alloys. Current and future applications include resistance heating wire and strip, foil for car exhaust catalyst substrates, electronic printed circuit board substrates, furnace cladding panels and other insulation systems, tubes for fluidised bed components, co-extruded clad tubing for power generation particularly for high sulphur fuels, and aerospace components.

ACKNOWLEDGEMENTS

Discussions and collaboration with many people in Harwell and in the steel industry are gratefully acknowledged with particular reference to the contributions and assistance of Dr. D.T. Livey, Dr. M.J. Bennett, Dr. B. Hudson, Dr. A.C. Roberts, Dr. E.A. Feest, Dr. P.T. Mosely, Mr. R. Ainsley and Mr. M. Lofting of Harwell, Mr. J. Hornbuckle of Stocksbridge Precision Strip, Mr. A.L.P. Wood, Mr. P. P. Wood, Dr. K.C. Barraclough, Mr. S. Coghlan and Mr. N. Woodcock of Resistalloy, Dr. F.B. Pickering and Dr. L. Taylor.

REFERENCES

1. J. E. Antill, D. R. Harries and H. Lloyd, AERE-M1608,1965.

2. J. E. Antill, H. Lloyd and A. C. Roberts, JNPC/FEWP(CPSG)/P(67)26,1967.

3. R. W. Barnfield and R. S. Nelson, Unpublished Information, 1980

4. Fecralloy brochure, Resistalloy Ltd.,1986.

5. J. Cook and A. C. Roberts, Unpublished Information, 1976.

6. C. S. Wukusik, (referenced in 7) GEMP-414,1966.

7. D. M. Poole, Unpublished Information, 1969.

8. A. C. Roberts, D. R. Harries, D. R. Arkell, M. A. P. Deway and J. D. H. Hughes, ASTM STP 457,312,1969.

9. E. A. Feest, Unpublished Information, 1977

10. J. E. Antill and K. A. Peakall, JISI,205,1136,1967.

11. P. T. Moseley, K. R. Hyde, B. A. Bellamy and G. Tappin, Corrosion Science,24,547,1984.

12. D. P. Whittle and J. Stringer, Phil.Trans.Roy.Soc.,A295,309,1980.

13. A. S. Pratt and J. A. Cairns, Platinum Metals Review,21,74.1977.

14. J. M. Davies and C. Steer, AERE-R6032,1969.

15. J. M. Davies, M. L. Noakes and J. C. Purchas, Unpublished Information, 1977.

16. J. Cook and A. C. Roberts, Unpublished Information,1979.

17. A. C. Roberts, Private communication.

18. R. M. Boothby and A. C. Roberts, Unpublished Information,1978.

19. R. C. Lobb and R. B. Jones, J.Nuc.Mat.,91,257,1980.

20. J. E. Antill, Stainless Steel Industry J.,7,13,1979.

TABLE 1. Typical Fecralloy specifications

	Steel Composition			
	C	Cr	Al	Y
General Range	<0.03	15-23	4-5.5	0.05-0.5
P.Wood & Co,1978 Car Exhaust Strip	0.03	15.5-16.5	4.7-5.0	0.4
Stocksbridge Prec.Strip 1987 Car Exhaust Strip	0.02	20-23	5.0-5.5	0.1
Resistalloy 1987 Fecralloy Wire	0.02	20-22	4.8-5.2	0.1
Resistalloy 1987 Fecralloy 'A' Wire	0.01	20-22	4.8-5.2	0.3

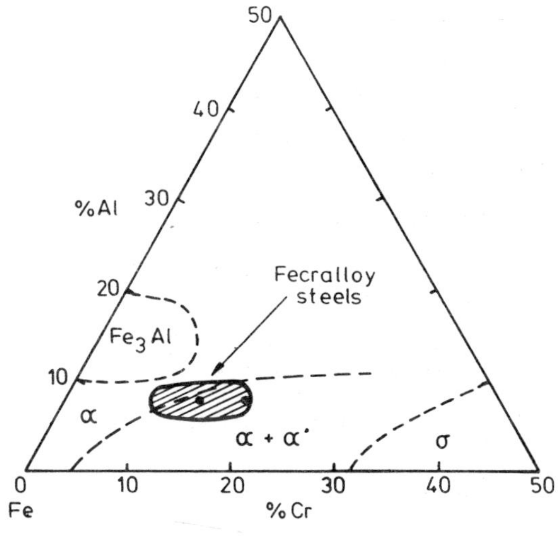

FIG. 1. SCHEMATIC REPRESENTATION OF FECRALLOY STEEL COMPOSITION RANGE ON Fe-Cr-Al TERNARY SECTION.

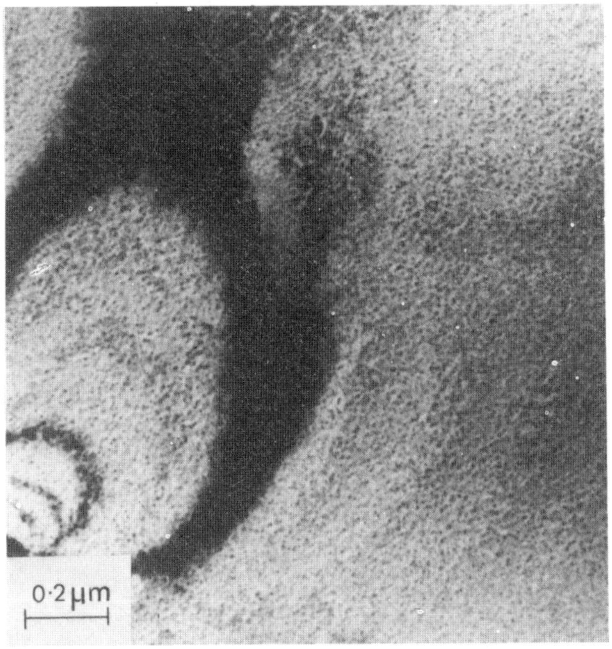

FIG.2 TRANSMISSION ELECTRON MICROGRAPH OF CLUSTERS IN FECRALLOY STEEL AGED AT 475°C

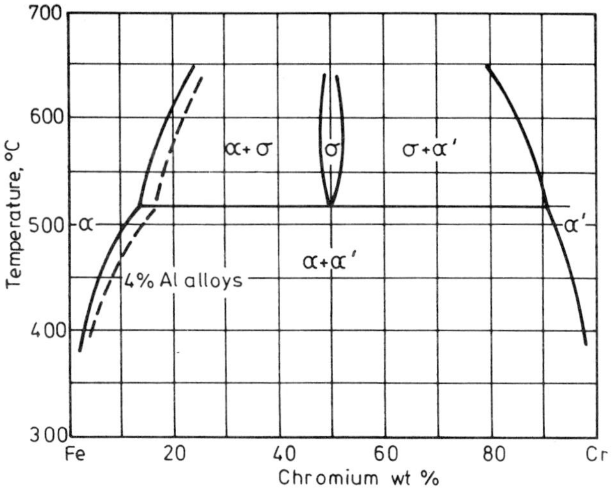

FIG. 3. IRON-CHROMIUM BINARY DIAGRAM AND THE EFFECTS OF 4% ALUMINIUM ADDITION.

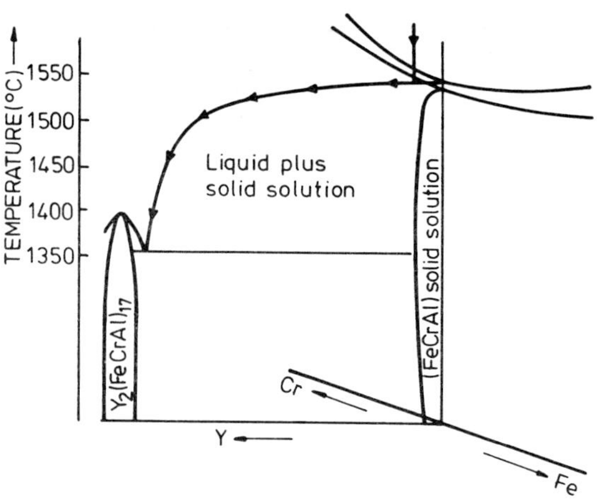

FIG. 4. SCHEMATIC DIAGRAM OF THE PHASE RELATIONSHIPS GOVERNING THE INITIAL STAGES OF FECRALLOY SOLIDIFICATION.

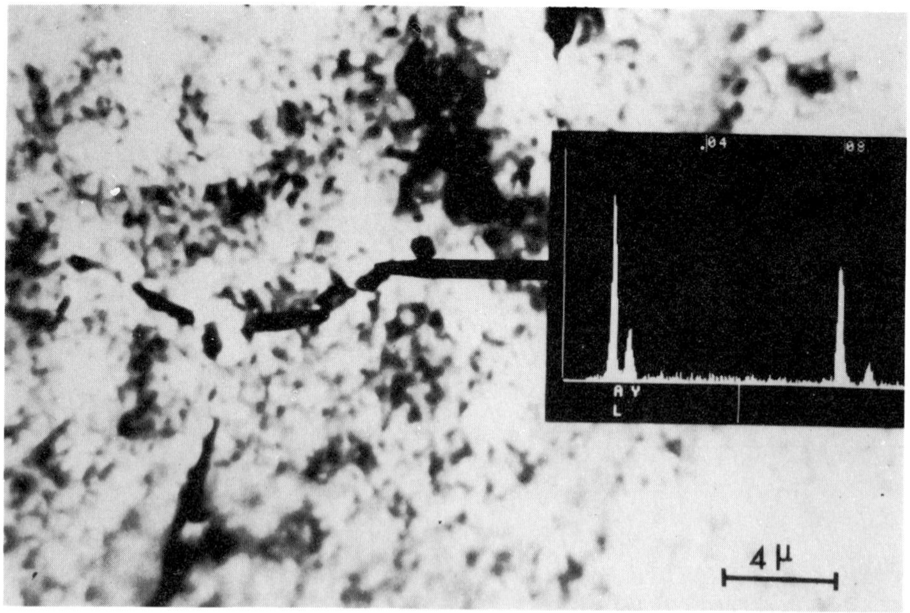

FIG.5 TRANSMISSION ELECTRON MICROGRAPH AND MICROANALYSIS TRACE OF THIN FRAGMENT OF SCALE OBTAINED ON FECRALLOY STEEL AGED AT 1100°C. THE SCALE IS α ALUMINA AND THE DARK RIBBONS ARE YTTRIUM-RICH AND COINCIDE WITH THE INTERSECTION OF THE STEEL GRAIN BOUNDARIES AND THE BASE OF THE SCALE i.e. GRAIN BOUNDARY MACRO-PEGS.

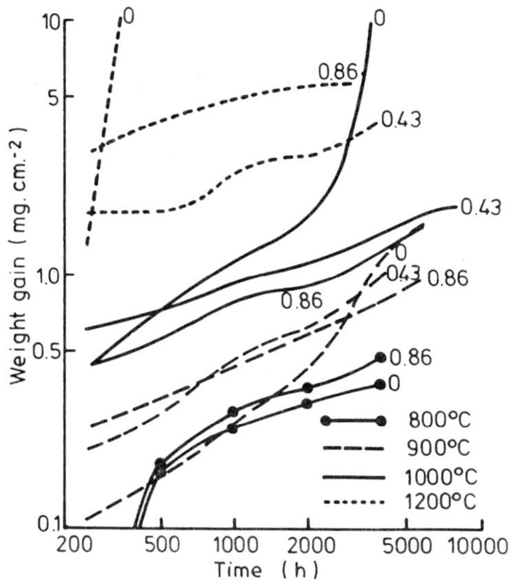

FIG. 6. INFLUENCE OF YTTRIUM ADDITION TO FECRALLOY STEELS UPON THE VARIATION OF WEIGHT GAIN WITH OXIDATION TIME IN AIR. VALUES AT END OF CURVES DENOTE YTTRIUM CONCENTRATION IN wt. %.

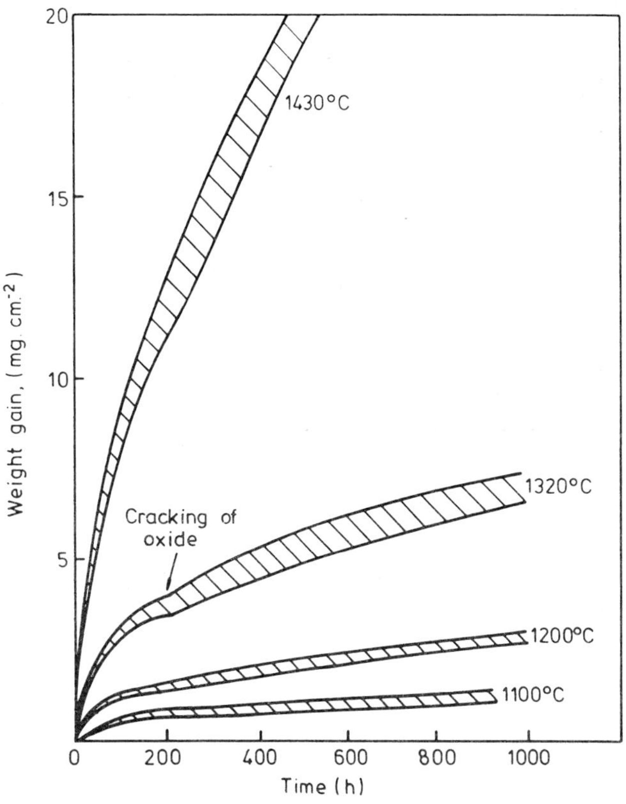

FIG. 7. VARIATION OF WEIGHT GAIN DURING THE OXIDATION OF 25% Cr. 4% Al. 1% Y. FECRALLOY STEEL IN AIR AT 1100-1430°C

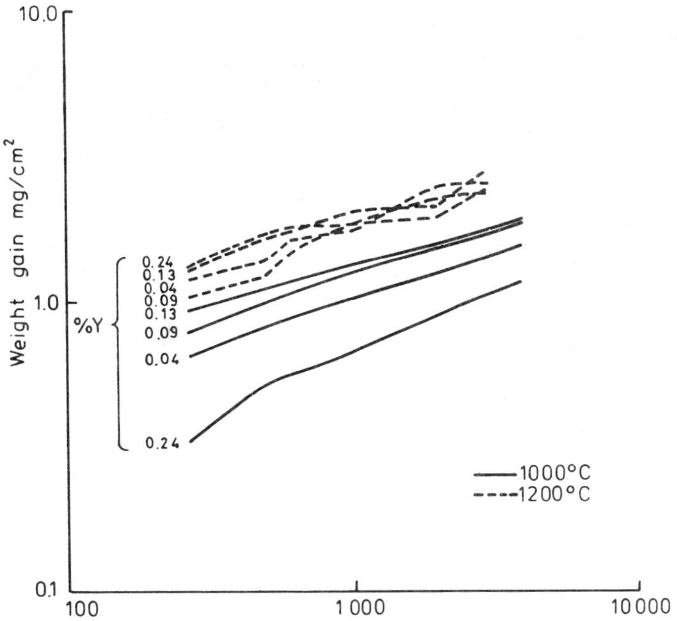

FIG. 8. VARIATION OF WEIGHT GAIN WITH TIME DURING THE OXIDATION OF FECRALLOY STEELS CONTAINING 0.04 TO 0.24% Y IN CARBON DIOXIDE AT 1000-1200°C. VALUES AT END OF CURVES DENOTE YTTRIUM CONCENTRATION.

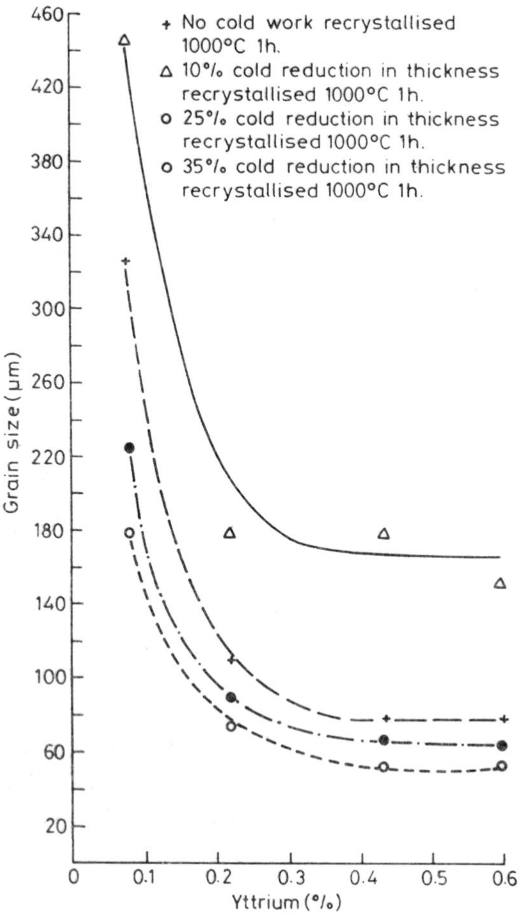

FIG. 9. INFLUENCE OF YTTRIUM CONTENT AND COLD WORK ON THE GRAIN SIZE OF FECRALLOY STEEL RECRYSTALLISED AT 1000°C

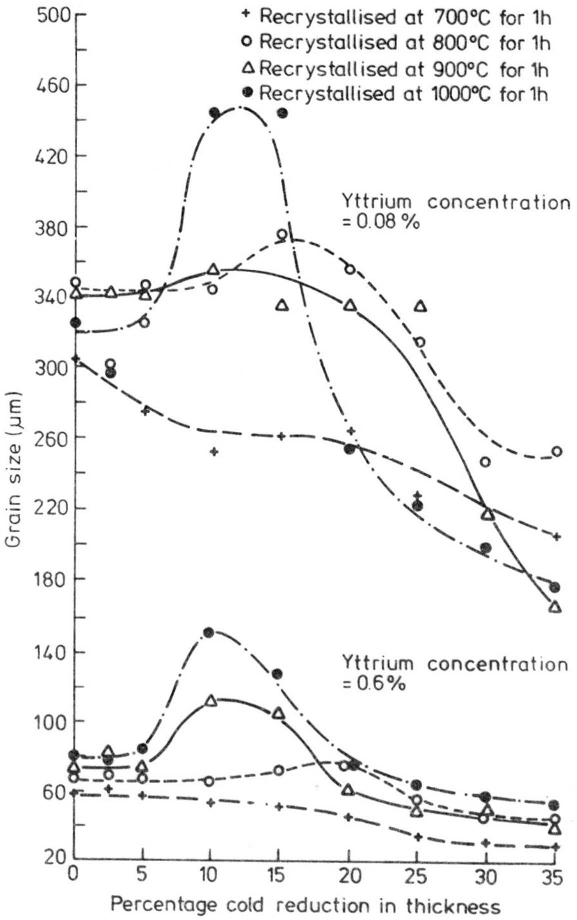

FIG. 10 EFFECT OF COLD WORK AND RECRYSTALLISATION TEMPERATURE ON THE GRAIN SIZE OF FECRALLOY STEELS CONTAINING 0.08 AND 0.6 % YTTRIUM

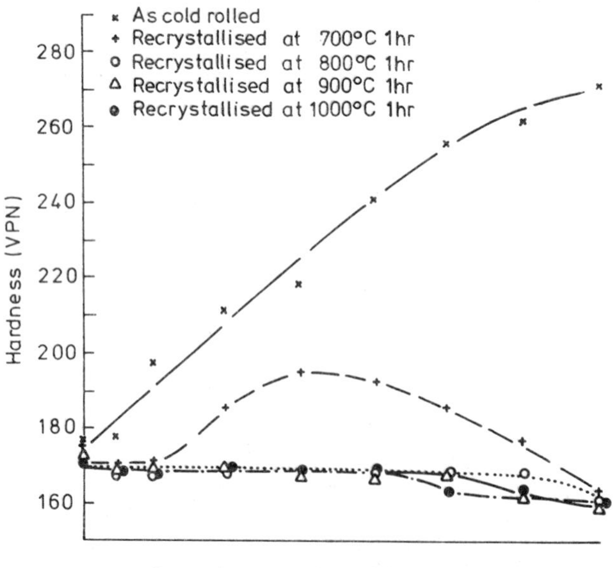

FIG. 11 EFFECT OF COLD WORK AND ANNEALING ON THE HARDNESS OF FECRALLOY STEEL CONTAINING 0·6% YTTRIUM

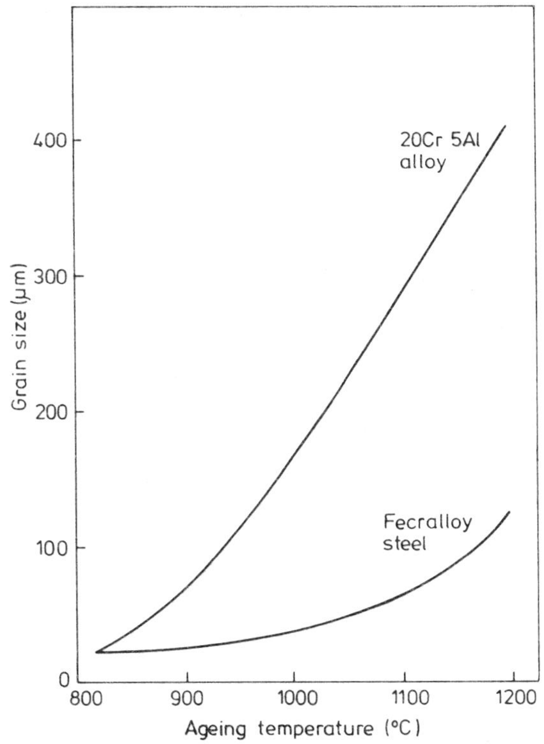

FIG. 12 GRAIN GROWTH OF 20 % Cr, 5 % Al STEEL AND FECRALLOY STEEL DURING 500h ANNEALING AT TEMPERATURES BETWEEN 800 AND 1200 °C

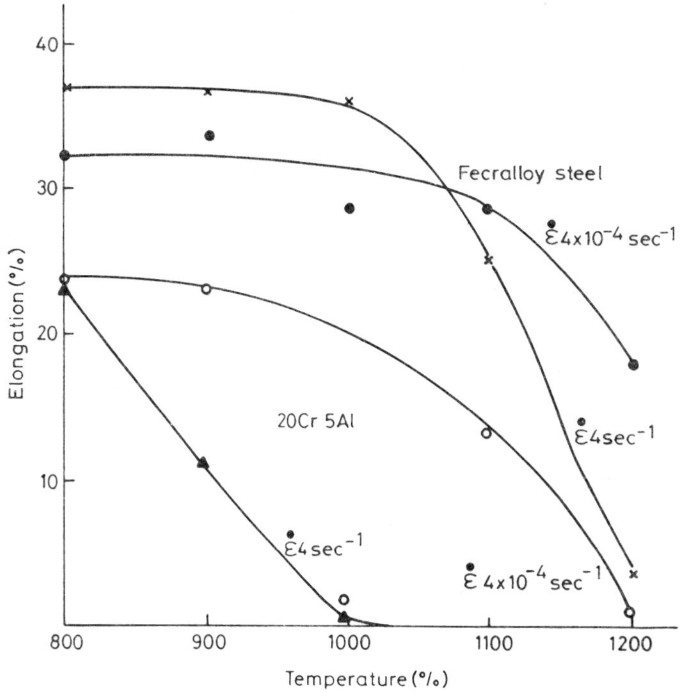

FIG. 13 TENSILE DUCTILITY AT STRAIN RATES OF 4 AND 4×10^{-4} SEC^{-1} FOLLOWING 1000h AGEING AT TEMPERATURES BETWEEN 800 AND 1200 °C.

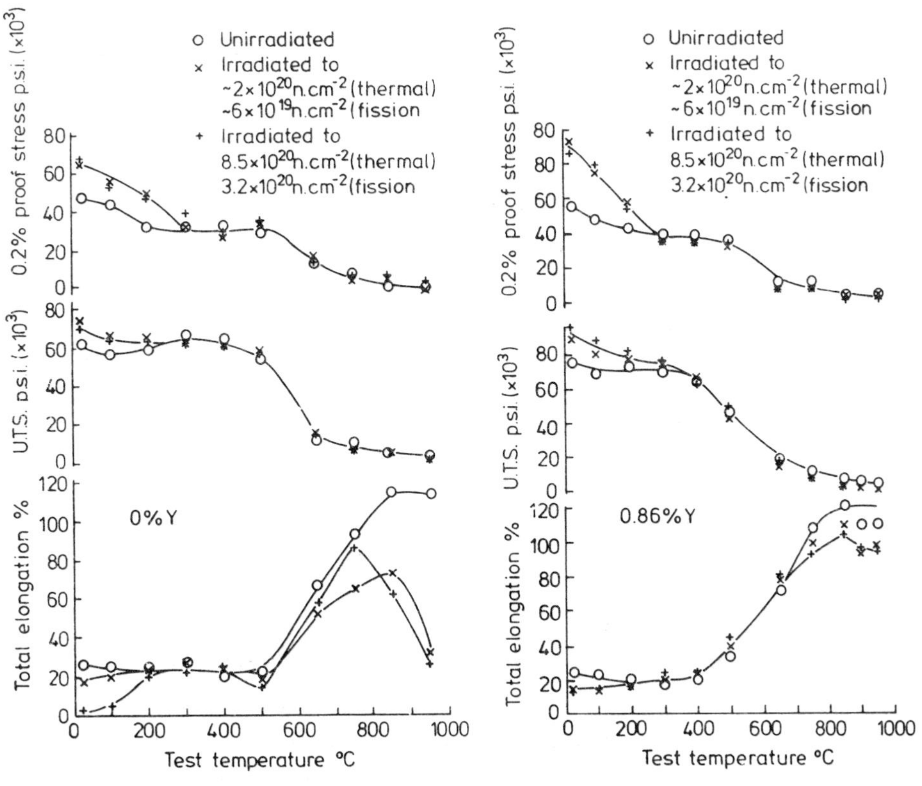

FIG. 14 TENSILE PROPERTIES OF YTTRIUM-FREE 14 % Cr, 4 % Al STEEL AND 0.86 % Y FECRALLOY STEEL

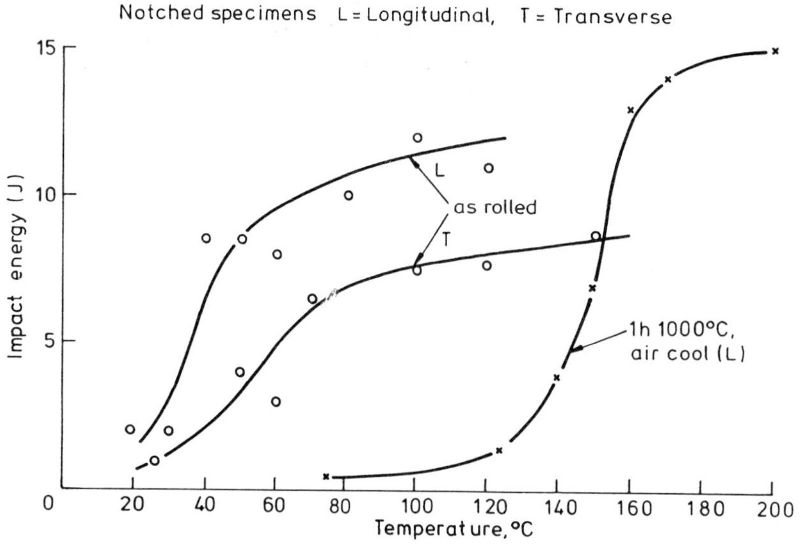

FIG. 15 CHARPY IMPACT PROPERTIES OF COMMERCIAL FECRALLOY STEEL IN THE HOT ROLLED AND ANNEALED CONDITIONS

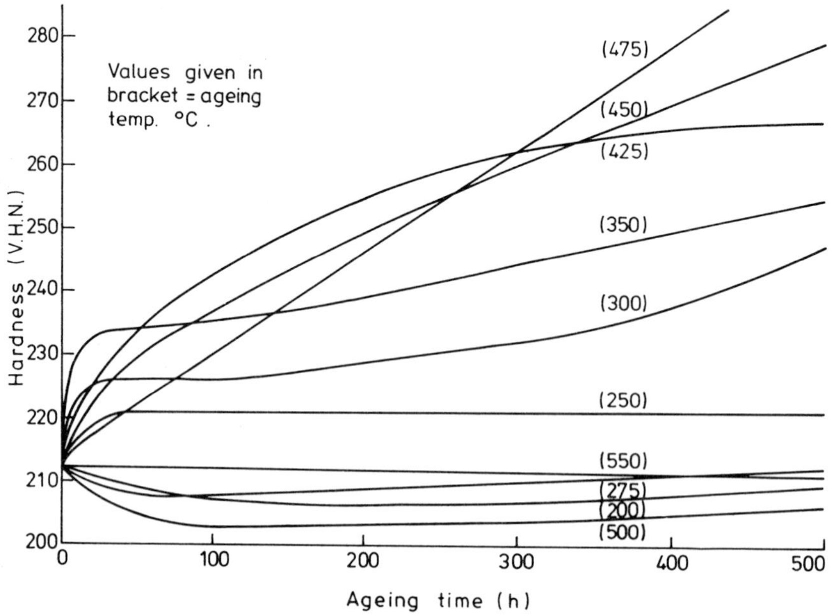

FIG. 16. AGE HARDENING OF FECRALLOY STEEL

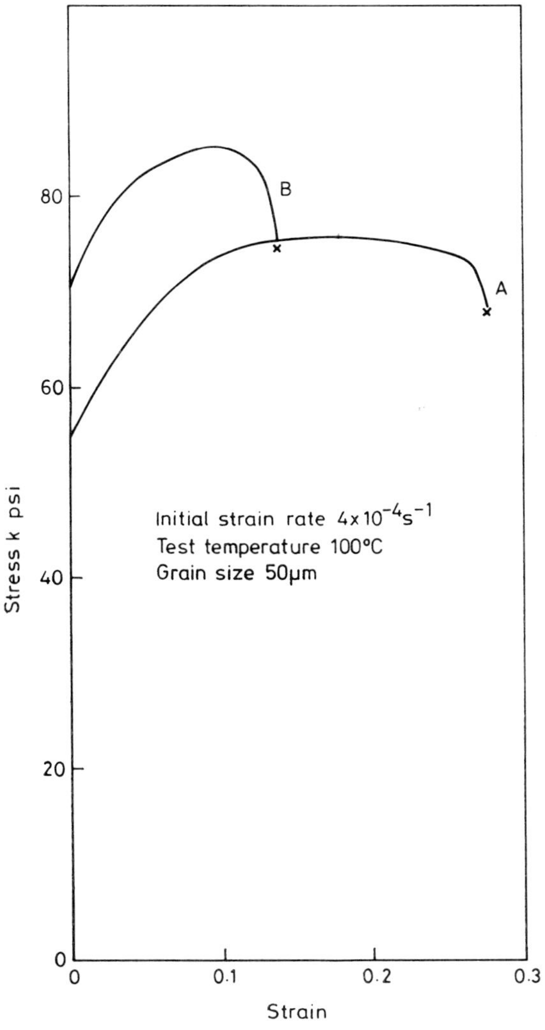

FIG.17 TENSILE CURVES FOR FECRALLOY STEEL SOLUTION TREATED(A) AND SOLUTION TREATED AND AGED 500h AT 475°C (B)

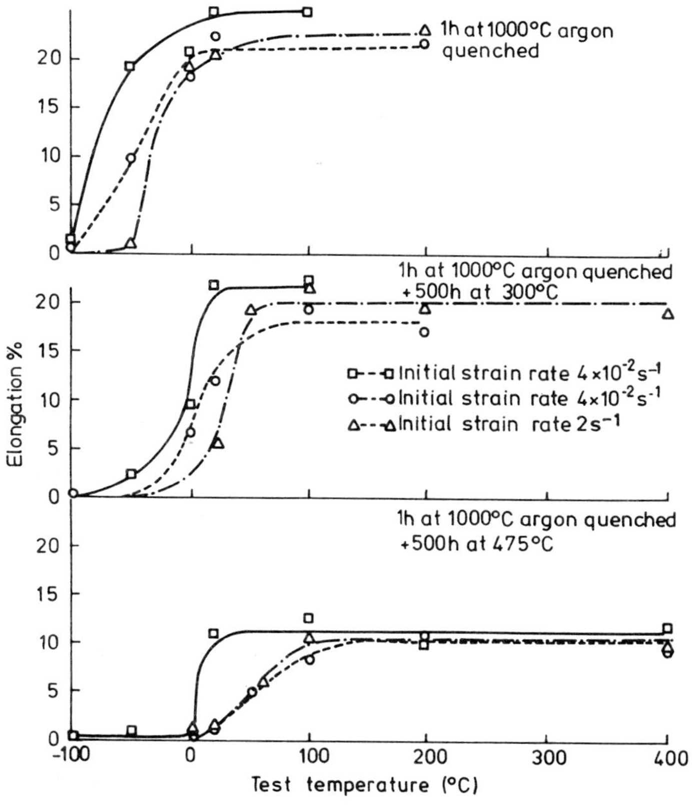

FIG. 18 EFFECTS OF AGEING, TEST TEMPERATURE AND STRAIN RATE ON THE TENSILE DUCTILITIES OF FECRALLOY STEEL

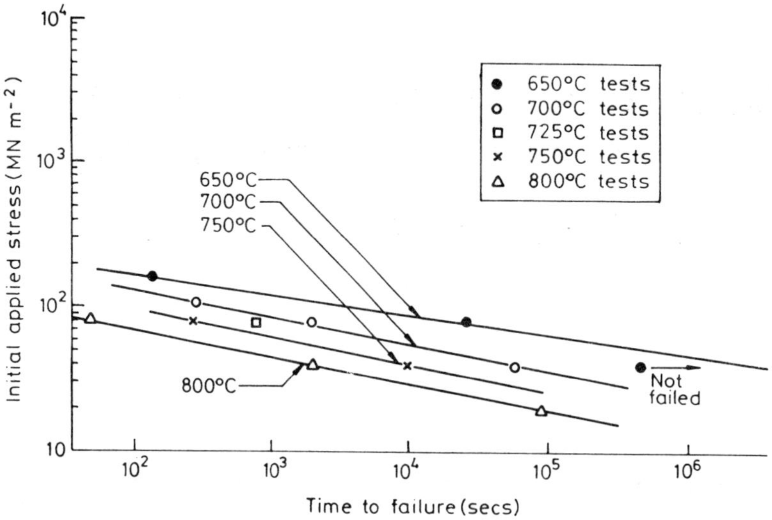

FIG. 19 VARIATION OF TIME TO FAILURE WITH INITIAL APPLIED STRESS FOR FECRALLOY STEEL

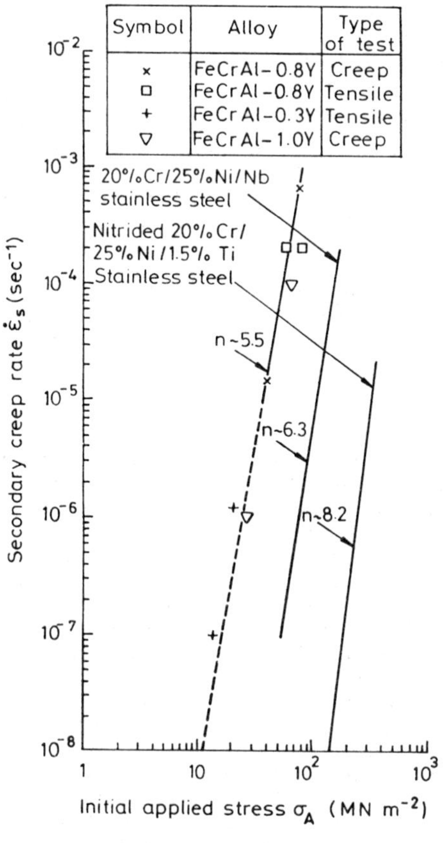

FIG. 20 COMPARISON OF CREEP RATES OF FECRALLOY STEEL, 20% Cr, 25% Ni/Nb STAINLESS STEEL AND NITRIDED 20% Cr/25% Ni/Nb/1.5% Ti STAINLESS STEEL AT 750°C

Near-threshold fatigue growth in a microduplex stainless steel

J Wasén, E U Engström and B Karlsson

JW and BK are in the Department of Engineering Metals, EUE the Department of Physics, Chalmers University of Technology, Göteborg, Sweden.

SYNOPSIS

In the present investigation the fatigue crack growth properties of a duplex, ferritic/austenitic stainless steel is studied. Cold rolling of the originally hot rolled, banded microstructure increases the fatigue crack growth threshold ΔK_{th} by 25 %. A similar increase is found by high temperature annealing where the banded structure is broken up. These improvements are caused by increases both in the closure level K_{cl} and in the effective crack growth resistance $\Delta K_{th,eff}$. The results are interpreted in terms of changes in the fracture surface topography and in internal changes in the flow properties. Newly developed fractographic techniques are applied and the results of these evaluations are related to the crack growth data.

INTRODUCTION

There is evidence in the literature that duplex ferritic/austenitic stainless steels exhibit very high fatigue strength as related to either the yield or the ultimate strength (1). However, very little has been published about the influence of the duplex microstructure and prior mechanical treatments like cold deformation on the near threshold fatigue crack growth properties. The aim of this paper is to demonstrate how the distribution of the austenite phase as well as cold rolling affect the near threshold fatigue crack growth in a microduplex, ferritic/austenitic stainless steel (Sandvik SAF 2205). It is a first report on the fatigue properties of this alloy system and uses newly developed techniques to investigate fracture surfaces and to relate fractographic information to the fatigue crack growth in a quantitative way. These techniques have been applied in other systems by the authors.

MATERIALS

The material employed is a high-alloyed duplex ferritic/austenitic (α/γ) stainless steel (SAF 2205 from Sandvik AB, Sweden) with the nominal chemical composition shown in Table 1. By controlled annealing in the ($\alpha+\gamma$)-region this material allows different amounts of the two phases to be obtained. In this investigation the proportion of the phases ferrite and austenite was intentionally chosen as 60:40, a composition often found in commercial grades. Water quenching from the high temperature annealing suppresses the precipitation of embrittling secondary phases such as σ, χ, α', and carbides (e.g. ref. 2). Microanalysis in SEM as well as thermodynamical calculations on this material (2,3) have revealed a slight segregation of Ni to austenite (from 5.5 to 7 %) and of Cr and Mo to ferrite (from 22 to 24 and from 3.0 to 3.5 % respectively). Being an austenite stabilizer nitrogen is strongly enriched in the austenitic phase (ca. 0.25 %; ref. 4), whereas ferrite does not contain more than 0.04 % of this element.

The material was tested in three different conditions:

(a) As received, hot rolled and quenched.
(b) As above and cold rolled 50 %.
(c) Annealed at 1060°C, 1h and quenched (quench annealed, "Q A").

The microstructures of the different conditions (Fig. 1) were characterized by point counting and linear intercept measurements. Table 2 shows mean values, each one based on a number of test points or intercepts corresponding to a relative standard deviation in the results of less than 5 %. As seen in Table 2 the volume fractions of the two phases in the three different conditions are identical within the experimental accuracy, leaving out the volume fractions of the phases as a variable in the experiments. The microstructures of the hot rolled and cold rolled conditions were heavily banded in the longitudinal (L) rolling direction and slightly banded in the long transverse (T) direction,(Figs. 1a and 1b). To ensure unambiguous data the mean intercept lengths were recorded in the short transverse (S) direction. During the annealing grain coarsening and partial spheroidization took place, producing a more particulate microstructure (Figs. 1c and 1d).

The tensile testing was performed in the longitudinal direction whereas the fatigue cracks were grown in the long transverse direction with the main crack plane perpendicular to the rolling direction (S-T orientation). In both cases, therefore, the nominal unidirectional stress is parallel to the fibres in the microstructure.

The static yield strengths are shown in Table 2. It is evident that the cold rolling causes a considerable strain hardening along with the geometrical thinning down of the fibre-shaped phase regions. On the other hand, the high temperature anneal with the accompanying grain coarsening results in a slight reduction of the yield stress level.

EXPERIMENTAL PROCEDURE

Fatigue crack growth

The fatigue tests were performed with single-edge notched sheet specimens (width 15 mm, thickness 1.5 mm). All tests were conducted in plane strain with fixed end displacements in tension mode with $R = K_{min}/K_{max} \leq 0.05$ using an Instron servohydraulic testing machine. The specimens were tested at room temperature in laboratory air (relative humidity 50-55 %) at a frequency of 50 Hz with a sinusoidal wave shape. The crack length was recorded on the polished specimen surface with a travelling microscope (resolution 0.01 mm). At least three specimens were tested for each structure. In the determination of the threshold stress intensity range (ΔK_{th}) the following procedure was used. The crack was initially grown approximately 1 mm from the starter notch, and the load was then reduced initially with 10 % and later with 5 % decrements. At any given load level the crack was allowed to grow at least a distance corresponding to five times the calculated plastic zone associated with the previous load level. This procedure was followed until the fatigue crack growth rate (da/dN) had slowed down to less than 10^{-11} m/cycle. The stress intensity amplitude (ΔK) corresponding to $da/dN \leq 10^{-11}$ m/cycle was designated as the threshold level (ΔK_{th}). In order to determine the growth rates above ΔK_{th} the load was raised slightly from that corresponding to ΔK_{th} and a ΔK-increasing test was performed under constant load amplitude. The crack propagation rates were then calculated using successive regression of the a-N data and ΔK was calculated by the aid of an equation derived by the finite element method (6).

The crack closure load (P_{cl}) was defined as the intersection of the two branches of the compliance curve recorded via COD-signals. This load lies between the conventionally defined opening and closure loads. The advantage of this procedure is that scatter due to evaluation problems is reduced.

Experimental evidence in this work and in related investigations in the authors' laboratory indicates an ΔK-independence of the closure level K_{cl}. This can conveniently be used to check whether the experimental conditions are acceptable during the crack growth test.

Fractography

The fracture surfaces of the fatigued specimens were geometrically quantified by the aid of profile analysis. The profiles were taken from vertical sections through the interior of the specimens parallel to the main crack propagation direction. High resolution optical micrographs were then used for computer aided digitizing of the profile. In the measurements the "real" profile line is replaced by a chain of chords with an individual length given by the resolution of the digitalization, in this case 0.2 μm.

The parameters used to describe the profile should describe the direction and the location of the approximating chords related to some suitable reference line. Parameters that are preferably used in order to describe the fracture surface (7) are for example the height distribution, the angular distribution of the approximating chords, and the angular distribution of surface elements on the curved surface. The linear and areal angular distributions are analytically related, and the average values from each of them can be described by corresponding roughness values (7,8).

A problem in profile analysis (and also in direct surface topography studies) is that a long-waved irregularity sometimes appears in the profile (7,9) and overlays the local fracture geometry. This waviness results in a non-averaged height distribution when the total length of measure is finite. A method to suppress this waviness in order to average the characteristics of the fracture surface over a limited interval is to use a high pass filter technique (9). This method creates a mathematically well-defined dividing line formed by the low frequencies of the frequency spectrum of the profile. With this dividing line (satisfying the geometrically necessary requirement that the height distribution should be symmetric around it) as a reference it is easy to determine the height distribution (7).

The digitized lengths of the different crack profiles were about 5 mm corresponding to typically 1000 grains. Together with the filtering technique used this secures a very high precision in the fracture profile data.

EXPERIMENTAL RESULTS AND DISCUSSION

Fatigue crack growth

The fatigue crack growth rate (da/dN) as a function of the stress intensity amplitude (ΔK) is shown in Figs. 2 and 3 for the conditions tested. It is obvious that both the phase morphology and the degree of deformation prior to testing affect the fatigue crack growth behaviour in the whole stress intensity range tested. In all conditions the Paris' law is closely followed in region II of the da/dN - ΔK plot. The banded, hot rolled condition exhibits a slope m of 3.1 (Figs. 2 and 3). Cold rolling by 50 % results in an increase of m to 4.2 (Fig. 2). Breaking up of the banded structure by quench annealing also leads to a larger slope, m = 3.8 (Fig. 3). Technically more important, the two treatments raise the nominal crack growth threshold (ΔK_{th}) by approximately 25 % from 4 to 5 $MNm^{-3/2}$ (Table 3). The increase in ΔK_{th} and m results in a cross-over point in the da/dN - ΔK curves as seen in Figs. 2 and 3. Therefore, the improvements found in the near threshold region by the two treatments are accompanied by slight impairments at $\Delta K \geq 15$ $MNm^{-3/2}$. Such cross-over points caused by variations in the m-values are frequently found in iron based alloys as well as in other alloys (10).

The nominal fatigue crack growth threshold is affected by many interlinked factors, which can be divided into mainly two groups: one intrinsic part describing the internal resistance of the material to crack extension ($\Delta K_{th,eff}$) and one extrinsic part taking the crack closure effects into account (K_{cl}), i.e.:

$$\Delta K_{th} = \Delta K_{th,eff} + K_{cl} \qquad (R = 0) \qquad (1)$$

As seen in Table 3 the crack closure level K_{cl} is increased from 1.1 to 1.6 by either the cold rolling or the annealing treatments. This difference explains half of the improvements found in the ΔK_{th}-values. The remaining part is due to changes in the microstructurally determined internal resistance $\Delta K_{th,eff}$. According to Table 3 $\Delta K_{th,eff}$ increases from 2.7 to 3.3 and 3.2 $MNm^{-3/2}$ respectively by the two treatments. As discussed below the dislocation cumulation upon straining and grain size effects caused by the heat treatments will affect the internal resistance to crack growth as well as the fractographic geometry. Both factors are keys to the changes in the threshold values found.

The microhardness of the present phases ferrite and austenite are reported to be fairly equal (11,12), although the experimental difficulties to obtain reliable values in these fine-grained materials are pronounced; thus the published values vary between 270 (11) and 330 (12). A more established fact is that austenite hardens more than ferrite upon monotonic straining (e.g. 13).

The cyclic straining behaviour in the process zone ahead of the advancing crack is of primary interest for the understanding of the fatigue crack growth behaviour. In the present case ferrite has a wavy slip character while the austenite exhibits a planar slip intensified by the high nitrogen content present (11). The cyclic properties of ferrite and austenite with chemical analyses and grain sizes as in the present material do not seem to be available in the literature. As a first approximation it can be assumed that the hardening is similar in monotonic and in cyclic straining, in analogy with experimental findings in low-alloyed ferritic steels (14). The ability of the wavy slip ferrite to rearrange the dislocation substructures upon cyclic straining (15) means that prior deformation does not play any role for the saturation flow stress in the crack tip zone. On the other hand, austenite with its planar slip characteristics exhibits a history dependence in the cyclic deformation behaviour (15). As a result the effective flow stress in austenite will be higher in prestrained (cold rolled) than in virgin condition. Such a difference will not be experienced in the ferrite phase. These arguments, although based on rather qualitative information, indicate a larger difference between the 'effective' flow stresses of austenite and ferrite in the cold rolled than in the undeformed conditions.

The cold rolling does not change the degree of banding in the originally hot rolled material. In both cases the crack is forced to pass both ferrite and austenite grains. The change in the 'effective' flow stresses upon cold rolling, as discussed above, results in an increase in the intrinsic resistance of the material to crack extension ($\Delta K_{th,eff}$). This behaviour is in contrast to that of low-alloy steels which often exhibit lower ΔK_{th} and $\Delta K_{th,eff}$ in prestrained conditions (5).

As shown in Table 3 both the nominal and the effective threshold values improve after annealing of the original banded structure. This heat treatment dissolves the banding and changes the morphology of the constituents. Since the crack is run perpendicular to the banding the crack will experience this change of morphology as an increase of the grain size by almost a factor four (cf. Table 2). Such a positive effect on ΔK_{th} and $\Delta K_{th,eff}$ caused by the increased grain size has also been found in single-phase, low-alloyed ferritic steels (6).

Compared to a fine grained ferritic-martensitic steel (DOCOL 800 from SSAB, Sweden) (5) the present steel in its quench-annealed condition exhibits similar $\Delta K_{th,eff}$ (Table 3).

Fractography

The fatigue crack is forced to intersect both phases in the banded structures. Furthermore, the fracture profiles in the quench annealed, partly spheroidized structure did not exhibit any significant preferential fracture path in the ferritic phase. This is probably due to the relatively small hardness difference between the austenite and ferrite.

The topography of the fracture surfaces was studied by the quantitative analysis technique described above. The profile line elements for each condition exhibit a symmetric, Gaussian shaped height distribution with a mean standard deviation of $\bar{H}$ as shown in Table 4. The angular distributions of the profile elements are also Gaussian in character and likewise described by its mean standard deviation $\bar{\theta}_L$ (Table 4).

The fracture surfaces in the different conditions investigated here are rather flat and less affected by the microstructure than is often the case in low-alloyed steels. The mean standard angular deviation $\bar{\theta}_L$ of the linear profiles increases from 12 to approximately 18 degrees upon cold rolling or annealing of the original hot rolled structure (Table 4). Thus the effective 'hardening' of the austenite in the banded structure causes an increased angular deviation of the advancing crack from the main crack plane. Spheroidization (high temperature annealing) causes a similar effect.

The profile line element angular distribution can easily be transformed to a surface element normal distribution. Such an analysis shows that the mean surface element normal direction, θ_{Sm}, is 15 degrees for the hot rolled and about 22 degrees for the other structures. This surface normal distribution is much narrower than those found in ferritic low-alloy steels, which exhibit mean surface normal directions about 45 degrees, i.e. in the direction of maximum shear (6).

Earlier investigations (6,16) reveal that the surface height deviations are determining for the degree of crack closure. As seen in Table 4 the mean standard deviation $\bar{H}$ increases from 0.7 to 1.4 μm for the cold rolled condition. The corresponding value is 1.8 μm for the quench annealed condition with a particulate microstructure, where the larger grain size also contributes to the increased surface roughness (6). In recent investigations by the authors (6,16) the following quantitative relation between K_{cl} and $\bar{H}$ has been established:

$$K_{cl} = 1.2\ \bar{H} \qquad (MNm^{-3/2};\ \bar{H} \text{ in } \mu m) \qquad (2)$$

Comparisons between the experimental data and predicted values from this equation show very good agreement (Table 5).

CONCLUSIONS

1. Dissolving of microstructural banding by high temperature annealing increases both the nominal and the effective threshold stress intensity amplitudes.

2. Heavy cold-rolling (50%) of initially banded microstructures increases both the nominal and the effective threshold values.

3. The crack closure level is relatively low for all conditions tested.

4. No significant preferential fracture path could be seen in quench annealed non-banded structures.

5. All conditions tested exhibit relatively flat fracture surfaces as compared to fracture surfaces in low-alloy ferritic steels.

6. The crack closure stress intensity (K_{cl}) approximately follows an earlier established relation between K_{cl} and the fracture surface roughness ($\bar{H}$): $K_{cl} = 1.2\ \bar{H}^{1/3}$.

ACKNOWLEDGEMENT

The supply of materials from Sandvik AB is gratefully acknowledged.

REFERENCES

1. H.W. Hayden, and S. Floreen: Metall. Trans., 1973, 4, 561.
2. T. Thorvaldsson, et al.: In Stainless Steels '84, Proceedings, The Institute of Metals, London, 1985, p. 101.
3. F.H. Hayes: J. Less Common Metals, 1985, 114, 89.
4. R.E. Johansson, and J-O. Nilsson: In Stainless Steels '84, Proceedings, The Institute of Metals, London, 1985, p. 446.
5. J. Wasén, and B. Karlsson: To be published, 1987.
6. J. Wasén, et al.: To be published, 1987.
7. B. Karlsson, et al.: Fatigue '87, Proceedings, EMAS, Warley, UK, 1987, p. 1479.
8. K. Wright, and B. Karlsson: J. Microscopy, 1983, 130 pt 1, 37.
9. J. Wasén, et al.: Acta Stereologica, 1987, 6, 199.
10. B. Karlsson, and K. Hamberg: Proceedings, 2nd Risö Intern. Symp. on Metallurgy and Materials Science, 1981, p. 431.
11. G. Wahlberg, and G.L. Dunlop: This conference.
12. W.B. Hutchinson, et al.: Mater. Sci. Tech., 1985, 1, 728.
13. M. Blicharski: Metals Science, 1984, 18, 92.
14. R.W. Landgraf: In Work Hardening in Tension and Fatigue (ed. A.W. Thompson), AIME, New York, 1977, p. 240.
15. C. Laird: Metall. Trans. A, 1977, 8A, 851.
16. K. Hamberg, et al.: Fatigue '87, Proceedings, EMAS, Warley, UK, 1987, p. 135.

Table 1 Nominal chemical composition of the material.

w/o						
Cr	Ni	Mo	N	C (max)	Si (max)	Mn (max)
22	5.5	3.0	0.14	0.030	0.8	2.0

Table 2 Metallographic and static mechanical characterization of the structures investigated.

Condition	α		γ		$R_{p0.2}$
	(%)	Mean intercept length (μm)	(%)	Mean intercept length (μm)	(MPa)
Hot rolled	60	2.4*	40	1.6*	650
Cold rolled 50%	63	1.8*	37	1.0*	1100
Q A 1060°C, 1h	59	8.9	41	6.2	570

* Heavily banded structures, cf. Fig. 1. The grain size measured perpendicular to the rolling plane.

Table 3 Fatigue crack growth threshold conditions.

Condition	ΔK_{th} ($MNm^{-3/2}$)	K_{cl} ($MNm^{-3/2}$)	$\Delta K_{th,eff}$ ($MNm^{-3/2}$)
Hot rolled	4.0	1.1	2.7
Cold rolled 50%	5.0	1.6	3.3
Q A 1060°C, 1h	4.9	1.6	3.2
DOCOL 800**	5.4	2.2	3.1

** Fine-grained (≈ 3μm) ferritic/martensitic low-alloy steel (42 % martensite); data from ref. 5.

Table 4 Fractographic data.

Condition	$\bar{H}$ (μm)	$\bar{\theta}_L$ (deg)	θ_{Sm} (deg)
Hot rolled	0.7	12	15
Cold rolled 50%	1.4	18	22
Q A 1060°C, 1h	1.8	19	23

Table 5 Comparison between the mean standard height deviation ($\bar{H}$) and experimental and predicted closure values (K_{cl}).

Condition:	$\bar{H}$ (μm)	K_{cl} ($MNm^{-3/2}$)	K_{cl}^* ($MNm^{-3/2}$)
Hot rolled	0.7	1.1	1.1
Cold rolled 50%	1.4	1.6	1.3
Q A 1060°C, 1h	1.8	1.6	1.5

* Calculated via $K_{cl} = 1.2\ \bar{H}^{1/3}$ (6,16).

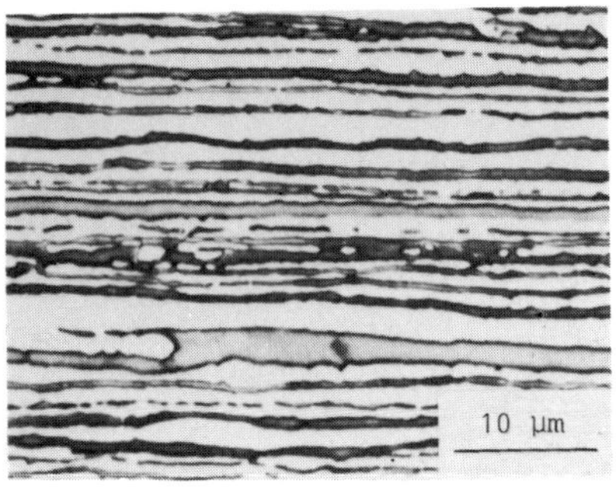

Figure 1a. Hot rolled (as received) condition, longitudinal section.

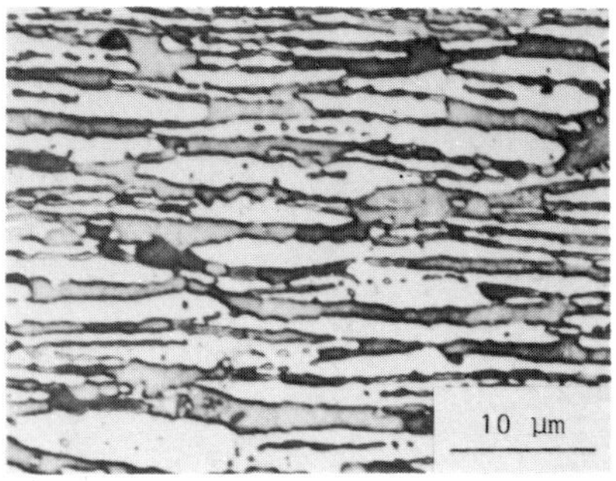

Figure 1b. Hot rolled (as received) condition, transverse section

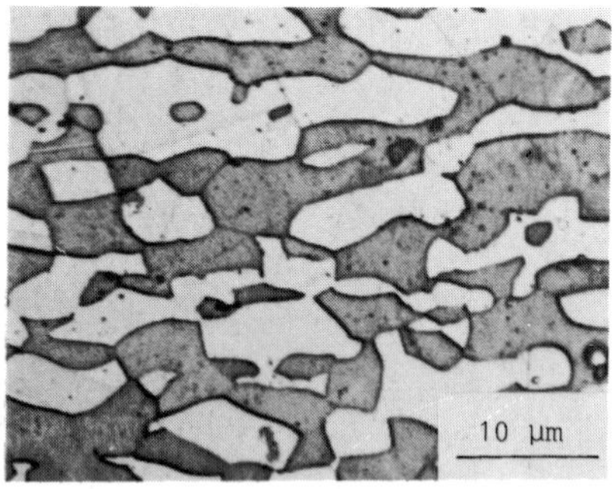

Figure 1c. Quench annealed (1h, 1060°C), longitudinal section.

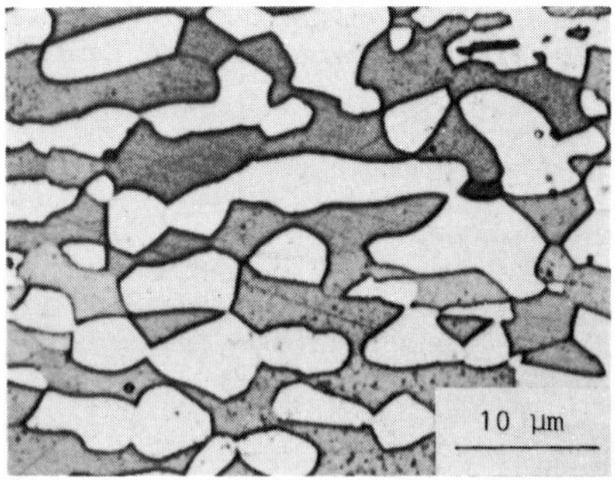

Figure 1d. Quench annealed (1h, 1060°C), transverse section.

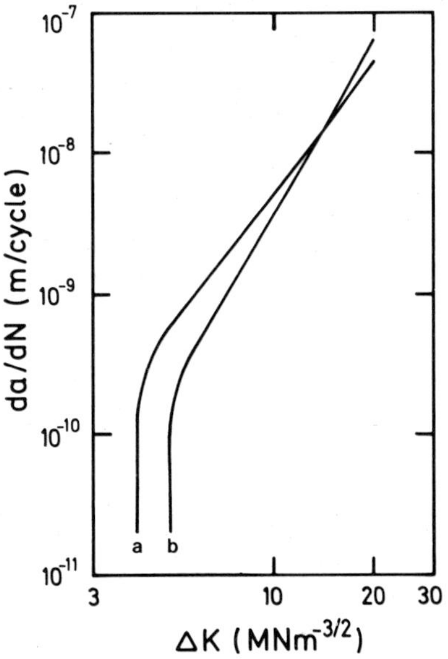

Figure 2. da/dN vs. ΔK for the hot rolled, as received (a) and the cold rolled (50%) conditions (b).

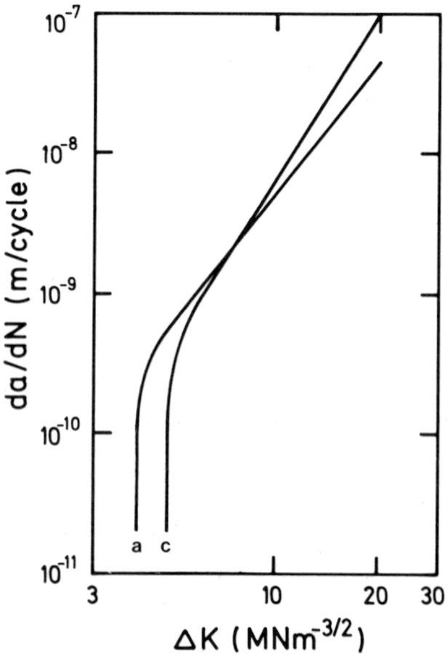

Figure 3. da/dN vs. ΔK for the hot rolled, as received (a) and the quench annealed (1h, 1060°C) conditions (c).

□

Influence of cathodic hydrogen on microstructure of duplex stainless steels

L J R Cohen, J A Charles and G C Smith

The authors are in the Department of Materials Science and Metallurgy, University of Cambridge, Pembroke Street, Cambridge, UK.

SYNOPSIS

The effects of cathodic hydrogen on the microstuctures of two commercial duplex stainless steels (Sandvik SAF 2205 and Sumitomo DP3) and a "model alloy" have been studied.

Hydrogen charging has been shown to produce microstructural changes in both the ferrite and the austenite phases of duplex systems. Lenticular features were visible in the ferrite phase, which were often slipped or cracked. Prolonged charging produced fine striations in the austenite phase. In addition, cracks were formed which exhibited a mixture of transgranular, intergranular and interphase morphologies and were microstructurally dependent.

T.E.M. work revealed that uncharged DP3 specimens were free from strain-induced twins and martensite. Charging thin foil specimens produced plates in the austenite phase, some of which were austenite twins and some which could also be identified as epsilon (h.c.p. martensite) and others which may have been a b.c.c. martensite phase. The lenticular features in the ferrite phase were identified as ferrite twins.

X-ray work on the commercial alloys showed that changes occurred in the lattice parameters of both phases after charging. There was also an indication of epsilon (h.c.p. martensite) formation. Epsilon was also identified on cold working the DP3 steel at 77 K.

A model is suggested to explain the behaviour of these materials, based on the surface stresses created by the different hydrogen diffusivities and occlusivities of austenite and ferrite.

INTRODUCTION

Duplex stainless steels contain both ferrite and austenite and combine the strength of ferritic stainless steels with the corrosion resistance of austenitic stainless steels. They are therefore candidate materials for use in oil and gas production. Oil and gas deposits can contain chloride ions, H_2S, CO_2 and brines, and may be at a temperature of up to 500 K. Many electrochemical reactions can therefore occur.

In the present investigation one of these electrochemical processes, namely the cathodic reaction evolving hydrogen, which could lead to hydrogen embrittlement, has been studied. It is known that cathodic hydrogen charging of fully ferritic or fully austenitic steels can induce cracking and void formation in ferritic steels *(1-3)* and phase transformation and cracking in austenitic steels *(4-8)*. The presence of external stresses in combination with such hydrogen ingress may then lead to accelerated cracking.

Previous work has shown that austenite and ferrite have very different hydrogen solubilities and diffusivities at room temperature; the ferrite having a high diffusivity and low solubility and the austenite having a low diffusivity and high solubility *(2,9,10,11)*. This can result, during cathodic charging, in higher surface concentrations of hydrogen and steeper concentration gradients, in austenite compared with ferrite. It is therefore of interest to study the response of ferrite and austenite to hydrogen charging when they are in close proximity, as in duplex stainless steels, and also when an additional microstructural feature is present, namely a large interfacial area between austenite and ferrite, which could act as a hydrogen trap.

Chan, Martinez-Madrid and Charles *(2)*, working on mild steel, found evidence for hydrogen-induced cracking at high-angle interfaces, such as high-angle ferrite grain boundaries and the ferrite-pearlite interface. This cracking susceptibility was shown to be associated with a high hydrogen concentration at the high-angle interfaces. Bentley and Smith *(4,8)*, working on austenitic stainless steels, found that hydrogen-induced cracking was associated with the generation of hydrogen-rich surface layers, and the subsequent effusion of hydrogen from epsilon martensite, produced by cathodic charging.

EXPERIMENTAL PROCEDURE

Materials

The chemical compositions of the as-received alloys tested are shown in *Table 1*. A model alloy was made to eliminate the effect of trace elements on the microstructure. This alloy was prepared by vacuum melting electrolytically pure ingredients and subsequent homogenisation (1200°C for 48 hours followed by water quenching). The alloy was then heat-treated by holding the material in the single-phase ferrite region for a short time in a vacuum furnace and slow cooling, so as to achieve a large-grained prior ferrite structure with grain boundaries decorated with allotriomorphic and widmanstätten austenite. Compositions of the phases in all three materials, as measured by E.D.S., are given in *Table 2*.

Hydrogen charging

Small coupons, 7 mm x 7 mm x 2 mm were cut from the bulk material and ground on 180 paper on all surfaces. One of the major square surfaces was subsequently reground and then electropolished in Struers A2 electropolishing solution (20V for 10 seconds at room temperature) and etched electrolytically in 20% NaOH at 5V for 20 seconds before hydrogen charging for subsequent examination. Insulated wires were spot welded onto the specimens which were then charged galvanostatically, on all surfaces in a solution of 4% H_2SO_4 in water with 200 mg l^{-1} As_2O_3 as a hydrogen recombination poison.

Examination of charged specimens

Specimens were examined optically immediately after charging and during room temperature outgassing. Acetate replicas were taken from some specimens during outgassing to monitor the microstructural changes more closely. Several specimens were examined, after outgassing, on a Camscan S4 scanning electron microscope at 30 kV.

Transmission electron microscopy

T.E.M. specimens were made from DP3 by grinding a small specimen, then spark eroding or punching out discs which were subsequently ground and then electropolished at 40 V in a solution of 7% perchloric acid and 23% glycerol in alcohol at 269 K. Prior to hydrogen charging the specimens were examined in a Philips EM300 at 100 kV or EM400T at 120 kV to check that specimen production had not created strain-induced twins or martensite. Specimens were then hydrogen charged and re-examined.

Hydrogen measurements

Diffusible hydrogen was measured by the standard technique of immersing a charged specimen in a mercury-filled Y-shaped tube immediately after charging. After all the diffusible hydrogen had been evolved - no change in the reading (usually after two weeks) - the experiment was terminated. Samples were then tested in a "Strohlein H-mat 250" machine to measure residual hydrogen contents.

X-ray work

Samples of as-received SAF 2205 and DP3 were examined before charging, immediately after charging (at 100 mA cm^{-2} for various times) and during outgassing on a Philips 1730 vertical diffractometer with Cu K_{alpha} radiation. Spectra were calibrated against a silicon standard.

RESULTS

Materials

Optical micrographs of the uncharged materials are shown in *figure 1*. Both of the as-received commercial alloy microstructures consist of austenite elongated parallel to the surface in a ferrite matrix. The cross-section of elongated austenite regions is in each case approximately 10 x 10 µm. The austenite regions in the SAF 2205 are of varying size, with some of the larger particles, which are polycrystalline, containing annealing twins. The austenite regions in the DP3 are more uniform in size and appear to contain just one grain. The model alloy microstructure consists of large prior ferrite grains decorated at the grain boundaries with austenite allotriomorphs and widmanstätten austenite plates.

Hydrogen charging - optical microscopy and S.E.M.

model alloy

On charging at 100 mA cm^{-2} for 20 minutes or more, small plates formed in some of the ferrite grains. The plates, which were visible immediately after charging, were lenticular, with a midrib, and some were slipped or cracked in a crystallographic manner *(figure 2a)*. The plate formation and damage had occurred during charging as these features were evident in the first replicas obtained. There was a profusion of plates around the widmanstätten austenite*(figure 2e)*. In a specimen charged for 2 hours at 100 mA cm^{-2} features became visible in the allotriomorphic austenite after about 15 minutes' outgassing. These features were later identified as slip steps and cracks. At this stage the widmanstätten austenite also showed some cracking *(figure 2e)*. In another sample with a similar microstructure some ferrite grains, which were free from plates, cracked on outgassing, the cracks becoming visible after about 15 minutes. The cracks were wavy, yet frequently appeared to be along specific planes and some appeared to emanate from the tips of widmanstätten austenite plates *(figure 2b)*.

S.E.M. examination allowed differentiation between cracking and slip. Features were observed with secondary electrons from both sides by rotation of the specimen through 180°: cracks will appear dark from both sides and slip steps will appear dark from one side and bright from the other. An example of how this was used is shown in *figures 2c and 2d*. This technique revealed that the majority of the features in the lenticular plates were slip steps, but a few cracks were also present. The allotriomorphic austenite was slipped and cracked, the widmanstätten austenite cracked, and the ferrite grains cracked. In addition, taper section examination of one sample showed that both the lenticular plates and the long wavy features in the ferrite extended a little way below the surface, of the order of 20 µm for the plates and 100 µm for the ferrite cracks.

SAF 2205

A range of charging conditions, for example 100 mA cm^{-2} for 20 minutes or more and 10 mA cm^{-2} and 1 A cm^{-2} for 2 hours produced similar plates in the ferrite phase to those in the ferrite phase in the model alloy. The plates were lenticular, with a midrib and were constrained by grain and interphase boundaries *(figure 3a)*. Some of the plates (1% to 5%, depending on the charging conditions) were damaged in a similar manner to those in the model alloy. As with the model alloy, some ferrite grains contained plates and others did not. Several plate variants (orientation directions) were sometimes present within a grain.

Charging at 100 mA cm^{-2} for 8 hours or more (or 8 hours or more at 10 mA cm^{-2} or at 1 A cm^{-2} for 2 hours) produced fine striations in the austenite phase after outgassing for about 30 minutes*(figure 3b)*. S.E.M. evidence, and that obtained by bending the specimen, whereby the features "opened up" yet remained closed at their ends, suggested crack formation. Bending the outgassed specimen also resulted in the opening up of some of the incipient microcracks in the ferrite phase parallel to the length of the plates. The cracks in the austenite phase formed on outgassing, appearing first at austenite grain boundaries and then at the interphase boundaries and profusely as straight transgranular cracks, possibly associated with annealing twin boundaries *(figure 3c)*.

DP3

Hydrogen charging of DP3 at 100 mA cm^{-2} for twenty minutes or more produced lenticular plates in the ferrite matrix, as in both the other materials. The plates were of similar orientation (direction) over several grains in the ferrite which suggests a high degree of texture, presumably arising from working the material *(figure 4a)*. Charging for longer periods (8 hours or more at 100 mA cm^{-2}) caused features to develop in the austenite phase on outgassing. Some features, obviously cracks from their morphology, formed initially at the interphase boundaries and then transgranular features appeared, which were easily visible after 30 minutes' outgassing *(figure 4b)*. Two types of transgranular feature were visible: straight-sided lines and wavy non-crystallographic veins. The former, which were opened up on bending the specimen while remaining closed at their ends were cracks and the latter, which were increased in number on bending were slip steps. This is illustrated in *figure 4c*. The presence of both slip steps and cracks was further verified by rotation of the specimen in the S.E.M. The transgranular austenite crack morphology in this material was different to that in the SAF 2205 alloy. In addition, small cracks were associated with some of the ferrite plates, parallel to the length of the plates after charging at 100 mA cm^{-2} for 12 hours *(figure 4d)*.

Transmission electron microscopy

The uncharged material is shown in *figure 1* and is free from twins and martensites. Charging the thin foils at 50 or 100 mA cm^{-2} for 5 minutes or more produced plates in the austenite phase. It was possible to identify a number of these as austenite twins. It must be stressed that it is not easy to differentiate twins from epsilon (h.c.p. martensite) *(12)*. Bowkett, Keown and Harries *(13)* have suggested tilting to either the <110> or the <114> zones in austenite, where unambiguous identification is possible. These authors suggested that it may be possible to see diffraction effects from epsilon and austenite twins, each producing, nominally, a separate set of diffraction spots at the <210> austenite zone. However, in this instance double diffraction can occur, and so the presence of *either* epsilon *and / or* austenite twins can generate both sets of spots. A similar problem occurs at the <111> austenite zone: only the epsilon spots *should* be visible, but double diffraction from twins, if they are present, may generate double diffraction spots in the same place as the epsilon spots. As all the possible diffraction spots for austenite twins and epsilon were visible at the <210> and <111> austenite zones, it can be inferred that epsilon and / or austenite twins were present in the DP3 as a result of the hydrogen charging.

There was unambiguous diffraction evidence at <110> and <114> for hydrogen-induced austenite twin formation *(figure 5)*, but the diffraction evidence at the <111> and <210> austenite zone axes could also support the presence of epsilon. Body-centred cubic martensite was looked for as previous workers had found that hydrogen charging could cause its formation in metastable austenitic stainless steels *(4,5)*. Only one piece of evidence was found for b.c.c. plates in the austenite phase of DP3.

While it was fairly easy to observe the microstructural changes in the austenite phase, difficulty was experienced with the ferrite phase as preferential electrodeposition of impurities occurred onto ferrite during the cathodic charging process. To observe the features in the ferrite phase some specimens were thinned further after charging using an ion beam thinner to expose the subsurface structure. The lenticular plates evident in the ferrite phase proved to be twins, and diffraction evidence supported their having the normal twin-matrix orientation for b.c.c. materials *(figure 6)*. Some cracking was visible around the twin plates, which could not be confused with slip, since any surface relief due to slip lines would have been removed by the ion beam thinning procedure.

Diffusible hydrogen and residual hydrogen measurements

Diffusible hydrogen contents are shown in *figure 7* for various charging conditions in all the materials. It should be noted that about 12 ml 100 g $^{-1}$ at S.T.P. corresponds to the hydrogen occupying an equal volume to the steel.

It can be seen from *figure 7* that the diffusible hydrogen content for a given charging time was highest in the model alloy, followed by the SAF 2205 and then the DP3. The higher hydrogen content of the model alloy as compared to the two commercial alloys could explain the cracking of the model alloy after charging at 100 mA cm^{-2} for 2 hours, but not the commercial alloys.

Residual hydrogen levels in all the materials are given in *figure 8*. All had a similar level of residual hydrogen, of the order 10% to25% of the diffusible hydrogen for a given charging time.

The model alloy contained the most total hydrogen for a given charging time (total hydrogen = diffusible hydrogen + residual hydrogen), possibly because the materials were not saturated after the relatively short charging times used. This being the case, the model alloy containing the most ferrite would have the highest overall diffusivity and so hydrogen could penetrate more easily into the bulk of the material. There are two further pieces of evidence supporting this idea. First, the amount of hydrogen taken up increases with charging time in the range of charging times studied, suggesting that not all the cross-sections are saturated. Second, the hydrogen content of a sample a quarter the thickness of a standard-sized sample, compared to that in a standard-sized sample and charged with the same conditions, was proportional to surface area ratios rather than volume ratios of the specimens, indicating that hydrogen was predominantly in the surface layers.

X-ray work (on commercial alloys)

X-ray studies were made on the DP3, mainly, with several experiments on the SAF 2205.

For both alloys, spectra taken before hydrogen charging showed characteristic sharp peaks for both the austenite and ferrite phases. Immediately after charging (usually within 2 minutes of terminating charging) a rapid scan was commenced in the region around the austenite {111} and ferrite {110} peaks, which was repeated (at progressively slower rates) during the outgassing process. Immediately after charging (for any time tested over 1 hour at 100 mA cm^{-2}) the ferrite peak remained sharp but the austenite peak became diffuse, indicating a spread in lattice parameter. After charging for a short time (1 hour) both phases expanded slightly. The ferrite reverted to its original dimensions quickly (within about half an hour's outgassing), but the austenite took considerably longer, the reversion not being complete within a day. Longer charging time tests were restricted to DP3 where charging caused the ferrite to contract after the initial expansion and the austenite to expand continuously. Subsequent reversion of ferrite to its original dimensions was much faster than the reversion of austenite.

With both alloys there was an indication of epsilon martensite formation in some of the charged specimens. Epsilon was also observed on cold-working DP3 at liquid nitrogen temperature (77 K). Examples of typical x-ray traces showing the above effects in DP3 are presented in *figure 9*.

DISCUSSION

Duplex stainless steels can take up significant quantities of hydrogen from cathodic charging as seen in this and in earlier work *(14)*. In the present investigation the specimens were 2 mm thick and the diffusible hydrogen results suggest that charging up to the maximum time used of 12 hours did not produce saturation. An estimate can be made of the hydrogen penetration distances in this time from the values of diffusion coefficient in ferrite and in austenite. Taking $D_{ferrite}$ as 10^{-11} m^2 s^{-1} and $D_{austenite}$ as 3×10^{-16} m^2 s^{-1} *(10)*, the penetration distances ($x^2 = Dt$) would be, repectively, of the order of 660 μm and 3.6 μm. In both materials there is marked directionality of the austenite, which is present as particles elongated parallel to the surface of approximately 10 μm width and varying length. Except at the edges, the specimens were charged through the 7 mm x 7 mm faces, at 90° to the elongation direction and it is therefore clear that only the austenite regions at a specimen surface would have approached saturation, whereas diffusion to considerably greater depths could have occurred in the ferrite. Damage created by the hydrogen could therefore have been present to a greater depth in the ferrite than in the austenite.

Microstructural changes are generated by the hydrogen and subsequent room temperatutre outgassing. Charging gives rise to lenticular features in the ferrite phase which are ferrite twins. The twins are visible immediately after terminating charging and are frequently slipped or cracked across their width.

Charging for longer times may cause cracking along the length of the twins. Charging for longer periods of time also induces damage in the austenite phase, which becomes apparent on outgassing. The austenite damage can take the form of slip or cracking: cracking and slip were visible in the DP3 and model alloy and cracking alone was visible in the SAF 2205 alloy, with intergranular and straight transgranular morphologies. In addition, interphase (austenite-ferrite) cracks were observed in the SAF 2205. Longer charging times are necessary to initiate damage in the austenite phase than in the ferrite phase. This might be explained by a higher local hydrogen concentration being required to generate damage in the austenite *(11)* (higher solubility), or non-uniform entry of the hydrogen into the duplex structure (see later). T.E.M. work has also shown the presence of austenite twins and possibly epsilon (h.c.p.) martensite in the austenite phase.

The observed microstructural changes in austenite are consistent with previous work on austenitic stainless steels *(4-6)*. Hydrogen can produce a high density of stacking faults and hence twins and epsilon martensite (which is equivalent to the presence of a stacking fault on every other close-packed plane). These features may be associated with the lowering of stacking fault energy by hydrogen, as suggested by Whiteman and Troiano *(15)* . They may also play a part in austenite damage by providing weak interfaces, or, in the case of epsilon, a region of inherently low ductility.

In the present work x-ray evidence showed that epsilon was formed from austenite during low temperature (77 K), rather than room temperature deformation. A reduction in the stacking fault energy by hydrogen could allow this to happen during cathodic charging at room temperature. Wakasa and Nakamura found both epsilon martensite and b.c.c. martensite in the austenite phase (composition, wt% 19.7 Cr, 5.9% Ni) of a deformed, relatively low alloy duplex stainless steel, with epsilon forming at higher temperatures than b.c.c martensite *(16)*. There has not been any systematic observation of b.c.c. martensite in hydrogen-charged DP3 in this investigation. Examination of the composition of the austenite phase *(table 2)* in the DP3 reveals segregation of nickel and manganese. It is also known that nitrogen segregates to the austenite phase in duplex stainless steels and is added as a potent austenite stabiliser. The higher concentration of austenite-stabilising elements in the austenite phase will make it more stable with respect to b.c.c. phase formation. In previous work Bentley and Smith *(7,8)*, comparing the effects of cathodic hydrogen charging on 18/8 and 18/12 austenitic stainless steels, showed that cathodic hydrogen produced greater transformation to b.c.c. martensite in the less stable austenite.

A possible model for hydrogen uptake in the surface layers of duplex stainless steels is given in *figure 10*. Initially both phases expand slightly as they take up hydrogen. The ferrite (low solubility) soon becomes saturated at a hydrogen level which depends on the fugacity of the hydrogen generated by cathodic charging. The austenite (high solubility) continues to expand as more hydrogen is dissolved. An explanation for the diffuse austenite x-ray peak is that the hydrogen concentration range and gradient is greater than in the ferrite, due to the lower hydrogen diffusivity, and hence a range of lattice parameter values is detected. Strains generated as the hydrogen is dissolved can give rise to plastic flow.

On outgassing, hydrogen can effuse from the ferrite phase quickly (high diffusivity), and this is manifest by a fast reversion to original lattice parameter and a sharp x-ray diffraction peak. The austenite, when it outgasses, will contract away from the ferrite slowly (low diffusivity) and this is shown by the diffuse austenite diffraction peak and a slow reversion to original lattice parameter. This contraction away from the ferrite will create tensile stresses, both within the austenite and at the ferrite-austenite interface, causing the damage observed in the austenite phase: twinning, slip, cracking (possibly associated with thin layers of epsilon) and the cracks at the interphase boundaries.

Twinning of ferrite is not normally observed when pure iron or ferritic/pearlitic steels are cathodically charged. However, hydrogen-induced twins have been observed in ferritic stainless steels by Tähtinen, Nenonen and Hänninen *(17)* and hydrogen-induced twinning in iron-titanium alloys has been observed by Hwang and Bernstein *(18)*. Desestret and other workers have associated the formation of "kinks" on a stress-strain curve with mechanical twin formation in duplex and ferritic stainless steels that were being strained in air or in various solutions. Cracking was sometimes associated with the "twins" and depassivation of the alloy was thought to be associated with the twin formation *(19-24)*. Magnin and Moret *(24)* suggested that nickel helped to promote twin formation in a Fe-Cr matrix, although the mechanism was not understood. There is evidence in the present work for cracking across and along the length of twins in addition to slip: possibly the cracks form as a result of the local stresses due to twinning in the presence of hydrogen, giving cleavage cracking. Slip in the twin could occur by a similar mechanism. In the model alloy, the appearance of twins appears to retard bulk matrix cracking: possibly by providing a local sink for the hydrogen and although this may lead to the formation of slip and incipient microcracks within the plates, this at least prevents bulk cracking of the matrix, which would obviously be more deleterious. The twinning in the ferrite may be a result of two effects; first, the generation of local plastic strains in the surface layers due to the hydrogen charging, and second, the possible reduction of twin boundary energy by the presence of hydrogen.

There is no evidence of blistering or internal porosity in the ferrite phase, as occurs when pure iron or plain carbon steels are charged *(1-3)*. The presence of austenite regions of higher hydrogen solubility may prevent an internal stress build-up, although this would not apply in the model alloy, where there is less, and more localised austenite.

The actual distribution of hydrogen in duplex stainless steels is not known (i.e. the concentration in each of the two phases). The discharge of H^+ ions may not occur at the same rate on austenite and ferrite, which might lead to non-uniform hydrogen uptake. The overall take-up rate may vary with relative amounts of austenite and ferrite and their composition.

The cracking observed in the present investigation is on a fine scale, but could be exploited as crack initiation sites by an applied external stress in the presence of further hydrogen charging. Tests on stressed samples are therefore important and are in progress.

CONCLUSIONS

Duplex stainless steels can take up considerable amounts of hydrogen from cathodic charging.

Hydrogen causes microstructural changes in duplex stainless steels. Lenticular twins, which may be slipped or cracked, form in the ferrite phase. Fine features form in austenite during outgassing after prolonged charging, which exhibit a mixture of cracking and slip. Electron diffraction evidence has also shown the presence of austenite twins and possibly epsilon martensite in the austenite in thin foils.

Hydrogen-induced cracking can occur on outgassing after prolonged charging, appearing in, and at the boundaries of austenite regions, with the allotriomorphic and widmanstätten austenite in the model alloy seemingly particularly susceptible. Cracking is also observed in the ferrite matrix, sometimes particularly associated with the twins.

The behaviour of these materials in relation to hydrogen charging can be interpreted in terms of the stresses set up as a result of the different hydrogen diffusivities and solubilities at the specimen surface.

ACKNOWLEDGEMENTS

The authors wish to thank Professor D. Hull for provision of laboratory facilities and Shell Research Limited for financial support. Thanks are also due to staff at the Shell Research Centre at Thornton, in particular, Dr T.D.B. Morgan, and to the T.E.M. group at Cambridge for helpful discussions.

REFERENCES

1. S. L. I. Chan, M. Martinez-Madrid, J. A. Charles: *Metals Technology,* 1983, (10), 464-470

2. M. Martinez-Madrid, S. L. I. Chan, J. A. Charles, *Materials Science and Technology*, 1985, (1), 454-460

3. A. S. Tetelman, W. D. Robertson, *Trans. Met. Soc. AIME*, 1962, (224), 775-783

4. A. P. Bentley, G. C. Smith, *Materials Science and Technology,* 1986, (2), 1140-1148

5. M. L. Holzworth, M. R. Louthan, *N.A.C.E. Corrosion*, 1968, (24), 110-124

6. A. Szummer, A. Janko, *Corrosion (N.A.C.E.)*, 1979, (35), 461-464

7. A. P. Bentley, G. C. Smith, *Met. Trans. A*, 1986, (17A), 1593-1600

8. A. P. Bentley, G. C. Smith, *Scripta Met.*, 1986, (20), 729-732

9. J. K. Tien, *Effect of Hydrogen on the Behaviour of Metals*, Proc. Int. Conf., Wyoming, September, 1975. Eds. A. W. Thompson, I. M. Bernstein, A.I.M.E., 1976, 309

10. T.-P. Perng, C. J. Altstetter, *Met. Trans.*, 1987, (18A), 123-134

11. K. Farrell, M. B. Lewis, *Scripta Met.*, 1981, (15), 661-664

12. H.-J. Kestenbach, *Metallography*, 1977, (10), 189-199

13. M. W. Bowkett, S. R. Keown, D. R. Harries, *Metal Science*, 1982, (16), 499-517

14. P. R. Rhodes, G. A. Welch, L. Abrego, *J. Materials for Energy Systems*, 1983, (5), 3-18

15. M. Whiteman, A. Troiano, *Phys. Status Solidi*, 1964, (7), 109

16. K. Wakasa, T. Nakamura, *J. Mat. Sci.*, 1978, (13), 807-811

17. S. Tahtinen, P. Nenonen, H. Hanninen, *Scripta Met.*, 1987, (21), 315-318

18. G. Hwang, I. M. Bernstein, *Scripta Met.*, 1982, (16), 85-90

19. A. Desestret, J. C. Colson, E. Mirabel, R. Oltra, *National Research Council, Canada, International Congress on Metallic Corrosion, Toronto,* June 3-7, 1984, (3), 578-583

20. A. Desestret, R. Oltra, *Corrosion Science*, 1980, (20), 799-820

21. A. Desestret, E. Mirabel, D. Catelin, P. Soulignac, *N.A.C.E. Corrosion '85*, Paper 229

22. R. Oltra, A. Desestret, E. Mirabel, J. P. Bizouard, *paper presented at conference on "Hydrogen Sulphide Induced Environment Sensitive Fracture of Steels" (European Federation of Corrosion),* Amsterdam, 12-14 September, 1986.

23. T. Magnin, J. Le Coze, A. Desestret, Duplex Stainless Steels, A.S.M., Ed. R. A. Lula, 1983, 535-551

24. T. Magnin, F. Moret, *Scripta Met.*, 1982, (16), 1225-1228

TABLE 1 - COMPOSITIONS OF MATERIALS (WEIGHT%)... BALANCE IRON

	DP3	SAF 2205	Model Alloy
Cr	25.1	21.7	31.1
Ni	6.85	5.4	6.6
Mo	3.2	3.0	
Mn	0.92	1.56	
Cu	0.48	<0.08	
Si	0.39	0.35	
W	0.27	<0.01	
N	0.13	0.14	
C	0.02	0.015	
P	0.027	0.025	
S	0.002	0.002	
Sb	<0.002	<0.002	
As	<0.002	<0.007	

TABLE 2 - COMPOSITIONS OF PHASES IN THE ALLOYS AS MEASURED BY E.D.S. (WEIGHT%)...BALANCE IRON

	DP3		SAF 2205		Model Alloy	
	γ	α	γ	α	γ	α
Cr	24.5	28.3	21.2	24.5	31.1	31.2
Ni	8.2	5.19	7.04	4.4	6.8	6.7
Mo	2.61	3.65	2.21	3.85		
Si	0.46	0.47	0.38	0.45		
Mn	1.28	1.1	1.85	1.51		
Cu	0.65	0.53	0.12	0.13		

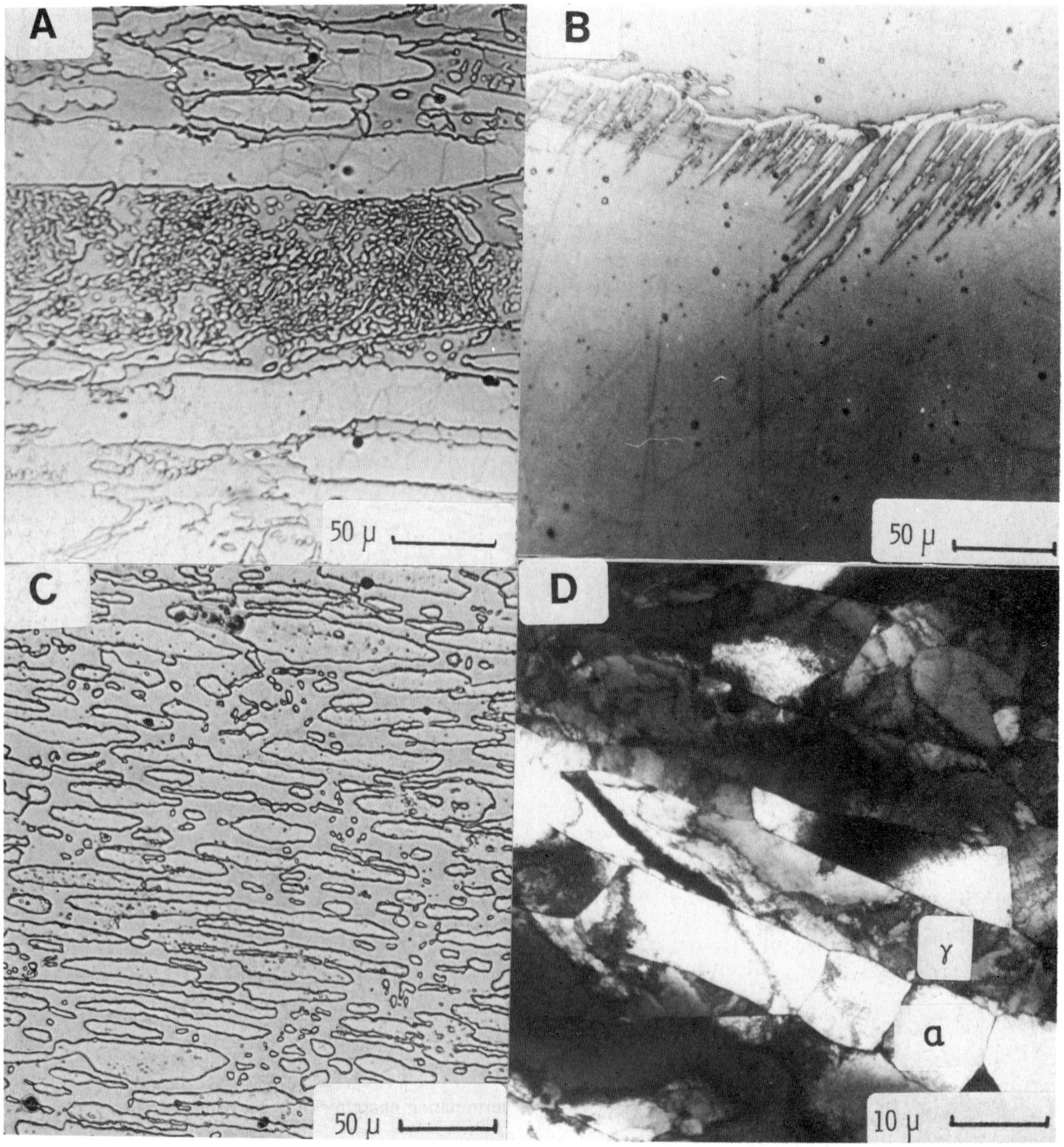

Figure 1. Uncharged materials:
a) SAF 2205, showing banding of austenite particles.
b) Model alloy grain boundary.
c) DP3. d) T.E.M. micrograph of DP3.

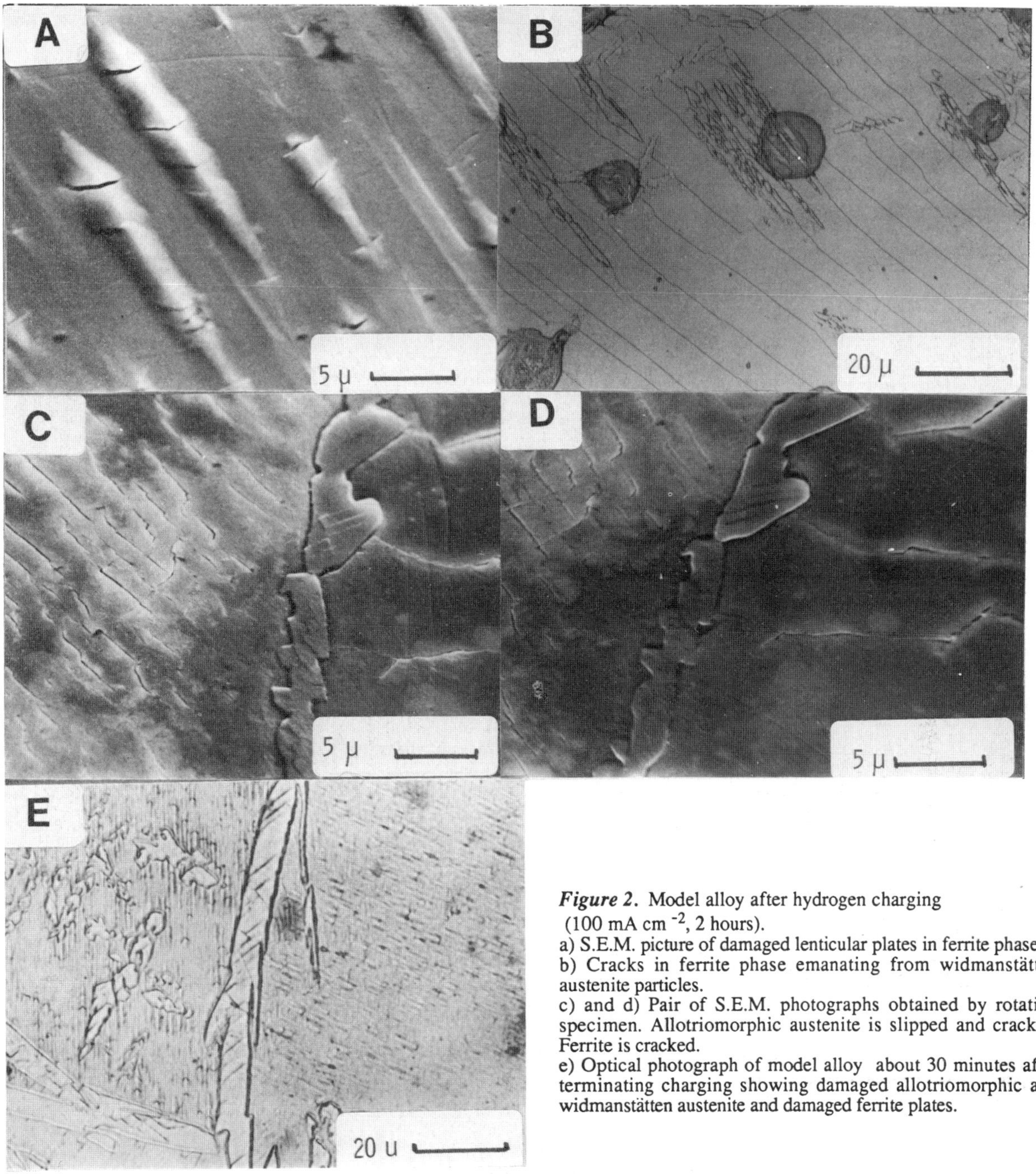

Figure 2. Model alloy after hydrogen charging (100 mA cm $^{-2}$, 2 hours).
a) S.E.M. picture of damaged lenticular plates in ferrite phase.
b) Cracks in ferrite phase emanating from widmanstätten austenite particles.
c) and d) Pair of S.E.M. photographs obtained by rotating specimen. Allotriomorphic austenite is slipped and cracked. Ferrite is cracked.
e) Optical photograph of model alloy about 30 minutes after terminating charging showing damaged allotriomorphic and widmanstätten austenite and damaged ferrite plates.

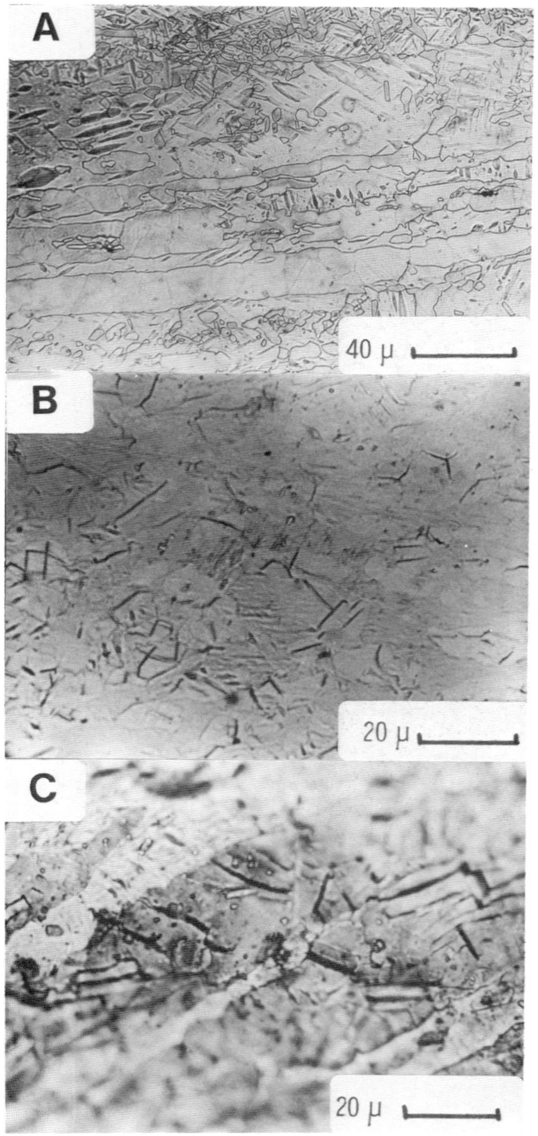

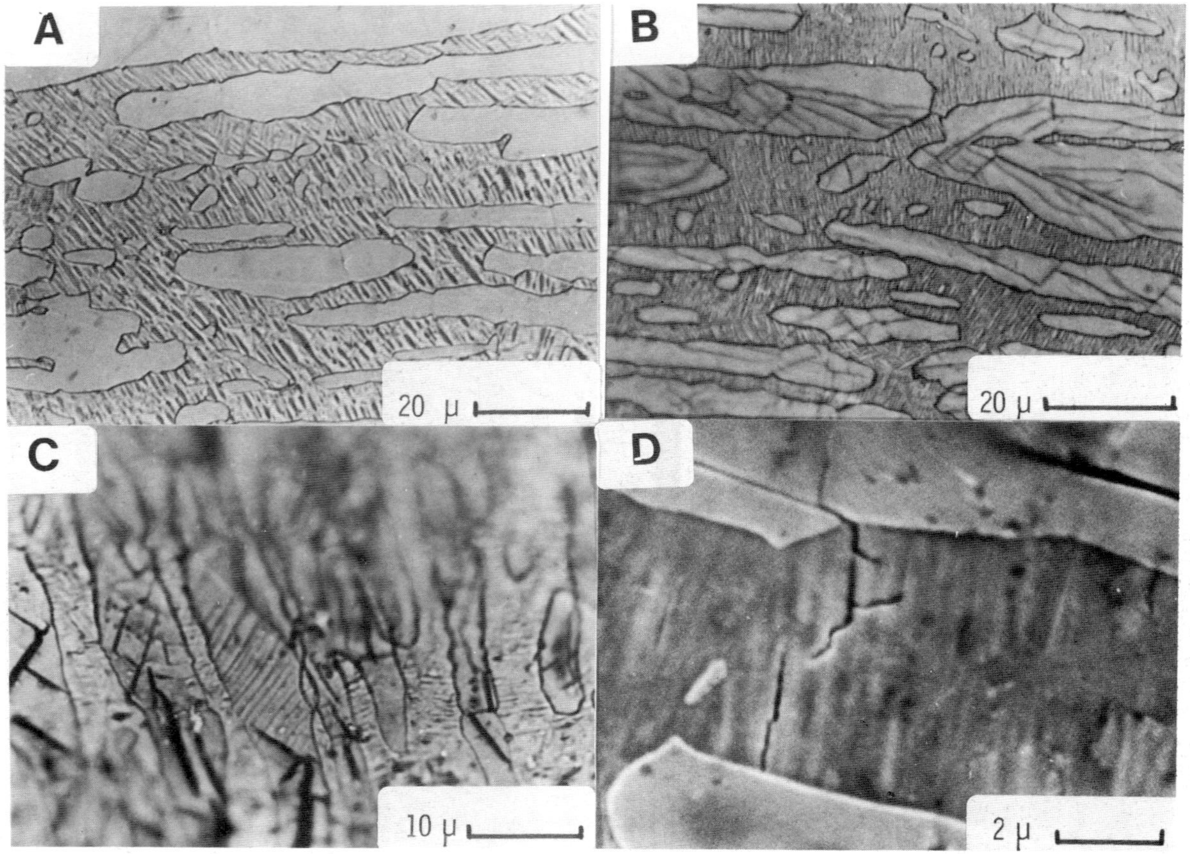

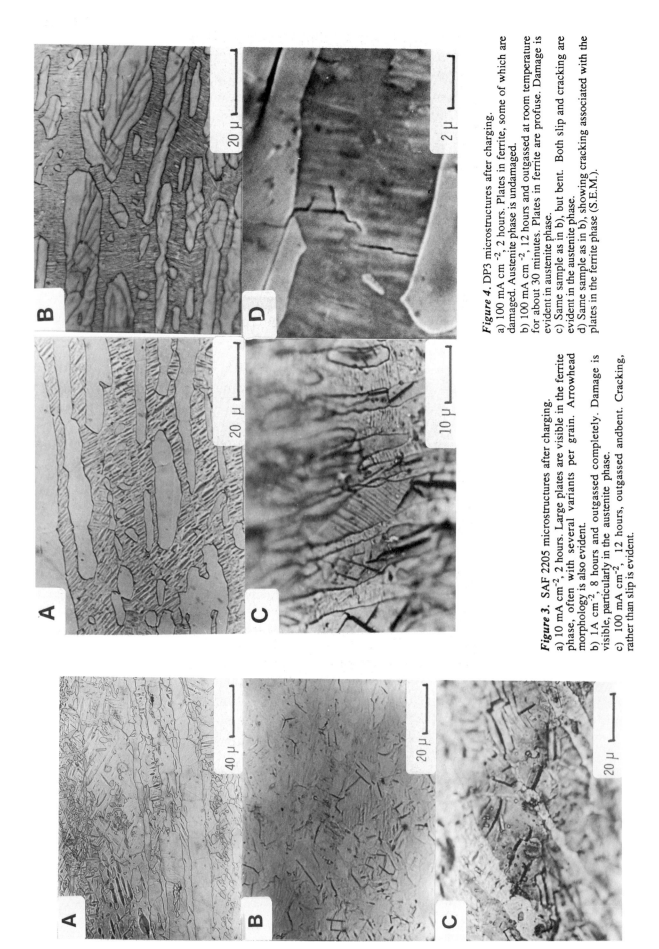

Figure 3. SAF 2205 microstructures after charging.

a) 10 mA cm^{-2}, 2 hours. Large plates are visible in the ferrite phase, often with several variants per grain. Arrowhead morphology is also evident.

b) 1A cm^{-2}, 8 hours and outgassed completely. Damage is visible, particularly in the austenite phase.

c) 100 mA cm^{-2}, 12 hours, outgassed andbent. Cracking, rather than slip is evident.

Figure 4. DP3 microstructures after charging.

a) 100 mA cm $^{-2}$, 2 hours. Plates in ferrite, some of which are damaged. Austenite phase is undamaged.

b) 100 mA cm $^{-2}$, 12 hours and outgassed at room temperature for about 30 minutes. Plates in ferrite are profuse. Damage is evident in austenite phase.

c) Same sample as in b), but bent. Both slip and cracking are evident in the austenite phase.

d) Same sample as in b), showing cracking associated with the plates in the ferrite phase (S.E.M.).

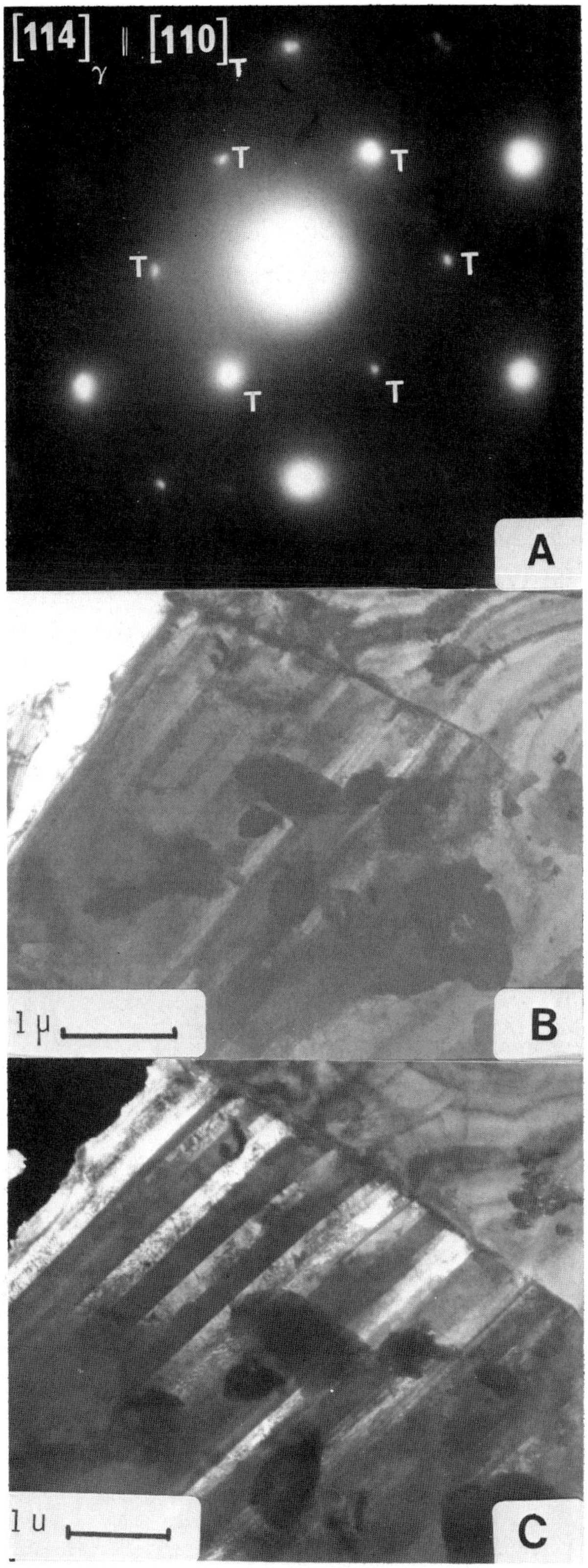

Figure 5. Diffraction evidence for austenite twins in austenite phase. Sample was charged at 50 mA cm^{-2} for 5 minutes.
a) Diffraction pattern: [114]austenite || [110]twin.
b) Bright field.
c) Dark field using twin spot ("T"). Plates are illuminated.

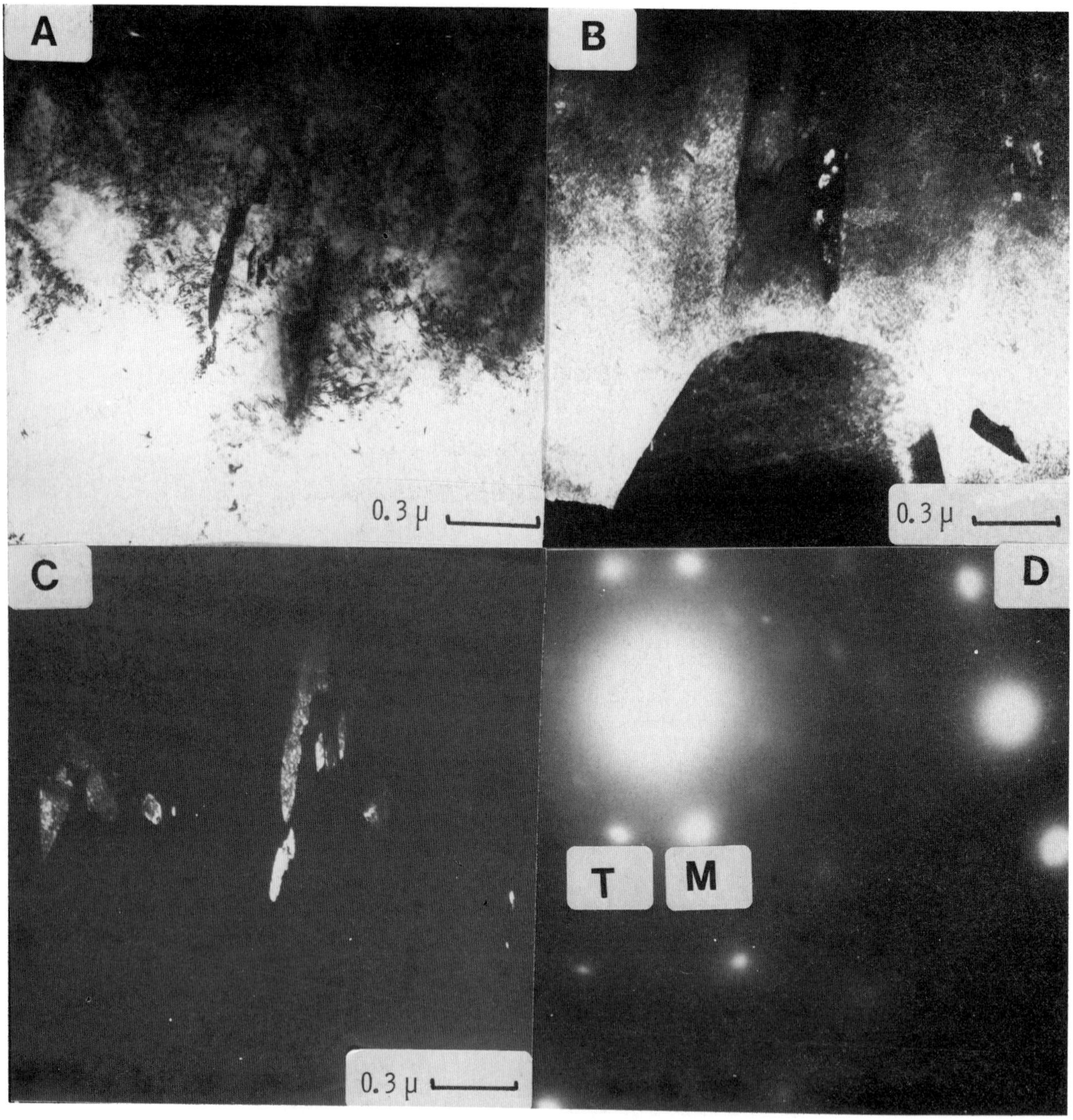

Figure 6. Diffraction evidence for ferrite twins in ferrite phase. Sample was charged at 100 mA cm^{-2} for 2 hours and then thinned.
a) Bright field.
b) Dark field using matrix spot "M".
c) Dark field using twin spot "T".
d) Diffraction pattern with [112]matrix || [121]twin.

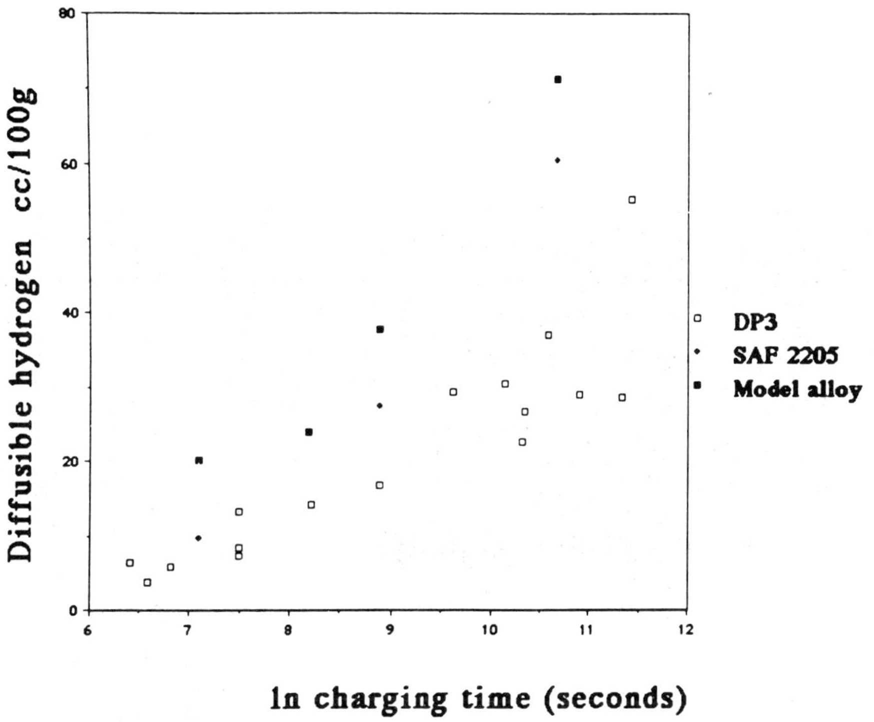

Figure 7. Graphs of diffusible hydrogen content vs. ln charging time at 100 mA cm $^{-2}$ for all three materials.

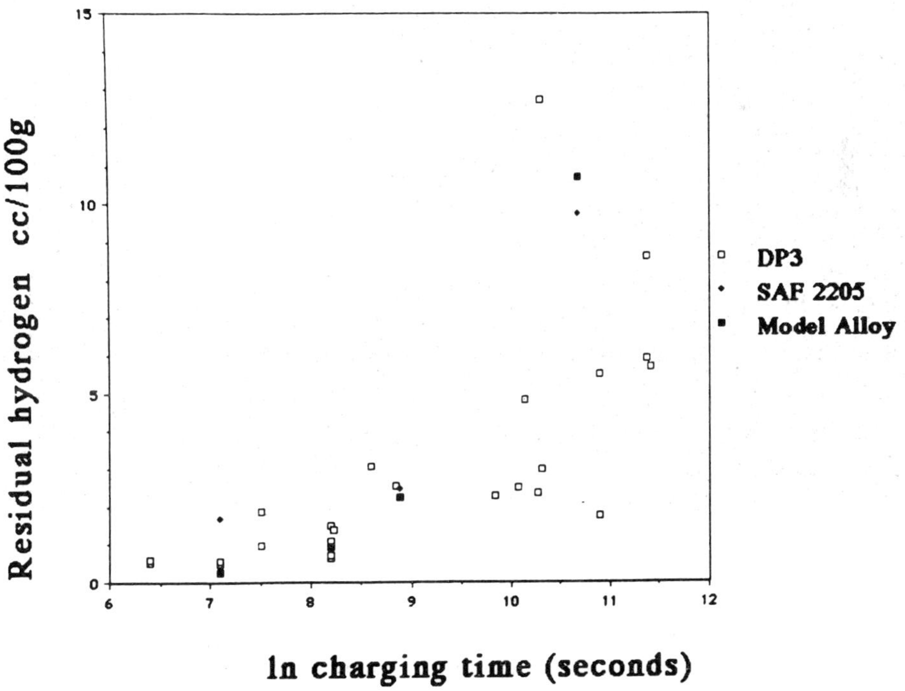

Figure 8. Graphs of residual hydrogen content vs. ln charging time at 100 mA cm^{-2} for all three materials.

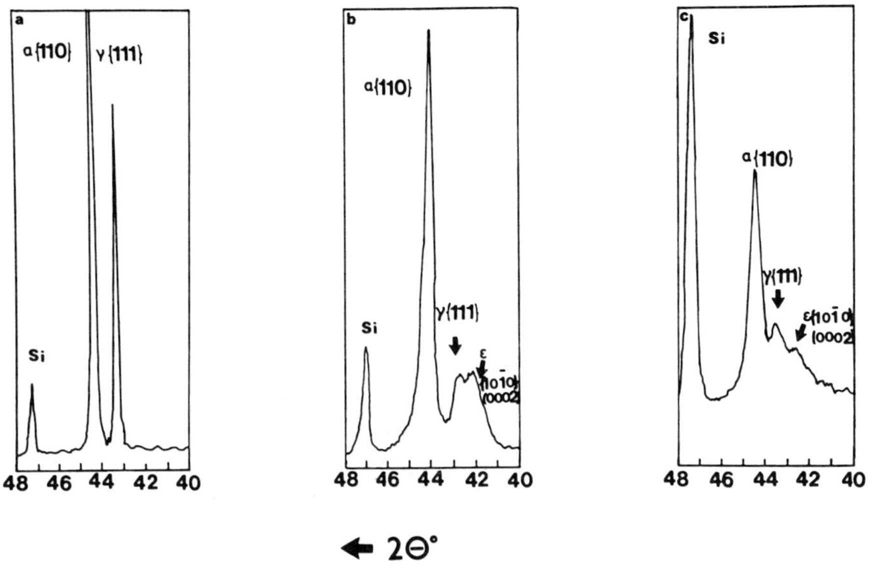

Figure 9. Typical x-ray traces for DP3.
a) Uncharged material.
b) Material charged at 100 mA cm $^{-2}$ for 3 hours 50 minutes. Spectrum taken after 10 minutes' outgassing.The change in lattice parameter in the sample shown corresponds to an approximate expansion of 0.5 % in the austenite and a contraction of 0.2% in the ferrite.
c) Material cold-worked to 48% reduction in thickness at liquid nitrogen temperature.

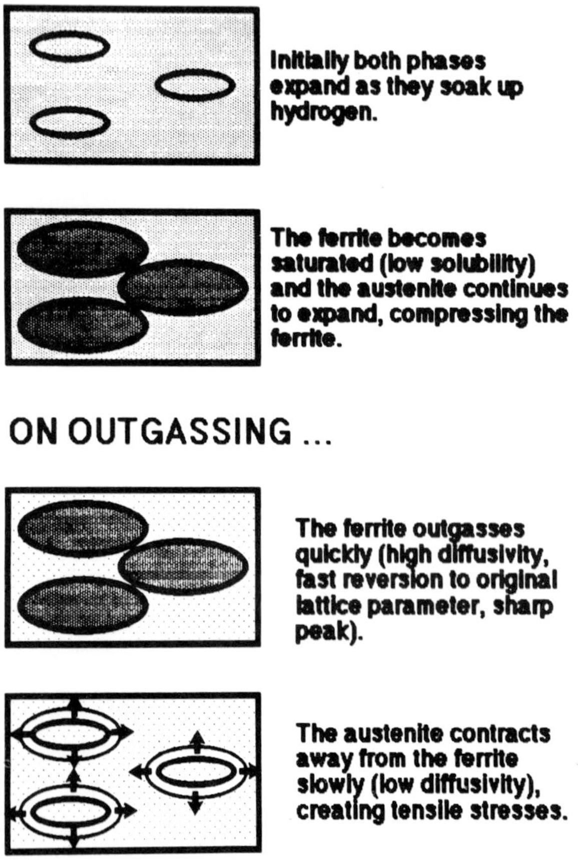

Initially both phases expand as they soak up hydrogen.
The ferrite becomes saturated (low solubility) and the austenite continues to expand, compressing the ferrite.
ON OUTGASSING ...
The ferrite outgasses quickly (high diffusivity, fast reversion to original lattice parameter, sharp peak).
The austenite contracts away from the ferrite slowly (low diffusivity), creating tensile stresses.

Figure 10. Suggested model for hydrogen charging and outgassing in the surface layers of duplex stainless steels to account for observed changes in lattice parameter and cracking.

SESSION VI
ELEVATED TEMPERATURE EFFECTS

High temperature aspects of stainless steels

T Gladman

The author is with BSC's Swinden Laboratories, Moorgate, Rotherham.

SYNOPSIS

In this introductory paper, the deformation and fracture characteristics of stainless steels at elevated temperatures are reviewed broadly. Similarities in deformation and fracture behaviour are seen at creep strain rates and at hot working rates in the appropriate temperature ranges. The effects of grain size and second phase particles are considered. The adverse effect of increased grain sizes on ductility in intergranular creep failure and hot tensile ductility are noted. The use of extremely fine-grained structures in duplex stainless steels enables superplastic deformation to be developed at practicable hot working strain rates. Attention is drawn to the adverse effects of delta ferrite on the hot ductility of austenitic steels.

A necessarily brief consideration of the finite element modelling of hot working processes is given, and suggestions for future developments are made.

1. INTRODUCTION

Stainless steels are commonly used at elevated temperatures because of their load carrying capacity (creep resistance), oxidation resistance, and fabricability. During manufacture, most stainless steels are subjected to substantial hot deformation in the shape of conversion process from ingot or concast product to plate, strip, tube or bar. Basic theoretical studies have indicated similarities in behavioural mechanisms in the two regimes of creep and hot working. In this introductory paper, these similarities are explored, together with the basic deformation and failure mechanisms. Developments arising from such studies are described covering the areas of elevated temperature deformation and failure in creep, microstructural developments in hot working, and the occurrence of superplastic behaviour.

2. DEFORMATION

2.1 Deformation Mechanism Maps

Ashby[1] produced deformation mechanism maps for a range of materials which included Type 304 austenitic steels. The basic constitutive equations developed earlier[2,3,4] were used to predict the strain rates that would be sustained by mechanisms such as dislocation creep, Coble creep (grain boundary diffusion) and Nabarro-Herring creep (bulk diffusion). These equations are shown below:-

Dislocation Creep[4]

$$\dot{e} = \frac{A\,G\,b.}{kT}\left(\frac{\sigma}{G}\right)^{n}.\ D_o.\exp(-Q_D/RT) \quad (1)$$

where e = strain rate
n = stress index
σ = applied tensile stress
T = absolute temperature
k = Boltzmann's constant
b = Burger's vector
G = shear modulus
Q_D = activation energy for dislocation creep
R = gas constant
A, D_o = constants

Coble Creep[3]

$$\dot{e} = \frac{C_1\,D_B\,\delta\,\Omega\,\sigma}{kT\,d^3} \quad (2)$$

where D_B = coefficient of grain boundary diffusion
Ω = atomic volume
δ = grain boundary width
d = grain diameter
C_1 = geometric constant

Nabarro-Herring Creep[2]

$$\dot{e} = \frac{C_2\,D_v\,\Omega\,\sigma}{kT\,d^2} \quad (3)$$

where C_2 = geometric constant
D_v = coefficient for volume diffusion

A typical deformation mechanism map for a Type 304L austenitic stainless steel is shown in Fig. 1. By normalising stresses with respect to modulus, and temperature with respect to melting point, many materials are found to show similar maps. The maps show the regimes of stress and temperature where a particular deformation mechanism gives the highest strain rate. Available experimental data were used to supplement the theoretical approach in establishing the actual deformation rates.

A particular point of interest in these maps is the extensive range of temperature(and strain rate) over which dislocation creep is the dominant mechanism. This range extends from modest temperatures in the creep range up to the melting point, and the creep rate increases dramatically with increasing temperature at a given stress level. Strain rates of 10^{-8} s^{-1} at 500 °C increase to 10^{0} s^{-1} at hot working temperatures.

In hot deformation processes, the Zener-Holloman parameter, Z, which defines the steady state deformation stress, σ, relates the strain rate and deformation temperature[5] by:-

$$Z = \dot{e} \exp(-Q/RT) = A \exp \beta\sigma \quad (4)$$

- and also shows that strain rates of 10^{0} s^{-1} at hot working temperatures are equivalent to strain rates of 10^{-6} to 10^{-8} s^{-1} at temperatures in the creep range, for a given stress level.

2.2 Stress and Temperature Effects

Experimental measurements[6] of creep rates of a Type 304L steel at temperatures and stresses across the deformation mechanism boundary illustrated in Fig. 1, are shown in Figs. 2 and 3. The change in stress dependency from a value of about 7 to a value of about 1 can clearly be seen in Fig. 2. The temperature dependence also changes, Fig. 3, from a value of 300 kJ/mole at high stresses (consistent with volume diffusion) to a value of 170 kJ/mole at low stress (consistent with grain boundary diffusion). Such changes are clearly in accord with the deformation mechanism map, and the constitutive equations, indicating a change in deformation mechanism from dislocation creep at high stresses to Coble creep at the lower stresses. A more detailed map for the Type 304L steel is shown in Fig. 4. The computed strain rates shown are based on the constitutive equations given earlier. Appropriate values of n, Q, and the various constants required in these equations were derived from the experimental data. The boundaries between regions where the various mechanisms are dominant are indicated together with the data points identifying the creep strain rate characteristics. The coincidence between the data and the map can clearly be seen.

2.3 Effects of Second Phase Particles

The effects of second phase particles on creep properties are complex. The impedance of dislocation movement by precipitates is well established[7] and creep rates in the dislocation creep range are observed to decrease. On this basis, the domain of Coble creep dominance in the mechanism map would be expected to expand as indicated in Fig. 5 for a Type 347 steel. However, the transition in the experimentally observed stress dependency was very different from that seen in the Type 304L steel. The Type 347 steel showed that the transition to low n values occurred at stress levels characteristic of the change from dislocation creep to Nabarro-Herring creep. Also, the activation energy for the deformation process was typical of volume diffusion at both high and low stresses. Similar effects were observed for Type 304 ($M_{23}C_6$ precipitation) Type 321 (TiC precipitation) and Type 316 ($M_{23}C_6$ precipitation) steels. The formation of precipitates in these cases (including Nb(CN) in Type 347 steels) is not confined to the matrix. Extensive precipitation is observed on grain boundaries, and the particles formed on the grain boundaries are generally larger than those formed intragranularly. The occurrence of Coble creep is dependent on diffusion along grain boundaries. The blocking of this diffusion path by grain boundary particles will therefore inhibit Coble creep.

The absence of specific effects of precipitates in the constitutive equations is seen as one of the shortcomings of the deformation mechanism maps.

The effects of precipitation during hot working produce a similar effect in impeding dislocation movement. During hot working, a fixed strain rate is imposed on the material, and therefore the precipitates increase the stress required to maintain the strain rate[7]. An additional effect is observed during hot working processes. The presence of a sufficiently fine dispersion of particles can retard not only dislocation recovery, but also recrystallisation. The retardation of recrystallisation can also have important effects on deformation stresses.

2.4 The Effect of Grain Size: Superplastic Behaviour

There is no effect of grain size in dislocation creep, as indicated by equation 1. On the other hand, in diffusion creep, the strain rate increases dramatically as the grain size is reduced as indicated in equations 2 and 3 which show grain size indices of -3 and -2 respectively.

The effect of grain size on both Coble and Nabarro-Herring creep has definite implications for the deformation mechanism map. The construction of the map involves equating the dislocation creep rate with the creep rates for the other mechanisms. As the latter are dependent on grain size, it is clear that any deformation mechanism map relates only to the grain size on which the calculations were based. A refinement of the grain size would increase both of the diffusion creep rates and would thus reduce the range of stress and temperature over which dislocation creep was the dominant mechanism.

The need for a fine grained structure to promote superplastic behaviour particularly in ferrous materials is well recognised[8]. Brophy and Miller proposed the following constitutive equation for superplastic deformation

$$\dot{e} = \frac{C.D_v \sigma^2}{T d^2} \qquad (5)$$

This equation shows many similarities with those for diffusion controlled creep, the creep strain rate having a relatively low stress index, and a strong inverse dependency on grain size.

The low stress index is imperative for superplastic behaviour; localised strain is much less likely to lead to a fully developed neck if the stress dependency index is low, as this promotes a more uniform strain distribution[9]. The use of a material with a very fine grain size is necessary to extend the range of stress or strain rate over which superplastic deformation can be maintained. This is particularly important in extending the upper limit of strain rate for superplastic deformation in order to accommodate commercially acceptable deformation rates. Equation 5 was used to evaluate superplastic deformation rates for the Type 304L and Type 347 steels used in the preparation of Figs. 4 and 5 respectively. The limits of stress and temperature over which superplastic deformation would dominate are superimposed on the deformation mechanism maps. With grain sizes of 75-100 µm, the maximum strain rate that would produce superplastic behaviour is of the order of 10^{-7} (s^{-1}) at 1000 °C. The strong inverse dependency of strain rate on grain size therefore requires very small grain sizes (below 5 µm) to increase the maximum limit of superplastic deformation rate to commercially acceptable forming rates.

An important practical aspect of superplastic materials is that the extremely fine grained material should be resistant to grain coarsening at the deformation temperature. It is probably this aspect which prevents the common austenitic stainless steels from exhibiting superplasticity. Duplex stainless steels, because of the extreme stability of their grain structures, have been used for superplastic forming processes.

It is interesting to note that even in the absence of a specific constitutive equation for superplastic deformation, superplastic behaviour would still be expected in these same domains of stress and temperature. Both Coble grain boundary diffusion creep and Nabarro-Herring volume diffusion creep show a low stress index and a strong inverse dependence on grain size.

3. FRACTURE

3.1 Fracture Mechanism Maps

The mapping of deformation mechanisms by Ashby[1] has been applied also to failure mechanisms[10]. A typical failure mechanism map for a Type 304 austenitic stainless steel is shown in Fig. 6. The mechanisms of failure are not based on constitutive equations but are derived from experimental observation.

At low temperatures, failure occurs exclusively by normal ductile transgranular fracture. Cavities are nucleated on non-metallic inclusions or other second phase particles, grow as a result of local strain intensification, and subsequently coalesce to give failure. At temperatures in the creep range, the same ductile transgranular fracture is observed at high stresses, but at the lower stresses, failure occurs mainly by intergranular cavitation and cracking. At high temperatures (above 0.7 Tm) failure by rupture is indicated. The term 'rupture' implies that the material will undergo extensive plastic deformation and will neck virtually to a point.

3.2 Intergranular Creep Failure

Intergranular creep cavitation is associated with the nucleation and diffusion controlled growth of voids. The volume fraction of cavities, f_v, can be related[11] to creep strain, ϵ, time, t, stress, σ, and temperature, T, by the following equation -

$$f_v = C \epsilon t \sigma^n \exp(-Q/RT) \qquad (6)$$

where C and n are constants
and Q is the activation energy for grain boundary diffusion.

More recent examination[12] has indicated that in most steels, both cavity nucleation and cavity growth are controlled by the maximum principal stress. Cavity growth can be calculated from the Hull and Rimmer[13] model provided that continuous nucleation of cavities is taken into consideration. The cavity nucleation rate although known to depend on maximum principal stress, σ_P, by -

$$N = K \sigma_P^m \qquad (7)$$

- has so far escaped any successful theoretical treatment. Nucleation rates in general cannot be calculated, but can be established by experimental observation. Such experiments have shown that the cavity nucleation rate is increased by

(a) increasing the grain size of the structure.
(b) the presence of second phase particles on the boundary; sulphide

particles are particularly detrimental in this respect.
(c) high residual element contents.

The effects of grain size and grain boundary sulphides are particularly important in welded structures where the heat affected zone can show both of these features, which result in intergranular fracture (Type III cracking) after much shorter exposure times than are observed for parent plate material. This has led to controlled heat input multipass welding to avoid the occurrence of coarse grained heat affected zones. It should be pointed out that much of this work was carried out on ferritic high temperature steels but would apply equally to the martensitic and other stainles steels.

3.3 Remanent Life

The need to replace expensive elevated temperature plant on reaching its design life has been questioned. Methods have been sought to give an estimate of the remanent life. Data on plant exposure to stress and temperature cycles have been used in conjunction with the Robinson life rule. Plastic strains have been monitored using pips or markers on the components. Post exposure creep testing of samples cut from the component have also been used to compare the rupture life with that of unexposed material. In the latter procedure, tests are usually accelerated by using test temperatures in excess of the service temperature; the use of higher stresses for accelerated testing could cause complications with a change in both deformation mechanism, Fig. 1, and failure mechanism, Fig. 6.

A further method of remanent life assessment, which is particularly useful for low ductility intergranular creep failures, derives from the cavitation studies. It has been shown [14] that many materials, including Type 347 austenitic steels exhibit a common relationship between the area fraction of grain boundaries covered by cavities, and the fraction of life expended, Fig. 7. Electron microscopical evaluation of the degree of cavitation can be carried out on 'boat' samples cut from the component, and the upper bound of life fraction expended can be derived from Fig. 7. The lower bound remanent life calculation is then simple, provided that the prior exposure time is known.

3.4 High Temperature Ductility

Although the failure mechanism map in Fig. 6 shows 'rupture' to be the common failure mechanism at hot working temperatures, these observations were generally based on long gauge length tensile tests. The failure mechanism and ductility can be influenced quite drastically by the application of constraints, such as will occur in commonly used short gauge length hot ductility tests. In the latter type of test, other failure mechanisms have been observed, including both transgranular ductile fracture, and intergranular cavitation and cracking. Theoretical developments in the quantification of hot ductility are lacking. Suitable hot working ranges are commonly selected by observing the temperature range of high ductility failures in hot ductility testing. The high ductility temperature range is commonly bounded on the high temperature limit by liquation problems, particularly if the steel has low melting eutectics, e.g. boron steels, and the Type 347 niobium grade of austenitic stainless steel, Fig. 8. The lower temperature is commonly limited by intergranular cavitation effects not dissimilar to those observed in creep fracture, or by the high strength levels which cause overloads in the hot mill.

Metallurgical factors which control hot ductility include the austenite grain size, second phase particles, and the recrystallisation characteristics.

The hot ductility decreases as the prior grain size is increased, Fig. 8. The nature of this effect may be similar to that observed in intergranular creep fracture, but there are additional features to consider at the generally high strain rates, high strains, and high temperatures associated with hot working. A common feature of the high temperatures associated with hot working is the dissolution of second phase particles, such as chromium carbide, giving relatively clean boundaries. The high strain rates and high strains are also conducive to recrystallisation. It has been suggested[7] that the formation or propagation of cavities on the transverse grain boundaries may be inhibited by recrystallisation, because the cavity then becomes isolated from the grain boundary. Any cavities formed prior to the onset of dynamic or static recrystallisation, cease to grow, and behave as isolated voids which contribute to normal transgranular ductile fracture in much the same way as sulphides or other non-metallic inclusions. Subsequent deformation of the sessile intergranular cavities and their subsequent contribution to the ductile fracture process has been observed directly. One of the important effects of prior austenite grain size on hot ductility may arise because of a major effect of the prior grain size on the kinetics of recrystallisation. A coarse grain size retards both dynamic and static recrystallisation, thereby allowing continued growth of the intergranular cavities.

The observed effect of niobium on the hot ductility of austenitic stainless steels may illustrate one method by which second phase particles affect hot ductility. Niobium carbide can be dissolved during high temperature soaking, and produces a marked retardation of recrystallisation, either because of the niobium in solution or because of strain induced precipitation of fine niobium carbides at lower hot working temperatures. Such a retardation of recrystallisation gives the opportunity for enhanced intergranular cavitation and lower hot ductility values.

A very different effect of second phase particles is observed when a duplex austenite-ferrite structure is considered. In this case, the delta-ferrite particles are usually large, and do not inhibit recrystallisation. The ferrite-austenite interfaces are weak, and decohesion occurs under a tensile stress[15]; the cavities thus formed cause a serious loss of hot ductility, Fig. 9. The effect of delta-ferrite on hot ductility can be related to the area of the austenite-ferrite interfaces per unit volume of material which are normal to the stress axis.

The effect of delta-ferrite on hot ductility is complicated by the constitutional relationships between delta-ferrite content, steel composition, and temperature. Most of the common grades of austenitic stainless steel show the presence of small amounts of delta-ferrite in cast steels. The Schaeffler[16] and DeLong[17] diagrams for predicting these delta-ferrite contents, from chromium and nickel equivalents are well established. The equilibrium delta-ferrite contents for many of the common 300 Series steels increase as the soaking temperature is increased above 1150 °C, thus giving the ferrite contents (about 5-10%) necessary to inhibit liquation cracking during welding. The design of many of the 300 Series in this way will necessarily introduce delta-ferrite at high hot working temperatures, and may give rise to hot ductility problems. It should be pointed out that many of the duplex stainless steels show a marked increase in ferrite content at high temperatures, and this can in fact reduce the number of austenite-ferrite interfaces when the equilibrium ferrite content exceeds 50%, as indicated in Fig. 10.

3.5 Stress State during Hot Deformation

The marked effects of second phase particles on hot ductility shown in hot tensile tests cannot be used as any more than a qualitative indicator for commercial hot working operations. Hot rolling is predominantly a compressive deformation, and the cavities developed in creep or hot tensile testing largely depend on the maximum principal tensile stress. Cavity nucleation is suppressed in compressive forming operations. Even when cavity nuclei are present, cavity growth is dramatically reduced when compressive forming operations are used. Relatively high bulk deformations can be sustained during hot rolling even though hot tensile testing may show marked reductions in ductility. The usefulness of hot tensile testing is probably most pertinent to edge cracking where the unsupported edges have to elongate in order to maintain continuity with the bulk (compressed) material during hot rolling passes.

Even so, the limited hot ductility of certain complex highly alloyed steels (with high hot strength) can cause problems in hot working, and it is interesting to see the development of sintering processes using rapidly solidified powders for the manufacture of substantial components for 'difficult' alloys[18]. The powder process route, however, does offer other advantages and is not used solely to overcome hot working difficulties.

4. HOT DEFORMATION MODELLING

Considerable attention is currently being given to the development of quantitative models for a range of processes, including solidification and hot working, and for the behaviour of structures in service e.g. the creep behaviour of welded joints. These models are frequently complex, in order to give due consideration to stress, strain, and temperature distributions, which affect the behaviour of material undergoing the solidification or straining operation. These models rely extensively on the availability of advanced computers capable of dealing with finite element or finite difference calculations. The models are not only capable of yielding important processing information such as rolling mill or forging loads, but can give important information on the evolution of microstructure.

Extensive studies of hot working have been carried out[19] over a considerable period of time to provide the necessary constitutive equations for hot deformation and the required material parameters. The roll separation force in a given hot rolling pass is dependent not only on the control variable such as temperature, strain rate, and rolling reduction, but also on the metallurgical state of the material, particularly that relating to the degree of recrystallisation in multipass operations. The constitutive equations for recrystallisation rates are available for Type 304 and Type 316 austenitic stainless steels, and cover the effects of prior grain size, temperature, strain and strain rate[20,21], and inter-pass hold times. Considerable success has been achieved in predicting rolling forces and microstructural evolution in plate and strip rolling where the strain is uniformly distributed. More recent developments however are aimed at a complete description of microstructural evolution in hot deformation processes where the strain distribution may be heterogeneous.

The relationship between the hot deformation schedule and the microstructural evolution will have important consequences for the control of hot band microstructure. Control of hot band microstructure in austenitic stainless steels is capable of yielding high strength products, although this has not been exploited seriously because of the present practice of giving a subsequent high temperature solution treatment. Direct quenching from the hot mill is a possibility.

The hot band grain size is also an important feature in controlling both the recrystallisation kinetics and the recrystallised grain size of cold rolled and annealed strip. In ferritic stainless steels, the control of hot band structure has important consequences for the riding and roping behaviour of these steels during cold forming of the final product, whether this is carried out on hot rolled product or cold rolled and annealed product. There is relatively little basic information on microstructural evolution of ferritic stainless steels at the present time.

5. SUMMARY

The basic deformation and failure mechanisms of stainless steels at elevated temperatures have been considered in relation to both creep applications and hot working practices. Changes in deformation mechanism with stress and temperature have been illustrated, and the low stress index power laws of deformation observed at low stresses are important to superplastic behaviour. The effect of grain refinement can change the limits of strain rate over which superplastic behaviour occurs, to include practicable deformation rates.

Important effects of grain size, both on creep cracking, and on ductility in hot working, are described. The grain size effect has particularly important implications for the welding practices used on components for high temperature services. The assessment of remanent life of components from cavitation studies is also described. In considering the effects of second phase particles, particular attention was given to the effects of delta-ferrite on ductility during hot working.

Finally, attention was drawn to finite element modelling methods which can provide useful information on hot deformation loads and on the evolution and control of microstructure during hot working. These models are currently being developed, and further basic information is yet required particularly on ferritic grades of stainless steel.

6. ACKNOWLEDGMENTS

The author would like to thank Dr. R. Baker, Director of Research, British Steel Corporation for permission to publish this paper.

7. REFERENCES

1. Ashby, M.F. Acta Met. 1972, 20, p887.
2. Weertman, J. J. App. Phys. 1957, 28, p196.
3. Coble, R.L. J. App. Phys. 1963, 34, p1679.
4. Mukherjee, A.K., Bird, J.E. and Dorn, J.E. Trans. ASM 1969, 62, p155.
5. Hughes, K.E., Nair, K.D. and Sellars, C.M. Metals Technology 1974, 1, p161.
6. Beckitt, F.R., Banks, T.M., and Gladman, T. 'Creep Strength in Steel and High Temperature Alloys', Iron and Steel Institute Conference, Sheffield University, Sept. 1972, p71.
7. Bywater, K.A. and Gladman, T. Metals Technology 1976, 3, p358.
8. Hayden, H.W. and Brophy, J.H. Trans. ASM, 1968, 61, p542.
9. Hart, E.W., Acta Met., 1967, 15, p351.
10. Fields, R.J., Weerasooriya, T. and Ashby, M.F., Met. Trans. A., Met. Soc. AIME, 1980, 11, P333.
11. Needham, N.G., Wheatley, J.E. and Greenwood, G.W., Acta Met. 1975, 23, p23.
12. Needham, N.G. and Gladman, T., Advances in Physical Metallurgy and Applications of Steels, London, Metals Society, 1981, p309.
13. Hull, D. and Rimmer, D.E., Phil.Mag. 1966, 14, p673.
14. Needham, N.G. and Gladman, T. to be published. Met.Sci. and Technology.
15. Decroix, J.H., Neveu, A.M. and Castro, R.J.' Deformation under Hot Working Conditions' Iron and Steel Institute, Special Report No. 108, London, 1968, p135.
16. Schaeffler, A.L. Metal Progress, 1949, 56, p680.
17. DeLong, W.T., Ostrom, G.A. and Szumachowski, E.R., Welding J. 1956, 35, p521s.
18. Lindenmo, M., Stainless Steels '87 - This conference.
19. Sellars, C.M., Materials Science & Technology 1985, 1, p325.
20. Towle, D.J. and Gladman, T. Metals Science 1979, 13, p246.
21. Barraclough, D.R. and Sellars, C.M. Metals Science 1979, 13, p257.
22. Gladman, T., Burke, P., Ingham, P. and Watts, G. to be presented at Institute of Metals Conference Sutton Coldfield, 1987.

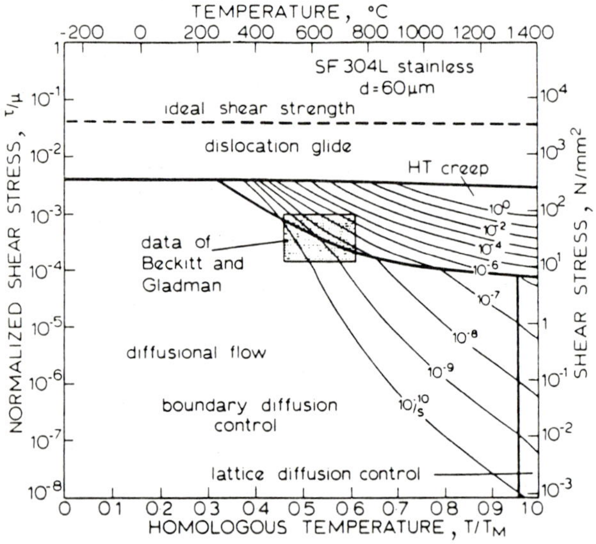

FIG. 1

Deformation mechanism map for SF 304 L austenitic stainless steel after Ashby[15]

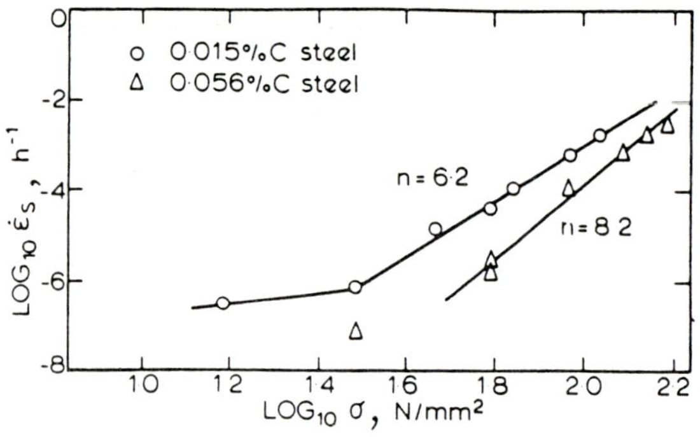

FIG. 2 **Stress dependence of the minimum creep rates**

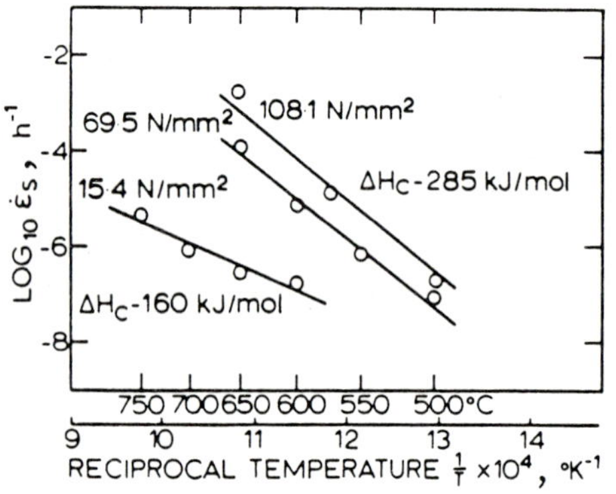

FIG. 3

Temperature dependence of the minimum creep rate for the low-carbon (0·015%) steel

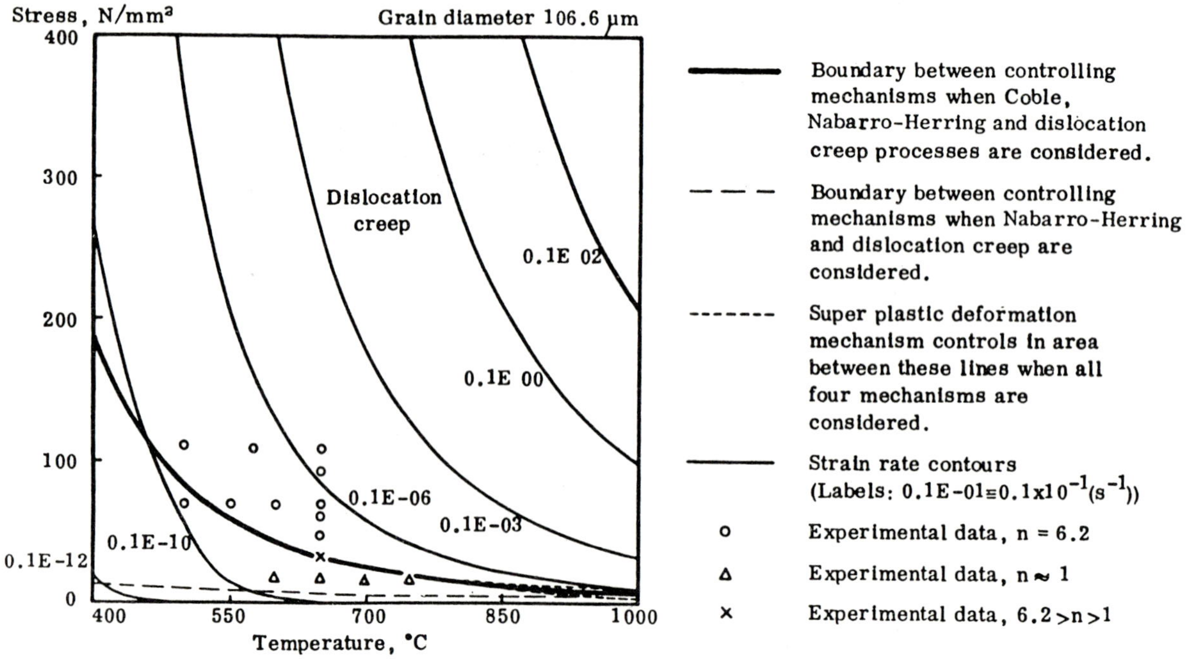

FIG. 4 DEFORMATION MECHANISM MAP FOR TYPE 304L STEEL

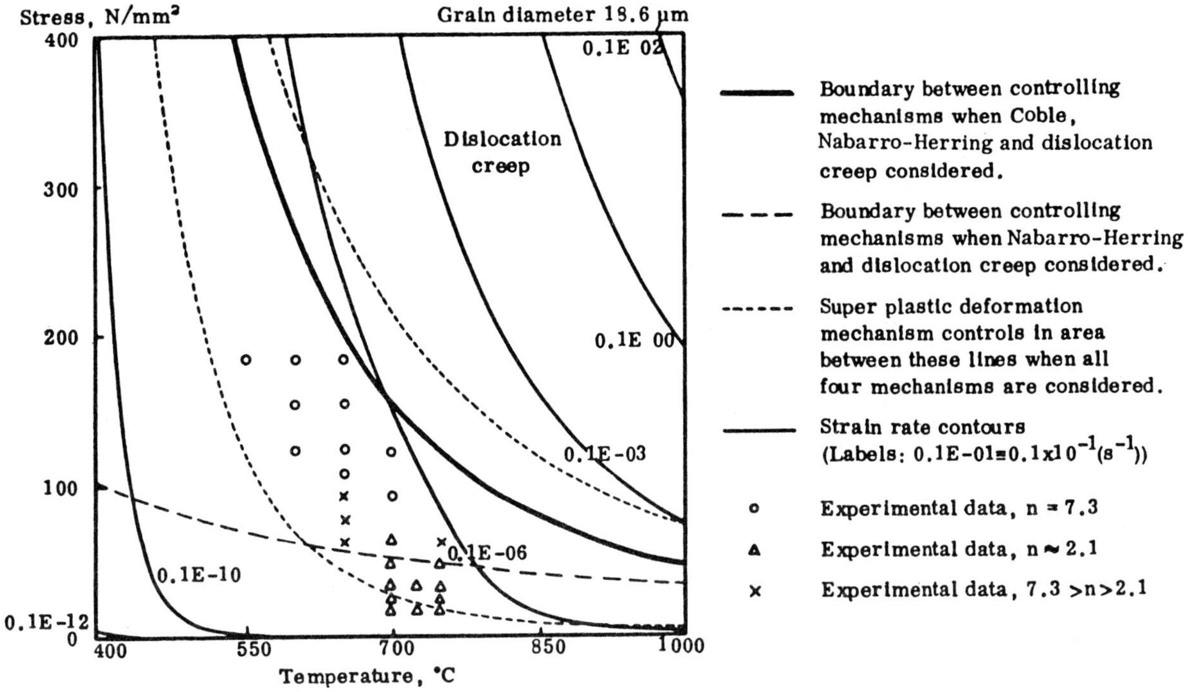

FIG. 5 DEFORMATION MECHANISM MAP FOR TYPE 347 STEEL

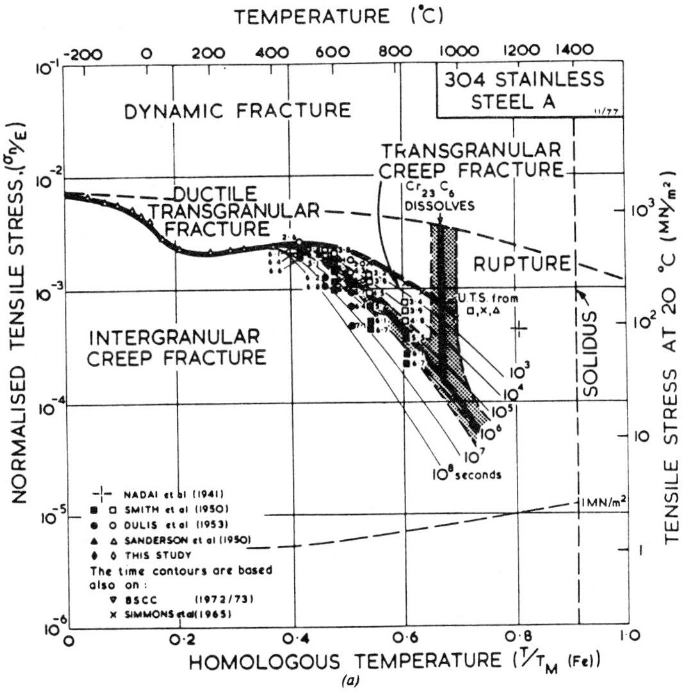

FIG. 6 Failure Mechanism Map

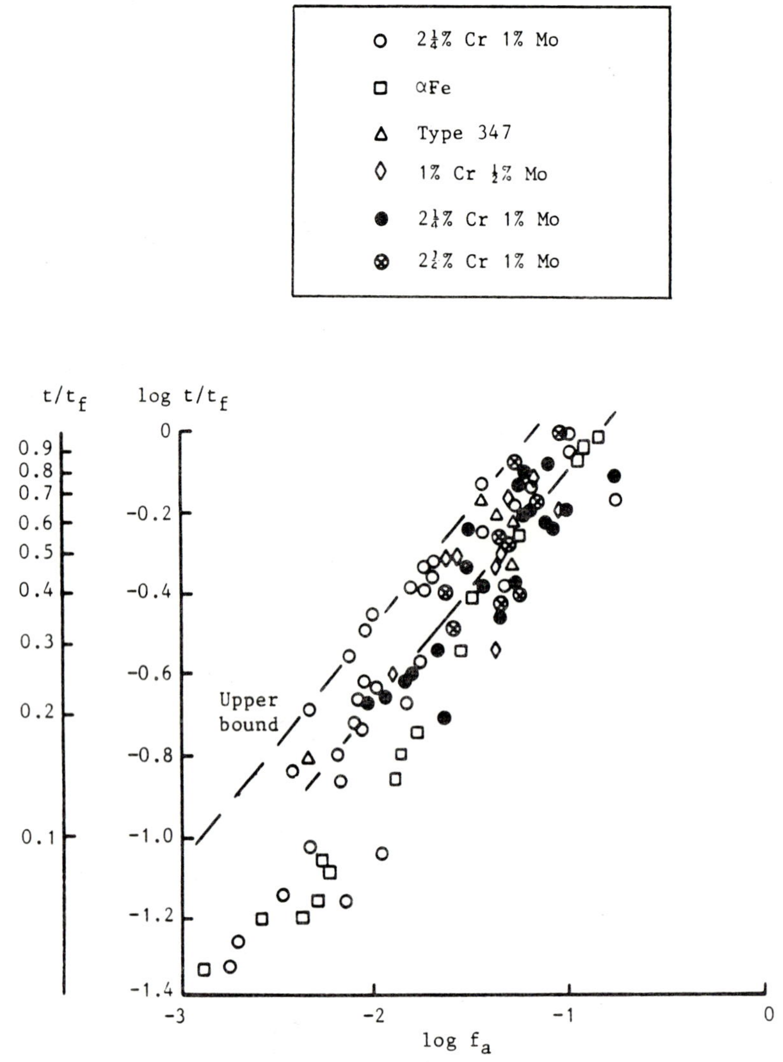

FIG. 7 THE DEPENDENCE OF THE LIFE FRACTION EXPENDED ON THE AREA FRACTION OF BOUNDARY CAVITATED

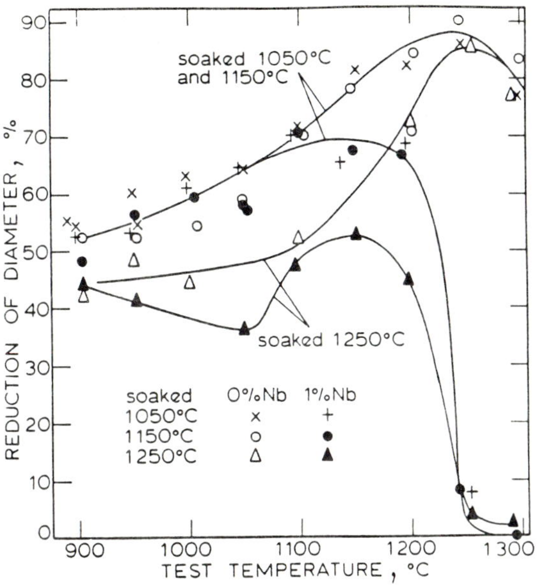

FIG. 8

Variation of hot ductility of 18Cr–10Ni steels, with and without 1% niobium, with soaking and test temperatures

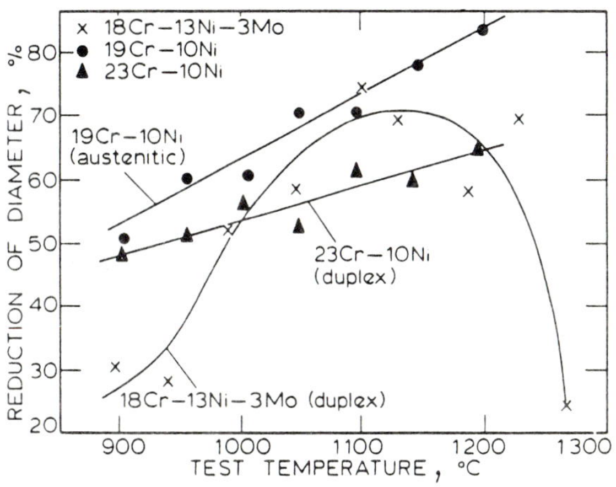

FIG. 9

Variation of hot ductility of 19Cr–10Ni, 23Cr–10Ni, and 18Cr–13Ni–3Mo steels with test temperature

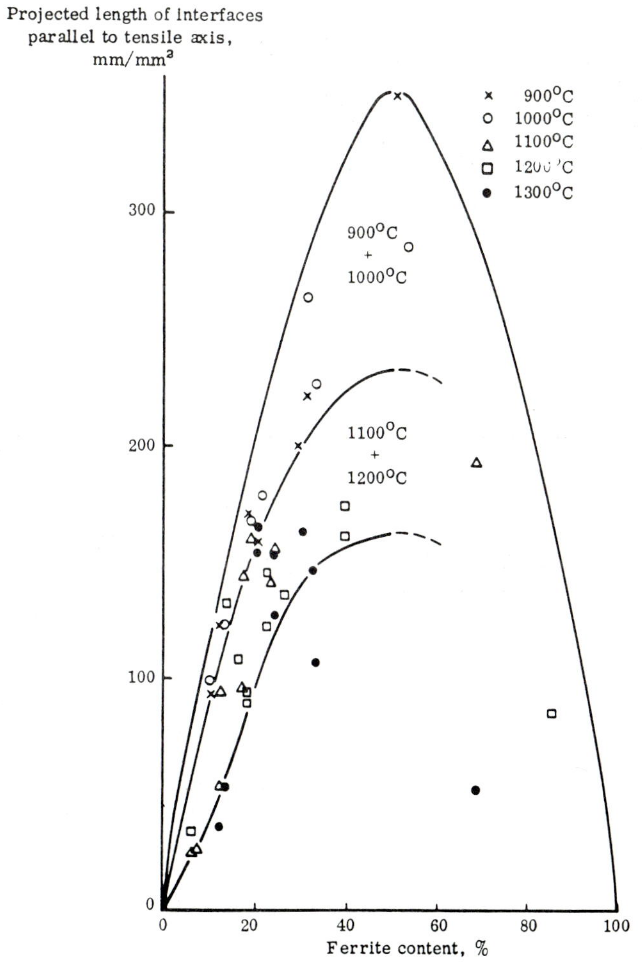

FIG. 10 THE VARIATION OF INTERFACE PROJECTED LENGTH PARALLEL TO THE TENSILE AXIS WITH FERRITE CONTENT

Formation of deformation band and recrystallization behaviour of fully ferritic stainless steels under hot rolling process

Y Uematsu and K Yamazaki

The authors are in the Shunan R & D Laboratory of Nisshin Steel Co. Ltd., Japan.

SYNOPSIS

The hot-rolling tests were carried out in a temperature range of between 600 and 1250°C with 40% reduction of thickness and the following annealing at the same temperatures, using specimens of the columnar and the equiaxed crystals of fully ferritic stainless steels. The hot-rolled microstructural changes and the following recrystallization behaviour were investigated by a Nomarsky's phase-contrast microscope, and the dislocation substructures were observed in detail by a transmission electron microscope. The orientation analysis of deformed matrix was performed by an electron channelling pattern.

The possibility of refinement of ferrite crystal grains under hot-rolling conditions was discussed from the point-of-view of deformation banding and the following recrystallization behaviour.

INTRODUCTION

Fully ferritic stainless steels are now widely used in the fields of drawing works, hot-water tanks, automotive moles and heat-resisting utilization, and so on, because of their sufficient corrosion resistance, formability, and heat resistance, in addition to cost-efficiency. However, on deep drawing, these steels sometimes bring about a ridging phenomenon and consequently lose surface beauty.

Generally, in a hot deformation, it is considered that ferritic stainless steels have a tendency to recover dynamically and hardly recrystallize,[1-3,6] therefore coarse, polygonized grains remain in a central layer of hot-rolled bars. These coarse grains are thought to cause a plastic anisotropy. In order to refine ferrite grains, it is important to make substructures non-uniform and to introduce many deformation bands as a result. However, few papers have investigated the occurrence of deformation banding in hot rolling.[6,7]

In this paper, microstructure changes developed in hot rolling, and the following recrystallization behaviour on annealing, were observed and the characteristics of deformation bands crystallographically studied, using fully ferritic stainless steels with the initial structures of the columnar or the equiaxed crystals. Finally, the roll of deformation bands to refinement of ferrite grains under hot-deformation process was discussed.

EXPERIMENTAL PROCEDURE

The chemical compositions of materials used are shown in Table I. Fe–16%Cr–0.3%Nb and Fe–13%Cr–0.2%Ti were melted in a 40-ton electric furnace and continuously cast. Specimens of both steels used for hot-rolling tests were cut from the columnar and the equiaxed parts of slabs in order to study the effect of initial crystal structures on the microstructures developed in the hot rolling. The average sizes of columnar and equiaxed grains were about 2.5 mm dia. x 14 mm length and 2.8 mm dia., respectively. Fe–19%Cr was laboratory-melted and cast in a 30 kg ingot. This ingot was hot-forged to 20 mm thickness and annealed at 1200°C for 30 min to obtain the average grain size of 1 mm. The specimen used for hot rolling was 20 mm thickness x 40 mm width x 50 mm length. The hot-rolling temperature range was between 600 and 1250°C and the reduction was 40%. The hot-rolling tests were carried out on the hot-rolling machine with a roll diameter of 330 mm at 50 rpm. The hot-rolled specimens were water-quenched immediately ($t < 5$ s) to examine the exact hot-rolled substructures and some of them were successively annealed at the soaked temperatures for 0.5~10 min to measure the fraction of recrystallized area.

Microhardness measurement was performed at the centre layer of hot-rolled specimens by a micro-vickers, 500 g in weight. The hot-rolled substructures were observed by a Nomarsky's phase-contrast microscope and the dislocation substructures were observed by an electron microscope. The crystal orientation of hot-rolled substructures was investigated by an electron channelling pattern analysis.

RESULTS AND DISCUSSION

Hardness of hot-rolled specimens

The hot-rolling tests were carried out at temperatures of between 600 and 1200°C, by one-pass hot rolling of 40% reduction for Fe–16%Cr–0.3%Nb, Fe–13%Cr–0.2%Ti, and Fe–19%Cr. The hardness of hot-rolled specimens was measured.

Figs. 1a and 1b show the variations in hardness of hot-rolled specimens with the hot-rolling temperatures for the columnar and the equiaxed crystals of Fe–16%Cr–0.3%Nb and Fe– 13%Cr–0.2%Ti respectively, compared with the forged and annealed Fe–19%Cr. The behaviour of stabilized steels was almost the same for the columnar and the equiaxed crystals, and a little different from that of Fe–19%Cr. At temperatures above 1000°C, the hardness curve of Fe–19%Cr was almost constant; those of stabilized steels on the other hand were recognized to increase slightly with increasing temperature. The reason why the hardness curves of Nb and Ti-stabilized steels slightly increased at above 1000°C, might be that NbC and TiC were starting to dissolve for soaking at these temperatures and reprecipitate during cooling after hot rolling. At temperatures below 950°C, all hardness curves were markedly increasing with decreasing temperature, because of work-hardening.

At those temperatures, the hardness of Fe–13%Cr–0.2%Ti was the smallest of the three steels, and that of the other two steels was nearly the same.

Microstructural observation of hot-rolled specimens

Figure 2 shows one of the typical microstructures observed in the hot-rolled Fe–16%Cr–0.3%Nb, the columnar crystals. It was shown that in specimens hot-rolled at 1200°C, well-developed equiaxed subgrains were formed. The subgrain could be observed down to about 1000°C, and with decreasing temperature, the subgrain size decreased. For the columnar crystals, deformation bands, indicated by arrows in Fig. 2, were recognized to form at temperatures below 1125°C and these deformation bands became clearer as the temperature decreased. Additional observation of equiaxed crystals indicated that deformation bands were formed at a slightly higher temperature, 1150°C.

Figure 3 shows the transmission electron microscope observations for the substructures of Fe–19%Cr, hot-rolled at 800, 900, 1050, and 1200°C. At 1200°C, dislocation networks were observed in subgrain boundaries. As the temperature was lowered, the subgrain size obviously decreased and the dislocation density increased: at 800°C, on the other hand, only tangled dislocation substructures could be observed without subgrain formation. These observations were thought to be in good agreement with the hardness changes (Fig. 1).

Recrystallization

Figures 4 and 5 show how the recrystallized fraction varies with temperature; and dependence of the recrystallized fraction with holding time, respectively, for Fe–16%Cr–0.3%Nb and Fe–13%Cr–0.2%Ti hot-rolled to 40% reduction and successively annealed. From these figures, it can be seen that the fraction of recrystallized area for Fe–13%Cr–0.2%Ti was higher than that for Fe–16%Cr–0.3%Nb and also that the equiaxed crystals recrystallized faster than the columnar crystals in both steels.

Figure 6 shows the nucleation site of recrystallization observed for columnar crystals of Fe–16%Cr–0.3%Nb. Figure 6 indicates that the recrystallization occurred preferentially at grain boundaries at 1150°C and at deformation bands at 1000°C.

These results indicated that recrystallization was strongly dependent on the chemical composition and the initial structure. Firstly, recrystallization might be retarded by addition of Nb, and secondly it might also be related to the surface area of grain boundary and deformation band. Therefore it is important to study the nature of deformation bands and so the deformation band was studied crystallographically.

Crystallographic observation

Figures 7 and 8 show the electron channelling patterns of the deformed matrix bands of the columnar and the equiaxed crystals of Fe–16%Cr–0.3%Nb, hot-rolled at 1050°C with 40% reduction, and their stereographic projections of <111> poles, respectively. From Fig. 7a and Fig. 8a, it is seen that all orientations of matrix bands in the columnar crystals, where ND means a normal direction to the rolling plane, were near (100). On the other hand, those of matrix bands in the equiaxed crystals separated into (100) and (111). The misorientation between subgrains in a matrix band was at most a few degrees and the rotation angle between two bands reached up to 10~50 degrees. The rotation angle between matrix bands was higher in the equiaxed crystals than in the columnar crystals. Columnar crystals have a <100> growing axis which is considered to be one of stable orientations under compression and this may be the reason why the orientation of the matrix bands was close to (100). The fact that the matrix bands in the equiaxed crystals were (100) and (111) indicated that a matrix might be divided into these stable orientations[4,5] in hot rolling.

Figure 9 shows the electron micrograph and the diffraction patterns of the deformed substructures of Fe–19%Cr hot-rolled at 1050°C. It is recognized from Fig. 9 that the diffraction pattern suddenly changed at 'b' boundary where the subgrains were relatively small. Therefore 'a' and 'c' regions had to be matrix bands and 'b' had to be a deformation band.

Finally the results mentioned above suggest the following refining mechanism of ferrite as shown schematically in Fig. 10. At higher temperatures only the equiaxed subgrains were formed and misorientation was very small. In this case, the following recrystallization might occur by grain-boundary migration. On the other hand, with decreasing temperature, the obvious deformation banding occurred with large misorientation. In that case, the following recrystallization might occur at deformation bands.

CONCLUSION

The characteristics of substructure changes of fully ferritic stainless steels hot-rolled in a temperature range between 600 and 1250°C were studied. The main results were as follows.

Substructures of fully ferritic stainless steels hot-rolled at temperatures above 1150°C were composed of only equiaxed subgrains and those below 1150°C contained the deformation band. Misorientation in a band was at most a few degrees; on the other hand, rotation angle between two bands reached up to 10~50 degrees. The equiaxed crystals had deformation bands with larger misorientation than the columnar crystals. Recrystallization at higher temperatures occurred preferentially at the grain boundary; at lower temperatures on the other hand it occurred in deformation bands. These results suggest that deformation bandings have an essential role to play in the refinement of ferritic stainless steels by hot-rolling process.

ACKNOWLEDGMENTS

The authors thank Prof. emeritus I Tamura and Assistant Prof. T Maki in Kyoto University for their advice and discussion of the results. Nisshin Steel is thanked for permission to publish this paper.

REFERENCES

1 A T English and W A Backofen: Trans. Metall. Soc. AIME, 230 (1964), 396
2 R Wusatowski: JISI, 204 (1966), 727
3 J P Sah and C M Sellars: 'Hot working and forging process', (ed. C M Sellars and G J Davis, Suppl.), 1979, 62, The Metals Society
4 I L Dillamore and H Katoh: Met. Sci., 8 (1974), 73
5 Y Inokuti and R D Doherty: Acta Metall., 26 (1978), 61
6 Y Uematsu, K Hoshino, T Maki, and I Tamura: Tetsu-to Hagane, 70 (1984), 2152
7 K Yamazaki and Y Uematsu: Trans. ISIJ, 26 (1986), B-281

Table 1 Chemical composition (wt%)

	C	Si	Mn	P	S	Cr	N	Nb	Ti	Fe
Base	0.001	0.27	0.24	0.001	0.004	18.75	0.002	—	—	Bal.
Nb	0.006	0.58	0.24	0.023	0.003	16.17	0.012	0.29	—	Bal.
Ti	0.011	0.40	0.23	0.021	0.008	12.93	0.010	—	0.20	Bal.

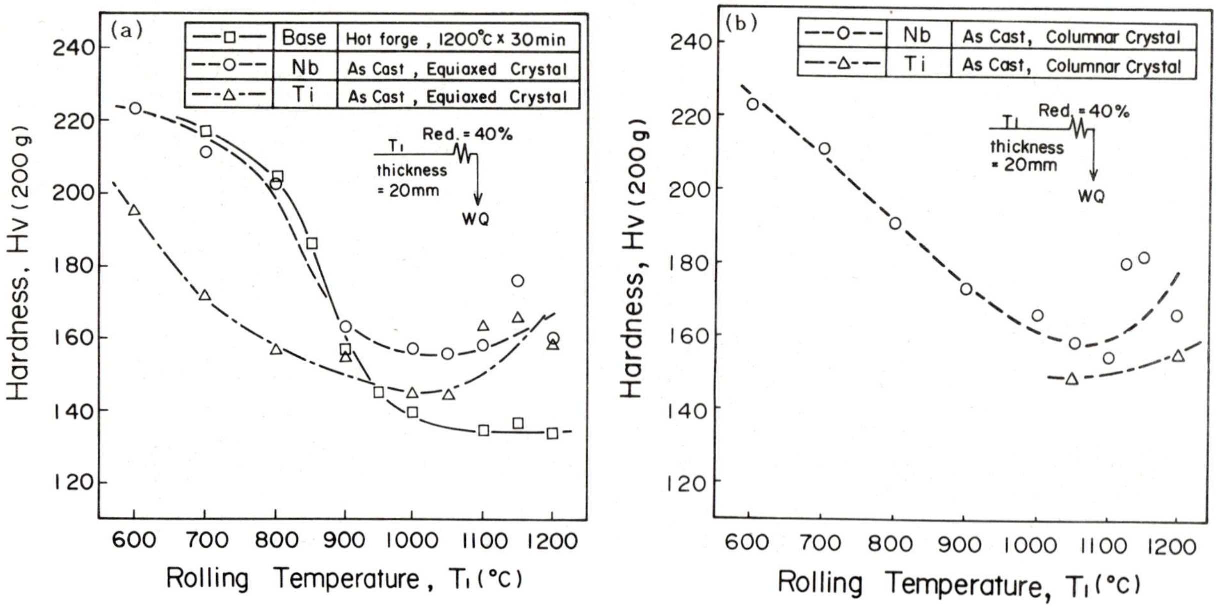

Fig. 1 Variation of hardness with hot-rolling temperature for Fe–16%Cr–0.3%Nb, Fe–13%Cr–0.2%Ti, and Fe–19%Cr: (a) for equiaxed crystals; (b) for columnar crystals

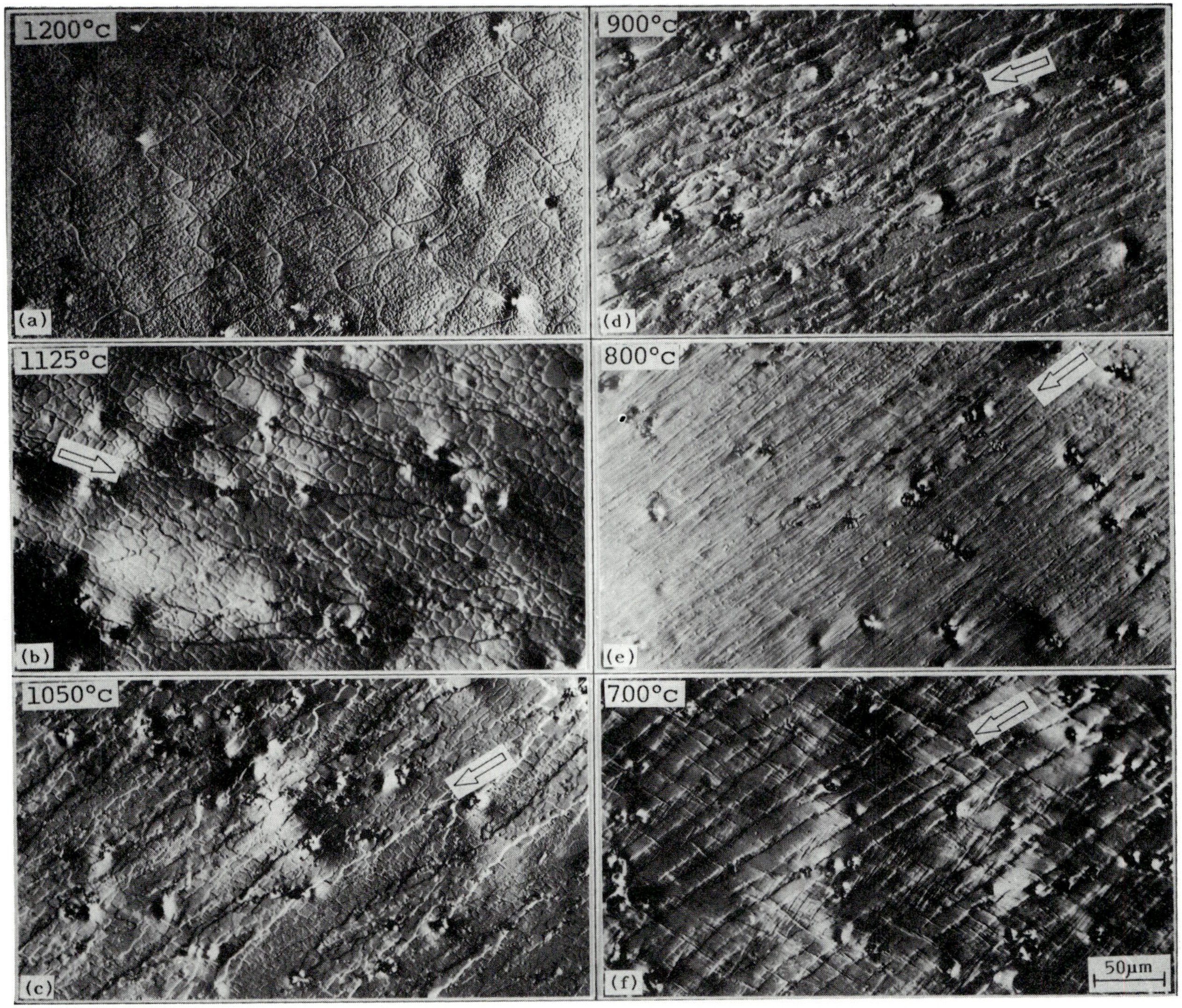

Fig. 2 Microstructures of hot-rolled Fe-16%Cr-0.3%Nb columnar crystals. Arrows indicate the deformation bands: (a) 1200°C, (b) 1125°C, (c) 1050°C, (d) 900°C, (e) 800°C, (f) 700°C

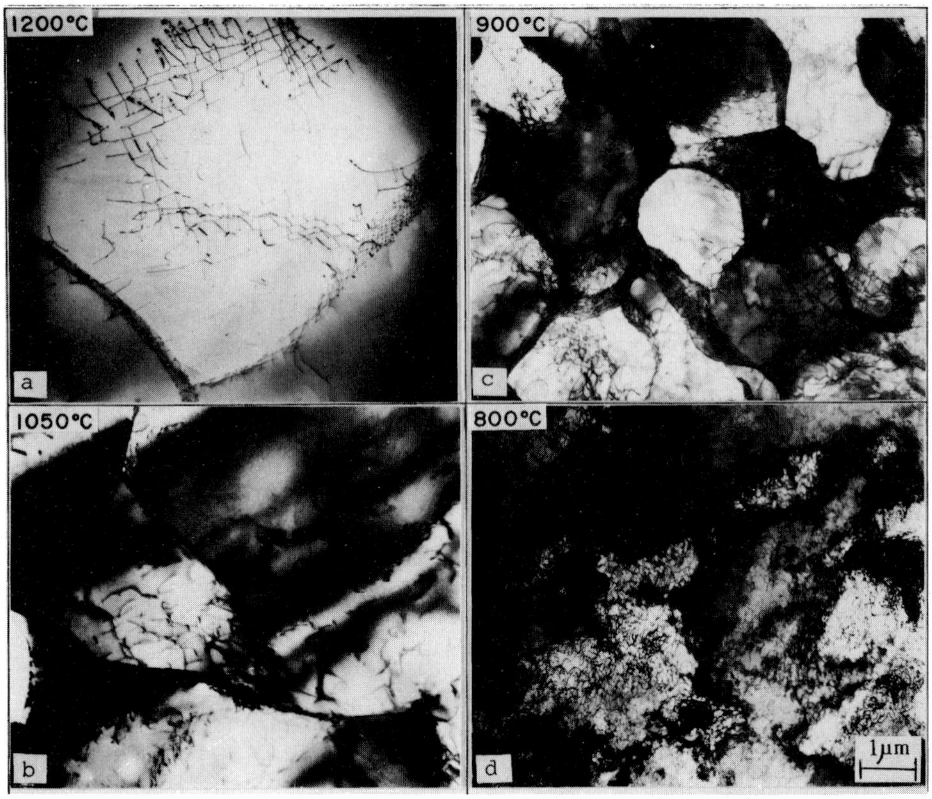

Fig. 3 Transmission electron micrographs of hot-rolled Fe–19%Cr: (a) 1200°C, (b) 1050°C, (c) 900°C, (d) 800°C

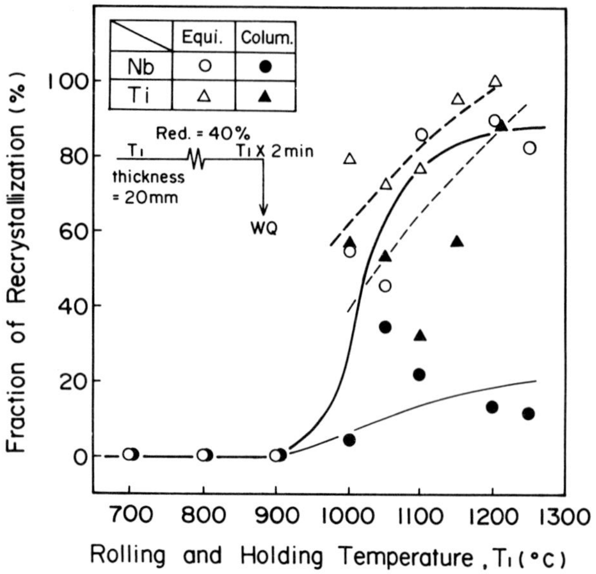

Fig. 4 Fraction recrystallized after annealing at the soaking temperatures for 2 min of hot-rolled Fe–16%Cr–0.3%Nb and Fe–13%Cr–0.2%Ti

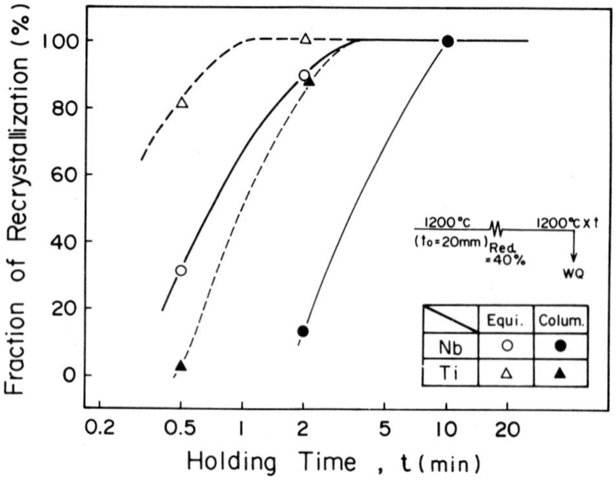

Fig. 5 Variation of fraction of recrystallization with annealing time for Fe–16%Cr–0.3%Nb and Fe–13%Cr–0.2%Ti

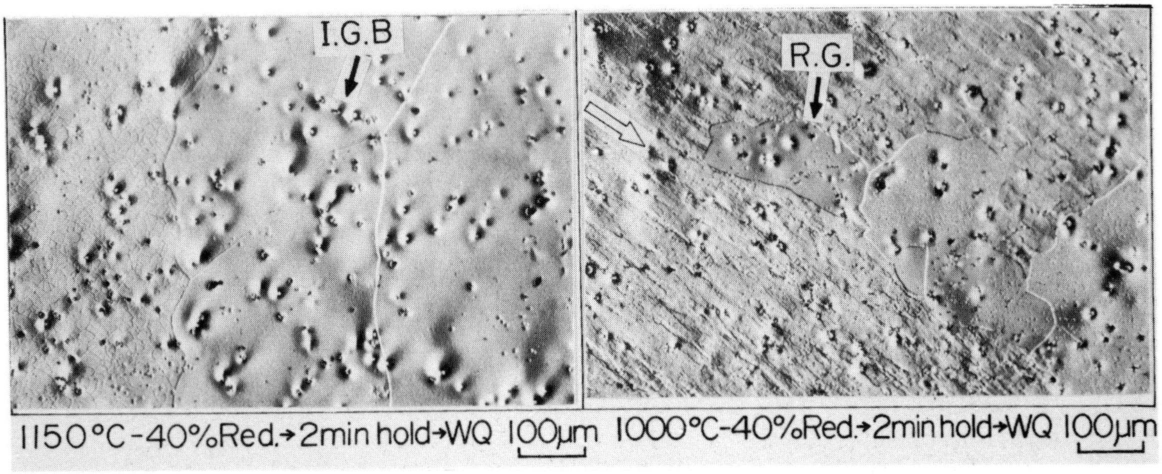

Fig. 6 Optical micrographs showing the nucleation site of recrystallization for Fe–16%Cr–0.3%Nb: (a) 1150°C at 40% hot-rolled and annealed at 1150°C for 2 min; (b) 1000°C at 40% hot-rolled and annealed at 1000°C for 2 min

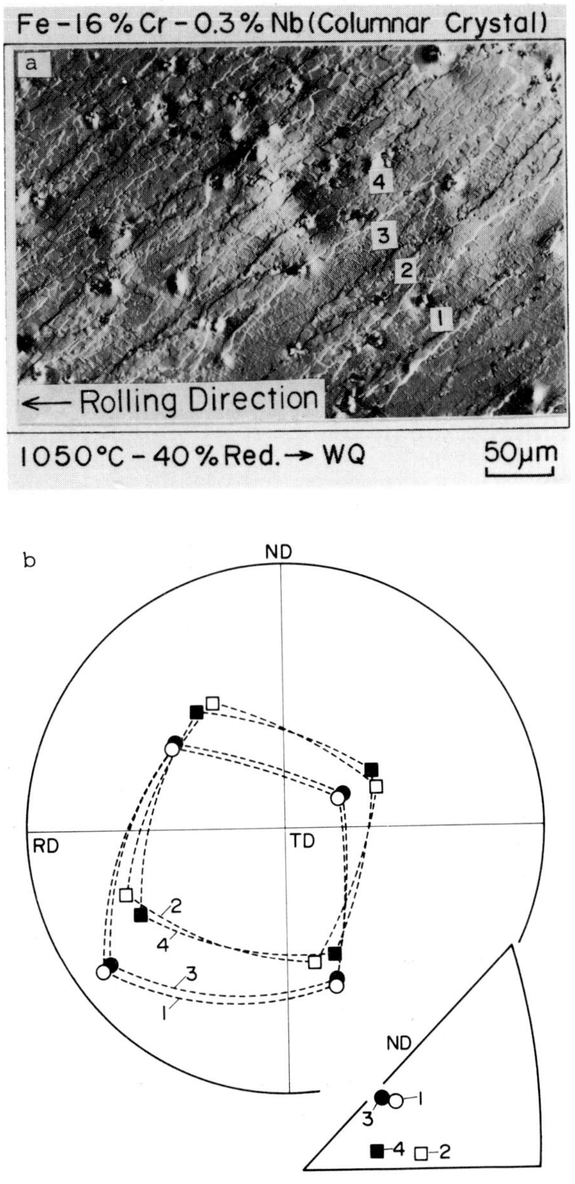

Fig. 7 Optical micrographs, stereographic projection of <111> poles and N D inverse pole figure for Fe–16%Cr–0.3%Nb, columnar crystals, hot-rolled at 1050°C at 40% reduction, using electron channelling pattern technique

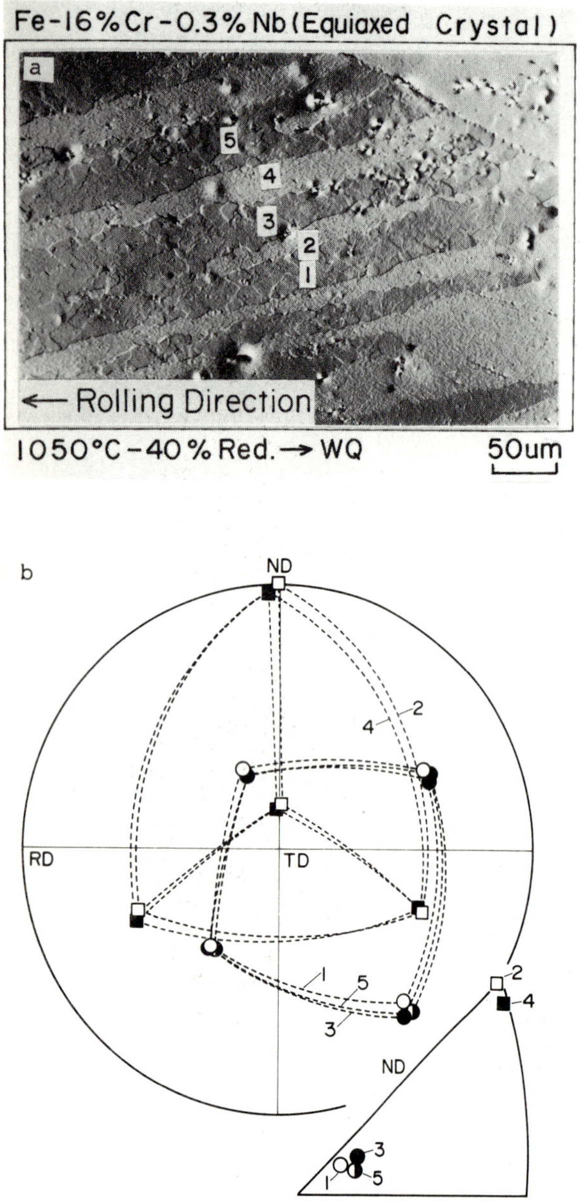

Fig. 8 Optical micrographs, stereographic projection of <111> poles and N D inverse pole figure, for Fe-16%Cr-0.3%Nb, equiaxed crystals, hot-rolled at 1050°C at 40% reduction, using electron channelling pattern technique

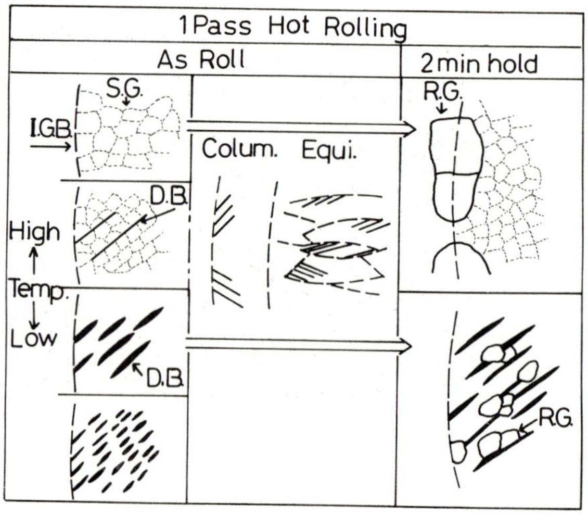

Fig. 10 Schematic diagram of refinement model of ferritic stainless steels under hot-rolling conditions

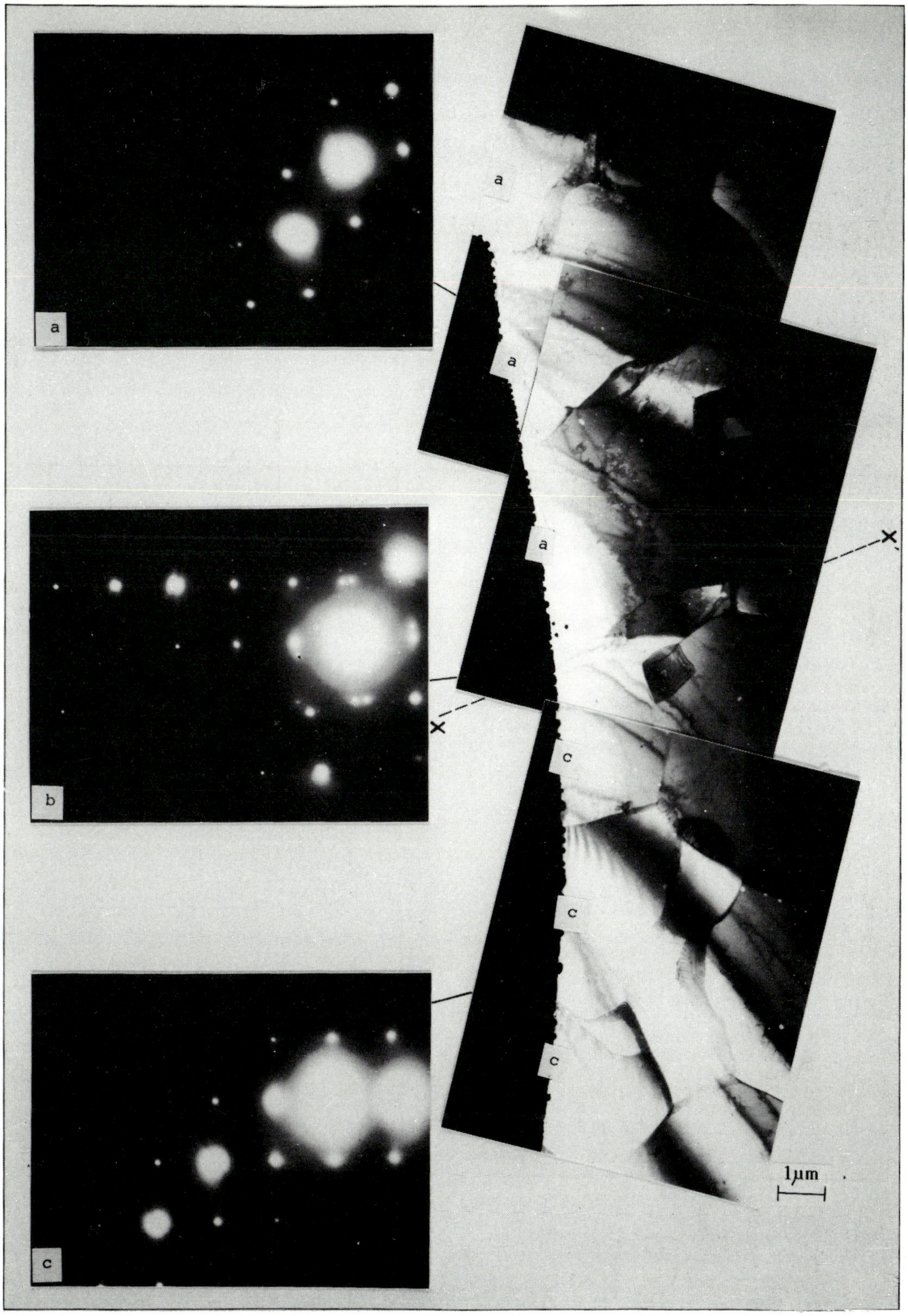

Fig. 9 Transmission electron micrographs and diffraction patterns of deformed substructures, showing the deformation band, observed for Fe-19%Cr, hot-rolled at 1050°C at 40% reduction

Effects of strain-rate history on strain-ageing phenomena in AISI 316 stainless steel

A M Eleiche, C Albertini and M Montagnani

AME is in the Department of Mechanical Design and Production at Cairo University, Egypt; CA and MM are with the Applied Mechanics Division of the CEC Joint Research Centre in Ispra, Italy, where AME was a visiting scientist when this work was performed.

SYNOPSIS

The present paper discusses the influence of strain-rate history on strain ageing behaviour in AISI 316 stainless steel observed in the temperature range 400-650°C. The material was tested monotonically in tension, and also under interrupted loading involving a strain-rate change of six orders of magnitude, from 0.004 to 500 s^{-1}, and vice versa, at seven temperatures ranging from the ambient to 650°C. Dynamic strain ageing clearly manifested itself at 550 and 650°C in all tests at the quasi-static strain rate. All symptoms of DSA were exhibited, but the normal level of ductility was maintained. Discontinuous loading was shown to produce effects which simple theory ignores. Thus, based on atomistic models for serrated yielding, a simple equation was derived to calculate the onset of serration in tests involving pre-straining as well as a change in strain rate. The flow stress for plastic deformation was also found to be sensitive to the entire mechanical history, and not a simple function of the instantaneous values of strain, strain rate and temperature.

INTRODUCTION

Austenitic stainless steels are frequently used in the construction of components of both nuclear and non-nuclear energy systems, where performance, safety and economy are of great concern. This makes it essential to describe, understand and rationalize their flow and fracture characteristics under all thermomechanical histories, simulating as much as possible those encountered during manufacture and service, and which may include large excursions in strain rate and temperature, reverse and cyclic loadings and multiaxial stress excitations.

Previous work by the authors /1-4/ has established new data base for AISI type 316H SS, particularly regarding its behaviour, in a wide range of temperature from the ambient up to 650°C, under a strain-rate loading history of constant strain rate tension interrupted and followed by a change of about six orders of magnitude from a quasi-static ($\sim 10^{-3}$ s^{-1}) to an impact ($\sim 10^{3}$ s^{-1}) strain rate, and vice versa. The deformation process was found to be rich in various types of instabilities, arising from physical phenomena such as adiabatic temperature rise and softening, recovery, and strain ageing, which exhibited themselves in specific domains of temperature, strain and strain rate.

The present paper examines in some detail the strain ageing phenomena, as affected by the strain-rate history imposed. Strain ageing is often accompanied by an increase in yield or flow stress, by unstable plastic flow (the well-known Portevin-Le Chatelier effect), by negative strain-rate sensitivity and by a decrease in ductility. The strengthening effect can be advantageous /5/, but other manifestations may prove detrimental in some large nuclear components, such as pressure vessels.
The paper further attempts to rationalize some particularities of the observed behaviour in the light of atomistic theories proposed in the literature.

EXPERIMENTAL PROCEDURE

All quasi-static tests were performed at a strain rate of about 10^{-3} s^{-1} in air on a Hounsfield tensometer. The impact tests at about 500 s^{-1} were conducted on a stored-energy type tensile split Hopkinson bar apparatus /6/ (Fig.1) which was properly modified to also allow dynamic specimen prestraining in the interrupted dynamic-to-quasi static strain-rate change tests. Full accounts of the apparatus design and instrumentation, as well as its performance were reported elsewhere /3,4/.

Heating in both setups was achieved by means of a small furnace surrounding the specimen gauge length. Temperature was monitored with a thermocouple attached to the specimen, and separately controlled by means of a Variac to ± 5°C. Thermal gradients along the bars surrounding the heated specimens were found to be small, and their effects neglected in impact test data analysis.

The material examined was AISI type 316H (nuclear grade) stainless steel (JRC reference material), supplied by Uddeholm in Sweden in the form of

50 mm thick hot-rolled plate. This nuclear grade, compared to its normal commercial variety, supposedly has more closely-controlled composition, lower concentration of residual elements and a lower inclusion content. The chemical composition (in wt%) was: 16.9 Cr, 12.4 Ni, 1.65 Mn, 2.45 Mo, 0.05 C, 0.35 Si, 0.008 S, 0.020 P, 0.023 Co, 0.001 B, 0.082 N and 66.066 (balance) Fe. The grain size of the material as received was about 0.09 mm. Tensile specimens were machined with their axes coinciding with the rolling direction of the supplied plate stock, and were not subjected to any post-machining heat treatment. No difference in characteristics was found between specimens taken from the outer or the inner layer of the plates. The specimens (Fig.2) were of such shape and dimensions that homogeneous stress distribution could be achieved, during the impact tests, by successive reflections of the elasto-plastic waves within the gauge length.

Procedures for the test performance and data reduction are quite straightforward /4/ and will not be repeated here. However, it is worth mentioning that, on the average, the time elapsed in the interrupted tests between the end of prestraining in a given setup and the start of reloading in the other did not exceed 20 minutes; specimens were usually left to cool in air down to a reasonable temperature, before transfer. Load and displacement signals during quasi-static testing were enregistered independently versus time on a chart recorder paper, whereas the incident-reflected and transmitted propagating pulses during impact testing were recorded on Polaroid film, and also in the memory of a Nicolet digital storage oscilloscope for subsequent data reduction.

RESULTS

Typical Test Records

Examples of test records collected from interrupted quasi-static to impact strain-rate change tests have been presented and discussed before /1/. Of particular interest in the present paper is the phenomenon of DSA, of which symptoms serrated flow represents an important visual feature. As will be discussed in a following section, this takes place for austenitic steels at low strain rates and at homologous temperatures in the neighbourhood of 0.5, which has also been confirmed in our tests. Therefore, typical quasi-static test records of load and elongation collected at 550 and 650°C are only shown in Fig.3(a) and (b), for monotonic loading to fracture as well as for various degrees of impact prestraining. Under monotonic loading, serrations in the load trace are observed to occur after some critical strain, smaller at 650 than at 550°C, i.e. in the neighbourhood of a homologous temperature of 0.5, as reported by early investigations /7,11/. More importantly, at both temperatures, this critical strain for serrated-flow initiation is seen to depend on the previous straining history, thus uniformly decreasing with increasing impact prestraining.

The type of observed serrations can be described according to their shape, using standard classification /12,13/, as follows. Under monotonic loading, type A serration started at 550°C, then changed to type B serration at larger strains. At 650°C, and also in quasi-static reloading after impact prestraining, serrations exhibited were always of type B.

Constant-Strain-Rate Response

Stress-strain curves. Figure 4 shows the stress-strain curves for the quasi-static and impact strain rates at various temperatures. The yield and flow stress are seen to be dependent on both temperature and strain rate, while the strain hardening is relatively dependent only on temperature. The low-rate results show the occurrence of DSA at 550°C, and more clearly at 650°C where the yield stress seems to have no rate sensitivity, while the flow stress, with continuing deformation, becomes higher than the corresponding dynamic one. Inside this DSA range, the material seems to maintain its normal level of ductility, unlike the blue-brittleness behaviour of other metals /5,14/.

Stress-temperature curves. Figures 5(a) and (b) show the flow stress, normalised with respect to the modulus, as a function of temperature and strain level, for the low and high strain rates, respectively. In Fig.5(b), the temperatures are initial not instantaneous values, thus neglecting the temperature rise due to adiabatic deformation. The dependence of the flow stress upon temperature and strain rate seems independent of the amount of strain. Also, the increment of stress due to the increase of strain seems larger at small strains, and independent of strain rate. The influence of dynamic strain ageing at the low strain rate is clearly seen in the curves of Fig.5(a); although a general trend exists for the flow stress to decrease with increasing temperature, a hump develops at high strain levels in the neighbourhood of 600°C (873 K) beyond which the stress presumably decreases with further increase in temperature, as reported in previous investigations /10,15/.

Figure 6 shows the variation of the initial work-hardening rate normalised with respect to the modulus, as a function of temperature and strain rate. Following a slight decrease, a consistent increase in the rate is observed, particularly at $\dot{\varepsilon} \sim 0.004\ s^{-1}$, probably reaching a maximum at 650°C before decreasing upon a further increase in temperature. This is again another symptom of the occurrence of DSA /9,10, 15/.

Strain-rate sensitivity of flow stress. From the curves of Figs.5(a) and (b), the strain rate dependence of the flow stress may also be deduced. Since tests were not carried out in the present work at intermediate strain rates, the continuous variation of the flow stress with strain rate could not be determined. However, based on previous tests conducted at the ambient temperature over a wider range of strain rates /16/, and assuming that the same thermally-activated deformation mechanism is governing in the range of strain rates from 10^{-3} to $10^{3}\ s^{-1}$, the flow stress may be related to the strain rate by:

$$\sigma = \sigma_0 + \mu_{12} \ln \dot{\varepsilon}, \qquad (1)$$

where σ_0 and μ_{12} are functions of temperature and strain. It follows that the mean apparent strain-rate sensitivity is defined by the parameter

$$\mu_{12} = (\partial\sigma/\partial \ln\dot{\varepsilon})_{\varepsilon} = (\sigma_2-\sigma_1)/\ln(\dot{\varepsilon}_2/\dot{\varepsilon}_1). \qquad (2)$$

Calculated values of μ_{12} are plotted against stress in Fig.7. μ_{12} decreases linearly with increasing stress. At 650°C, where DSA occurs in the quasi-static test, μ_{12} assumes negative values starting from a value of strain very close to the critical strain for onset of serrations, in agreement with previous observations on other alloy systems /17,18/. Negative values of μ_{12} are important in metal forming or other high-temperature high-rate processes, but they give no indication of the intrinsic rate sensitivity of the material which can only be measured by strain-rate-change testing.

Incremental Strain-Rate Change Response

A complete set of the stress-strain curves obtained at the temperatures of 20, 100, 200, 300, 400, 550 and 650°C for interrupted tests involving strain-rate changes from quasi-static to impact, and vice versa, at different values of prestrain, has been presented before /4/. Repeated here in Figs.8(a) and (b) are only those for 550 and 650°C where DSA is strongly exhibited, and also for 20°C as reference. Monotonic stress-strain curves to fracture at the quasi-static and impact strain rates are denoted in these figures by A and B, respectively. The following observations can be made.

For quasi-static-to-impact strain rate change. (a) The dynamic reloading curve is generally characterised by a well-defined yield point whose level is higher than the corresponding impact flow stress (on curve B) at the same strain, this positive difference is seen to increase with increasing prestrain, especially at 550 and 650°C. This behaviour is in contrast to that exhibited by copper and titanium, under incremental strain-rate jumps in shear /19,20/, but qualitatively similar to that obtained with mild steel /14/; (b) The initial jump in stress is followed by a yield drop only at 550°C, but not at the other testing temperatures; (c) At each temperature, the initial stress jump increases with increasing prestrain. The subsequent hardening rate, however, tends to decrease with increasing straining most probably due to thermal softening accompanying adiabatic deformation; (d) Although the retained ductility decreases with increasing prestrain, no dramatic change in total ductility (as expressed by total strain to fracture) seems to occur. At any given prestrain, the intrinsic strain-rate sensitivity parameter can be defined as follows:

$$\bar{\mu}_{12} = (\sigma_{1\rightarrow 2}-\sigma_2)/\ln(\dot{\varepsilon}_2/\dot{\varepsilon}_1), \qquad (3)$$

where $\sigma_{1\rightarrow 2}$ is the stress exhibited at $\dot{\varepsilon}_2$ following the strain-rate change. The variation of $\bar{\mu}_{12}$ with strain is shown in broken lines in Fig.9, from which values for a prestrain of 0.2 have been estimated and separately plotted in Fig.10; for 550°C, points are plotted corresponding to both upper and lower yield points. $\bar{\mu}_{12}$ is always positive, and generally decreases with increasing temperature. Also, comparing values of μ_{12} and $\bar{\mu}_{12}$ in Fig.9, it is seen that $\bar{\mu}_{12}$ is always greater than μ_{12}. Thus, if an impact straining follows a quasi-static prestrain, an increased strain-rate sensitivity should be accounted for. In Fig.10, radial lines are also shown corresponding to constant values of V/b^3, where V $(=kT/\bar{\mu}_{12})$ is the activation volume calculated on the assumption of a constant pre-exponential factor during the stress increment. From these, it follows that the activation volume increases rapidly with temperature.

For impact-to-quasistatic strain rate change. The quasi-static reloading behaviour in this case is complex and governed by strain-ageing, both static and dynamic, which depend on dynamic prestrain and temperature. (a) The reloading curve is always characterised by a not-so-well-defined yield point (rounded elastic-plastic transition) of a stress level which lies in general above that attained at the same strain in the exclusively quasi-static test (on curve A), except at 650°C where the level of the reloading curve lies always below the monotonic one, which may be attributed to overageing; (b) This difference in stress level, whether positive or negative, increases in all cases with increasing prestrain. Fig.11 illustrates this point at 650°C, where the reduction in quasi-static flow stress, $\Delta\sigma$, as defined in the inset, is plotted as a function of impact prestrain; (c) At 550 and 650°C (where DSA is exhibited phenomenologically by jerky flow after some critical strain, as discussed above), the reloading quasi-static curves exhibit also jerky flow after a total "critical" strain which generally decreases with increasing impact prestraining; (d) The retained ductility progressively decreases with increasing prestrain, but the total ductility remains relatively constant.

DISCUSSION

Serrated-Flow Characteristics

The change in the form of serrations under monotonic loading from type A at initiation to type B at larger strains at 550°C and to type B throughout jerky flow at 650°C, agrees with previous findings on stainless steel /11/ and other fcc alloys /13/. Following the conclusions of Mannan et al. /11/, and according to recent concepts of diffusion-controlled models of dynamic strain ageing /21,22/, type A serrations result from the diffusion of solute atoms to mobile dislocations temporarily arrested at penetrable obstacles lying in the slip path, and thereby the attainment of a critical composition of solutes to lock the dislocations, thus causing a step rise in stress. Unlocking of the dislocations from the solute atmosphere causes a subsequent load drop. Repetitions of the process cause a serrated curve. On the other hand, type B serrations, which oscillate about the general level of the stress-strain curve and occur in quick succession, result from locking of mobile dislocations in the propagating type A band. This is generally due to a rapid solute diffusion and a long time taken for the stress to recover, thus causing the dislocations in the band front to age and subsequent band motion to be discontinuous. Switching to type B serrations, of large amplitudes and wider strain spacings, with increasing deformation can be rationalized in terms of increase in the mobile dislocation density with strain /8,13/. This dependence on strain is usually expressed by /23/

$$\rho = N \varepsilon^{\beta}, \quad (4)$$

where the exponent β has a value close to 1, independent of strain rate and temperature /24/, whereas N is a constant which depends on both strain rate and temperature. At the same temperature, but a higher strain rate, Eq.(4) takes the form

$$\rho = n \varepsilon^{\beta}, \quad (5)$$

where n(> N) is the corresponding factor at the higher rate. Using Eqs.(4) and (5) together with experimental data of Michel et al. /15/ and that used by Schmidt and Miller /25/ for 316 SS, the ratio n/N at 650°C was calculated /26/ to be about 1.8.

Effect of Test Conditions on the Critical Strain for Onset of Serrations

The onset of serrations, observed experimentally, was found to depend on temperature, strain rate and strain rate history. Thus: (a) under monotonic loading, serrations were observed to start for the quasi-static strain rate at a critical strain smaller at 650°C than at 550°C; (b) serrations were not observed at any of the testing temperatures up to 650°C for the impact strain rate of about 500 s^{-1}; and (c) for the strain-rate history involving interrupted impact to quasi-static strain rate change, total strains for onset of serrations at 550 and 650°C were found to be smaller, the larger the impact pre-strain. These observations can be rationalized, at least qualitatively, as follows.

Atomistic models for the onset of serrated yielding, whether the solute drag model /27/ or the solute diffusion to arrested dislocations model /28/, predict the critical condition for dynamic locking of mobile dislocations in terms of applied strain rate, for a fixed grain size, as given by:

$$\dot{\varepsilon} = A \rho C_v \exp(-Q/kT), \quad (6)$$

where $\dot{\varepsilon}$ is the applied strain rate while serrations are observed, A is a constant, Q is the effective activation energy for solute migration, the mobile dislocation density ρ is as in Eq.(4) a function of strain, the vacancy concentration C_v is similarly a function of strain, k is Boltzmann's constant and T is the absolute temperature. The vacancy concentration is expressed as /29/:

$$C_v = B \varepsilon^{m}, \quad (7)$$

where B is a constant which depends on temperature and strain rate, whereas m is a constant. Substituting Eqs.(4) and (7) into Eq.(6) yields

$$\dot{\varepsilon} = ANB \varepsilon_c^{\beta+m} \exp(-Q/kT). \quad (8)$$

The critical strain can then be explicitly expressed by

$$\varepsilon_c = D \dot{\varepsilon}^{q} \exp(q Q/kT), \quad (9)$$

where D = 1/ANB, and $q = 1/(\beta+m)$.

Effect of temperature. Although jerky flow strongly exhibited itself at only 550 and 650°C during quasi-static testing, it is clear from Figs.3 and 8 that serrated yielding starts earlier at the highest temperature (at a strain of about 0.4 and 0.8 at 650 and 550°C, respectively). Similar observations have been made by other investigators /7-11/. Using data from a wider elevated-temperature range, plots of $\log \varepsilon_c$ against 1/T can be used to determine the apparent activation energy Q, as suggested by Eq.(9). Mannan et al. /11/ thus reported a value of ∿ 255 kJ/mol for 316 SS, and using the results of Jenkins and Smith on 330 SS calculated a corresponding value of Q in the range 190 to 290 kJ/mol. These values support diffusion, and hence an associated pinning effect, of substitutional solute atoms, probably chromium, as the mechanism responsible for DSA in austenitic steels. This also explains the increasing initial strain hardening rates at about 0.5 T_m, as shown in Fig.6, although Michel et al. /15/ attributed this increase to the transition from cell to subgrain formation, while Almeida and Monteiro /10/ associated the increase to interactions of dislocations with interstitials or interstitial pairs.

Effect of strain rate. According to Eq.(9), a plot of $\log \dot{\varepsilon}$ against $\log \varepsilon_c$ should be linear with a slope of $(\beta+m)$. Previous results on AISI 316 over wide ranges of quasi-static strain rates and elevated temperature inside the DSA region /11/, show that this linear relationship is observed, and give a value for the slope $(\beta+m)$ of about 2.3. Thus as the strain rate is increased, the critical strain for the onset of jerky flow is also increased; at a strain rate of 500 s^{-1} and a temperature of 600°C, therefore a critical strain of 1.8 is predicted, which is far beyond fracture. Diffusion-controlled theories of DSA also predict such dependence. In terms of the Cottrell solute drag model /27/, jerky flow occurs when the velocity of moving solute atoms becomes nearly equal to that of the mobile dislocations. Since the latter increases with strain rate, a higher velocity for diffusion is needed to start jerky flow at higher strain rate. This means that for constant deformation temperature, a higher vacancy concentration is needed, and thus a larger value of strain. On the other hand, in McCormick's solute diffusion to arrested dislocations model /28/ the waiting time of dislocations at obstacles decreases with strain rate, and consequently a higher velocity of diffusing solute atoms is required, which means a larger strain to initiate jerky flow. Since a linear increase in ρ with ε is known to be valid for most metals /23,24/, corresponding to β=1 in Eqs.(4) and (5), a reasonable value for m which can be adopted in calculations is 1.3.

Effect of strain-rate history. Equation (8) can be rewritten for constant temperature in the form

$$\dot{\varepsilon} = (\text{constant})\, NB \varepsilon_c^{\beta+m}. \quad (10)$$

If a specimen is pre-strained at a given strain rate $\dot{\varepsilon}_p$, then unloaded and rested, the resulting accumulation of vacancies will be lost, while the accumulation of dislocations will not, as was first noted by Ham and Jaffrey /23/. When deformation is resumed at the same temperature but at a different strain rate $\dot{\varepsilon}_r$, the vacancy concentration C_v, and mobile dislocation density ρ will start to build up at new rates corresponding to the new rate of restraining; however, the starting value of C_v will be zero, whereas the starting value of ρ will be the final value reached during prestraining at the initial strain rate. The situation is schematically represented in Fig.12. With some simplified assumptions, it may be further proved that the first jerk will occur at some total strain ε_t (including the prestrain ε_p), which is different from the critical strain with zero prestraining, ε_o. At this first jerk, C_v and ρ assume the critical values

$$C_v)_c = (B)_{\dot{\varepsilon}_r} (\varepsilon_t^m - \varepsilon_p^m), \quad (11a)$$

$$\rho)_c = (n)_{\dot{\varepsilon}_p} \varepsilon_p^\beta + (N)_{\dot{\varepsilon}_r} (\varepsilon_t^\beta - \varepsilon_p^\beta), \quad (11b)$$

so that a modified form of Eq.(10) now applies, namely:

$$\dot{\varepsilon}_r = (\text{constant})(B)_{\dot{\varepsilon}_r} (\varepsilon_t^m - \varepsilon_p^m)\left[(n)_{\dot{\varepsilon}_p} \varepsilon_p^\beta + (N)_{\dot{\varepsilon}_r} (\varepsilon_t^\beta - \varepsilon_p^\beta)\right] \quad (12)$$

When $\varepsilon_p = 0$, then $\varepsilon_t = \varepsilon_o$ and Eq.(12) must reduce to Eq.(10), so that we have finally /26/:

$$\varepsilon_o^{m+\beta} = (\varepsilon_t^m - \varepsilon_p^m)\left[\varepsilon_t^\beta - \varepsilon_p^\beta\left(1 - \frac{(n)_{\dot{\varepsilon}_p}}{(N)_{\dot{\varepsilon}_r}}\right)\right]. \quad (13)$$

Note that, when $\dot{\varepsilon}_r = \dot{\varepsilon}_p$, Eq.(13) reduces to

$$\varepsilon_o^{m+\beta} = (\varepsilon_t^m - \varepsilon_p^m)\varepsilon_t^\beta, \quad (14)$$

which is the same form derived by Ham and Jaffrey /23/. Equation (13) can be further manipulated to take the final form:

$$1 = \left[\left(\frac{\varepsilon_t}{\varepsilon_o}\right)^m - \left(\frac{\varepsilon_p}{\varepsilon_o}\right)^m\right]\left[\left(\frac{\varepsilon_t}{\varepsilon_o}\right)^\beta - \left(\frac{\varepsilon_p}{\varepsilon_o}\right)^\beta\left(1 - \frac{(n)_{\dot{\varepsilon}_p}}{(N)_{\dot{\varepsilon}_r}}\right)\right]. \quad (15)$$

To use Eq.(15) for the present case of interrupted impact to quasi-static strain rate change at 650°C, various constants are substituted by their experimental values. Thus, putting $\dot{\varepsilon}_p = 500\ s^{-1}$, $\dot{\varepsilon}_r = 0.004\ s^{-1}$, hence $(n)_{\dot{\varepsilon}_p}/(N)_{\dot{\varepsilon}_r} = 1.8$, $m = 1.3$ and $\beta = 1$, reduces Eq.(15) to:

$$1 = \left[\left(\frac{\varepsilon_t}{\varepsilon_o}\right)^{1.3} - \left(\frac{\varepsilon_p}{\varepsilon_o}\right)^{1.3}\right]\left[\left(\frac{\varepsilon_t}{\varepsilon_o}\right) + 0.8\left(\frac{\varepsilon_p}{\varepsilon_o}\right)\right]. \quad (16)$$

For various values of $(\varepsilon_p/\varepsilon_o)$, Eq.(16) may be solved for $(\varepsilon_t/\varepsilon_o)$, thus allowing the calculation of the new critical strain for jerky flow. Generalization of Eq.(15) to include temperature change in addition to the strain-rate change, as well as basic assumptions involved in the derivations are reported elsewhere /26/.
Using Eq.(15) and within the accuracy of the numerical values of various constants involved, the calculated critical strains for the onset of jerky flow in quasi-static testing following various impact prestrains are found to satisfactorily correlate with experimental observations.

CONCLUSIONS

The flow stress for plastic deformation is not a simple function of the strain, strain rate and temperature, but it is sensitive to the entire thermal-mechanical histories. This was confirmed in the present work by examining the behaviour of AISI type 316H SS, over a wide range of temperatures, under a strain-rate history consisting of tensile loading interrupted and followed to fracture by a strain-rate change of six orders of magnitude, from 0.004 to 500 s^{-1} and vice versa. Strain ageing dominated the behaviour at 550 and 650°C, i.e. in the neighbourhood of a homologous temperature of about 0.5. At the quasi-static rate, dynamic strain ageing manifested itself, as with other metallic solid solutions, through increase in flow stress, unstable plastic flow, negative strain-rate sensitivity; the material, however, maintained its normal level of ductility.

When quasi-static deformation was preceded by impact prestraining of various magnitudes, it was found that the onset of unstable flow, as well as other symptoms of dynamic strain ageing, was accelerated. Using commonly-accepted atomistic models for serrated yielding, a simple equation was derived to account for the amount of prestraining and its strain rate in the form:

$$1 = \left[\left(\frac{\varepsilon_t}{\varepsilon_o}\right)^{1.3} - \left(\frac{\varepsilon_p}{\varepsilon_o}\right)^{1.3}\right]\left[\left(\frac{\varepsilon_t}{\varepsilon_o}\right) - \left(\frac{\varepsilon_p}{\varepsilon_o}\right)\left(1 - \frac{(n)_{\dot{\varepsilon}_p}}{(N)_{\dot{\varepsilon}_r}}\right)\right]$$

where ε_o is the critical strain for onset of jerky flow under continuous loading, ε_p is the amount of prestrain, $(n)_{\dot{\varepsilon}_p}$ and $(N)_{\dot{\varepsilon}_r}$ are constants describing the rate of increase of mobile dislocation density with strain during prestraining and restraining, respectively, and ε_t is the critical total strain for onset of serrations under discontinuous loading at two different strain rates, namely $\dot{\varepsilon}_p$ followed by $\dot{\varepsilon}_r$.

Future work will concentrate on modelling and explicitly calculating the change in flow stress introduced due to the ageing process. Theoretical guidelines are available in the literature, e.g. /18,30/, which should be accounted for in any unified constitutive equation adopted for modelling material behaviour with reasonable accuracy.

REFERENCES

1. Eleiche, A.M., Albertini, C. and Montagnani, M. Nuclear Engineering and Design, 88, 131 (1985).
2. Eleiche, A.M., Albertini, C. and Montagnani, M. J. de Physique, Colloque C5, Supplément au no.8, Tome 46, 495 (1985).
3. Albertini, C., Montagnani, C. and Eleiche, A.M., Proc. SMiRT 8, Paper L8/4, 383 (1985).
4. Albertini, C., Eleiche, A.M. and Montagnani, M., in Metallurgical Applications of Shock-Wave and High-Strain-Rate Phenomena, ed. by L.E. Murr, K.P. Staudhammer and M.A. Meyers, 583, Marcel-Dekker, Inc. (1986).
5. Baird, J.D., Metals and Materials, 5, Metallurgical Review No.149, 1 (1971).
6. Albertini, C. and Montagnani, M., Tech. Rep. No. EUR 5787 EN, JRC, Ispra, Italy (1977).

7. Tamankar, R., Plateau, J. and Crussard, C., Rev. Met., 55, 383 (1958).
8. Naybour, R.D., Acta Met., 13, 1197 (1965).
9. Jenkins, C.F. and Smith, G.V., Trans. A.I.M.E., 245, 2149 (1969).
10. Almeida, L.H. and Monteiro, S.N., in Proc. 2nd Int. Conf. on Mech. Beh. of Mat., 1697, Fed. Mat. Sci. ASM (1976).
11. Mannan, S.L., Samuel, K.G. and Rodriguez, P., Trans. Ind. Inst. Metals, 36, 313 (1983).
12. Russel, B., Phil. Mag., 8, 615 (1963).
13. Brindley, B.J. and Worthington, P.J., Metall. Rev., 15, 101 (1970).
14. Eleiche, A.M., Exp. Mech., 21, 285 (1981).
15. Michell, D.J., Moteff, J. and Lovell, A.J., Acta Metall., 21, 1269 (1973).
16. Albertini, C., Cenerini, R., Curioni, S. and Montagnani, M., Proc. SMiRT 7, Paper L2/4 (1983).
17. van den Brinck, S.H., van den Beukel, A. and McCormick, P.G., Phys. Stat. Sol. (a), 30, 469 (1975).
18. Mulford, R.A. and Kocks, U.F., Acta Metall., 27, 1125 (1979).
19. Eleiche, A.M., TMS Technical Paper No. A86-62, The Metallurgical Society of AIME, Warrendale, PA, USA (1986).
20. Eleiche, A.M., in Titanium '80 Science and Technology, ed. by H. Kimura and O. Izumi, 831, The Metallurgical Society of AIME (1981).
21. McCormick, P.G., Acta Metall., 20, 351 (1972).
22. van den Beukel, A., Phys. Stat. Sol. (a), 30, 197 (1975).
23. Ham, R.K. and Jaffrey, D., Phil. Mag., 15, 247 (1967).
24. Gilman, J.J., Micromechanics of Flow in Solids, McGraw-Hill (1969).
25. Schmidt, C.G. and Miller, A.K., Res. Mechanica, 3, 175 (1981).
26. Eleiche, A.M., to be published.
27. Cottrell, A.H., Phil. Mag., 44, 829 (1953).
28. McCormick, P.G., Acta Metall., 20, 351 (1972).
29. van Bueren, H.G., Imperfections in Crystals, North-Holland (1960).
30. van den Beukel, A. and Kocks, U.F., Acta Metall., 30, 1027 (1982).

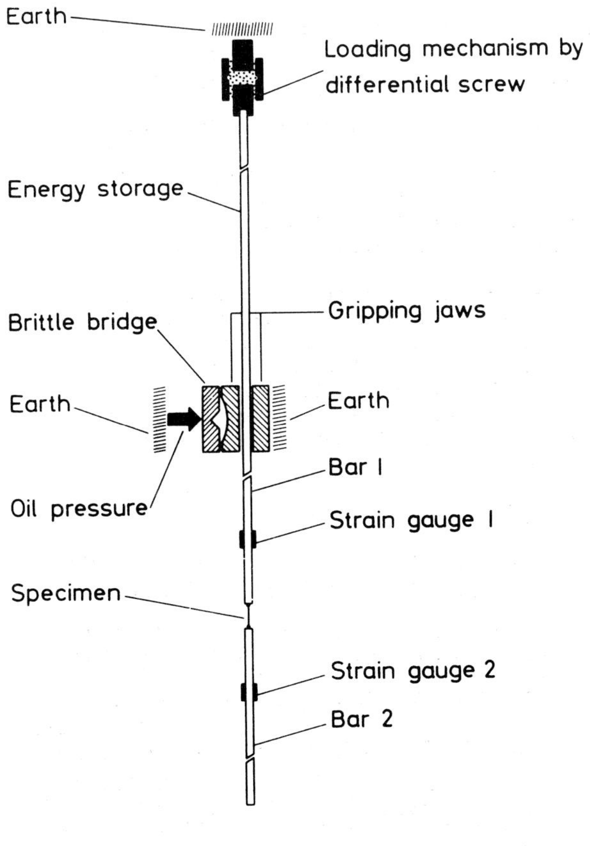

1 Basic configuration of the tensile split Hopkinson bar apparatus.

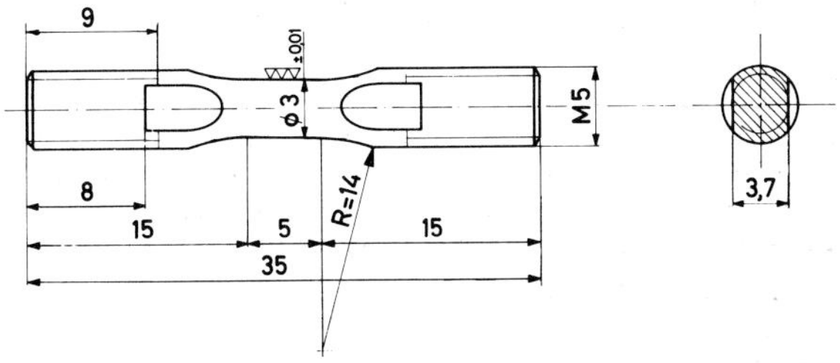

2 Test specimen configuration and dimensions (in mm).

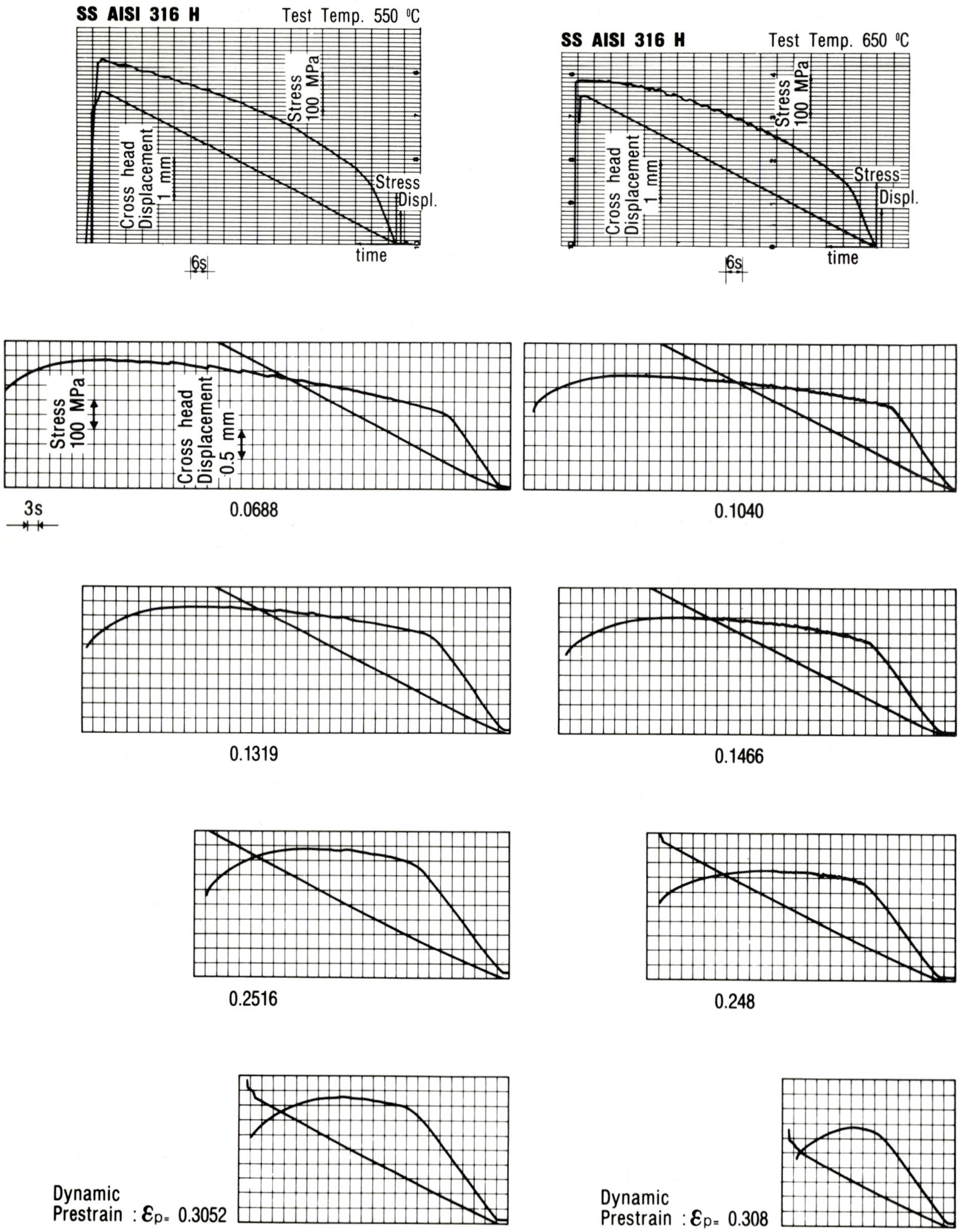

3 Typical chart records from quasi-static tests following impact prestraining: (a) at 550°C; (b) at 650°C.

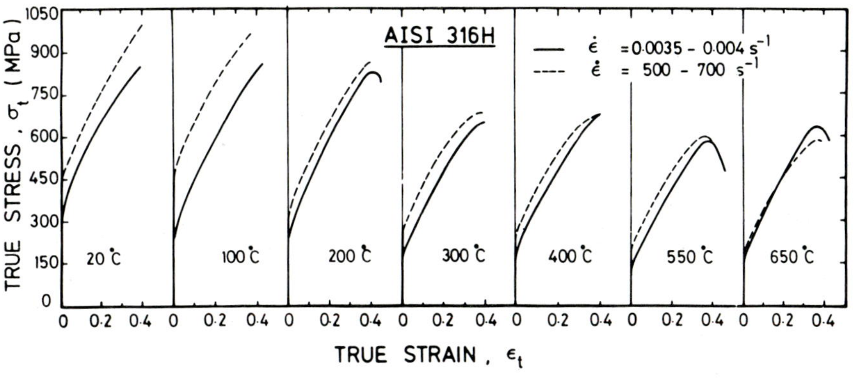

4 Monotonic quasi-static and impact stress-strain curves at various temperatures.

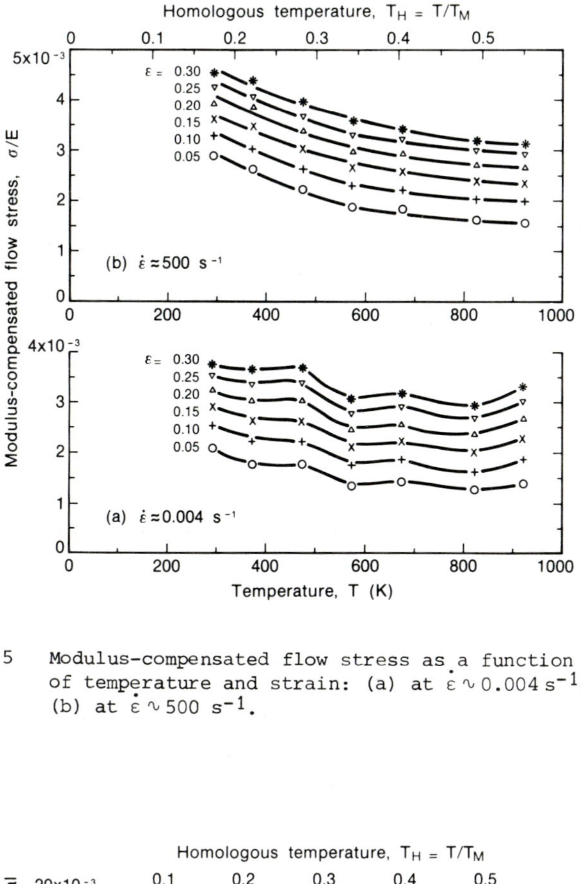

5 Modulus-compensated flow stress as a function of temperature and strain: (a) at $\dot{\varepsilon} \sim 0.004\,s^{-1}$; (b) at $\dot{\varepsilon} \sim 500\ s^{-1}$.

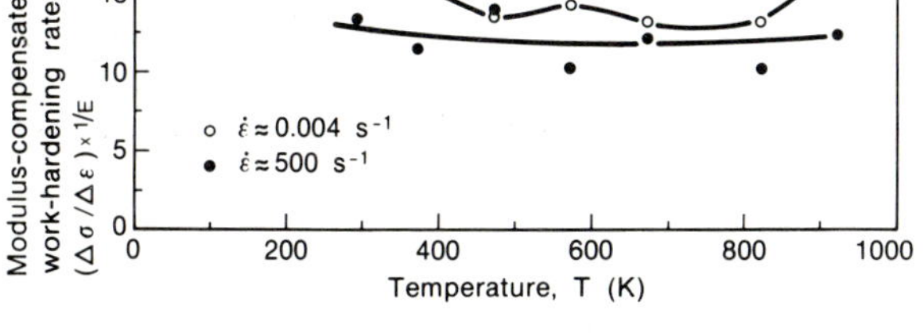

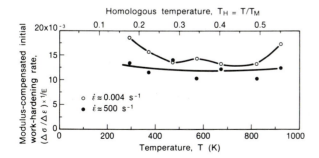

6 Modulus-compensated initial work-hardening rate as a function of temperature and strain rate.

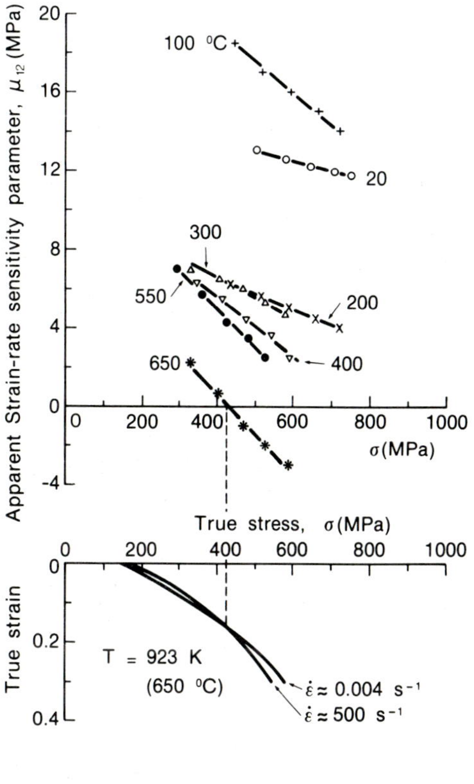

7 Variation of apparent strain-rate sensitivity parameter, μ_{12}, with stress and temperature. Negative values of μ_{12} occur at 650°C at stresses and strains very close to initiation of jerky flow at the low strain rate.

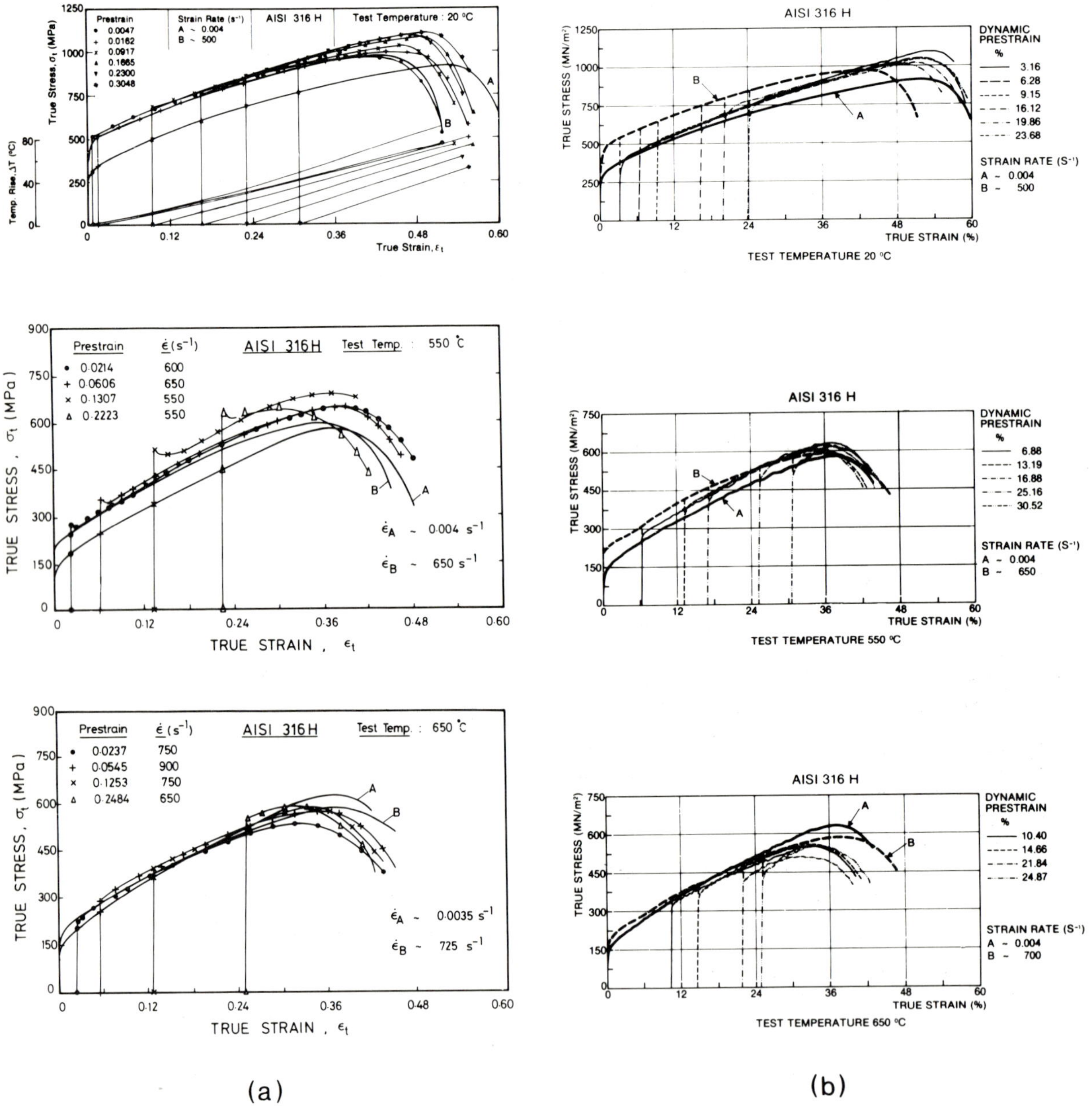

8 Stress-strain curves from interrupted strain-rate change tests at 20, 550 and 650°C: (a) quasi-static to impact; (b) impact to quasi-static.

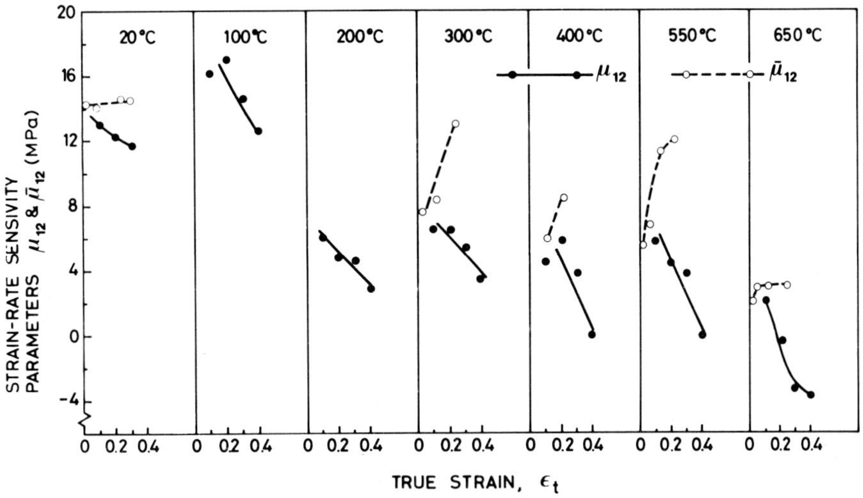

9 Strain dependence of apparent and intrinsic strain-rate sensitivity parameters, at constant temperatures.

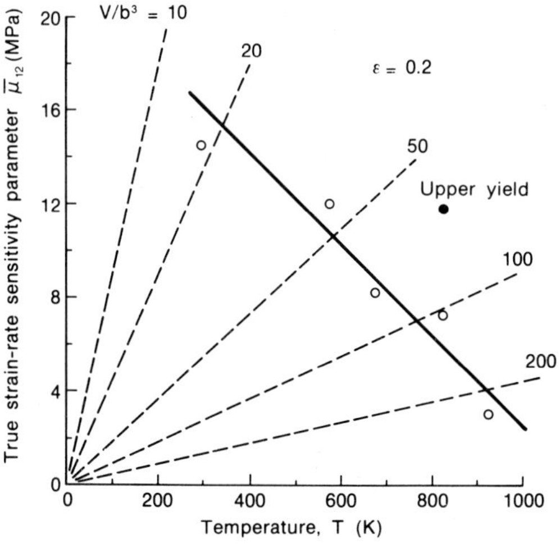

10 Temperature dependence of intrinsic strain-rate sensitivity parameter, at ε = 0.2.

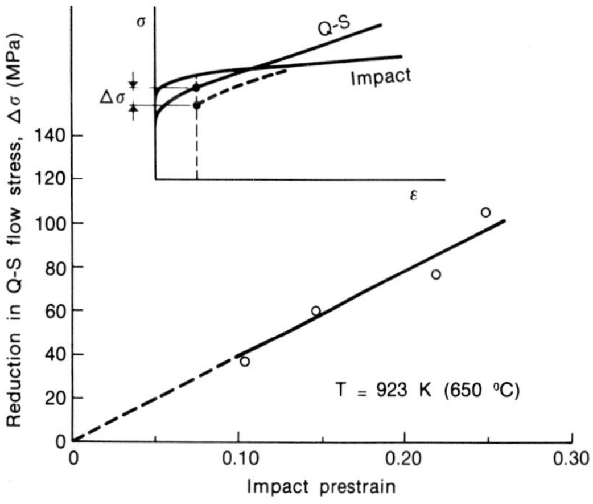

11 Variation of the reduction in quasi-static flow stress, $\Delta\sigma$, at 650°C, with impact prestrain.

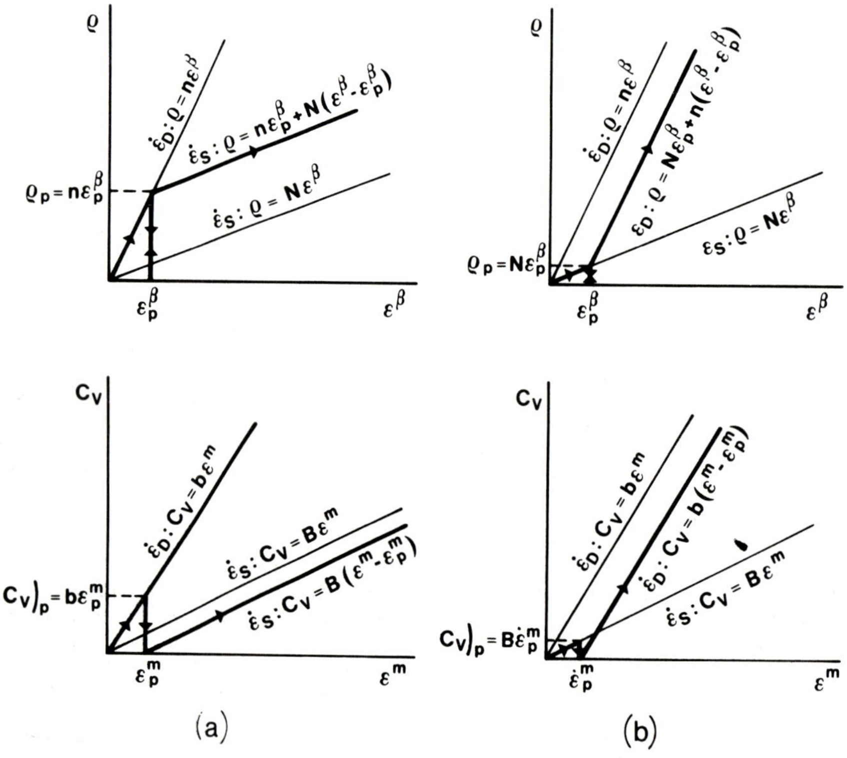

12 Schematic representing theoretical variations of the dislocation density, ρ, and the vacancy concentration C_V, as a function of strain, for monotonic loading and for interruped strain-rate change loading: (a) impact to quasi-static; (b) quasi-static to impact.

Creep properties of duplex 316 weld metals: role of residuals

J J Smith, R A Farrar and J Myers

RAF and JJS are in the Department of Mechanical Engineering, University of Southampton; JM is at Marchwood Engineering Laboratories, CEGB.

SYNOPSIS

Several 316 manual metal arc deposits, produced with commercial and experimental consumables, have been creep tested at 600°C, in both the as-welded and 700 and 850°C heat treated conditions. Deposits of nominally the same compositions were shown to exhibit very large differences in the minimum creep rate at the same stresses.

Additional creep tests on selected deposits, were interrupted in secondary creep by cooling under load. Thin foils from these interrupted tests have been examined using transmission electron microscopy.

A deformation mechanism has been identified which is controlled by the residual content, and a correlation has been established between the minimum creep rate and the composition for all the deposits. The model implies that variations in rupture properties are controlled by the residual levels of Nb, Ti and V present in the deposits.

INTRODUCTION

Type 316 steel is used extensively in both conventional and nuclear power stations for high temperature applications. It is normally welded with electrodes depositing metal of a similar composition, but containing 3-8% δ-ferrite which prevents solidification cracking [1,2,3]. Although the parent steel possesses excellent high temperature properties, the creep strength and ductility of the weld metal can vary considerably and even deposits of the same nominal composition can show marked differences in these properties [4,5,6,8]. The weld metal may also exhibit poor room temperature impact properties after ageing at high temperatures [3,9]. It is however, the poor creep ductility, which can be less than 1% in long term tests, which is of greatest concern.

Although a number of factors for the large scatter in creep rupture data have been suggested, such as type and distribution of precipitates, ferrite levels, welding process and general inhomogenity of the weld metal, the rôle of residuals and the mechanism via which these can affect the properties has not been successfully identified.

This present work considers several commercial and experimental type 316 manual metal arc weld deposits. The deposits are basic and rutile coated varieties of 17Cr, 8Ni, 2Mo type consumables with the exception of one fully austenitic deposit. This paper is mainly concerned with the effect of residual Nb, Ti and V on the secondary creep levels rates and the properties of the weld metal.

Binkley, Goodwin and Harman [4] studied the effect of electrode coating on the creep rupture properties of AISI 308 SMA weld metals. These authors reported that the use of lime-titania or titania (rutile) electrodes compared to lime-fluorspar (basic) electrodes, significantly improved the creep rupture life of the weld metal. No explanation was put forward to explain the variations in properties.

Binkley, Berggren and Goodwin [5] concluded that small residual additions of titanium, boron and phosphorous produced significant improvements in creep strength and ductility. Further work by Cole et al [6] claimed that the creep properties were optimised when approximately 0.007%B, 0.42%P and 0.06%Ti was present in the weld metal.

Thomas [7] concluded from studies on 17/8/2 weld metal, the superior strength of rutile deposits were attributable to differences in silicon and phosphorous, which arose from the changes in slag basicity. It was thought these elements induced solid solution strengthening of the weld deposit.

The influence of residuals on the properties of 308 weld metal has been recently reviewed by Marshall and Farrar [8]. They concluded the rupture life was dominated by the niobium present in the rutile deposits, whilst rupture ductility was controlled by the silicon content which was related to flux basicity.

EXPERIMENTAL PROCEDURE

The chemical analyses of the weld metals and the as deposited ferrite contents are shown in Table 1. The weld metals, BW13, 14, 15 and 19 were deposited using rutile coated 17/8/2 Metrode R.C.F. electrodes. In BW 13 and 14, further additions of B and Ti + B were added to the coatings in order to assess the effects of the controlled residual elements on the mechanical properties and transformation behaviour. Deposit BW19 was a low carbon version of BW15 and was incorporated into the programme in order to investigate the effects of carbon on transformation behaviour. BW16 was a fully austenitic deposit based on a 19Cr 15Ni 3Mn 2Mo consumable. BW17 was deposited using a new metal powder electrode, based on a pure iron core wire. The BW18 consumable was a rutile version of a 17/8/2 Armex GT consumable which has been widely used in the power generating industry. The weld metal BW20 was deposited with a consumable which has been developed by Soudometal Belgium, and was thought to optimise the mechanical properties. Some of the trace elements listed in Table 1 are near to the limit of accurate 'Quantavac' analysis and to ensure that errors were similarly weighted, all deposited, with the exception of BW20, were analysed in the same batch using the same calibration standards. Several analyses were carried out on each deposit and a mean value reported.

CREEP TESTING

The creep specimens were machined with their axes normal to the direction of columnar growth. The specimens were subsequently tested in the as welded condition, after 15 hours at 700°C or 8 hours at 850°C. The specimens had a parallel gauge length of 58 mm long and 50 mm^2 round cross section.

All creep testing was under constant load conditions. After allowing the temperature to stabilise, the specimens were loaded at 600°C. Specimen strain was monitored using extensometers fitted with linear voltage transducers and this allowed minimum deformations of 0.01 mm to be measured. The temperature was monitored using Pt/Pt-Rd thermocouples, and was maintained constant to better than ± 0.5°C. All voltage outputs were logged at regular intervals. When rupture occurred the furnace was switched off automatically within 10 or 20 minutes in most instances. Interrupted tests were cooled slowly under load in order to 'freeze' in the mobile dislocation structures. In the interrupted tests, stresses were chosen to be below the 80% proof stress to avoid rapid yielding. A summary of these tests is shown in Table 2. Initially strong and weak deposits were tested so as to give approximately the same minimum creep rate; tests 1 and 2 were performed at 156 and 226 MNm^{-2} respectively. They were interrupted after similar times and strains.

A further test, 3, was carried out on a strong deposit after a post weld heat treatment of 850°C/8 hours to give a minimum creep rate similar to tests 1 and 2.

SPECIMEN PREPARATION AND EXAMINATION

Rods 3 mm diameter were machined from the gauge length of the creep specimens. After slicing and mechanical polishing to a thickness of 110-130 μm, the discs were electropolished using a Struers 'Tenupol' twin jet electro-polisher in a solution of 5% perchloric in acetic acid. Microstructural and micro-compositional analyses were carried out on a JEOL 100 CX scanning transmission electron microscope, STEM, operating at 100 kV. Suitable thin foil calibration and correction procedures were adopted as recommended by Farrar and Thomas [12].

Estimates of the dislocation densities in the creep specimens were also obtained using transmission electron microscopy. Several representative areas in the tests were photographed using two beam conditions. The operating reflection g giving rise to dislocation contrast was recorded and correction factors reported by Hirsch [11] for the proportion of invisible dislocations as a result of a particular g vector were employed. Foil thickness measurements were obtained using either extinction distances or by a carbon deposition technique. The foil thickness varied between 240-320 nm. The dislocation density was measured using a technique reported by Keh [12]. Ham [13] has observed that up to 50% of the dislocations may be lost during electropolishing. Despite these limitations in the measurement of dislocation density values reported here are believed to reflect trends although they may not be absolute.

RESULTS

Creep Test Properties

Creep tests were performed on specimens at 600°C for failure times up to 36000 hours. The variation in rupture life t_f, with nominal applied stress σ, is linear over the test stress range and is shown in Figure (1). The rutile deposits BW13, 14, 15 and 18 are consistently stronger than the basic deposits. The rupture life for the fully austenitic weld metal BW16 is greater than the ferrite containing welds. The creep strength of all deposits was reduced by heat treatment for 15 hours at 700°C and was further reduced by 8 hours at 850°C.

The stress dependence of the minimum creep rate $\dot{\varepsilon}_s$ is shown in Figures (2) (3) and (4) for specimens in the as welded, the 700°C and 850°C heat treated conditions respectively. All the minimum creep data could be approximately related to the stress by the expression:-

$$\dot{\varepsilon}_s = A\,\sigma^n \qquad \dots (1)$$

where σ is the nominal stress, n is the stress exponent and A is a constant. The stress exponent (n) varied between 10 and 24 depending on the material under consideration and the prior heat treatment. Very large differences in creep strength were observed in the deposits. Comparison of the creep rate at one stress in the as welded and 700°C heat treated condition spanned nearly three orders of magnitude. After heat treatment at 850°C the spread was greater than two orders of magnitude.

Microstructure Prior to Creep Testing

The as deposited weld metals which contained 3-8% δ-ferrite were typically vermicular in morphology. On increasing the base levels of δ-ferrite in the deposits to above 8% a gradual transition in ferrite morphology from the vermicular to the lacy type was observed. Both these structures were characterised by columns of aligned δ-ferrite, the alignment being along the heat flow axes, and thus the primary dendrite growth direction.

The fully austenitic weld deposit BW16, was typified by very large columnar grains, approximately 100 μm wide and 1000 μm long in the growth direction. In the duplex deposits, the presence of the δ-ferrite refined the grain size of the weld metal, the primary dendrite spacing being approximately 10-15μm.

In all the weld deposits, the dislocation densities in the austenite and ferrite were constant. These were approximately 8 x 10^{13} m^{-2} and 6 x 10^{13} m^{-2} in the austenite and ferrite respectively.

The effect of prior heat treatments at 700°C and 850°C on the δ-ferrite content and transformation product are summarised in Table 3. After 15 hours at 700°C and 8 hours at 850°C approximately 50% and 90% respectively of the δ-ferrite transformed to a mixture of austenite, $M_{23}C_6$ and intermetallic phases.

The 850°C/8 hours post weld heat treatment resulted in the rapid conversion of the δ-ferrite to intermetallic σ phase. The distribution of the intermetallics was discontinuous and partial spheroidization had occurred. Microcompositional traverses across a number of intermetallics showed the boundary region to be significantly depleted in molybdenum and to a lesser extent in chromium. This depleted zone was approximately 0.5-1μm wide. Some residual δ-ferrite, typically 1%, was present in all the deposits. The dislocation densities in the austenite were reduced by an order of magnitude relative to the as welded deposits.

Interrupted Test 1 - BW11/as-welded

The dislocation density in the austenite was approximately 4 x 10^{13} m^{-2}, a little lower than in the as welded condition. The dislocation structure in the δ-ferrite varied considerably. Some laths were heavily dislocated with densities of up to 10^{14} m^{-2}, but the majority were almost free of dislocations with densities of 10^{11} m^{-2}. Some δ-ferrite laths showed bowed fronts of dislocation free grains sweeping down heavily dislocated laths. Other dislocation-free laths were crossed by high angle grain boundaries, Figure(5). This indicated that re-crystallisation of the δ-ferrite was occurring during creep. A narrow band of dislocations approximately 0.2 μm wide lay on the austenite side of most of the δ/γ boundaries, Figure (5). Dislocations within this region were so densely packed that no separation could be detected and it appeared as a black band. Occasional constrictions in the band appeared to be associated with the migration of the δ/γ boundary. These dislocation bands were not observed in the normally fractured creep specimens and it is thought that they anneal out during cooling. This was verified by reheating a portion of the interrupted specimen to the test temperature of 600°C and recooling. No near boundary dislocation bands were observed after this heat treatment.

These observations indicate that very heavy deformation was occurring in the ferrite and in the neighbouring δ/γ grain boundary region. Recovery by recrystallisation appeared to occur almost simultaneously in the ferrite, but recovery in the austenite in the region of the boundary took place more slowly. It was however unclear whether the deformation at the boundaries was due to grain boundary shear or was peripheral damage associated with the deformation of the ferrite.

Interrupted Test 2 - BW15/as welded

This specimen had a loose cellular dislocation structure in the austenite with a density of approximately 3 x 10^{14} m^{-2} which was higher than that observed in the as welded condition. Some small precipitates were observed on dislocations in the austenite which were thought to be M(C,N) carbonitrides. These were very sparsely distributed with a typical spacing of 300 - 900 nm.

Discontinuous precipitates of $M_{23}C_6$ carbides had formed at the δ/γ boundaries and these frequently extended across the ferrite. Dislocation densities within the ferrite were less extreme than in the previous specimen. The formation of low angle boundaries was again indicative of recovery processes operating within the δ-ferrite. Significantly there was no evidence of shear bands or localised deformation at the δ/γ interface and the absence of any high angle boundaries inferred that recrystallisation was not a dominant mechanism in this material.

In regions where the δ-ferrite had transformed to austenite a fine precipitate was resolvable. The precipitates were thought to be M(C,N) carbonitrides but could not be positively identified, Figure (6).

This specimen, which had been tested at a higher stress, had sustained the same total strain at approximately the same strain rate as test 1, yet unlike test 1 it showed no regions of intense strain accumulation. Therefore the metallographic evidence suggested strain had been more uniformly distributed throughout the austenite matrix.

Interrupted Test 3 - BW18 - 850°C/8 hrs

After 741 hours at 600°C a semicontinuous σ network had developed along the prior δ/γ boundaries. The distribution and morphology of the intermetallics was similar to that observed prior to creep in the 850°C heat treated condition. The intermetallics were coarse typically 2-3 μm in size. Magne Gage measurements indicated that approximately 1% residual δ-ferrite remained untransformed in the specimen. In the δ-ferrite, recovery occurred by the formation of low angle boundaries. The boundary regions were essentially featureless although local strain accommodation had occurred at the σ/γ interfaces which were denuded in molybdenum and chromium.

The austenite dislocation density was approximately 8 x 10^{12} m^{-2} and there was no evidence of a cell substructure. A sparse distribution of M(C,N) carbonitrides was observed in the austenite, these were of the order of 20 nm in size and a typical spacing of 700 nm.

DISCUSSION

In support of a number of studies [7,8,9,10] the rupture data from the present series of weld metals showed that the basic coated deposits were consistently weaker in creep rupture testing compared to the rutile variety.

Although the effect of different ferrite levels on the creep properties has not been specifically studied in this report, this has been examined by other authors [14]. Berggren, et al [14] were able to show that increasing the ferrite levels from 2-15% produced approximately half an order of magnitude decrease in the minimum creep rate (m.c.r.). The weld deposits studied in this work exhibited very large differences in creep strength minimum creep and the rates for the same applied stress differed by up to 3 orders of magnitude. The changes in m.c.r. could not be explained by differences in the level of δ-ferrite in the weld metal and it was concluded that microcompositional differences between the deposits were responsible.

The bulk analyses in Table (1) inferred that carbon, niobium, titanium and vanadium were the major strengthening elements, the only exception being weld BW16, the fully austenite deposit which was low in all these elements. A plot of [Nb + Ti + V][C] against the m.c.r. at an applied stress of 180 MNm^{-2}, gave a good correlation for all the ferrite containing deposits, Figure(7). similar relationship was also obtained for the 700°C and 850°C heat treated deposits, Figure (8) and (9).

The curves in Figures (7), (8) and (9) exhibit a number of very interesting features.

1. Very small quantities of Nb, Ti and V have a marked effect on the m.c.r. of the duplex deposits.

2. The minimum creep rates for the fully austenitic deposit were equivalent to the stronger duplex deposits, despite it having a low residual content.

3. The expression [Nb + Ti + V][C] has the form of a solubility product which suggests that precipitation in either the austenite or the ferrite was controlling the m.c.r.

4. As the residual content was increased the minimum creep rate of the duplex deposits approached that of the fully austenitic deposit. No further strengthening was achieved above 0.04 at %. precipitate. It is thought that saturation is achieved at these residual content levels.

5. The curves level out at a m.c.r. similar to a number of highly alloyed austenitic weld metals containing up to 1.5 wt% Nb, Ti and V. (Essehete 1250 weld metals Clark and Myers, private communication).

Extensive transmission electron microscopy on the duplex deposits in the aged and crept conditions revealed the presence of a sparse distribution of M(C,N) carbonitrides in the austenitic matrix. Some estimate of the contribution of the matrix particle spacing to the creep strength can be derived from Orowan equation:-

$$\sigma = \frac{Eb}{\lambda} \qquad \ldots (2)$$

where Young's modulus (E) of austenite at 600°C $\simeq 1.7 \times 10^5$ MNm^{-2} [15] Burgers Vector (b) $\simeq 3 \times 10^{-10}$m and particle Spacing (λ) $\simeq$ 300 - 900 nm. However from estimates of 50-145 MNm^{-2} for the stress necessary to move dislocations through the austenite, it was concluded that the carbonitride particles were too widely spaced to result in any effective dispersion strengthening.

Tensile tests at 600°C had exhibited discontinuous yielding in the fully austenitic deposit BW16, suggesting that solute strengthening could be controlling creep rate. However BW16, although among the stronger deposits was low in residuals and carbon content. This anomaly suggested solute strengthening was not a controlling factor. This specimen, after creep testing for 10,000 hours, was found to contain a profuse dispersion of intragranular $M_{23}C_6$ carbides which would contribute significantly to the matrix strength.

Microcompositional analyses in the rutile deposits indicated that small quantities of Nb, Ti and V had segregated to the δ-ferrite during solidification. The Nb, Ti and V remained in solid solution in the δ-ferrite during ageing or creep due to the low solubility of carbon and nitrogen in the δ-ferrite. However, during creep or prior heat treatment at 700 or 850°C, M(C,N) carbonitrides precipitate in the new austenite formed by the migration of the δ/γ interface. Precipitation in this region might be expected to inhibit grain boundary shear and this would strengthen deposits in which boundary shear was the dominant mode of creep deformation.

In the duplex weld deposits the δ-ferrite forms a semicontinuous network, thus the δ-ferrite refines the effective grain size, typically 10-15 μm in the duplex deposits compared with 1000 μm in the fully austenitic weld metal. Barrett et al [16] proposed that the increase in creep rate with decreasing grain size was attributable to the importance of grain boundary sliding at small grain sizes. The refinement of the austenite grain size by the δ-ferrite weakens the deposits due to the increased grain boundary area and the predominance of grain boundary sliding at the test stresses under consideration.

In the basic deposits BW11 and 17 where the ferrite was low in Nb, Ti and V, heavy deformation occurred in the δ-ferrite with accompanying shear in the grain boundary region. Since the near creep strength of the δ-ferrite was low compared to the austenite, the localised deformation caused the δ-ferrite to undergo dynamic recrystallisation. Some estimate of the relative creep rates of the δ-ferrite and austenitic phases can be derived from the ratio of the diffusion coefficients at the creep temperature of 600°C.

$$\frac{\dot{\varepsilon}_\delta}{\dot{\varepsilon}_\gamma} \approx 100 \qquad \ldots (3)$$

This is based on evidence of the dependence of the secondary creep rates of austenite and ferrite on their self diffusion coefficients

[17]. It was unclear whether the deformation at the δ/γ boundary was a consequence of deformation in the neighbouring ferrite or of grain boundary shear or a combination of both. However, the presence of shear at the δ/γ boundary is of importance since it may lead to the nucleation and growth of grain boundary cavities or cracks, and hence to the eventual creep rupture. In particular, under conditions where continuous grain boundary precipitates immobilize boundaries then the localised deformation will result in low elongations to rupture.

In the rutile deposits BW13, 14, 15 and 18, the decrease in minimum creep rate appeared to be related to the precipitation of M (C,N) carbonitrides in the near boundary region. The amount of precipitation [MC] will be determined by the carbon or nitrogen concentration [C] and the residual element concentration [M] in the new austenite formed by the migration of the δ/γ interface.

$$[C] \times [M] = k[MC] \qquad \ldots (4)$$

where k is the solubility product constant in the austenite. The grain boundary shear strength might be expected to increase with [MC] and the values of [C] and [M] are expected to be proportionally related to the bulk contents.

Significantly there was a notable absence of any localised shear bands at the δ/γ boundaries. The higher dislocation density in the austenite compared to the basic deposits inferred that deformation occurred predominantly in the matrix. Recrystallisation did not occur in the δ-ferrite, although sub boundary formation was evidence of some strain recovery. This suggested that the δ-ferrite in these deposits was either precipitation or solution strengthened by residuals, and deformation in the ferrite was minimal.

In the post weld 700 and 850°C heat treated conditions approximately 50% and 90% respectively of the original δ-ferrite transformed to austenite and or intermetallic phases. The ferrite was therefore replaced by a high angle boundary network. The curves at 700 and 800°C, Figures (8) and (9) respectively, have a similar form to the as welded curve, Figure (7), and consequently similar deformation mechanism is implied. This suggests that grain boundary shear and not deformation of the ferrite is the dominant deformation mode.

The curves in the as welded and heat treated conditions, however, do show small differences. Heat treatment of deposits at 850°C reduces the creep rate span by nearly an order of magnitude. This might be interpreted as removing the contribution due to deformation in the ferrite, but there is no accompanying decrease in creep rate to support this interpretation. In the 850°C heat treated condition it was not known what effect the soft zone would have on the total creep strain. It is proposed that some strain accommodation within the grain boundary region occurs as the flow stress is much lower than the matrix.

Previous workers Thomas [7], Farrar and Marshall [8] have suggested that differences in silicon levels between the basic and rutile deposits were sufficient to account for variations in rupture lives. The strengthening of silicon has not been specifically studied in this report, due to the limited variations in silicon content between the deposits. However Smith, Farrar, Myers [19] have found that silicon tends to segregate to the δ-ferrite during solidification. Consequently it is possible that silicon variations can account for some difference in rupture life between the basic and rutile deposits, although evidence from this work suggests that this is only a second order effect and rupture times are primarily controlled by residual Nb, Ti and V contents.

In the low stress regime the contribution of grain boundary shear to the total creep strain is large and consequently the model of residuals controlling the m.c.r. via grain boundary strengthening is valid. However, at stresses above the macroscopic yield stress, deformation is considered to occur predominately via the grains themselves with a smaller contribution from grain boundary shear and thus a different strengthening mechanism exists. Above the tensile proof stress, creep strength is derived from dislocation locking due to interstial carbon and nitrogen.

CREEP MODEL

A schematic model summarising the effect of residual Nb, Ti and V and prior heat treatment on the minimum creep rate or mode of deformation is shown in Figure (10). In the as welded condition the steep part of the curve represents a regime where deformation occurs primarily via grain boundary shear with some contribution from the δ-ferrite. As the carbon and or residual content is increased the precipitation of M (C,N) carbonitrides in the rear boundary regions impede grain boundary slip and thus promote matrix deformation. The curve levels out as the optimum grain boundary strengthening is achieved. The primary effect of the heat treatments at 700 and 850°C was to reduce the strength of the austenite by factors of 2 and 10 times respectively.

SUMMARY AND CONCLUSIONS

1. Duplex 316 manual metal arc deposits of nominally the same composition exhibited very large differences in minimum creep rate at 600°C. A fully austenitic deposit exhibited equal strength with the stronger duplex deposits.

2. Post weld heat treatment at 700 or 850°C reduces the creep strength of the deposits but the large differences in minimum creep rate remain.

3. In the duplex deposits, the presence of the δ-ferrite refined the grain size of the weld metal to 10-15 μm. The grain size in the fully austenitic deposit was approximately 100 x 1000μm.

4. Thin foils from interrupted tests of low creep strength revealed evidence of deformation at δ/γ boundaries and within the δ-ferrite. No evidence of deformation was observed in these regions in deposits of high creep strength.

5. The creep strength of the duplex deposits below the yield stress was governed by Nb, Ti and V derived from the electrode coating. These elements were observed to segregate to

the δ-ferrite during solidification and during creep or post weld heat treatment precipitate in the form of M(C,N) carbonitrides in the new austenite formed by the migration of the δ/γ interface.

6. A correlation between minimum creep rate at constant stress and composition was obtained for all duplex deposits in the as welded, 700°C and 850°C heat treated conditions. The compositional factor has the form of a solubility product relating carbon and total residual content.

7. The dominant deformation mechanism in the weaker deposits was thought to be by shear at the δ/γ boundaries, but deformation of the δ-ferrite also made a contribution. In the stronger deposits deformation of the austenite appeared to be the dominant mechanism.

ACKNOWLEDGEMENTS

The authors wish to acknowledge the provision of excellent technical facilities by the Laboratory Manager, CEGB Marchwood. Acknowledgement is made of financial support by SERC Case Award No. SB101, for one of the authors (JS).

REFERENCES

[1] Delong, W.T. (1974) 'Ferrite in austenitic stainless steel weld metal'. Welding Journal 53 p.273.

[2] Honeycombe, J. and Gooch, T.G. (1970) 'Microcracking in fully austenitic stainless steel weld metal'. Metal Construction 9 p.375.

[3] Hull, C.F. (1967) 'Effect of δ-ferrite on the hot cracking of stainless steel'. Welding J. 46 p.399.

[4] Binkley, C. Goodwin, G.H. and Harries, S. (1973). 'Effect of electrode covering on elevated temperature properties of austenitic stainless steel weld metal'. Welding J. 52 p.306.

[5] Binkley, C, Berggren, R.G., Goodwin, G.H. (1974) 'Effect of slight compositional variation on type E308 electrode deposits'. Welding J. 53 p.91.

[6] Cole, N.C., Berggren, R.G., Goodwin, G.R.M. (1974) American Welding Soc. Texas.

[7] Thomas, R.G. (1981) 'The effect of electrode coating on the high temperature mechanical properties of AISI 316 austenitic weld metals CEGB report No. RD/M/1191R81.

[8] Marshall, A.W., and Farrar, J.C.H. (1984). 'The influence of residuals on properties of austenitic stainless steel weld metal with particular reference to energy industries' Stainless Steel 1984 Institute of Metals.

[9] Emmanuel, G.N. (1950). 'Sigma phase and other effects of prolonged heating at elevated temperature on 25% chromium 20 nickel steel". ASTM 110 p.82.

[10] Farrar, R.A. & Thomas,R.G. (1983). 'Microstructure and phase transformations in duplex 316 submerged arc weld metal'. J. Material Sci. 18 3461-3474.

[11] Hirsch, P.B. (1967). 'Electron microscopy of thin crystals' Butterworth Publications p.423.

[12] Keh, A.S. (1960). J. Applied Physics 31 p.1501.

[13] Ham, R.K.(1961) Phil mag 6 p.1183.

[14] Berggren, R.G., Cole, N.C., Goodwin, G.M. (1978). 'Structure and elevated temperature properties of type 308 stainless weld metal with varying ferrite contents'. Welding J. p.167.

[15] Eldridge, E.A., Deem,H.W.(1981). Handbook of Stainless Steel.

[16] Barrett, C.R., Lytton, J.L., Sherby, O.D. (1967). 'Effect of grain size and annealing treatment on steady state creep of copper. AIME 170 p.239.

[17] Sherby, O.D., Lytton, J.L.,(1956) 'Possible rôle of diffusion in the creep of alpha and gamma iron'. Journal of Metals p.928.

[18] Smith,J., Farrar, R.A., Myers,J. (1987). 'Microstructure and the effect of ageing on duplex type 316 weld metal'. CEGB report to be published.

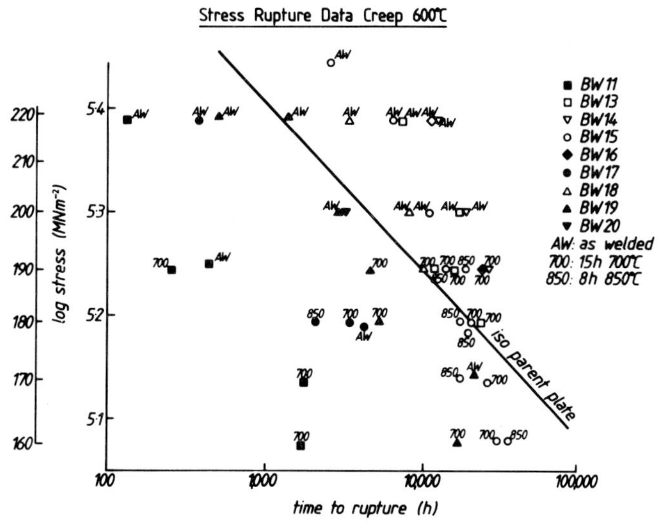

FIGURE 1. STRESS RUPTURE PROPERTIES.

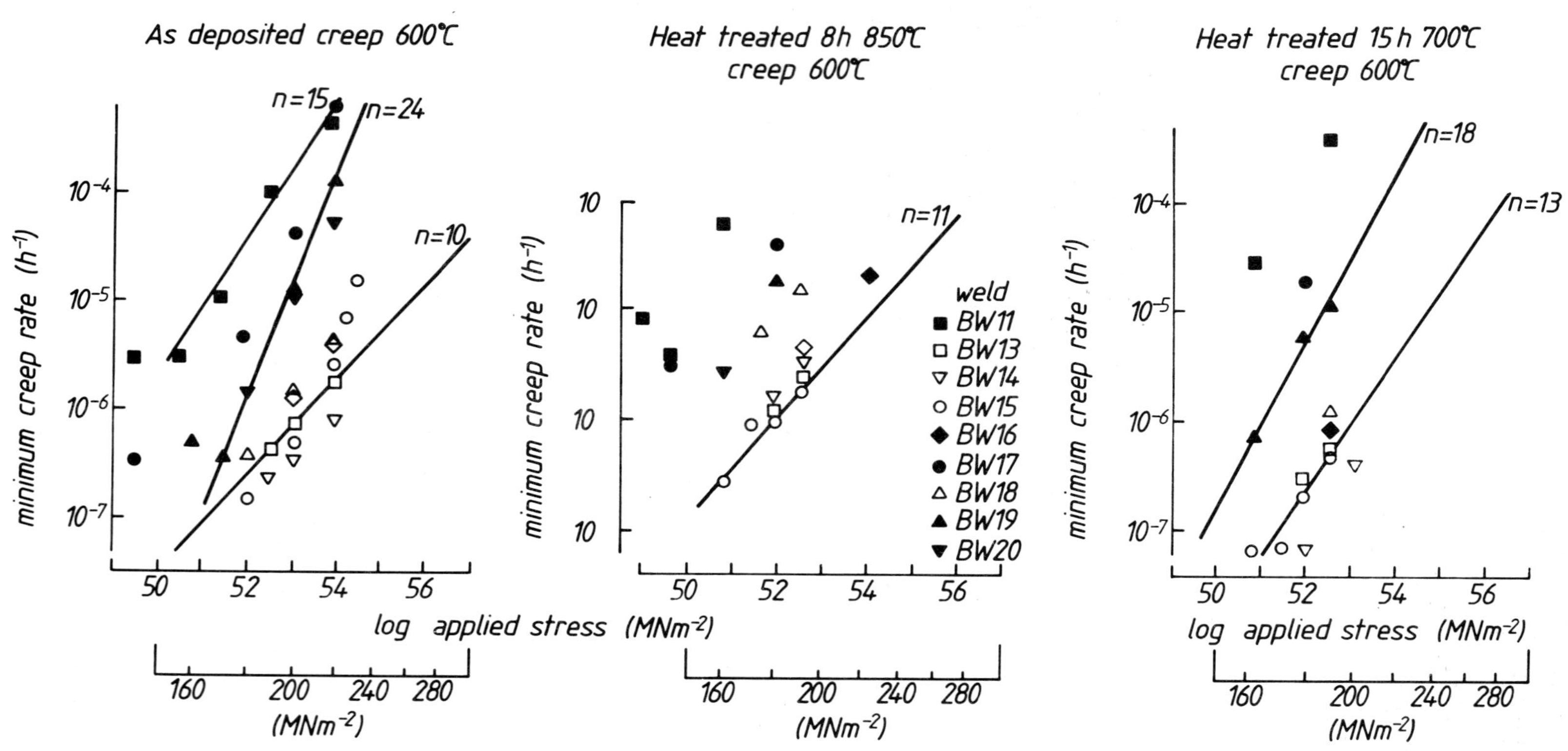

FIGURES 2, 3 AND 4. THE STRESS DEPENDENCE OF MINIMUM CREEP RATE AT 600°C FOR DEPOSITS IN THE AS WELDED, 700°C AND 850°C POST WELD HEAT TREATED CONDITIONS.

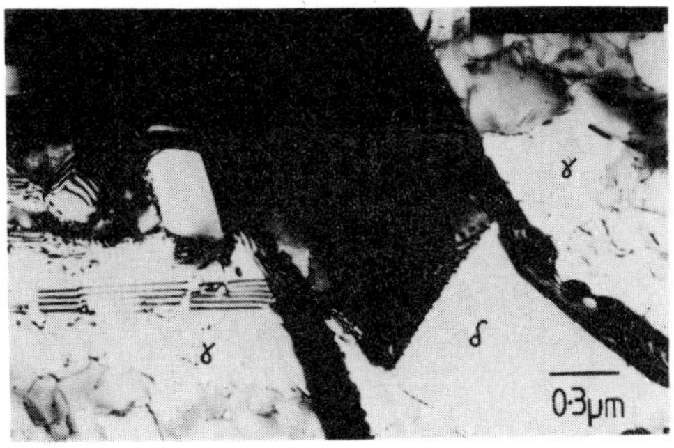

FIGURE 5. BW11/AW INTERRUPTED TEST 1 - HIGH ANGLE BOUNDARY IN THE δ-FERRITE WITH HEAVY SHEAR AT THE δ/γ BOUNDARY.

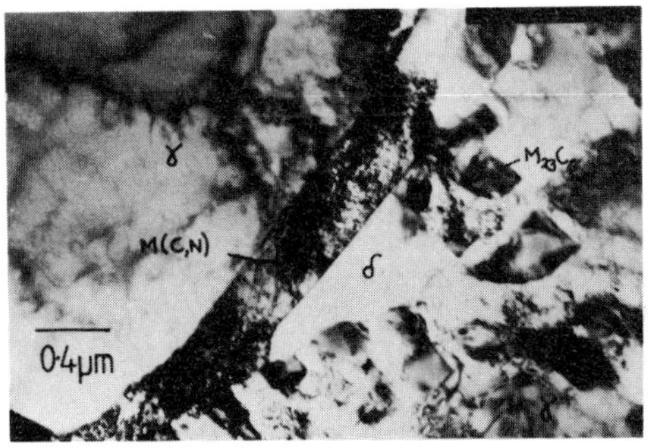

FIGURE 6. BW15/AW INTERRUPTED TEST 2 - PRECIPITATION IN NEW AUSTENITE FORMED BY THE MIGRATION OF THE δ/γ INTERFACE.

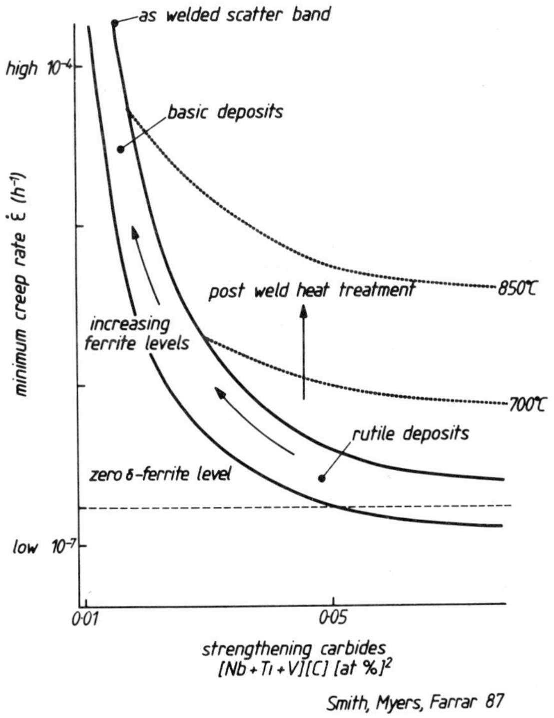

FIGURE 10. SCHEMATIC MODEL SHOWING EFFECT OF RESIDUAL Nb, Ti AND V AND PRIOR HEAT TREATMENT ON THE MINIMUM CREEP RATE.

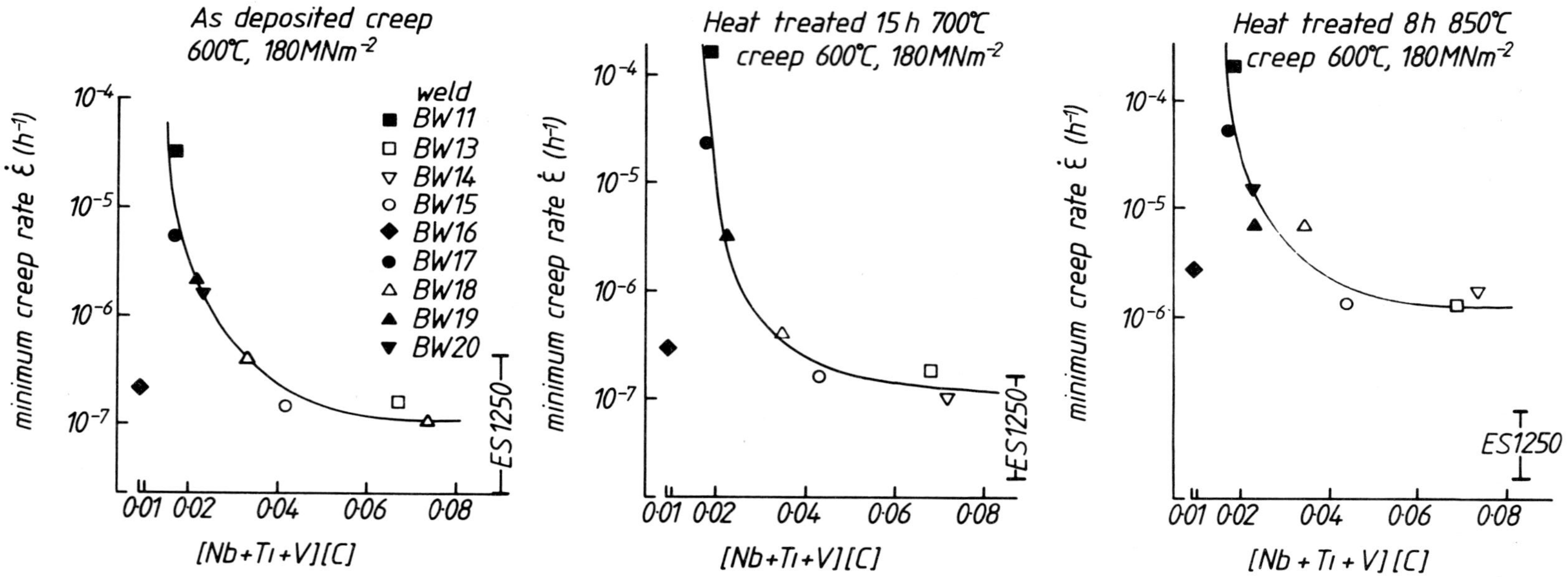

FIGURES 7, 8 AND 9 SHOWING VARIATION OF MINIMUM CREEP RATE 600°C AT $180MNm^{-2}$ WITH THE RESIDUAL CARBON PRODUCT FOR THE AS WELDED 700°C AND 850°C POST WELDHEAT TREATED DEPOSITS.

Transformation characteristics and microstructure of 12%CrMoV weld metal

C S Wright and T N Baker

The authors are in the Department of Metallurgy, University of Strathclyde.

Synopsis

12% CrMoV steels have applications in steam pipework and superheater tubing in both fossil fuelled and nuclear power stations. The alloy is also under consideration as a first wall material in fusion reactors.

The present work has studied the transformation characteristics of nominally 12% Cr steel electrodes at two carbon levels, 0.05 and 0.2%, together with wrought material for comparison. Samples were taken from multirun MMA welds which were given standard post-weld heat treatment, solution treated at 1135°C and cooled between 430°C min^{-1} and 3°C min^{-1}.

A modified Linseis dilatometer was used to determine CCT diagrams. The microstructure was studied using transmission electron microscopy.

M_s and M_f transformation temperatures for the wrought alloy and the 0.2%C consumable were similar, but for the 0.05% weld metal were significantly higher.

The microstructure consisted of lath martensite with retained austenite films between the laths, and delta ferrite, whose volume fraction increased with decreasing carbon content.

A complex precipitation of carbides occurred. The paper discusses the effect of cooling rate on the type and morphology of the precipitates, and also the modifications associated with the presence of delta ferrite.

Introduction

12% CrMoV steels have applications in steam pipework and superheater tubing in both fossil fuelled and nuclear power stations. For many applications martensitic stainless steels are fusion welded and whilst data is available on the properties and microstructure of the parent material [1-4], far less is known about the modifications introduced as a result of welding [5].

Carbon levels of about 0.2% are still used in the heat resistant grades containing Mo and V and welding practice has remained virtually unchanged, using preheats of 250 to 450°C which are higher than the M_s temperature of about 200°C [5]. This level of preheat has been chosen to avoid the risk of hydrogen induced cracking, but requirement of such a high temperature has been questioned [6]. To ascertain more appropriate preheat/interpass temperatures, a knowledge of the M_s-M_f temperature range for the weld fusion zone is required.

The present work is concerned with the effects that variations in composition, particularly carbon content, microstructural features such as grain size and delta ferrite content, and cooling rate have on the transformation characteristics of CrMoV weld metal.

Experimental Procedure

Multirun MMA welds, of composition shown in Table 1, were laid down onto a 25mm thick plate of 12 CrMoV steel. A preheat temperature of 170°C was maintained between the welding passes. A post-weld heat treatment of 750°C for 2hrs, followed by furnace cooling was then carried out. Electrodes with different carbon contents were used to prepare the deposits. These were made at Babcock Power plc, Renfrew, where the welding and heat treatment were undertaken.

Phase transformation characteristics of the wrought material and the two electrode steels were determined using a modified Linseis dilatometer. A considerably higher solution temperature than usual in phase transformation work prior to cooling was incorporated into the procedures to model as close as possible the weld cycle. Samples were heated at 50°C min^{-1} to 1135°C and held for 10 min before cooling in argon or helium at rates between 450°C min^{-1} and 3°C min^{-1} (measured at 700°C).

Metallographic observations were made using optical and electron microscopes, and precipitates were identified by a combination of electron diffraction and EDX.

Results

(i) Phase Transformation Data:

The transformation data is presented in the form of CCT diagrams, Figs. 1 and 2. It was found that the wrought alloy and weld metal A gave comparable transformation characteristics, while the M_s and M_f temperatures for the low carbon weld metal B were somewhat higher than for weld metal A.

(ii) Metallography:

Optical micrographs of the wrought alloy and weld metal B in the 'as received' conditions are shown in Figs. 3 and 4, and following cooling at 200°C min^{-1} in Fig. 5. The as-received wrought alloy consisted of tempered martensite, and was martensitic after cooling at 200°C min^{-1}. No delta ferrite was seen in this alloy in either condition; on the other hand, weld metal B possessed a duplex structure of tempered martensite and delta ferrite (>0.23).

Transmission electron microscopy of thin foils showed similar features in the martensite present in all three materials; parallel laths grouped in packets, irregular dovetail laths, Fig. 6, and laths containing microtwins. All three materials contained some retained austenite which was present as thin films separating individual martensite laths, Fig. 7.

No precipitation was observed in the wrought material when the cooling rate exceeded 200°C min^{-1}. At this cooling rate, autotempered laths containing M_3C precipitates were found, the number and size increasing with decreasing cooling rate. Cooling at 17°C min^{-1} or slower, resulted in localized precipitation at prior austenite grain boundaries, Fig. 8. This precipitation was more pronounced at slower cooling rates which led to a coarsening of precipitates characterized as mainly $M_{23}C_6$.

The precipitation in weld metal A was comparable to that of the wrought material, but with weld metal B, autotempered martensite was first seen after cooling at 480°C min^{-1}. However, unlike the other two materials, decreasing the cooling rate did not lead to a substantial increase in the number of autotempered laths. Cooling at rates of 170°C min^{-1}, or less gave a localized precipitation of $M_{23}C_6$ carbides in the delta ferrite adjacent to the martensite regions, and of M_2X carbides at delta ferrite interfaces, Fig. 9. At rates of less than 10°C min^{-1} another precipitate morphology was occasionally observed at delta ferrite grain boundaries. This consisted of a regular network of precipitates nucleated on dislocations, Fig. 10. Electron diffraction and microanalysis indicated that the precipitates were MC carbides containing vanadium and substantial amounts of chromium.

Discussion

The transformation data is given in Figs. 1 and 2 for material of differing carbon level. Data for the wrought material is in agreement with that previously published [7,9]. The presence of delta ferrite in weld metal B and also the possibility of a different austenite grain size prior to transformation (it was not possible to etch the prior austenite grain boundaries in weld metal B) may account for the higher M_s and M_f temperatures than in the wrought alloy and weld metal A.

Fenn and Jordan [9] have shown by electron probe microanalysis, that the presence of delta ferrite in martensitic stainless steel results in the partitioning of alloying elements, with the delta ferrite being enriched in Cr, Mo and V, and depleted in C, compared with austenite. An increase in the delta ferrite volume fraction, would therefore be expected to lead to an enrichment in carbon of the remaining austenite.

On the other hand, the lower carbon in weld metal B and the consequent presence of delta ferrite should markedly influence the austenite grain size at the solution temperature, as in the case of 9%Cr steels [3].

In the 'as received' condition, weld metal A contained a small volume fraction (0.02) of delta ferrite. However, after solution treatment and cooling, no delta ferrite was detected, which would suggest, in agreement with Fenn and Jordon, that it had transformed to austenite during the solution treatment.

Microstructural examination of thin foils revealed that the austenite-martensite transformation was not complete at room temperature. All three materials studied contained small amounts of retained austenite in the form of thin films between individual martensite laths, as also reported in low alloy steels [10].

The decomposition of these films during low temperature tempering (300°C) to a lath boundary cementite can lead to tempered martensite embrittlement, but by heating to 700°C, retained austenite decomposes to $M_{23}C_6$ which has been claimed to confer microstructural stability, particularly under creep conditions [11].

Most of the precipitates identified in the continuously cooled samples studied in the present work, are parallel with those observed after tempering [3].

It is interesting to note the presence of polygonized boundaries in weld metal B when cooled at rates of less than 10°C m^{-1}. However, in contrast to tempering above 500°C [3] which also produced low angle boundaries pinned by M_2X or $M_{23}C_6$, the present studies identified the dislocations as sites for a fine network of precipitation which was found to be MX containing substantial amounts of chromium and also vanadium.

Conclusion

The transformation characteristics of 0.2%C wrought alloy and weld metal A were very similar, but for weld metal B (0.05%C), the M_s and M_f temperatures were significantly higher after solution treatment of 10 mins at 1135°C and cooling between 430 and 20°C min^{-1}.

For all three alloys, the as received material consisted of lath martensite with thin films of retained austenite between the laths. Both weld metals contained delta ferrite, weld metal B having a substantial volume fraction.

After solution treating and cooling the precipitates which developed were similar to those present after tempering.

Acknowledgements

The authors would like to express their gratitude to Mr Iain Hamilton and Mr Bruce Brown of Babcock Power plc for supplying material, and for their guidance during this work. An SERC grant to support CSW is also acknowledged.

References
1. K J IRVINE, D J CROWE and F B PICKERING, J Iron Steel Inst 1960, 195, 43-62.
2. K J IRVINE and F B PICKERING, J Iron Steel Inst 1960, 195, 386-405.
3. E A LITTLE, D R HARRIES and F B PICKERING, Ferritic Steels for Fast Reactor Steam Generators, 1978, BNES:London, 120-127.
4. C LOIER and C LEYMONIE, Mat Sci & Eng, 1984-5, 68, 165-174.
5. T G GOOCH, Stainless Steels '84, 1985, Inst Metals:London, 249-261.
6. R CASTRO and J J de CADENET , Welding Metallurgy of Stainless and Heat Resisting Steels, 1974, Cambridge Univ.Press:Cambridge.
7. J Z BRIGGS and T D PARKER, The Super 12%Cr Steels, Climax Molybdenum Co. 1965.
8. R PETRI, L SCHNABEL and P SCHWAAB, Mannesmann Forchungsberichte 1979, 799.
9. R FENN and M F JORDAN, Metals Tech. 1982, 9, 327-337.
10. N C LAW, P R HOWELL and D V EDMONDS, Metal Sci 1979, 13, 507-515.
11. T LECHTENBERG, Ferritic Steels for high Temperature Applications (Ed A A Khare) 1983, ASM:Ohio, 163-177.

Table 1. Chemical compositions of experimental materials.

Material	C	Cr	Mo	V	P	S	Ni	Mn	W	Si
Wrought	0.2	10.9	1.34	0.33	0.0156	0.007	0.31	0.67	-	0.34
Weld Metal A	0.17	9.8	1.37	0.27	0.022	0.003	0.31	0.87	0.38	0.45
Weld Metal B	0.05	11.1	1.1	0.12	0.02	0.001	0.46	0.57	0.33	0.51

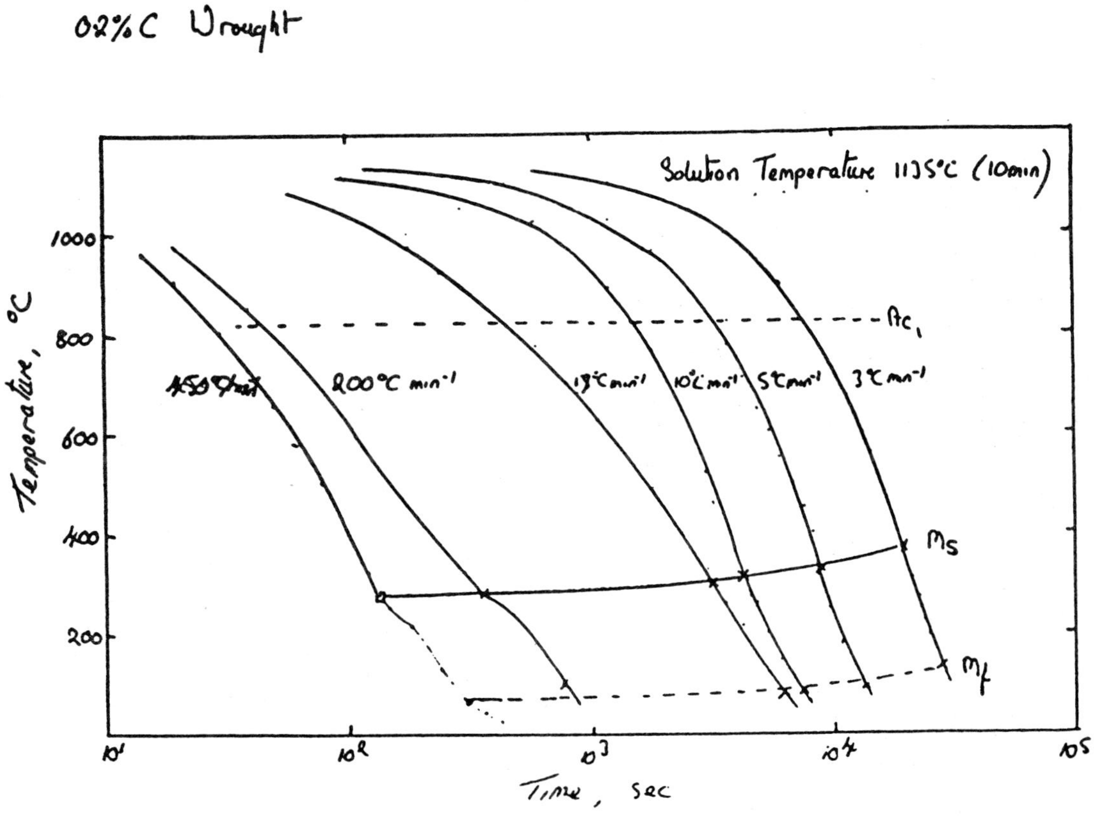

1. CCT diagram for the 0.2%C wrought material, solution treated at 1135°C.

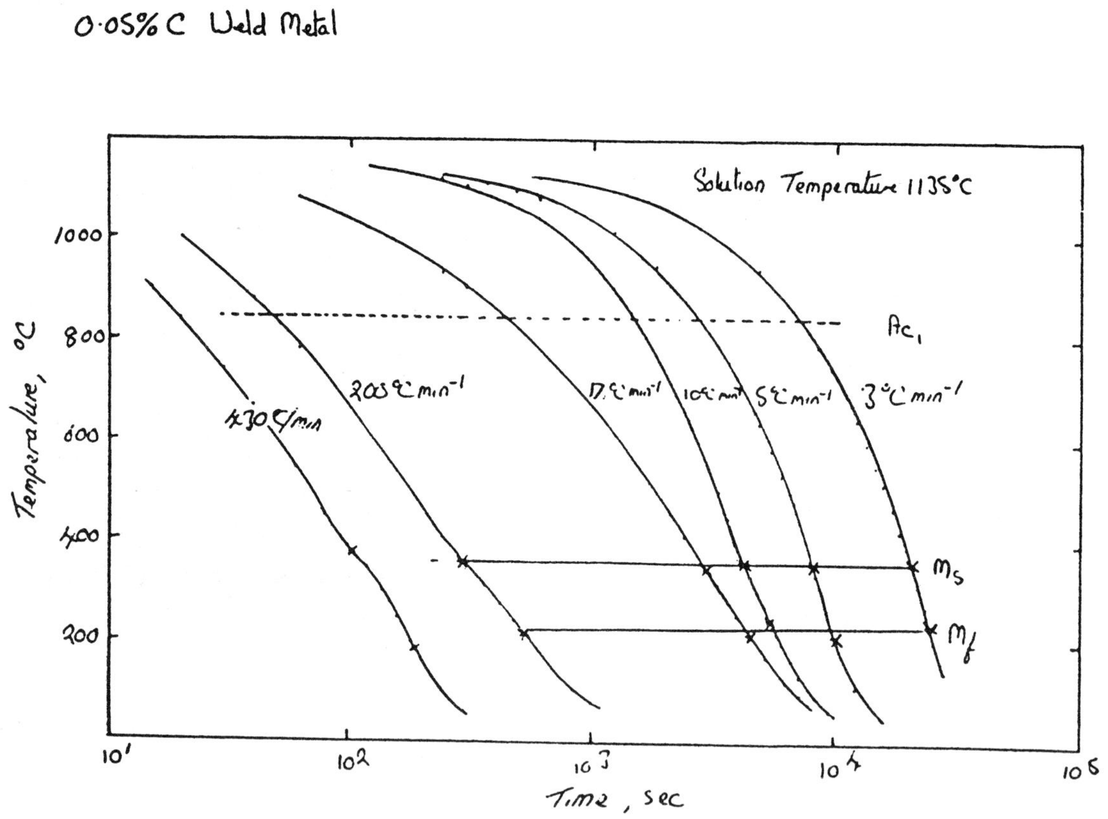

2. CCT diagram for the 0.05%C weld metal B solution treated at 1135°C.

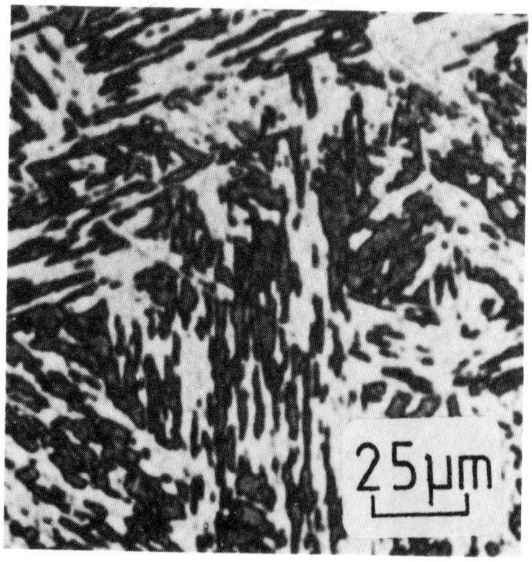

3. Optical micrograph of as received wrought material.

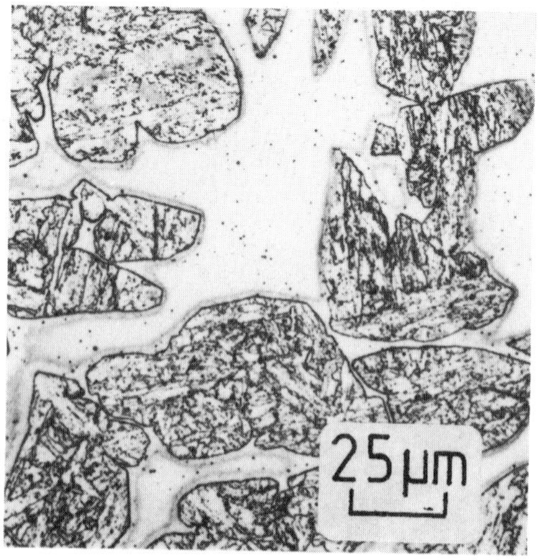

4. Optical micrograph of as received 0.05%C weld metal B.

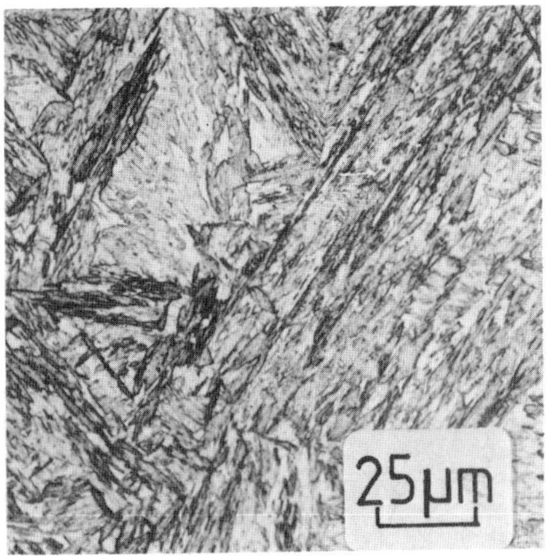

5. Optical micrograph of wrought material cooled at 200°C min^{-1}.

6. Thin foil micrograph of irregular dovetail laths.

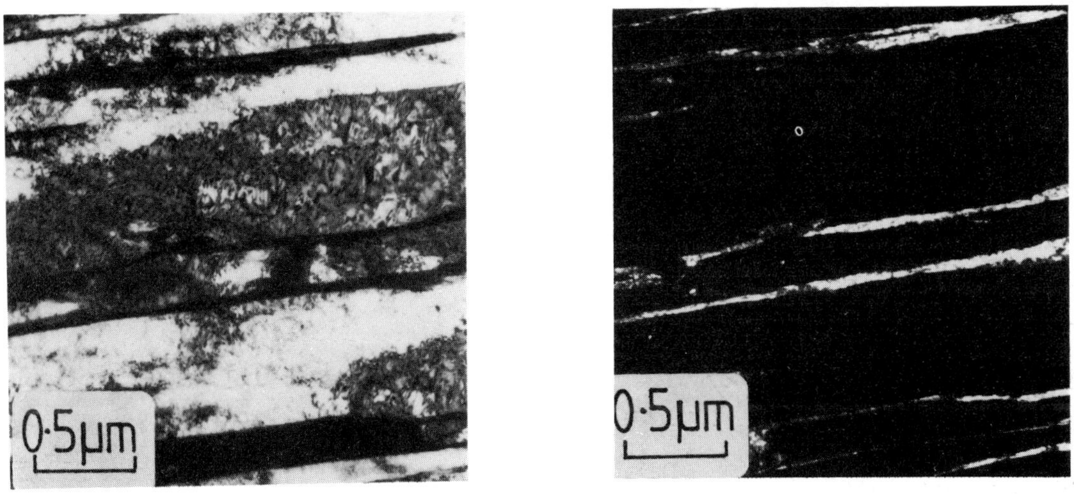

a

b

7. Thin foil micrographs of retained austenite films
(a) bright field (b) dark field.

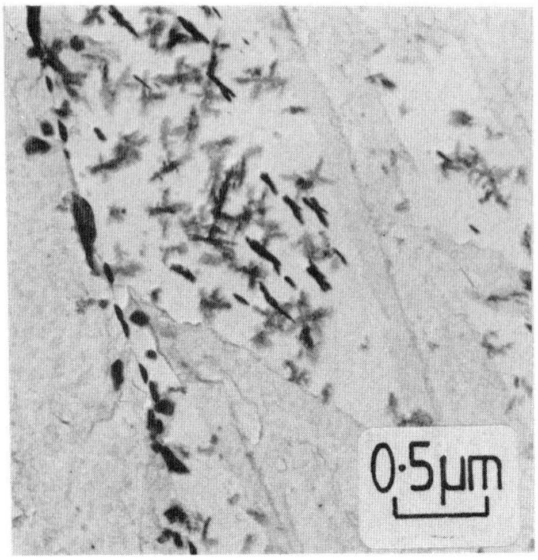

8. Carbon extraction replica of autotempered martensite in wrought material cooled at 17°C min^{-1}.

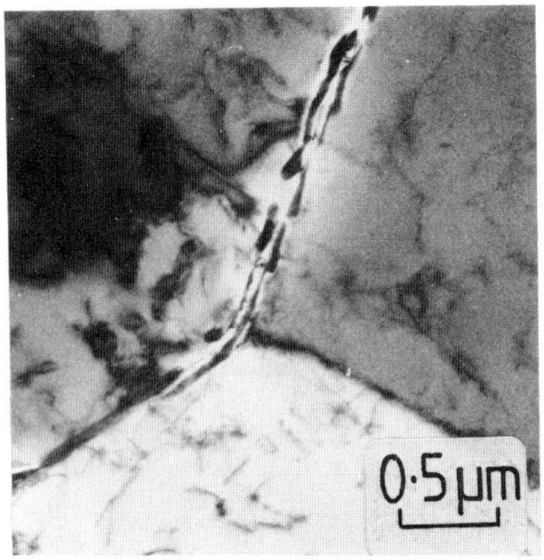

9. Thin foil micrograph of M_2X carbides in the delta ferrite interfaces in weld metal B cooled at 10°C min^{-1}.

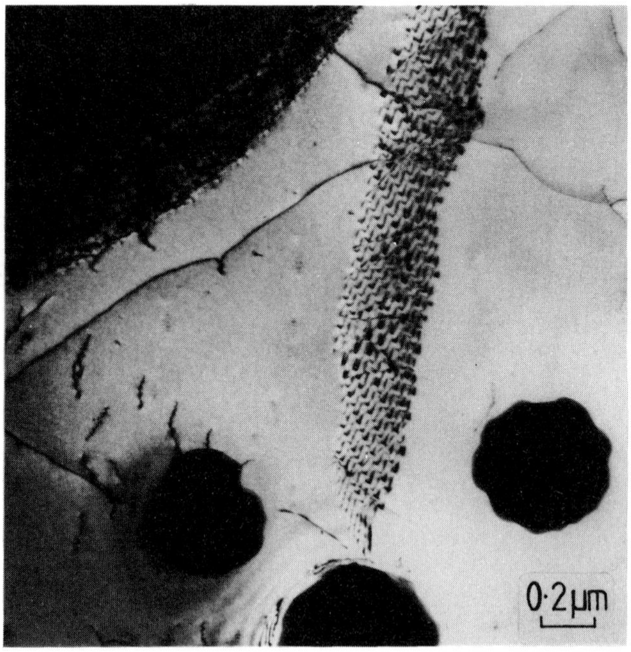

10. Thin foil micrograph of MX carbides precipitated on dislocation networks (sub boundaries) in delta ferrite.

POSTERS

Duplex stainless steel as modern material: corrosion resistance of base metal and welded joints

J M Lardon, J Charles, F Dupoiron, P Pugeault and D Catelin

Dr. Lardon is a Research Metallurgist at the Research Centre of Le Creusot (CRMC) at Creusot-Loire Industrie. Dr. Charles is Head of the Stainless Steel and Corrosion Department of the CRMC. Dr. Dupoiron and P. Pugeault are Research Metallurgists at the CRMC. D Catelin is Head Manager of Stainless Steel and Clad Products at Creusot-Loire Industrie.

SYNOPSIS

The purpose of this paper is to present the corrosion resistance of the duplex stainless steel family in relation to its structure. The latest commercialized molybdenum-free duplex stainless steel, which will supply the AISI 304 and 316 austenitic stainless steels in many applications, is particularly investigated.

The corrosion resistance (pitting and crevice) of welded joints at different levels of heat input is presented.

Results are discussed by means of temperature-time precipitation diagrams. Optimal welding parameters are defined from these results in order to obtain an equivalent corrosion resistance for welded joints and base metal.

INTRODUCTION

Austenitic stainless steels are in very common use in many processes of the chemical industry. Presenting both high ductility and good resistance to general and localized corrosion, such as pitting and crevice, they have been chosen for many applications. However, conventional austenitic stainless steels (ie 304, 316) are very prone to stress-corrosion cracking in hot chloride environments, and in this case the use of higher nickel-content stainless steels or ferritic stainless steels is recommended. This has several drawbacks. Alloying austenitic stainless steels with nickel makes them much more expensive and ferritic stainless steels suffer from a lack of ductility and are difficult to weld.

Between these two different kinds of stainless steel (austenitic and ferritic), a family of duplex austenitic–ferritic stainless steels has been developed over many years by Creusot-Loire Industrie.[1-3] Both interesting mechanical properties and corrosion resistance (including stress-corrosion cracking) can be obtained with these alloys. Moreover, the wide range of composition of duplex stainless steels now available makes adjusting the choice of materials for given service conditions easier and cost-efficient.

The aim of this paper is therefore to examine this family of duplex stainless steels, paying attention to both base metal and to welded joints. The microstructure, mechanical properties, and corrosion resistance are presented. The behaviour of the latest commercialized duplex stainless steel, lower alloyed and molybdenum-free, URANUS 35N, is emphasized.

STRUCTURE AND COMPOSITION OF DUPLEX STAINLESS STEEL

The average chemical compositions of the different duplex stainless steels developed by Creusot-Loire Industrie are shown in Table 1. The ferrite contents of quenched structures are also given.

Duplex stainless steels solidify first on α-phase (Fig. 1). The austenitic γ-phase appears only during cooling. Composition is adjusted in order to obtain a 50%α-50%γ structure after quenching. The microstructure of such a duplex stainless steel plate is presented in Fig. 2.

The chromium and molybdenum contents may vary from 22 to 25% and from 0 to 3% respectively, depending upon the corrosion properties required. The low carbon and relatively high chromium contents (see Table 1) avoid sensitization effects due to carbide precipitations $Cr_{23}C_6$. Nitrogen content is increased up to 0.15% in order to stabilize the austenitic phase and improve localized corrosion resistance. Moreover, alloying with nitrogen delays the first appearance of brittle intermetallic compounds such as σ-phase (usually present in Fe–Cr–Mo alloys) during expositions in the range of temperatures 600–950°C. Another beneficial effect of nitrogen addition is that segregation of chromium between the two phases is reduced. Compositions of both α-phase and γ-phase determined by microprobe analysis are shown in Table II both for base metal and for welded joints of several duplex stainless steels.

MECHANICAL PROPERTIES OF DUPLEX GRADES

The mechanical properties of the duplex grades are listed in Table III. Particularly high yield strength can be seen; about twice that of conventional austenitic stainless steels (Fig. 3).

The ultimate tensile strength (UTS) lies between 700 and 900 MPa for most of the duplex grades, with an elongation to rupture (A%) always greater than 20%. These particularly good results have to be related to the microstructure of the duplex stainless steels, which can be compared to that of composite materials. The small size of the ferritic and austenitic grains, the dynamic strengthening of the γ-phase due to nitrogen additions, and the mechanical coupling effects occurring between the two phases, can all account for the high resistance of duplex stainless steels.

Moreover, it is important to notice that these alloys have good mean toughness values at room temperature (Table III) and down to very low temperatures, as shown in Fig. 4.

Dynamic strengthening of duplex stainless steels is possible; very high properties can then be achieved, with UTS of more than 1000 MPa and an elongation to rupture superior to 10%.[4]

CORROSION RESISTANCE OF DUPLEX STAINLESS STEELS (BASE METAL)

General corrosion resistance

The corrosion resistance of duplex stainless steels is strongly dependent upon their chemical compositions (chromium and molybdenum content). Most of the duplex grades are suitable alloys for very aggressive environments. Figure 5 shows the good behaviour of these alloys in some acidic media. The better performance of URANUS 45N grade, compared to standard 304L and 316L stainless steels, has to be related to its higher chromium and molybdenum content. But it can be seen that the lower alloyed steel URANUS 35N has similar corrosion resistance to 316L stainless steel.

In phosphoric acids (Fig. 6), duplex grades UR45N and UR47N can be compared to high-alloyed, super-austenitic grades such as URANUS B6 (25%Ni–20%Cr–4.5%Mo).

The most alloyed duplex grade URANUS 52N has even better corrosion resistance in aggressive environments. Figure 8 shows, for instance, its behaviour in sulfuric media contaminated with chlorides.

These results show that duplex stainless steels present a high resistance to general corrosion in very aggressive conditions. They are potential choices in many applications where conventional austenitic stainless steels (304, 316) are not resistant enough and where higher-alloyed nickel or chromium steels are no more cost-efficient.

Pitting and crevice corrosion resistance

It is well-known that austenitic stainless steels (304 or 316 grades) are very prone to pitting or crevice corrosion in chloride media. These types of corrosion are closely related to chloride content and temperature.

It is now well-established that increasing chromium, molybdenum, and nitrogen contents decrease the pitting (or crevice) susceptibility of stainless steels. A pitting index is generally used in order to compare the different stainless steel grades' resistance to localized corrosion. Table I shows different pitting indexes for austenitic and duplex grades. It appears that the duplex family provides a wide range of pitting indexes and thus a wide range of corrosion resistance. The URANUS 35N grade can be compared to 316L grades while the higher-alloyed duplex grades URANUS 47N and URANUS 52N should be compared to super-austenitic stainless steels URANUS B6 or UR SB8 (Fig. 8).

Many other tests have been performed in industrial media. For instance, URANUS 47/52N proved to be suitable material for sea-water applications, the pulp and paper industry, and in scrubbers and depollution equipment (FDG systems).

Thus duplex grades provide a wide range of pitting or crevice-resistance materials. The choice of such a stainless steel can be optimized according to the corrosion properties required.

Stress corrosion cracking (SCC)

Austenitic stainless steels are very prone to SCC, especially in hot chloride environments, whereas duplex stainless steels are quite resistant to SCC in many chloride environments. Figure 9 presents laboratory test results in an aerated 300 g/l NaCl at 110°C. The duplex grades URANUS 47/52N can withstand SCC at up to 90% of the yield strength, as opposed to only 75% for the 316L stainless steel. Moreover, the much higher yield strength of the duplex steel, compared to 316L, results in much higher threshold stresses for SCC, as can be seen in Fig. 9.

The good performances of duplex stainless steels have been explained by both mechanical and electrochemical coupling between the two phases.[5,6] At low stresses, only the austenitic phase is deformed (due to the higher yield strength of the ferritic phase) and that phase is cathodically protected by the ferrite phase in chloride media, so that the occurrence of SCC is delayed.

WELDING DUPLEX STAINLESS STEELS: METALLURGY AND CORROSION RESISTANCE OF WELDED JOINTS

Primary solidification structures are accurately predicted using the RH ESPY and SCHAEFFLER diagram (Fig. 10) which presents γ and α primary solidification areas.[7]

Solidification as δ ferrite prevents the duplex stainless steels from hot-cracking risks in heat-affected zones (HAZ) or weld metal. Figure 11 shows the hot-cracking sensitivity of a duplex stainless steel compared to that of some austenitic grades using varestraint tests.

For base metal the composition is adjusted in order to obtain after solution annealing a near 50%α–50%γ structure. For weld metal and HAZ the ferrite/austenite ratio is modified since cooling occurs from a near α region and cooling rates are generally too high to allow a complete $\alpha \rightarrow \gamma$ transformation.

Ferrite contents are thus dependent on welding conditions. Figure 12 shows, for instance, the HAZ ferrite content, determined by optical metallographic counting, of several single-pass weld beads as a function of the cooling rate of the HAZ (measured at 700°C). Figure 13 illustrates the evolution of the structure from base metal to weld deposit metal. Two different zones are pointed out: the first one, near the base metal, presents an increasing ferrite content without recrystallization, as the second one, near deposit metal, is characterized by a coarsening of the ferrite grains and the recrystallization of γ plate-like grains – called Widmanstätten structure – during cooling.

Corrosion resistance : influence of welding conditions

We shall restrict ourselves to the pitting corrosion resistance of duples stainless steel welding joints.

Table IV presents the welding parameters of the welds investigated. Two welding processes were used: SMA (coated electrode) welding process for the UR 35N and UR 52N alloys, and electron-beam (EB) process for the UR 45N and 52N duplex grades.

Pitting potentials were determined by means of polarization curves. Experimental conditions were: 30 g/l NaCl solution, argon deaeration, 900 mV/h scanning rate, 22°C, 100 $\mu A/cm^2$ as critical pitting current.

SMA welding process: UR 35N alloy

Figure 14 shows the pitting potential measures obtained with SCE for the less-alloyed duplex grade. Results are compared with AISI 304 and 316 pitting potentials (base metal). Pits were initiated in HAZ and particularly in the high ferrite-content zone. Pitting potential remained nevertheless always greater than 500 mV/SCE, which is higher than that obtained for base metals of AISI 304 and 316 grades.

SMA welding process: UR 52N alloy

For the highest alloyed duplex stainless steel, pits were initiated in the base metal. The pitting potentials presented in Fig. 14 were thus equivalent to those of base metal for all heat input tested. Pitting potential is very high (about 950 mV/SCE). Optical and MEB examinations showed that pits resulted mainly from equivalent dissolution of the two phases. Oxide inclusions were clearly identified as possible pit initiation sites.

Electron-beam process: UR 45N and UR 52N alloys

Both alloys presented very high pitting potentials (about 950 mV/SCE). Most of the pits were initiated in the weld metal for the UR 45N alloy; in this case, a preferential dissolution of the ferrite was observed. Pitting potential remained nevertheless equivalent to that of base metal. No difference has been observed between welds and base metal for the UR 52N duplex grade, pits initiating in both weld and base metal.

Discussion

The duplex structure results from the transformation during cooling of ferrite in austenite. The composition of each phase and the α/γ ratio are dependent on the average chemical composition of the alloy and of the annealing heat treatment. Due to rapid cooling from temperatures higher than 1100°C, weld metal and HAZ consist of large ferrite grains with transformed austenite grains of Widmanstätten morphology. With a high rate of cooling, elements such as Cr, Ni, Cu, Si, and Mo are less divided between the two phases and the α/γ ratio can be modified (by increase of ferrite content).

This results in an increase of average nitrogen content in the austenite of the HAZ, compared to the austenite of the base metal, and thus increases its corrosion-resistance properties.

In the electron-beam welding process, weld metal and HAZ seemed to be somewhat more sensitive to pitting than for SMA welding joints. This probably results from higher cooling rates and from the absence of external input of nitrogen in electron-beam welds. In addition to the partitioning effect and the higher α/γ ratio, this induces a less beneficial nitrogen effect on the pitting corrosion resistance of the γ-phase and on the carbide precipitation kinetics. Corrosion resistance remained nevertheless very high.

CONCLUSIONS

The duplex austenitic-ferritic alloys are a new family of stainless grades, combining the high mechanical properties of the ferritic grades and the ductility of the austenitic stainless steels. Moreover, their high chromium content makes them a potential choice for many applications where high corrosion resistance is required, including stress-corrosion cracking. Thus it has been shown that:

1) The duplex stainless steels, presenting a 50%α -50%γ microstructure, offer high yield strength (twice as high as austenitic AISI 316) and good ductility (A% > 20%).

Despite the differences in structures and partitioning of the alloying elements between HAZ and base metal, no difference in pitting potential has been found between weld structures and base metal for the heat input energies investigated. Even for very aggressive conditions, UR 35N, the lowest alloying duplex stainless steel, remained more corrosion-resistant than 304 and 316 alloys. The UR 45N and 52N presented very high corrosion-resistance properties in the chloride-containing solutions for both base metal and welded joints. This explains why they are to be taken into consideration for off-shore applications.

2) The chemical composition of the alloy can be adjusted according to the aggressivity of the environment. The wide range of duplex grades now available provides a wide range of corrosion resistance.

3) In welding duplex stainless steels, avoiding too fast a cooling rate makes it possible to optimize the corrosion resistance of weld metal and heat-affected zones. Thus heat input should preferentially be in the range of 10 to 20 kJ.

4) With optimum welding conditions, HAZ and weld metal present equivalent resistance to pitting corrosion in chloride solutions at room temperature to the base metal.

5) Nitrogen additions in the welding gas (GTAW or PAW processes) or in the filler metal (SMA process) are beneficial to the corrosion resistance of welded structures by reducing the α/γ ratio,

increasing the pitting resistance of the γ-phase (solid solution effect), and increasing the corrosion resistance of the α-phase by reducing the carbide precipitation kinetics.

REFERENCES

1 J Charles, P Soulignac, J P Audouard, D Catelin: 25èmes journées des aciers spéciaux, St-Etienne, March 1986, 32, 1-9

2 J Charles, B Bonnefois, F Dupoiron, D Catelin: Proc. Conf. in 'Eurocorrosion', Karlsruhe, 1987, 635

3 A Desestret, J P Audouard, D Catelin, P Soulignac: Proc. Conf. in 'Corrosion 85', March 25–29, Boston, Nace, 1985, No. 48

4 P Bourgain, Y Grosbety, Ph Demarez, J C Bavay: 26èmes journées des aciers speciaux, Toulouse, May 1987, 17, 1–13

5 R Oltra, A Desestret, J C Colson: Mem. Sci. Rev. Metall., Nov 1980, 1003

6 T Magnin, J Le Coze, A Desestret: ibid., (3), 535

7 B Bonnefois, R Blondeau, D Catelin: ibid., (1), 14, 1–8

TABLE I – CHEMICAL COMPOSITION

	Si	Mn	Ni	Cr	Mo	N	Cu	FERRITE	*P.I.
URANUS 35N	0.6	1.5	4.0	23.0	0.2	0.12	–	45	26.9
URANUS 45N	0.4	1.7	5.7	22.0	2.8	0.12	–	45	33.6
URANUS 47N	0.3	1.2	6.5	25.0	3.0	0.17	0.2	45	38.3
URANUS 52N	0.3	1.2	6.5	25.0	3.0	0.17	1.5	45	38.3
ICL 472 BC	0.5	1.5	10	18.5	0.5	0.07	–	2	21.5
UR B6	0.2	1.5	25	20	4.3	0.08	1.5	0	35.8
UR SB8	0.20	1	25	25	4.8	0.2	1.6	0	44.8

* P.I. : Pitting Index
Cr + 3.3 Mo + 20 N

Table II – Microprobe analysis of α and γ phases of UR 35N Duplex stainless steels – base metal and H.A.Z

UR 35 N		Si	Mn	Ni	Cr	Mo	Cu
BASE METAL	α	0.562	1.697	3.05	25.75	0.145	0.104
	γ	0.483	1.903	5.031	21.65	0.088	0.154
H.A.Z	α	0.536	1.812	3.968	22.636	0.109	0.135
	γ	0.520	1.851	4.383	22.025	0.084	0.150

Table III - Mechanical Properties of Duplex stainless steels

GRADES	YS 0.002 (MPa)			UTS (MPa)	EL (%)	KCV Transverse + 20°C (J/cm2)	
	+ 20°C	+ 100°C	+ 200°C	+ 20°C	+ 20°C	Mini	Average
UR 35N	≥ 400	≥ 330	≥ 280	600-800	≥ 25	-	≥ 120
UR 45N	≥ 480	≥ 360	≥ 310	680-880	≥ 25	≥ 90	≥ 120
UR 47N	≥ 500	≥ 370	≥ 330	700-900	≥ 25	≥ 90	≥ 120
UR 52N	≥ 540	≥ 400	≥ 360	740-920	≥ 20	≥ 75	≥ 100

Table IV - Welding parameters

REPERE	I (A)	U (V)	V	E	% α HAZ
35 V 10	100	29.5	7.7	22.98	77
35 V 30	100	30	12	15	78
35 V 60	87	28.5	14.2	10.47	81
35 V 100	80	27	16.96	7.64	84
35 V 150	73	26.5	18.65	6.22	86
52 V 10	110	26	10.86	15.8	76
52 V 30	90	26.5	15.07	9.49	79
52 V 60	80	27	18.27	7.09	84
52 V 100	73	27	20.16	5.86	88
52 V 150	70	27.5	23.16	4.98	91

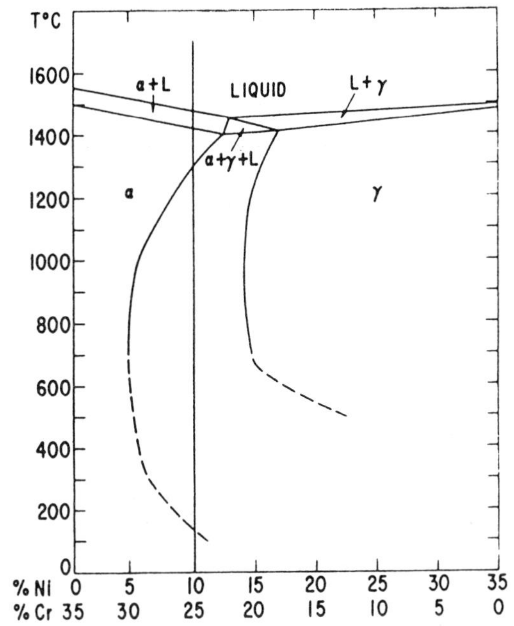

1 Pseudo binary diagram for 65%Fe–Cr–Ni alloys

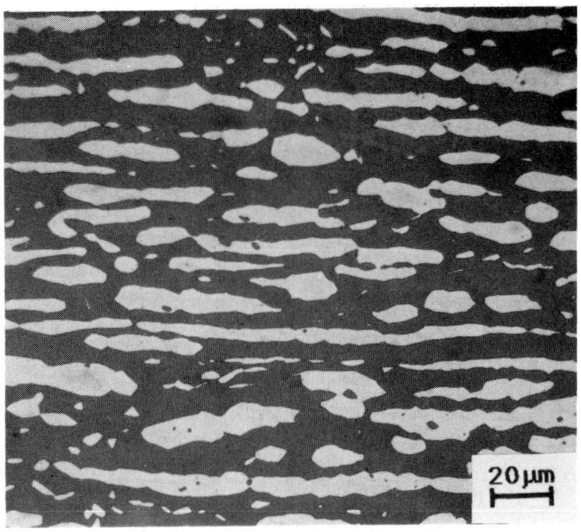

2 Optical micrograph of a duplex stainless steel (white: γ dark: α)

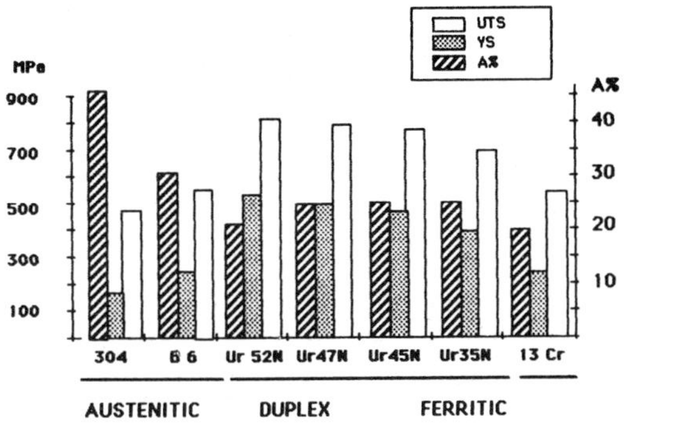

3 Mechanical properties of several stainless steels

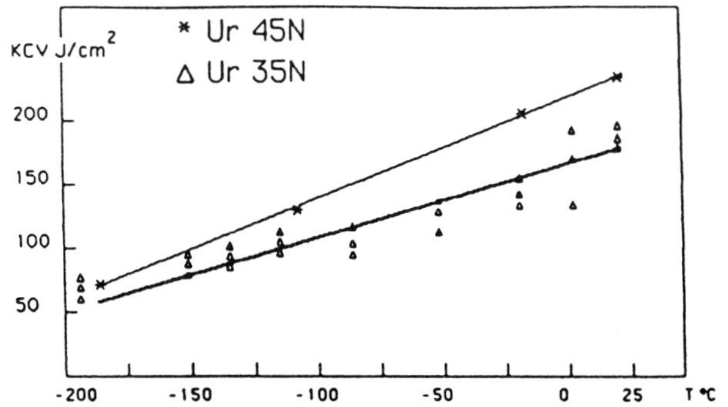

4 Evolution with temperature of the thoroughness of two duplex stainless steels

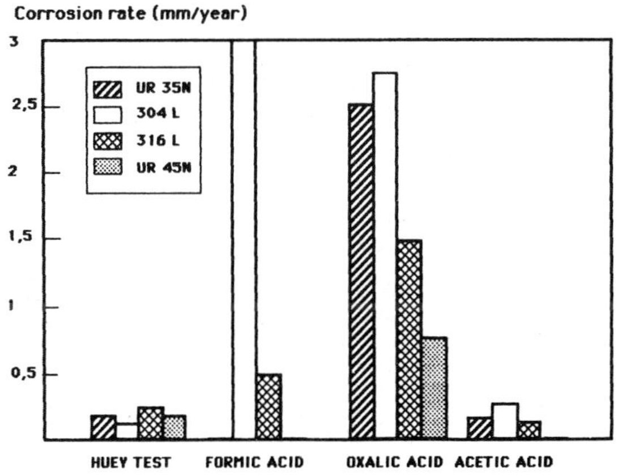

5 Corrosion rate of duplex UR 35N and UR 45N alloys compared to 304L and 316L alloys for some acidic media

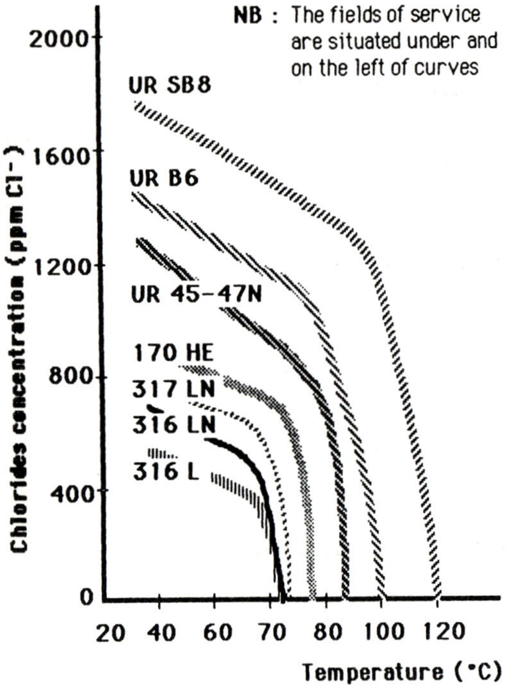

6 Maximum chloride contents allowed according to temperature for an industrial 54% P_2O_5 solution (H_2SO_4 < 4%; F^- < 1% and HF < 0.2%)

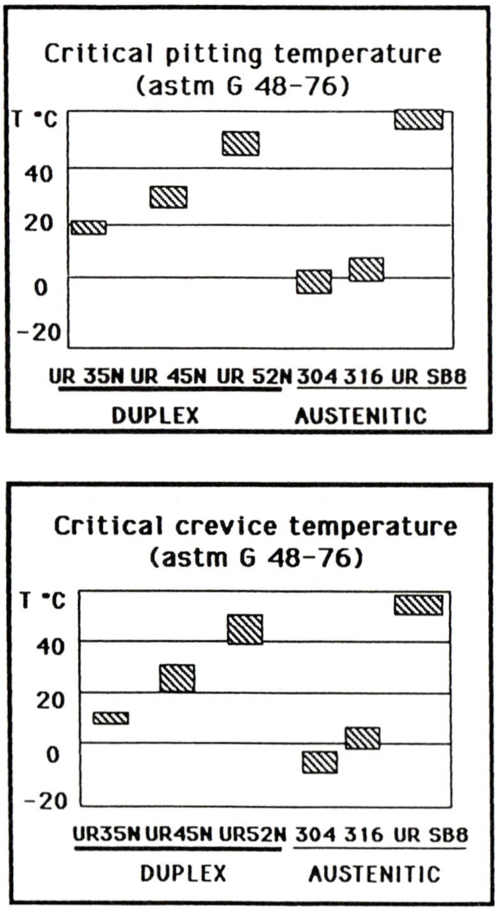

8 Critical pitting and crevice temperatures determined for several stainless steels following ASTM G48 specifications

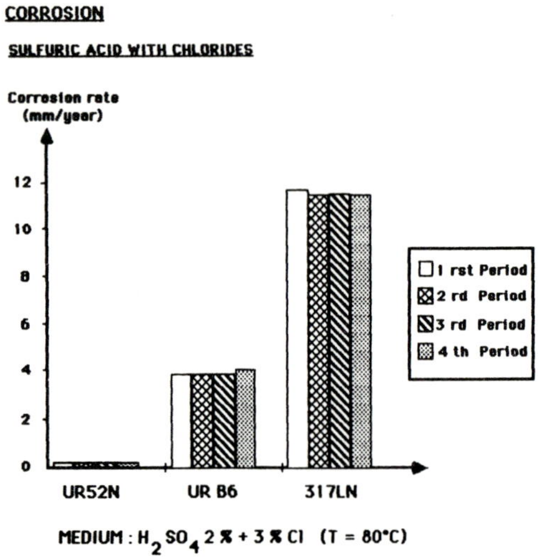

7 Corrosion rate of several stainless steels in a 2% H_2SO_4 + 3% Cl^- solution tested at 80°C

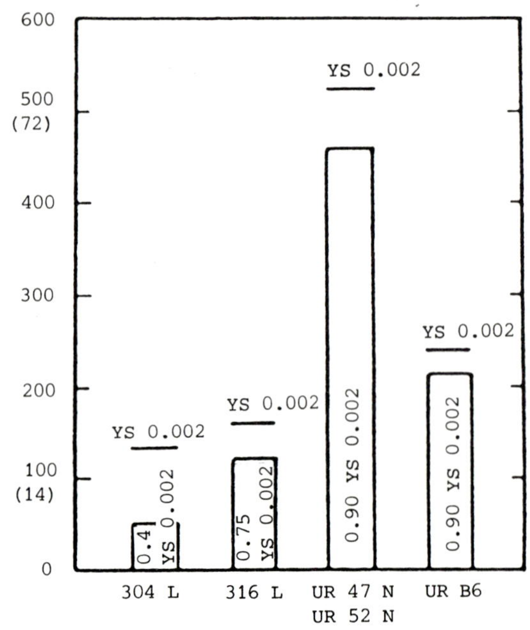

9 SCC resistance of several stainless steels in hot chloride medium (300 g/l NaCl, 110°C)

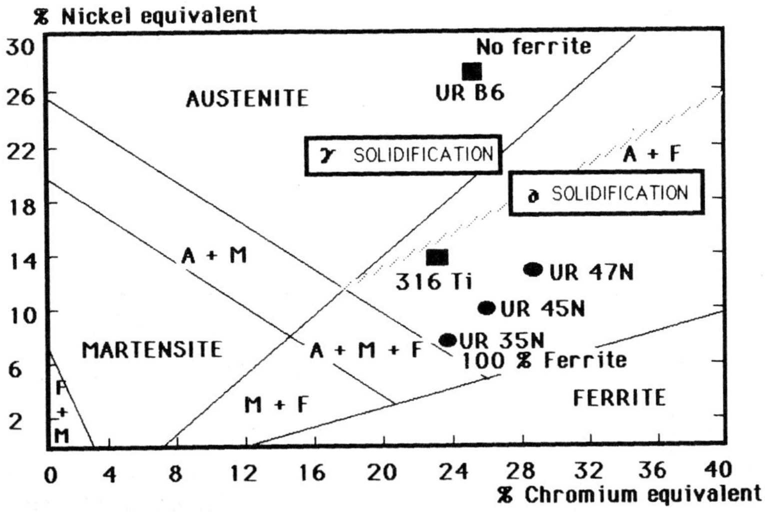

10 RH ESPY and SCHAEFFLER diagram showing the primary solidification structure areas

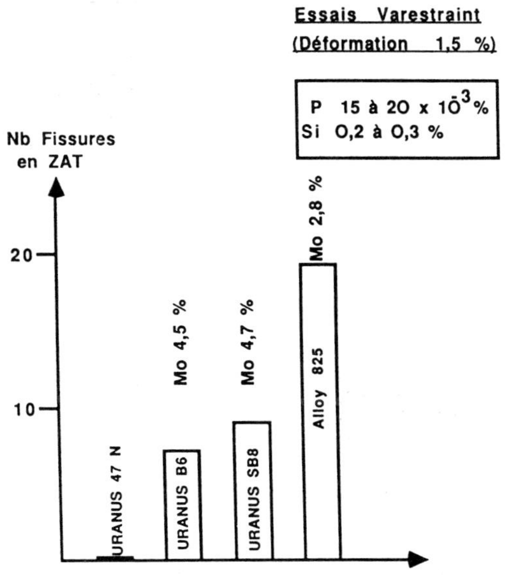

11 Hot-cracking sensitivity of duplex stainless steels compared to austenitic structures

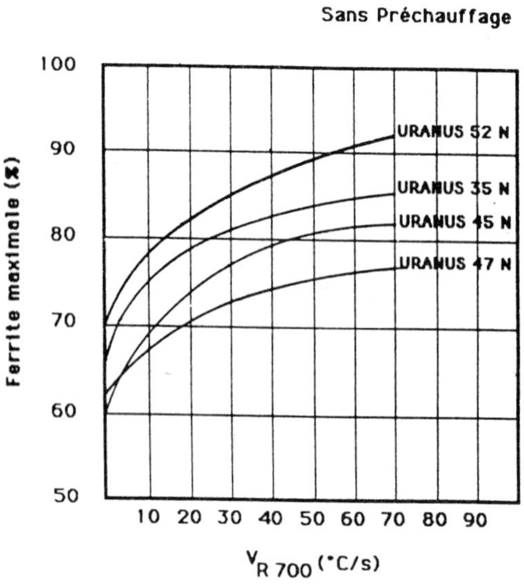

12 Ferrite content determined by optical metallographic examinations of several single-pass weld beads as a function of the cooling rate of HAZ (measured at 700°C)

13 Optical micrograph of a welded joint

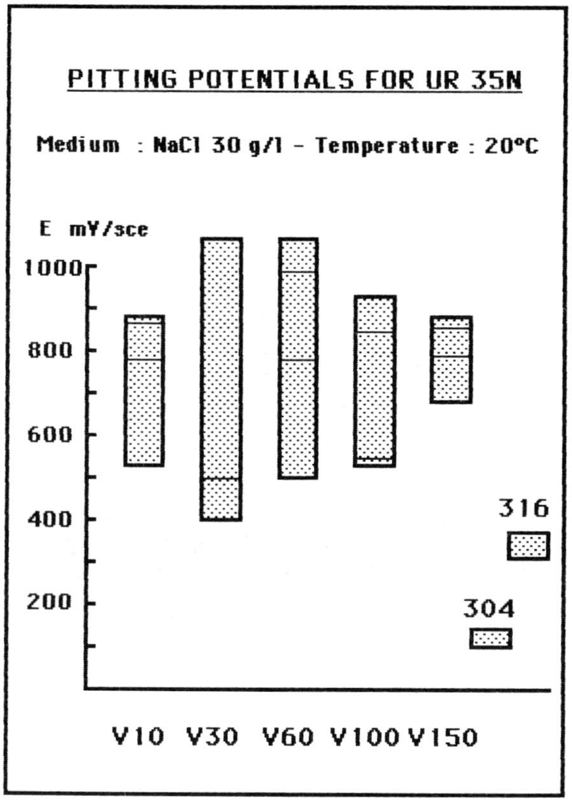

Figure 14 (a)

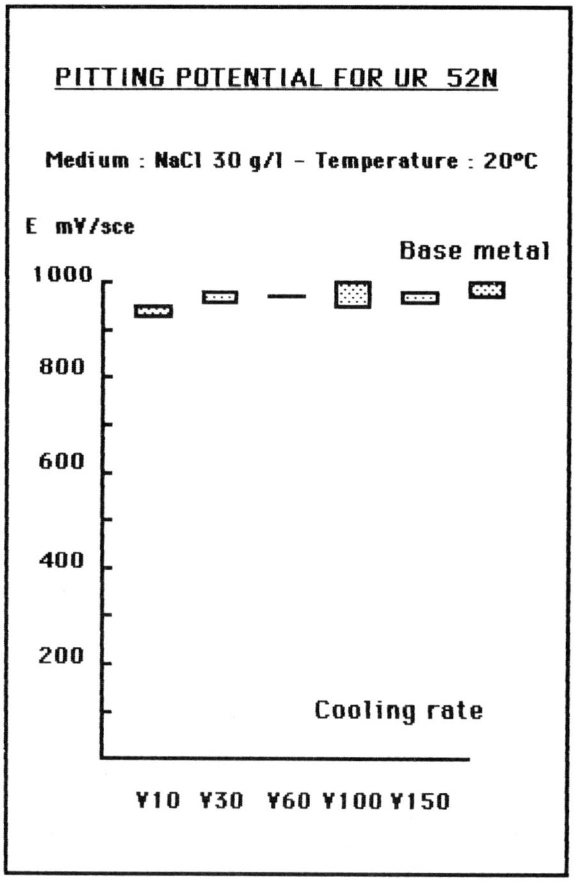

Figure 14 (b)

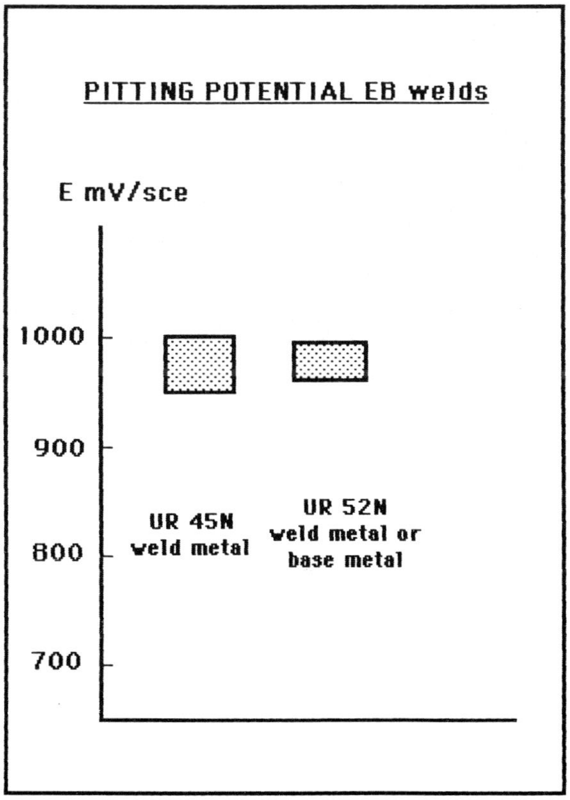

Figure 14 (c)

14 Pitting potential determined by polarization curves in a NaCl 30 g/l medium (HAZ and base metal):

a) UR 35N alloy compared to 304 and 316 alloys UR 35N SMA welds

b) UR 52N - SMA welds

c) UR 45N and UR 52N - EB welds

Role of secondary phase precipitation in intergranular corrosion of differently stabilized ELI ferritic stainless steels

M Barteri, S Fortunati, L Sassetti, L Scoppio and A Tamba

The work was carried out at the Centro Sviluppo Materiali SpA, Rome, Italy; SF is now with Acciaieria Foroni SpA, Gorla Minore, Varese, Italy.

SYNOPSIS

Critical time intervals and temperature ranges which render Ti, Nb and Ti+Nb stabilized steels of 18Cr2Mo and 21Cr3Mo grades sensitive to intergranular corrosion (IC) are identified and correlated with grain boundaries precipitates.

Time-temperature-IC diagrams are obtained and the role of the stabilizing elements is discussed.

The role of molybdenum-rich $M_{23}C_6$ microprecipitates and of the Laves phases in promoting intergranular corrosion susceptibility has been examined.

INTRODUCTION

Susceptibility of ferritic stainless steels to intergranular corrosion (IC) can be lessened by reducing the C+N interstitials content and by stabilizing the interstitials still present by appropriate alloying with Ti, Nb or Zr.

Nevertheless, fully stabilized extra-low interstitials (ELI) ferritic stainless steels, though immune to IC as detected by the Strauss test (ASTM A262-E), suffer severe IC in highly oxidizing media such as nitric acid (ASTM A262-C) when heated in the 500-800°C temperature range/1/.

There are interesting industrial prospects for Ti-and Nb-stabilized ELI ferritic stainless steels whose composition is Cr 18-21%, Mo 2-3%, C+N $\simeq$ 400 ppm, because of their immunity to stress corrosion and excellent resistance to localized corrosion in chloride media (pitting, crevice corrosion), as well as their high thermal conductivity and good hot formability/2-6/.

These steels are especially suitable for many uses, but their wide-scale application is limited by their low toughness at room temperature and low intergranular corrosion resistance in media with a high redox potential. This latter characteristic emerges as a result of their being held at certain critical temperatures during welding or particular heat treatments and the subsequent cooling rate.

The purpose of the work reported here was thus to identify the critical time intervals and temperature ranges which render Ti, Nb and Ti+Nb stabilized ELI steels sensitive to intergranular corrosion, and to correlate the sensitivity revealed by the corrosion tests (ASTM A262-C) with such secondary phases as may be present on the grain boundaries.

EXPERIMENTATION

Materials

The materials utilized were 18Cr2Mo and 21Cr3Mo ELI ferritic stainless steels in which the C and N interstitials were stabilized with Ti, Nb and Ti+Nb. A non-stabilized 18Cr2Mo steels was used for comparison. Table I lists the relevant chemical compositions.

All the steels had been cold rolled to a thickness of 1.5 mm and all had been subjected to solubilization heat treatment at 1050°C followed by water quenching.

Intergranular corrosion tests

The Huey test (ASTM A262-C) was used as the indirect method for obtaining quantitative measurement of the degree of heterogeneity created on the grain boundary as a result of isothermal heat treatments; the test was adopted because of the manner in which it selectively corrodes intermetallic phases (sigma, chi and Laves phases), Cr-and/or Mo-rich complex carbides ($M_{23}C_6$, M_3C, M_7C_3) or attacks the grain boundaries when these are affected by segregation.

The test consists of immersing the sample in boiling HNO_3 for five 48 hour periods and then calculating the corrosion rate after every period. The average corrosion rate after each (expressed in mm/month) provides precise information on the susceptibility of the grain boundary to corrode in oxidizing acid media.

In order to establish the time-temperature-intergranular corrosion (TTC) diagrams, the average intergranular corrosion rate threshold value (v_c) was fixed at 0.08 mm/month, it being considered that v_c values below this limit are typical of intergranular corrosion resistant steels.

Microstructure investigations

Electron microscope extraction replicas and thin films were observed. The extraction replicas were etched with a solution of Nital 2%. The thin films were prepared for chemical thinning in perchloric acid, followed by electrolytic thinning in ethylene glycol with 11% percloric acid solution. A JEOL 200CX scanning transmission electron microscope was employed (acceleration voltage 200 kV). This was equipped for microanalysis with a LINK 860 energy-dispersion X-Ray analyser (EDXA).

The phases were identified by electron diffraction on a selected area, completed by EDXA microanalysis on extraction replicas.

RESULTS

Intergranular corrosion

The TTC curves for the 18Cr2Mo and 21Cr3Mo steels are as follows: Non-stabilized 18Cr2Mo steel (Fig. 1); 18Cr2Mo steel stabilized with a) Ti, b) Nb and c) Ti+Nb (Fig. 2); 21Cr3Mo steel stabilized with a) Ti, b) Nb and c) Ti+Nb (Fig. 3).

The following information can be derived regarding the steels. The non-stabilized 18Cr2Mo steel becomes susceptible to intergranular corrosion very rapidly (t = 3s at T = 750°C), but resistance can again be restored by adopting a sufficiently long treatment time (Fig. 1). This fact is in line with the theory which postulated the formation of de-Cr zones near chromium-rich carbides, even with very short holding times. However, with longer treatment times there is steady recovery of intergranular corrosion resistance due to the increase in chromium around the precipitates as a result of diffusion.

The sensitization range of Ti-stabilized 18Cr2Mo steel is between 750 and 550°C with short holding times (t = 3s at T = 750°C) (Fig. 2a). Unlike what happens with 18Cr2Mo steel, lengthening the holding time does not promote recovery of intergranular corrosion resistance. The sensitization range of Nb-stabilized 18Cr2Mo steel occupies the same temperature interval but times are longer (t = 20s at T = 750°C) (Fig. 2b). The sensitization range of the Ti+Nb-stabilized steel extends up to 850°C even for very short holding times.

In the 21Cr3Mo steels, sensitization occurs at more or less the same temperatures as the 18Cr2Mo steels but with longer times (Figs 3a, 3b and 3c). The Nb-stabilized 21Cr3Mo steel (Fig. 3b) has the best intergranular corrosion resistance properties. In fact the nose position is at T = 700°C and t = 5 minutes.

Analysis of grain boundary precipitation

Non-stabilized steel

Examination of non-stabilized 18Cr2Mo steel reveals constant precipitation of $M_{23}C_6$ carbides fairly rich in chromium. Even after only five minutes heat treatment at T = 550°C quite fine precipitates (about 5 to 15 nm in diameter) start to decorate the boundaries (Fig. 4). The presence of carbides results in local de-Cr of the matrix near the grain boundaries and is thus the cause of the intergranular corrosion observed. An example of this de-Cr phenomenon is provided in Fig. 5. The chromium concentration profile was measured at right angles to the boundary where there is an intergranular precipitate 100 nm in diameter. The de-Cr zone extends for a width of about 0.5 mm and causes a decrease of over 2.5% in the chromium content in the area closest to the precipitates.

In actual fact, to explain the observed susceptibility to intergranular corrosion it would be necessary to assume de-Cr of such intensity as to lower the local concentration of chromium to below 12%. It must be assumed, therefore, that the size of the 12% Cr zone is of the order of magnitude of the spatial resolution of the EDXA microanalyser (30-50 nm).

The effect of chromium diffusion is evident in the sample treated at 550°C for a much longer time (t = 20 h) (Fig. 6); indeed, as can be seen from Fig. 1, the steel recovers its intergranular corrosion properties. There is fair growth of precipitates between 30 and 50 nm in diameter which decorate the entire boundary. These are $M_{23}C_6$ precipitates whose composition (Tab.II) corresponds fairly closely to the formula$(Cr_{14}Fe_{7.5}Mo_{1.5})C_6$. The chromium concentration profile near the intergranular precipitates indicates that the de-Cr condition has completely disappeared.

Titanium-stabilized steels

Observations made on the sample treated at T = 675°C for t = 30 minutes - a condition located very close to the nose of the TTC diagram - reveal widespread microprecipitation on the grain boundaries (Fig. 7) in the form of iron-chromium carbide particles of the $M_{23}C_6$ type measuring less than 5 nm.

Examination of various samples treated at lower temperatures or for shorter times reveals no trace of precipitation on the grain boundaries.

The extreme fineness of the precipitates and the ensuing difficulty in putting them into the Bragg condition owing to the low intensity of the electron beam diffracted by them may explain the apparent absence of intergranular precipitation on the boundaries of the nose of the Fig. 2a sensitization curve.

Niobium-stabilized steels

It is evident from the TTC diagram that there is a quite narrow sensitization zone between 800 and 600°C. Typical of the sensitization region in Fig. 8 is the sample treated for 30 minutes at 725°C; when subjected to the Huey test this fails after the first cycle due to intergranular corrosion. Examination under the STEM reveals abundant precipitation mainly of Type A_2B Laves phases at the grain boundaries. The quantity of this precipitation which is continuous along the whole boundary surface could well account for the poor results in the corrosion test, assuming that the phase of the type mentioned are themselves subject to attack in the Huey test.

Table II indicates the composition of the Laves phases as determined by averaging the values for nine particles. It can be ascertained that the A_2B formula typical of these phases is respected quite closely: in fact, considering that the chromium and the silicon can enter into these compounds together with the iron (A) and that B includes niobium, molybdenum and titanium, the formula obtained is of the type $(Fe, Cr, Si)_{2.1}(Nb, Mo, Ti)$.

Ti+Nb-stabilized steels

From the TTC diagrams it ensues that there is a sensitization zone between 850 and 550°C. The time needed to cause sensitization is quite small at the nose of the curve, situated around 750°C. Various samples in the 600-850°C temperature range have been examined. It ensues that:

- No precipitation is observed on the grain boundaries for holding times of 1 to 2 minutes at any of the temperatures examined: indeed, for T = 600°C no precipitation is observed even after t = 30 min, despite the fact that all the samples examined come from the IC sensitization zone.

- For longer times (t $\geqslant$ 30 min), carbide precipitation is seen on the quite thickly decorated boundaries. Laves phases were always present in all the samples examined.

- In the sample treated at T = 850°C, in addition to the Laves phases there is also a fair quantity of sigma phase present; in the others the Laves phase is sporadically accompanied only by M_6C carbides.

DISCUSSION

There is a marked difference in the intergranular corrosion resistance of the various steels examined.

Non-stabilized 18Cr2Mo steel is sensitized intensely as a result of short holds at temperature of T < 800°C, but recovers with long holding times. With the stabilized steels, sensitization increases with holding time. To be more precise, sensitization of the Nb-stabilized steels is the lowest of all, while that of the Ti-steels is comparable with that of the non-stabilized steel and there is the further disadvantage of lack of any recovery with longer holding times at the treatment temperature. The behaviour of the Ti+Nb-stabilized steels, instead, is somewhere between that of the other two, the nose being shifted quite definitely towards higher temperatures (T = 600 to 850°C). The markedly beneficial effect of a higher alloy content must be stressed: in fact, the shift in the start of sensitization condition towards much longer times for the same temperature is more noticeable in the case of 21Cr3Mo steels than for the similar 18Cr2Mo variety.

To sum up, the steel most resistant to intergranular corrosion is Nb-stabilized 21Cr3Mo steel which becomes sensitized only after heat treatments lasting more than three minutes at 700°C.

To explain these differences in behaviour, which depend on the type of stabilizing element, it must be considered that titanium and niobium added individually or together to 18Cr2Mo and 21Cr3Mo steels differ basically in their tendency to form second phases. In fact steels stabilized with titanium alone do not form intermetallic compounds at holding times of less than 1 h at the sensitization temperature, while niobium-stabilized steel exhibits closely-knit decoration of the grain boundaries with Laves phases after only thirty minutes at 700°C. The behaviour of Ti+Nb stabilized steel is similar, but a greater variety of second phase is to be seen: Laves, sigma, and carbides, as well as chi phases, too, sometimes.

The favourable effect of the addition of Ti and Nb on the thermodynamic stability of the Laves phase is the result of atom size and electron concentration. As regards this latter point, it must be stressed that an average concentration of d and s electrons of less than

8 per atom favours the formation of the Laves phase/7/. As titanium has only four s+d electrons, niobium 5 and molybdenum 6, these elements help reduce the average electron concentration, thus favouring the formation of Laves phases.

Redmond/8,9/ has plotted TTP diagrams for 18Cr steels containing molybdenum in quantities ranging from 0 to 5%. The noses for the Laves, chi and sigma phases appear more or less clearly separated. The sigma especially tends to form at lower temperatures (T = 550-650°C in the case of 18Cr2Mo steel). Also according to Redmond, molybdenum accelerates the precipitation processes of all the intermetallic phases.

During the research reported here no detailed study was made of sigma and chi phase precipitatations which are not responsible for the sensitization phenomena, but observation of the sigma phase (at T = 850°C) in the Ti+Nb-stabilized steel is in contrast with Redmonds findings.

In point of fact, in Nb+Ti-stabilized 25Cr3Mo4Ni steels, Brown et al/7/ observed almost simultaneous precipitation of Laves, chi and sigma phases in a temperature range of 700 h to 950°C with the maximum precipitation rate at around T = 800°C (start of precipitation for t = 0.5-1 h).

These data are in good agreement with what is reported here, at least as far as the formation of Laves phases is concerned, considering that the greater alloying of the steels studied by Brown causes a shift in the TTP curves towards higher temperatures.

Moreover, the work performed by Sawatani et al./10/ on 19Cr2Mo steels stabilized with Ti or Nb indicates maximum precipitation of Laves phases around 700°C.

A recent paper on Ti-stabilized 22Cr3Mo2.5Ni stees/11/ reports that for treatment times of t $\geqslant$ 1h at various temperatures, the precipitated phases are Laves phases at T = 650-750°C, while at higher temperatures (T = 750-900°C) they are sigma or chi.

A critical point that no one appears to have cleared up, however, is the disagreement between susceptibility to intergranular corrosion for short sensitization times and the apparent absence of precipitation on the grain boundaries.

The assumption that some "invisible" phase plays a role in the intergranular corrosion of Ti-stabilized ferritic stainless steels has also been put forward by Steigerwald/12,13/.

Lizlov & Bond have postulated the formation of intergranular carbide "precursors" to interpret the behaviour of 18Cr2Mo steels containing titanium/14/. However, there are shortcomings in this reasoning too, because they observe that intergranular attack in Streicher solution (ASTM A262-B) after treatments at 620°C is more severe for long holding times, up to 64 h. If it were only local de-Cr that was responsible for susceptibility to intergranular corrosion, then diffusion of matrix chromium should be capable of re-establishing the average chromium level for such long times, thus reducing the corrosion rate. This kind of recovery in resistance to intergranular corrosion, after long treatment times, was clearly apparent, in fact, in the non-stabilized 18Cr2Mo steel, where the de-Cr of areas next to the grain boundary, caused by the precipitation of chromium-rich $M_{23}C_6$ carbides (chromium concentration 60%) is the only mechanism that can adequately explain the sensitization.

The documentation on the formation of extensive microprecipitation of $M_{23}C_6$ carbides on the boundaries of the grains even in Ti-stabilized steels, permits a possible explanation to be put forward for the sensitization of stabilized steels.

For holding times that are short in relation to the treatment temperature, the formation of $M_{23}C_6$ chromium carbides and the consequent local de-Cr in the vicinity thereof, is sufficient to explain the susceptibility to corrosion encountered in the nitric acid tests. Subsequently there is massive precipitation on the grain boundaries of corrosion-susceptible Laves phase particles/15,16/. The presence of such extensive precipitation could account for the susceptibility to corrosion at long treatment times. It is likely that the presence of $M_{23}C_6$ carbides quite rich in molybdenum as well as chromium (Mo = 12%) favours the subsequent formation of Laves phases in which the molybdenum content exceeds 25%.

The hypothesis postulated here could explain the intergranular corrosion shown in Figs 1 to 3 and is in good agreement with the numerous data and observations reported by other workers regarding the behaviour of steels subjected to long sensitization treatments.

The very fine nature of the $M_{23}C_6$ precipitates and the difficulty of detecting these in treated samples near the noses of the TTC curves may well explain why so little mention is made of these in published research, especially since there are no reports which indicate the performance of particularly painstaking electron microscope analyses, though the papers generally refer to situations involving holding times of the order of a few hours at sensitization temperature.

To sum up, according to the proposed model, sensitization occurs owing to poor stabilization, since despite the fact that compared with carbon and nitrogen the titanium concentration is more than three times stoichiometric, there is still a sufficient quantity of free carbon which gives rise to the precipitation of chromium-rich $M_{23}C_6$ carbides.

This phenomenon does not occur, however, with niobium stabilization. In fact in 18Cr2MoNb steel, where compared with the interstitials the niobium concentration is only 1.5 times higher than stoichiometric, no sensitization occurs at the shorter treatment times. It would thus appear that the greater affinity of niobium for carbon ensures such a marked reduction in the concentration of carbon in solid solution at the sensitization temperature that the formation of chromium carbides is prevented. Viewed in this light it is also easier to understand the superior performance of 21Cr3MoNb steel. In fact the greater chromium alloying renders this steel less subject to de-Cr beneath the Cr = 13% threshold. According to the hypothesis which ascribes sensitization solely to the formation of intermetallic phases which are anyway susceptible to corrosion, a steel with a higher chromium and molybdenum content should be very subject to the formation of intermetallic compounds (e.g. Laves, sigma or chi phases) and is therefore more markedly inclined to sensitization.

CONCLUSIONS

The following conclusions can be drawn from the work done:

1) The choice of stabilizer in ELI ferritic stainless steels can have a great influence on intergranular corrosion resistance. More precisely, the addition of niobium guarantees a minimum tendency towards sensitization, while titanium steels may be subject to intergranular corrosion at much shorter holding times in a temperature range extending from 750 to 450°C. The behaviour of steel with mixed Nb+Ti stabilization is fairly similar to that of titanium-stabilized steel, with a slight upward shift in the sensitization temperature.

   For a given stabilizer the alloy content appears to have a marked effect. For instance the TTC diagrams of 21Cr3Mo steel are similar to those of 18Cr2Mo steel as regards temperature but much longer times are necessary for sensitization to occur.

2) Intergranular corrosion sensitization of the non-stabilized steel occurs at temperatures between 750 and 550°C. Sensitization is a reversible phenomenon bound up with the treatment time: as this rises above a given value, which depends on temperature (the lower the temperature the longer the time), the steel again becomes immune to intergranular corrosion. This behaviour is due to the general intergranular precipitation of Cr-rich $M_{23}C_6$ carbides which, right from the initial stage of precipitation, results in marked de-Cr near the boundaries, making them susceptible to corrosion in environments with a high redox potential. This situation remains until the amount of de-Cr falls off, owing to the diffusion effect.

3) The addition of stabilizers does not always overcome this phenomenon, even when the additions, compared with the carbon and nitrogen content, are superstoichiometric. In fact, while the addition of niobium ensures that there is no formation of intergranular $M_{23}C_6$ carbides and guarantees that the steels have optimum intergranular corrosion resistance, titanium stabilization does not significantly reduce susceptibility to intergranular corrosion. This behaviour is the result of the rapid precipitation of intergranular $M_{23}C_6$ microcarbides which form despite the addition of Ti in quantities that are decidedly greater than stoichiometric for the formation of Ti(C,N) carbonitrides.

4) In stabilized steels titanium and niobium favour rapid formation of intermetallic phases, especially Laves phases of the $(Fe,Cr,Si)_2(Mo,Nb,Ti)$ type. This phenomenon causes another sensitization mechanism which occurs when intergranular precipitation of the intermetallic phases is sufficiently abundant to form virtually continuous decoration of the boundaries. In that case intergranular corrosion is the result of lower stability compared with the matrix in strongly oxidizing media. Intergranular precipitation of Laves phases may be favoured by the pre-existence of Mo-rich $M_{23}C_6$ microprecipitates and this could well explain the susceptibility to intergranular corrosion that occurs with long treatment times.

REFERENCES

/1/ R.J. HODGES - Corrosion, 27, 119 (1971).

/2/ S.FORTUNATI, L.SASSETTI, A.TAMBA - Arch. Eisenhuttenwes., 54, 323 (1983).

/3/ C. ÅSLUND - Proceed. of "Stainless Steel 77" Conf., London 26-28 Sept. 1977, 173.

/4/ H.J. DUNDAS, A.P. BOND - NACE CORROSION/82, paper no. 192 (1982).

/5/ K. FÄSSLER, H. SPÄHN - Anticorrosion, 29 (8), 4 (1982).

/6/ K. YOSHIOKA et al. - Kawasaki Steel Giho 17, 249 (1985).

/7/ E.L. BROWN, M.E. BURNETT, P.T. PURTSCHER, G. KRAUSS - Metall. Trans. 14A, 791 (1983).

/8/ J.D. REDMOND - MICON 78 Conf., Houston, April 1978.

/9/ J.D REDMOND, P.J. GROBNER, V. BISS - J. Metals, 33, 19 (1981).

/10/ T. SAWATANI, S. MINAMINO, H. MORIKAWA - Trans. ISIJ, 22, 172 (1982).

/11/ G. RONDELLI, B. VICENTINI, M. MALDINI - La Metallurgia Italiana, 77, 857 (1985).

/12/ R.F. STEIGERWALD, H.J. DUNDAS, J.D. REDMOND, R.M. DAVIDSON - Proceed. of "Stainless Steel 77" Conf., London 26-28 Sept. 1977, 57.

/13/ R.F. STEIGERWALD, A.P. BOND, H.J. DUNDAS, E.A. LIZLOVS - Corrosion, 33, 279 (1977).

/14/ E.A. LIZLOVS, A.P. BOND - J. Electrochem. Soc., 122, 589 (1975).

/15/ R.F. STEIGERWALD - Corrosion, 33, 339 (1977).

/16/ H.BRANDIS, H.KIESHEYER - Proceed. of "Stainless Steel 84" Conf., Göteborg, 3-5 September 1984, 217.

TABLE I - Chemical composition (% wt)

	C	N	S	P	Cr	Mo	Mn	Si	Ti	Nb
18-2	0.0098	0.0090	0.007	n.d.	18.70	2.50	0.20	0.39	---	--
18-2 Ti	0.0230	0.0170	0.020	0.020	17.90	2.00	0.20	0.52	0.47	--
18-2 Nb	0.0230	0.0220	0.007	0.026	18.47	2.11	0.29	0.36	0.004	0.50
18-2 NbTi	0.0180	0.0250	0.003	0.026	18.10	2.08	0.17	0.53	0.23	0.21
21-3 Ti	0.0170	0.0126	0.008	0.013	19.60	3.02	0.42	0.44	0.65	--
21-3 Nb	0.0180	0.0165	0.009	n.d.	21.45	3.11	0.21	0.41	--	0.29
21-3 NbTi	0.0180	0.0160	0.008	0.015	21.20	3.10	0.20	0.38	0.14	0.18

TABLE II - Composition of $M_{23}C_6$ carbides and of M_2B Laves phases: average values and 95% confidence interval

	Si	Cr	Fe	Nb	Mo
Carbides	-	56.0 ± 0.8	31.8 ± 1.1	-	12.2 ± 1.3
Laves phases	8.2 ± 2.1	12.3 ± 0.5	47.4 ± 1.7	13.5 ± 0.9	18.4 ± 1.3

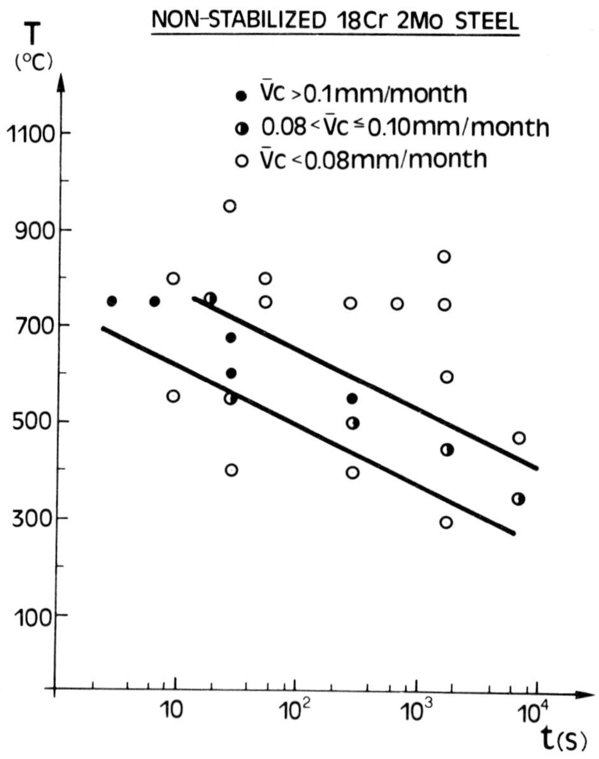

1) Non-stabilized 18Cr2Mo steel. TTC curve.

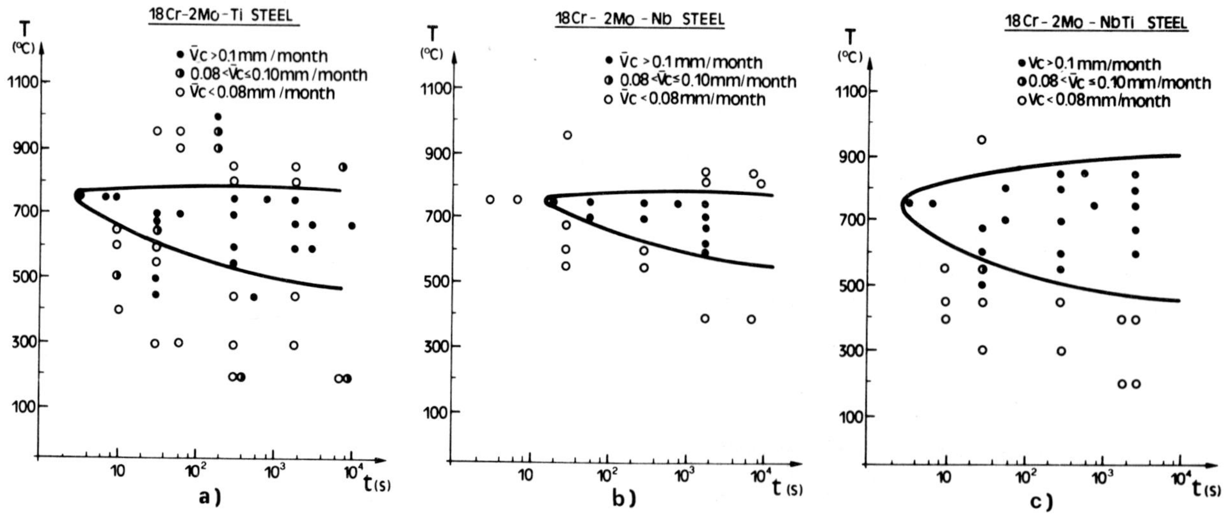

2) 18Cr2Mo steel. TTC curve.
a) Ti-stabilized
b) Nb-stabilized
c) Ti+Nb-stabilized

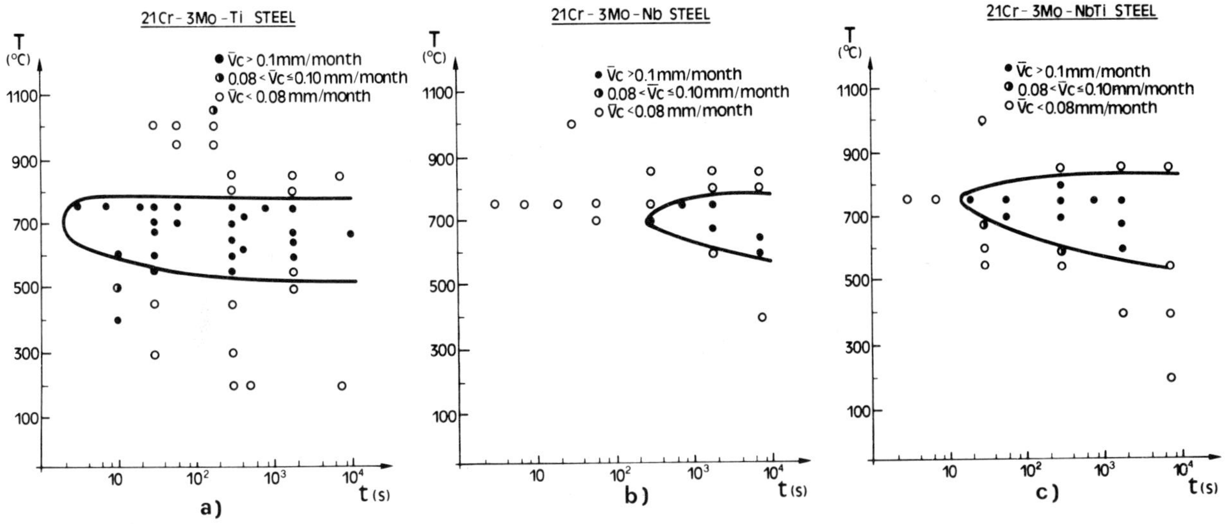

3) 21Cr3Mo steel. TTC curve.
a) Ti-stabilized
b) Nb-stabilized
c) Ti+Nb-stabilized

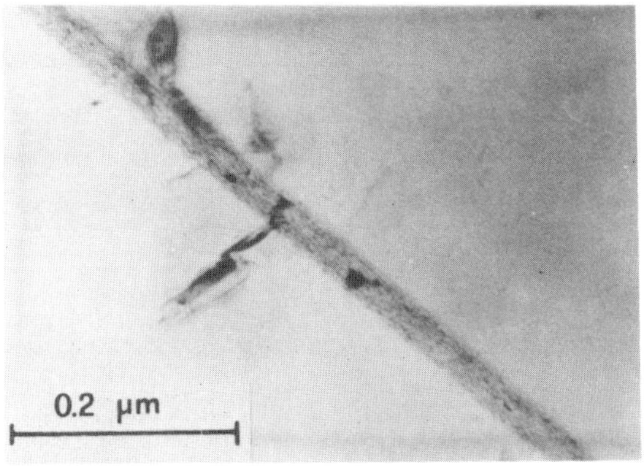

a

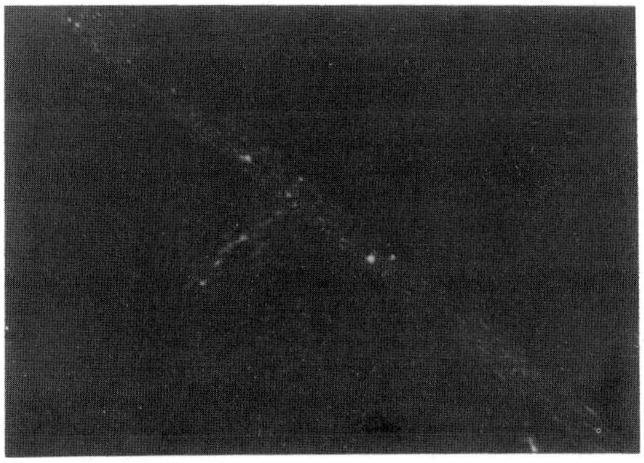

b

4) Non-stabilized 18Cr2Mo steel.
Heat treatment: t=5 min; T=550°C.
Fine $M_{23}C_6$ carbide precipitates on grain boundaries.
a) bright field;
b) dark field

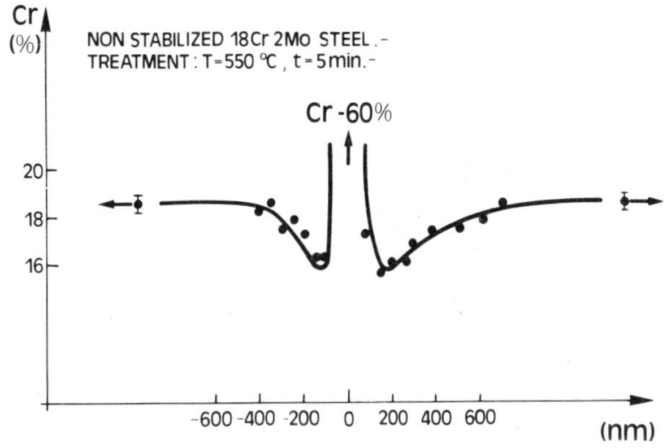

5) Chromium depletion near an $M_{23}C_6$ precipitate (≃ 100 nm)

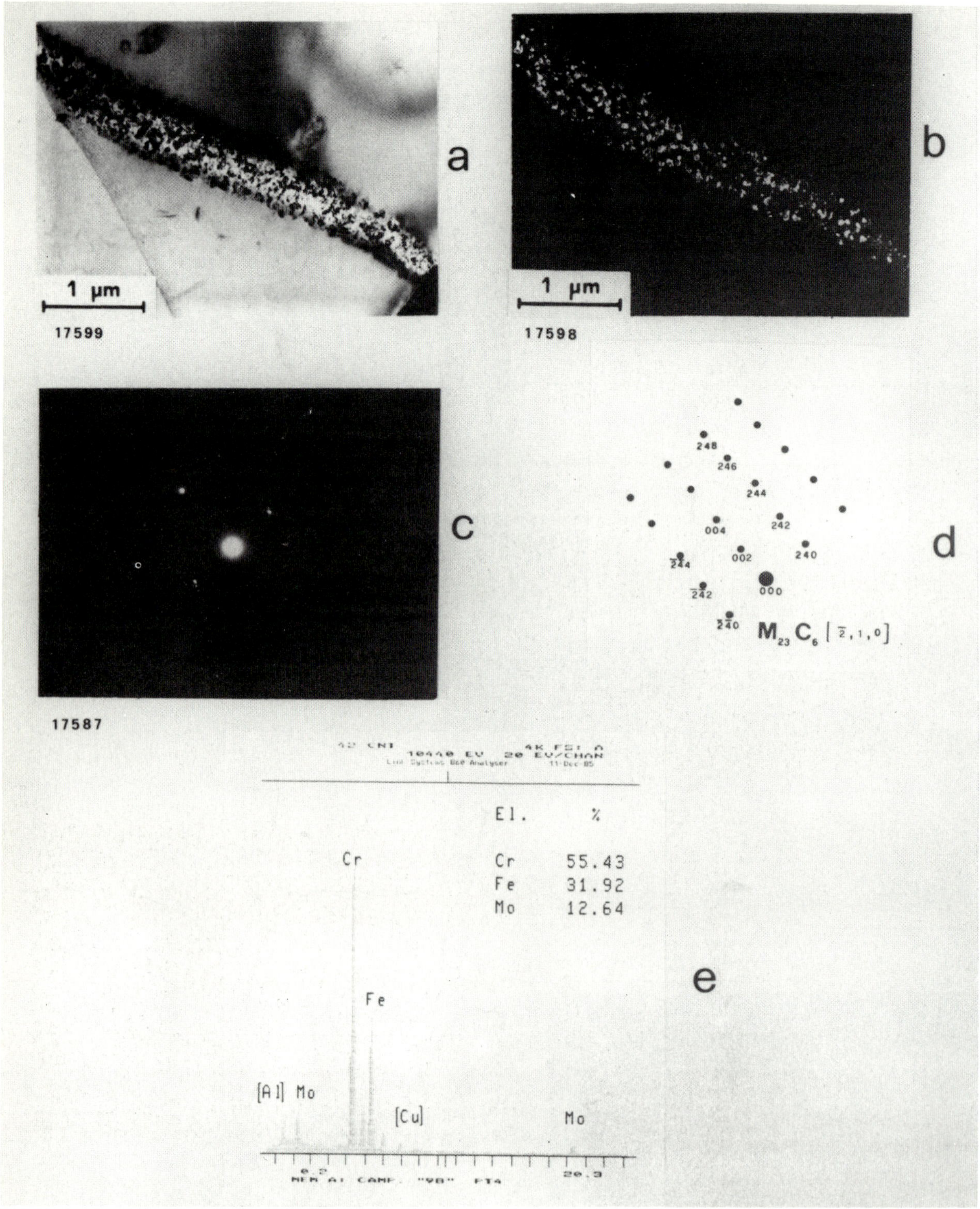

6) Non-stabilized 18Cr2Mo steel.
Heat treatment: t=20 h; T=550°C.
a) bright field
b) dark field
c) diffraction diagram
d) interpretative scheme
e) EDXA microanalysis of replica extracted precipitates
The peaks for elements Al and Cu come from the replica support grid and the anticontaminator respectively.

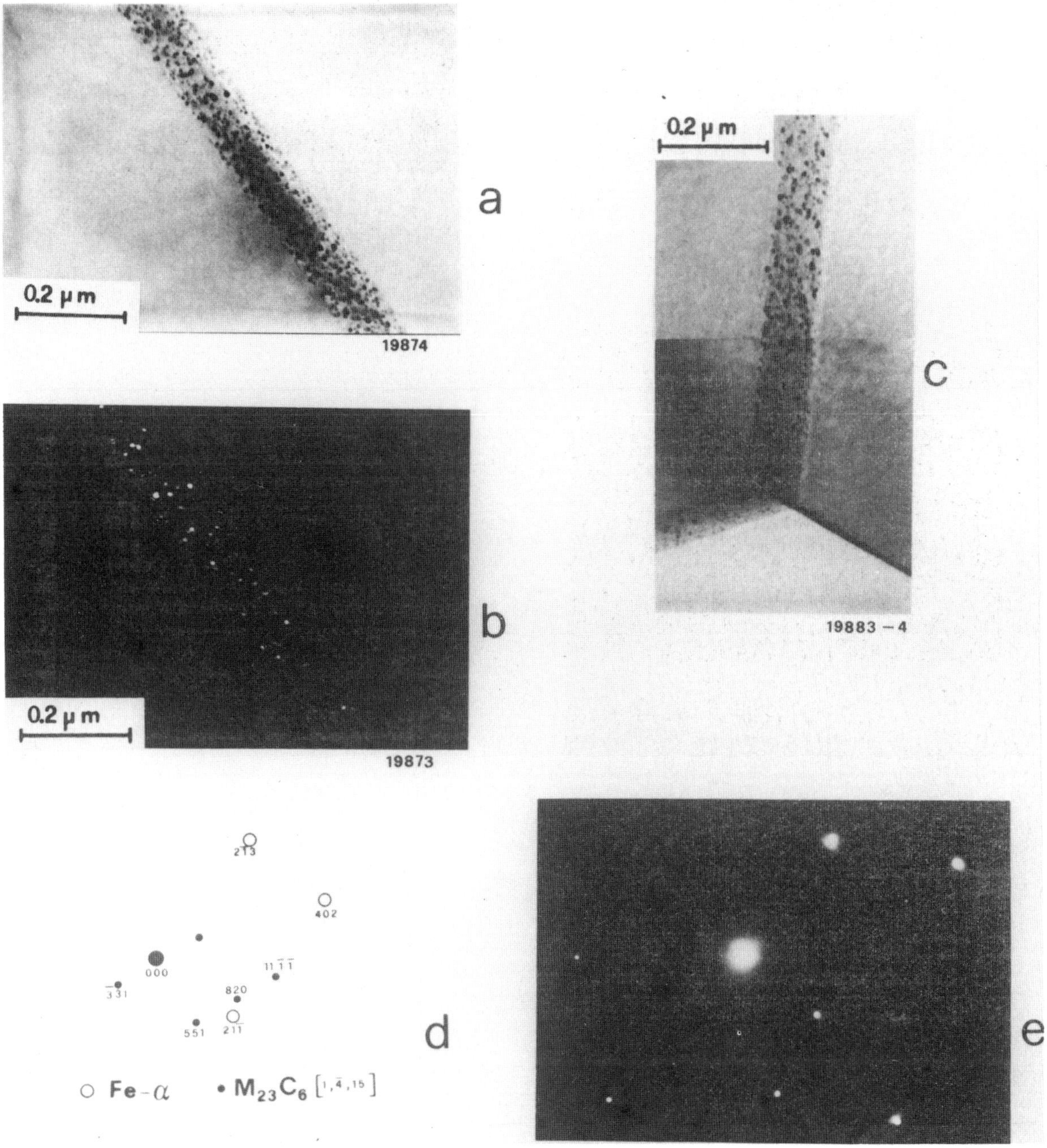

7) Ti-stabilized 18Cr2Mo steel.
Heat treatment: t=30min; T=675°C.
a) $M_{23}C_6$ precipitation on grain boundaries
b) dark field
c) bright field
d) interpretative scheme
e) diffraction diagram.

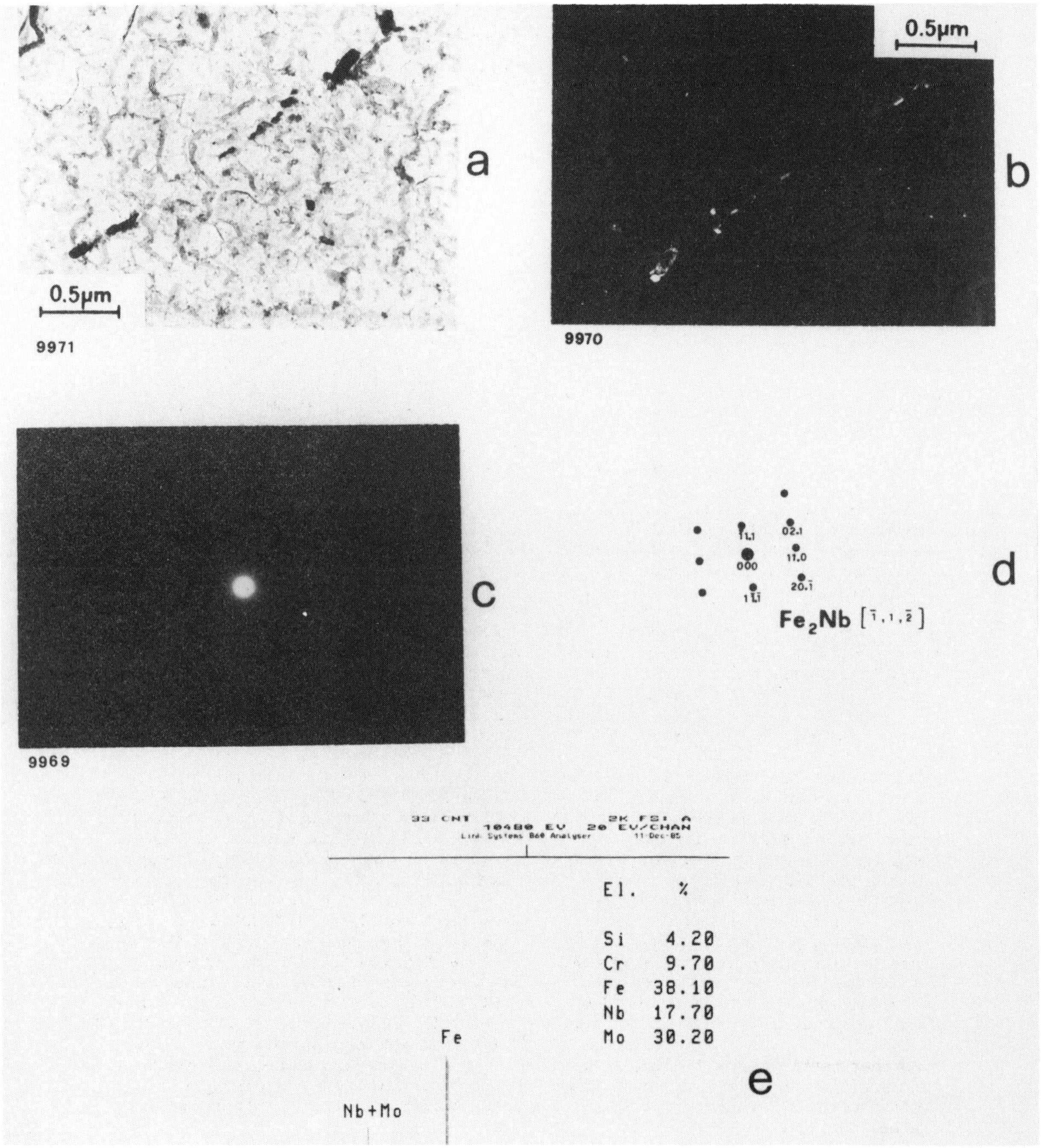

8) Nb-stabilized 18Cr2Mo steel.
Heat treatment: t=30min; T=725°C
Laves-phases precipitation on grain boundaries
a) bright field
b) dark field
c) diffraction diagram
d) interpretative scheme
e) EDXA microanalysis.

Post-fabrication painting systems for stainless steels and their performance in marine atmospheres

K Johnson and P G Stone

KJ: BSC Swinden Laboratories
PGS: BSC Stainless

SYNOPSIS

A range of proprietary primers which exhibit good adhesion to properly prepared mild steel surfaces have been tested on Type 430 and 304 stainless steels in various surface finishes. Both accelerated and natural weathering tests have shown that some of these primers are quite unsuitable for stainless steels. However, others give extremely good results and are suitable for this use.

Unfortunately, the comparison of these results with similar tests carried out elsewhere indicates that, at the present time, it is not possible to make recommendations based upon the usual generic descriptions of primer-type. Therefore, until there is a more general understanding of the reasons for these differences, each proprietary primer, of specific composition, will have to be assessed individually. However, accelerated testing appears to be adequate for this purpose.

Further testing is to be carried out to identify the basic causes of loss of adhesion of primers on stainless steel surfaces.

INTRODUCTION

Stainless steel has historically been mainly used for its corrosion resistance, hygenic or aesthetic properties, but its potential as a structural engineering material is becoming increasingly appreciated. This is particularly evident in applications which call for long term structural integrity in aggressive environments, as illustrated by its use for topside architectural components on offshore oil and gas platforms. The background to this usage was described in a paper presented to Stainless Steel '84 (1) where it was mentioned that in response to a demand for colour, as well as structural integrity, a research programme was being initiated on the weathering performance of painted stainless steel.

Obviously stainless steels do not require painting for corrosion protection, and any suggestion to paint them is alien to many people. However, colour can sometimes be required in the offshore market for identification reasons, or to comply with an oil company's house style. There is also an additional demand in the onshore market for coloured stainless steel for purely aesthetic reasons. For instance, the use of coil painted stainess steel roofing in Japan is a good example of stainless steel meeting the aesthetic, low maintenance and long life requirements of Japanese planning authorities, architects and property owners. Some of the Japanese work has been reported (2, 3), but nothing appears to have been published on the post fabrication painting of stainless steel. Never the less, it is known that techniques have been developed to paint both road and rail vehicles in the United States, and elsewhere. However, in some instances, it appears that colour has been achieved by a laminated film technique rather than painting. (4)

A common feature of these applications is that stainless steel was chosen for life cycle cost reasons. Maintenance costs obviously play a crucial role in this type of analysis (5) and it is vital that any paint system adopted should perform well. This is particularly important in oil and gas platform applications in view of the high cost of offshore painting (6). A programme of work was therefore undertaken to examine the factors influencing the weathering performance of painted stainless steels in marine atmospheres.

STAINLESS STEELS.

Before describing this work, it is perhaps useful to consider some major aspects associated with the painting of stainless steel.

Steel Types

For the purpose of this paper, it is proposed to concentrate on the more common types used for architectural components.

Offshore, Type 316 is usually used for unpainted stainless steel architectural applications. However, one platform has used Type 304 successfully for nearly fifteen years (7). On the other hand, a steel with a higher molybdenum content than Type 316 has been adopted for a more northern platform (8). Onshore, Type 304 has been predominantly selected for architectural applications, except for coastal locations where Type 316 is specified in order to give increased resistance to pitting corrosion.

It appears that the additional protection given by painting has been utilised to reduce the initial cost of material. Certainly most onshore painted stainless steel applications have used a Type 304 substrate, rather than Type 316. Trials are also being undertaken with a ferritic material, Type 430 (9) in order to determine whether a cheaper overall system could be used.

Surface Variation

An important aspect which must be considered in relation to the "paintability" of stainless steel is the range of surface finish which can be encountered. In general, there are two broad groups of finishes.

(a) Mill Finishes

As the name implies, these finishes are produced as a result of the steelmaker's production route.

The British Standard classification (10), and its relationship to surface roughness are given in Table I.

In practice, most stainless steel fabrications are produced from cold rolled material in the 2B condition with perhaps some heavier sections, eg angles, being produced from hot rolled material in the No. 1 condition. Sometimes, however, it may be required to paint the much smoother, bright annealed material. The development of these surfaces during processing, and further details of their topography have been given elsewhere (11). The important feature to note is the marked differences in roughness.

(b) Secondary Finishes

Most secondary finishes involve mechanically deforming the mill surface and, on economic grounds, it is not usual to use this type of finish as a substrate. However, another form of local deformation, namely blasting, is one of the accepted methods to remove remnant slag and heat tinting from welded areas of a fabrication. In addition, experience has shown that paint manufacturers sometimes suggest that the complete surface of a stainless steel fabrication should be blast-cleaned before painting. This obviously increases the effective surface area of the component; a practice widely used in the post fabrication painting of mild steel.

It was therefore felt useful also to examine the effect of blast cleaning systems when considering the experimental design.

THE ADHESION OF PAINTS TO METAL SURFACES

The prime requirement of a paint film is that it must adhere to its substrate throughout the useful life of the coating system.

The predominant bonding forces in the adhesion of surface coatings to metals are normally attributed to secondary or Van der Vaal's forces. These are weak attractive forces which arise from either permanent or temporary dipoles in the organic compounds. The forces are essentially short range in nature. It follows that, for them to be operative, the substrate metal must be free of surface contamination and furthermore, the applied coating should readily wet the substrate surface. Because of this, when paints are to be applied to mild steel, the component is invariably subjected to some form of surface preparation. This will involve degreasing, removal of surface oxides and occasionally some form of chemical modification eg phosphating.

The problems arising from the painting of steels are essentially concerned with adhesion between the organic paint film and the steel substrate. In stainless steel, this problem is accentuated by the presence of the passive oxide film. As with many metals and alloys, stainless steels owe their excellent corrosion properties not to the inherent corrosion resistance of the metal itself, but to the properties of the oxide on its surface. This oxide film is very thin; about 30-50 A°, i.e. 15-25 atomic layers. It consists essentially of oxy-hydroxides of iron and chromium and is normally formed in air during the steel production process.

In service, the passive film protects the steel surface from most corrosion environments. However, this is not a static process. If the oxide layer is damaged, further oxidation of the exposed metal will occur, to heal the passive film. If the steel is exposed to a corrosive environment, oxide dissolution occurs. In oxidising environments, further oxidation of the metal then occurs to repair the protective film.

EXPERIMENTAL DESIGN

Paint Systems and Materials

Possibly the first response of a paint manufacturer who is asked for a painting system for use on stainless steel would be to consider the use of a two-pack etch primer. Unfortunately, BSC experience shows that two-pack etch primers do not adhere well on stainless steels. This is, perhaps, not surprising since stainless steels are very resistant to phosphoric acid, the etchant on which these primers are based. Paint systems applied on top of etch-primers eventually show a loss of adhesion on most stainless steel surfaces, Fig 1.

An experimental design was therefore adopted to evaluate primer systems, using a two-pack etch primer as a reference. The design consisted of:-

* A two-pack, polyvinyl butyral etch primer.
* A zinc phosphate, vinyl copolymer primer.
* A zinc phosphate, vinyl copolymer primer, but with an added organo-silane adhesion promoter.
* A two-pack polyamide-cured epoxy primer.
* A two-pack polyamide-cured epoxy primer, but with added organo-silane adhesion promoter.
* A maleic acid modified vinyl copolymer primer.
* A moisture cured urethane primer.
* An alkyd/vinyl primer.

These primers were selected for evaluation largely on the basis of the excellent adhesion they exhibit on properly prepared mild-steel surfaces.

Organo-silanes of the general formula $R^1\text{-}Si(OR^2)_3$ have been examined by several research establishments and found to be beneficial on surfaces where it is difficult to maintain good paint adhesion, such as copper and zinc. It was considered that their use on stainless steel surfaces could be beneficial.

All of the primers were selected for evaluation in conjunction with a paint manufacturer. They were therefore either proprietary products or modifications of proprietary products.

The primers were applied to two grades of stainless steel:-

* Type 430 (17% Cr)
* Type 304 (18% Cr 10% Ni)

Four surface finishes were tested:-

* Bright Annealed
* 2B, (cold rolled, softened, descaled and planished).
* No 4, (polished).
* Blast cleaned, using 2 different abrasives, ie: fine and coarse copper slags.

Each primer was overcoated with two finishes:-

* Alkyd finish.
* Vinyl finish.

Test Methods

Large composite test panels were initially subjected to an accelerated corrosion test. This was a cyclical salt spray test which involved spraying for one hour with a test solution containing 0.5g/l sodium chloride and 3.5g/l ammonium sulphate at a temperature of 35°C. This was followed by a period of two hours during which the test chamber was exhaust ventilated and the panels were allowed to dry. The three hour cycle was then repeated for a total of 3,000 hours when cross-hatch adhesion tests (BS 3900 Part E6) were carried out to assess the adhesion of the paint finishes to the stainless substrates.

Subsequently, the same combination of primers, stainless steels, and surface finishes were subjected to three years atmospheric exposure testing in a severe marine environment on a beach at Rye on the South Coast of England (Fig 2). The individual test panels were cross-scribed before exposure and adhesion was assessed periodically using the cross-cut test described in ASTM D3359, Method 'A'. This test was considered to be more realistic in terms of adhesion-demand during service than the cross-hatch tests used to assess the accelerated tests.

RESULTS

The results obtained from the cyclical salt-spray accelerated tests indicated that:-

i) All the stainless steel surfaces tested could be primed to give satisfactory adhesion with either the polyamide-cured epoxy primer or the alkyd/vinyl primer. Fig 3 shows the good cross-hatch adhesion results obtained on both of these primers on bright annealed 430.

ii) None of the other primers tested gave satisfactory adhesion and in particular, the moisture-cured urethane and maleic-modified vinyl primers showed a total loss of adhesion.

iii) There were no significant differences between the performance of painted ferritic and austenitic steels.

iv) The addition of an organo-silane to two of the primers produced no improvement in adhesion.

v) Consideration of the adhesion classification ratings obtained from the cross-hatch test (as described in BS 3900, E6) indicated that, in general, adhesion improved with increasing surface roughness (Fig 4).

vi) Blast cleaning improved the adhesion of most of the primers. The test panel which was blast-cleaned with a fine copper-slag expendable abrasive produced a good result but exhibited localised rust-spotting on some of the primed-only areas. (Fig 5) In the case where a coarse copper-slag abrasive was used, extensive rusting of the painted 430 surface occurred (Fig 6).

The results obtained after three years atmospheric exposure testing in a severe marine environment indicated that:-

i) Only the polyamide-cured epoxy primer and the alkyd/vinyl primer showed good adhesion to all the stainless steel substrates. Fig. 7 illustrates the good adhesion obtained with these primers in 430 BA, 430 2B and 304 2B.

ii) Two primers, the moisture-cured urethane and the maleic-modified vinyl exhibited loss of adhesion and stripping from the cross-cut after only 6 months (Fig 8).

iii) In general, the results obtained from atmospheric testing were in good agreement with those obtained from the accelerated tests.

DISCUSSION

The experience of the paint industry over many years is that etch primers, vinyl primers, polyamide-epoxy primers, urethane primers, and alkyd/vinyl primers all show excellent long-term adhesion when applied to properly prepared mild steel surfaces. Both accelerated and natural weathering tests have now shown that some of these primers do not adhere to many stainless steel surfaces.

At the present moment, there is no understanding of why some primers adhere to stainless steel and some do not. Indeed there is some indication that the properties required of primers for stainless steel may be quite different to those required for use on mild steel. For example, in the case of the latter, low oxygen and water permeability are required to provide maximum corrosion protection. However, in the case of stainless steels, high oxygen and water permeability may be advantageous in order to facilitate repair of the passive oxide film under the paint coating.

Analysis of the test data indicates that, in general, the adhesion of primers to stainless steel substrates improves as the surface roughness increases as shown in Fig. 4. Of the standard mill finishes that are available, Table 1, it can therefore be assumed that bright annealed surfaces will be the most difficult to prime and that coil and sheet produced in 2B and 2D and plates in No. 1 finish will be easier.

Blast-cleaning improves the adhesion of primers to stainless steel, but, in practice, the associated extra cost must be taken into account, particularly as the selection of a suitable primer may make this preparatory work unnecessary. In addition, blast-cleaning also produces crevices on the steel surface which can provide sites for localised corrosion. This effect was particularly apparent when a coarse irregular abrasive was used, which produced a peak-valley surface profile of 75-100 microns. This surface exhibited heavy rusting, even on those areas treated with both a primer and top coat. This point may have relevance not only in the specification of original surface finishes but also with respect to the treatment of site welds.

As mentioned earlier, welded areas are normally cleaned to remove flux residues, weld spatter and localised surface colouring. Methods used include acid pickling pastes, glass-bead blast-cleaning, and the use of needle-guns. In the absence of tests, it is suggested that the first two methods would probably be satisfactory before painting but that the surface produced by needle-gunning may be too coarse.

Both accelerated and atmospheric exposure testing have demonstrated that, of the primers tested, the polyamide-cured epoxy and alkyd/vinyl primers can achieve satisfactory long term adhesion on all the surface finishes in which the common ferritic and austenitic stainless steels are supplied.

It must be emphasised that this conclusion relates entirely to the specific proprietary products which were tested. For each of these generic types, there is scope for formulation variations which could have either beneficial or adverse effects on adhesion to stainless steel. To illustrate this point, Fig 9 shows the results of tests carried out on a polyamide-cured epoxy primer which originated from another paint manufacturer, evaluated elsewhere but using a similar cyclical salt spray method. The primer, which was applied to a 430 2B surface showed an early and total loss of adhesion.

Clearly, until there is a more general understanding of why some primers adhere to stainless steels and others do not, it will not be possible to recommend generic types of primer.

Consequently, it is currently necessary for each proprietary primer to be evaluated individually. Fortunately, there appears to be excellent correlation between the accelerated test method used in this work and the results of natural weathering so that such evaluations can be carried out rapidly.

ACKNOWLEDGEMENT

The authors would like to thank Mr. A. P. Pedder, Chief Executive BSC Stainless and Dr. R. Baker, Director of Research - BSC for approval to publish this paper.

REFERENCES:

1. P G Stone, R Hudson, D R Johns, Performance of Stainless Steel in Topside Constructions Offshore, Stainless Steels '84. The Metals Society 1985, p 478.

2. R. Kato et al. Development of Painted Stainless Steel Sheet. Nisshin Steel Tech. Rep. Dec. 1983.

3. R. Kato et al. Development of Fluoro-Carbon Resin Paint Coated Stainless Steel Sheet. Nisshin Steel Tech. Rep. Dec. 1984.

4. Private communication.

5. R Kiessling - Thirty First Hatfield Memorial Lecture, Sheffield 1983.

6. R. A. E. Hooper. Cost Effective Use of Stainless Steel in Topside Module Cladding and Structural Applications. Conference on Corrosion and Marine Growth on Offshore Structures. Society of Chemical Industry. Aberdeen 1982.

7. Private communication.

8. North Sea Observer - 27th May 1983.

9. 'Stainless' 19. BSC Stainless, 1984.

10. BSC 1449 part 2.

11. P G Stone, C V Honess - Metallurgical interpretation of Surface Requirements in the Cold Rolled Stainless Steel Market. Advances in Cold Rolling Technology. Institute of Metals Conference 1985.

TABLE 1 **SURFACE FINISH INFORMATION**

FINISHES	DESCRIPTION	SURFACE ROUGHNESS TRANSVERSE CLA um RANGE
MILL		
1	Hot Rolled Softened & Descaled	4.0 - 7.0
2D	Cold Rolled Softened & Descaled	0.4 - 1.0
2B	CRS & D followed by Planish	0.1 - 0.5
2A	Cold Rolled and Bright Annealed	0.05 - 0.1
SECONDARY		
4	Polished with Fine Grit	0.2 - 1.5
7	Bright Buffed	0.1
8	Bright Polished	0.2

Fig. 1. Loss of adhesion of etch primer on stainless steel.

Fig. 2. Exposure tests at marine site at Rye.

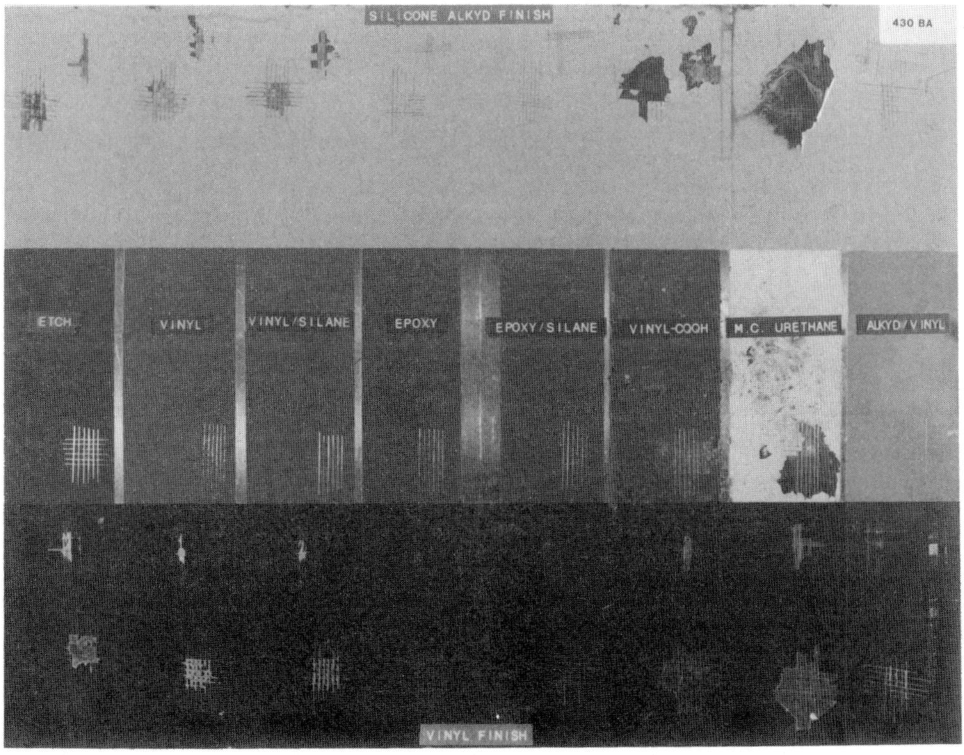

Fig. 3. Cyclical salt spray results on 430BA.

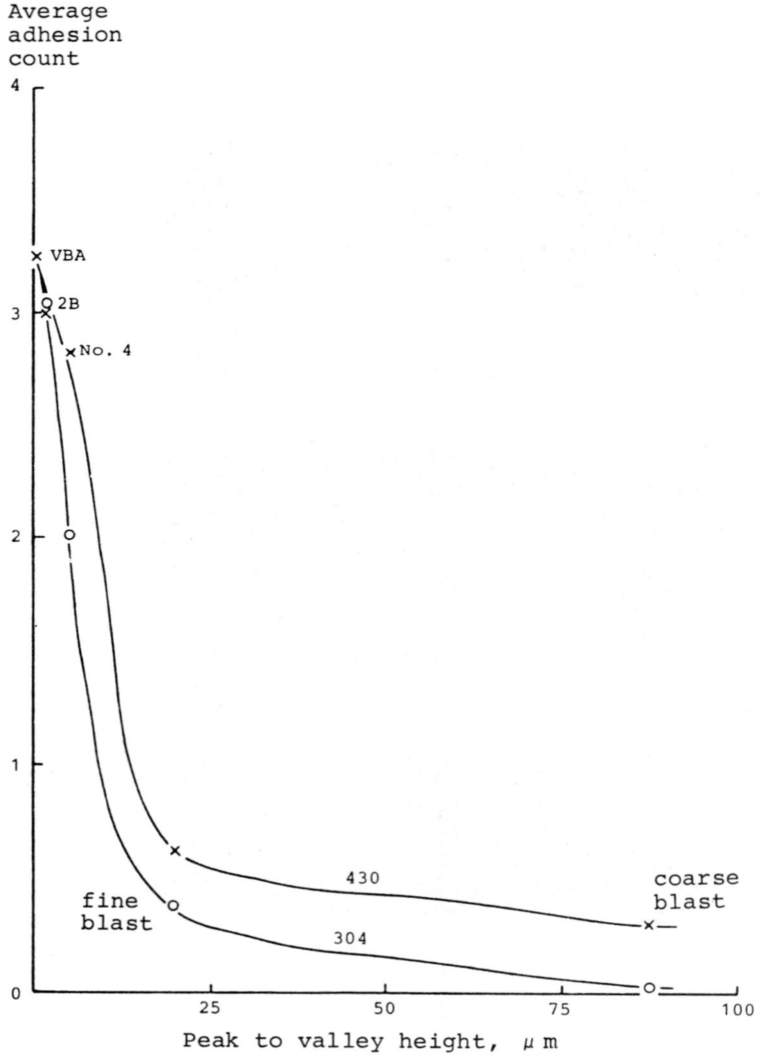

Fig. 4. Variation of adhesion with surface finish

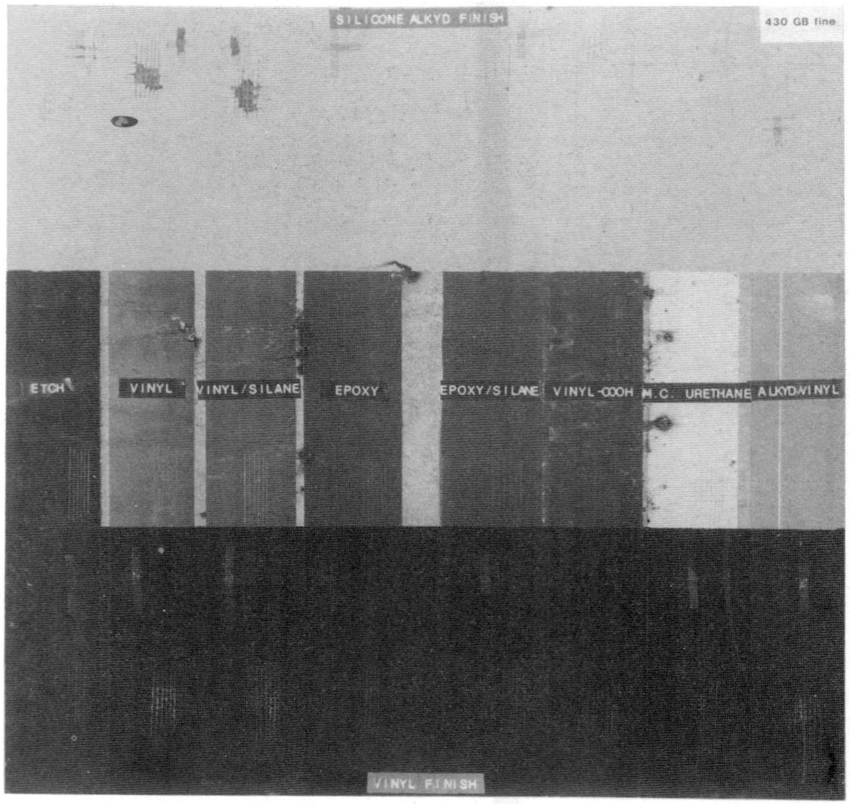

Fig. 5. Cyclical salt spray test results on 430, Fine Grit Blast.

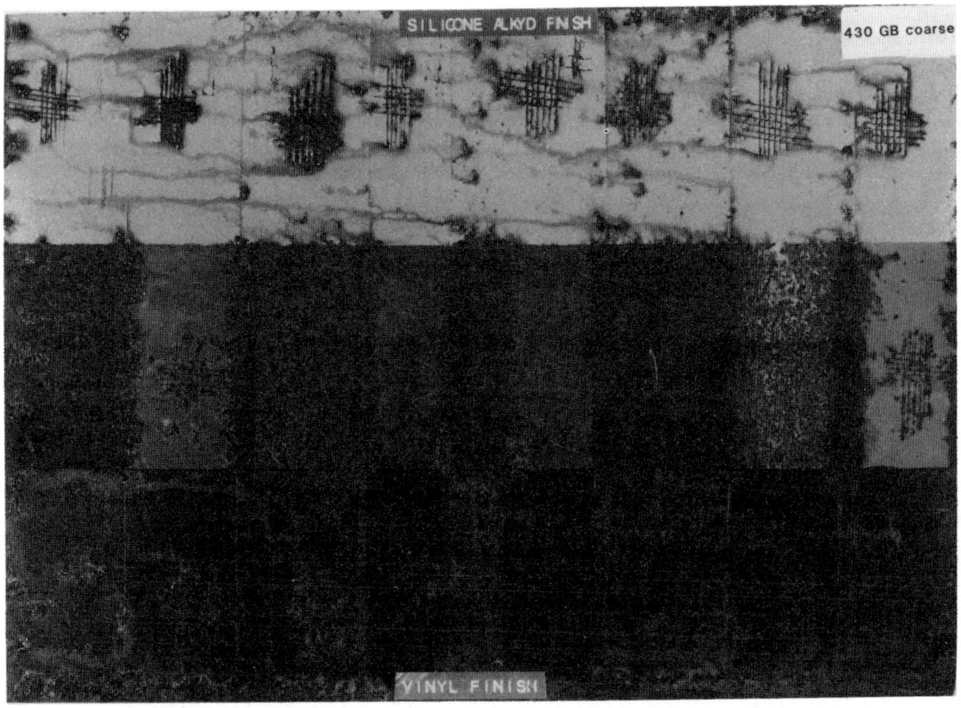

Fig. 6. Cyclical salt spray test results on 430, Coarse Grit Blast.

430 BA　　430 2B　　304 2B

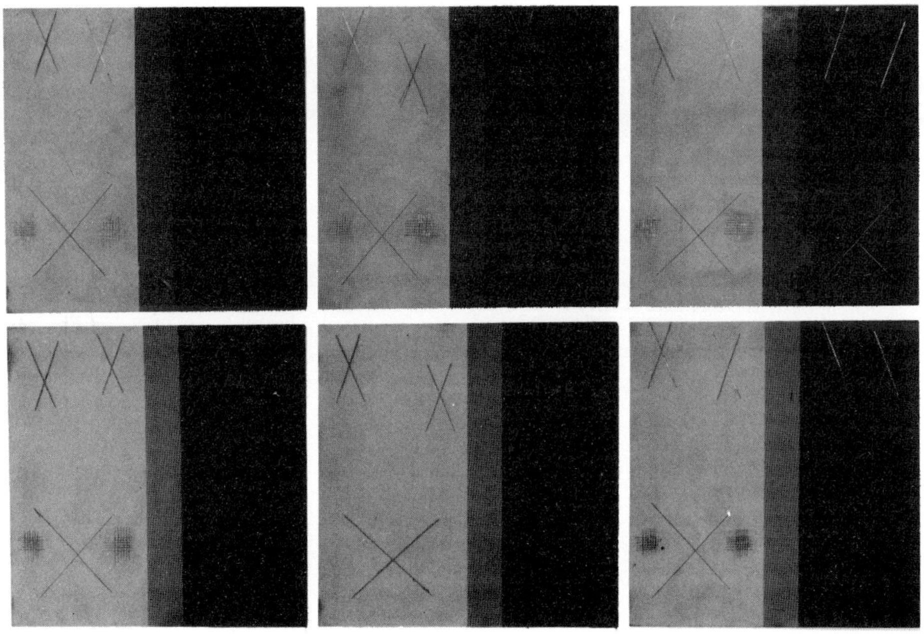

Polyamide Epoxy

Alkyd/ Vinyl

Fig. 7. Good adhesion of polyamide-epoxy and alkyd vinyl primers after 3 years exposure at Rye.

Etch M.C. Urethane Maleic/Vinyl

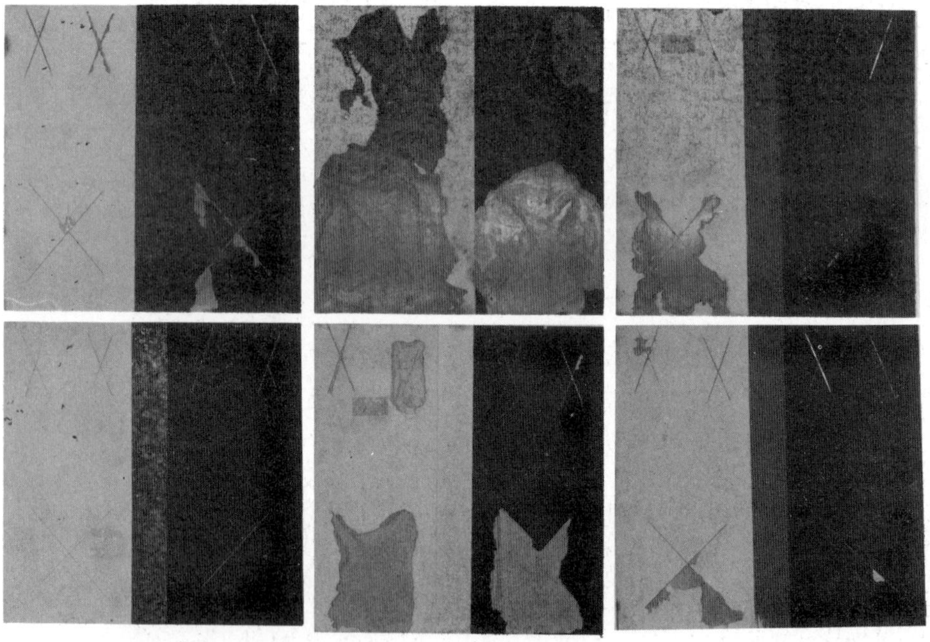

430 BA

304 2B

Fig. 8. Loss of adhesion of the moisture-cured urethane and maleic-modified vinyl primers after 3 years exposure at Rye.

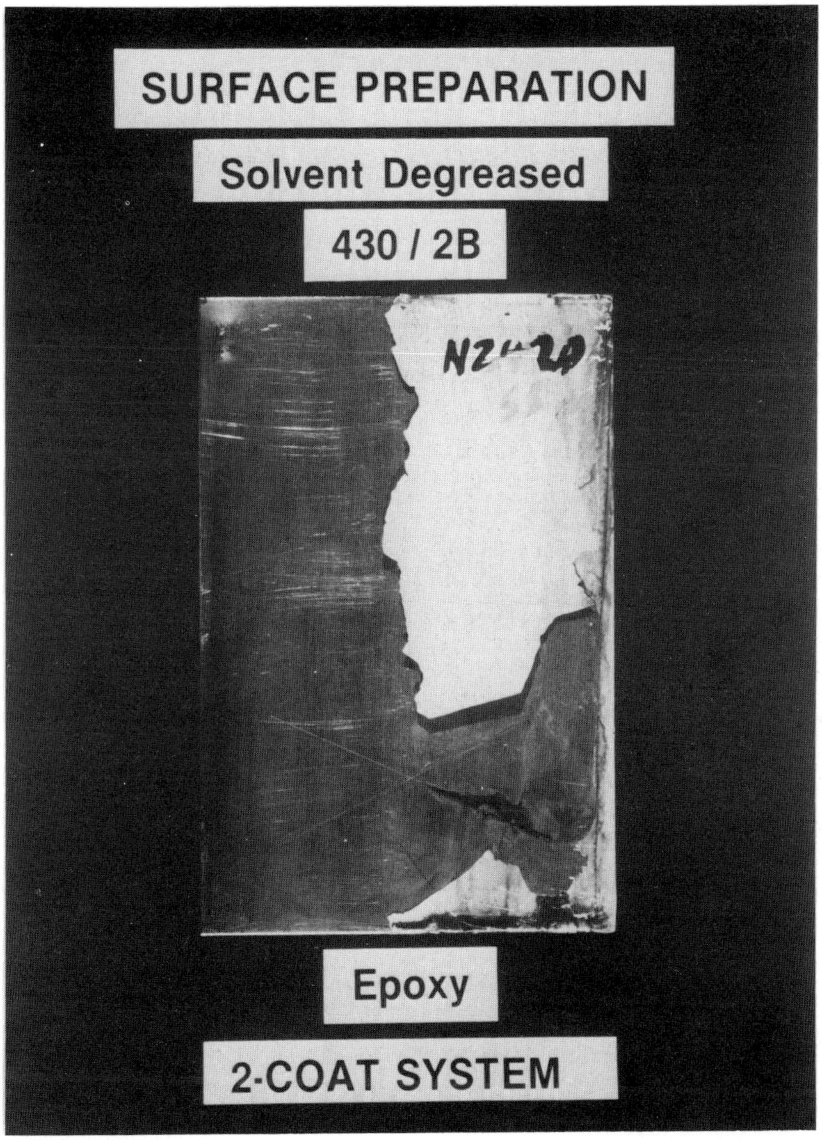

Fig. 9. Cyclical salt spray test result on a polyamide-epoxy primer from another manufacturer.

Formation of intermetallic phases and their influence on mechanical properties and corrosion behaviour in 25Cr—4Mo—4Ni—Ti ferritic stainless steel ('MONIT')

H L Cao, S Hertzman and W B Hutchinson

The authors are with the Swedish Institute for Metals Research, Stockholm, Sweden.

SYNOPSIS

Systematic investigations have been carried out on specimens of a ferritic stainless steel (MONIT) aged in the temperature range from 600°C to 900°C for times up to 2060h. Sigma (σ) phase forms over the whole temperature region investigated and chi (χ) phase forms in the region 700°C-900°C. Chi phase forms faster than σ, and acts as nuclei for σ phase formation. At lower temperatures (600°C and 650°C) Laves (η) phase particles appear to have nucleated at pre-existing titanium carbide particles.

The morphologies and distributions of the intermetallic phases have a strong effect on both the mechanical and corrosion properties of the steel. A dense amount of fine precipitation of χ and η raises the ultimate strength although having little influence on yield stress, whereas the network of χ along the grain boundaries together with σ brings about a serious loss of almost all the mechanical properties.

Corrosion testing in 10% oxalic acid indicates that the σ-phase formation causes heavy intergranular attack, but the presence of χ-phase is not revealed in this test.

INTRODUCTION

Since the end of the 1970s, a new family of high Cr-Mo ferritic stainless steels possessing excellent resistance to pitting, crevice and general corrosion have been developed (1). These steels, containing nominally 25-28%Cr, 2-4%Mo and 2-4%Ni (wt%) are expected to have excellent potential for application in seawater, pulp and paper production and chemical industry (2). They are generally ferritic at all temperatures below the melting point, so that grain size and mechanical properties are controlled by cold working and annealing.

Two types of metallurgical transformation may occur in high Cr ferritic stainless steels; both cause embrittlement and a reduction in corrosion resistance. One consists of the precipitation of alpha prime (α') and is responsible for the phenomenon referred to as 475°C embrittlement (3). Alpha prime is characterized by coherent Cr-rich clusters. The second type of transformation consists of the precipitation of intermetallic phases, such as sigma, chi and Laves phase (2). These phases form when highly alloyed stainless steels are heated into or slowly cooled through the 500°C to 1000°C temperature range. The formation of these intermetallic precipitates could be encountered commercially during cooling after hot working or during elevated temperature exposure, such as welding or severe service conditions.

Many investigations showed that sigma, chi and Laves phases should form in ferritic stainless steels during aging in the 600°C to 1000°C range. Redmond et al (4) studied 18 pct Cr ferritic stainless steel with 0 to 5%Mo and showed that increasing Mo additions shift the T-T-P curves for all three intermetallic phases to shorter times. Also Ni additions accelerated the intermetallic phase formation (5). The study of the effects of heat treatments upon mechanical properties and microstructures of high purity 20~28%Cr, 0~5%Mo ferritic stainless steels (6) indicated that chi phase precipitates faster than sigma, and that sigma rarely appears without chi phase. Vicentini (7) and Sinigalia (8) investigated two experimental ELI (extra-low-interstitial) ferritic stainless steels (22Cr-2.5Ni-3Mo and 22Cr-2.5Ni-3Mo-Ti) and compared these with a commercial alloy (21Cr-3Mo-Ti). The results showed that the unstabilized steel has better ductility and toughness than the Ti stabilized one, and the latter, after heat treatment at 600°C, 850°C and 1250°C followed by water quenching, was immune to intergranular corrosion. Lately, Rondelli et al (9,10) have studied the ELI ferritic stainless steels, one of which was similar to that used in the present work, while the other differed by being Nb stabilized. They showed that embrittlement of the alloys occurs most rapidly at 800~ 850°C and that samples heat treated in the high temperature range (850~900°C), for times between 20 min. and 5 hours, had low corrosion rates. However, specimens heat treated in the lower temperature range (650°- 750°C), even for only 5 min., were subject to severe intergranular corrosion in oxidizing media. This was caused by the presence of a continuous network of precipitates of a phase along the grain boundaries which is preferentially attacked in these environments.

MONIT is an example of a newly developed highly alloyed stainless steel with 25Cr, 4Mo, 4Ni, low contents of C and N and stabilized with Ti to ensure best possible resistance to intergranular attack after welding (11,12). It is expressly developed for application in seawater cooled heat exchangers and condensers. The purpose of this investigation was to identify the intermetallic phase morphologies, microstructural distribution and their influence on both the mechanical and corrosion properties in commercial MONIT.

EXPERIMENTAL PROCEDURE

Material and heat treatments

The material used in this investigation was obtained as a 3 mm thick sheet in the as-cold rolled condition with composition shown in table 1. Specimens were solution treated at 1050°C for 10 minutes followed by water quenching to provide a recrystallized ferrite. The specimens were then isothermally aged at 600,650,700,800 and 900°C for times of 1,3,10,30,90 minutes, 2,5,24 hours respectively. Some additional long time aging treatments were performed to assess the effect of extreme conditions on the precipitation sequence at those temperatures.

Phase identification

Metallographic specimens were prepared and then etched using two different etchants, namely Groesbecks (9) and Murakamis (14) reagents in order to conduct a systematic metallographic investigation a of the different phases present after heat treatment.

Transmission electron microscopy (TEM) and analytical scanning transmission electron microscopy (ATEM) were used to provide a quantitative analysis of small particles, which could not be identified using the optical microscope. Carbon extraction replicas were prepared for these analyses. Semiquantitative EDS analyses were also carried out on selected specimens using a scanning electron microscope (SEM) together with back-scattered electron imaging (BSE) to distinguish the σ and χ or σ and η using atomic number contrast. Electron microprobe analysis was carried out to identify accurately the chemical composition of the mixed microstructures.

X-ray diffraction was performed on solid specimens containing relatively large fractions of precipitates for the purpose of phase identification at room temperature.

Quantitative image analysis was performed, using an automatic image analysis system (IBAS) in order to correlate the amount of intermetallic phases to the mechanical and corrosion properties.

Mechanical testing

Charpy V-notch impact testing was performed at room temperature using subsize samples (55x10x3 mm) cut parallel to the rolling direction with the notch lying perpendicular to the sheet plane. Room temperature tension tests were also performed using specimens parallel to the rolling direction.

Corrosion tests

The mechanically polished specimens were tested in 10% oxalic acid at room temperature according to ASTM A 262-79 practice A in order to determine the type of grain boundary attack (i.e. step, slight boundary attack, ditch present and continuous ditch) as one of the evaluations of corrosion resistance.

EXPERIMENTAL RESULTS

Intermetallic phase formation

Results of systematic metallographic investigations and the phase identifications conducted, are presented in the Time-Temperature-Precipitation (TTP) diagram shown in Fig.1. It can be seen that the noses of the individual C-shaped curves displace to times longer than 1 minute in the temperature region 800°~1000°C and >3 minutes at 700°C. At lower temperatures (600°C and 650°C), intermetallic phases form after times longer than 10 minutes. The Laves phase, which was identified by ATEM with the composition $Fe_{0.45}$ $Mo_{0.30}$ $Cr_{0.10}$ $Ni_{0.07}$ $Ti_{0.08}$, (almost the same as found by Weiss and Stickler (16) in an alloy of type 316) was observed as intra-granular particles (Fig.2) and appeared to have nucleated at pre-existing titanium carbide particles (Fig.3). After long aging times some sigma phase also appeared at these low temperatures.

The results from quantitative image analysis of chi phase formed at 700°C and 800°C clearly showed that at both temperatures it passes through a maximum (Fig.4), which is displaced towards shorter times for higher temperatures. The chi phase formed faster than sigma and it appeared within 3 minutes, for example, at 800°C (see Fig.1). After 10 minutes sigma phase formed around chi phase and after longer times it became more widespread both along the grain boundaries and inside the grains (Fig.5). After 24 hours, most of the chi phase had disappeared and only in some local regions could a few chi phase particles be found (Fig.6). After 90 hours chi phase had almost completely disappeared.

These new results show that the chi phase is metastable and acts as a transitional phase in the transformation from ferrite to sigma, verifying the suggestion by Koh (19), see discussion below. From the metallographic results it can be concluded that:

(i) σ-phase forms over the whole temperature region investigated.
(ii) σ-phase and chi phase form in the temperature region from 700-900°C, however, at 900°C the chi phase is very rare.
(iii) the formation rate of chi phase is higher than that of sigma phase and sigma phase usually forms around chi phase.
(iv) it is experimentally verified that chi phase is metastable and can be characterised as a transition phase in this alloy.
(v) at lower temperatures, 600°-650°C, Laves phase forms mostly as intra-granular particles, which appear to have nucleated at pre-existing titanium carbide particles.

Mechanical properties

Fig.7a and b show the longitudinal room temperature tensile properties of the solution treated steel and those after subsequent ageing for various times at different temperatures. At lower temperatures (600°C-700°C), there is an increase in tensile strength (Fig.7a) without accompanying change in yield strength (Fig.7b). The ultimate strength increased with aging time at first and then decreased. The lower the temperature the longer was the time to reach the peak, and the higher was the maximum strength. From the results of impact tests as shown in Fig.8, it can be seen that brittleness occurred more rapidly with increasing temperature (Fig.8a). The aging times after which the impact energy decreased to less than 15 Joules were about 12, 20 and 35 minutes at 900, 800 and 700°C respectively, which approximately coincide with total intermetallic phase area fraction of 1%. With 10% intermetallic phase the steel became extremely brittle, having almost zero impact energy (Fig.8b).

Corrosion behaviour

The results of oxalic acid etching tests are plotted in the form of a time-temperature-corrosion (TTC) graph (Fig.9a) and a typical etching structure is illustrated in Fig.9b. From a comparison with Fig.1 it is clear that the σ phase caused the formation of ditches and heavy boundary attack, but the presence of χ was not revealed in this test.

DISCUSSION

Sigma and chi phase transformation

As shown above the chi phase is a transition phase in this alloy. Two years after Andrews discovered chi-phase (17), he suggested that this may be low-temperature modification of σ phase (18). It was also suggested by Koh (19), that chi phase might occur as a very transient intermediate phase between, say a body-centered cubic structure and sigma itself. In a literature review of sigma phase Lena (20) pointed out that the chi phase has so far been found only in steels containing Mo. This behaviour may be associated with the fact that (i) the σ-phase formation is a sluggish and largely nucleation controlled process and (ii) the chi phase is a Mo-favoured intermetallic structure. In the competition between the two phases the one with the lower nucleation barrier starts the precipitation sequence (χ phase) and the most stable phase (σ phase) dominates after long times. Molybdenum is enriched in the σ-phase and to an even greater extent in χ-phase. Inhomogeneous Mo distributions caused for example by a segregation of atoms to dislocations or grain boundaries could then promote the nucleation of χ-phase in the manner that was observed. Generally σ-phase formation is promoted by prior precipitation of phases (21), such as χ which also can act as reservoirs for chromium. Sequentially the prior χ-phase acts as nuclei for σ-phase formation, transforms to σ and finally disappears.

Another aspect concerns the morphologies of intermetallic phases formed at different temperatures. At low temperatures (600-650°C), the intermetallic phase appeared to be finely dispersed in all of the specimens (Fig.2) while at higher temperatures (800-900°C) they were present primarly as wide grain boundary phases, Fig.5. This behaviour can be explained on the basis that at higher temperatures the available driving force for transformation is lower, leading to a smaller nucleation tendency so that only very favourable sites such as grain boundaries are active. At the same time, there is a higher diffusion rate for the alloy elements so that growth is rapid. For lower temperatures the larger driving force permits profuse nucleation at many sites but subsequent growth of the finely dispersed precipitates is limited by the low diffusion rate.

Mechanical behaviour

The high Cr content stainless steel are prone to precipitation of σ as well as χ phase in the 500-900°C range (22). From the present results one could say that both σ and χ phases result in loss of toughness, and this is especially true if both σ and χ phases are present and form continuous networks along the grain boundaries (Fig. 10a). The work by Aggen et al (23) demonstrated that the fracture surfaces of brittle impact tested specimens showed the Mo content to be three times that of the matrix, indicating that much of the fracture occurred at grain boundaries through a Mo-rich phase. This appears to agree well with the coexistence of grain-boundary phases σ and χ, especially Mo-rich χ, observed by SEM in the impact test specimens aged at 700 and 800°C (Fig.10b, 800°C/2h).

In order to understand the variation in tensile properties in the present material, it is useful to consider some of the basic mechanisms which are known to affect the strength behaviour of alloys. These include (i) solid solution effects; (ii) precipitation hardening involving closely spaced, usually coherent, second phase particles; (iii) dispersion hardening involving relatively coarse, widely spaced particles; (iv) composite reinforcement due to continuous networks of hard phase around grain boundaries.

The present tensile test data show almost no change in yield stress with aging except for the very longest times at 700°C and above, where a marked decrease occurs. By contrast, the tensile strengths pass through maxima on aging in the range 600-700°C. Of the mechanisms described above, only dispersion hardening is capable of explaining such behaviour and this is fully compatible with the scale of the dispersion of intra-granular particles observed. The decreases in yield and tensile stress after long aging times can be attributed to premature fracture associated with the brittle intermetallic networks.

Corrosion behaviour

By matching the TTP curve (Fig.1) and the TTC curve (Fig.9a), one can readily identify the origin of etching structures. Light boundary attack was caused by presence of chi phase (at 700°C and 800°C) and Laves phase (at 600°C and 650°C) along the grain boundaries during the earlier aging stage. The σ phase is responsible for the ditch structure and caused also continuous boundary attack when coarse networks of σ phase formed along the boundaries.

The fact that σ-phase itself is attacked rather than the adjacent matrix can be explained on the basis of the electrochemical conditions used. The OAET was performed at a potential in the transpassive region. This region is characterized by a chromate forming reaction which implies that structure elements with high Cr contents, such as chromium carbides, chromium nitrides and in the present case σ-phase are preferentially attacked, i.e. the phases with higher Cr content have a higher dissolution rate. Accordingly it is quite reasonable that chi phase, Laves phase and ferrite, which have lower Cr contents than σ, were not heavily attacked in this test.

The morphology and the distribution of the intermetallic phases have thus a strong effect on both the mechanical and corrosion properties of the steel. However, the incubation times leading to these detrimental effects are fairly long, even at high temperatures, which indicates that welding procedures adequately performed should not influence the HAZ properties adversely.

SUMMARY AND CONCLUSIONS

Microstructure, mechanical and corrosion properties were examined in a 25Cr-4Mo-4Ni-Ti ferritic stainless steel MONIT, after aging in the temperature range 600-900°C. The following conclusions can be drawn concerning formation and effects of intermetallic phases.

- at high temperatures σ-phase dominates, in the low temperature region Laves phase precipitates and in the intermediate range it was established that χ acts as a transition phase in the ferrite to σ transformation.
- mechanical properties were strongly affected by intermetallic phase formation such that one area percent of intermetallic phase reduced the impact energy to half its original value. However, the incubation time leading to these detrimental effects are fairly long, even at high temperatures.
- a fine distribution of Laves and chi phase formed at low temperatures increased the UTS by a dispersion mechanism without an accompanying increase in yield stress.
- sigma phase caused heavy intergranular attack in oxalic acid testing while chi and Laves phases were almost unaffected.

ACKNOWLEDGEMENTS

This work has been carried out as a part of the basic research programme of the Swedish Institute for Metals Research. The authors wish to thank Avesta AB for providing the material and Jan Wåle and Rachel Jargelius for assistance and advice. The microprobe analyses were performed by Connie Westman and Börje Lehtinen carried out the ATEM measurements.

REFERENCES

1. J.D.Redmond, Chemical Engineering, July 25, (1983), 93-96.
2. R.F.Steigerwald, H.J.Dundas, J.D.Redmond and R.M.Davison, Stainless Steel' 77, Climax Molybdenum Company, Greenwich, CT, (1977), 57-72.
3. D.Peckner and I.M.Bernstein, Handbook of Stainless Steel, N.Y. (1977), Section 14-14 and 44-47.
4. J.D.Redmond, P.J.Grobner and V.Biss, J.of Metals, 33 (1981), 2, 19-25.
5. M.A.Streicher, Corrosion, 30 (1974),4, 115-124.
6. H.Brandis, H.Kiescheyer und G.Lennartz, Arch. Eisenhüttenwes., 4b (1975),12, 799-804.
7. B.Vicentini, D.Sinigaglia, R.Taceani, G.Rondelli, F.Cherardi and P.L.Ortali, Werkstoffe und Korrosion, 33 (1982), 132-143.
8. D.Sinigaglia, G.Taceani, B.Vicentini, G.Rondelli and L.Gallelli, Werkstoffe und Korrosion, 33 (1982), 591-601.
9. G.Rondelli, D.Sinigaglia, B.Vicentini and G.Taceani, Proceedings of International Congress on Metallic Corrosion, 3 (1984), Toronto, June 3-7.
10. G.Rondelli, B.Vicentinit and D.Sinigaglia, Proc. of 16th Annual Technical Meeting of IMS, 12 (1985), 74-87.
11. G.Gemmel, Stainless Steel'84, Göteborg, Sept.3-4 (1984), 166-172.
12. G.Gemmel and G.Molinder, Int.Coll.; Choice of materials for condenser tubes and plates and toughness testing, Avignong (1982), 251-258.
13. G.Gemmel and S.Nordin, Proc.Conf.of advanced stainless steels for seawater applications, Pracenza, Italy, Fe.1980 1981, 69-80.
14. G.Petzow, Metallographic Etching, ASM, Ohio (1978).
15. ASTM A262-79, practice A, 84-88.
16. B.Weiss and R.Stickler, Metallurgical Transactons, 3 (1972), 851-866.
17. K.W.Andrews, Nature, 164 (1949), 1015.
18. K.W.Andrews and P.E.Brookes, Metal treatment and drop forging, July (1951), 301-311.
19. P.K.Koh, Trans.Amer.Inst.Min.Met.Eng., 197 (1953), 339.
20. A.J.Lena, Metal progress, Sept. (1954), 122-128.
21. E.L.Brown, M.E.Burnett, P.T.Purtscher and G.Grauss, Met.Trans A, 14A (1983), 791-800.
22. R.N.Wright, ASTM STP 706 (1980), 2-33.
23. G.Aggen, M.E.Deverell and T.J.Nichol, Micron 78, ASTM, STP 672 (1979), 334-366.
24. J.B.John et al., Proceedings of 12th Saganiore Army Materials Research Conference, 1966.

Table 1. Chemical composition of MONIT (wt%)
Charge number - G6275.

Fe	Cr	Mo	Ni	Ti
65.4	25.0	3.85	4,13	0.54

Si	C	N	S	P
0.26	0.015	0.025	0.008	0.024

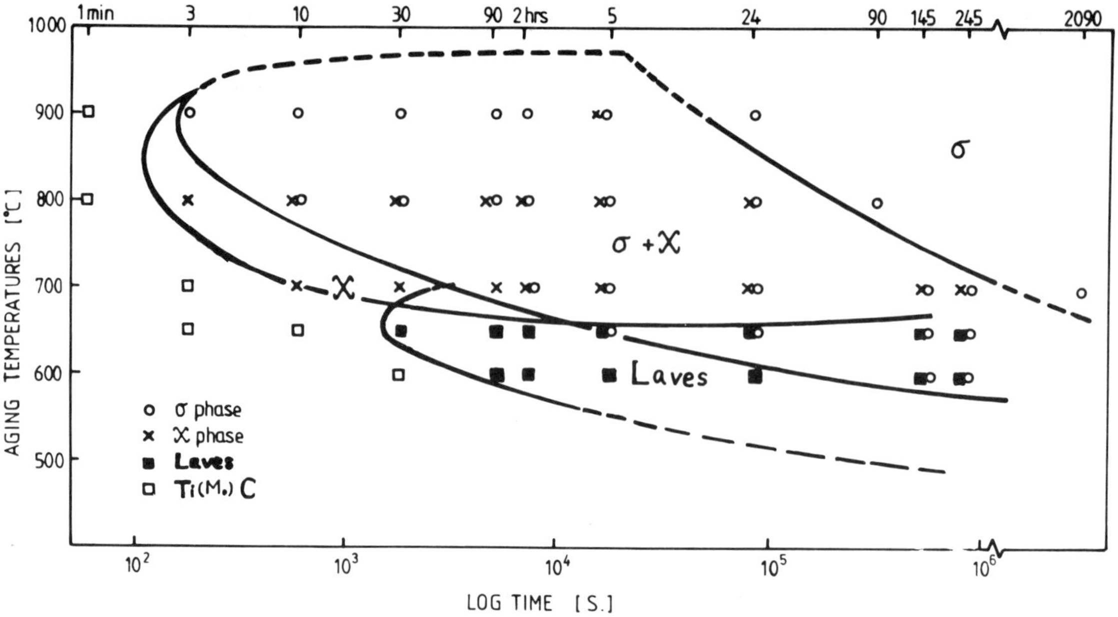

Fig.1 Time-Temperature-Precipitation curve, aged after solution treatment at 1050°C for 10 minutes, water quenched.

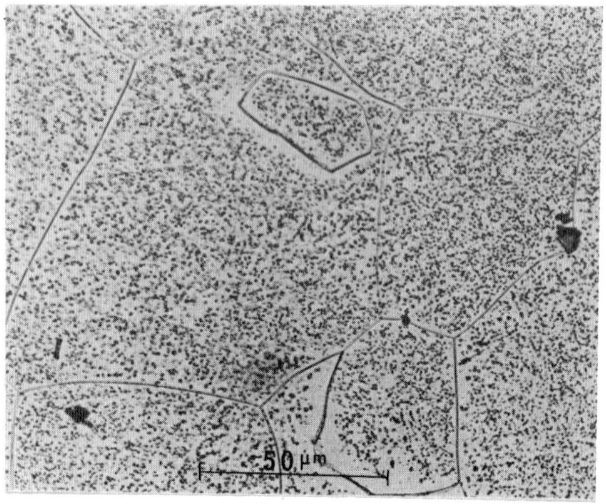

Fig.2 Dense precipitation of fine particles formed in the specimen aged at 600°C/145h.

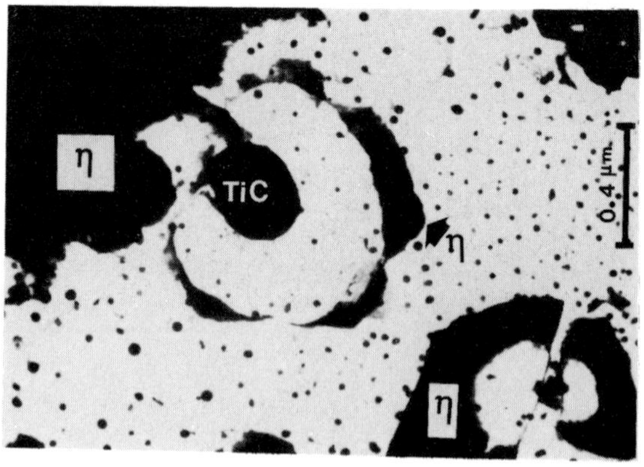

Fig.3 Laves phase (η) nucleated at pre-existing titanium carbide particles ($Ti_{0.89}Mo_{0.11}C_{0.92}$).

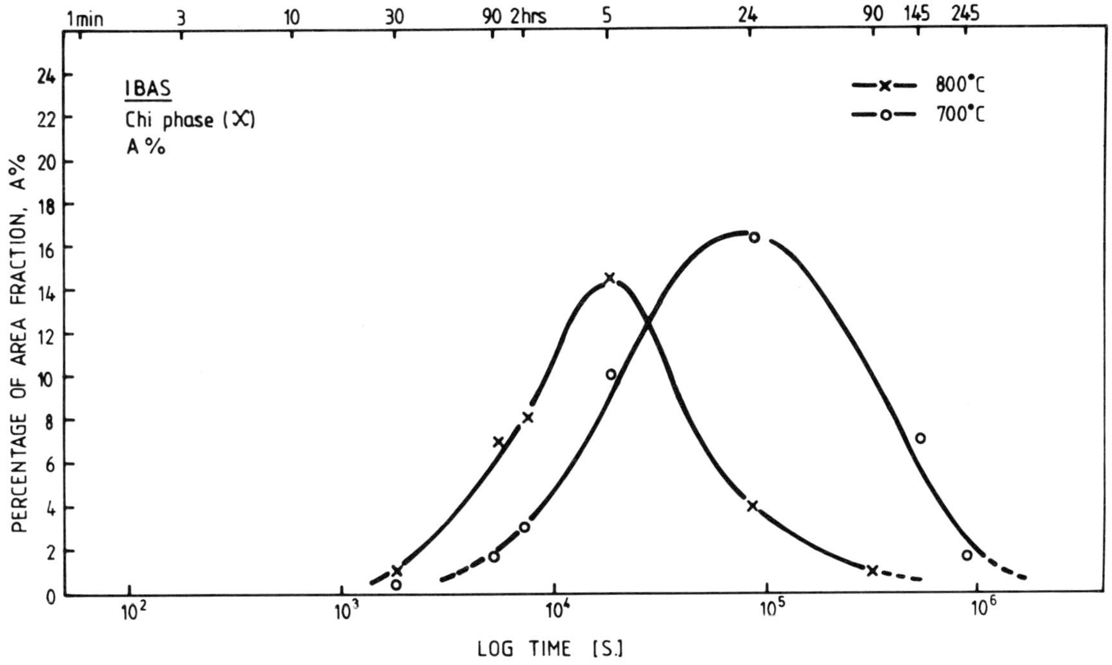

Fig.4 Area fraction of chi phase formed at 700°C and 800°C, IBAS analysis.

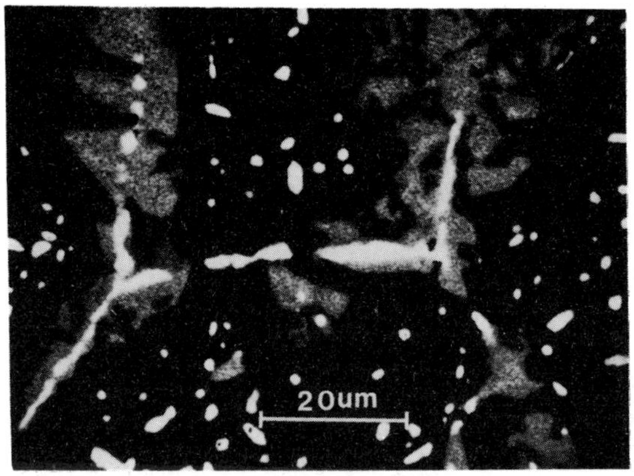

Fig.5 BSE image, aged at 800°C/5h, bright-chi, grey-sigma, black-ferrite (background).

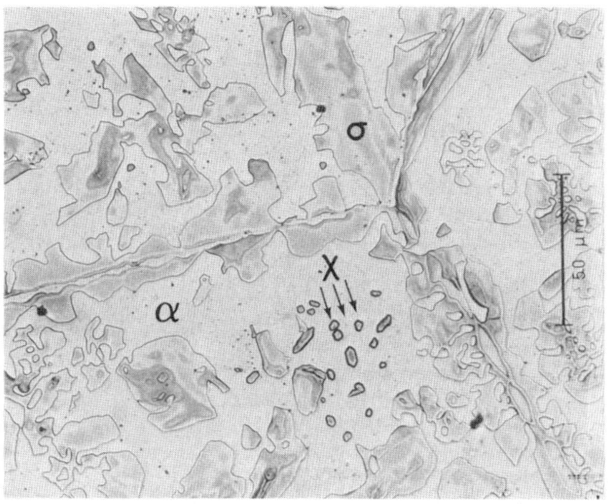

Fig.6 Aged at 800°C/24h, ferrite and sigma-phase, only in some local regions can a few chi phase particles be found.

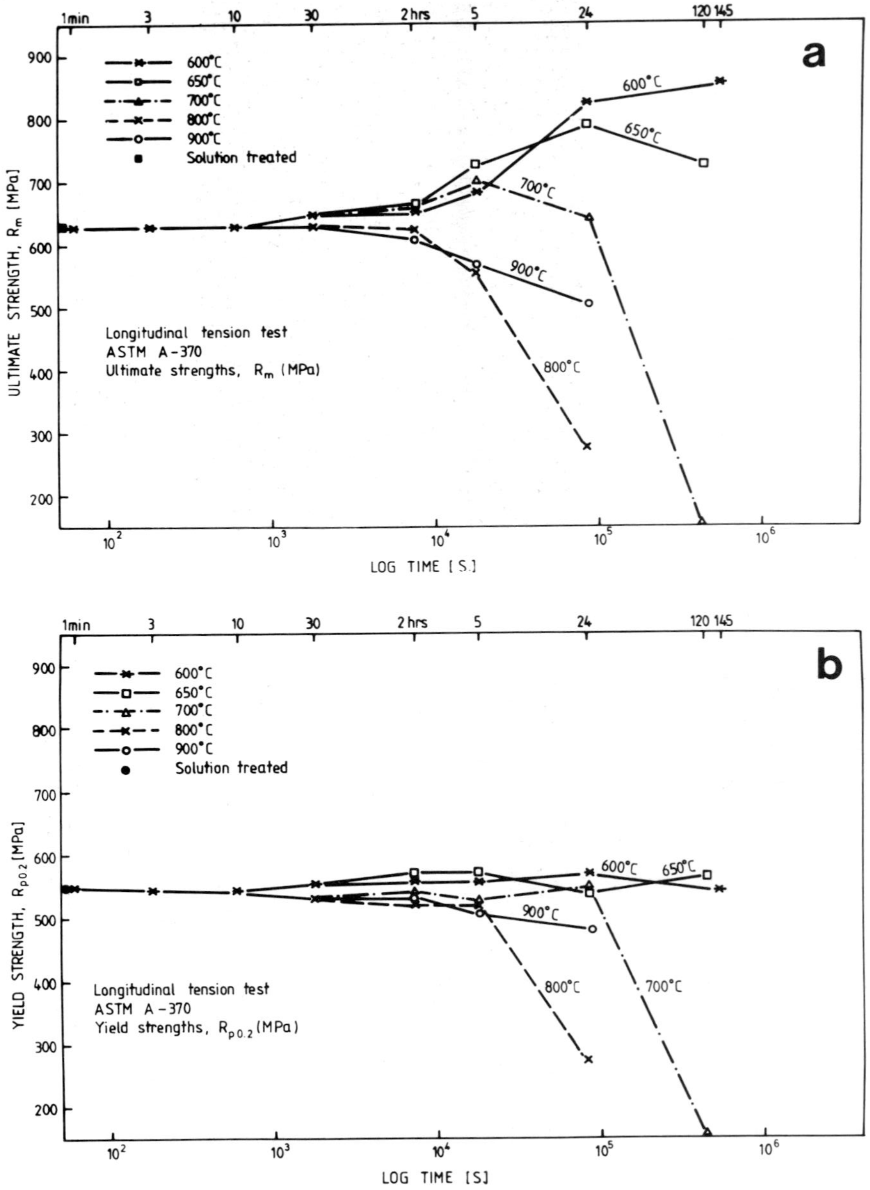

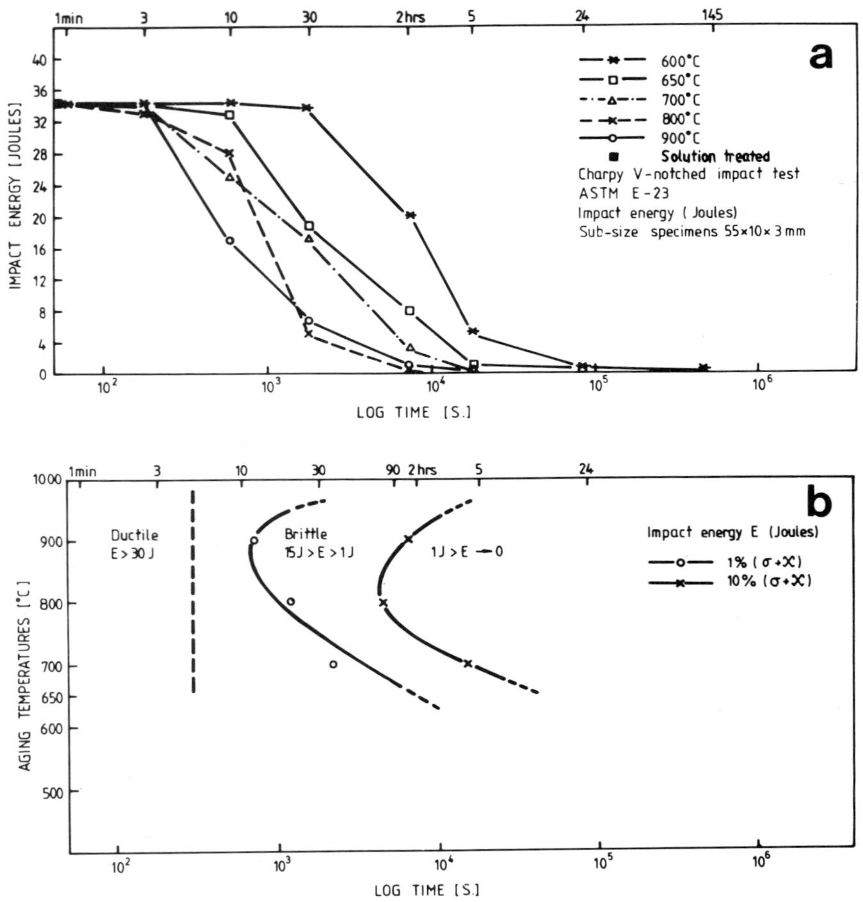

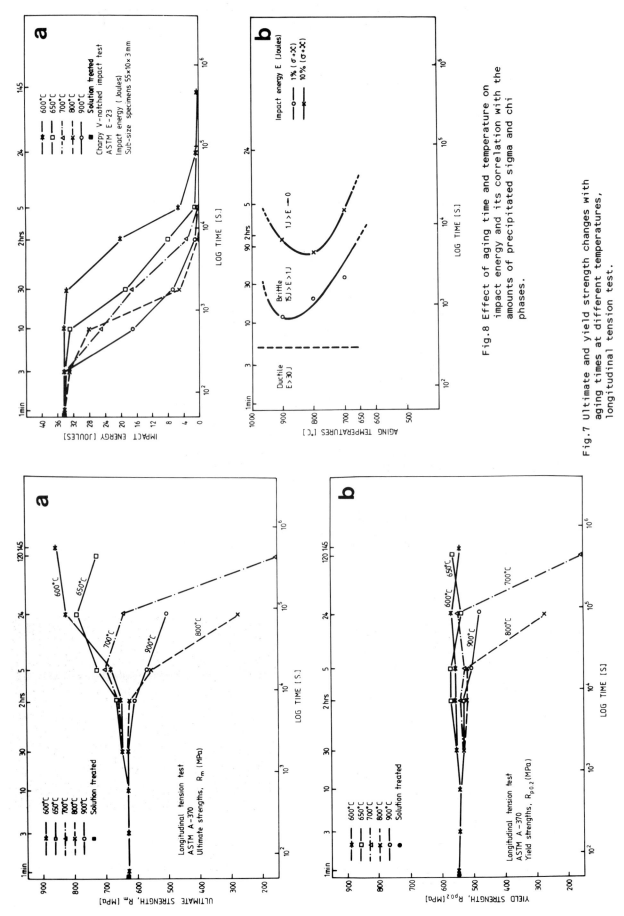

Fig.8 Effect of aging time and temperature on impact energy and its correlation with the amounts of precipitated sigma and chi phases.

Fig.7 Ultimate and yield strength changes with aging times at different temperatures, longitudinal tension test.

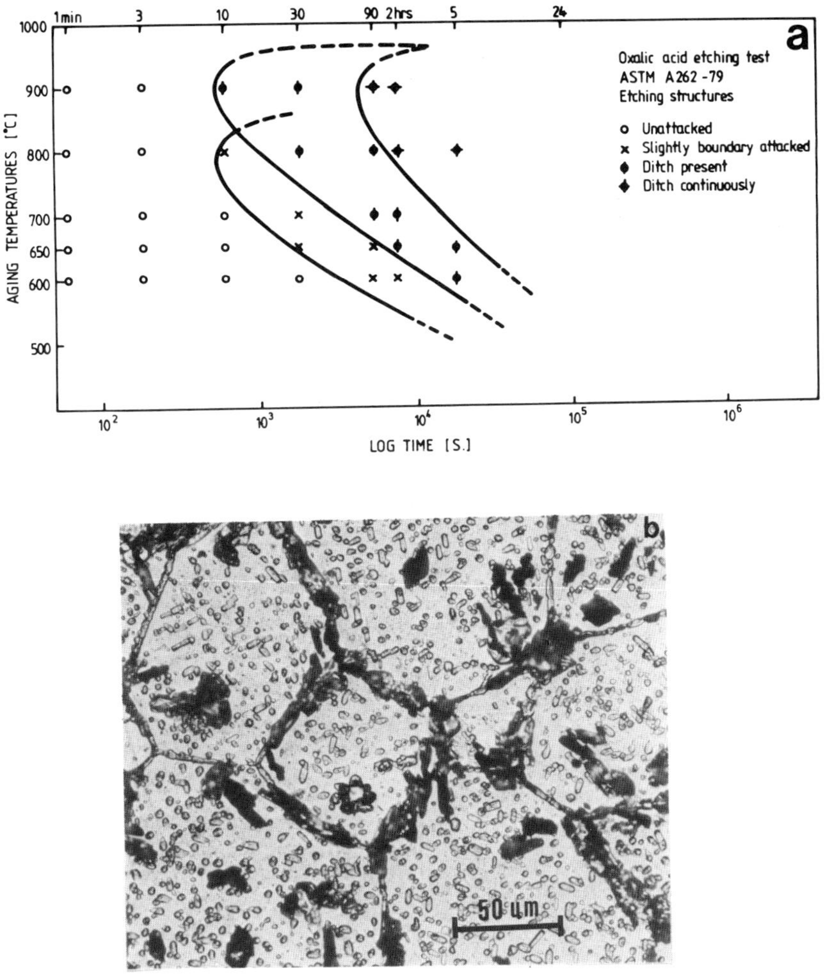

Fig.9 Time-Temperature-corrosion curve (a) and etching structure of specimen aged at 800°C/5h (b) by oxalic acid etching test.

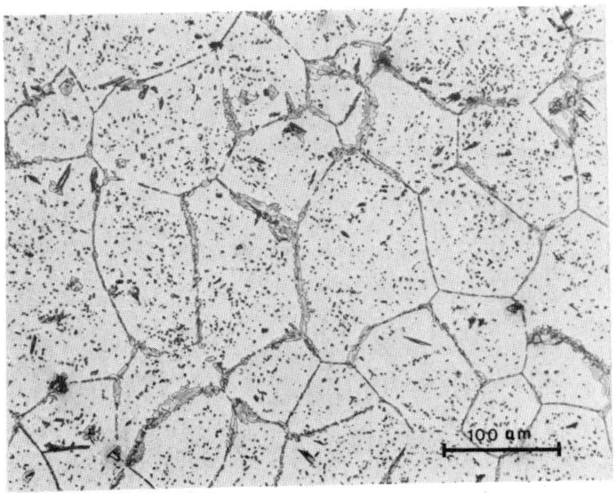

Fig.10a Grain boundary network of fine chi phase surrounded by sigma phase formed at 800°C/90 min; and

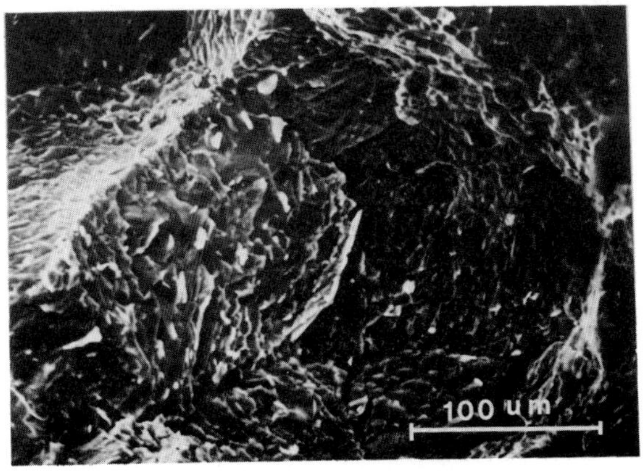

Fig.10b Scanning electron micrograph showing fracture surface of the steel embrittled by coexistence of sigma and chi phases.

□

Machinability of stainless steels with controlled oxide inclusions

O Bletton, R Duet, B Heritier and P Pedarre

The authors are in the Research Center of UGINE SAVOIE, France.

SYNOPSIS

Stainless steels containing malleable oxides form the latest generation with improved machinability. The influence of the oxide nature on the machinability with carbide tools of AISI 316L grades is outlined. The best results are obtained with anorthitic inclusions.

1 - INTRODUCTION

Over half of the stainless "long products" (billets, bars, wire rods) undergo at least one machining operation in the fabrication of final products. Thus any improvement in a stainless steel's machinability will significantly affect the cost of manufacture, especially in the free machining industry.

Among the different methods used for this purpose, the addition of sulfur is best known ; however, the good machinability is obtained at the expense of other properties (hot and cold ductility, weldability, corrosion...).

The control of oxide inclusions constitutes an alternative approach. By a suitable control of steelmaking, oxide inclusions which are malleable at the cutting temperatures obtained with carbide tools can be produced.

The chemical composition of these oxide lies in the SiO_2 - CaO - Al_2O_3 ternary phase diagram.

The relative proportion of the three components in the oxides and the composition of sulfides influence to a large extend machinability. Three variations of steelmaking applied to an AISI 316L stainless steel are presented here.

2 - COMPOSITION OF INCLUSIONS -

The different types of inclusions are listed below :

N°	STEELMAKING	MAIN TYPE OF OXIDE	TYPE OF SULFIDES
1	référence 316L	Alumina	MnS
2	316L with controlled oxide inclusions	Gehlenite pseudowollastonite	MnS (Mn, Ca)S
3	316L with controlled oxide inclusions	Anorthite	MnS only

Their positions in the SiO_2 - CaO - Al_2O_3 ternary phase diagram are shown in figure 1.

3 - MACHINABILITY -

The machinability of these three types of stainless steels was evaluated by various methods.

* **Turning with HSS tool**

The criterion Vm is the cutting speed at tool failure. With HSS tools, similar results are obtained with the three types of stainless steels (fig 2). Because of the low cutting speeds, the chip temperatures are not high enough to take advantage of the oxides.

* **Turning with carbide tool**

- Fig. 3 and fig 4 show that the lowest flank and crater wear are obtained with steel 3 containing anorthitic type oxides.

- The flank wear of steel 2, with gehlenite and pseudowollastonite, displays a behaviour intermediate between 1 and 3 ; with respect to the reference steel 1, the crater wear is not improved.

- Under identical cutting conditions (speed, feed, depth of cut, tool...) steels 1 and 3 show the best chip breakability (fig. 5)

PRODUCTIVITY IN SCREW MACHINING

Parts, shown in fig. 6a, were screw-machined with the 3 types of stainless steels.
The highest productivity (fig. 6b) is obtained with steel 3 (30 % over reference steel 1) while steel 2 offers a low improvement only, with respect to reference steel 1, steels 2 and 3 show the best surface aspect, especially steel 2 after the reaming operation.

4 - DISCUSSION - CONCLUSIONS

- The oxide control in stainless steels can improve machinability by the formation of a layer at the tool/chip interface which acts as a lubricant, reduces friction and heat and limits heat diffusion in the tool.

- The type of oxide in the SiO_2 - CaO - Al_2O_3 ternary phase diagram influences greatly machinability. Anorthitic inclusions lead to the best results, as long as sulfides do not contain calcium in carbide tool machining.

- The beneficial influence of anorthitic inclusions does not appear with HSS tools due to lower chip temperatures, this limitation is lessening with the continuing progress in tool materials.

- Stainless steels with controlled oxide inclusions are very attractive in difficult operations and unlike the sulfur bearing grades do not suffer any deterioration of properties such as cold and hot ductility, corrosion resistance, weldability this generation of stainless steels is marketed by UGINE SAVOIE under the IMA (improved machinability) generic name.

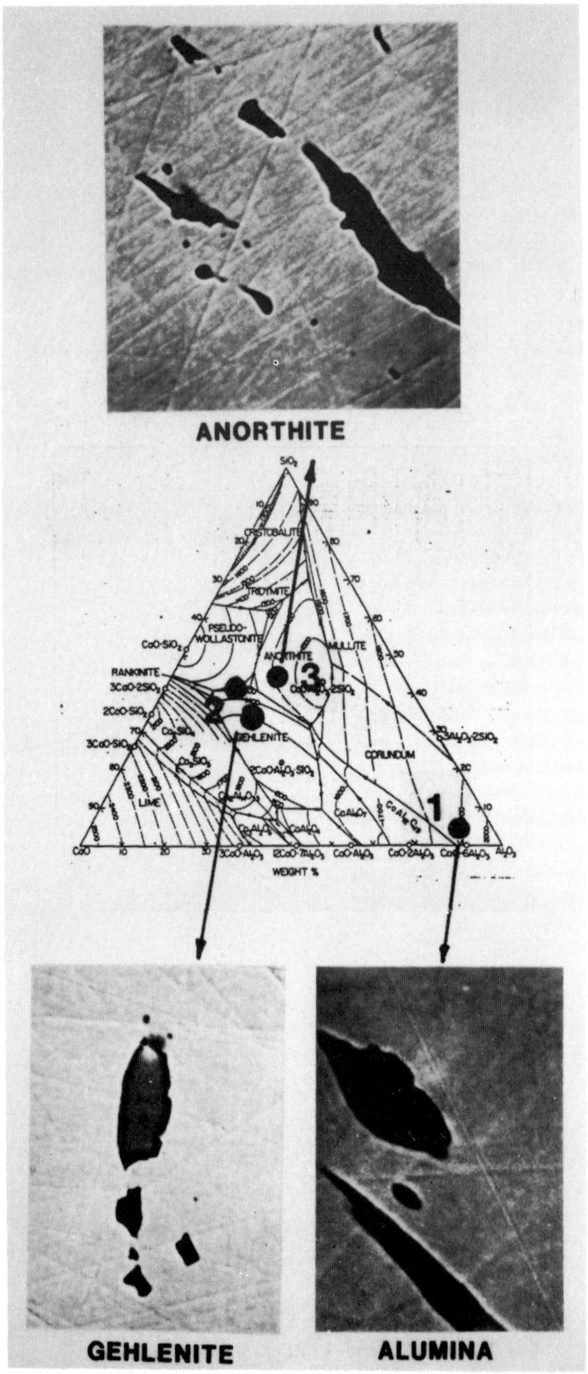

Fig. 1 Position in the SiO2-CaO-Al2O3 ternary phase diagram

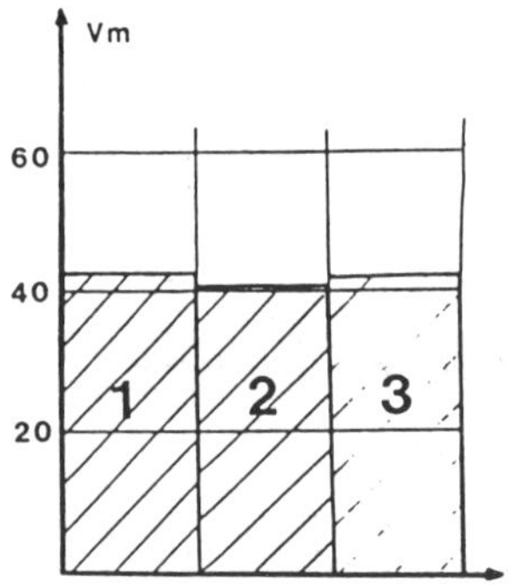

Fig. 2 Turning with HSS tools

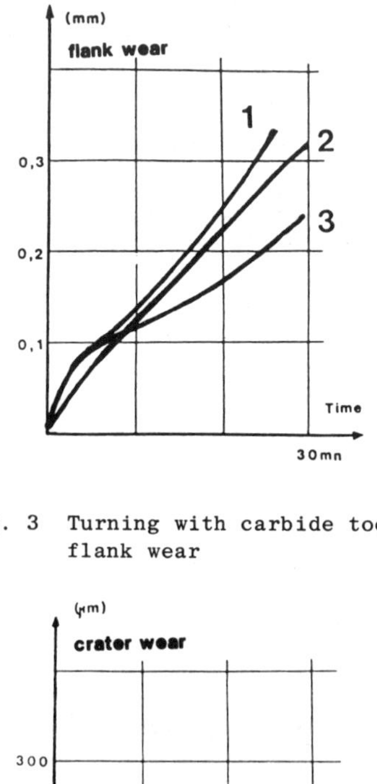

Fig. 3 Turning with carbide tools - flank wear

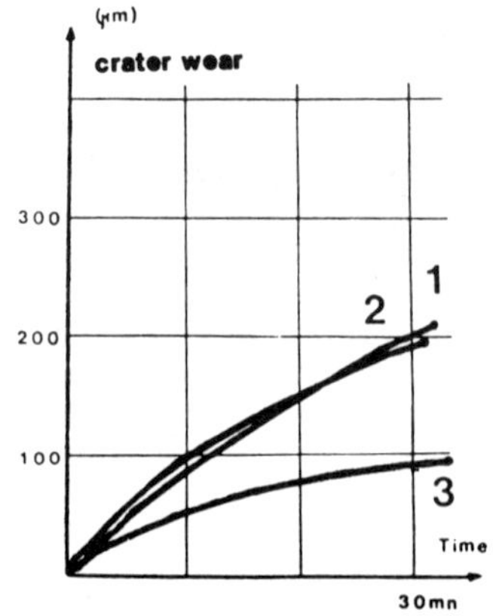

Fig. 4 Turning with carbide tools - crater wear

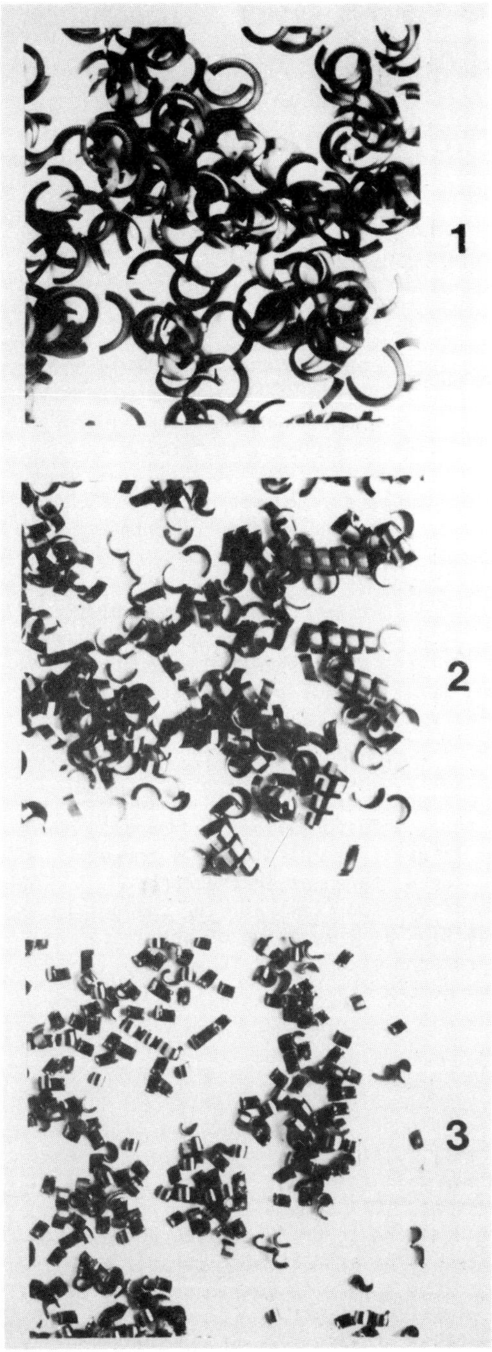

Fig. 5 Turning with carbide tools - chip breakability

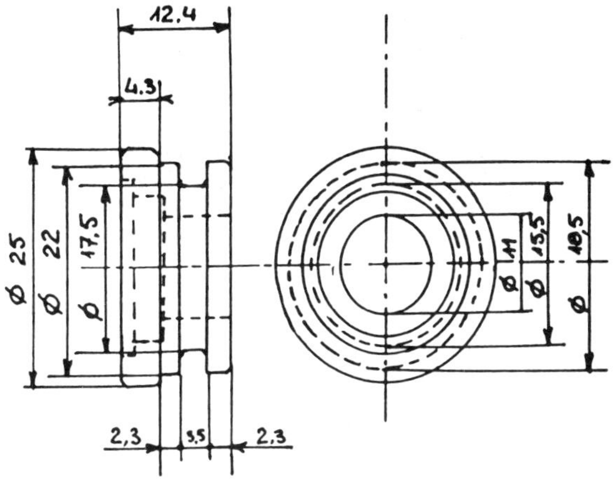

a

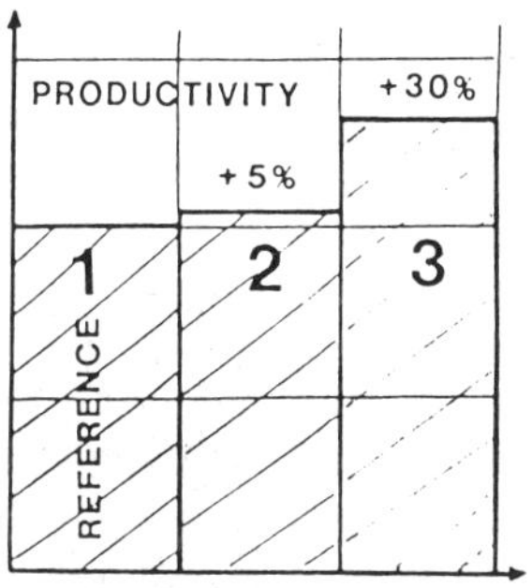

b

Fig. 6 Screw machining
(a) screw machined part
(b) productivity

Effects of ageing on creep behaviour of two Type 347 stainless steel castings

J C Lawrenson, R Pilkington and D A Miller

JCL and RP are in the Department of Metallurgy and Naterials Science, University of Manchester/ UMIST, Grosvenor Street, Manchester; DAM is with the Operational Engineering Division, Central Electricity Generating Board, Bedminster Down, Bridgwater Road, Bristol.

SYNOPSIS

The creep behaviour of two centrifugally cast type 347 stainless steels containing 3% and 13% delta-ferrite has been investigated. In the as-received materials, the effect of increasing the δ-ferrite content from 3% to 13% is to raise the minimum creep rate, whilst lowering the creep ductility, with a resultant shorter time to failure. Ageing the high δ-ferrite cast, prior to creep testing, resulted in increasing both the minimum creep rate and rupture ductility but had little influence on the rupture life compared with that observed in the as-received material. On pre-ageing the low δ-ferrite cast, a similar trend was observed in that both the minimum creep rate and rupture ductility increased, however in this case the rupture life was reduced compared to the as-received material. This behaviour is discussed in terms of the microstructural transformations and the development of fracture damage leading to creep rupture.

INTRODUCTION

AISI Type 347 is a niobium stabilised grade of 18/8 austenitic stainless steel in which the formation of intergranular niobium carbide prevents the development of intergranular, chromium rich $M_{23}C_6$ during exposure to elevated temperatures. The niobium carbide also precipitation strengthens the austenite grains. These properties have led to the use of type 347 steel for high temperature components in nuclear power stations. In applications where complex shapes are required, this alloy is used in the cast rather than wrought form. In common with austenitic stainless steel weld metals the composition of castings is designed to produce a limited amount of primary δ-ferrite during solidification. This alleviates the problem of hot tearing during casting or welding (1), but results in the retention of metastable δ-ferrite in the room temperature microstructure. During subsequent ageing or exposure to service temperatures above ~550°C the δ-ferrite may be partially or completely transformed to intermetallic phases and austenite, often accompanied by the precipitation of carbide particles. Extensive studies of the creep properties of δ-ferrite containing austenitic stainless steel weld metals have found that low ductility failures are associated with the presence of intermetallic phases (2, 3, 4). However, the number of similar studies for cast austenitic steels appears to be very limited (5, 6).

The present work is a study of the microstructural transformations and creep properties of two Type 347 stainless steel castings. The differences in the creep properties of the two casts and the changes in these properties as a function of pre-ageing at 650°C are rationalised with reference to comparative matrix strengths and initial δ-ferrite contents.

EXPERIMENTAL

The material studied in this work was obtained from two centrifugally spun cast tubes (OD = 235 mm, wall thickness = 32 mm) of Type 347 stainless steel. The compositions of these two casts were designed to produce different amounts of δ-ferrite in the as-cast microstructures. The compositions and δ-ferrite contents of the two casts are given in Table 1.

Characterisation of the as-received aged and crept microstructures has been achieved using a combination of optical and electron microscopy. Optical metallography has been carried out on the cast and aged material and on longitudinal sections from the gauge length of creep specimens. Several etches have been employed to reveal the phases present in the microstructures. Electrolytic etching in 10% KOH followed by 10% oxalic acid, or Kegley's modification to Murakami's etch followed by an electrolytic etch in 10% oxalic acid were found to be the most useful techniques to distinguish between the phases, the latter producing more colour contrast. The creep specimens have also been examined using a Philips 505 SEM. To avoid confusion between creep damage and preferential etching of certain phases (mainly σ-phase) a very short electrolytic etch in 10% HCL-90% methanol was employed prior to examination, and the absence of etch pitting was confirmed by optical examination.

Thin foil specimens and extraction replicas have been examined using a Philips EM301 TEM and a Philips EM400T STEM. The foil specimens were produced using a twin jet-polishing apparatus with an electrolyte of 5% perchloric acid in glacial acetic acid. Extraction replicas were prepared by employing a light electrolytic etch in 10% oxalic acid, depositing a thin carbon film, then removing the film by electrolytically etching in 10% HCL/90% methanol solution. A Zeiss TGA10 particle size analyser was used to measure the matrix NbC particle size distributions, in the as-received and 10000h aged conditions, from photographs of extraction replicas of these specimens. Approximately 1000 particles were analysed in each case.

Creep specimens (gauge lengths 50.8 mm, diameters 6.5 mm) were machined from the tube walls with the gauge lengths perpendicular to the columnar grains and the tensile axis along the longitudinal axis of the tube. Constant load creep tests were carried out, in air at 650°C, on material from both casts. The tests were performed on as-received material and on material which had been aged at 650°C for 2000h and 10000h. Specimen temperatures were measured using thermocouples attached to the specimen gauge length and were maintained to within ± 2°C. Strain measurements were made using capacitance transducers sensitive to 2.54 x 10^{-4} mm.

RESULTS

The microstructure of as-received Cast L consisted of isolated islands of δ-ferrite in an austenite matrix. Niobium carbide was present both as fine intragranular precipitates and within a NbC/austenite eutectic phase (Fig 1a). The niobium carbide in this phase had a coarse rod-like morphology and the austenite associated with it was distinguishable by the absence of intragranular particles. The microstructure of Cast H (Fig 1b) consisted of a nearly continuous network of δ-ferrite in an austenite matrix which again contained intragranular niobium carbide precipitates. Only a very small proportion of NbC/austenite eutectic phase existed in the microstructure of this cast.

Ageing at 650°C produced significant microstructural changes and in both casts phase transformations were only observed at δ-ferrite regions. In the case of Cast L the precipitation of a small amount of $M_{23}C_6$ at the δ-ferrite/austenite boundaries, was followed by the transformation of δ-ferrite to σ-phase. After 2000h at 650°C small amounts of σ-phase (Fig 2a) and $M_{23}C_6$ were present. After 10000h at this temperature the amount of σ-phase had increased slightly, but untransformed δ-ferrite remained in the microstructure. The transformations which occurred in Cast H were similar to those for Cast L. However, this sequence was accompanied by the formation of new austenite within the δ-ferrite regions. After 2000h ageing at 650°C the formation of $M_{23}C_6$ and the transformation of δ-ferrite to σ-phase and austenite was effectively complete (Fig 2b). No other phase transformations were detected on ageing up to 10000h.

In both casts the matrix niobium carbide particles and the NbC/austenite eutectic phase were unaffected in size or morphology by ageing treatments at 650°C. The results of particle size analysis for the matrix carbides are shown in Table 2.

CREEP TESTING

The creep properties of the two casts in the as-received and aged conditions were characterised in terms of minimum creep rates, rupture ductilities and rupture lives. A comparison of the creep properties of the two casts and a summary of the influence of preageing at 650°C on the above parameters are given in Figures 3, 4 and 5 and Tables 3 and 4.

DISCUSSION

The two casts of type 347 examined in the as-received and aged condition exhibit large differences in creep behaviour. These differences can be rationalised by consideration of the relationship between mechanical and microstructural properties. For clarity it is appropriate to consider the creep parameters in the following sequence: minimum creep rate; rupture ductility and rupture life.

The order of magnitude increases in the minimum creep rate as a result of pre-ageing both casts is considered to be a consequence of a softening mechanism occurring in the austenite matrix. Three mechanisms have been considered, these are: particle coarsening, solid solution strengthening and changes in dislocation substructure.

Time dependent coarsening of the matrix niobium carbide particles was initially considered to be the most likely mechanism. Intragranular niobium carbides are known to be particularly effective in strengthening austenite[7] and have been observed to coarsen in other alloys[8]. This coarsening would lead to a reduction in the Orowan stress which is considered to be the factor controlling strength in particle strengthened alloys[9]. Comparison of the results of particle size measurements carried out on as-received and on 10000h aged material, Table 2, shows that the matrix particles do not coarsen during the preageing treatment. This niobium carbide size stability has also been observed in type 347 weld metal[10]. Therefore the reduction in creep strength cannot be attributed to a carbide coarsening mechanism.

Consideration of solid solution strengthening in this type of steel suggests that the most important substitutional elements are likely to be Ni(weakening) and Mo (strengthening). Lattice parameter measurements and micro-analysis of the matrix compositions, as a function of ageing and of creep testing did not show any significant compositional variations of these two elements. Thus it is felt that changes in solid solution strengthening can be discounted in terms of explaining the observed order of magnitude changes in the creep rate.

The third mechanism for matrix softening is related to the dislocation substructure. The austenite in the as-received microstructures of both casts contains extensive regions of stacking faults, Figure 6. The presence of stacking faults in austenitic stainless steels is to be expected in view of their relatively low stacking fault energies[11]. The stacking faults observed in the micro-structure of Cast L show a much larger separation of the partial dislocations than those in Cast H. This suggests that the stacking fault energy (SFE) of Cast L is lower than that of Cast H and could account for the lower minimum creep rates of Cast L in the as-received condition, Figure 3.

Ageing at 650°C leads to a change in the dislocation structure in the austenite of both casts. Faulted regions become much fewer and less extended but many more unit dislocations are observed, Figure 7. It is thus suggested that ageing reduces the number and size of the stacking faults and that this is probably due to an increase in the stacking fault energy of the austenite matrix. The main influence on the value of SFE is composition and during ageing the transformations which occur in the δ-ferrite will alter the concentration of alloying elements in the austenite. This effect of alloying element segregation, or partitioning between δ-ferrite and austenite on the SFE has been observed in 304L and 308L weld metals[12].

The effect of changing the dislocation structure from widely separated partial dislocations to unit dislocations is to soften the austenite matrix. Both the deformation and recovery processes become easier in the aged condition because the interactions between unit dislocations are much less complex than those between extended dislocations where multiple jogs and complex stacking sequences are formed[11]. This concept can be applied to the minimum creep rate results for both casts, in the as-received and aged conditions. Cast L in the as-received condition contains widely separated partial dislocations. During ageing it is suggested that the SFE is reduced and hence the minimum creep rate of the 2000h aged material becomes substantially greater than that of the as-received material. On ageing for 10000h at 650°C, the continued transformation of δ-ferrite to σ-phase may further reduce the SFE leading to the further increase in minimum creep rate. Indeed, the microstructure of the 10000h aged material of this cast contains very few stacking faults compared to the 2000h aged material.

Cast H in the as-received condition contains less widely separated partial dislocations than Cast L. This infers that the SFE of Cast H is higher than that of cast L and explains the higher minimum creep rates for as-received Cast H. On ageing Cast H for 2000h at 650°C, the SFE would be expected to increase as consequence of the transformation of δ-ferrite to σ-phase. Since this is complete after this time no further change in SFE would be expected after 10000h ageing. This view is in agreement with the microstructural observations and the creep test results where mainly unit dislocations are observed after 2000h ageing and the minimum creep rate of the 2000h aged material is greater than that of the as-received material. Furthermore, no significant difference exists between the minimum creep rates of 2000h and 10000h aged materials, Figure 3. It is clear that the above comments for both casts of material suggest that stacking faults have a significant influence on the creep strength of this type of steel.

Comparison of material ductilities in the as-received condition, Figure 4, shows that Cast L is significantly more ductile than Cast H. This is explained by consideration of the δ-ferrite distribution, which in Cast L consists of isolated islands, whereas in Cast H consists of a continuous network. During creep testing at 650°C the δ-ferrite transforms to σ-phase (<500 hours). This phase is brittle in character and during creep deformation easily cracks. The continuous network of cracked σ-phase in Cast H would provide an easy path for failure, Figure 8a. In Cast L these cracked σ-phase particles are present as isolated islands in an austenite matrix. The observed higher ductility in Cast L is thus a consequence of the difficulty of linking the cracked σ-phase regions which are surrounded by the austenite matrix, Figure 8b.

Consideration of the material ductility changes as a function of ageing shows that Cast L has an increased rupture ductility after ageing but the similar increase observed for Cast H is much larger in magnitude. To explain this observation it is necessary to consider not only the change in the matrix strength but also the formation of σ-phase particles during ageing. It is thus suggested that in Cast L the austenite fails at a critical value of strain, which is modified only slightly by ageing resulting in only a small increase in rupture ductility.

The rupture ductility of Cast H is significantly increased by preageing. In the aged condition, this cast contains a nearly continuous network of σ-phase. At first sight it is therefore surpising that the aged material exhibits a higher ductility than the as-received material. However, this observation is explained by the lower creep strength of the matrix in the pre-aged condition, Figure 3. Therefore in the aged condition the matrix accommodates more strain thus inhibiting crack nucleation and propagation at the σ-phase.

In the as-received condition the rupture life of Cast L is significantly better than for Cast H, Figure 5. This is a simple consequence of the greater creep strength and better creep ductility for Cast L. The rupture life results for aged Cast L show that the large increase in the minimum creep rate, as a result of pre-ageing, has a dominant influence on the life compared to the influence of the small increase in rupture ductility. This qualitative correlation between increase in creep rate and decrease in rupture life is true for both aged conditions. The observed insensitivity of the creep rupture life to material condition for Cast H is a consequence of the simultaneous increase in creep rate and increase the rupture ductility after ageing.

CONCLUSIONS

1. Differences in the as-received creep strength of two casts of 347 stainless steel has been attributed to the presence of different amounts of stacking faults in the austenite matrix.

2. Pre-ageing at 650°C for 2000h and 10000h results in a reduction in the creep strength. This is a consequence of a reduction of the number of stacking faults in the austenite matrix.

3. Creep fracture of Cast L occurs by transgranular failure of the austenite together with small amounts of creep damage at isolated σ-phase particles. Preageing produces a large increase in creep rate significantly reducing the rupture life.

4. Creep failure of Cast H occurs by crack interlinkage along the continuous σ-phase network. Pre-ageing reduces the creep strength of this cast but significantly increases its ductility. Hence the rupture life is apparently unaffected by pre-ageing.

ACKNOWLEDGEMENTS

The paper is published with the permission of the Central Electricity Generating Board.

REFERENCES

1. F C Hull. Welding Journal Res. Suppl. 46 p339 (1967).

2. R G Berggren, N C Cole, G M Goodwin, J O Stiegler, G M Slaughter, R J Gray, R T King. Welding Journal Res. Suppl. 57 p167 (1978).

3. R G Thomas, R D Nicholson, R A Farrar. Metals Technology 11 p61 (1984).

4. C A P Horton, P Marshall, R G Thomas. CEGB Report RD/L/2045/N/81 (Sept 1981).

5. V K Sikka and S A David. Met. Trans 12A p883 (1981).

6. G Bernasconi, W Nicodemi, R Roberti. Proc. Conf. Mechanical Behaviour and Nuclear Applications of Stainless Steel at Elevated Temperatures ASM, Varese, Italy 1981, p182, The Metals Society.

7. G Knowles. Metal Science 11 p117 (1977)

8. D J Powell, R Pilkington, D A Miller. Proc. Conf. Stainless Steels '84 Gothenburg 1984 p382 Inst. of Metals, London (1985).

9. R C Ecob. Acta Met. 32 p2149 (1984).

10. B A Senior. Private Communication.

11. L E Murr. 'Electron optical Applications In Materials Science' McGraw-Hill (1970).

12. P Marshall. 'Austenitic Stainless Steels, Microstructure and Mechanical Properties; Elsevier Applied Science Publishers (1984).

TABLE 1

Cast Identification	C	Si	Mn	S	P	Ni	Cr	Mo	V	Co	Ti	Nb	Cu	N	δ-ferrite
Cast L	0.076	0.99	0.52	0.0081	0.009	8.13	16.87	0.48	0.38	0.025	0.003	0.74	0.044	0.082	2.7
Cast H	0.075	0.963	0.62	0.0069	0.0035	8.08	19.63	0.23	0.38	0.025	0.002	0.569	0.036	0.060	13.4

Table 2

Comparison of Matrix NbC Particle Sizes

Cast Identification	Mean Particle Size As-received	Mean Particle Size 10000h Aged at 650°C
Cast L	30.08 nm	31.12 nm
Cast H	35.59 nm	31.20 nm

Table 3

Comparative Synopsis of Creep Properties of Casts L and H

	As-received	Aged 2000h at 650°C	Aged 10000 at 650°C
Minimum Creep Rate	L << H	L < H	L < H
Rupture Ductility	L >> H	L > H	L ~ H
Rupture Life	L >> H	L > H	L ~ H

Table 4

Comparative Synopsis of Creep Properties as a Function of Ageing

	Cast L	Cast H
Minimum Creep Rate	AC << 2K < 10k	AC << 2K ~ 10K
Rupture Ductility	AC < 2K ~ 10K	AC << 2K < 10K
Rupture Life	AC >> 2K > 10K	AC ~ 2K ~ 10K

AC-As-received; 2K-2000h aged at 650°C; 10K-10000h aged at 650°C

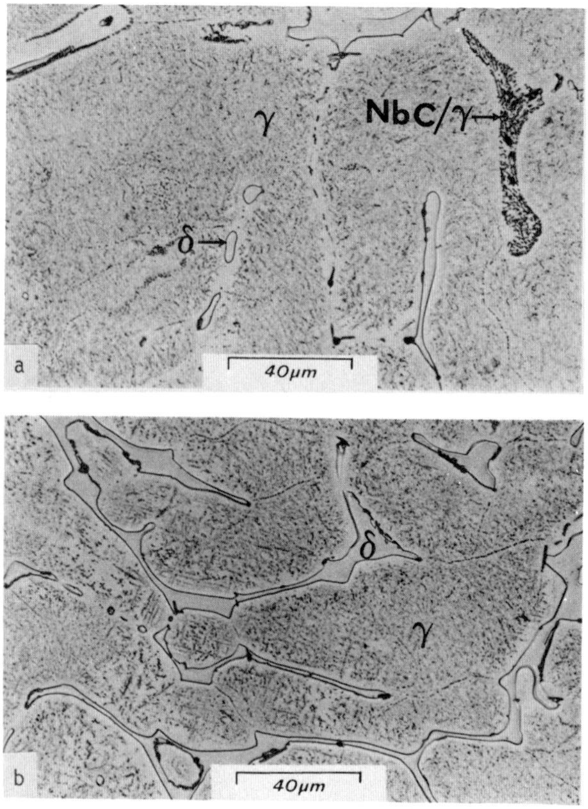

Figure 1. Optical micrographs of the as-received microstructures showing δ-ferrite distributions for: a) Cast L b) Cast H

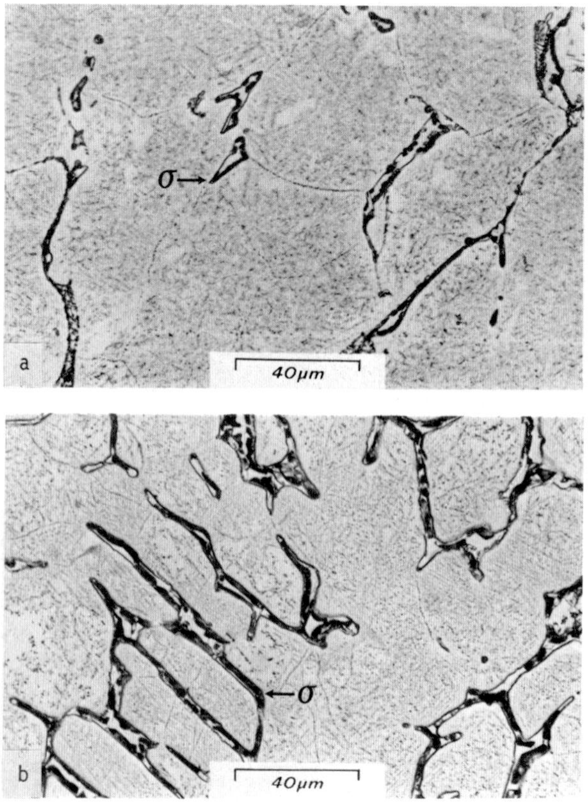

Figure 2. Optical micrographs of 2000h aged at 650°C microstructures showing development of σ-phase for: a) Cast L b) Cast H

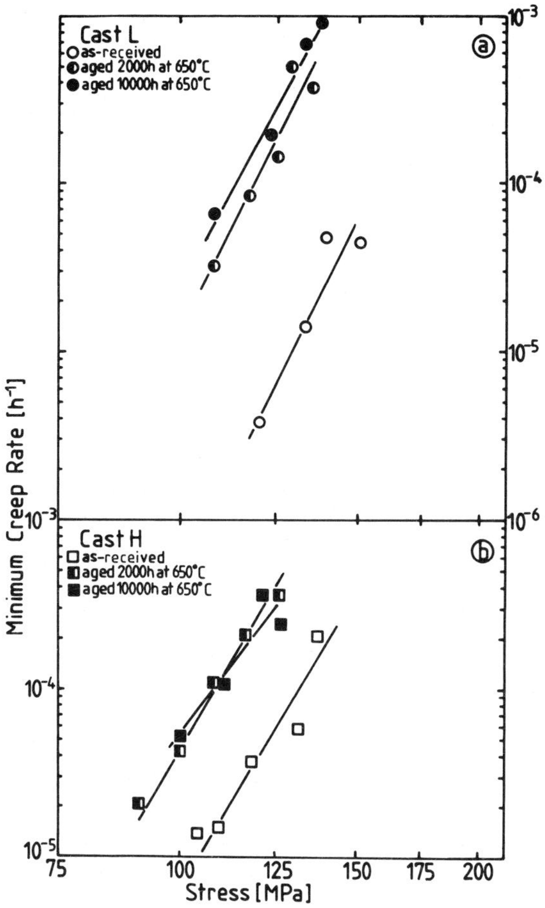

Figure 3. Variation of minimum creep rate with stress for: a) Cast L b) Cast H

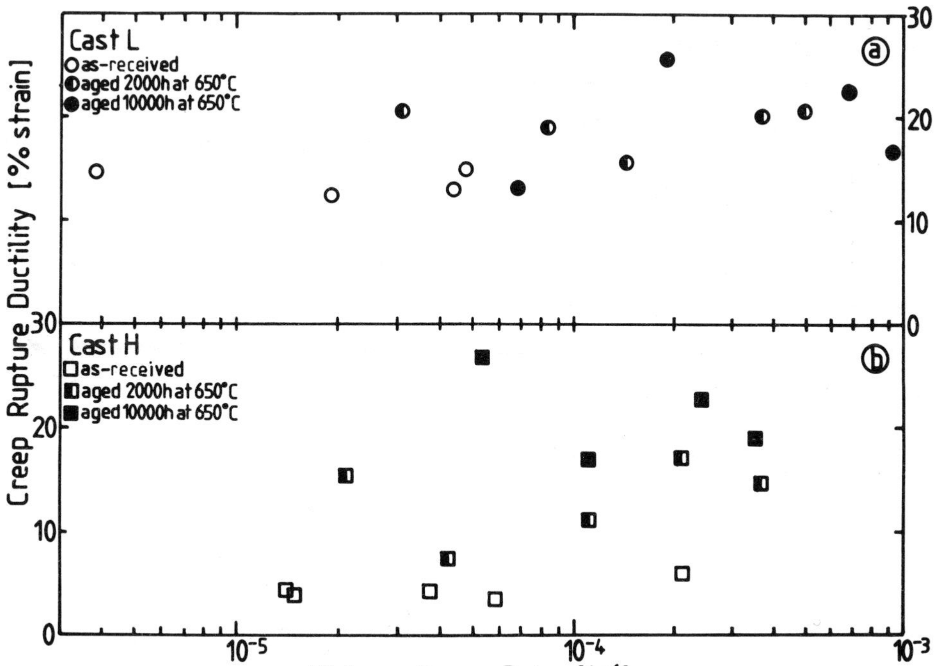

Figure 4. Variation of creep rupture ductility with minimum creep rate for: a) Cast L b) Cast H

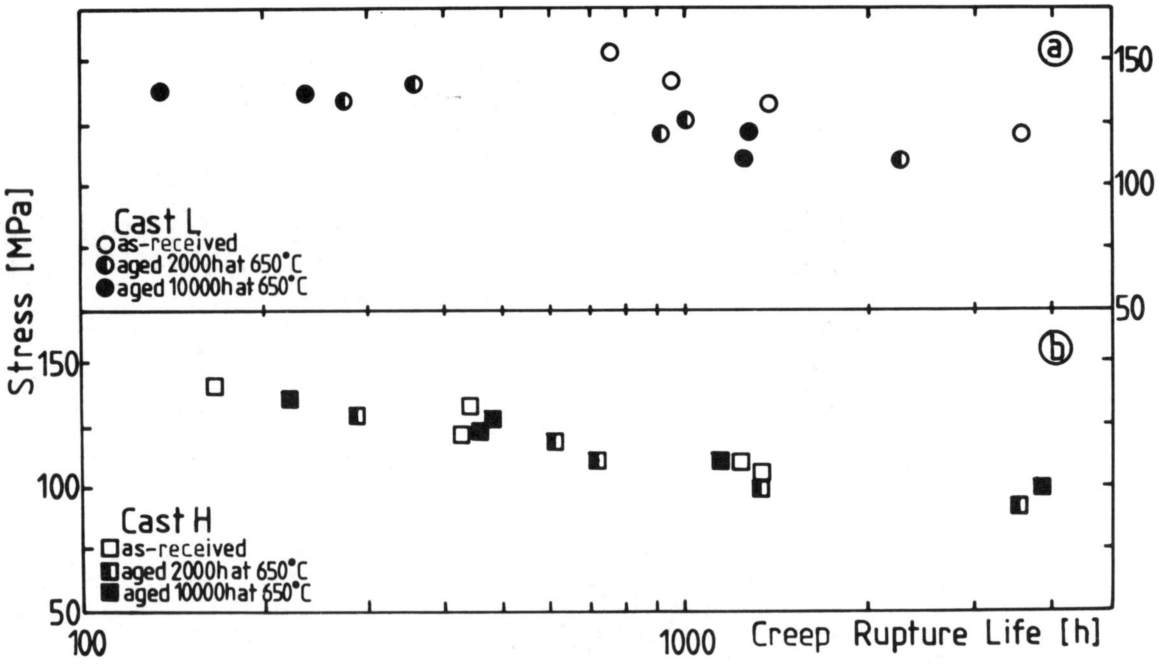

Figure 5. Variation of creep rupture life with stress for: a) Cast L b) Cast H

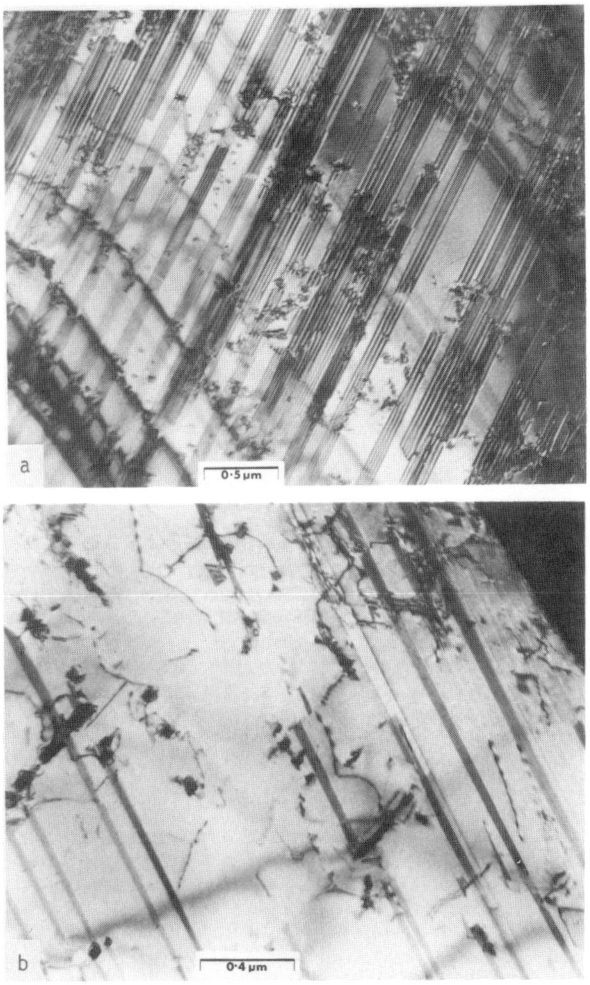

Figure 6. Transmission electron micrographs of thin-foil specimens showing stacking faults in as-received microstructures of: a) Cast L b) Cast H

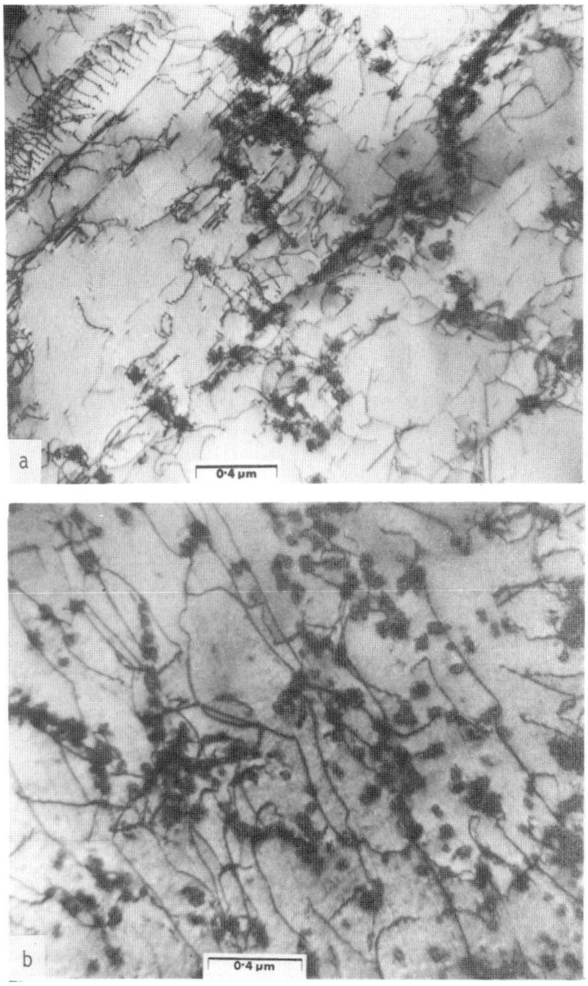

Figure 7. Transmission electron micrographs of thin foils from material aged 10000h at 650°C, showing unit dislocations in: a) Cast L b) Cast H

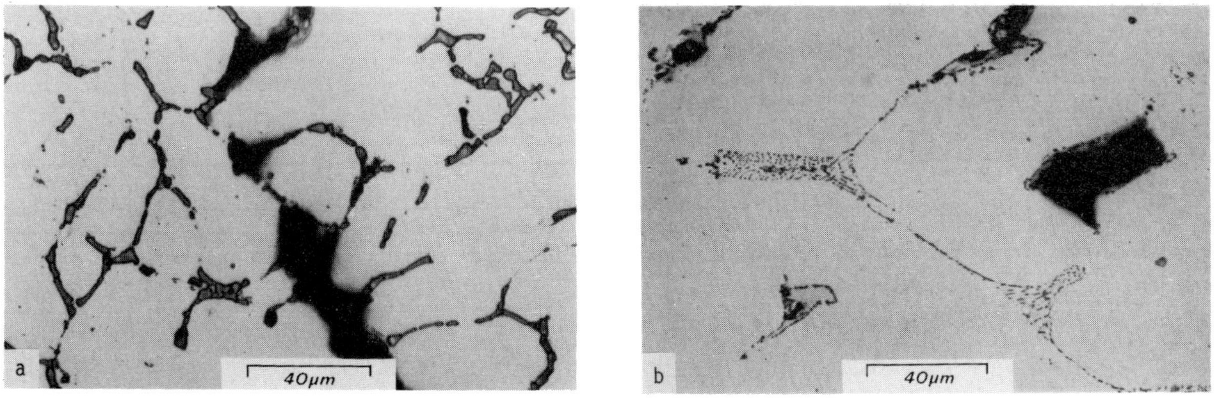

Figure 8. Optical micrographs of creep damage in specimens of as-received material: a) Cast H b) Cast L

□

Mechanical behaviour of a 29Cr—4Mo—2Ni superferritic stainless steel

M Anglada, M Nasarre and J A Planell

The authors are in the Department of Metallurgy, Polytechnical University of Catalonia, Barcelona, Spain.

SYNOPSIS

The tensile and cyclic deformation behaviour of a 29Cr-4Mo-2Ni superferritic stainless steel has been studied, the tensile deformation in a temperature range from 21 to 500°C, and the cyclic deformation at room temperature in specimens in either the annealed or the 475°C embrittled condition. It is shown that the existence of deformation by twinning and dynamic strain ageing above 230°C is associated to the presence of molybdenum and nickel. The cyclic stress-strain curves have been determined at several strain rates, and the dislocation substructure generated has been observed by transmission electron microscopy. Examination of the surface of the fatigued specimens of the steel in the annealed condition has shown that cracks nucleate mainly at the grain boundaries, while in the 475°C embrittled specimens they form mainly at the slip bands and twin-matrix interfaces.

INTRODUCTION

The superferritic stainless steels have been developed in the last decade as a response to the need of steel with high pitting and corrosion resistance in aggressive environments such as polluted sea water (1,2). In spite of the high chromium content, these alloys do not suffer from the common low ductility of the traditional ferritic stainless steels with high Cr content. This is achieved by keeping the amount of interstitial elements in solution at very low levels and by stabilizing the concentrations of these elements with the addition of small amounts of Nb and Ti to form carbonitrides. However, there are two ranges of temperature centred around 475 and 800°C where some embrittlement reactions occur, namely, the spinodal decomposition which produces a phase rich in Cr and a phase rich in Fe, and the precipitation of the sigma phase, respectively (3,4). Therefore, the annealing treatments must be carried out at high temperature (about 1000°C) and the specimens must be quenched in water in order to ensure that no precipitation occurs during cooling.

The study of the mechanical properties of the superferritic stainless steels has not received much attention, and the few papers published so far on this matter, have mainly dealt with the measure of the energy absorbed in the impact test and the tensile behaviour of the steel in the embrittled condition (5-9). In the present paper, the tensile and cyclic deformation behaviour of the superferritic stainless steel 29Cr-4Mo-2Ni is studied. Attention is focussed on the presence of twinning at above room temperature, the dislocation structure generated during fatigue, the influence of the 475°C embrittlement on the cyclic deformation behaviour, and the sites of nucleation of the fatigue cracks. The only works known to us on the cyclic deformation behaviour of superferritic stainless steels have concentrated on the study of the alloy Fe-26Cr-1Mo in the annealed (10) and 475°C embrittled condition (11), while the occurrence of twinning during the deformation in tension has been investigated by Magnin and Moret (12) in various superferritic stainless steels with variable amounts of Cr, Mo, Cu and Ni.

EXPERIMENTAL PROCEDURE

The stainless steel investigated was kindly supplied by Allegheny Ludlum Steel Corporation in the annealed condition, in the shape of plates of 13 mm thickness. The grains had rather a pancake structure with grains of very large size at about half depth of the plates and grains of much smaller size near the surfaces. In order to decrease and homogenize the grain size, the plates were cold reduced to about 7 mm in thickness, recrystallized at 1030°C in an argon atmosphere and quenched in water. The final grain size was 30 μm. The chemical composition of the alloy as specified by the manufacturer is given in Table 1.

Table 1. Composition, in wt.%, of the steel studied.

Cr	Mo	Ni	P	Mn	Si	N	C	Fe
28.3	3.92	2.43	0.010	0.10	0.14	0.012	0.002	bal.

The tensile specimens were machined in a rectangular shape of 2 mm thickness, 10 mm width and 20 mm length. The tensile tests were carried out in an Instron testing machine in a temperature range from 21°C to 500°C and at strain rates between 6.0×10^{-5} and 3.3×10^{-2} S-1.

The fatigue specimens were machined with a dumb bell shape of 3mm diameter and 5mm gauge length and were electropolished in a chemical solution of acetic acid and perchloric acid (9:1) until a specular surface finish was achieved. The fatigue tests were conducted at room temperature in tension -compression with R=-1 by controlling either the total strain amplitude or the plastic strain amplitude.

In order to study the influence of the 475 °C embrittlement on the tensile and cyclic deformation behaviour, several specimens were heat treated in a tubular furnace in air at 475 ± 2 °C and quenched in water. Before testing, a surface layer about 0.1 mm deep was eliminated by electropolishing the specimens.

The surface of the fatigued specimens was observed by optical microscopy using a Nomarski interferential contrast attachment, and by scanning electron microscopy.

The preparation of the thin foils for observation by transmission electron microscopy was carried out by cutting foils of 0.3 mm thickness perpendicularly to the axis of the fatigued specimens. The disc faces were ground to make them parallel in a special attachment and by using a fine grain silicium carbide grinding paper. Final thinning was accomplished by jet electropolishing at 13 °C in the solution mentioned above. The thin foils were examined in the electron microscopes Jeol CMX-100 and Philips 300 both operated at 100 kV.

EXPERIMENTAL RESULTS

The main feature of the tensile behaviour is the occurrence of twinning, which was detected by the typical sounds and load drops, and by examining several sections of the tensile specimens after electropolishing and etching. At room temperature, twinning was far more frequent in specimens in the as received condition which had a larger grain size, and at the higher strain rates used. Close examination of the drops in load showed that twinning first appeared at stresses well below the conventional elastic limit.

The tension tests were conducted from room temperature to 500°C in recrystallized specimens and the tensile curves are shown in Fig. 1. It may be noticed that the load drops begin to appear at temperatures above 230°C and that the maximum amount of serrated flow occurs at around 280°C and disappears at 450°C. At the highest temperature where twinning still occurs, the load drops with audible noises appeared only after reaching the maximum load in the tensile curves.

A detailed examination of the load drops showed two distinct types of serrations, as depicted in Fig. 1. Most of the more regular drops in load took place without any audible noise and they were associated to dynamic strain ageing. Since a negative strain rate sensitivity of the flow stress is one of the main characteristics of dynamic strain ageing (13), the sign of this parameter was determined by changing the strain rate during the tensile tests and measuring the variation in flow stress. It was found that the strain rate sensitivity of the flow stress was negative at temperatures where there were regular drops in load. In addition, it may also be noticed in Fig.1 that when serrated flow occurs, the tensile strength increases, which is another well known phenomenon associated to dynamic strain ageing (14). These two observations strongly suggest that the serrated flow that appears without any audible noise may be caused by dynamic strain ageing.

In order to examine whether the presence of twinning above room temperature could have been induced by some modification of the structure taking place while heating up the steel to the test temperature, two specimens were aged at 283 °C and 480 °C for 7 and 27 hours respectively. These times are much longer than those required to reach and stabilize the temperature, which are typically about 2 hours. Then, the specimens were deformed at 150°C, temperature at which twinning does not occur in the untreated specimens. The result was that the tensile curve of the specimens treated at 283 °C did not show any load drop. This seems to indicate that the appearance of twinning at 283 °C is not caused by the heat treatment that the specimens undergo while heating them up in the testing machine. On the other hand, the specimen treated at 480°C deformed partly by twinning at 150°C, which is a well known phenomenon of the Fe-Cr alloys (15) treated for long times at 475 °C and is caused by the spinodal decomposition. The fact that twinning was not detected in the specimens deformed at 500 °C indicates that, even at this temperature, the time spent in heating up the specimens in the testing machine was too short to induce twinning.

The response of the steel to cyclic deformation in tension-compression is given in Fig.2 where the cyclic stress-strain curves have been plotted for specimens deformed at different strain rates. In these curves, the average of the peak stresses in tension and compression at saturation has been plotted versus the logarithm of the amplitude of the plastic strain, which was taken as half the width of the hysteresis loops. The average peak stress changed with the number of cycles only during the first 20 cycles approximately, in which some small amount of cyclic work hardening took place.

The hardening curves of specimens aged for different times at 475 °C and cyclically deformed at constant plastic strain amplitude of 0.1 %, are shown in Fig.3. It can be seen that besides an increase in the elastic limit measured at 0.1 % plastic strain, which is given by the peak stress corresponding to the first cycle, there is strong hardening during the first 15 cycles,approximately. During these initial cycles there were drops in load in the hysteresis loops which were accompanied by audible noises characteristic of twinning. Later on, the alloy softened to a considerable extent and, in the case of specimens aged for long times, softening persisted until fracture.

The observation by transmission electron microscopy of thin foils of undeformed specimens of the steel, either in the recrystallized or in the as received condition, showed the existence of a monophasic structure of very low dislocation density and with very few precipitates (Fig.4). A comparative chemical analysis by EDAX of the composition of the matrix and of the precipitates gave as the main difference a much higher Mo content in the precipitates. The composition of these was found to be close to that of the chi-phase, $Cr_{12}Fe_{30}Mo_{10}$.

The observation of the dislocation structure generated by cyclic deformation was mainly carried out in the recrystallized steel and in specimens deformed in a range of strain rates between 10^{-4} and 10^{-2} S^{-1}, only few observations were made in the steel in the as received condition. In both cases, the same basic dislocation structures were found. At plastic strain amplitudes of 0.4 %, the dislocation structure at saturation consisted of tangles and cells (Fig.5) and walls perpendicular to the primary Burgers vector. This last structure was more abundant at high plastic strain amplitudes (Fig.6). The dislocation structure in the initial cycles was examined after 15 cycles and it was found that at plastic strain amplitudes of 0.94 % dislocation walls had already begun to form form, while at plastic strain amplitudes of 0.1 %

the dislocation density was much lower and there were dislocation loops and dipoles.

In the case of the specimens aged at 475°C and fatigued at a plastic strain amplitude of 0.1 %, the observation by TEM confirmed the existence of twinning (Fig.7). Twins were also observed by optical microscopy on the surface of the fatigued specimens.(Fig.8).

Microscopic examination of the specimens after cyclic deformation showed the existence of a relief caused by changes in shape of the surface grains. Cracks nucleated at the surface grain boundaries and at deformation bands (Fig.9), the former sites predominated at the highest strain rate.In specimens heat treated at 475°C, cracks nucleated at deformation bands were much more abundant (Fig. 10).

DISCUSSION

The most interesting observation about the tensile behaviour of the steel studied is the occurrence of deformation by twinning at temperatures as high as 400°C, that is, at about 0.4 T_m where T_m is the melting point of the alloy studied. In b.c.c. metals, deformation by twinning is usually restricted to low temperatures or high loading rates (16), although the presence of high concentrations of substitutional elements in solid solution, as in the present case, also favours deformation by twinning (17). To our knowledge, twinning in the high temperature regime of b.c.c. metals has only been reported recently by Magnin and Moret (12) in alloys with about 25% Cr and smaller amounts of Ni, Mo and Cu. These authors have suggested that the existence of twinning at high temperature is caused by an intrinsic effect of Ni and by the spinodal decomposition. However, this idea is difficult to reconcile with the observation that ageing at 283°C, even for periods of time much longer than those required to heat up the specimens and stabilize the temperature in the testing machine, does not induce deformation by twinning at 150°C, while there is profuse twinning during deformation at 283°C. We suggest that the twinning observed at high temperature is related to dynamic strain ageing by substitutional elements. It should be noticed that the temperature range where twinning occurs is roughly the same as that where dynamic strain ageing takes place in binary substitutional alloys of α-Fe (230-500°C) (14). As the mobile dislocations become locked by the diffusion of solutes during dynamic strain ageing, the stress required for glide increases and in some places it may reach the value necessary for the onset of twinning at the test temperature.

Twinning has not been observed at high temperature in Fe-Cr and Fe-26Cr-1Mo alloys, and this may be related to the fact that Cr does not produce dynamic strain ageing (14), and that in the latter alloy the amount of Mo is too small. According to Leslie (14), the minimum concentrations of Mo and Ni required to have serrated flow in binary iron alloys are between 0.6 and 1.8% for Mo, and between 3.0 and 5.0% for Ni. In addition, twinning occurs in the alloy Fe-29Cr-4Mo (9) which does not contain Ni. Therefore, since Mo is more effective than Ni in causing dynamic strain ageing, it seems very likely that Mo had even a stronger influence than Ni in promoting deformation by twinning.

The low work hardening exhibited by the alloy studied is a well known characteristic of the ferritic stainless steels in both unidirectional and cyclic deformation (10,18). The increase in the saturation stress with the plastic strain amplitude displayed in the cyclic stress-strain curves of Fig. 2 , arises mostly from the increase in the peak stress of the first cycle at each strain amplitude. The contribution to the saturation stress from cyclic hardening during the subsequent cycles is relatively small.

The fact that the saturation stress is larger at high strain rates may be caused by an increase in the thermal component of the stress. In b.c.c. metals, this is the major component of the total stress and is very sensitive to the temperature and the strain rate at temperatures below T_0 (10,19-21). This is the temperature above which the mobility of edge and screw dislocations become similar and athermal behaviour is approched. Magnin et al.(22) have shown that the addition of nickel to Fe-26Cr alloys induces high friction stresses which increase T_0 and change the plastic deformation mode. For example, the deformation mode of the Fe-26Cr-5Ni alloy correspond to the low temperature regime even at low strain rates.

As can be seen in Fig.2, at the highest strain rate studied, the saturation stress increases more slowly with the plastic strain amplitude than at lower strain rates. This may be related to a relative decrease of the athermal component of the saturation stress caused by a diminution in the dislocation density, as has been observed in α-iron single crystals (20) and in the alloy Fe-26Cr-1Mo (10).

The dislocation structures developed under cyclic deformation in the steel studied are characteristic structures which have also been observed in other b.c.c. alloys. For example, Sesták et al. (23) have recently shown that in Fe-3%Si single crystals with crystallographic orientation near the centre of the unit stereographic triangle, the veins formed at high strain amplitudes during the initial cycles, give way to walls and cells at saturation. Therefore, the different dislocation structures developed in distinct grains of the same specimen of the steel studied, could correspond to different stages of the evolution of the dislocation structure. However, if this were the case, the change from veins to walls and cells would have taken place without a change in the saturation stress, since in the present steel this remains constant after the initial hardening stage until propagation of macroscopic cracks sets in. The change from veins to cells has also been reported in the alloy Fe-26Cr-1Mo (11), but it is accompanied by an increase in the saturation stress. In the present case, the dislocation structure in the initial hardening stage has been examined after 15 cycles at plastic strain amplitudes of 0.1% and 0.9%. In the former case, only dislocation dipoles and loops are found, which evolve to dislocation tangles without hardly any increase in the peak stress, while in the latter a dislocation wall structure is already present after 15 cycles.

It is now well known that in Fe-Cr alloys with high chromium content, ageing around 475°C causes the spinodal decomposition of the alloy in zones rich in Cr and others rich in Fe. The cyclic deformation behaviour of spinodally decomposed Fe-Cr alloys has been hardly studied. Park et al. (11) have recently investigated the evolution of the composition fluctuations during cyclic deformation in the Fe-26Cr-1Mo steel by using small angle neutron scattering. They found that the amplitude of the composition fluctuations decrease with the number of cycles but the wavelength remains essentially unchanged.

The fact that the strain rate sensitivity of the saturation stress of the as quenched material is not altered by ageing (24), shows that the increase in the cyclic stress produced by this treatment (Fig.3) arises from an increase in the athermal component of the total stress.

The softening observed during the cyclic deformation of the aged material could arise from the irreversible to-and-fro motion of dislocations in tension and compression which disorders the composition waves (11), decreases the composition amplitudes and reduces the coherency strains.

In the present work, the final hardening stage observed by Park et al. (11) and which mark the change from a vein to a cell structure, has not been observed. The reason may lie in that in our material the embrittlement produced by ageing is more severe and fracture occurs before this stage is reached as it happens in Fe-26Cr-1Mo aged for long times.

The tendency for cracks to nucleate at grain boundaries at high strain rates can be rationalized in terms of the slip asymmetry (21). The operation of different slip planes in tension and compression may induce a net change in shape of the surface grains which deform more freely than the interior grains (25). If neighbouring surface grain suffer incompatible changes in shape, cracks may nucleate at the grain boundaries. At high strain rates the slip asymmetry increases, so that is more likely that cracks nucleate at grain boundaries as has been observed in Fe-26Cr-1Mo (10). Similarly, the incompatible changes in shape of the twins and of the surrounding matrix can also induce the nucleation of cracks at the twin-matrix interface in the aged steel.

ACKNOWLEDGEMENTS

The authors wish to thank the Allegheny Ludlum Steel Corporation for providing the material, the "Servei de Microscopia Electrònica" of the University of Barcelona and the Department of Materials of Queen Mary College of the University of London for use of their TEM facilities. This work has been supported by the CAICYT of Spain under Grant No. 2696/83.

REFERENCES

1. M. A. Streicher, Corrosion, 30, 77 (1974).
2. R. F. Steigerwald, A. P. Bond, M. J. Dundas and E. A. Lizlovs, Corrosion, 33, No,8, 279 (1977)
3. R. F. Steigerwald, H. D. Dundas, J. R. Redmond and R. M. Davison, Proc. of Stainless Steels '77, p. 284, Climax Molybdenum Company,London London, 1979.
4. J. J. Demo, Structure, Constitution and General Characteristics of Wrought Ferritic Stainless Steels, ASTM STP 619, American Society for Testing and Materials, 1977.
5. T. J. Nichol, A. Datta and G. Aggen, Met. Trans. 11A, 573 (1980).
6. H. Kiesheyer and H. Brandis, Z. Metallkunde, 67, 258 (1976).
7. N. Ohashi, Y. Ono, N. Kinoshita and K. Yoshioka, Toughness of Ferritic Stainless Steels, ASTM STP 706, Ed. by R. A. Lula, American Society for Testing and Materials, 1980
8. T. J. Nichol, Metallurgical Trans. 8A, 229 (1977).
9. G. Aggen, H. E. Deverell and J. J. Nichol, Proc. of "MiCon 78: Optimization and Processing Properties and Service Performance Through Microstructural Control", Ed. by H. Abrams, G. N. Maniar, D. A. Nail and H. D. Solomon, ASTM STP 672, American Society for Testing and Materials, 1979.
10. T. Magnin and J. H. Driver, Mater. Sci. Engng, 39, 175 (1979).
11. K. H. Park, J. C. LaSalle and L. H. Schwartz, Acta Metall. 33, 205 (1985).
12. T. Magnin and F. Moret, Scripta Met. , 16, 1225 (1982).
13. R. A. Mulford and U. F. Kocks, Acta Metall., 27, 1125 (1979).
14. W. C. Leslie, The Physical Metallurgy of Steels, Tokyo, 1982.
15. M. J. Blackburn and J. Nutting, Journal of The Iron and Steel Institute, July 1964, p.610.
16. S. Mahajan and D. F. Williams, Int. Met. Reviews, 18, 43 (1973).
17. J. O. Stiegler and C. J. McHargue, Deformation Twinning, Ed. by R. E. Reed-Hill, J. P. Hirth and H. Rogers, Gordon and Breach, New-York, 1964.
18. F. B. Pickering, Physical Metallurgy and the Design of Steels, Applied Science, London, 1983.
19. A. Seeger, Philos. Mag., Ser. 7, 45, 771 (1954).
20. H. Mughrabi, K. Herz and X. Stark, Acta Metall., 24, 659 (1976).
21. M. Anglada and F. Guiu, Philos. Mag. A, 44, 499 (1981).
22. T. Magnin, L. Coudreuse and A. Fourdeux, Mater. Sci. Engng, 63, L5 (1984).
23. B. Sesták, V. Novák and S. Libovicky, to be published.
24. M. Anglada and M. Nasarre, unpublished results.
25. H. Mughrabi, Z. Metallkd., 66, 719 (1975)

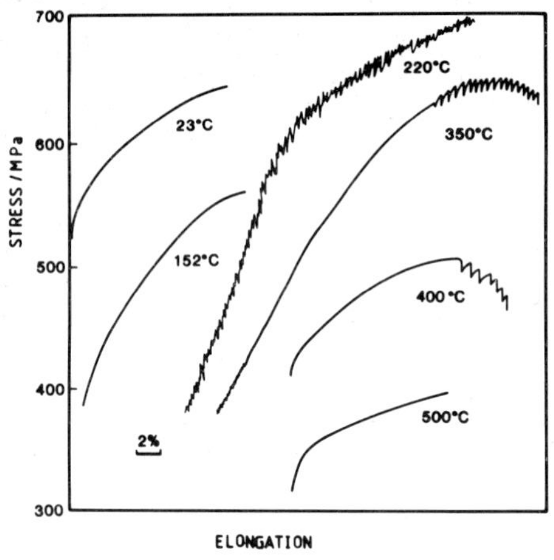

Fig. 1. Tensile curves at different temperatures and at a strain rate of 1.33×10^{-3} S^{-1}.

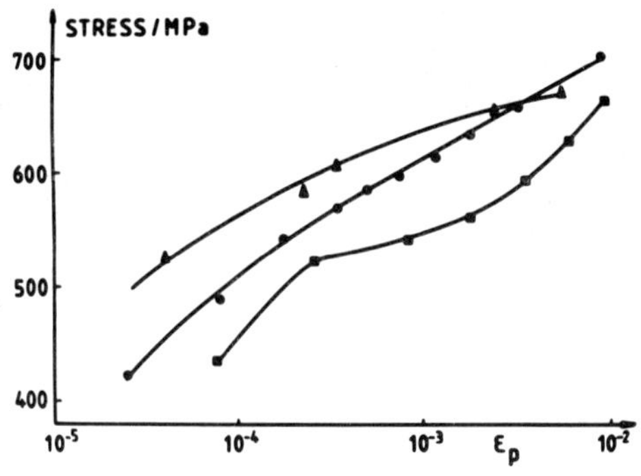

Fig. 2. Cyclic stress-strain curves at room temperature and at different strain rates: ■ 10^{-5} S^{-1}, ● 10^{-3} S^{-1}, ▲ 2×10^{-2} S^{-1}.

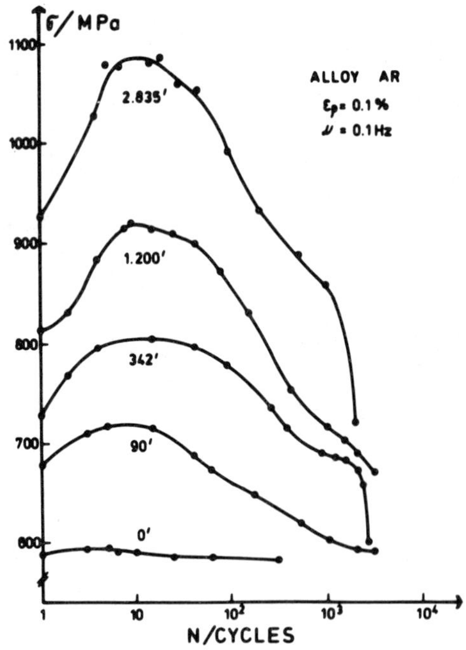

Fig. 3. Hardening curves of the steel aged at 475°C.

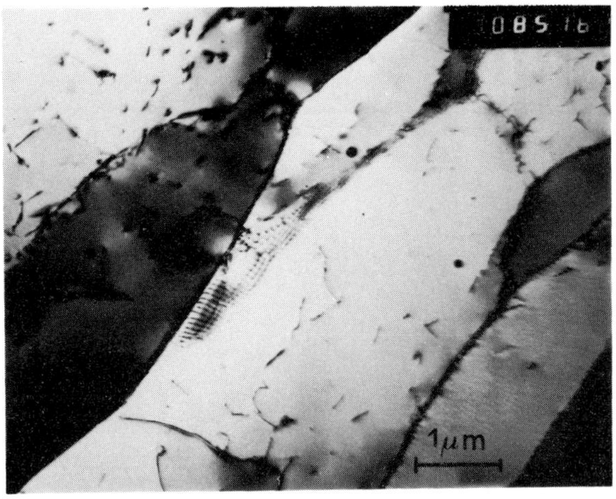

Fig. 4. Structure of the undeformed annealed steel.

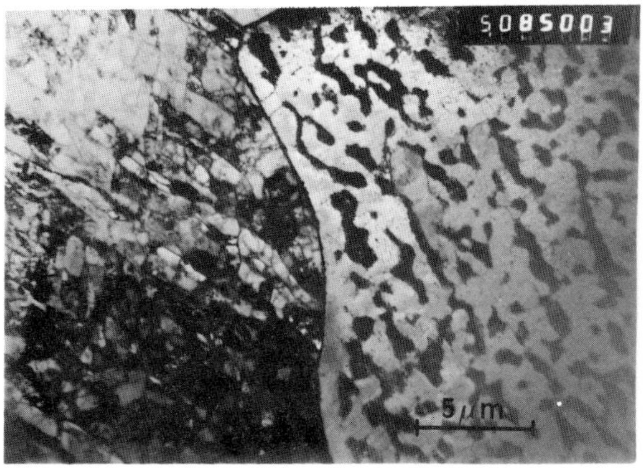

Fig. 5. Dislocation structures at saturation of the steel fatigued at $\epsilon_p = 0.4$ %, $\dot{\epsilon} = 4x10^{-4}$ S^{-1} for 356 cycles.

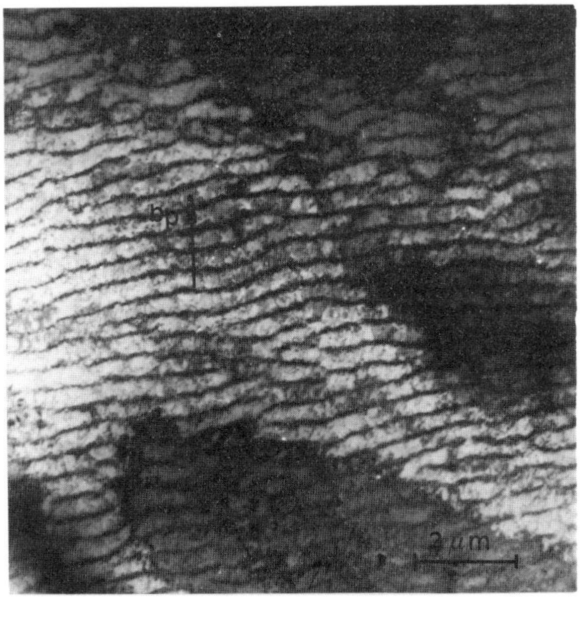

Fig. 6. Dislocation structure at saturation, $\epsilon_p = 0.68$ %, $\dot{\epsilon} = 4x10^{-4}$ S^{-1}, 157 cycles.

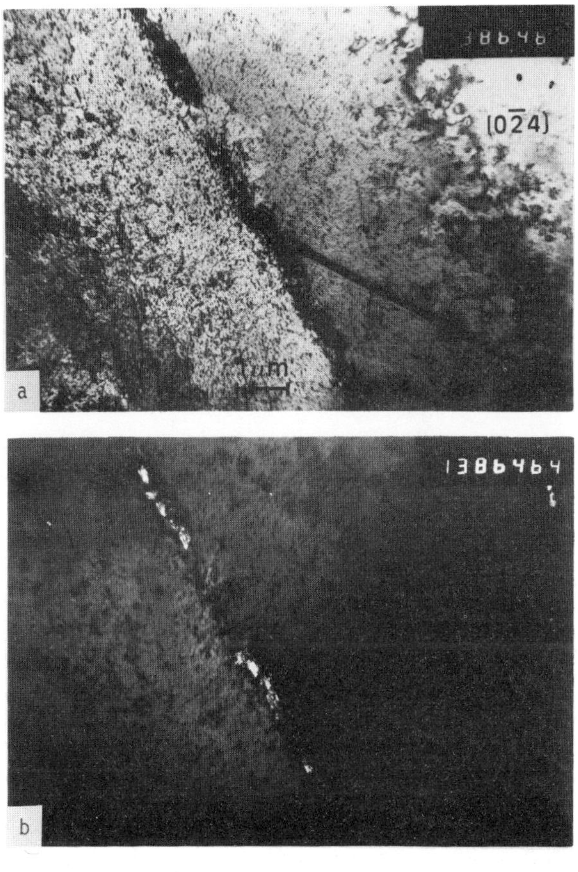

Fig. 7. Twin in the steel aged for 47 hours at 475 °C and fatigued at $\epsilon_p = 0.1$ %, $\dot{\epsilon} = 2.2x10^{-3}$ S^{-1}: (a) Bright field image; (b) Dark field image using $\bar{g} = 11\bar{2}$.

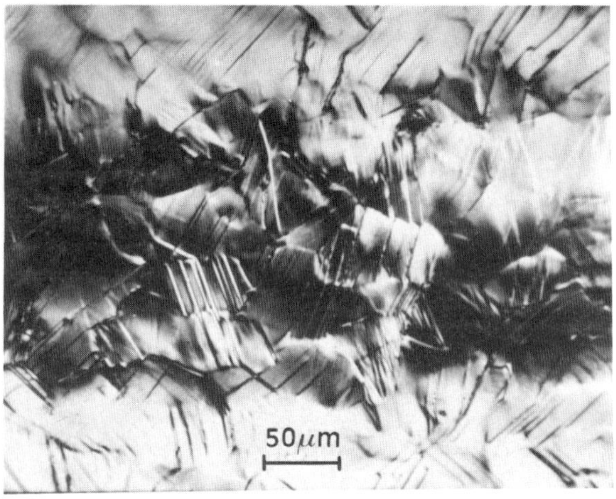

Fig. 8. Aspect of the surface of a specimen of the steel aged for 47 hours and fatigued at ϵ_p = 0.1 % and $\dot{\epsilon}$ = $2.2x10^{-3}$ S^{-1} for 44 cycles.

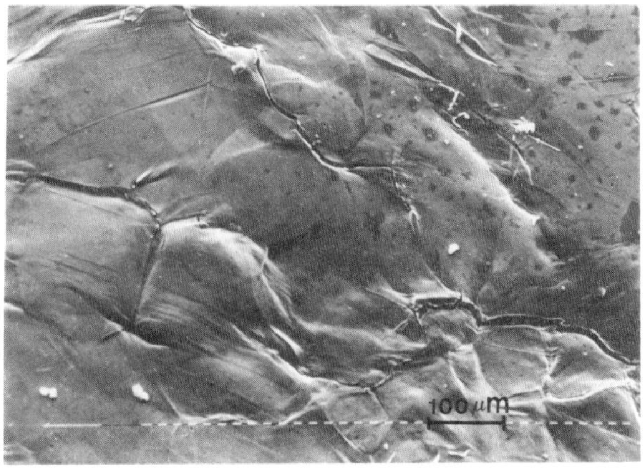

Fig. 9. Cracks in a specimen fatigued at high strain rate ($\dot{\epsilon}$ = $2x10^{-2}$ S^{-1}) and at a saturation plastic strain of $0.4x10^{-2}$.

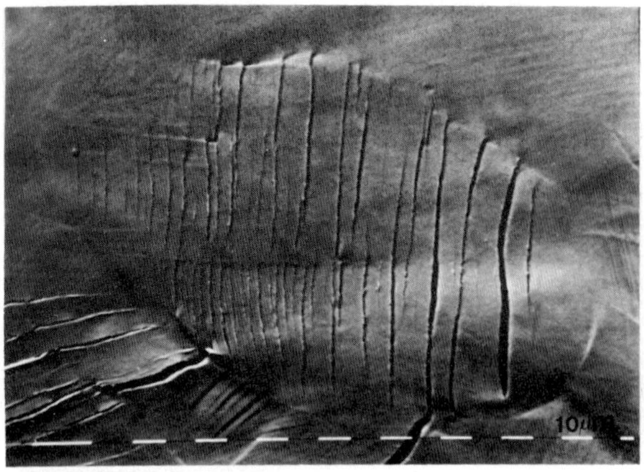

Fig.10. Cracks at deformation bands in the steel aged for 47 hours at 475°C and fatigued at ϵ_p = 0.1 % and $\dot{\epsilon}$ = $2.2x10^{-3}$ S^{-1} for 2000 cycles.

Structure stability of two highly-alloyed nickel-base alloys

A M Wardle and H Eriksson

The authors are in the Research and Development Centre, AB Sandvik Steel, Sandviken, Sweden.

SYNOPSIS

The aim of this work was to obtain detailed microstructural information of two highly alloyed nickel-base alloys, Alloy C and Alloy 625, and to correlate the findings to mechanical properties.

It is well known that inproper welding and heat treatments can cause precipitation in these types of material. The precipitates have been shown to have deleterious effects on the corrosion resistance and the toughness.

It has been shown that the mechanical properties can be directly correlated to the precipitation of carbides. In Alloy C both $M_{23}C_6$ and M_6C were observed, M_6C being the most critical factor on the toughness. In Alloy 625 only $M_{23}C_6$ was observed.

1 INTRODUCTION

Highly alloyed nickel base alloys such as the Alloy C series and Alloy 625 have proved to be of major importance in the oil and gas and chemical process industries, where extremely harsh corrosion conditions exist. The corrosion resistance of the nickel-base alloys is primarily due to the high chromium, molybdenum, tungsten and niobium contents.

These alloys are complex and consequently difficult to manufacture by conventional means. The Osprey spray deposition process offers a method to more readily produce such materials [1].

The high alloying contents unfortunately reduce the structure stability. Hence, precipitation may occur at temperatures which cannot be avoided, for example during cooling after deposition and subsequently in connection with welding or annealing operations. These precipitates have been shown to have deleterious effects on the corrosion resistance and the toughness [2-4]. In order to avoid these unwanted precipitates it is, therefore, important to have information concerning the kinetics and types of precipitates that occur during different heat treatment conditions.

2 EXPERIMENTAL

Alloy C and Alloy 625 were produced as tubular forms by the Osprey spray deposition method. These then underwent a small amount of machining in order to obtain a finished tube with an outside diameter of 200 mm and a wall thickness of 15 mm.

The chemical composition of the two alloys are given in Table 1.

The tubes were heat treated at 1225°C for 2.5 hours and water quenched in order to obtain a precipitate free (ie no carbides or intermetallics) microstructure.

Specimens were heat treated at different temperatures and for various times. These heat treatments were selected in order to cover the ranges over which the precipitation reactions are known to occur in conventionally produced alloys. The temperature range involved was from 700 to 1100°C and the time duration from 1 to 60 minutes, see Table 2.

As can be seen in Table 2, approximately half of the heat treatments were prepared as tensile specimens whilst the others were only for microstructural analyses. Two tensile tests were performed at room temperature for each of the test treatments.

Impact strength tests were also performed on standard Charpy V specimens on Alloy C samples which had been heat treated at 900°C. This temperature is known to cause rapid precipitation in conventionally produced Alloy C and Alloy C-276 [2].

Thin foil specimens were prepared from the various heat treatments for analytical transmission electron microscopy. The foils were electrolytically polished in a solution of 15 % perchloric acid in methanol using a Struers Tenupol; the approximate polishing conditions were -25°C, 15V and 0.20A.

A number of precipitates from each foil were analysed in a JEOL 2000FX analytical transmission electron microscope, ATEM. The analyses were both chemical using a Link Systems AN 10000 Analysis System, and structural via electron diffraction. On each precipitate tilting experiments were performed in order to obtain a series of diffraction patterns and thus reliably determine the structure.

3 RESULTS

3.1 Mechanical properties

The results of the tensile tests for Alloy C and Alloy 625 are summarised in Tables 3 and 4 respectively. The results given are the average of two performed tests.

The effect of the heat treatment, within the test ranges on the tensile properties was observed to be more significant in Alloy C than Alloy 625.

3.1.1 Tensile properties of Alloy C

The largest ductility in Alloy C was exhibited by the as solution treated specimens. In this condition the elongation obtained was 71 % and the reduction in area 66 %. With the shortest heat treatment of 1 minute (which was performed in the range 800-1000^oC) there was a continual reduction in ductility. At 1000^oC the elongation was down to 48.5 % and the reduction in area was 37 %. With increasing time little effect was observed up to 850^oC. At all temperatures from 850 to 1050^oC increasing time corresponded to a significant loss in ductility, this was most significant at 1000^oC. After 30 minutes at 1000^oC the elongation had been reduced to 21.5 % and reduction in area had decreased to 20.5 %. From 1000 to 1100^oC increasing the temperature corresponded to an increase in ductility.

The tensile strengths of Alloy C exhibit less significant and consequently less obvious trends. The initial as solution treated samples had the lowest observed 0.2 % yield strength, YS, of 341 Nmm^{-2} and an ultimate tensile strength, UTS, of 723 Nmm^{-2}. With increasing temperature, at the three primary test times of 1, 5 and 30 minutes, there was initially an increase in the tensile properties which appeared to maximise at 900,850 and 800^oC respectively. The maximum 0.2 % YS was 369 Nmm^{-2}. At temperatures above the maximum 0.2 % YS value increasing time corresponded to a decrease in yield strength. The effect was most significant at 1000 and 1050^oC. The lowest tensile strength combination was observed after a 30 minute heat treatment at 1000^oC which resulted in a 0.2 % YS of 353 Nmm^{-2} and a UTS of 583 Nmm^{-2}.

3.1.2 Tensile properties of Alloy 625

The elongation of as solution-treated Alloy 625 was 67 % and the reduction in area 58.5 %. The shortest time/low temperature heat treatment of 1 minute at 800^oC resulted in a slight increase in the ductile properties, ie 68.5 % and 59 % respectively. Other than this marginal increase the ductility followed exactly the same trend as Alloy C. Although, in the case of Alloy 625 the reduction in area showed much more significant changes than the elongation. As with Alloy C the effect was most pronounced at 1000^oC. After 30 minutes at temperature the elongation was down to 52.5 % and the reduction in area had decreased to 39.5 %.

The 0.2 % YS of as solution treated Alloy 625 was 323 Nmm^{-2} and its UTS was 732 Nmm^{-2}. Increasing time and/or temperature up to 1050^oC had the effect of increasing the tensile strengths. The maximum observed 0.2 % YS was 359 Nmm^{-2} which was obtained after 5 minutes at 1050^oC and the maximum UTS was 756 Nmm^{-2}, obtained after 30 minutes at the same temperature.

3.1.3 Impact resistance of Alloy C

The impact test results, carried out on Alloy C specimens which had been treated at 900^oC, are given in Table 5. The impact strength is highest in the as solution annealed condition ie 194 Nm and subsequently decreases with increasing time to 30 Nm after a 1 hour heat treatment.

3.2 Microstructural results

3.2.1 Solution treated specimens

The solution treated samples of both Alloys as expected were free of all carbides and intermetallic phases. However, there were small amounts of oxides present. These were slag inclusions which occurred during the production of the tubular pre-forms. Alloy C contained Fe_3O_4, SiO_2 and Al_2O_3, primarily the former. Alloy 625 contained primarily $MnCr_2O_4$, with some SiO_2 and $MnSiO_3$. These oxides were analysed by electron diffraction and examples of micrographs and diffraction patterns for the common oxides are given in Figure 1.

The oxides were observed in all specimens.

3.2.2 Precipitation

The precipitation which occurred after the various heat treatments for the two alloys are summarized in Figures 2 and 3.

3.2.2.1 Precipitates in Alloy C

It can be seen from Figure 2 that Alloy C contained $M_{23}C_6$ even after the shortest time - lowest temperature heat treatments. Following increased time/temperature combinations M_6C precipitates were observed. Hence with further increases in time/temperature M_6C becomes the more dominant carbide, forming an increasingly more continuous layer of coarser precipitates.

The precipitates were primarily nucleated at the grain boundaries, Figure 4 shows typical examples of micrographs, diffraction patterns and EDS spectra of the carbides. When compared with the matrix the precipitates were high in molybdenum and tungsten content, but also contained chromium, nickel and iron. The specimens which contained the largest amounts of M_6C tended to have precipitates which were even higher in molybdenum and tungsten content and substantially lower in chromium.

3.2.2.2 Precipitates in Alloy 625

Alloy 625 was not as susceptible to carbide precipitation as Alloy C (compare Figure 3 with Figure 2). The temperature at which the $M_{23}C_6$ carbides were first observed was 800^oC, after a 1 hour treatment. The shortest time of 1 minute first produced carbide precipitation at 950^oC. The precipitates were nucleated at both the grain boundaries and at the $MnCr_2O_4$ inclusions (Figure 5). With increasing time and temperature up to approximately 1100^oC there was an increase in the proportion of $M_{23}C_6$.

When compared with the matrix these carbides were high in niobium, molybdenum and chromium (Figure 5c). With increasing time and temperature to approximately 1 hour at 900^oC or 2 minutes at 1100^oC there was a greater tendency for the carbides to be nucleated at the oxides. These carbides had a higher molybdenum content and also contained silicon (Figure 5d).

4 DISCUSSION

4.1 Alloy C

4.1.1 Precipitate Formation

The precipitate fields observed for Alloy C are shown in Figure 2 along with the curves given in the literature [2] for conventionally produced Alloy C and Alloy C-276. It shows that in the conventionally produced Alloy C-276 material intermetallics were precipitated out, ie P-phase or Ni_7Mo_6. In Alloy C although it was not specified by Leonard what the precipitates were it has been shown [3,4] that σ and A_7B_6 type intermetallics along with M_6C, $M_{23}C_6$ and Mo_2C_6 and Mo_2C carbides can form following the heat treatment of Alloy C. It is suggested that this line represents the onset of carbide formation with the intermetallics forming after longer/higher heat treatment conditions.

The Osprey spray deposited Alloy C resulted in only the precipitation of carbides, and possibly nitrides of the same structure as the carbides. The diagram shows distinct boundaries of the two overlapping precipitate fields. The first to precipitate out was the $M_{23}C_6$ type which occurred even after the shortest times and lowest temperatures, where carbides from about 20 nm diameter upwards were observed. Following higher temperature heat treatments precipitates of the M_6C type were obtained, which with increasing heat treatment became the more dominant carbide.

The curves for the precipitation of carbides in Osprey produced Alloy C appear to be comparable with the conventionally produced alloy. The carbide precipitation occurs in Alloy C and not Alloy C-276 because the carbon content was slightly higher.

No intermetallic phases were observed in the Osprey produced Alloy C.

4.1.2 Tensile properties and their correlation to the precipitation

The trends of the tensile properties are most clearly seen on contour maps plotted on temperature versus log time graphs. Figure 6 shows the Alloy C results along with the precipitate fields from Figure 2.

Figure 6a shows the trend towards the minimum ductility, which was observed after a 30 minute heat treatment at 1000°C. This plot is of the reduction in area results, although it could equally be a plot of the elongation results which show the same trend. It is well established in the literature [eg 5] that the precipitation of $M_{23}C_6$ reduces ductility, this is also apparent from Figure 6a. With an increasing amount of precipitation the ductility decreases. It also appears that the line which indicates the onset of M_6C precipitation corresponds approximately to a contour of 50% reduction in area (or 58-60% elongation); compared to the as solution treated value of 66% (or 71%). Thus it can be deduced that the precipitation of $M_{23}C_6$ has a greater effect on the reduction in area than the elongation. This is because the $M_{23}C_6$ precipitation at the grain boundary hinders the plastic deformation and results in a more homogeneous elongation ie it reduces necking.

The loss in ductility is related to the coarsening of the carbides, particularly the M_6C precipitates. As the M_6C is higher in molybdenum and tungsten than the $M_{23}C_6$ this continual decrease in ductility towards the minimum obtained value must therefore be dependent on the diffusion of molybdenum and tungsten as well as the carbon.

The tensile strengths are illustrated in Figure 6b, which is a contour map of the 0.2% YS; a similar trend is exhibited by the UTS results. This figure demonstrates that with the precipitation of $M_{23}C_6$ the strength increases slightly, ie the 0.2% YS increases from 341 Nmm^{-2} (as solution treated) to approximately 369 Nmm^{-2}. This is a consequence of grain boundary strengthening due to the nucleation of intergranular precipitation. A contour line of 369 Nmm^{-2} coincides approximately to the line demonstrating the onset of M_6C precipitation. From this onset there is a reduction in strength towards a minimum at 1000 to 1050°C after longer heat treatments. Thus, the indication is that with the increasing formation of M_6C, where the molybdenum and tungsten are diffusing out of the matrix, the solution strengthening decreases. At the highest temperatures the strength is higher again; this is because more of the molybdenum and tungsten are retained in solution at the higher temperature and the number of nucleation sites is decreased.

4.1.3 Impact strength

The impact results for Alloy C treated at 900°C showed a decrease with increasing time. This corresponded to an increasing amount of M_6C precipitates at the grain boundary and a subsequent embrittlement.

4.2 Alloy 625

4.2.1 Precipitation

Figure 3 shows that the carbide precipitation reaction is more retarded in Alloy 625 than it was in Alloy C with only $M_{23}C_6$ being observed.

The difference in the susceptibility to carbide formation for these two Osprey spray deposited tubes is most probably due to the different alloying contents. Alloy C contains a much higher alloying content. The most important factor appears to be the much lower molybdenum content in Alloy 625.

4.2.2 Tensile properties and their correlation to precipitation

Figure 7a shows the reduction in area contour map of Alloy 625. After a heat treatment of 1 hour at 800°C there appears to be a slight increase in ductility probably because of some stress annealing. Beyond this increase in ductility the reduction in area behaves similarly to Alloy C, ie decreases towards a minimum observed ductility after 30 minutes at 1000°C. There is a slight reduction of ductility before the observed carbide precipitation as the chromium, niobium, molybdenum and carbon diffuse nucleating carbides at nearby grain boundaries. Then the growth of carbides begins which corresponds approximately to a reduction in area contour of 52% (or an elongation contour of 62% compared with an initial value of 58.5% (or 67%). Following the precipitation of $M_{23}C_6$ there is a continual reduction in ductility, again the change within the $M_{23}C_6$ field is more significant on the reduction in area than the elongation because of the precipitation hindering plastic deformation.

The precipitates which contained high silicon and niobium which were primarily nucleated at $MnCr_2O_4$ oxide inclusions, occurred after slightly longer - higher temperature heat treatments, when more diffusion, particularly niobium, had been possible. This corresponded to a reduction in area of approximately 50% (or a 60% elongation). Thus, the primary factor influencing the ductility of Alloy 625 is the diffusion of niobium and carbon, with niobium diffusion being the rate controlling parameter.

The reduction of ductility following the heat treatments is less in Alloy 625 than in Alloy C because the carbide precipitation reaction is retarded, as explained above, and due to there being a greater tendency for the carbide to precipitate on sites other than the grain boundaries, ie at the oxides.

The yield strength properties illustrated in Figure 7b show that there is a tendency to increase strength with an increased heat treatment up to about 1050°C, which is a consequence of the embrittlement discussed above. The onset of carbide precipitation corresponds to a 0.2% yield strength of approximately 342 Nmm^{-2} (or UTS of 735 Nmm^{-2}) compared to the as solution treated value of 323 Nmm^{-2} (or 732 Nmm^{-2}).

The onset of the high molybdenum, silicon containing $M_{23}C_6$ particles corresponded to a 0.2% YS of 350 Nmm^{-2} (or UTS of 745 Nmm^{-2}). Therefore, it again appears that the precipitation can be linked to the mechanical properties.

CONCLUSIONS

1. No intermetallic precipitates were observed in either Alloy C or Alloy 625, produced by the Osprey process.
2. The carbide precipitation of Alloy C produced by the Osprey process is comparable to that of the conventionally produced alloy.
3. Two distinct carbide precipitate fields were observed in Alloy C.
4. In both Alloy C and Alloy 625 precipitation of carbides is deleterious to the ductility.
5. Precipitation of $M_{23}C_6$ gives some increase in the tensile strengths.
6. $M_{23}C_6$ precipitation markedly decreases the amount of reduction in area whilst having less effect on the elongation.
7. The most important factor with regard to the tensile properties and impact strength of Alloy C is the amount of M_6C present, which is controlled primarily by the diffusion of molybdenum and tungsten.
8. The tensile properties of Alloy 625 can be correlated to $M_{23}C_6$ precipitates, where niobium diffusion is the rate controlling parameter.

REFERENCES

1. R W Evans, A G Leatham and R G Brooks; Powder Metallurgy, 28, 13, (1985).
2. R B Leonard; Corrosion, 25, 222, (1969)
3. I Class, H Grafen and E Scheil; Zeit für Metallkunde, 53, No 5, 283-293, (1962)
4. C H Samans, A R Meyer and G F Tisinai; Corrosion, 22, 336-345, (1966)
5. R W K Honeycombe; "STEELS Microstructure and Properties", Edward Arnold, (1981)

Table 1

Chemical composition (wt %)

	C	Si	Mn	Cr	Ni	Mo	W	Nb	Fe
Alloy C	0.011	0.33	0.73	15.7	57.8	15.2	3.5	-	6.6
Alloy 625	0.014	0.39	0.38	21.5	64.6	8.2	-	3.5	1.3

Table 2

Osprey produced Alloy C and Alloy 625

Heat treatment program

T^oC	1 min	2 min	5 min	15 min	30 min	60 min
700				o	x	x
750		o	x	o	x	x
800	x	o	x	o	x	o
850	x	o	x	o	x	o
900	x	o	x	o	x	o
950	x	o	x	o	x	o
1000	x	o	x	o	x	o
1050		o	x	o	x	x
1100			x	o	x	x

o - material for microstructural analyses

x - material for both microstructure and tensile properties

Table 3

Tensile properties of Alloy C
(results also illustrated in figure 6)

Temperature (^oC)	Ultimate Tensile Strength (Nmm^{-2}) 0.2% Yield Strength					% Elongation % Reduction in Area				
	Tim/ mins	1	5	30	60	t	1	5	30	60
As Solution Treated	723 341					71 66				
700			734.5 356.5	753.5 354					66 63	64.5 57.5
750			736.5 362	731 358.5	733 360			63 58	65 55	65 54
800		736.5 357.5	733 360	735.5 367			66 59	63 53.5	60.5 46.5	
850		736 364	736 366.5	698.5 360.5			64.5 54	58 45	44 29	
900		737.5 369	725.5 358.5	669.5 340.5			56.5 45	56 40.5	33.5 28	
950		726 360	705 360.5	651.5 363.5			53.5 39	46.5 34	30.5 23.5	
1000		715.5 361.5	687.5 360	583 353			48.5 37	37.5 28	21.5 20.5	
1050			694.5 357.5	616.5 353	613.5 351.5			38.5 30	28.5 22	23.5 20.5
1100			724.5 360.5	718 356.5	725.5 356			53 38.5	47 37	53 40.5

Table 4

Temperature (^{o}C)	Ultimate Tensile Strength (Nmm^{-2}) 0.2% Yield Strength Tim/ mins	1	5	30	60	% Elongation % Reduction in Area t	1	5	30	60
As Solution Treated	732 322.5					67 58.5				
700			734 328	739.5 341					67.5 56	64.5 56
750			738 333	744 341	735.5 336			64 58.5	62.5 56.5	65 56.5
800		736 337	733 333.5	737 342			68 59	61.5 51	62 50.5	
850		738.5 336.5	734 339.5	739.5 347			64 53.5	64.5 56	59.5 49.5	
900		733.5 339.5	740.5 343	744.5 348.5			65.5 56.5	62.5 52.5	58 46.5	
950		741.5 344.5	744.5 351	753 352.5			62.5 52.5	60 49.5	54.5 42	
1000		746 345	742 348	747 353			59.5 49.5	58.5 48	52.5 39.5	
1050			747.5 358.5	755.5 356.5	745 348.5			61 48	55.5 46.5	51 45
1100			749 355	745.5 352	750 351.5			61.5 53.5	59.5 52	60 51

Table 5

Impact strength of Alloy C heat treated at 900°C for various times

Treat ment	Solution annealed	1 min	2 min	5 min	15 min	30 min	60 min
Impact strength (Nm)	194	122	110	94	69	50	30

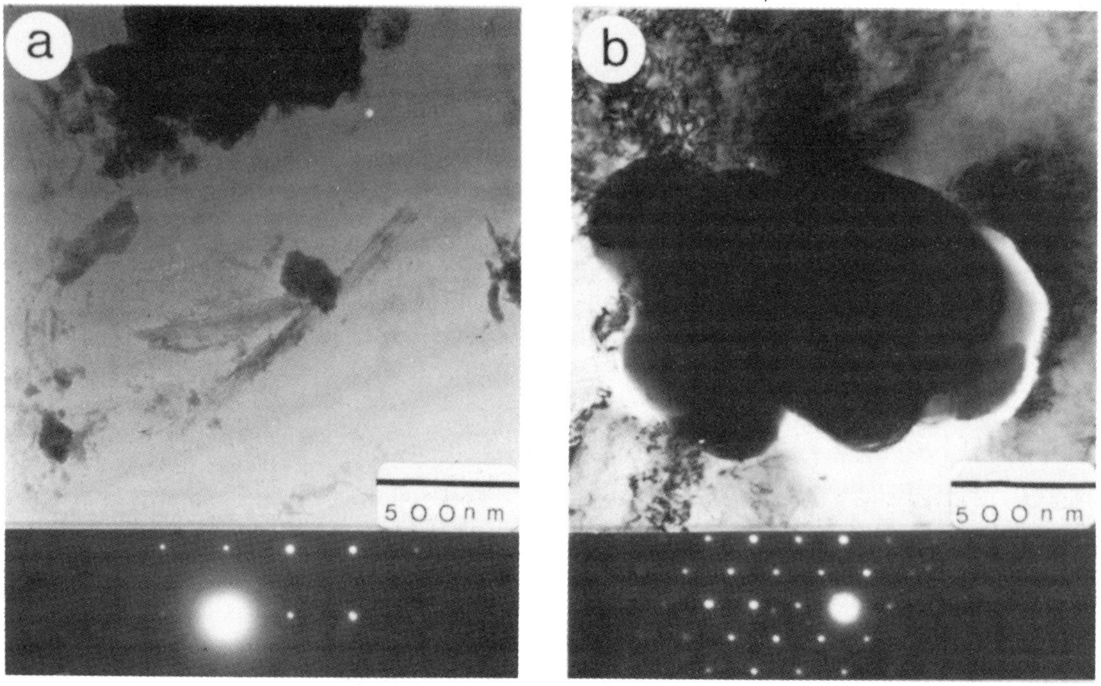

1 Oxide inclusions in as solution treated specimens showing (a) Fe_3O_4 in Alloy C and its 100 pole and (b) $MnCr_2O_4$ in Alloy 625 and its 110 pole.

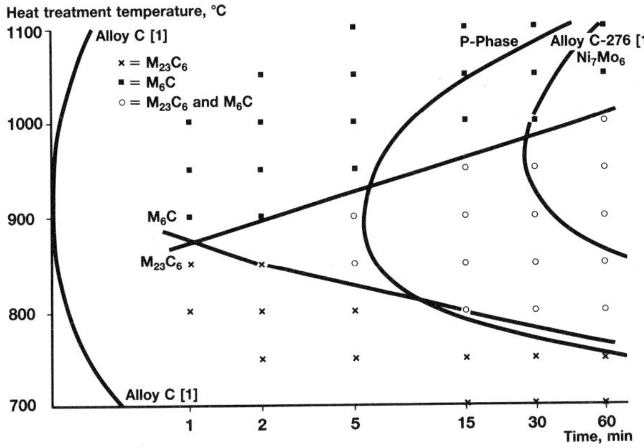

2 Time-Temperature-Transformation diagram for Alloy C showing the observed precipitation fields and curves from the literature [1].

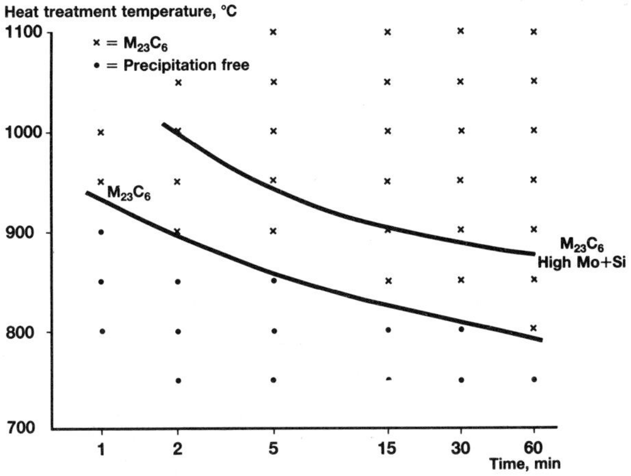

3 Time-Temperature-Transformation diagram for Alloy 625.

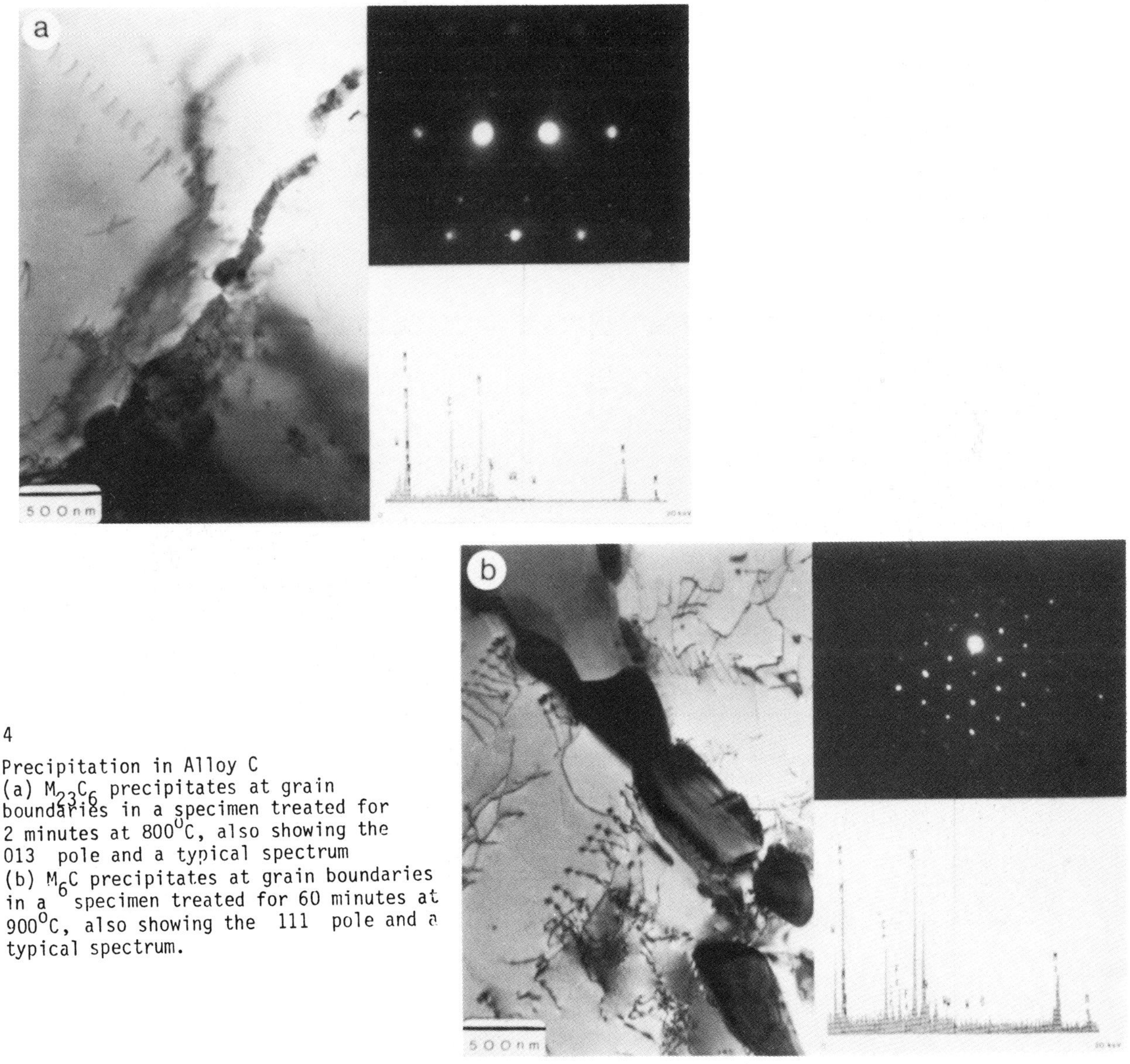

4

Precipitation in Alloy C
(a) $M_{23}C_6$ precipitates at grain boundaries in a specimen treated for 2 minutes at 800°C, also showing the 013 pole and a typical spectrum
(b) M_6C precipitates at grain boundaries in a specimen treated for 60 minutes at 900°C, also showing the 111 pole and a typical spectrum.

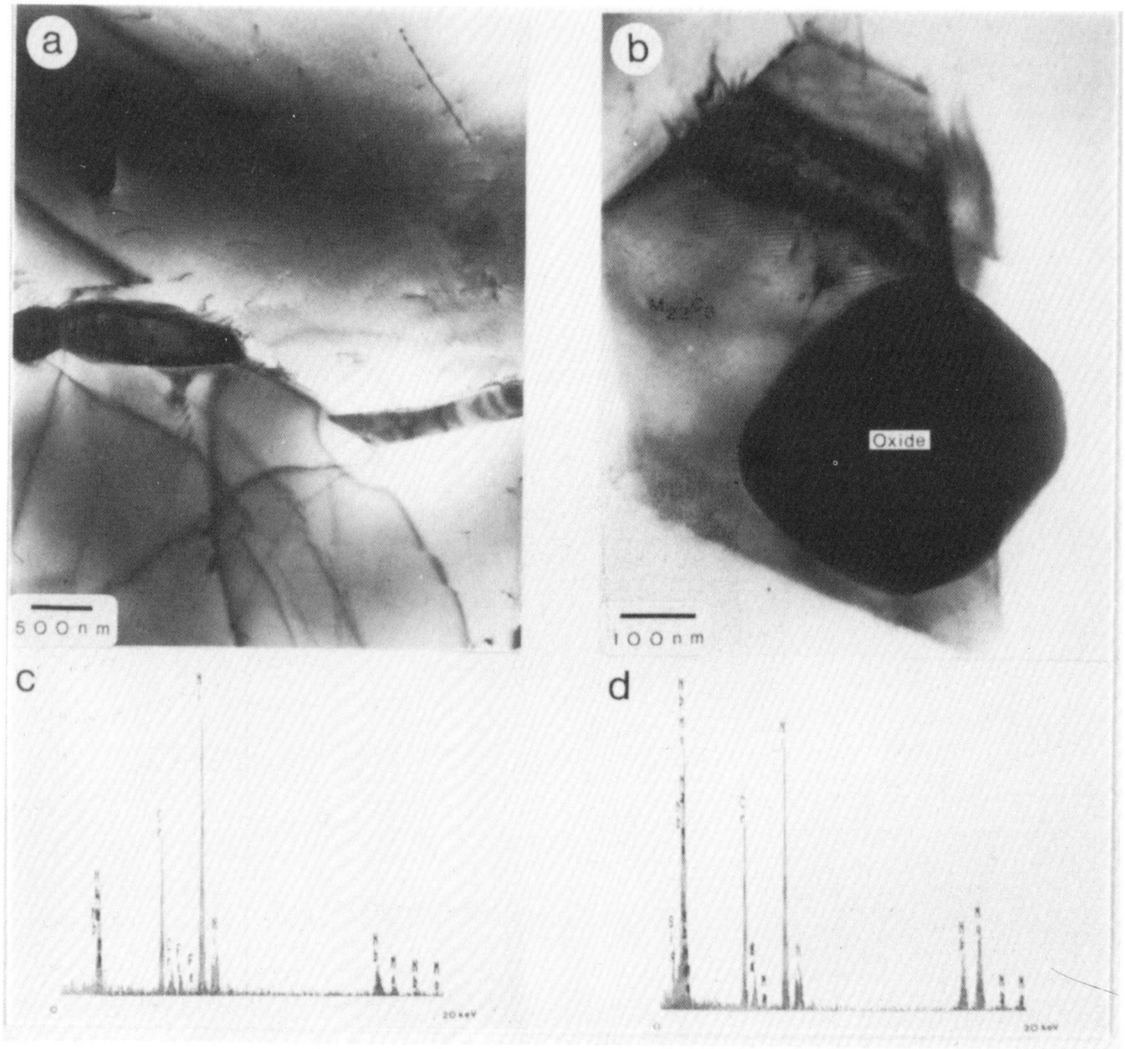

5 Precipitation in Alloy 625 treated for 15 minutes at 1050°C.

(a) $M_{23}C_6$ precipitates at a grain boundary
(b) $M_{23}C_6$ precipitates at a $MnCr_2O_4$ inclusion
(c) Typical spectrum of a grain boundary precipitate
(d) Typical spectrum of a precipitate nucleated at an inclusion

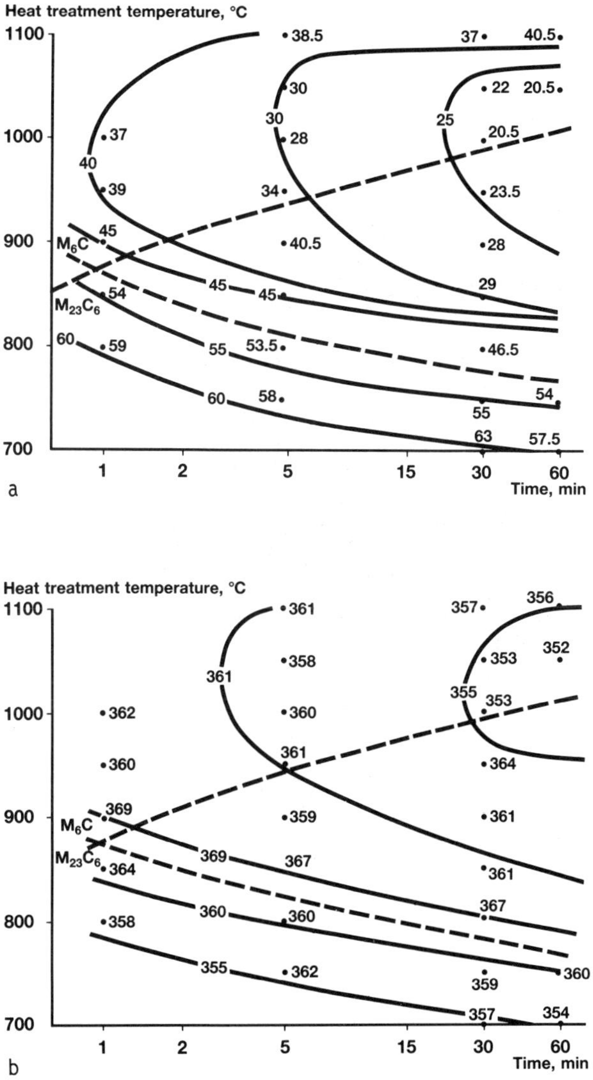

6 Illustration of the Alloy C tensile results on time-temperature plots (a) the reduction in area contours (b) the 0.2% yield strength contours. The carbide curves are also shown.

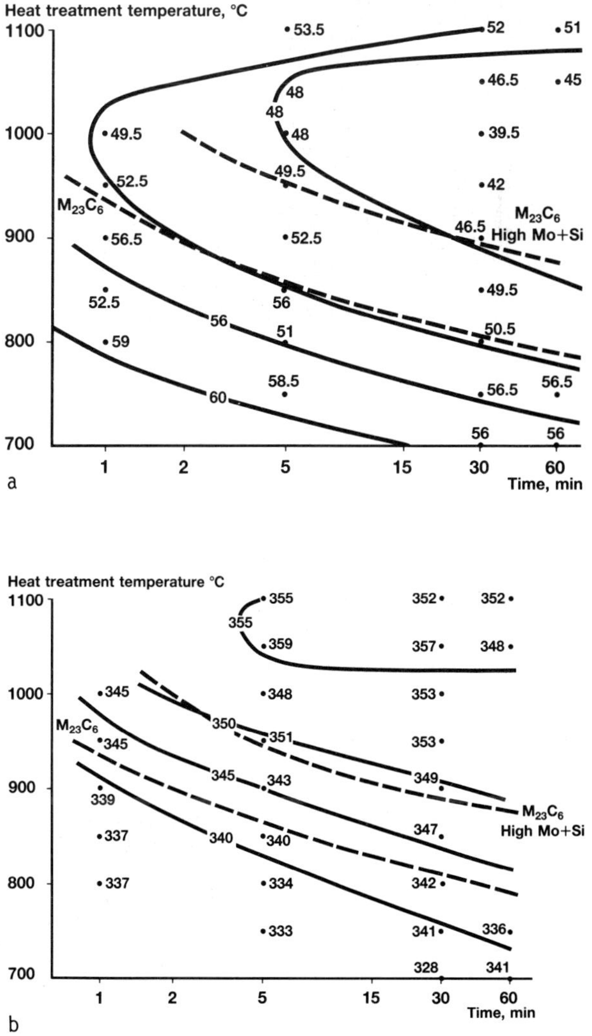

7 Illustration of the Alloy 625 tensile results on time-temperature plots (a) the reduction in area contours (b) the 0.2% yield strength contours. The $M_{23}C_6$ curve is also shown.

Superplasticity of a duplex stainless steel

K Osada, S Uekoh and K Ebato

The authors are in the Technical Research Centre, Nippon Yakin Kogyo Co. Ltd., Kawasaki, Japan.

SYNOPSIS

This paper investigates the superplasticity of duplex stainless steel with coarse grains elongated towards hot-rolling direction. Despite the microstructure, even the as-hot rolled steel showed more than 300% elongation over a wide range of strain rates. The effect of cold rolling on the as-hot rolled strips was investigated, and consequently it was found that while the deformation behaviour of an as-hot rolled specimen was controlled mainly by the recrystallization of primary austenites, the deformation of as-cold rolled specimens depended on the precipitation in the ferrites. According to the microstructural changes, ductility maxima were observed at relatively higher strain rates. The relation between log (stress) and log (strain rate) indicated that superplasticity in the steel derives from two microstructural conditions at least. Variation of the m-value with strain, and microstructural changes during deformation, supported this view. The process by which the steel with coarse, elongated grains became superplastic was discussed.

INTRODUCTION

Superplasticity of duplex stainless steels has been studied extensively by many authors. Interest has been concentrated on the superplasticity of IN744, and the manufacturing processes which make duplex stainless steels superplastic.[1-3] As a result the microstructures were adjusted until they were very fine. It has also been found that in order to make duplex stainless steels superplastic, the phase ratio should be kept at approximately 50:50, and that precipitates like Ti(C,N) limit superplastic ductility.[4,12] However, few authors have studied superplasticity of the duplex stainless steel with coarse, elongated ferrite/austenite grains.

We developed a new generation of duplex stainless steel and reported its properties at 'STAINLESS STEEL '84'.[5] This duplex stainless steel is characterized by a phase ratio of austenite to ferrite ranging from 35:65 to 40:60 after annealing. It also features very low carbon and sulphur contents; an alloying of Mo (about 3.3wt%); and an addition of nitrogen up to 0.2wt%. The mechanical properties of the steel at high temperatures predicted a superplasticity which surpassed the capability of superplastic IN744. We have already published a paper related to the superplasticity of the steel in which more than 3000% elongation was reported.[6] According to our previous experiments sigma phase was found to play an important role in superplasticity in the duplex stainless steel. With respect to this role, N Ridley compared the superplasticity of 3RE60 with that of IN744 and reported that sigma phase in 3RE60 was harmful to its superplastic ductility.[7]

In this paper we will deal with the duplex stainless steel which shows unforeseen ductility even in an as-hot rolled plate. The microstructure of the as-hot rolled plate consists of coarse, elongated austenite/ferrite grains towards hot-rolling direction. The effect of cold-rolling reduction on the superplasticity of the as-hot rolled steel will be studied in terms of microstructure. Also the reasons which explain this characteristic superplasticity will be discussed.

EXPERIMENTAL PROCEDURE

Chemical composition

The duplex stainless steel used in this investigation had the composition Fe–25%Cr–6.5%Ni–3.2%Mo–0.5%Si–0.7%Mn–0.001%S–0.008%C–0.12%N. The phase ratio of austenite to ferrite ranges from 35:65 to 40:60 after various heat treatments.

Manufacturing process

The manufacturing process is summarized as a flow chart in Fig. 1. The melt was refined by the AOD process with the aid of a simulation program followed by casting into a continuous-casting machine. A PL hot-rolling mill that provided a very high rolling speed was used to roll the slab down to a strip of 4.0 mm thickness within seconds. A water jet was applied to quench the strip immediately after the rolling operation. As-hot rolled sheets were cut from the strip for later use. As-hot rolled specimens were rolled in a four-high rolling mill with reductions of 10, 20, 40, 50, and 60% without prior annealing. Those were supplied as as-cold rolled specimens.

High-temperature tensile test

Tensile test specimens were machined from the rolled sheets with the tensile axis perpendicular to the cold-rolling direction which was similar to hot-rolling direction. The gauge length was 10 mm. High-temperature tensile tests were carried out in a tensile-test machine equipped with a furnace by which the temperature was controlled to an accuracy of 5 K. Specimens were kept for 5 min at deformation temperature and pulled at constant cross-head speeds. Tests were conducted in the air. A strain rate was represented by an initial strain rate calculated from a constant cross-head speed and the gauge length.

Microstructures

Test pieces taken from near fractured tips were etched by a 5N KOH solution electrolytically for optical metallographic observations. Microstructures of as-received and aged specimens were also examined.

RESULTS

Variations in elongation with strain rate are shown in Figs. 2 and 3. Almost every specimen shows more than 400% elongation between 10^{-3} and $10^{-2}s^{-1}$ of strain rate. The elongation at 1173 K does not depend on the amount of cold-rolling reduction. In a 10% cold-rolled specimen the elongation obtained at 1223 K shows an improvement over the elongation at 1173 K. Although slight improvement by temperature is observed on the elongation of the other specimen with rolling reductions of 20, 40, 50, and 60%, that is not so large as that in the 10% cold-rolled specimens. Characteristic maximum peaks of elongation are observed. It seems that there is no relation between the position of these maxima and the rolling reduction. However, especially in the 60% cold-rolled specimen, even at 1173 K a peak begins to be enhanced at a higher strain rate.

Figs. 4 and 5 illustrate the strain rate dependence of flow stress at 1173 and 1223 K respectively. Flow stresses of 10 and 20% rolling reduction show larger values than those of as-hot rolled specimens at both 1173 and 1223 K. This result is noticeable when deformed under a higher strain rate at 1173 K. It is said that one of the distinctive properties of superplastic materials is deformation with low flow stress. In the experiment a noticeable decrease in flow stress is observed in the specimens with 40, 50, and 60% cold-rolling reduction, compared with those with 0, 10, and 20% reduction. When the specimen is rolled with reduction of 40%, its flow stress begins to decrease, especially at higher strain rates at 1173 K. This suggests that the effect of cold rolling works at higher strain rates, and the amount of rolling reduction has to exceed a certain level, in this case 40%. Usually m-values are evaluated from the slopes of log (stress)–log (strain) relations; however, at 1223 K the curves do not maintain good linearity and show irregularities at a higher strain-rate range. This was thought to be caused by the same mechanism which controls the behaviour of flow stress of cold-rolled specimens with 40% or higher reductions at 1173 K.

A m-value of a material is defined as the slope of a regression line which is a portion of sigmoidal curve of log (flow stress) vs. log (strain-rate) relationship. When the m-value shows a larger value than 0.3, the material is regarded as a superplastic material. But in the experiment it was supposed that microstructure continued to change during deformation. Hence m-values were calculated from peak flow stresses taken at the same strain in the stress–strain curves. Variation of m-value with strain at 1173 and 1223 K is plotted in Figs. 6 and 7 respectively. The m-value of the as-hot rolled specimen at 1173 K gradually increases and then exceeds 0.3 at a strain rate of 0.6. At 1223 K the m-value of the as-hot rolled specimen showed more than 0.3 at an early stage of straining. This result means that thermal energy is important for the as-hot rolled specimen to become superplastic. The cold-rolled specimens show high m-values at the beginning of their deformation and gradually decrease at 1173 K. There seems to be no relation between cold reductions and the m-values at 1173 K. All the specimens tested at 1223 K indicate larger m-values than 0.3 at less strain compared with the specimens tested at 1173 K. As straining continues the m-value starts to drop significantly. The maximum m-value taken from the relation between the m-value and strain is plotted as a function of rolling reduction in Fig. 8. The variations of maximum m-value with rolling reduction at 1173 and 1223 K are almost identical up to 10% rolling reduction. But once rolling reduction exceeds about 40% or less, the maximum m-value at 1223 K has a bigger value than that at 1173 K. This phenomenon is consistent with the results in the behaviour of flow stress with strain rate. The maximum m-values both at 1173 and 1223 K seem to keep their value below 0.5.

The microstructures of as-hot and as-cold rolled specimens are shown in Fig. 9. While in the as-hot rolled specimen, coarse, elongated grains are observed with a small quantity of sigma on the boundaries of austenites/ferrites, in the as-cold rolled specimens, straining induced by cold rolling is observed as displacement of inside grains. Figure 9 shows the effect of heat treatment at 973 K. By that heat treatment sigma-phase precipitates on slip lines in the coarse ferrite grains as well as on the austenite/ferrite grain boundaries. The microstructures deformed superplastically are shown in Figs. 10 and 11. There are three types of microstructure. The first type, which is observed mainly at 1173 K, consists of sigma and austenite phases. The second type of microstructure consists of three phases – sigma, austenite, and a small quantity of ferrite. The last type of microstructure contains very fine precipitates in the austenite matrix. In the stress–strain curve for the last type of microstructure, work-hardening effect was observed followed by a fracture. The volume fraction of sigma phase in the austenite matrix is very large at 1173 K. The individual size of sigma grains varies with strain rate, ie when the specimen deforms at a higher strain rate, the sigma grains are small. In all cases the average grain size of particles in the matrix is measured at less than 5 micro-metres. The specimens with a lot of sigma grains in the austenite matrix did not yield large elongation.

DISCUSSION

According to extensive work on the precipitation behaviour of a similar type of duplex stainless steel by Maehara et al.[8] sigma-phase precipitation plays an important role in the superplasticity of the duplex stainless steels. It is also well-known that residual strain significantly accelerates the precipitation of sigma.[9] It was proposed by Maehara et al. that sigma precipitation was effective in the dynamic crystallization of the austenites.[8] Zhang et al. studied the superplasticity of three different duplex stainless steels and concluded that the steel which kept an equal phase ratio showed excellent superplasticity.[12] In the present investigations our results showed the characteristic behaviour in the elongation vs. strain rate, the irregularities in the strain-rate dependence of flow stress, interesting variation of the m-value with strain, the effect of rolling reduction on the maximum m-value, and the microstructural change after superplastic deformation. All of those phenomena seem to be explained by recrystallization of the steel during heating.

There are some papers which deal with the recrystallization of duplex stainless steels.[10] Cooke et al.[11] concluded that the recrystallization behaviour of a ferrite depends on the phase volume fraction of the steel, and that the steel with 75% ferrite recrystallizes discontinuously. In our experiment the as-hot rolled duplex stainless steel was used as a starting material and this as-hot rolled steel contains as much as 70 to 75% ferrite due to high-speed hot rolling from a cast slab followed by a rapid water quench. By metallographic observations in the 50% cold-rolled steel, it was confirmed that ageing caused the rapid precipitation of the sigma phases on the slip lines. This phenomenon controls the superplasticity of the duplex stainless steel under investigation. To explain the elongation maxima seen in the higher strain-rate range, a process which accommodates high-speed diffusion must be provided. The slip lines induced by more than 40% cold reduction may work as high-angle reaction fronts in the ferrites and may provide high-diffusivity paths for sigma nucleation. So if sigma is not the first phase, secondarily precipitated austenite may play this role with somewhat different behaviour. However in as-hot rolled and 10% cold-rolled specimens, recrystallization of primary precipitated austenites is a rate-controlling process for superplasticity and the effect can be expected in the lower strain-rate range, ie below $1.00 \times 10^{-3}s^{-1}$. The former process begins to function at the temperature where sigma precipitation can be expected; the latter has to be expected at higher temperatures where recrystallization of primary austenite occurs. The reaction of sigma-phase precipitation takes place very fast. So once conditions reach equilibrium, few diffusion paths will be left; in other words, as deformation continues, the m-value begins to decrease. Finally voids originate from the interfaces between the sigma and the austenite matrices, and then the voids coalesce to failure in a pseudo-brittle manner. If recrystallization of primary austenites succeeds, however, superplasticity based on the fine ferrite/ austenite duplex structure may be expected. In that case grain-boundary sliding, supported by volume diffusion of an element, is regarded as an accommodation process for superplasticity at lower strain rates, ie below $1.0 \times 10^{-3}s^{-1}$.

The microstructure which has very fine precipitations in the austenite matrix with cracks, is due to nitrogen pick-up from the air during a long deformation time at high temperatures. Y Zhang et al. reported the same microstructure in superplastically deformed duplex stainless steels.[12,13] It was also suggested that the fine precipitation might be Cr_2N.

CONCLUSIONS

The results of the present investigation into the superplastic behaviour of as-rolled duplex stainless steels with coarse grains indicated that:

1 The characteristic maximum peaks in relation to elongation vs. strain-rate are accentuated by cold reduction.

2 According to analysis of the relation between peak stress and strain rate, the drop of flow stress begins to appear in the 40% cold-rolled specimen at 1173 K. By raising the temperature to 1223 K, this was significantly emphasized. It is assumed that at least two kinds of microstructural change during heating and deformation are connected to the superplasticity of the steel, and one of the changes is accelerated by cold reduction.

3 From the study of the m-value an as-hot rolled steel needs a strain of 0.60 before becoming superplastic.

4 The effect of cold rolling on superplastic properties begins to be observed at a reduction of 40%.

5 The three types of microstructure after deformation were observed. The austenite-enriched structure is thought to be caused by nitrogen pick-up from the testing atmosphere.

6 Superplasticity of the duplex stainless steel can be explained by the precipitation of sigma mainly in the ferrites, combined with the recrystallization of the primary austenites.

REFERENCES

1 H W Hayden, R C Gibson, H F Merrick, and J H Brophy: Trans. ASM, 1967, 60, 3

2 H W Hayden and S Floreen: ibid., 1968, 61, 489

3 C I Smith, B Norgate, and N Ridley: Met. Sci., 1976, 10, 182

4 N Ridley and C W Humphries: in 'Grain boundaries', 1976, E29, London, Institution of Metallurgists

5 R Nemoto, K Osozawa, K Osada, and M Tsuda: 'Stainless steel '84', 1984, 148

6 K Osada, S Uekoh, and K Ebato: Trans. Iron Steel Inst. Jpn, 1987, Vol. 27, No. 9, 713

7 N Ridley and L B Duffy: in 'Strength of metals and alloys', (Ed. H J McQueen, J P Bailon, J I Dickson, J J Jonas, M G Akben), 1985, 853, Oxford, Pergamon Press

8 Y Maehara and Y Ohmori: Metall. Trans. A., 1987, Vol. 18A, 663

9 Y Maehara, M Koike, N Fujino, and T Kunitake: Trans. Iron Steel Inst. Jpn, 1983, Vol. 23, 240

10 B A Cooke, A R Jones, and B Ralph: Metal Sci., March—April 1979, 179

11 T R Parayil and P R Howell: Met. Sci. Tech., 1986, Vol. 2, 1131

12 Y Zhang, F Dabkowski, and N J Grant: Met. Sci. Eng., 1984, 65, 265

13 Y Zhang, F Dabkowski, S Kang, and N J Grant: Powder Met. Intn'l, 1985, Vol. 7, No. 1, 17

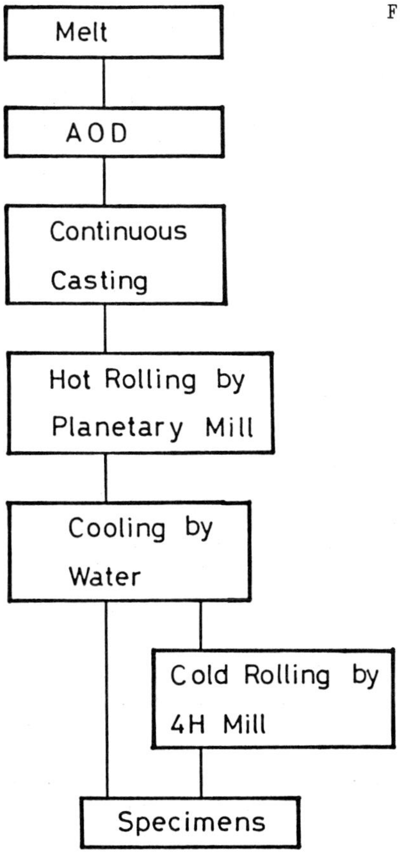

Fig. 1 Process flow chart for duplex stainless steel used for experiments. This process characterizes a high-speed rolling operation from a cast slab to a strip within seconds

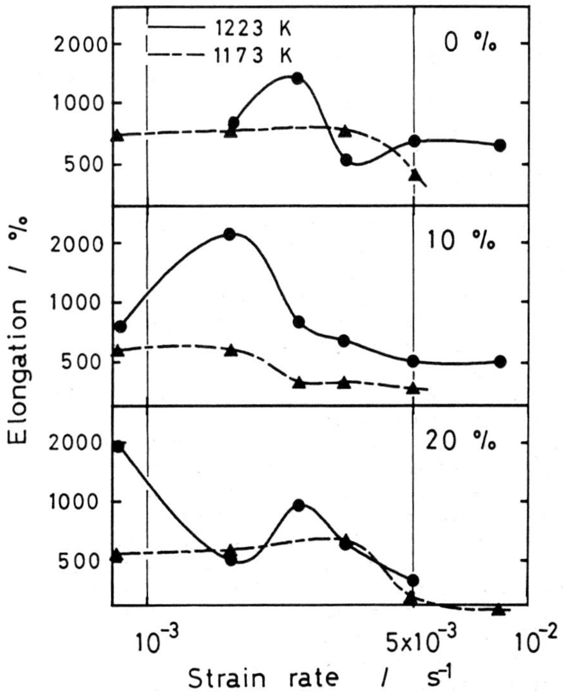

Fig. 2 Variation of elongation with strain rate on specimens cold-rolled at 10, 20% and as-hot rolled (designated as 0%). A solid and a broken line show the results at 1223 and 1173 K respectively

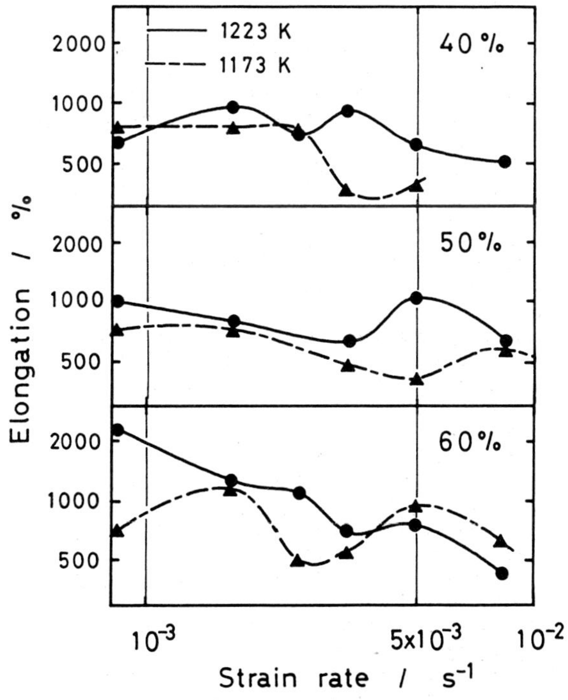

Fig. 3 Variation of elongation with strain rate on specimens cold-rolled at 40, 50, and 60%. A solid line shows the results at 1223 K and a broken line those at 1173 K

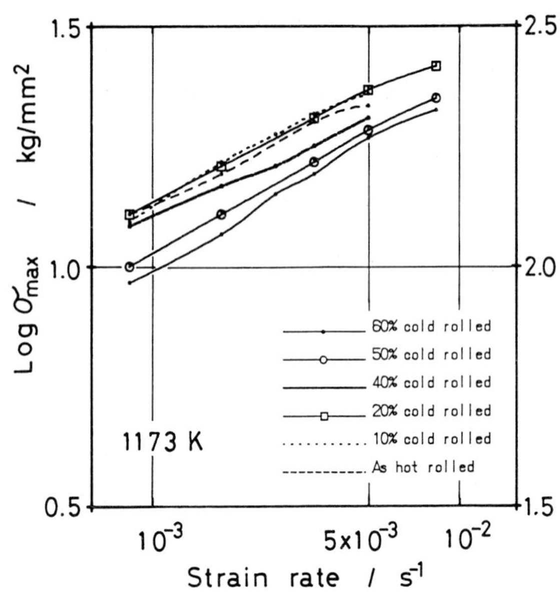

Fig. 4 Relation between peak flow stress and strain rate at 1173 K

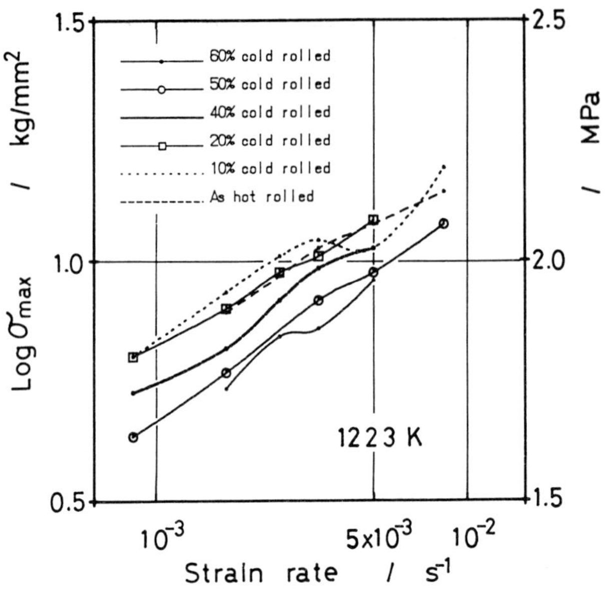

Fig. 5 Relation between peak flow stress and strain rate at 1223 K

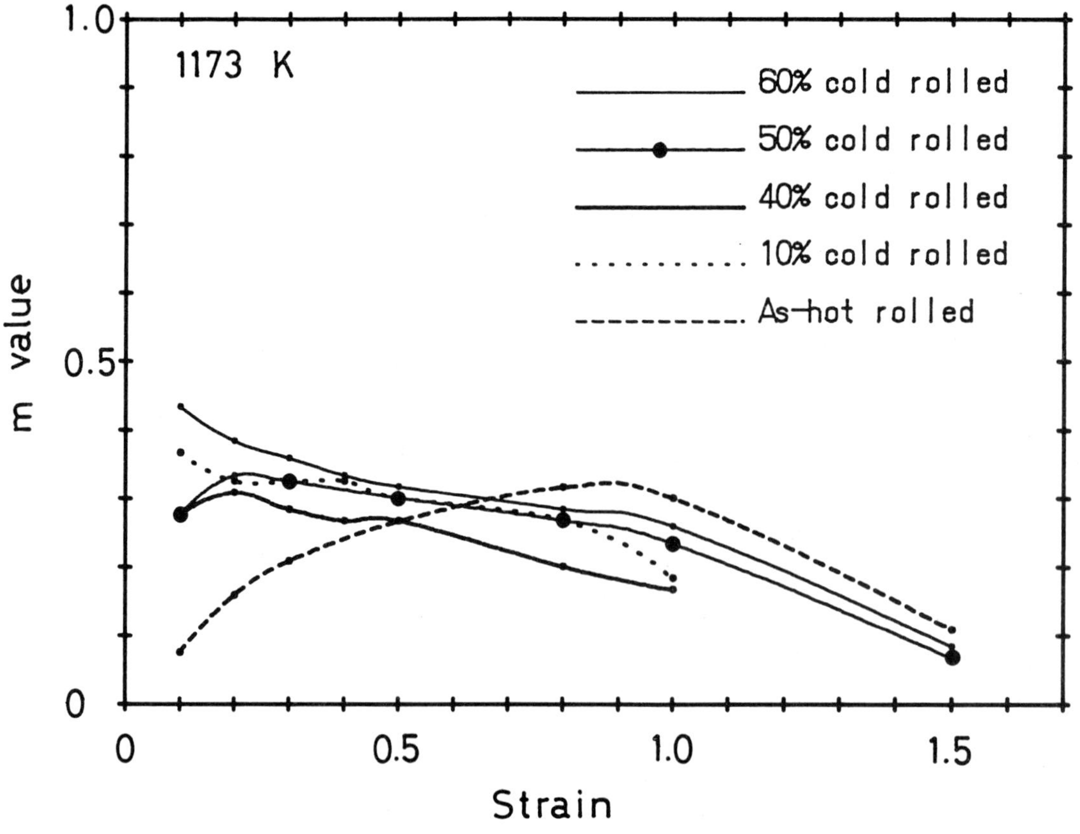

Fig. 6 Variation of m-value with strain at 1173 K

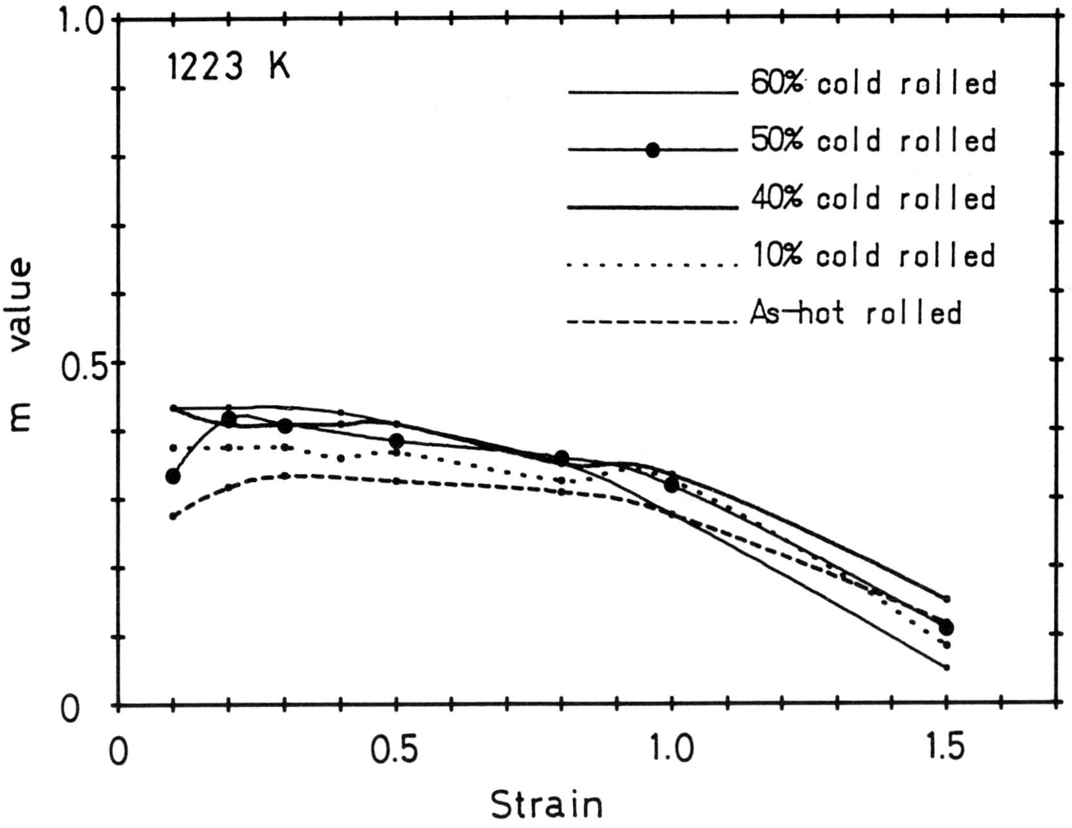

Fig. 7 Variation of m-value with strain at 1223 K

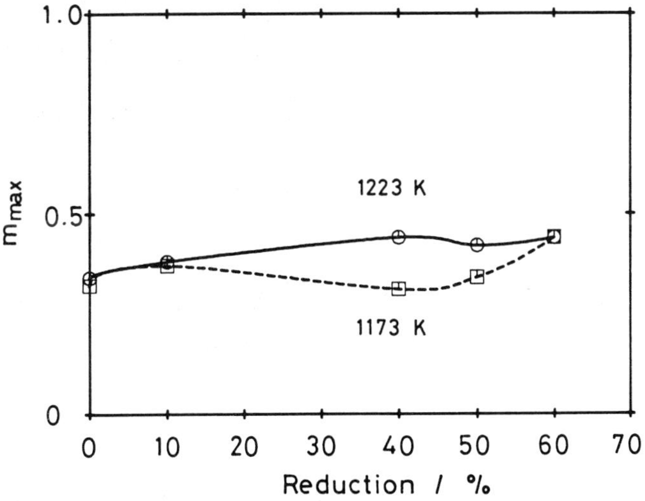

Fig. 8 Relation between maximum m-value and cold-rolling reduction

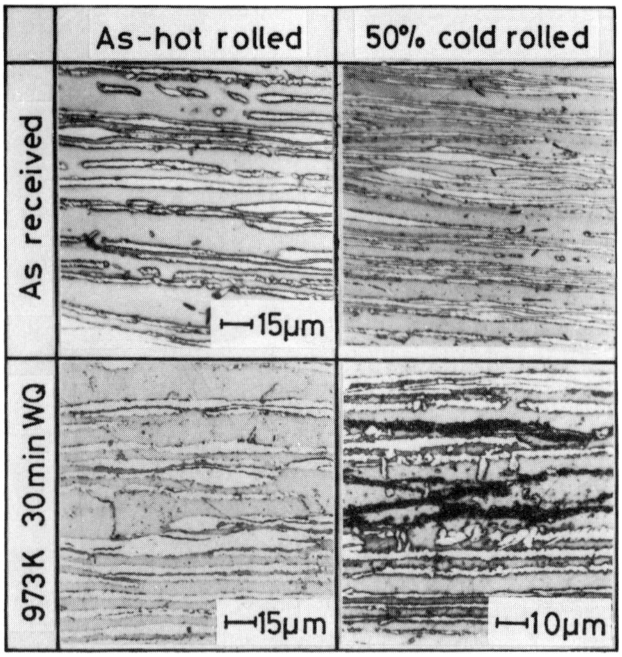

Fig. 9 Microstructures of an as-hot rolled and a 50% cold-rolled specimen. Also the changes by heat treatment at 973 K are shown

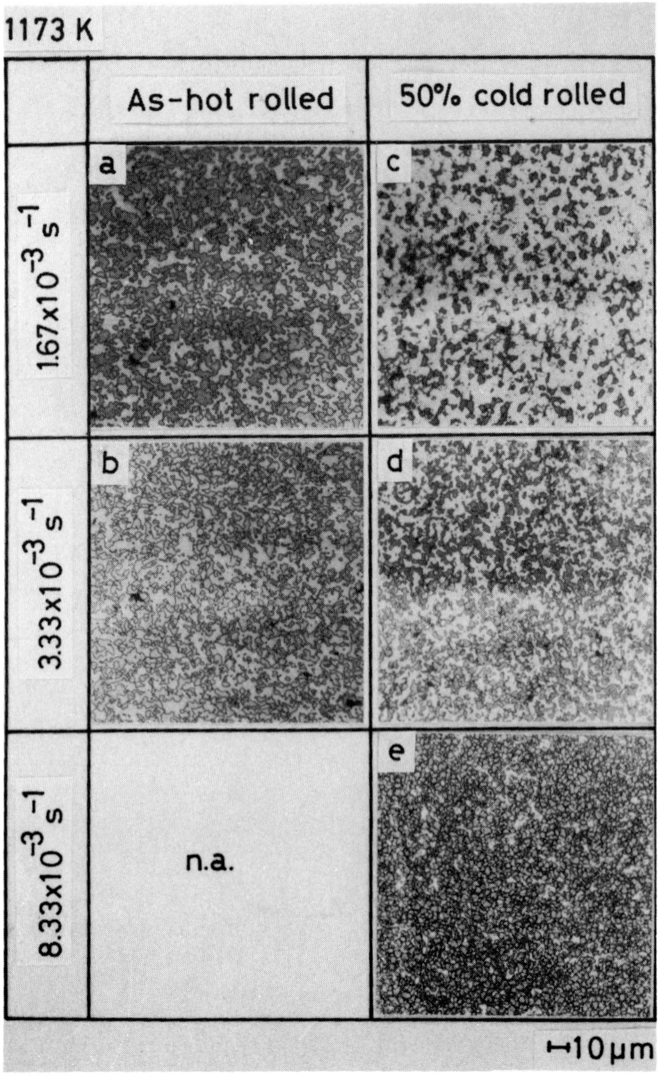

Fig. 10

Microstructures superplastically deformed at 1173 K. Cross-sections are perpendicular to the tensile axis

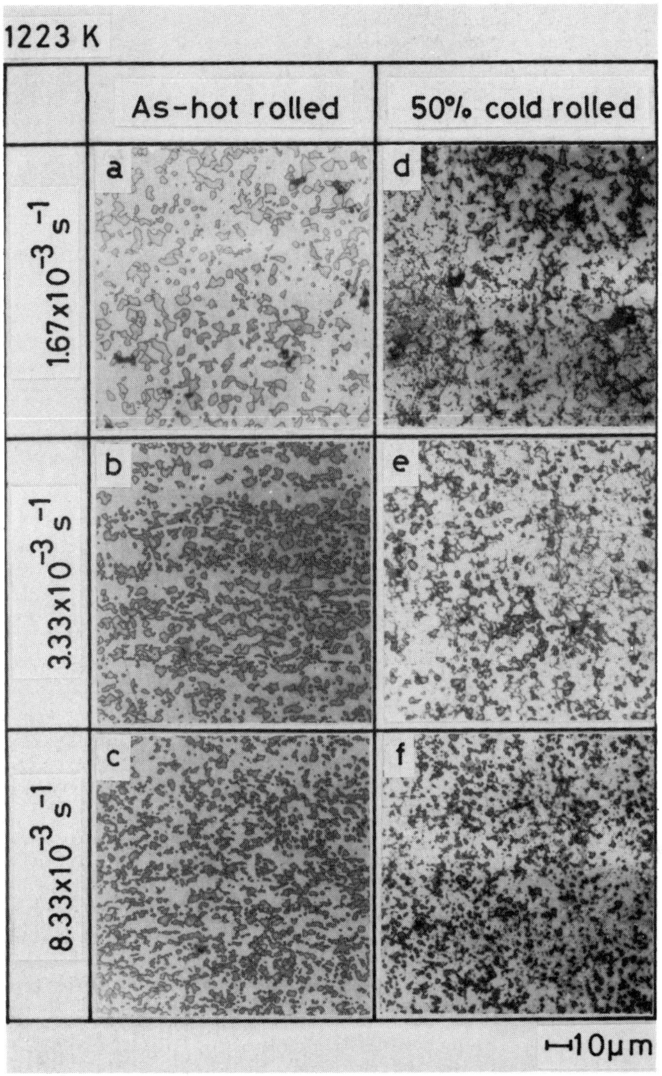

Fig. 11 Microstructures superplastically deformed at 1223 K. Cross-sections are perpendicular to the tensile axis

Strain rate and temperature dependence of hot strength in 301, 304, 316 and 317 steels in as-cast and worked conditions

N D Ryan and H J McQueen

NDR is Research Associate and HJMcQ is Professor in the Department of Mechanical Engineering, Concordia University, Montreal, Canada.

SYNOPSIS

The torsion data on the four grades of as-cast and worked stainless steels have been analyzed by the relation

$$Z = \dot{\varepsilon} \exp(+Q/RT) = A(\sinh \alpha\sigma)^n$$

over the ranges T=1200-900°C and $\dot{\varepsilon}$=0.1-5 s^{-1}(Z is the Zener-Hollomon parameter). The activation energies Q_{HW} for the worked steels were 399, 393, 454 and 496 kJ/mol respectively. The values of stress exponent n are about 4.5 for the worked but were lower for the as-cast. In addition Q_{HW} and n values compiled from 70 reports on various modes of testing published in the literature have a scatter of about ±6% for Q_{HW} and ±20% for n. There were good agreements between the present results and the averages of published values. The values of Q_{HW} for all the steels gave a good correlation with total metallic solute content. The variations in hot strengths were related to specific alloy contents with reasonable success. For the 304 steels, the strain to the peak stress, which gives the limit of the strain hardening for calculation of mean rolling stresses, is proportional to $Z^{0.125}$ and to (initial grain size)$^{0.75}$. The flow curves and constitutive behavior are consistent with a dynamic recrystallization mechanism which is retarded by additional solute or carbonitride precipitates.

INTRODUCTION

Although the hot working behavior of austenitic stainless steels has been reviewed extensively in the recent past [1-5] the current objective is to examine in greater depth the temperature and strain rate dependence of the hot strength and ductility. This has grown out of efforts by the authors to compare results on 301, 304, 316 and 317 grades in both as-cast and as-hot-worked-homogenized conditions.[6-15] The mechanistic framework for the analysis is the softening during deformation by dynamic recovery (DRV) and recrystallization (DRX).[1-19] Post-deformation softening by static recovery (SRV) and recrystallization (SRX) is only cursorily discussed. These mechanisms are impeded by increasing concentrations of solutes or of precipitates.[1-18] The traditional constitutive equations are examined with emphasis on the variation of the empirical constants in these equations for steels within the normal composition limits. The paper opens with presentation of the σ-$\dot{\varepsilon}$-T data fitted to the hyperbolic sine and Arrhenius functions for each of the alloys 301, 304, 316 and 317 for both cast and worked materials.[8-15 20 21] In each case, a limited amount of comparative published data has been added, whereas for the bulk of the extensive literature surveyed, the constitutive constants are tabulated for each alloy (Tables 1-4).[22-68] Although a $(d\sigma/d\varepsilon)$-σ analysis has been performed with success [8-14 69-70] it is not discussed as it does not clarify the present approach.

EXPERIMENTAL TECHNIQUES AND RESULTS

The four grades of austenitic stainless steel were continuously cast as 127mm slabs by Atlas Steels of Tracy, Québec. Three grades, 304, 316 and 317 were studied in the as-cast condition with specimen axes parallel to the casting direction.[6] Different heats of the above grades and 301 were rolled to 19mm plate from which specimens were machined with their axes in the rolling direction.[7-14] The specimens were deformed on a hydraulically-powered, computer-controlled, hot-torsion machine, being heated in a radiant furnace protected by argon.[6 7 14] The tests were conducted in the range of temperatures, T, 1200-900°C and strain rates, $\dot{\varepsilon}$, 0.1-5 s^{-1}. The equivalent flow stress σ corrected for ε-$\dot{\varepsilon}$ gradient and the surface strain were calculated by standard techniques.[6-15] Some representative stress strain curves for each grade, as-cast and worked, (Fig. 1) are given for 3 conditions: 1000°C, 5 s^{-1}; 1000°C, 0.1 s^{-1} and 1200°C, 5 s^{-1}, illustrating the influence of T and $\dot{\varepsilon}$ on the flow curves and the hot ductility. The curves exhibit the peak (ε_p, σ_p) and work softening to steady state characteristic of DRX. The details of these mechanisms and their confirmation by metallography have been described previously.[7 9-12 14] The ductility increases with rise in T, $\dot{\varepsilon}$ (301,304) and homogeneity or with decline in $\dot{\varepsilon}$ (316,317) and alloying additions.[7-14]

The constitutive equations relating σ_p-$\dot{\varepsilon}$-T were selected from the following:[3 6 7 21]

$$Z = \dot{\varepsilon} \exp(+Q'_{HW}/RT) = A' \sigma^{n'} \quad (1)$$

$$Z = \dot{\varepsilon} \exp(+Q''_{HW}/RT) = A'' \exp(\beta\sigma) \quad (2)$$

$$Z = \dot{\varepsilon} \exp(+Q_{HW}/RT) = A (\sinh \alpha\sigma)^n \quad (3)$$

where A, A', A", α (= 1.2 x 10^{-2} MPa), $\beta = \alpha n$, n, n' and Q_{HW} are material constants (R = 8.32

kJ/°K mol). The sinh law (Eqn. 3) was found most suitable giving a linear fit across the entire range examined (Figs. 2 to 5).[7-14] The stress exponents n and n' are almost equal being close to 4.5 for the present steels. The activation energy Q_{HW} lies in the range 390-510 kJ/mol, increasing with alloy content. With Q_{HW} derived, it is possible to prepare a graph of Z vs sinh $\alpha\sigma$ which draws the present data for each grade into a single line (Fig. 6a). In order to simplify comparison where differences are greatest, the Z scales have been normalized to 900°C, 1.0 s^{-1} with the temperatures given for a strain rate of 1.0 s^{-1}.[20] The limited suitability of Eqns. 1 and 2 for the present steels is shown in Fig. 6bc where it is seen that the power law (Eqn. 1) breaks down for high stresses and the exponential law (Eqn. 2) for low stresses.[6 7] The stress at which the transition from the power to exponential laws takes place is approximately the inverse of α.[71]

For each alloy there are two graphs: 301, Fig. 2[8 9 54 55]; 304, Fig. 3[5 7 9 14 29 42]; 316, Fig. 4[7 13 55 59 60 63 64] and 317, Fig. 5.[7 10-12 27 53 68] The first graph is a plot of log $\dot{\varepsilon}$ vs. log sinh $\alpha\sigma$ which consists of a series of parallel lines for differnt T with slopes equal to n, having different values for cast and worked. The second graph is a plot of log (sinh $\alpha\sigma$) vs. (1/T) consisting of parallel lines for various $\dot{\varepsilon}$. From the slopes of these lines, Q_{HW} are calculated and are usually higher for the cast than for the worked. The graphs show that Eqn. 3 is a good representation of the σ-$\dot{\varepsilon}$-T relationship not only for the present alloys but for published data, the comparison with which will be carried out in the discussion. The values of Q_{HW} derived from Figs. 2-5 are tabulated in Tables 1-4 along with results of other researches.[6-14 22-68] The values of log Q_{HW} have been plotted against log (solute content, S) (Fig. 7). The data fall in a fairly narrow band of ±25 kJ/mol which is approximately ±6%. Q_{HW} increases by about 13.5 kJ/mol per 1% solute.

The constitutive Eqn. 3 has been inverted and altered by Tanaka et al.:[71]

$$\sigma = (1/\alpha)\sinh^{-1}\left[\dot{\varepsilon}/\dot{\varepsilon}_n \exp\{Q/R[(1/T)-(1/T')]\}\right]^{1/n} \quad (4)$$

where $\dot{\varepsilon}_n = 1\ s^{-1}$ and a characterizing temperature T' is given by

$$T' = Q_{HW}/R\ \ell n\ A \quad (5)$$

which replaces the pre-exponential constant A. In the normal form of Eqn. 3, A has to be determined for each material; however, T' is a constant for the material[71] as illustrated in Fig. 8 where a plot of Q vs ln A for the steels in Tables 1-4 are presented. The lines for each steel pass through the origin and show that T'= 1486 K for 304 and 317; for 316, T'= 1522 K and for 301, T'= 1403 K. Tanaka et al.[71] further reported that T'= 0.78 T_m which is confirmed in the present case with T'/T_m ranging from 0.788 to 0.855.

In association with rise in σ_p as Z increases (Fig. 1), the peak strain ε_p also becomes larger as described earlier.[4 6 7 9 20] For the cast material, ε_p is smaller (σ_p larger) than the worked as a result of the the large δ ferrite particles producing a localized fine substructure which enhances both strain hardening and DRX.[6-14] Values of ε_p are shown in Fig. 10 for 304 stainless steel from publications listed in Table 2.[5 6 7 14 32 37 41 48 52] It can be seen that ε_p increases not only with Z but also with grain size because nucleation occurs principally at the elongated boundaries.

DISCUSSION

For the present homogenized-worked steels it is clear that σ_p and ε_p increase as the alloying additions increase. These effects become more pronounced as Z increases ($\dot{\varepsilon}$ rises, T falls) which is reflected in the increased Q_{HW}.[6-14] These effects are fully supported by the averages of the published data (Tables 1-4). Partial cause of the mounting difference as Z augments arises from the increased solute hindrance to GB migration, but partially because of the increased likelihood of precipitation of Mo or Cr carbonitrides.[1-4] This appears in the 301 which becomes stronger (and less ductile) than the 304 only at 900°C because of the greater C content.[8 9] The as-cast materials show a divergence from this pattern insofar as the δ ferrite raises σ_p but lowers ε_p; the increase in substructure density and the enhancement in nucleation of DRX at δ particles has been confirmed by metallography.[10-14] Moreover because of higher ferrite levels, 304C has lower ε_p and ductility than 316C.[6]

The 304 grade is clearly the most studied with 40 entries in Table 1. The mean value of Q_{HW} is 410 kJ/mol and of n is 4.3 which confirm the present results for 304C and for 304W when account is taken of the lower solute level. There are only 4 cases of Q_{HW} below the -6% margin, three of which have low solute levels (e,f,i,n, Table 2)[22 23 26 30] and 3 above the +6%, 2 of which have high solute levels (K,R,S, Table 2).[26 53] In Fig. 3a, the three comparison steels have similar n values but are considerably stronger than the present 304W (m,q,D,Table 1).[5 29 42] They are less strong than the present 304C which indicates that the differences cannot be considered substantial since less than the strength variation due to segregation. The three alloys have greater solute content; the order is affected by additions of Cu and Co to the steel of McQueen et al. (q)[5] and reduction of Cr in that of Rossard (D).[42] The results of Hashizume (m)[29] are in compression which always tends to give higher values than torsion[73] even though the compositions are almost identical. In Fig. 3b, McQueen et al. (q)[5] show a high activation energy (equal to the mean) whereas Hashizume (m)[29] is the same as the present since the compression has raised the strength uniformly across the range. In analyzing the hot torsion results from an 18-8 steel, McShane and Sheppard[32] corrected for adiabatic heating to obtain Q_{HWC}= 483 kJ/mol. This compares closely with Q_{HWC}= 479 kJ/mol for the present 304.[14] From the 22% factor of increase over the uncorrected value,[14] a mean Q_{HWC} has been calculated (Table 2).

For 301, the mean value of Q_{HW} is considerably lower than that of the 304 (378 vs. 410 kJ/mol).[29] This appears consistent with the decrease in total metallic solute (26.3% vs. 30.3%). However, the limited results give rise to a great deal of scatter with 3 out of 4 results beyond the ±6% limit (Table 2). As seen in Fig. 3, the 301 of Suzuki et al. (d)[53] has high values of strength, n and Q_{HW} because it is in compression and has a high level of Al and N. The extrusion results of McCallum and Cockcroft (b)[54] are at low T and high $\dot{\varepsilon}$ and hence more affected by deformational heating which results in apparently lower stress and Q_{HW}; correction for adiabatic heating gives a value of Q_{HWC} of 508 kJ/mol compared to 459 kJ/mol for the present results.[7-9 14]

The present 316W (A',Table 3) is only slightly different from the mean which has

Q_{HW}= 460 kJ/mol and n = 4.3[7 13] which are derived from about 20 reports. There are only two cases (C',T',Table 2)[6 57] in which Q_{HW} fall more than 6% below the mean and only three (G',K',L')[63 66 67] which exceed it by more than 6%. In the latter cases the solute contents are very high; this also leads to higher strength as is illustrated for Teodosiu et al. (G')[63] in Fig. 4a,b.[13] In Fig. 4a, Suzuki et al. (Z)[55] exhibits higher stresses because of compression. In Fig. 4b, Hughes et al. (B')[60], Bywater and Gladman (E')[47] and Donadille et al. (H')[64] have higher strengths than the present 316W because of considerably higher Mo and Cr contents although their Q_{HW} are near the average. The very low Q_{HW}, 402 and 410 kJ/mol, are both observed in as-cast material (C',T')[6 57] with about 20% δ ferrite. This is also observed for other cast segregated steels in Fig. 7 but not for the present 304C or 317C.[6] In the present 316C, the value of Q_{HW} below that of 304C (r, 31% δ) was associated with higher ductility and ε_p but lower σ_p.[6 7] This could arise because the δ phase is acting as a lower strength component rather than as particles which harden the γ phase. However, there is insufficient knowledge of the relative morphologies of δ in 316C and 304C. Correction for adiabatic heating gives Q_{HWC}= 562 kJ/mol for 316W.[14]

For the 317 steels in Table 4 and Fig. 5, everyone has Q_{HW} differing by less than ±2% from the mean of 503 kJ/mol.[6 10-12 27 53 68] In Fig. 5, the three results from the literature do have higher flow stresses than the present 317W (but only one above 317C) because of higher Mo contents (M',N',P').[27 53 68] Correction for adiabatic heating gives Q_{HWC}= 605 kJ/mol for 317W.[14] There are, however, two values of n which are about 10% less (M',Q') which is a smaller variation than for 304 or 316. For those two steels combined (60 alloys), there are 7 below and 5 above the ±20% limits (indicated by arrows in Table 2). This large scatter in n, the range being 2.1 to 5.8, is unrelated to solute content or to Q values as noted before.[71]

The reason for determining the flow stresses and the dependencies on T and $\dot{\varepsilon}$ is the development of the capability of estimating the forces during a stage of processing. Several methods of increasing rigor for estimating σ_p will be described. With the average T and $\dot{\varepsilon}$, the graphs in Figs. 2-5 can be used for the compositions near one of the included steels. Since it is difficult to interpolate it is possible to draw up graphs of log Z vs. sinh $\alpha\sigma$ (Fig. 6a) which provide a single line for all T and $\dot{\varepsilon}$. Alternately one can use Eqn. 4 with $\alpha = 1.2 \times 10^{-2}$ MPa^{-1}, with 4.3 for n since it exhibits so much random scatter and with the T' value for the grade from Fig. 8. For Q_{HW}, the suitable mean value can be utilized or one selected on the basis of composition from the Tables or Fig. 7. These uncorrected Q_{HW} values are satisfactory approximations when the deformational heating in the processing has not been determined.[32 52] Estimated mean Q_{HWC} are given in Tables 1-4 having been calculated from the mean Q_{HWC} and the correction factors for the present worked alloys.[14] In calculating rolling forces, the value of interest is not the peak stress, which is an upper limit, but the mean stress for the pass strain in question. In order to estimate the stress at a given strain, one needs to know the peak strain ε_p as well as σ_p. From a knowledge of ε_p and σ_p, one may use a common flow curve equation to calculate the mean flow stress for any desired strain.[3] One may also employ the principle that the flow curve has constant shape for a given Z value independent of T or $\dot{\varepsilon}$.[20 41] Additional principles of such calculations are given for a planetary mill with no softening between passes of the small rolls[74] and for schedules with long interstage intervals in which there is considerable softening.[4 6 21 75]

An important constraint on forming operations is limited hot ductility. In the early passes where T is high, the main problem is the segregated cast microstructure which generally has fracture strains ε_f extending from about 2.0 at 1200°C to 0.4 near 900°C (Fig. 1).[4 6 7] The present continuously cast steels are superior to many reported which were ingot cast.[1-3] The 316 has better ductility than 304 and 317 because of less δ phase.[4 6 7] The other critical condition for cracking is the finishing stages where the temperature becomes very low; the edges are most susceptible because T is lowest and secondary tensile stresses are highest. Because the steel is much more uniform from earlier working, the dominant mechanism is intergranular fissuring initiated at triple junctions as a result of grain boundary (GB) sliding.[1-3 16 18] The failure strain is about 4 near 900°C which is generally better than the as-cast near 1200°C (Fig. 1).[7-14] ε_f rises rapidly with T (reaching 10-20 at 1200°C) as a result of increasingly rapid DRX which moves the GB away from the fissures inhibiting their propagation. The 304 now appears to be the most ductile because it has the lowest amount of solute and carbides which slow down GB migration. In the same way that DRX slows down fissure propagation, SRX during the intervals between stages of deformation also increases the ductility.[18 21]

CONCLUSIONS

The present hot torsion peak flow stresses in the range 1200-900°C and 0.1-5 s^{-1} can be adequately represented by hyperbolic sine and Arrhenius functions. Results compiled from the literature show similar behavior and mean constitutive constants quite close to the present values. The mean value of n applicable to the 4 alloys is about 4.3. The mean values of Q_{HW} are: for 301, 378 kJ/mol; for 304, 410 kJ/mol; for 316, 460 kJ/mol; and for 317, 503 kJ/mol. The values of Q for the seventy steels increase as a function of total metallic solute content. The inverted sinh formula was presented and values of the constant T' were found for each of the 4 alloys. The peak strain was shown to increase as Z and grain size rise; however, it is generally lower for cast material containing δ ferrite.

REFERENCES

1. W.J. McG. Tegart and A. Gittins, Hot Deformation of Austenite, J.B. Ballance ed., Met. Soc. AIME, Warrendale, PA, (1977), 1-46.
2. B. Ahlblom and R. Sandstrom, Int. Metall. Rev., 27 (1982), 1-27.
3. C.M. Sellars and W.J. McG. Tegart, Int. Metall. Rev., 17 (1972), 1-24.
4. H.J. McQueen and N.D. Ryan: Stainless Steels '84, Inst. of Metals, London, (1985), 50-61.
5. H.J. McQueen, R.A. Petkovic, H. Weiss, and L.G. Hinton, Hot Deformation of Austenite, J.B. Ballance, ed. Met. Soc. AIME, Warrendale, PA, 1977, 113-139.
6. N.D. Ryan, H.J. McQueen, and J.J. Jonas, Can. Metall. Q., 22 (1983), 369-378.

7. N.D. Ryan and H.J. McQueen, New Developments in Stainless Steel Technology, R.A. Lula, ed., Am. Soc. for Metals, Metals Park, OH, (1985), 293-304.
8. N.D. Ryan and H.J. McQueen, Strength of Metals and Alloys, ICSMA 7, Vol. 2, H.J. McQueen et al. eds., Pergamon Press, Oxford, (1987), 935-940.
9. N.D. Ryan and H.J. McQueen, In Intnl. Symp. on Plasticity and Resistance to Metal Deformation, S. Blecic ed., Ferrous Metall. Inst., Niksic, Yugoslavia, 1986, 11-26.
10. N.D. Ryan and H.J. McQueen, Mat. Sci. Eng., 81 (1986), 259-272.
11. N.D. Ryan, H.J. McQueen, and E. Evangelista, Annealing Processes, Recovery, Recrystallization and Grain Growth, H. Hansen et al. eds., Riso Natl. Lab., Roskilde, DK, (1986), 527-534.
12. E. Evangelista, N.D. Ryan, and H.J. McQueen, Teksid, 3 1987, in press.
13. N.D. Ryan and H.J. McQueen, Czech. J. Phys., B37 (1987), in press.
14. N.D. Ryan and H.J. McQueen, Can. Met. Q., 27 (1988), in press.
15. L. Fritzmeier, M.J. Luton, and H.J. McQueen, Strength of Metals and Alloys, ICSMA 5, Vol. I, P. Haasen et al. eds., Pergamon Press, Oxford, (1979), 95-100.
16. H.J. McQueen and J.J. Jonas, Treatise Mat. Sci. Tech., Plastic Deformation of Materials, Vol. 6, R.J. Arsenault ed., Academic Press, New York, (1975), 393-493.
17. H.J. McQueen and J.J. Jonas, J. Appl. Metalwork, 3 (1984), 233-241.
18. H.J. McQueen and J.J. Jonas, J. Appl. Metalwork, 3 (1985), 410-420.
19. H.J. McQueen and D.L. Bourell, Inter-Relationship of Metallurgical Structure and Formability, A.K. Sachdev ed., Met. Soc. AIME, Warrendale, PA, (1987), in press.
20. N.D. Ryan and H.J. McQueen, J. Mech. Working Tech., 12 (1986), 279-296.
21. N.D. Ryan and H.J. McQueen, J. Mech. Working Tech., 12 (1986), 323-349.
22. T. Ikeshima, Nihon Kinsoku Gakkai, 17 (1953), A25-29, A55-58.
23. R. Hinkfoth and H.D. König, Neue Hütte 12 (1967), 212-214.
24. P.N. Cook, The Properties of Materials at High Rates of Strain, Inst. Mechanical Engineers, London, (1957), 86-97.
25. G. Wallquist and J.C. Carlen, Mem. Sci., Rev. Metal., 56 (1959), 268-274.
26. C. Gavrila, Cercetari Metal., 14 (1973), 213-222.
27. J.M. Dhosi, L. Morsing, and N.J. Grant, Proc. 4th Int. Conf. on Mechanical Working of Steel, Vo. 44, D.A. Edgecombe ed., Gordon and Breach, New York (1965), 265-283.
28. L. Zela, J. Fuxa, M. Holler, and L. Maly, Intnl. Symp. on Plasticity and Resistance to Metal Deformation, Ferrous Metall. Inst., Niksic, Yugoslavia (1986), 179-193.
29. S. Hashizume, Sosei-To-Kako, 6 (1965), 71-75.
30. J. Elfmark, Hutnické Listy, 26 (1971), 107-113.
31. T. Maki, K. Akasaka, K. Okuno, and I. Tamura, Trans. ISIJ., 22 (1982), 253-261.
32. H.B. McShane and T. Sheppard, J. Mech. Working Tech., 9 (1984), 147-160.
33. B. Ahlblom, Studies of the Deformation of FCC Materials at Both High and Low Temperatures, IM-1208, Royal Inst. Tech., Stockholm (1977).
34. V.I. Zyuzin, M.Y. Brovman, and A.F. Melynikov, Soprotivenie deformacii, Izdatel'stvo, Metallurgiia, Moscow, (1964), 153-182.
35. V.S. Zotyeyev, Novyie metody ispitanii metallov, Tecknicheskoe Izdatel'stvo Metallurgiia, Moscow, (1962), 342-348.
36. K. Inouye, Tetsu-To-Haganê, 41 (1955), 15-23.
37. C. Ouchi and T. Okita, Trans. ISIJ, 22 (1982), 543-551.
38. A. Nadai and M.J. Manjoine, J. Appl. Mech., 8 (1941), A77-91.
39. L.D. Sokolov, Chernaia Metal., 7 (1964), 59-63.
40. T.L.F. Müller, Ph.D. Thesis, University of Sheffield (1967).
41. D.R. Barraclough, Ph.D. Thesis, University of Sheffield (1974).
42. C. Rossard, Metaux, Corrosion, Industrie, 35 (1960), 102-115, 140-153, 190-205.
43. C.M. Sellars and W.J. McG. Tegart, Mem. Sci. Rev. Metall., 63 (1966), 731-746.
44. I.J. Tarnovoskij, Chernaia Metalurgiia, 4 (1961), 82-90.
45. A.T. Cole and G.J. Richardson, Hot Working and Forming Processes, Metals Society, London, (1979), 128-132.
46. G. Carfi, C. Perdix, D. Bouleau, and C. Donadille, Strength of Metals and Alloys, ICSMA 7, Vol. 2, H.J. McQueen et al. eds., Pergamon Press, Oxford, (1987), 929-934.
47. K.A. Bywater and T. Gladman, Met. Tech., 3 (1976), 358-365.
48. Y. Ohtakara, T. Nakamura, and S. Sakui, Trans. ISIJ, 12 (1972), 36-44, 207-216.
49. T. Nakamura, M. Ueki and Y. Ohtakara, Tetsu-To-Hagane, 58 (1972), A87-90.
50. D. Hengerer, Radex-Rundschau, 1 (1977), 72-82.
51. M. Hildebrand, Neue Hütte, 17 (1972), 724-727.
52. S.L. Semiatin and J.H. Holbrook, Metall. Trans., 14A (1983), 1681-1695.
53. G. Radu, D. Moisescu, C. Vaida, and I. Ilca, Cercetari Metal., 18, (1977), 239-246.
54. R. McCallum and M.G. Cockcroft, Structure Properties of Warm Extruded γ Stainless Steels, NEL Rep. 517, Dept. Trade and Industry, London (1972).
55. H. Suzuki, S. Hashizume et al., Rep. Inst. Sci. Univ. Tokyo, 18 (1968), 1-240.
56. A. Gittins, J.R. Everett, and W.J. McG. Tegart, Met. Tech., 4 (1977), 377-383.
57. M. Zidek and B. Kubickova, Technical Digest (SNTL, Prague), 3 (1968), 154-161.
58. R. Barbosa and C.M. Sellars, to be published (1987).
59. R. Colas and C.M. Sellars, Strength of Metals and Alloys, ICSMA 7, Vol. 2, H.J. McQueen et al. eds., Pergamon Press, Oxford, (1987), 941-946.
60. K.E. Hughes, K.D. Nair, and C.M. Sellars, Met. Tech. 1 (1974), 161-169.
61. I. Dragan, C.G. Radu, and C. Vaida, Metalurgica, 30 (1978), 439-443.
62. C.M. Young and O.D. Sherby, JISI, 211 (1973), 640-647.
63. C. Teodosiu, V. Nicolae, E. Soos, and C.G. Radu, Rev. Roum, Sci. Techn. Mec. Appl., 24 (1979), 13-43.
64. C. Donadille, C. Rossard, and B. Thomas, Annealing Processes: Recovery, Recrystallization and Grain Growth, N. Hansen et al. ed., Riso Natl. Lab., Roskilde, DK (1986), 285-290.
65. R. Johansson, Clean Steel, 1 (1971), 201-216.
66. K. Nikkila, Scand. J. Metall., 1 (1972), 9-15.
67. A. Tokarz and J. Bik, Intnl. Symp. on Plasticity and Resistance to Metal Deformation, Ferrous Metall. Inst., Niksic, Yugoslavia (1986), 424-433.
68. H. Bodén, IM-1233, Swedish Institute for Metals Research, Stockholm (1977).

69. U.F. Kocks, J. Eng. Mat. Tech., 98(1976), 75-85.
70. H. Mecking, Dislocation Modelling of Physical Systems, M.F. Ashby ed., Pergamon Press, Oxford, (1981), 197-211.
71. K. Tanaka, T. Nakamura, Y. Hoshida and S. Hara, Res. Mech., 12 (1986), 41-57.
72. T. Nakamura and M. Ueki, Zairyo, 23 (1974), 182-188.
73. C. Tome, G.R. Canova, U.F. Kocks, N. Christodoulou, and J.J. Jonas, Acta Metall., 32 (1984), 1637-1654.
74. N.D. Ryan and H.J. McQueen, Proc. 4th Steel Rolling Congress, (Deauville 1987), B. Fazan et al. eds., IRSID, Mazieres-Le-Metz, France (1987), F.17.1-F.17-9.
75. H.J. McQueen, Can. Met. Q., 21 (1982), 445-460.

TABLE 1 COMPOSITIONS, GRAIN SIZES, STRESS EXPONENTS AND ACTIVATION ENERGIES FOR 304

Steel Condition	C	Mn	P	S	Si	Cr	Mo	Ni	N	O	MET+	D_0 x	Mode	n	Q_{HW}	REFERENCE		
			wt.	%					ppm	ppm	%	μm						
304LW	.100	.41	--	--	.45	18.30	--	8.00	-	--	27.16	--	TEN	4.6	381↓	Ikeshima	e	22
304W	.070	.48	--	--	.43	18.60	--	7.70	-	--	27.21	--	ROL	4.7	377↓	Hinkfoth, Konig	f	23
304W	.070	.48	--	--	.43	18.60	--	7.70	-	--	27.21	--	COM	4.4	424	Cook	g	24
304W	.047	.81	.033	.019	.28	18.10	--	8.50	-	--	27.69	--	TOR	4.3	398	Wallquist, Carl.	h	25
304C	.030	.91	.021	.030	.34	17.80	--	9.70	-	(δ=7%)	28.75	--	TOR	4.9	347↓	Gavrila	i	26
304C	.040	.80	.013	.008	.28	17.70	0.04	9.90	490	--	28.75	--	TEN	4.5	396	Dhosi et al.	i'	27
304LW	.030	.91	.021	.030	.34	17.80	--	9.70	-	--	28.75	--	TOR	3.8	404	Gavrila	j	26
304W	.110	1.17	.020	.013	.58	16.55	0.12	9.85	-	(Ti .57)	28.84	--	TOR	4.0	416	Zela et al.	k	28
304W	.062	1.72	.030	.008	.47	18.28	0.28	8.27	590	35	29.02	70	TOR	4.6	393	McQueen & Ryan	l	14
304W	.080	1.06	.037	.005	.49	18.37	--	9.16	-	--	29.08	--	COM	4.5	393	Hashizume	m	29
304W	.100	.92	.018	.014	.37	17.80	--	9.50	-	(Ti .57)	29.16	-	TEN	3.5	351↓	Elfmark	n	30
304W	.050	1.07	.029	.003	.53	18.42	0.09	9.14	380	(Cu .05)	29.30	150	TEN	4.6	435	Maki et al.	p	31
304W	.070	1.76	--	--	.82	17.60	0.27	8.52	(Cu .30,	Co .17)	29.45	--	TOR	4.3	410	McQueen et al.	q	5
304C	.069	1.76	.020	.002	.68	18.31	0.08	8.68	-	(δ=31%)	29.51	64	TOR	4.5	407	Ryan et al.	r	6
304W	.042	1.29	.008	.011	.53	18.32	0.01	9.34	240	(Cu .04)	29.56	28	COM	3.8	418	Ahlblom	s	33
304W	.070	.82	.030	.011	.47	17.80	--	10.50	-	--	29.59	--	COM	4.9	398	Zyuzin et al.	t	34
304W	.070	.82	.030	.011	.47	17.80	--	10.50	-	(Ti .20)	29.79	--	COM	3.5	422	Zotyeyev	u	35
304C	.030	.92	.021	.030	.34	17.70	--	11.00	-	(δ=4%)	29.96	--	TOR	4.0	429	Gavrila	v	26
304LW	.030	.92	.021	.030	.34	17.70	--	11.00	-	--	29.96	--	TOR	4.0	431	Gavrila	w	26
304W	.060	1.21	.032	--	.57	18.69	--	9.42	-	--	30.10	--	TEN	5.6↑	391	Inouye	x	36
304W	.060	1.56	.031	.008	.75	18.70	0.22	9.07	271	--	30.33	--	COM	4.4	393	Ouchi & Okita	y	37
304W	.054	1.19	.025	.016	.48	18.29	0.29	10.10	-	--	30.35	--	TEN	3.0↓	425	Nadai, Manjoine	z	38
304W	.090	.95	.018	.021	.64	18.00	--	10.40	-	(Ti .43)	30.42	--	COM	3.3↓	407	Sokolov	A	39
304W	.050	.92	--	--	.11	18.20	0.02	11.30	(Nb .02,	Ti .02)	30.67	--	TOR	4.0	402	Müller	B	40
304W	.050	.92	--	--	.11	18.20	0.02	11.30	(Nb .02,	Ti .02)	30.67	250	TOR	4.6	410	Barraclough	C	41
304W	.055	1.06	.036	.022	.59	18.20	0.50	10.60	-	(Cu .26)	30.95	--	TOR	4.3	406	Rossard	D	42
304W	.055	1.06	.036	.022	.59	18.20	0.50	10.30	-	(Cu .26)	30.95	--	TOR	4.3	414	Sellars, Tegart	E	43
304W	.110	1.07	.024	.012	.80	18.20	--	10.30	-	(Ti .65)	31.02	--	COM	2.9↓	404	Tarnoyskij et	F	44
304W	.062	1.02	--	--	.17	18.50	0.02	11.40	-	--	31.24	120	TOR	5.8↑	424	Cole, Richard	G	45
304LW	.018	1.20	--	--	.32	18.60	0.19	11.10	60	(Cu .16)	31.57	280	TOR	4.5	415	Carfi et al.	H	46
304LW	.020	1.57	--	--	.47	19.50	--	10.20	-	--	31.74	-	TEN	4.4	424	Bywater, Gladm.	I	47
304C	.030	.91	.021	.030	.35	17.60	--	13.00	-	(δ=0%)	31.86	-	TOR	4.2	431	Gavrila	J	26
304LW	.030	.91	.021	.030	.35	17.60	--	13.00	-	--	31.86	-	TOR	4.1	438↑	Gavrila	K	26
304LW	.026	1.72	.026	.005	.54	19.02	--	10.80	(Cu .07,	Nb .06)	31.21	300	TOR	4.7	418	Ohtakara et al.	L	48
304L	.026	1.72	.026	.005	.54	19.02	--	10.80	(Cu .07,	Nb .06)	32.21	-	TOR	5.8↑	427	Nakamura et al.	M	49
304W	.084	1.40	.036	.021	.67	18.85	0.47	10.92	-	--	32.32	200	TOR	4.5	398	Hengerer	N	50
304W	.084	1.40	.036	.021	.67	18.85	0.47	10.12	(Cu .26,	Ti .57)	32.34	-	TOR	3.4↓	409	Hildebrand	P	51
304L	.013	1.80	.012	.004	.64	19.60	0.03	10.60	380	(Co .07)	32.74	27	TOR	4.6	411	Semiatin, Holbr.	Q	52
304W	.040	1.70	.012	.010	.55	18.23	0.44	12.43	-	--	33.35	-	TOR	4.4	439↑	Radu et al.	R	53
304W	.040	1.90	.017	.017	.67	17.63	1.37	12.64	-	--	34.26	-	TOR	4.2	464↑	Radu et al.	S	53
MEAN	.073	1.15	.025	.016	.48	18.22	0.10	10.17	344	35	30.30	168	-	4.3	410	Q_{HWC} 500 kJ/mol		

+ In increasing order of metallic solute; x cast, dendrite arm spacing; worked, original grain size.
W=Worked, C=As Cast, L=Low Carbon, TOR=Torsion, TEN=Tension, COM=Compression, EXT=Extrusion, ROL=Rolling.

TABLES 2-4 COMPOSITIONS, GRAIN SIZES, STRESS EXPONENTS AND ACTIVATION ENERGIES: 2,301; 3,316; 4,317.

Steel Con-dition	C	Mn	P	S	Si	Cr	Mo	Ni	N	O	MET[+]	D_0[x]	Mode	n	Q_{HW}	REFERENCE		
	wt. %								ppm	ppm	%	μm						
Table 2: Values for 301																		
301W	.150	.48	.031	.016	.48	16.21	--	7.62	-	-	24.79	--	TOR	3.7	335↓	Elfmark	a	30
301W	.080	1.16	.013	.026	.32	17.44	--	7.37	-	-	26.29	--	EXT	4.4	340↓	McCallum,Cockc.	b	54
301W	.110	1.12	.036	.002	.54	17.12	0.20	7.92	190	43	26.90	66	TOR	4.4	399	Ryan & McQueen	c	89
301W	.080	1.10	.009	.014	.93	16.99	0.31	6.96	200	(Al .93)	27.24	--	COM	4.9	437↑	Suzuki et al.	d	55
MEAN	.105	.97	.022	.058	.57	16.94	0.26	7.47	195	(.07)	26.29	66	-	4.3	378	Q_{HWC} 435 kJ/mol		

Steel Con-dition	C	Mn	P	S	Si	Cr	Mo	Ni	N	O	MET[+]	D_0[x]	Mode	n	Q_{HW}	REFERENCE		
Table 3: Values for 316																		
316W	.070	.97	.020	.021	.46	17.10	2.48	11.40	-	--	32.41	--	TOR	4.9	444	Gittins et al.	T	56
316C	.068	1.15	.021	.013	.40	16.90	2.35	12.00	(δ=20%,	Ti .22)	32.02	--	TOR	4.0	410↓	Zidek, Kubicko	T'	57
316W	.068	1.15	.021	.013	.40	16.90	2.35	12.00	-	(Ti .22)	32.02	--	TOR	4.5	453	Zidek, Kubicko	U'	57
316W	.068	1.15	.021	.013	.40	16.90	2.35	12.00	-	(Ti .22)	32.02	--	TOR	3.4↓	455	Zela et al.	U	28
316W	.070	1.57	.039	--	.90	17.40	2.60	10.60	-	--	33.07	--	TEN	5.7↑	465	Inouye	V	36
316L	.024	1.50	--	--	.29	16.70	2.63	12.20	390	--	33.32	191	TOR	4.4	450	Barbosa,Sellar.	W	58
316LW	.024	1.50	--	--	.29	16.70	2.63	12.20	-	--	33.32	30		4.5	460	Colas,Sellars	X	59
316W	.090	1.00	.031	.017	.60	17.20	2.30	11.80	-	(Ti .58)	33.48	--	COM	3.8	448	Zyuzin et al.	Y	34
316W	.070	1.34	.030	.008	.67	17.29	2.26	12.04	-	--	33.60	--	COM	4.2	459	Suzuki et al.	Z	55
316LW	.010	1.87	.034	.007	.62	16.40	2.73	12.05	140	37	33.67	60	TOR	4.5	454	McQueen & Ryan	A'	13
316W	.070	1.27	.020	.023	.64	17.20	2.92	10.90	(Co .32,	Cu .49)	33.76	--	TOR	4.5	460	Hughes et al.	B'	60
316C	.017	1.68	.022	.007	.52	16.92	2.76	12.42	110	(δ=20%)	34.60	63	TOR	4.5	402↓	Ryan et al.	C'	6
316W	.040	1.68	--	--	.40	17.00	2.20	12.90	(δ=1.3%,	Ti .42)	34.60	50	TOR	4.8	458	Dragan et al.	D'	61
316LW	.030	1.76	--	--	.70	17.60	2.67	12.80	(δ=3%)	--	34.83	111	TEN	4.8	479	Bywater & Glad	E'	47
316W	.060	1.74	.024	.014	.43	17.93	2.50	12.38	--	--	34.96	50	TOR	5.8↑	481	Young & Sherby	F'	62
316W	.550	1.76	--	--	1.08	17.42	2.14	12.80	--	(Ti .05)	35.25	60	TOR	4.8	499↑	Teodosiu et al.	G'	63
316LW	.020	1.76	--	--	.35	17.40	2.35	13.70	600	--	35.52	80	TOR	4.6	460	Donadille etal.	H'	64
316LW	.030	2.13	.014	.013	.69	17.56	2.73	12.82	--	--	35.93	--	TOR	4.0	479	Radu et al.	I'	53
316LW	.013	1.72	.007	.011	.60	17.38	2.71	13.67	420	--	36.08	--	TEN	4.0	464	Johansson	J'	65
316W	.040	1.39	--	.010	.52	17.60	2.73	13.90	--	(Cu .22)	36.36	--	TOR	2.1↓	498↑	Nikkila	K'	66
316W	.070	1.22	.026	.030	.52	17.04	2.97	15.39	(Ti .57,	Cu .19)	37.90	--	TOR	3.1↓	491↑	Tokarz & Bik	L'	67
MEAN	.048	1.49	.024	.015	.52	17.17	2.32	12.47	332	(.30)	34.27	77	-	4.3	460	Q_{HWC} 570 kJ/mol		

Steel Con-dition	C	Mn	P	S	Si	Cr	Mo	Ni	N	O	MET[+]	D_0[x]	Mode	n	Q_{HW}	REFERENCE		
Table 4: Values for 317																		
317LW	.027	1.59	.017	.015	.53	15.80	4.30	14.00	1400	--	36.22	75	COM	4.0	504	Boden	M'	68
317LW	.030	1.60	.015	.015	.51	17.28	3.66	12.85	--	--	35.90	--	TOR	4.5	503	Radu et al.	N'	53
317C	.050	1.46	.015	.023	.51	16.70	4.06	15.00	360	(Cu .110)	37.73	--	TEN	4.5	502	Dhosi et al.	P'	27
317C	.035	1.73	.032	.004	.44	18.60	3.22	13.88	170	(δ=23%)	37.87	72	TOR	4.0	508	Ryan et al.	Q'	6
317W	.035	1.73	.032	.004	.44	18.60	3.22	13.88	170 (δ=5%)	61	37.87	57	TOR	4.5	496	McQueen & Ryan	R'	10
MEAN	.035	1.62	.022	.012	.49	17.40	3.69	13.92	525	61 (.11)	37.12	68	-	4.3	503	Q_{HWC} 615 kJ/mol		

+ In increasing order of metallic solute; x cast, dendrite arm spacing; worked, original grain size.
W=Worked, C=As Cast, L=Low Carbon, TOR=Torsion, TEN=Tension, COM=Compression, EXT=Extrusion, ROL=Rolling.

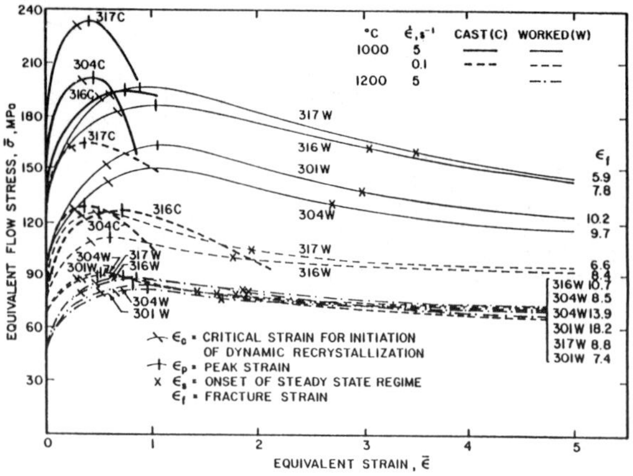

Fig. 1: Representative flow curves for the alloys 301W, 304W, 304C, 316W, 316C, 317W and 317C. The effects of T are shown by 1200 and 1000°C at 0.1 s^{-1} and of $\dot{\varepsilon}$ by 0.1 and 5 s^{-1} at 1200°C.

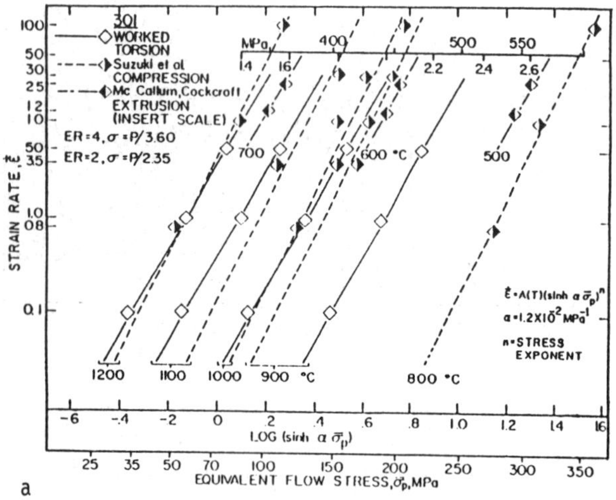

a

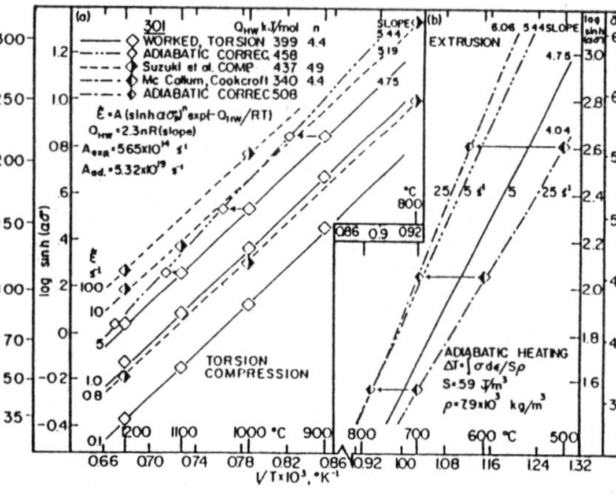

b

Fig. 2: Interdependence a) of σ_p - $\dot{\varepsilon}$ by the hyperbolic sine function and b) of σ_p - T by the Arrhenius function for 301W in torsion (C, Table 2)[8 9] compression (d)[55] and extrusion (b)[54].

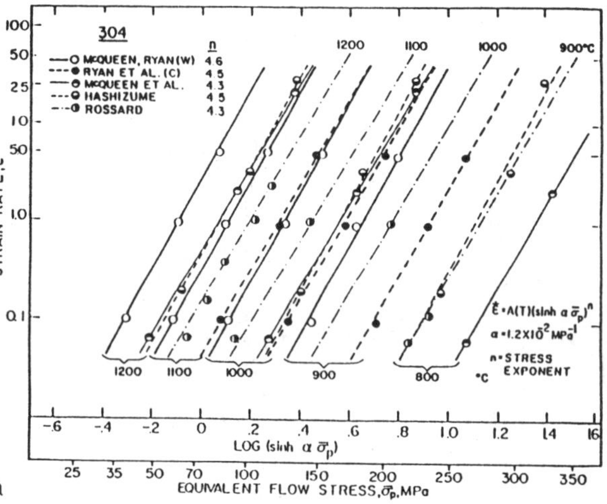

a

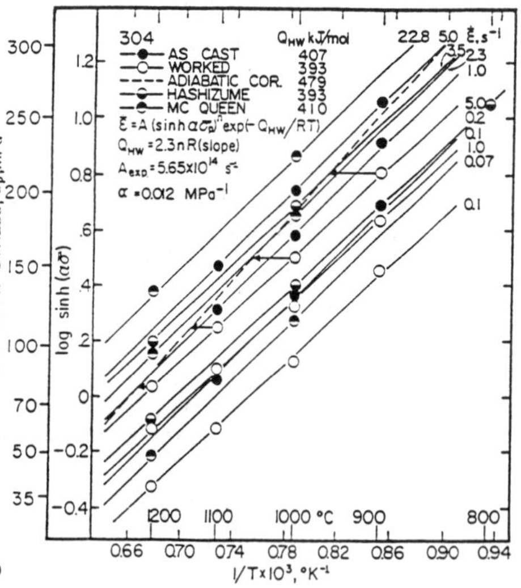

b

Fig. 3: Interdependence of σ_p - $\dot{\varepsilon}$ by hyperbolic sine function and b) of σ_p - T by the Arrhenius function for 304W and 304C (1,r)[7 9 14] with additional results form the literature for comparison (m,q,D, Table 2).[5 29 42]

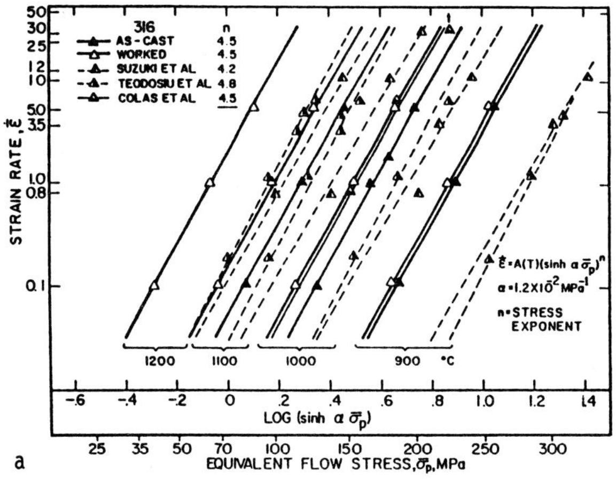

a

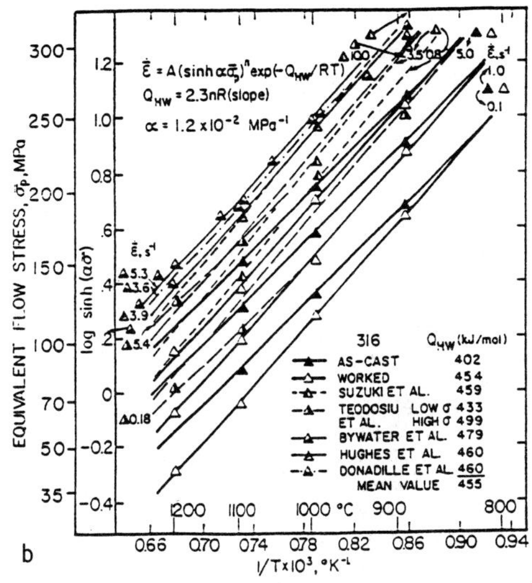

b

Fig. 4: Interdependence a) of σ_p - $\dot{\varepsilon}$ by the hyperbolic sine function and b) of σ_p - T by the Arrhenius function for 316W and 316C (A',C', Table 2)[7 13] with additional results from the literature for comparison (X,Z,B',E',G',H').[47 55 59 60 63 64]

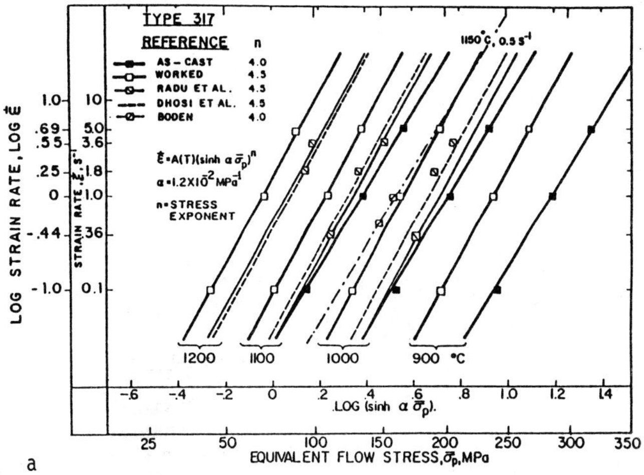

a

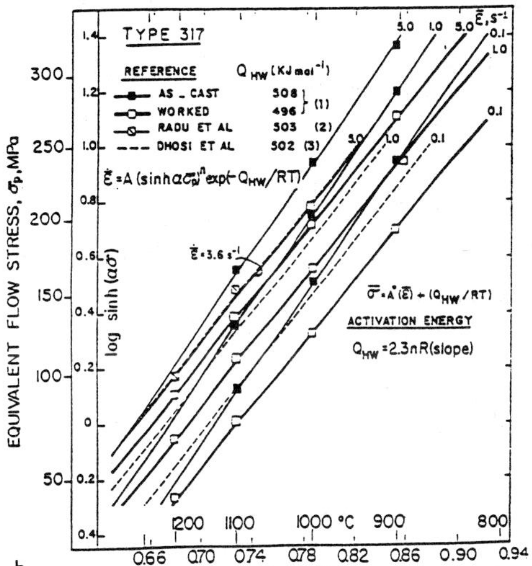

b

Fig. 5: Interdependence a) of σ_p - $\dot{\varepsilon}$ by the hyperbolic sine function and b) of σ_p - T by the Arrhenius function for 317W and 317C in torsion (Q',R', Table 2)[7 10 12] with other published data for comparison (M',N',P').[27 53 68]

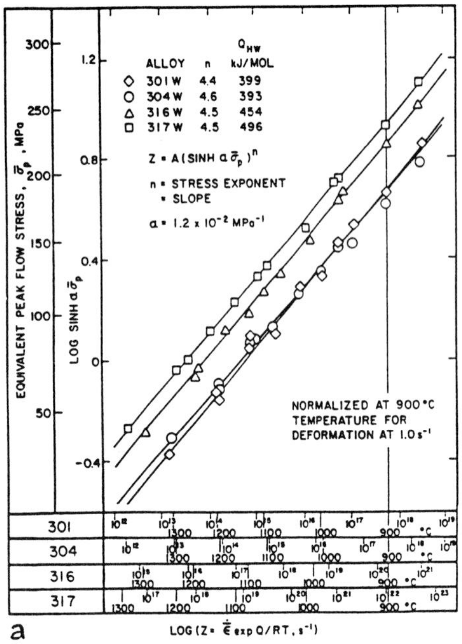

a

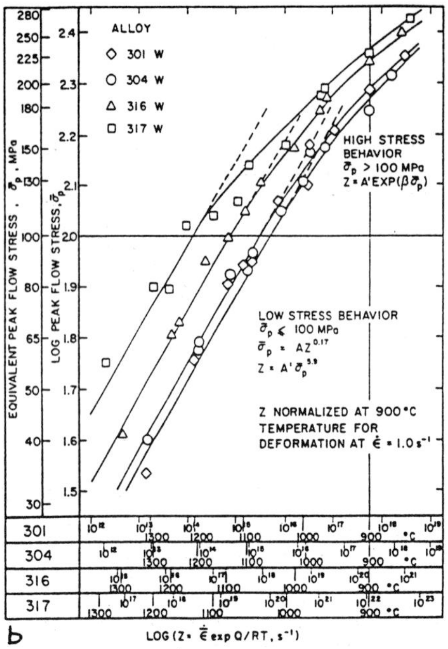

b

Fig. 6: The relationship of σ_p to Z by three different functions: a) hyperbolic sine (Eqn. 1) b) power law (Eqn. 2) and c) exponential law (Eqn. 3). It is evident that the power law breaks down above 100 MPa whereas the exponential has problems below 100 MPa.[6-15]

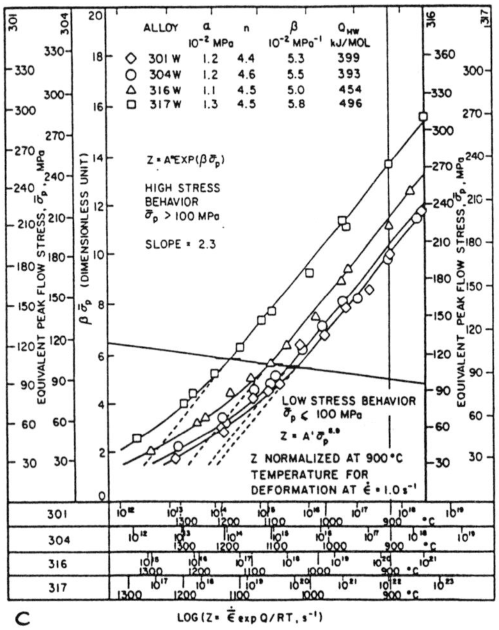

c

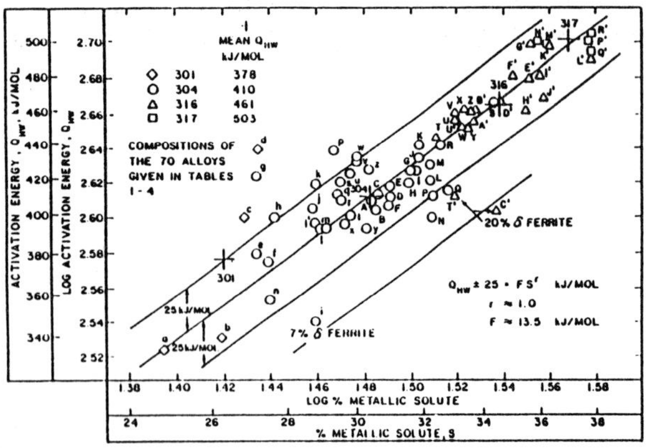

Fig. 7: The relationship of Q_{HW} to the solute concentration for 301, 304, 316 and 317 stainless steels.[6-15] The concentrations of the 70 alloys are given in Tables 1-4.[22-68] For some segregated cast alloys, Q_{HW} is unusually low.

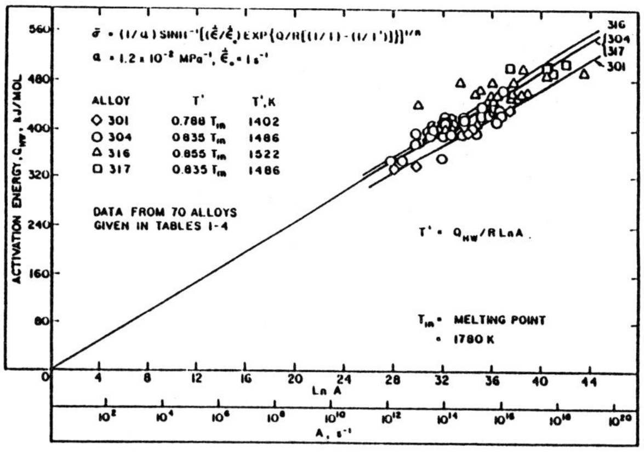

Fig. 8: This plot of the activation energy Q_{HW} against the constant A in Eqn. 1[6-13 22-68] allows derivation of the constant T' for the inverted sinh function of Tanaka et al.[71]

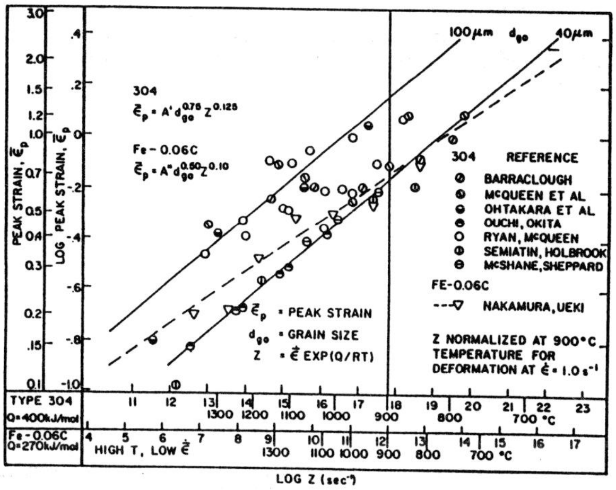

Fig. 9: The peak strain for 304W is shown to be a function of Z and to depend on the initial grain size.[5 7 14 32 37 41 48 52]

Rapid electrolytic descaling process for stainless steel wire rods

R H G Rau

The author is General Manager (R & D), Mukand Iron and Steel Works Ltd., Bombay, India.

SYNOPSIS

For smooth cold-working operations, including wire drawing of stainless steel rods, a scale-free surface is a prerequisite. Hot-rolled stainless steel wire rod is known to have an adherent tenacious complex oxide layer over its substrate, due to chromium and nickel presence in the base metal, coupled with high-temperature rolling. This paper critically reviews the available methods for stainless steel wire rod descaling, highlighting their advantages and disadvantages. It summarizes experience with the Rapid Electrolytic Descaling process, developed in the research laboratories of Mukand Iron & Steel Works Ltd, Bombay, India. The basic principles of the R.E.D. process are explained, together with successful laboratory studies and pilot plant experiments conducted with the aim of developing a single-stage process for descaling of stainless steel wire rods.

INTRODUCTION

Hot-rolled stainless steel wire rod has surface scale which is an adherent tenacious complex oxide layer over its substrate. This scale, besides wearing the dies during wire drawing, may get rolled into the base metal, thereby damaging the smooth and clean product and making subsequent operations like cold-forming extremely difficult. Furthermore, the resultant surface imperfections can act as nuclei to initiate and promote corrosion of stainless steel rods and in the cold-worked products. Descaling is therefore considered an essential step before cold working.

The scale on stainless steel wire rods is rich in oxides of chromium, iron and nickel. The characteristics and the structure of the scale depend on:

- heating temperature and time of billets
- atmosphere in the billet reheating furnace
- rate and type of cooling of rods

The scale formed on stainless steel contains tenacious oxide of chromium, like Cr_2C_3, and the spinel constituents (M_3O_4) like $FeCr_2O_4$, Fe_2NiO_4 and Fe_3O_4. These oxides, being quite stable, are difficult to remove by the usual pickling methods.

Various methods of descaling are presently being used for stainless steel wire rods, such as acid pickling, high-temperature fused salt baths, salt-bath treatment followed by acid pickling, etc. These methods are cumbersome, time-consuming and require handling of large volumes of concentrated acids and chemicals. In these conventional descaling methods the wire rods are kept in acid contact for a longer period, which results in pitting and substantial metal loss.

Salt-bath treatment is seldom employed alone for scale removal and normally the descaling operation is followed by quenching and dilute-acid pickling. The operating temperatures of salt baths are in the range of 700–1000°F. Moreover, these systems do not ensure complete cleaning since the density of coil turns inhibits satisfactory pickling action and results in inconsistent and selective gauge losses.

Though processes like 'Neutral Electrolytic Descaling' of hot- and cold-rolled stainless products are available for commercial exploitation, they are restricted to sheets and strips, and not available for wire rods. The present investigation is related to the design and development of a rapid electrolytic descaling process for stainless steel wire rods of austenitic grades, produced through hot-rolling process.

PRINCIPLES OF R.E.D. PROCESS

In the case of electro deposition, metal ions are dissolved from the anode and are deposited on the cathode. Similarly in rapid electrolytic descaling (R.E.D.), scale is caused to be detached from the anodes under controlled conditions.

The electrolytic bath of the R.E.D. process consists of a mixture of dilute acids and cleaning or complexing chemicals. During the passage of electric current at low voltage the acids get dissociated, liberating hydrogen which has a reducing effect on the scale. Due to the chemical pickling action and electro-chemical reduction, the scale is either dissolved in the form of a lower oxide or is detached by the mechanical action of hydrogen. This loosened scale is found relatively sticky in nature and is removed by passing the rod through a high-pressure water-cleaning system, which detaches the scale completely from the rod.

R.E.D. EQUIPMENT AND PROCESS

The R.E.D. process is electrochemical in nature and uses electrical and chemical energy modes for effective removal of scale. The equipment used consists of three major parts:

- mechanical system
- electrolytic tank assembly
- electrical system

The mechanical system includes an uncoiler-coiler assembly along with various rollers, guides, and straightening units. In the R.E.D. pilot plant set up, the uncoiler of wire rod is of rotating type while the coiler is of inverted type, with drum mounted on the top with a variable speed motor. The speeds can be controlled in the range of 0.75–1.25 m/second, which in turn is the speed of descaling.

Electrolytic tank assembly has been designed and fabricated with an acid-resistant, non-corrosive, and high-strength material which can be used at higher temperatures. The electrolytic tank consists of various components designed for electrolytic descaling, mechanical cleaning and passivation treatment.

During the electrolytic descaling process, continuous acid circulation is ensured using a seal-less and magnetic-coupled, centrifugal, miniature acid-resistance pump. This also helps in replenishment of fresh acid during the descaling process.

The temperature of the electrolyte is maintained using a carefully controlled heating system and the temperature of the electrolytic bath is constantly monitored. The power supply used in this descaling process is a continuously variable D.C. current rectifier of 1000 amps and 20 volts capacity. This rectifier is oil-cooled and operated on 3-phase A.C. input of 415 volts, 50 Hz supply. The power supply to anodes and cathodes is provided through a bus bar system of thick copper strips.

The hot-rolled stainless steel coils are loaded onto the uncoiler which is situated at one end of the tank assembly. The wire rod is passed through various rollers, guides, and finally through a straightening unit into the electrolytic tank assembly, followed by mechanical brushing or a high-pressure water-cleaning system and a passivation bath, and is finally wound on the drum situated over the inverted coiler, as shown in Fig. 1.

As the power is on, the moving wire rod passes through various electrolytic cells provided in the electrolytic tank assembly. The electrolyte has a relatively simple dilute hydrochloric acid as the base, with dilute hydrofloric acid and controlled amounts of citrates constituting other major elements. Here, the wire rod is subjected to electrochemical action and due to the dual action through chemical and electrical energies, the anodically treated wire rod loosens its scale and is transformed into a sludge form. The part of the scale which is left behind is then removed by passing it through a brushing or pressurized cleaning system. The wire rod is passed through a bath containing passivating or lime solution followed by a drying system before it gets wound over the drum, and continuously falls onto the inverted coiler.

PROCESS PARAMETERS

During the descaling process, concentrations of each constituent gradually change. As the concentration of acids decreases, so does conductivity. Regular checking and replenishment of acids and water is therefore essential during the descaling operation.

The hydrogen ion concentration of the electrolytic bath is another important factor which must be controlled for uniform and smooth descaling performance. In the present bath, best descaling results are obtained when the pH is maintained at between 2.5 and 4.5. When the pH falls below this range, an attack on the base metal is found to take place, coupled with slowing-down of the dissolution of the scale. When the pH value goes beyond this range, the pickling action is reduced and descaling efficiency decreases.

An increase in temperature is found to enhance the conductivity of the electrolyte. The efficiency of descaling increases with rise in temperature. However, the rise in temperature also results in an increase in evaporation loss in acids. Therefore the bath composition has to be selective. Keeping the concentration of the bath, pickling time, and other parameters constant, the results obtained on the effect of temperature on descaling performance are illustrated in Fig. 2.

The higher the current density, the better is the descaling performance. However, the optimum current density desirable to move the scale varies depending on the type of material being descaled. Higher current density does not significantly reduce the descaling time, and on the contrary it may have a negative effect as the electrodes get overheated. (In the present situation, as the wire rod is anodic, it may become hot and might even break at the contact junction.) As the temperature of the electrolyte goes up, current density can be brought down without adversely affecting the descaling efficiency, as can be observed from Table 1.

The duration of electrolytic descaling depends on the thickness of the scale layer, which is a function of annealing temperature and time. Thicker rod gauges require more time for annealing and as such have a thicker scale layer in comparison with thinner gauges, and can be effectively descaled only at lower current densities and at higher speeds.

QUALITY OF DESCALING

Figs. 3a and 3b illustrate the effectiveness of the R.E.D. process on the surface quality of 5.5 mm dia. stainless steel wire rods of 18/8 quality. It is apparent that the scale could be fully removed without any damage to the substrate when the rods are treated in line with the general principles described above.

ADVANTAGES OF R.E.D. PROCESS

The experience so far showed that the pickled surface of the R.E.D.-treated rod is uniformly descaled and is bright in appearance. It is possible to increase the pickling rate, thereby increasing productivity by modifying temperature and concentration of the bath and by varying current density in the system, making the process quite flexible.

No toxic fumes are generated in the process. The fumes which are produced can be taken care of by providing a proper exhaust system in the electrolytic tank assembly. The pollution problem is therefore minimal.

Due to the low concentrations of the acid solutions employed, the loss of base metal is significantly small in comparison with conventional pickling methods. It is possible to ensure a process yield of over 99.6%.

The whole R.E.D. system can be mechanized, automated and computerized to enhance its productivity and quality. The operational costs are lower, due primarily to only marginal loss in chemicals during the descaling operation.

CONCLUSIONS

The quality of wire rods obtained through the novel Rapid Electrolytic Descaling (R.E.D.) process is far superior with a clean, smooth, and bright surface without any attack on base metal. The innovative R.E.D. process is likely to emerge soon as one of the outstanding technological developments in the field of stainless steel wire rods.

Table 1 Effect of temperature on current density

Temperature of the electrolyte (°C)	Current density (amp/dm^2)
95	15
85	16
75	20
60	25
50	32
40	45
30	60

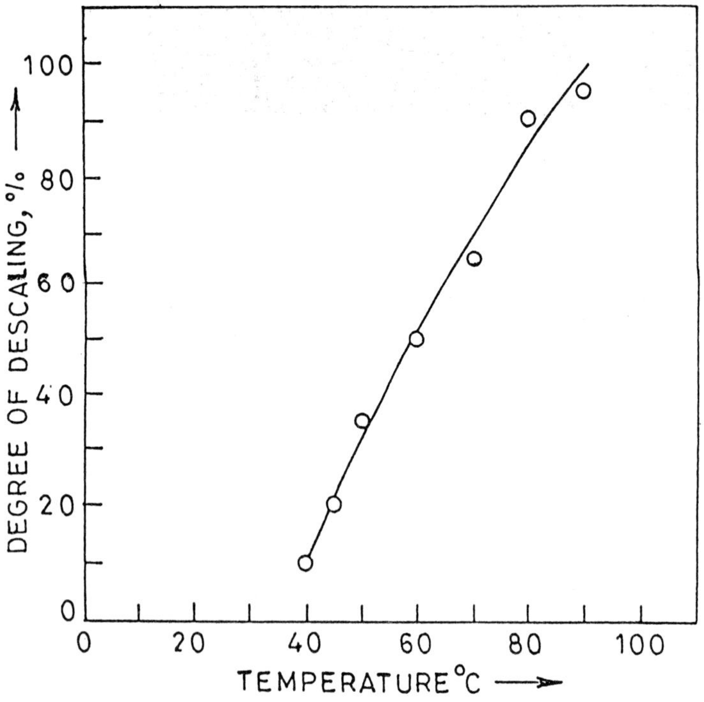

Fig. 1 Effect of bath temperature on degree of descaling

Fig. 2 Pilot plant of R.E.D. process

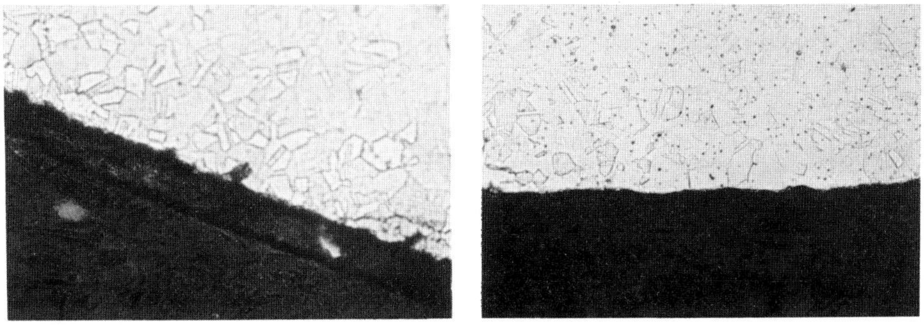

a   b

Fig. 3 Wire rod surface before and after descaling (x 375)

Review of physical metallurgy, properties and applications of dual-phase ferritic-martensitic steel designated '3CR12'

R D Knutsen, A Ball, J Hewitt, J P Hoffman and R Hutchings

RDK and AB are in the Department of Materials Engineering, University of Cape Town; JH and JPH are at Middelburg Steel and Alloys (Pty) Ltd., Transvaal; RH, formerly of the Department of Materials Engineering, University of Cape Town, is now at Beta R & D Ltd., Derby.

ABSTRACT

A steel containing approximately 12 wt. % chromium has been developed in order to fill the 'gap' between mild steel and stainless steels. The steel, designated 3CR12, has similar strength to mild steel, but possesses relative wear resistance and corrosion resistance far superior to that of conventional carbon steels. It therefore provides an effective replacement for many existing structural steels without invoking the higher costs of austenitic steels. This paper serves to review the physical metallurgy of 3CR12 which gives rise to a stable fine-grained microstructure. The excellent mechanical and corrosion properties are discussed and numerous applications are listed.

1. INTRODUCTION

Alternatives to the use of mild steel are continually being sought on the basis of improved mechanical efficiency and corrosion resistance. The cost effectiveness of attempting to protect the commonly used mild steel against corrosive attack is often questioned and this has led many development organisations towards the consideration of ferritic grade stainless steels. However, what ferritic grade stainless steels gain in their resistance to corrosive attack, they generally lose in terms of ductility, toughness and forming and welding properties. In a comprehensive review of the physical metallurgy of stainless steels[1], the deleterious effects of grain growth in ferritic grades are clearly outlined. It is noted that due to the greater atomic mobility in ferritic structures, grain-coarsening occurs readily and poor toughness is a consequence. Ferritic stainless steels, like mild steel, show a ductile-brittle cleavage transition, but the impact transition temperature is considerably higher than that for mild steel due to the embrittling effect of chromium dissolved in ferrite. This impact transition temperature can be reduced by decreasing the grain size, but, due to the non-transformable nature of ferritic grades, the coarse grain size is a major problem. Therefore, a corrosion resistant replacement for mild steel is required which contains approximately 12 wt. % chromium, and maintains a stable fine grained ferritic microstructure which will not be subject to excessive grain growth at elevated temperatures. A corrosion resistant martensite-ferrite dual phase steel, designated 3CR12, has been developed by Middelburg Steel & Alloys (Pty) Ltd, Transvaal, Republic of South Africa, which provides formability, weldability and mechanical properties comparable to that of mild steel, but possesses corrosion resistance far superior to that of conventional carbon steels. This paper serves to review the physical metallurgy, properties and applications of 3CR12 which allow it to fill the 'gap' between mild steels and stainless steels.

2. PHYSICAL METALLURGY

The composition of 3CR12 essentially arises from small but significant composition deviations from that of AISI 409. Ferrite and austenite forming elements have been carefully balanced in order to give a ferrite factor in the range 8 - 12 which is calculated from the formula given by Kaltenhauser[2]:

$$\text{ferrite-factor} = \%Cr+6(\%Si)+8(\%Ti)+4(\%Mo)+2(\%Al) -2(\%Mn)-4(\%Ni)-40[\%(C+N)]$$

This composition (Table 1) gives rise to a banded two-phase structure of martensite and ferrite after hot rolling, which, in view of the low carbon content of the martensite, produces a ferritic microstructure after tempering. The origin of the banded two-phase microstructure is due to hot-rolling being carried out in the duplex ferrite-austenite region. The dilatometrically determined $\alpha \rightarrow \gamma$ transformation temperature is approximately 800°C (depending upon the ferrite factor of the steel) and indicates that the composition of 3CR12 corresponds to the vertical line at approximately 12% Cr on the Fe-Cr phase diagram (fig. 1).

The response of the as-rolled material to heat treatment in the range 25°C to 1200°C is shown in fig. 2, which is taken from the work of Brink and Ball[3]. On reaching 780°C the as-rolled dual-phase martensitic and ferritic microstructure has become almost completely ferritic and shows a minimum in bulk hardness. Above 800°C a fine duplex microstructure is generated which is retained until at least 1200°C; the duplex structure becomes coarser with increasing temperature. A study of the microstructures by Ball et al [4] using transmission electron microscopy (TEM) indicates that after the heat treatment at 700°C, the martensite becomes highly tempered and starts to lose its lath character. The dislocations within the martensite form low-angle boundaries and the structure is reminiscent of a 'recovered' structure. At this temperature, the ferrite nucleation barriers are exceeded and if sufficient time is allowed, new ferrite can nucleate and grow. A further increase in temperature to 800°C provides sufficient thermal energy to induce substantial martensite decomposition to ferrite plus carbides. In this way, the hot rolled dual phase ferrite/martensite structure is essentially transformed to a stable fine-grained ferritic structure by annealing at temperatures just below the A_1 temperature. The stability of this fine-grained microstructure is undoubtedly due to the occurrence of a phase separation $\alpha \rightarrow \alpha + \gamma$ at temperatures of about 800°C. The presence of a duplex structure during hot-rolling not only restricts grain growth, but confers on the steel a very uniform state of plastic deformation[5]. This uniform plastic deformation of the duplex microstructure in turn produces a fine grain size during subsequent annealing.

Annealing above the A_1 temperature results in the nucleation of austenite (which forms martensite on cooling to room temperature) at the ferrite grain boundaries. The austenite content increases with increasing annealing temperature, thereby resulting in an increase in hardness of material quenched from these temperatures (fig. 2). Above 1000°C, the amount of ferrite increases and considerable grain growth occurs. This in turn causes a decrease in hardness as reflected in fig. 2. The kinetics of the transformation from ferrite to austenite at constant temperatures between 800°C and 975°C are shown in fig. 3. The reaction is very slow at temperatures below 850°C and at 915°C the transformation to austenite remains 90 per cent complete after extended times. This indicates that a small amount of ferrite remains in equilibrium with the austenite. When quenched from these duplex regions, the microhardness of the martensite is considerably greater than that of the ferrite and will in turn be determined by the amount of carbon in solution in the austenite prior to quenching. The higher the holding temperature and the longer the holding period, the greater will be the amount of carbon in solution in the austenite. The martensite formed on cooling will consequently be harder. The change in the macrohardness of the steel and the volume fraction of martensite as a function of annealing temperature is shown in fig. 4. The kinetics of the reverse reaction at temperatures below 800°C are indicated in fig. 5. Even at 750°C, a time in excess of 1 hour is required before the regions of ferrite in the austenite begin to grow. Complete transformation to ferrite and carbides takes about 16 hours. This result indicates that the dissolution of carbide and subsequent diffusion of carbon from the carbides into solution in austenite and the reverse are major factors influencing both the kinetics of reaction and in turn the microstructural condition and mechanical properties of the final product.

3. PROPERTIES

Mechanical Properties at Room Temperature

It is well known that refinement of the ferrite grain size not only increases the yield and tensile strengths according to the Hall-Petch relationship, but also increases the toughness. It is thus of little surprise that the stable fine-grained structure of 3CR12 obtained by annealing just below the A_1 temperature provides attractive strength, toughness and forming properties. In their initial study of 3CR12, Ball and Hoffman[5] demonstrated that the impact properties of 3CR12 were much improved as compared with those for AISI 409 type steel. They determined values of room-temperature impact energy, 0°C impact energy, and the ductile-brittle transition temperature (regarded as the temperature at which the impact energy is less than 30J) to be 85J, 65J and -20°C respectively for 3CR12, compared with values of 20J, 10J and 40°C respectively for samples of AISI 409 tested under identical conditions.

The room temperature tensile properties of a typical 3CR12 composition as a function of annealing temperature are indicated in fig. 6[6]. The ultimate tensile strength follows a similar pattern to the variation in hardness with annealing temperature depicted in fig. 2 and shows a minimum at around 800°C with maxima at approximately 400°C and 1050°C. The elongation to fracture traces an inverse pattern to the tensile strength showing minima at around 400°C and 1050°C and a maximum in the region of 800°C. It has also been shown that the impact strength of 3CR12 is greatly affected by the annealing temperature[7]. The impact strength follows a similar trend to that exhibited by the elongation to fracture curve and gives rise to a maximum in impact strength of 80J at around 800°C. This means that the formation of new austenite (transforms to martensite on cooling) at temperatures above 800°C results in an increase in hardness and tensile properties, but reduces impact strength and ductility. The most attractive combination of mechanical properties is therefore obtained from material annealed at a temperature just below the A_1 temperature. Typical mechanical properties of material in this condition (hot-rolled-annealed condition) are listed in Table 2.

Like all steels, the mechanical properties of 3CR12 are affected by the impurities in the steel. The effect of non-metallic inclusions on the impact toughness of 3CR12 has been studied by Knutsen and Hutchings[8] and they found that the presence of titanium carbonitride and stringered sulphide inclusions can have a deleterious effect on the impact values, and contribute towards delamination of the steel in the rolling plane. However, a more favourable inclusion morphology which gives rise to improved impact properties can be achieved by careful control of the impurity levels. The maintenance of low nitrogen and sulphur levels results in both the formation of fewer titanium carbonitride inclusions and a change in the stringered morphology of the sulphide inclusions. The occurrence of a lower volume fraction of titanium carbonitride inclusions means that there is more titanium available to form globular titanium sulphide instead of stringered manganese sulphide. The increase in impact toughness as the sulphide inclusions become less stringered is especially marked and this effect is illustrated in fig. 7. The increasing Ti/S ratio indicates a greater tendency for the formation of titanium sulphide as opposed to the formation of manganese sulphide and hence the impact values increase. These impurity controls, which ensure optimum properties, are easily accomplished in the production of 3CR12.

Mechanical Properties at Elevated Temperatures

The mechanical properties of 3CR12 at elevated temperatures have been extensively studied by Brink and Ball[3]. They found that the deformation behaviour at elevated temperatures can be divided into three temperature ranges, each with distinctive flow characteristics. These temperature ranges are marked A, B and C in fig. 8. In region A, the strength of the dual-phase structure declines with increasing temperature due to the progressive tempering of the martensite to ultimately form, at 780°C, pure ferrite and free carbides. However, the work hardening (Hollomon parameter, n) initially increases due to a dynamic strain ageing effect. Region B is characterised by a dramatic increase in elongation which can be attributed to the stable, fine-grained, duplex austenitic and ferritic structure intrinsic to 3CR12 in this temperature range. As mentioned previously, the austenite content increases with increasing temperature above the A_1 temperature and since austenite has a higher work hardening capacity than ferrite, the attenuation of the Holloman parameter to an undefined minimum between 800°C and 900°C is unexpected. However, a proposed explanation for this anomaly is that under straining conditions in the region between 800°C and 900°C, 3CR12 experiences a rapid increase in dynamic recrystallisation of ferrite together with an increase in the rate of nucleation of austenite grains induced by a straining matrix [3]. Both factors work together to retain a low work hardening condition. Since the nucleation of austenite above 900°C occurs rapidly even during static annealing, the work hardening of the austenite can no longer be retarded by the nucleation of new grains and therefore the work hardening of the material increases accordingly. Region C is characterised by a decrease in all three parameters due to the grain growth which occurs in 3CR12 above 1000°C. In terms of these elevated temperature mechanical properties, it is important to note that for hot-forming operations, minima in deformation energies occur in the temperature region between 800°C and 900°C.

Weldability

The successful weldability of 3CR12 is mainly due to the occurrence of the second phase austenite (transforms to martensite) in the heat affected zone (H.A.Z.) which restricts ferrite grain growth. The H.A.Z. in 3CR12 contains between 20% and 90% martensite, depending on the ferrite factor of the steel, the heat input per pass during welding and the number of passes. These factors all contribute towards determining the strength, ductility and toughness of the welded material[9]. The following paragraphs discuss these properties of weldments in turn.

The grain size in the H.A.Z. is determined by the heat input and the volume fraction of martensite. This means that steels with a high ferrite factor will have a low martensite content and therefore ferrite grain growth is less likely to be inhibited. On the other hand, if the steel is welded with a high heat input, the grains will be coarse with the occurrence of intergranular martensite. The weld H.A.Z. of a steel with a low ferrite factor will consist of 80-90% martensite and grain growth is inhibited. The increase in strength (and hardness) brought about by the low carbon martensite more than compensates for the loss in strength as a result of the coarser grain. The grain growth is also restricted to a band two to three grain diameters wide. It has been determined that the tensile strength of the H.A.Z. exceeds that of the parent material in all cases[9].

The presence of a low fraction of martensite phase on the grain boundaries in the H.A.Z., coupled with grain coarsening and high residual stresses, results in reduced ductility in the H.A.Z. However, if the size of the H.A.Z. is restricted (by employing low heat input procedures), the occurrence of reduced ductility is limited. Slow bend tests performed on welds, with weld root in compression and tension respectively, indicate that 90°-180° bends, depending on specimen thickness, can be successfully achieved (Table 3). Only in cases of poor penetration or slag inclusions are the bends unsuccessful.

The toughness of the H.A.Z. is determined by the volume fraction of the inherently tough, low carbon lath martensite formed. If the martensite content is maintained at approximately 45%, then this second phase effectively pins grain boundaries. The synergistic effect of low carbon lath martensite and restricted grain growth thus limits the increase in the ductile-to-brittle transition temperature of welded 3CR12.

Since many applications involve cyclic loading, welded structures must be designed for resistance to failure by fatigue. The fatigue values for 3CR12 fall within the confidence limit bands for C-Mn steels [10] and it has been shown that cruciform specimens tested give satisfactory lives with respect to weld metal failure in toe and throat of welds. This indicates that the through thickness properties are not detrimental to the fatigue strength of this type of joint. The fractures show, however, that the steel has poor resistance to crack development in the rolling plane. Although this apparently does not affect the fatigue life, caution is expressed when designing joints in which high cyclic stresses are transmitted through the plate thickness.

Corrosion and Abrasion Resistance

The use of 3CR12 as a structural steel is increasing mainly due to its superior corrosion resistance to galvanised and painted steels in marine, chemical and industrial environments. Although not labelled a stainless steel, 3CR12 exhibits a passive behaviour not unlike that of stainless steels and has very good atmospheric corrosion resistance, even when the atmosphere is heavily polluted by contaminants such as sulphur dioxide, chlorides and nitrous fumes. Extensive atmospheric corrosion tests conducted over a two year period(11) at various sites show that 3CR12 has a low corrosion rate compared with that of mild steel, Cor-Ten A, zinc, copper and aluminium. Results shown in Table 4 indicate that only in the case of a severe marine environment, is the corrosion rate higher than that of aluminium alloys.

Although the abrasion resistance of 3CR12 under dry sliding conditions is similar to mild steel, conditions in a real industrial situation where corrosion is present indicate that the relative wear resistance of 3CR12 is twice that of carbon and low alloy steels(12). Simulated laboratory corrosion-abrasion tests and tests performed in-situ on mine conveyors show that when the effects of corrosion and abrasion are combined, 3CR12 out-performs not only mild steel, but certain proprietary wear alloys as well (figs. 9 and 10).

Corrosion studies of welded 3CR12 indicate that the material is not prone to intergranular corrosion as tested according to ASTM A262(E) since the steel is properly stabilised(11). However, it has been shown that under certain environmental conditions tramline corrosion can occur if the weld oxide is not removed and the steel is not re-passivated after welding. Most metals suffer a loss in corrosion resistance in the H.A.Z. due to a combination of factors such as unfavourable microstructure and residual stresses. These can be removed by a proper post weld heat treatment. However, since 3CR12 has been developed to be a low cost steel which can be welded without a pre- or post-weld heat treatment, a stress relieving heat treatment is only considered if the steel is to be used in severe applications.

4. APPLICATIONS

The excellent mechanical properties combined with adequate corrosion resistance has resulted in 3CR12 being used in a wide spectrum of applications. Designing structures in 3CR12 is very similar to designing to existing mild steel practice in view of their similar mechanical properties. In fact, existing mild steel stresses can often be compared to the lower boundary design values for 3CR12(14). This means that, although designing with 3CR12 does not result in a significant mass saving compared to mild steel, a greater cost effectiveness is achieved due to the much lower maintenance and longer service life. However, an even greater mass saving, and hence the greatest minimisation of the cost of 3CR12, can be achieved with the use of an appropriate and efficient structural form. Therefore, the ability to design a section to suit the structural requirements at considerable mass saving allows 3CR12 to compete with conventional steel structures, even in a mildly corrosive environment.

The steel 3CR12 has found acceptance in numerous applications in the coal and South African gold mining industry. Ore and coal transport wagons, coal bunkers and washing plants, railway masts, storage tanks, chutes, spiral welded tube, square and round tubing, battery boxes, expanded or pressed flooring, have all been successfully fabricated from 3CR12 in various gauges. Furthermore, the use of this material has been extended into the agricultural, fishing and liquor and wine industries for application where abrasion-corrosion conditions are prevalent(15).

More recently, 3CR12 has found acceptance in the international market as a cost effective replacement for mild steel in many structural applications. In Europe, the steel is used extensively in the coal mining, sugar and transport industries. There is an increasing demand for 3CR12 and some of the applications include ash disposal systems, cross connection cabinets for telecommunications systems and chassis structures for the motor industry. Considerable tonnages of 3CR12 are being used in prototype milk bank units and it is expected that around two tonnes of 3CR12 will be used in each unit. Spiral wound tube is now being extensively used in the sugar refining industry and provides an enhanced life of typically five to ten times that of mild steel. It is found that the wall thickness of the pipe can be reduced by 50% and yet still give a projected life in excess of 12 years compared with only two years for mild steel. This is achieved at a cost of only two and a half times that of mild steel. The potential of the steel is still far from realised and cost saving applications abound.

ACKNOWLEDGEMENTS

This work has been undertaken as a collaborative research programme between Middelburg Steel and Alloys (Pty) Ltd and the Department of Materials Engineering of the University of Cape Town. Staff and student members of these organisations are thanked for their many and various contributions to the work. Special thanks are due to Mrs A C Ball and Mr B Greeves for their assistance in the preparation of this manuscript.

REFERENCES

1. Pickering, F.B. (1976), International Metals Reviews, 211, pp 227-268.

2. Kaltenhauser, R.H. (1971), Met.Eng., 11, (2), pp 41-47.

3. Brink, A.B. and Ball, A. (1984), Proc. of Conference "Stainless Steels '84"'Götenburg, pp 69-76.

4. Ball, A., Chauhan, Y. and Schaffer, G.B. (1987), Mat. Sci. and Technology, 3, pp 189-196.

5. Ball, A. and Hoffman, J.P. (1981), Metals Technology, 8, (9), pp 329-338.

6. Brink, A.B. (1983), MSc thesis, Univ. of Cape Town, South Africa.

7. Matthews, L.M. and DeMarsh, E.A. (1985), Mintek Report No. M213, Council for Mineral Technology, Randburg, South Africa.

8. Knutsen, R.D. and Hutchings, R. (1987), Mat.Sci. and Technology, in press.

9. Hoffman, J.P. (1984), ASM International Conference on New Developments in Stainless Steels Technology, Detroit, Michigan, paper 8410-014.

10. Tubby, P.J. (1984), Proceedings of the Inaugural International 3CR12 Conference, Johannesburg, South Africa, pp 263-280.

11. Hoffman, J.P. (1984), Proceedings of the Inaugural International 3CR12 Conference, Johannesburg, South Africa, pp 82-96.

12. Allen, C., Ball, A. and Protheroe, B.E. (1981-82), Wear 74, pp 287-305.

13. Allen, C. Protheroe, B.E. and Ball, A. (1981), J. S. Afr. Inst. Min. Metall., 81, (10), pp 289-287.

14. Swallow, J. (1984), Proceedings of the Inaugural International 3CR12 Conference, Johannesburg, South Africa, pp 304-312.

15. Market Development Division, Middelburg Steel and Alloys, Johannesburg, South Africa.

Table 1 : Compositional limits for 3CR12 steel.

C	N	Ni	Mn	Si	P	Cr	Ti
0.03 max	0.03 max	1.5 max	1.5 max	1.0 max	0.03 max	11.0-12.0	4(C+N) min

Table 2 : Mechanical properties of 3CR12 in the hot-rolled-annealed condition.

PROPERTY	UNIT	MIN.	TYPICAL
Tensile strength	MPa	460	530
Proof stress (0.2%)	MPa	280	380
Elongation (in 50mm)	%	20	26
Hardness	HB	220*	165
Charpy V Impact	J	50	80

* Maximum hardness

Table 3: Mechanical properties of welded 3CR12 as a function of plate thickness and welding parameters (Hoffman (9)).

TYPE	GAUGE	PROCESS/ ELECTRODE	PASSES	TOTAL HEAT INPUT (KJ/mm)	$RP_{0\ 2}$ (MPa)	Rm (MPa)	A $_{50}$ (%)	IMPACT (RT) CHARPY "V" (J)	BEND RC	BEND RT
3CR12	6.0mm			Parent	367	506	32	54	180°	
		MMA 308L	2	1.51	345	499	26	21	180°	180°
		MMA 316L	2	1.51	347	502	23	19	180°	180°
		MMA 310	2	1.51	360	496	27	21	180°	180°
		MMA 309	2	1.51	364	500	26	21	180°	180°
3CR12	12.0mm			Parent	376	490	36	78	180°	
		MIG 308L	4	5.99	353	489	31	38	90°	90°
		MIG 309	4	5.43	351	494	31	35	90°	90°
		MIG 310	3	4.85	299	465	26	31	90°	90°
		MIG 316L	4	4.98	336	488	27	43	90°	90°

RC Root in compression

RT Root in tension

Table 4: Comparison of the atmospheric corrosion rate of 3CR12 with some common alloys after two years exposure at various sites. Results indicate material loss in units of micrometres/year (Hoffman (11)).

METAL (Environment)	TRANSVAAL (Rural)	CAPE TOWN (Marine)	WALVIS BAY (Severe Marine)	PRETORIA (Industrial)
Mild Steel	10.9	46.7	110	31.2
'Cor-Ten' A	6.10	24.9	141	19.6
Zinc	0.203	3.20	63.1	0.279
Copper	0.838	1.40	4.34	3.48
Aluminium				
3S	0.038	0.478	0.757	0.399
M57S	0.028	0.051	0.500	
50S	0.051	0.142	0.483	
B51S	0.046	0.058	0.739	
Stainless Steel				
430	0.0025	0.0635	0.2210	
304	0.0025	0.0102	0.0330	
316	0.0025	0.0051	0.0229	
3CR12 (Pickled)	0.007	0.192	1.130	0.063

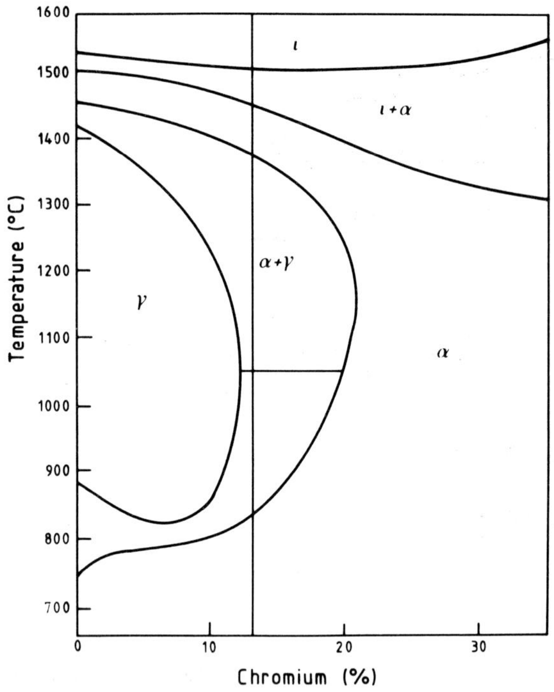

1. Fe-Cr phase diagram as a function of chromium content showing approximate position of 3CR12 (after Ball et al[4]).

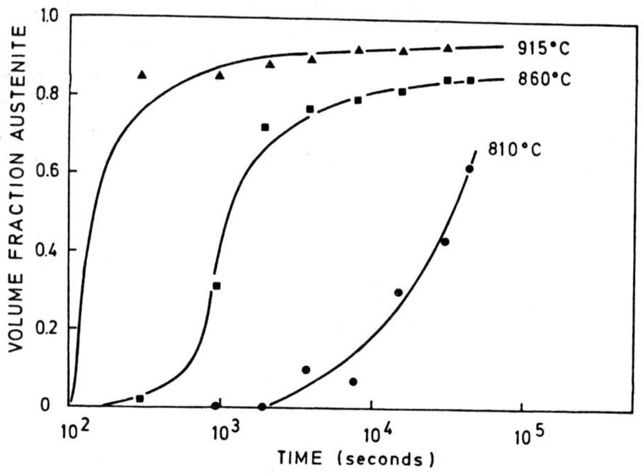

3. Volume fraction of austenite as a function of time at various temperatures for transformation of 3CR12 steel from ferrite structure (Ball et al[4]).

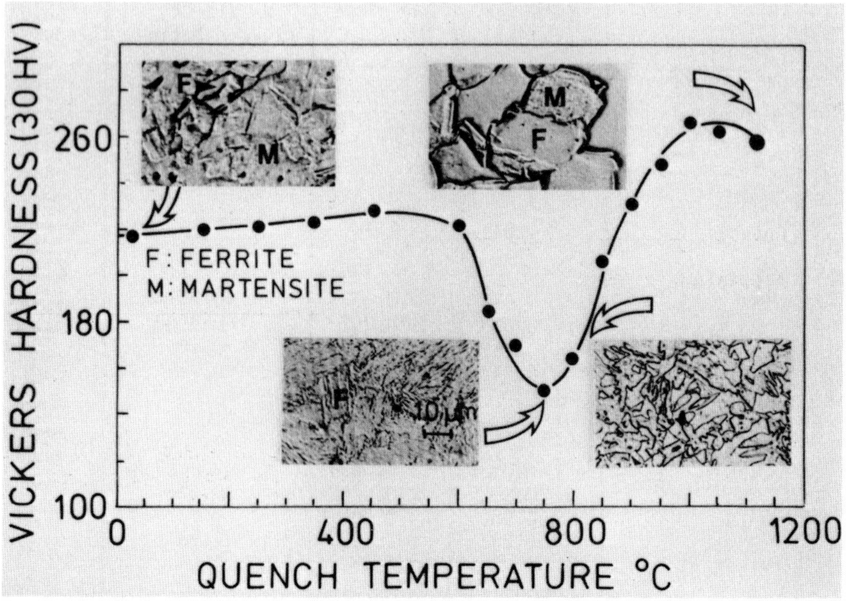

2. Variation of bulk hardness and microstructure of 3CR12 versus annealing temperature. Each specimen was soaked at indicated temperatures for 1h followed by oil quenching (Brink and Ball[3]).

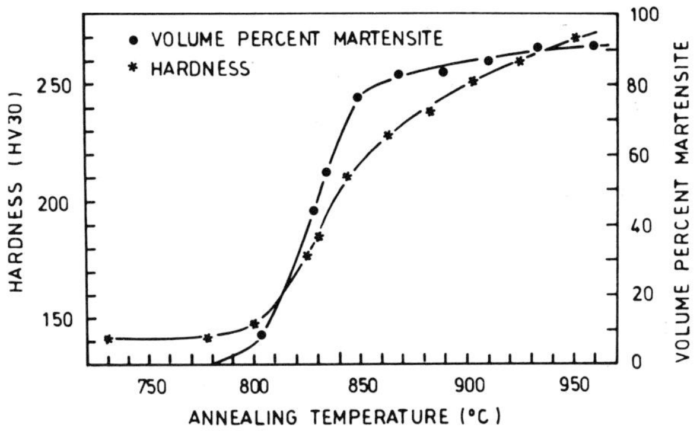

4. The effect of annealing temperature within the duplex region on hardness and martensite content for 3CR12 (Brink[6]).

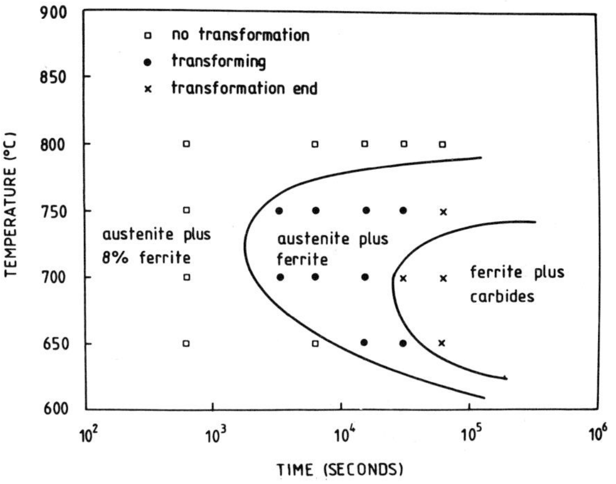

5. Isothermal transformation for decomposition of austenite to ferrite + carbides in 3CR12 steel (Ball et al[4]).

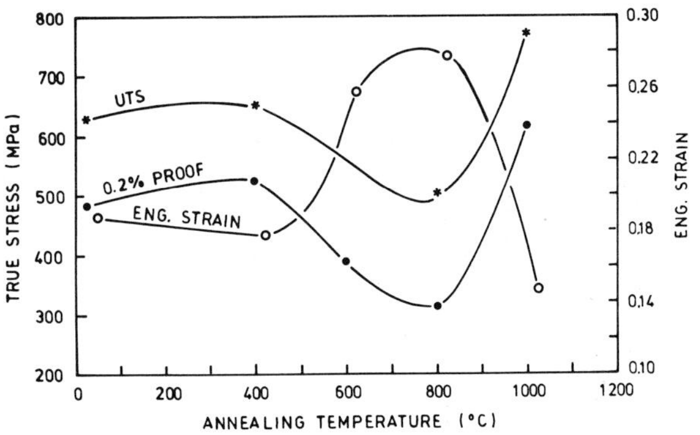

6. The effect of annealing temperature on the mechanical properties of 3CR12 (Brink[6]).

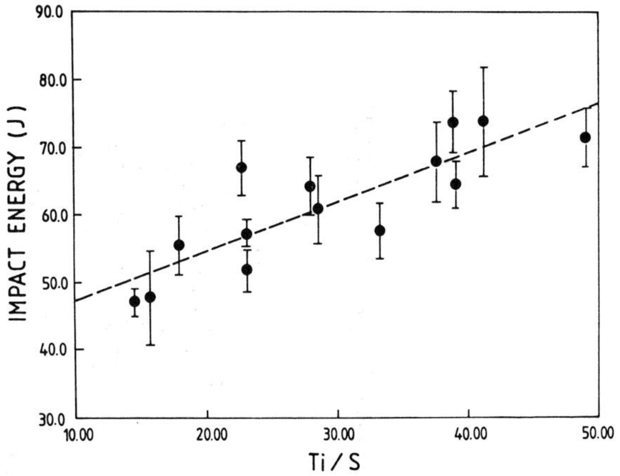

7. Impact energy of 3CR12 as a function of the Ti/S ratio of the steel (Knutsen and Hutchings[8]).

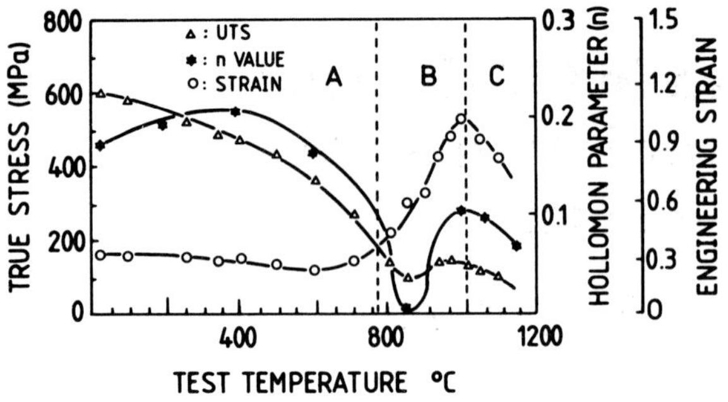

8. Deformation of 3CR12 at elevated temperatures showing three temperature ranges A, B and C, each with distinctive flow characteristics (Brink and Ball[3]).

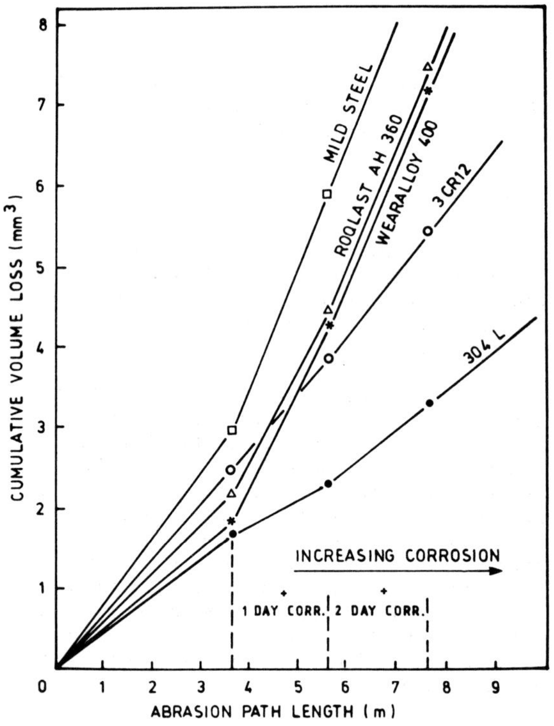

9. The effect of introducing corrosive periods on the rate of volume loss with abrasive path length in a simulated laboratory test (Allen et al[12]).

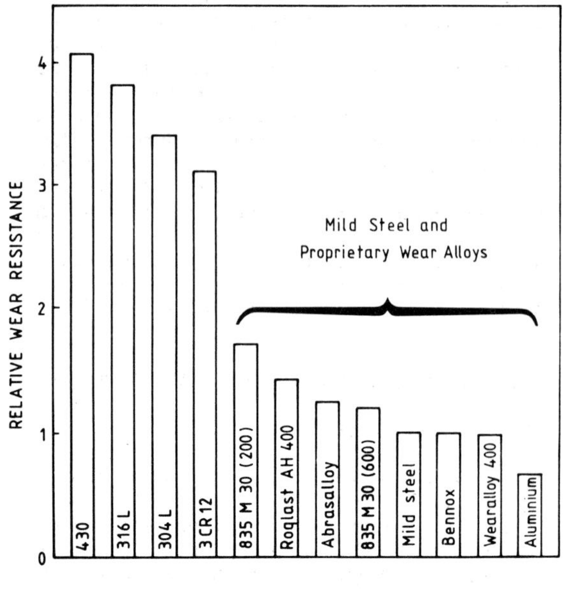

10. Relative wear resistance of 3CR12 compared to some common alloys during tests performed in-situ on mine conveyors (Allen et al[13]).

□

Production of cold-rolled stainless steel sheets for tower internals of heavy water plants

V P Sardana and B B Patnaik

VPS is General Manager and BBP Superintendent (Metallurgical Services), Steel Authority of India Ltd., Salem Steel Plant, Salem, India.

SYNOPSIS

Bulk of the heavy water required for nuclear reactors is produced by hydrogen sulphide and water dual temperature chemical exchange process. The reaction towers used in this process are provided with tower internals for facilitating exchange reactions. For satisfactory performance in this highly corrosive media, tower internals are fabricated from austenitic stainless steel sheets in AISI 304L and 316L grades due to their excellent corrosion resistance, mechanical properties and formability. Experience relating to development and industrial production of cold rolled austenitic stainless steel sheets in AISI 304L and 316L grades, as material of construction for tower internals of a heavy water plant in India, is briefly discussed in this paper.

1.0 INTRODUCTION

Heavy water is used as a moderator and coolant in pressurised heavy water type nuclear fission power reactors due to its high moderating ratio and cooling efficiency enabling the use of natural uranium as fuel.

Though heavy water can be commercially produced by a number of processes like chemical exchange methods, hydrogen distillation, water electrolysis, LASER separation process etc, most of the heavy water is produced by chemical exchange methods involving exchange of hydrogen sulphide and water or hydrogen and water or hydrogen and ammonia. Of these, the chemical exchange process involving hydrogen sulphide and water is the most commonly used.

In industrial production of heavy water by H2O-H2S isotopic exchange process, the exchange reaction takes place in counter current scrubbing towers. Feed water is fed from top and enriched H2S gas enters the bottom of the tower.

A separation stage consists of two cascading columns maintained at two different temperature ranges of 32-35 deg .C and 130-135 deg. C. In the cold section of the tower, water extracts deuterium from H2S and in the hot section, H2S extracts deuterium from water(1).Such a process is known as dual temperature process. After the first stage of processing, the concentration of heavy water is increased from about 0.015% to 0.15% which further increases to 15-20% after the second and third stages of processing. The concentration is further raised to 99.80% by fractional electrolysis and fractional vacuum distillation.

The tower columns for H2O-H2S system are provided with tower internals such as the bubble plates and sieve plates which facilitate intimate contact between the liquid and gaseous phases in counter current for exchange reaction to take place. The sieve plate is an efficient component which permits operation of the system at higher loading factors with minimum pressure drop. Besides foaming, the efficiency of the plates is also adversely affected by the deposition of corrosion products. It is, therefore, essential to ensure sufficient resistance to corrosion in the material of tower internals for efficient and durable performance of the tower columns. The components of bubble and sieve plates include downcomers, tray necks, inlet weirs, multi-downcomer trays, supports, single and double bends, bubble promotors etc. which are assembled in the tower in matching dimensions to permit desired conditions of fluid flow for maximum efficiency.

2.0 MATERIAL OF CONSTRUCTION FOR TOWER INTERNALS

2.1 Choice of material

For reliable performance of exchange towers, the material to be used in construction of tower internals plays an important role. The material should have precise chemical composition, high level of microcleanliness and desired mechanical properties coupled with adequate resistance to corrosion, particularly inter granular & stress corrosion, in addition to good weldability and formability. The components of tower internals are fabricated by a number of cold forming operations viz. roll forming, bending, pressing, drawing, punching (including perforating and louvering) and other operations including welding. The cold forming operations involve stretching, straining and drawing resulting in cold working of material. Since it is not possible to anneal the fabricated components without the risk of distortion, it is essential that the material of construction must have minimum work hardening tendency(2) besides other material characteristics. This ensures efficient and durable performance of tower internals in a highly corrosive media of H_2O-H_2S system with minimum rate of corrosion loss and risk of failure over a long period of use. To meet such demanding conditions, austenitic stainless steel in AISI 304L and 316L grades is used in sheet form in annealed condition.

2.2 Chemical Composition

The specification of chemical composition for AISI 304L and 316L grades is given in Table No.1.(3)

To minimise the risk of workhardening during cold forming operations, it is necessary to increase austenite forming and stabilising elements in the above ranges of chemical composition without sacrificing other characteristics. Though carbon, nitrogen, nickel and manganese are austenite formers, carbon content should be maintained at a low value to ensure good intergranular corrosion resistance and the nitrogen content preferably below 0.05% to achieve good formability. Hence, it is imperative to increase other austenite forming elements namely nickel and manganese towards higher side of the limits without disturbing the chemical balance. While manganese content is maintained between 1.5 to 2.0%, nickel content is generally maintained at a minimum of 9.0% for AISI 304L grade and 11.0% for AISI 316L grade. For load bearing components involving more cold working, it is desirable to increase nickel content further towards higher limits of the specification(4,5,6,7).

2.3 Cleanliness and surface condition

Cleanliness of steel used for fabrication of tower internals plays an important role in influencing their material characteristics and corrosion behaviour. Modern stainless steel-making practices involving EAF steel melting and AOD/VOD processes of secondary refining followed by argon purging before continuous casting of slabs ensure a high degree of steel cleanliness. The con-cast slabs are ground to remove surface imperfections, reheated in walking beam type furnaces with controlled parameters and hot rolled to coils in hot strip mills, equipped with automatic gauge control, rolling temperatures control and descaling facilities to achieve close dimensional tolerances and good surface. The original specification had stipulated use of solution annealed hot rolled stainless steel sheets as material for construction of tower internals. The cold rolled solution annealed stainless steel flat products have superior material characteristics resulting from fine grained uniformly recrystallised grain structure in addition to superior surface finish, gauge uniformity, strip shape and freedom from surface defects, particularly in lower range of thicknesses. It was, therefore, proposed to use cold rolled products in 2B finish subject to their meeting all quality stipulations.

3.0 QUALITY STIPULATIONS

Precision cut stainless steel sheet sections in 77 different sizes in AISI 304L grade with 9.0% min. Ni and AISI 316L grade with 11.0% min. Ni. were required in the thicknesses of 2.0, 2.5, 2.75, 3.0, 3.5 and 4.0 mm in different lengths and widths, with precise material characteristics, good corrosion resistance properties, minimum workhardening tendency and close tolerances on dimensions and flatness. The hardness of annealed material was specified to be RB 82 max as against RB 88 and RB 95 max specified in relevant ASTM specifications for AISI 304L and 316L grades respectively. The finished components should not workharden beyond RC 22 at the point of severest cold forming. For this purpose, microhardness test was to be conducted across thickness of the samples drawn in longitudinal and transverse directions of rolling and bent to 150 degrees through a mandrel of diameter equal to four times the thickness of the samples. The microhardness should be limited to VPN 250 at 1 Kg load at the point of maximum cold work.

The chloride content in pickling media and rinse water was specified to be maintained below 30 ppm to prevent risk of chloride contamination leading to pitting and chloride stress corrosion cracking. It was stipulated that the material must pass the corrosion tests as per the practices A and B of ASTM A 262 with the maximum rate of corrosion loss being limited to 0.004"/month for practice B.

The sheet sections were required to be supplied in length range of 361 to 4240 mm plus minus 2 mm, width range of 30 to 1250 mm plus minus 1 mm, out of squareness restricted to plus minus 3 mm and plus minus 6 mm, camber on longer side limited to 2 mm (max) and flatness restricted to maximum deviation of 2 mm from a flat reference surface.

4.0 EXPERIMENTAL

Depending on the severity of cold work involved in components during fabrication processes and stress conditions in actual application, tower internal components were categorised into two groups, viz. Group A and Group B. The components of Group A involved a lower degree of cold work and were a non load bearing type and those of Group B were load bearing critical components involving intricate fabrication processes. To assess the feasibility of producing base material with required characteristics particularly in respect of workhardening behaviour during cold forming, experiments were designed to arrive at specific chemical composition of material and operating parameters for their processing.

4.1 Initial Trials

Industrial trials were first conducted with standard manufacturing practices and operating parameters and evaluation tests were carried out for cold rolled, solution annealed austenitic stainless steel in AISI 304L and 316L grades so produced from available hot rolled coils.The test results are given in Table 2.

As can be seen in the table, though the material characteristics were satisfactory, the workhardening tendency was found to be significant in some cases. The microhardness values on bent samples exceeded the stipulated value of VPN 250 in certain samples at the point of maximum cold work, the tendency being relatively more in transverse samples. It was observed that for AISI 304L material with 8.6% and 9.33% Ni, the microhardness values exceeded the limit of VPN 250. In case of AISI 316L material, though the microhardness values exceeded the limit marginally for material with 11.10% Ni, the values were satisfactory with 12.60% Ni.

4.2 Laboratory and industrial trials

Laboratory annealing trials were conducted for cold rolled samples in AISI 304L and 316L grades with different annealing temperatures and soaking periods to explore the possibility of achieving lower annealed hardness and lower microhardness values on bent samples. Annealing temperatures of 1050 deg C and 1100 deg C were tried with normal and excess soaking periods. The results were satisfactory both for AISI 304L and 316L grades. With adequate soaking even at an annealing temperature of 1050 deg C, the annealed hardness could be brought down to RB 73 and microhardness values could be contained below VPN 250. There was no significant improvement in results at an annealing temperature of 1100 deg C or prolonged soaking period which on the contrary led to grain coarsening.

The effect of percentage cold reduction and number of stages of cold reduction on material characteristics were also examined. Typical test results are given in Table 3. Though the available data was not very exhaustive, it was observed that the material characteristics particularly the workhardening behaviour did not improve significantly with variation of percentage of cold reduction and increase in the number of stages of cold rolling with intermediate annealing.

Based on the observations and results of laboratory experiments, trial production was taken up with modified annealing parameters. The solution annealing temperature was maintained at 1050 deg.C and speeds of processing were reduced for each strip thickness so as to increase the annealing time by about 25% both for AISI 304L and 316L grades with around 9.0% and 11.0% Ni respectively. The annealed product exhibited a fully recrystallised structure with ASTM grain size of 7 (Fig .1)and the overall results were satisfactory. There was no significant change in material characteristics after skinpassing.

Extensive corrosion tests(8) were carried out on samples of products to ensure that corrosion resistance properties as stipulated could be achieved in a reproducible manner.

Oxalic acid etch test, as per practice A of ASTM A 262, was carried out for samples of AISI 304L and 316L grades with different nickel contents. The etched surface of the samples was examined under microscope at 500x. It was observed that the samples did not exhibit undesirable ditch structure.

Ferric sulphate - Sulphuric Acid test conforming to Practice B of ASTM A 262 was carried out on samples of AISI 304L and 316L grades in spite of the fact that the samples showed acceptable etch structure in Practice A - Oxalic acid test. The rate of corrosion was observed to be less than the specified value of 0.004"/month.

4.3 Trial fabrication of components

To evaluate further the suitability of cold rolled products in 2B finish as material of construction for tower internals, different components were fabricated from cold rolled trial products employing actual industrial fabrication processes. Micro hardness values measured on typical samples drawn from fabricated components of bubble promoter,single bend and double bend are shown in Fig. 2. It was observed that the microhardness values did not exceed the stipulated limit of VPN 250.

5.0 MATERIAL SPECIFICATION FOR INDUSTRIAL PRODUCTION

Based on results of the above trials, material specifications were drawn for chemical composition and product characteristics to be achieved in actual industrial production. Moreover, the sampling and inspection procedures were also formulated. Though it was possible to achieve specified characteristics for most of the components of tower internals in AISI 304L and 316L grades with 9% and 11% Ni respectively, it was considered prudent to increase nickel content to 10.5% minimum for AISI 304L and 13.5% minimum for AISI 316L grades for Group B components. Besides, the maximum annealed hardness and microhardness values for these components were also fixed at RB 77 and VPN 235 respectively. However, in case of AISI 316L grade, when nickel content was increased to 13.5%, the material was prone to exhibit slivers.This is because the delta ferrite content is almost nil at this concentration. In normal AISI 316L grade, nickel content is maintained around 11% which according to Delong diagram leads to a delta ferrite content of 4 to 5%. Thus for AISI 316L material with 13.5% Ni,it was necessary to grind the coils to remove sliver type of surface defects.

6.0 COMMERCIAL PRODUCTION

Industrial production was undertaken to produce 77 different sections in AISI 304L and 316L grades in 2B finish with stringent quality surveillance and a monitoring system involving systematic process control, inspection and testing of material at all stages of processing. The production cycle involved annealing, shot blasting and pickling of hot rolled coils in continuous annealing and pickling line and cold rolling in 20-high Sendzimir Mill to different thicknesses with close dimensional tolerances, followed by finish annealing and pickling in a continuous line. The possible source of chloride contamination was kept under check by maintaining the chloride content of pickling baths and final rinse water at a level below 30 ppm. The coils were skinpassed to improve flatness. To take care of slivers in AISI 316L material with Ni content of 13.5%, the coils were ground at an intermediate stage.

Samples drawn in 2B finish were tested for hardness, tensile properties, microhardness on bent samples, grain size and corrosion resistance to ensure conformity to specifications before release of material for finishing. Typical material characteristics as obtained in finished products in different thicknesses are shown in Table 4. The samples passed the corrosion tests for practices A and B of ASTM A 262 and the rate of corrosion was between 0.0013"/month and 0.0037"/month for both the grades of stainless steel, majority of them having a rate of corrosion less than 0.003"/month. The skinpassed material was slit to desired widths, precision levelled and cut to size to close tolerances.

Flatness of such products being an important quality criteria, it was checked for each section in the transverse and longitudinal directions with an elaborate flatness checking device assembled in the plant(Fig.3). A surface table with machined surface of precision tolerance was used as the reference surface for checking the flatness of sheets. The assembly consists of a bright straight steel rod which is mounted on two "V" blocks on either end with a magnetic base. Dial gauges were fixed to the rod so that flatness could be assessed in different locations across the width by changing the position of dial gauges. For measurements of flatness along the length of the sheet, the assembly was moved on the surface table along the length of the sheets. The sheets produced were marked for their identity sheet by sheet and adequate precautions were taken in handling and packing of the material to avoid possible risk of mix-up and damage.

Components of tower internals were fabricated successfully from sheet sections, employing different fabrication processes. Various field tests were conducted on samples drawn from these components. They have since been assembled in exchange towers.

7.0 CONCLUSIONS

i) Cold rolled stainless steel sheets in AISI 304L and 316L grades were successfully produced in thicknesses of 2.0, 2.5, 2.75, 3.0, 3.5 and 4.0 mm in 2B finish with desired material characteristics, and precision tolerances on dimensions and flatness to be used as material of construction for tower internals in H_2O-H_2S exchange towers.

ii) Components of Group A (non load bearing with a low degree of cold forming) fabricated from AISI 304L and 316L material with 9% Ni and 11.6% Ni respectively exhibited microhardness values below the specified maximum limit of VPN 250 on cold forming. Likewise components of Group B (load bearing with a higher degree of cold forming) showed microhardness values below VPN 235 with 10.5% and 13.5% Ni for AISI 304L and 316L grades respectively.

iii) The parameters in cold rolling & processing were so adjusted as to get annealed hardness of less than HRB 82 and HRB 77 for sheet sections of Group A and Group B respectively.

REFERENCES

1 Mc Graw Hill Encyclopaedia of Science and Technology,4th ed, V6, 1977, pp 432-434

2 Spray, P.H.G, Fabrication and Construction Experience - Bruce and Port Howkesbury Heavy Water Plants, presented at the CNEN Nuclear Symposium on Technical and Economic Aspects of Heavy Water production, Turin, Italy, 30 Sep - 1 Oct, 1970.

3 Standard specification for stainless steel and Heat Resisting Chromium Nickel Steel Plate, Sheet and Strip, A S T M A 167-82.

4 Novak, C.J., Structure and Constitution of Wrought Austenitic Stainless Steels (In: Peckner, B and Bernstein, I.M. (Eds), Handbook of Stainless Steels, McGrew Hill, 1977 pp 4.2-4.53)

5 Spahn,H., Performance Requirements for Stainless Steels in the Chemical Process Industry (In: Stainless Steels' 77, Climax Molybdenum Co., 1977, pp 161-172)

6 Pickering, F.B., The Metallurgical Evolution of Stainless Steels, ASM, 1979 pp 1-42

7 Pickering, F.B.,Physical Metallurgical Developments of Stainless Steels (In: Stainless Steels' 84, The Institute of Metals, 1985, pp 2-28)

8 Standard Practices for Detecting Susceptibility to Intergranular Attack in Stainless Steels, ASTM A 262-81

***

Table No.1 Chemical Composition of AISI 304L and 316L grades

Grade AISI	Chemical composition [%] C max	Mn max	P max	S max	Si max	Cr	Ni	Mo
304L	0.03	2.00	0.045	0.03	1.00	18.00-20.00	8.00-12.00	----
316L	0.03	2.00	0.045	0.03	1.00	16.00-18.00	10.00-14.00	2.00-3.00

Table : 2

EVALUATION OF MATERIAL CHARACTERISTICS OF STANDARD ANNEALED PRODUCTS IN AISI 304L AND 316L GRADES

Sample No.	Grade AISI	Thickness mm	Chemical Composition % C	Si	Mn	P	S	Cr	Ni	Mo	N	Properties in annealed condition Hardness RB	Y.S N/mm^2	UTS N/mm^2	% E (Lo: 50 mm)		Micro hardness, VPN on samples bent to 150°, 4t 1	2	3
1	304L	2.0	0.03	0.48	1.49	0.027	0.008	18.40	8.60	-	0.039	74	302	608	54	L	239	210	265
																T	220	180	260
2	304L	2.50	0.018	0.46	1.64	0.035	0.011	18.70	9.33	-	-	75	241	585	54	L	248	162	260
																T	248	195	263
3	304L	3.15	0.018	0.46	1.64	0.035	0.011	18.70	9.33	-	-	74	256	597	54	L	251	198	251
																T	286	163	301
4	316L	1.60	0.025	0.48	1.52	0.036	0.008	17.40	12.60	2.61	0.034	78	334	624	46	L	225	184	226
																T	205	165	232
5	316L	2.0	0.025	0.48	1.52	0.036	0.008	17.40	12.60	2.61	0.034	75	349	617	46	L	225	182	234
																T	221	167	248
6	316L	2.0	0.025	0.48	1.52	0.036	0.008	17.40	12.60	2.61	0.034	75	331	588	46	L	217	158	219
																T	219	156	234
7	316L	2.5	0.025	0.48	1.52	0.036	0.008	17.40	12.60	2.61	0.034	79	293	586	45	L	227	178	235
																T	230	184	235
8	316L	3.15	0.029	0.51	1.51	0.04	0.004	17.10	11.10	2.15	0.047	80	264	589	48	L	217	175	225
																T	257	219	269

Table : 3

EFFECT OF COLD REDUCTION AND NO.OF STAGES ON MATERIAL CHARACTERISTICS OF FINISHED PRODUCTS

Sl No	Grade AISI	Reduction in thickness mm	Chemical Composition %				Properties in annealed condition				Microhardness, VPN on Samples bent to 150°at 4t		
			C	Mn	Ni	Mo	Hard RB	Y.S N/mm²	U.T.S. N/mm²	% E (Lo:50mm)	1	2	3
1	304L	4.0 -- 2.0	0.020	1.60	9.24	-	76	300	612	50	228	149	230
2	304L	5.0 -- 2.0	0.020	1.60	9.24	-	76	315	617	52	223	174	241
3	304L	6.0 -- 2.0	0.020	1.50	9.10	-	73	284	584	52	224	168	233
4	304L	6.0 -- 3.15 -- 2.0	0.018	1.58	9.44	-	74	301	595	52	225	145	226
5	304L	6.0 -- 4.0 -- 2.0	0.020	1.50	9.10	-	76	321	596	51	242	153	241
6	304L	4.0 -- 2.5	0.018	1.64	9.33	-	78	228	583	55	216	156	234
7	304L	5.0 -- 2.5	0.020	1.60	9.24	-	75	256	589	53	239	162	245
8	316L	4.0 -- 2.0	0.021	1.50	11.50	2.55	76	313	596	49	214	178	217
9	316L	5.0 -- 2.0	0.024	1.53	11.60	2.59	78	333	612	46	202	175	216
10	316L	6.0 -- 3.15 -- 2.0	0.017	1.48	11.60	2.57	74	290	561	50	226	150	229

TABLE-4

Typical Material characteristics of finished products in AISI 304L and 316L grades for tower internals

Sl No	Grade AISI	Thickness mm	Chemical Composition [%] C	Si	Mn	P	S	Cr	Ni	Mo	N	Properties in annealed condition: Hardness RB	Y.S. N/sq.mm	U.T.S N/sq.mm	% E Lo= 50 mm	ASTM Grain Size No.	Corrosion rate inches/month	Microhardness VPN on samples bent to 150 deg	1	2	3
1	304L	2.0	0.020	0.41	1.47	0.028	0.006	18.20	9.10	-	0.034	75	285	581	51	7	0.0020	L	221	133	221
																		T	221	184	221
2	304L	2.0	0.022	0.42	1.46	0.033	0.011	18.30	9.10	-	0.04	74	305	603	53	7	0.0023	L	240	185	232
																		T	236	174	228
3	304L	2.0	0.022	0.42	1.46	0.033	0.011	18.30	9.10	-	0.04	72	302	605	53	6	0.0022	L	224	185	227
																		T	230	170	231
4	304L	2.0	0.018	0.46	1.64	0.035	0.011	18.70	9.33	-	0.04	76	313	618	49	7	0.0025	L	206	171	184
																		T	189	164	184
5	304L	2.5	0.018	0.46	1.64	0.035	0.011	18.80	9.33	-	0.04	78	228	583	55	7	0.0025	L	221	162	227
																		T	233	161	239
6	304L	3.0	0.018	0.46	1.64	0.035	0.011	18.80	9.33	-	0.04	79	233	570	56	7	0.0025	L	221	201	233
																		T	226	175	239
7	304L	2.0	0.023	0.28	1.47	0.027	0.003	18.50	10.09	-	0.04	76	302	600	47	7	0.0032	L	233	168	221
																		T	226	196	216
8	304L	2.5	0.023	0.28	1.47	0.027	0.003	18.50	10.09	-	0.04	79	245	559	56	7	0.0032	L	233	153	233
																		T	233	157	226
9	304L	3.0	0.023	0.28	1.47	0.027	0.003	18.50	10.09	-	0.04	79	254	585	53	7	0.0032	L	221	133	221
																		T	221	184	221
10	304L	2.0	0.015	0.40	1.73	0.029	0.009	18.70	10.60	-	0.03	71	272	563	51	6	0.0017	L	221	182	217
																		T	226	192	219
11	316L	2.0	0.024	0.47	1.53	0.037	0.008	17.20	11.60	2.59	0.06	77	333	612	56	6	0.0028	L	202	175	216
																		T	209	184	221
12	316L	2.0	0.020	0.48	1.53	0.036	0.003	17.20	11.60	2.55	0.04	77	328	589	44	7	0.0035	L	234	175	225
																		T	239	181	229
13	316L	2.5	0.018	0.49	1.69	0.026	0.001	17.60	13.60	2.14	0.03	74	300	570	46	7	0.0020	L	226	158	229
																		T	229	160	228
14	316L	2.5	0.019	0.46	1.63	0.027	0.001	17.70	13.70	2.14	0.02	73	299	567	44	7	0.0024	L	216	160	226
																		T	221	171	231
15	316L	2.75	0.018	0.43	1.58	0.032	0.001	17.70	13.60	2.13	0.04	74	304	562	47	6	0.0023	L	193	150	218
																		T	201	164	222
16	316L	2.75	0.016	0.47	1.61	0.028	0.001	17.80	13.60	2.15	0.03	75	304	564	46	7	0.0020	L	216	146	218
																		T	221	157	228
17	316L	2.5	0.025	0.45	1.60	0.027	0.001	17.80	13.50	2.17	0.03	73	314	597	45	7	0.0020	L	221	141	228
																		T	230	154	229
18	316L	3.0	0.018	0.43	1.58	0.032	0.001	17.70	13.60	2.13	0.04	75	305	563	47	7	0.0020	L	220	145	223
																		T	229	157	227

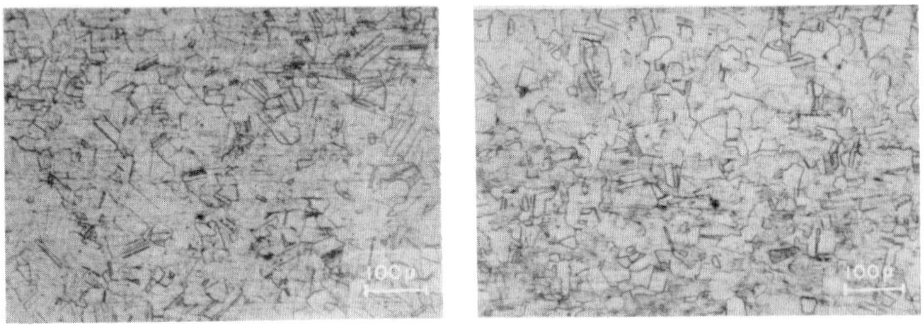

a b

Fig.1 Typical microstructure of cold rolled and annealed austenitic stainless steel of AISI grade (a) 304L and (b) 316L.

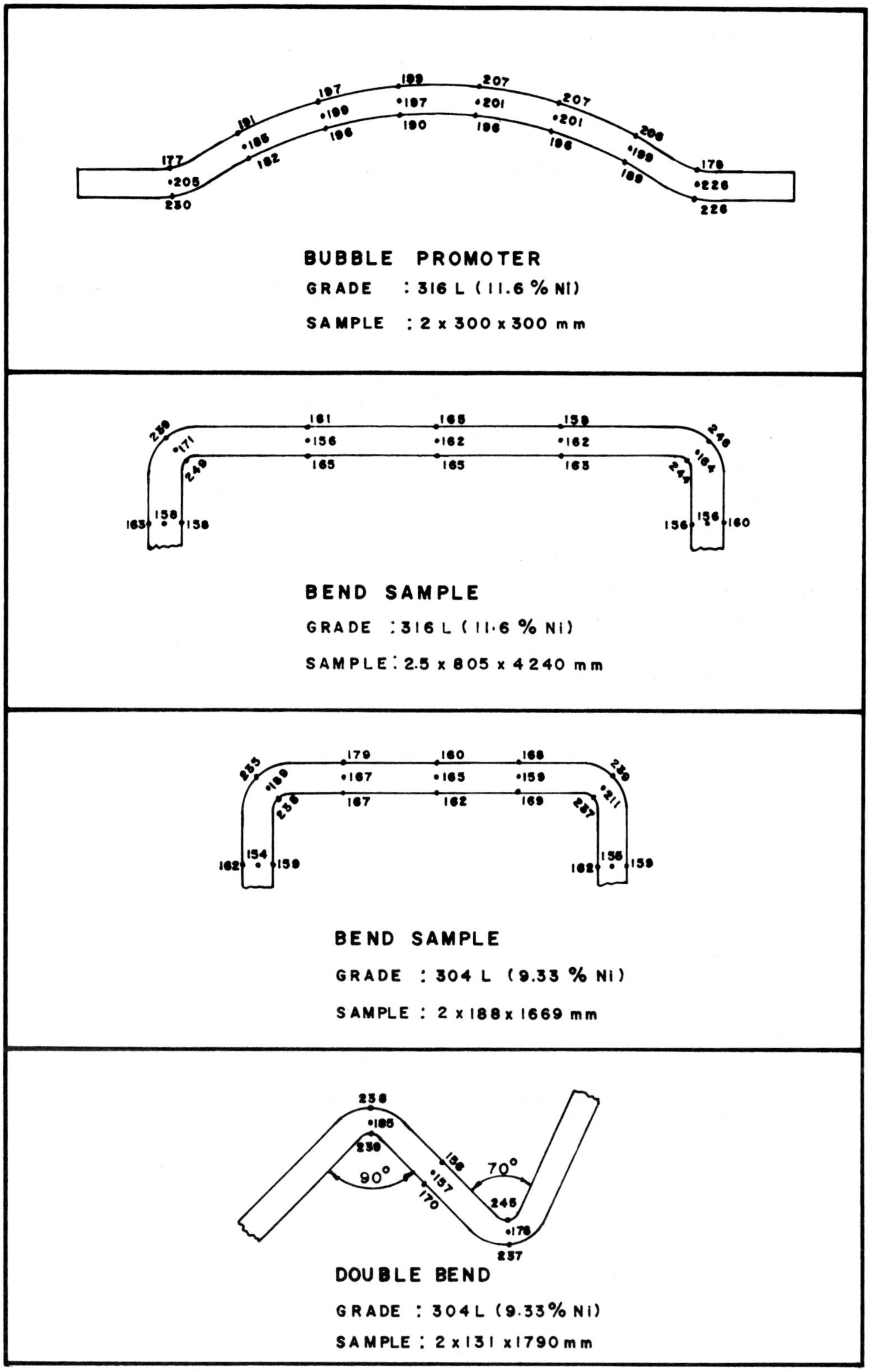

Fig.2 Micro hardness values for typical components.

Fig.3 Flatness measuring device for stainless steel sheets

Improved technique for control of NO_x gas emissions from pickling of stainless steel

I Dalin, P Andréasson and T Berglind

The authors are with Eka Nobel AB, Bleaching Chemicals Division, Surte, Sweden.

Summary

In Sweden, a dosage of hydrogen peroxide to reduce emissions of NOx gases from stainless steel pickling baths has been established in several steel plants.

Through recent development work at Eka Nobel, a new control system has been developed, which continuously adapts the hydrogen peroxide dosage rate to the time variations of the NOx production in each pickling bath. This is an essential improvement over the original method, since the NOx production varies a lot with time and since an excess of hydrogen peroxide has a limited stability in the pickling acid.

The new system is permanently installed at one of Sweden's largest band pickling units. Data will be presented from this full scale operation showing the system to be able to reduce NOx emissions by 70-85 % and also to save hydrogen peroxide due to the ability to dose the right quantity of hydrogen peroxide in each moment.

The new system is subject to patent applications in several countries.

Introduction

Pickling of stainless steels in nitric acid/hydrofluoric acid mixtures generates an environmental problem - NOx gases are formed. These gases contribute to the so-called acid rain. Due to its toxicity, one of the oxides, nitrogen dioxide (NO_2) also presents a problem in the working environment.

Starting in the late 1970s, several trials have been performed at Swedish steel plants to find a useful method to reduce the NOx emission. Of the alternatives investigated, an addition of hydrogen peroxide into the pickling bath was recommended as the most useful method. This method is now used in seven different sites in Sweden.

By combining the practical experience from those operations with a developed control technique that will be described in this paper, a more cost-efficient technique has been established.

Pickling Chemistry

A typical pickling bath contains 15-20 % nitric acid (HNO_3) and 2-4 % hydrofluoric acid (HF).

Dissolution of metal is the dominating chemical reaction taking place during pickling. Example for iron:

$$4Fe + 10HNO_3 + 8HF \rightarrow 4FeF_2^+ + 4NO_3^- + 6HNO_2 + 6H_2O \quad \ldots (1)$$

The nitrous acid (HNO_2) formed during pickling appears to have a reasonably high solubility in the pickling acid. Typical HNO_2 concentrations during steady state pickling that we have measured are in the range of 20-80 m moles/l (about 1-4 g/l) depending on pickling intensity, temperature and venting conditions.

The nitrous acid dissolved in the pickling acid is in equilibrium with the dominating nitrogen oxides (NO and NO_2) which tend to leave the system as gases:

$$2HNO_2 \rightleftharpoons N_2O_3 + H_2O \quad \ldots (2)$$

$$N_2O_3 \rightleftharpoons NO + NO_2 \quad \ldots (3)$$

Hydrogen peroxide chemistry in the pickling acid

When hydrogen peroxide is added to a pickling acid containing HNO_2, two competing reactions take place.

Oxidation of HNO_2:

$$HNO_2 + H_2O_2 \rightarrow HNO_3 + H_2O \quad \ldots (4)$$

Decomposition of H_2O_2:

$$H_2O_2 \rightarrow 1/2O_2 + H_2O \quad \ldots (5)$$

The result of a simple laboratory test shows which of the reactions is the fastest:

A pickling acid sample (Fe 60, Cr 9, Ni 5 g/l, HNO_3 194, HF 49 g/l) containing HNO_2 42 m moles/l (1,97 g/l) was treated with H_2O_2 42 m moles/l (equivalent to the HNO_2 content) at 50°C. An identical pickling acid sample was left untreated at the same temperature as a reference.

HNO_2 and H_2O_2 content was determined after 15 min. Result, see table 1.

Table 1. HNO_2 removal in pickling acid by equivalent H_2O_2 addition.

	Initial conc. HNO_2 m mol/l	H_2O_2 add. m mol/l	Conc.after 15 min. H_2O_2 m mol/l	Conc.after 15 min. HNO_2 m mol/l	HNO_2 decrease %
Sample	42	42	0,0	1,2	97
Reference	42	-	-	36,7	13

Table 1 shows that nearly all of the HNO_2 is eliminated in the sample treated with an equivalent amount of H_2O_2. The 13 % loss in HNO_2 in the reference is due to NOx gas leaving the sample.

The conclusion is that most of the hydrogen peroxide has reacted (as desired) according to formula (4) and only a minor part has decomposed according to formula (5).

In fact reaction (4) is extremely rapid (ref.1) and is essentially completed in less than a second.

Figure 1 shows the decomposition rate of excess hydrogen peroxide in a pickling acid free from HNO_2.

As seen in the figure the decomposition rate is such that an essential part of the H_2O_2 excess is wasted after 5 minutes.

The conclusion from table 1 and figure 1 are (a) that an equivalent or sub-equivalent dosage of H_2O_2 into the pickling bath has a good chance to give a good HNO_2 reduction rate (and hence a good NOx gas removal rate) and (b) that excess quantities of hydrogen peroxide will cause some decomposition which will waste hydrogen peroxide and hence add to the treatment cost.

HNO_2/H_2O_2 redox potential behaviour

The unusual behaviour of the redox potential of the system HNO_2/H_2O_2 in the pickling acid appears to be a useful tool when aiming at a H_2O_2 dosage that is related to the equivalent dosage. See figure 2.

Figure 2 shows (going from right to left) the gradual increase in redox potential that is a result of oxidation of HNO_2 by H_2O_2, the arrival at a redox potential maximum (corresponding to the equivalence point) and finally a rapid drop in redox potential as the result of an H_2O_2 excess.

The redox potential curve changes its level with variations in pickling acid composition and acid strength but the characteristic shape of the curve remains within normal acid compositions.

Full scale operation

For about 8 months the technique has been in successful large scale operation at the Avesta AB cold strip plant at Torshälla, Sweden. The system is laid-out as shown in figure 3.

Hydrogen peroxide dosage is put into a pickling acid pumparound line. The dosage is continuously regulated against the redox potential downstream of the dosage point. The redox potential set point is selected in relation to the potential maximum. Two different modes of regulation can be used, either a slight H_2O_2 excess (area III in figure 2) or a slight H_2O_2 deficiency (area II in figure 2).

Comparing full-scale results of these two modes gives interesting information. Table 2 contains steady-state H_2O_2 consumption rates and NOx reduction rates for the two different modes of regulation. The comparison is made for two different steel qualities (low and high NOx production).

Table 2. Comparison of H_2O_2 excess and H_2O_2 deficiency modes.

Steel type CrNi steel	Operational mode (cf fig.2)	Steady state data H_2O_2* flow l/h	Steady state data NOx red.%
Low carbon (low NOx-prod.)	Area III	24	82
	Area II	14	80
High carbon (high NOx-prod.)	Area III	61	87
	Area II	42	84

* 35 % H_2O_2 (400 g/l)

As seen, operation in the H_2O_2 deficiency area (area II in fig. 2) gave about 30-40 % lower H_2O_2 consumption with only a few percent lower NOx reduction rate.

The data (Table 2) are from a trial in one of the pickling baths. Based on these results, the operational mode with a H_2O_2 deficiency was selected for the permanent operation.

The cold rolling strip plant at Torshälla consists of two pickling lines: Line 55 that produces finish-product strips (two pickling baths in series) and line 60 that produces raw strips (one bath). The production rate is 75000 tons/year in line 55 and 85000 tons/year in line 60.

Table 3 summarizes the results achieved during a two-day recording period with hydrogen peroxide addition together with an extensive reference period. NOx production was recorded in the mutual venting line from all three pickling baths.

Table 3. Summary of NOx reduction data from controlled H_2O_2 addition. Avesta AB Torshälla, lines 55 och 60.

Pickling line no.	Pickling bath no.	Ref. NOx* prod. w/o H_2O_2 kg/h	Period 1 (8 hrs avg.) Prod. rate m^2/h	NOx residual kg/h	H_2O_2** consumption l/h	Ref. NOx* prod. w/o H_2O_2 kg/h	Period 2 (16 hrs avg.) Prod. rate m^2/h	NOx residual kg/h	H_2O_2 consumption l/h
55	1				27				20
	2				8				4
	1 + 2		2860		35		2360		24
60	5		676		33		715		31
Total		29		5,0-6,8	68	27		4,7-5,7	55

* Assumed mol wt of NOx = 38. Data based on reference NOx production per m^2 in each pickling line.
** 35 % H_2O_2 (400 g/l)

As can be calculated from table 3, an average NOx removal rate of 80 % was achieved.

Data in table 3 consist of operations with higher than average NOx production rates. With lower NOx production, the removal rate (%) tends to be somewhat lower. Over long term, typical reduction rates have been in the range of 70-80 %.

The environment protection authorities have allowed the NOx emission to be max. 6 kg/h on average. This requirement is well met by the installed NOx elimination unit.

Advantages of the new technique

The technique of controlled H_2O_2 dosage has shown to give essentially lower H_2O_2 consumption rates than can be achieved without the control system. The control system takes care of the time variations in NOx production rate that take place due to irregularities in the production and is therefore equally applicable to batchwise and continuous pickling. An important benefit is that the control is individual for each pickling bath.

The installed system requires some minor maintenance on a daily basis.

Acknowledgements

The contribution from personnel at Avesta AB is greatly acknowledged. Particularly the willingness of the production manager Torbjörn Andersson and his co-workers to provide the first full scale object has been very helpful. The central laboratory at Avesta (Sven Eric Lunner and Billy Skoglund) has contributed with discussions over pickling chemistry and also with extensive recording of the NOx emissions.

Reference

1. DE Damschen, LR Martin, Atmospheric Environment, vol 17, No 10 (1983) pp. 2005-2011.

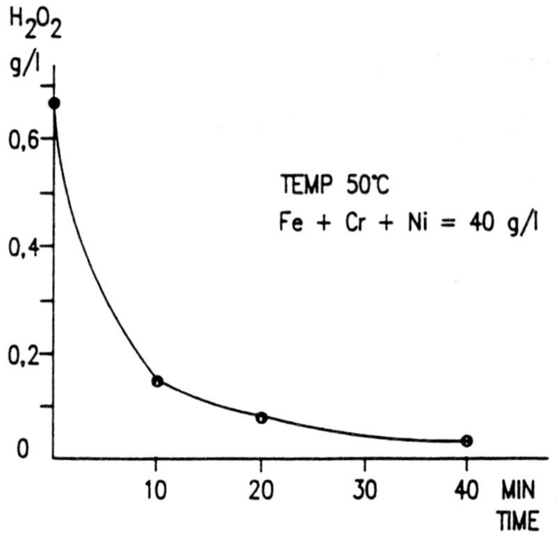

Figure 1. Decomposition of H_2O_2 in pickling acid.

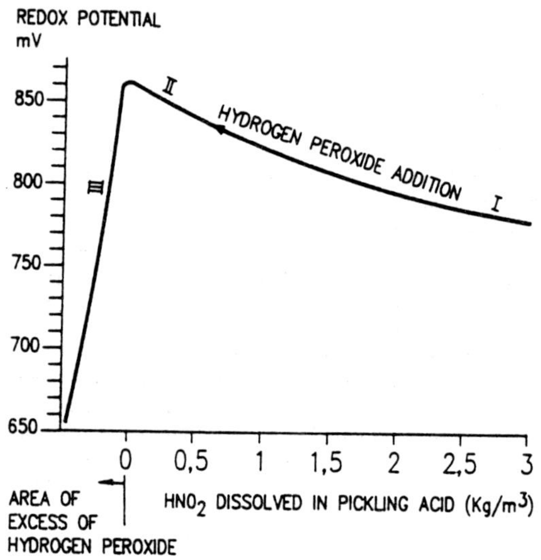

Figure 2 Redox potential behaviour of the system HNO_2/H_2O_2 in pickling acid.

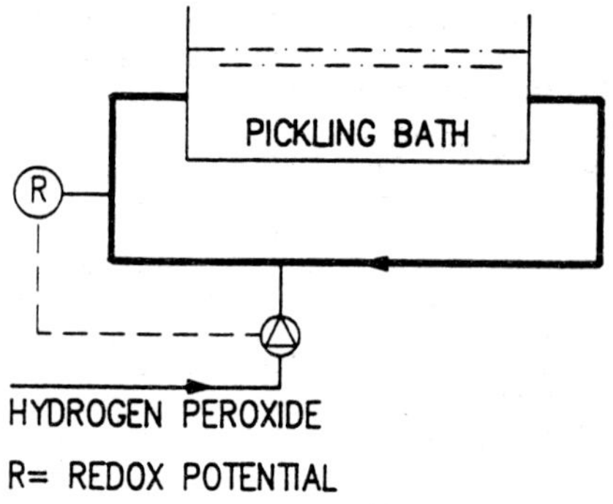

Figure 3. Outline of a NOx elimination unit.

Substitute nickel-free chromium, manganese, nitrogen austenitic stainless steels

B R Nijhawan

Dr. B. R. Nijhawan is currently Senior Consultant to UNIDO (United Nations Industrial Development Organization) and worked with UNIDO for about two decades as the Senior Interregional Advisor (Metallurgical Industries) and was Director, National Metallurgical Laboratory, Jamshedpur, India for many years before joining UNIDO.

SYNOPSIS

The development of substitute nickel-free austenitic stainless steels based on chromium, manganese and nitrogen systems, as distinct from ferritic stainless steels, has warranted much attention and concerted work during emergencies retarding or shutting off the nickel supplies and/or in countries that lack nickel resources and depend upon its imports.

During World War II, developments in substitute nickel-free stainless steels were significant in the USA, USSR and Germany. Equally considerable research and development work and also industrial scale production of these substitute austenitic stainless steels took place in the USA in the post war period leading to the production of "Tenelon" and "Century" trade mark austenitic stainless steels by US Steel and the Crucible Steel Co. respectively.

In the USSR, China and India also this subject received concerted attention during the last 2-3 decades both for studies on the inherent stainless properties of these substitute steels and for high strength properties achieved after their cold working; the industrial potentials and limitations of these substitute austenitic nickel-free stainless steels have been highlighted in this paper; a summary of R and D work done inter-alia by the author and in some other countries has been furnished along with industrial scale applications of such substitute austenitic nickel-free stainless steels.

Some of the recent work undertaken in the USSR on the substitute austenitic-ferritic stainless steels has been referred to as well as the subject of manganese, aluminium stainless steels, commonly known as the common man's stainless steels.

The above subject and its multiple facets present highly interesting academic and practical areas for research and development work, industrial studies and potential applications.

INTRODUCTION

RESEARCH AND DEVELOPMENT WORK on substitute austenitic nickel free stainless steels based on chromium, manganese and nitrogen system with optional additions of copper and investigations on production scale have been extensively undertaken and reported upon in many many countries; such work has tended to receive concerted attention during emergencies cutting off supplies and nickel and particularly in countries deficient in resources of nickel. Extensive technical papers, documents and publications have been published on the subject both in the developing countries and in advanced countries such as the USA, USSR and others. It is not intended in this paper to provide a Bibliography on the subject; however, some typical references will be made notable to the very recent publications (1985-86) by the manganese centre 1/ in France including recent developments of austenitic manganese steels for non-magnetic and cryogenic applications in Japan 2/ and to most recent work in the USSR and India.

Duplex austenitic - ferritic steels type 22-7-(2Mo) have during the last few years been commercially developed in the USSR (3) and are being produced as materials that are equal or even superior in corrosion resistance to traditional type 18-8 stainless steel; these steels are characterized by high resistance to stress corrosion. Their economic advantage lies in the savings of nickel and possibilities they offer for using thinner sections since duplex austenitic - ferritic stainless steels are 1-3 to 1-5 times stronger than austenitic 18-8 stainless steels.

Influence of nitrogen additions to Duplex stainless steels has been extensively covered in Japan by Akihiko Hoshino 4/ and by Asphahani et al 5/ in USA. Several other papers/publications/patent have been developed in USA by US Steel Corporation and Crucible Steel Co. and reported upon in the extensive technical literature on the subject.

In India, Nijhawan et al 7/,8/ has in a series of papers dealt extensively with the development of substitute nickel free stainless steels 6/ and with substitute high temperature creep resistance steels based on chromium, manganese, nitrogen systems; equally "Creep properties and micro-structure of austenitic steels containing manganese" has received concerted attention in India and reported upon by Gibbons et al 9/, by Choubey et al 10/ and by Nijhawan et al 11/. Most recently an Indian patent has been taken for the production of chromium - manganese - nitrogen - carbon - tungsten and molybdenum creep resistant steels by Kumar et al 12/. Nijhawan covered these developments concerning substitute Nickel-free stainless steels in several patents in different countries including the UK 13/. Research and development work and industrial scale applications of substitute stainless heat - resistant steels based on iron, manganese, aluminium system were investigated by Nijhawan and Khalaf 14/ and their main conclusions were the following:-

a) for successful hot forging, hot rolling and cold rolling of iron, manganese, aluminium steels, manganese and aluminium contents should be below 36% and 9% respectively. Austenitic steels were obtained in alloy compositions containing 0.3%C, 35% Mn, and 7% Aluminium. Steels containing 0.3%C, 28-35% Mn and 8-9% Al gave duplex austenitic - ferritic structures. Corrosion resistance of these steels under different media gave mixed results whilst the oxidation characteristics of these steels were satisfactory.

"DEVELOPMENT OF STAINLESS STEELS BASED ON CHROMIUM, MANGANESE AND NITROGEN COMPOSITION."

Following the basic themes of development of substitute alloys based on the indigenous alloying elements to the exclusion of those whose resources are deficient or totally non-existent in India, such as nickel, cobalt, molybdenum, tungsten, tin etc. Research and development work conducted in different countries and by various authors have resulted in the formulation of processes for the production of nickel-free austenitic stainless steel alloys from entirely indigenous raw materials, such as in India.

The research and development work on nickel-free austenitic stainless steels based on chromium, manganese, nitrogen and copper is very significant in the context that these substitute stainless steels have been made fully austenitic and non-ferro-magnetic despite their being wholly nickel-free, by virtue of their high optimum nitrogen contents-- and element readily available in nature. Comprehensive investigations were carried out on the determination of physical, mechanical and high temperature creep properties of the new steels developed. Tests were conducted to determine the low temperature and sub-zero tensile and impact properties, high-temperature tensile strength, weldability, oxidation resistance and ageing characteristics of different compositions of the nickel-free austenitic stainless steels.

A long range research project to develop nickel-free austenitic stainless steels processing significant corrosion resistance under specific media and adequate ductility and deep drawing characteristics for undergoing cold forming operations led to the development of a family of chromium-manganese-nitrogen austenitic stainless steels.

Some typical compositions of the four broad ranges of the austenitic stainless steels are given below. The addition of copper up to a maximum of 1% assists in somewhat improved corrosion resistance but affects the deep forming characteristics of the stinless steels and in the following ranges of chemical composition of the stainless steels, addition of copper has been excluded.

A.	Chromium	... 17.54%
	Manganese	... 12.57%
	Nickel	... Nil
	Carbon	... 0.054%
	Nitrogen	... 0.21%
B.	Chromium	... 16.68%
	Manganese	... 14.13%
	Nickel	... Nil
	Carbon	... 0.055%
	Nitrogen	... 0.49%
C.	Chromium	... 17.71%
	Manganese	... 17.55%
	Nitrogen	... 0.525%
	Carbon	... 0.08%
D.	Chromium	... 20.9%
	Manganese	... 13.8%
	Nitrogen	... 0.68%
	Carbon	... 0.12%
E.	Chromium	... 21.56%
	Manganese	... 18.47%
	Nitrogen	... 0.8%
	Carbon	... 0.06%

In each of the above composition ranges, sulphur, phosphorus and silicon contents are below 0.06%, 0.06% and 1% respectively.

Melting technique and production of ingots

The technique of making the austenitic low carbon nickel-free Cr-Mn-N austenitic stainless steels is very similar to the standard procedure of making 18-8 austenitic stainless steels. The important point to remember in melting these stainless steels is that loss of nitrogen occurs in the melt with longer holding time and in order to minimise the loss of nitrogen, the period after the additions of nitrogen bearing ferro-alloys to the melt should be as short as possible. In direct electric arc furnace melting, it is desirable to reduce the period after oxidation of the melt for reasons of carbon pick up in the manufacture of the 18-8 austenitic stainless steels, but in the case of Cr-Mn-N stainless steels this aspect is of still greater importance. Since the additions of cold nitrogen-bearing ferro-alloys will prolong the melt time, it is preferable to preheat the nitrogen bearing ferro-alloys up to 600°C separately. For the same reason, it is preferable to attain a high melt temperature before the additions of these ferro-alloys.

The heats of Cr-Mn-N stainless steels were made in high frequency induction furnace as also in electric arc furnace. Furances that have been actually used are:

1. Basic lined high frequency induction furnaces of 75 lb, 20 lb and 5 lb capacities.
2. 2-ton basic high frequency induction furnace.
3. 0.8-ton basic arc electric arc furnace.
4. 3-ton basic electric arc furnace.

Since no refining or carbon pick-up takes place in the high frequency furnace, it is indeed relatively easier to make the stainless steel in the induction furnaces during melt trials of which normal losses of various alloying elements were determined.

The basic high frequency induction

furnace heats are made by first charging the furnace with low carbon steel scrap or ferritic stainless steel scrap. The furnace is then started and on softening of the charge, low carbon ferro-chrome is gradually added. A part of the low carbon ferro-chrome is often replaced by nitrogen containing ferro-chrome to raise the nitrogen input of the charge. Balance of the charge consisting of either nitrided electrolytic manganese or Fe-Cr-MN-N master alloy is slowly added to the molten bath which is kept at a low temperature.

During such addition period, care should be taken to see that the bath is not suddenly cooled down by abrupt addition of large quantities of material at a time; this precaution is taken to avoid the formation of bridge on the top of the melt which may at times be difficult to break. After completion of melting, ferro-silicon is added to the bath and some time is allowed for the melt to attain the pouring temperature. The pouring temperature is controlled between 1500°C-1550°C. The melt is poured into preheated ladles and thence into the ingot moulds which have earlier been properly dressed with lime to facilitate stripping and dried beforehand. The ingots are immediately capped soon after the completion of pouring and jets of water are directed on the top of the "capped" ingots. Capping gives rise to a sound and defect-free ingot top as per the observations and experience gained.

At times, when available, Armco ingot-iron was also somewhat used along with the mild steel scrap but generally its use was avoided owing to its high cost.

The procedure of melting these stainless steels in direct arc electric furnace is similar to the procedure of melting normal low carbon stainless steel. For this reason, the melting procedure of normal stainless steels will be briefly described and any difference necessary in the melting procedure of the Cr-Mn-N stainless steels would be pointed out. The melting of normal stainless steels takes place in three stages.

(a) The oxidizing stage - The aims in this stage are:
   (i) to obtain a sufficiently low carbon value to make up for any carbon pick up during the later stage and
   (ii) to obtain a sufficiently high temperature to minimise or completely avoid arcing in later stages. The oxidation of the melt can be brought about either by oxygen lancing or iron ore additions; former is preferred on account of relatively shorter time required and ready high temperature obtained.

(b) The reducing stage - the oxygen-saturated melt must be deoxidised in order to take a substantial portion of chromium from the slag back to the steel melt, in case the starting material is straight chromium steel scrap; this deoxidation is also needed to reduce subsequent losses of alloying elements which would occur in case effective deoxidation has not been earlier accomplished. Ferro-silicon is used for deoxidation in the case of these nickel-free stainless steels. Since aluminium combines with nitrogen, the use of this element for deoxidation purposes is not possible in the manufacture of CR-Mn-N stainless steel.

(c) Finishing stage - after removing the first slag, a fresh finishing slag of lime and flurospar is prepared; requisite alloy additions are then made. In the manufacture of Cr-Mn-N stainless steels as pointed out earlier, the nitrogen-bearing ferro-alloys, such as nitrided electrolytic manganese, nitrided ferro-manganese or nitrogen-bearing ferro-chrome are added last after requisite preheating. These alloy additions are then thoroughly rabbled into the bath. In certain cases, these nitrogen-bearing alloys may be added even during the reducing stage. Having made the additions of the alloying elements and measurements of the pouring temperature which should be below 1600°C, the melt is ready for tapping. It is advisable to take a sample from the bath, quench it for immediate magnetic testing which should clearly show it to be totally non-magnetic. The melt is now poured in a ladle and a slag cover is maintained over it. The steel is poured from the bottom of the ladle through a magnesite nozzle into properly dressed ingot moulds as stated earlier.

Mechanical capping and water quenching of the ingot tops should be done in a similar way as referred to earlier.

Hot working operations

The ingots after being stripped from the ingot moulds are heated to 980°C in a gas fired furnace and soaked at this temperature for sufficient time to allow full homogenization. The temperature is then raised to the forging temperature of 1150°C and the ingots are forged by pneumatic hammers into flat bars suitable for rolling into sheets. During forging, reheatings have to be carried out due to loss of temperature of the bar.

The slabs on cooling are ground to remove surface defects and cut to proper size for rolling into sheets. The slabs are slowly reheated to a temperature of 1150°C and rolled into sheets. Reheating and rolling has to be repeated before obtaining the final thickness of the hot rolled sheets.

Annealing

For solution annealing the sheets are reheated in an electric furnace to 1050°C and quenched in oil.

Pickling

Pickling of the sheets is carried out by first degreasing the sheets in a suitable solvent and carrying out the pickling operation in two stages. The sheets are first pickled in a solution containing 6% H_2SO_4 and 2% HCl at 50-60°C whereby the scale on the surface of the sheets gets loose and dislodged. The second pickling solution is composed of 10% HNO_3 and 1% HF. Pickling is carried out at a temperature of 50-60°C whereby the scale still remaining on the sheet surface after the first pickling operation is completely removed.

Physical properties

Physical properties determined on specimens made from the forged and solution treated bars are given below.

The solution treatment was carried out by soaking at 1050°C followed by quenching in water.

Chemical composition		Young's modulus
Heat No. E-68/2		
Cr	... 21.5%	
Mn	... 13.2%	28.15×10^4 lb/sq(determined by resonance vibrating method)
N	... 0.78%	
C	... 0.06%	
Si	... 0.27%	

Chemical composition		Resistivity
Heat No. E-13		
Cr	... 21.4%	
Mn	... 13.4%	
N	... 0.71%	74.59 micro ohm cm.
C	... 0.11%	
Si	... 0.46%	

Chemical composition	Saturation induction
Heat No. E-13 given above	Less than 2500 gausses for magnetizing field of 2340 oersteds

Chemical composition	Average permeability
Heat No. E-13 given above	1.05

Chemical composition		Co-efficient of thermal expansion		
Heat No. A-87		°C		
Cr	... 22.67%	30-100	15.7×10^{-6}°C	
Mn	... 14.88%	30-200	16.47	do
N	... 0.67%	30-200	17.57	do
C	... 0.05%	30-500	18.08	do
Si	... 0.58%	30-1000	23.7	do
		100-500	18.5	do
		500-1000	29.00	do

Results of tensile tests in solution treated condition and after cold colling (Thickness of sheet - 0.050"-0.040")

Chemical composition	% reduction	Direction of test	Max. stress tons /sq.in	Elongation % on 1 each G.L.	Hardness V.P.N
Heat No. B_2					
Cr .. 20.9%	0%	..	58.65	45.3	270
Mn .. 13.8%	(solution treated)				
N .. 0.68%	10%	Longitudinal	70.78	34.4	332
C .. 0.12%		Transverse	67.47	37.5	328
Si .. 0.70%	20%	Longitudinal	76.17	26.6	375
		Transverse	74.42	25.0	387
-	30%	Longitudinal	85.98	18.8	429

Sub-zero tensile tests

A special fixture was made for sub-zero tensile testing. Specimens of 1 inch G.L. were tested immersed in petroleum ether cooled to the desired temperature by liquid air. Results of the tests are given below:

Results of sub-zero tensile tests

Heat No.	Temperature	Maximum stress Kg/mm^2	Elongation % on 25 mm G.L.
B_2	Room Temperature	97.01	45.3
	-10°C	114.8	46.8
	-20°C	99.90	43.7
	-30°C	114.2	43.7
B_4	Room Temperature	104.1	48.4
	-10°C	122.3	53.0
	-20°C	116.1	50.0
	-30°C	122.6	46.8
B_6	Room Temperature	94.82	50.0
	-10°C	102.8	Broken outside G.L.
	-20°C	104.7	50.0
	-30°C	109.7	Broken outside G.L.

Weldability, oxidation resistance and low temperature tensile properties of different compositions of chromium-manganese nitrogen stainless steels used in the investigations are given in Table I.

Table I - Chemical composition of Cr-Mn-N stainless steels

Heat No.	Cr %	Mn%	N%	C%	Si%
B_2	20.9	13.8	0.68	0.12	0.70
B_4	22.07	16.63	1.01	0.04	0.04
B_5	17.9	16.93	0.69	0.05	0.28
B_6	17.59	16.93	0.66	0.04	0.31

Table II - Tensile properties of the welded specimens

Heat No.	Filler metal	Tensile strength kg/mm^2	Elongation % on 25 mm G.L.	Location of fracture
B_2	Parent metal	95.4	50.0	Base metal
	18/8 type	84.4	31.0	At the weld
B_4	Parent metal	100.3	21.8	At the weld
	18/8 type	108.6	25.0	At the weld

Weldability tests

Welded specimens were prepared by argon arc welding. Strips cut from the annealed sheets of 0.035-0.040 in. thickness were welded using 18/8 type stainless steel welding rods and thin strips cut from the original sheets as filler metal. Single layer butt joints were produced. The welded strips were ground smooth at the weld and tensile specimens made from the strips were tested at room temperature. Rsults of the tests are given in Table II. Tensile properties of unwelded base metal are also given in Table III for the purpose of comparison. Results of low temperature impact tests (V-notch Charpy specimens) are given in Table IV.

Table III - Tensile properties of unwelded base metal

Heat No.	Tensile strength kg/mm^2	Elongation % on 25 mm.G.L.
B_2	94.2	45.3
B_4	102	50.0

Table IV - Results of low temperature Charpy impact tests

Chemical composition Heat No. E-28	Temperatures °C	Energy of fracture in ft. lb.
Cr ... 21.3%	25	204
Mn ... 14.1%		
N ... 0.64%	0	188
C ... 0.05%		
Si ... 0.30%	-20	170
	-40	152
	-50	104
	-57	76
	-60	72
	-70	58.60

Results of high temperature short time tensile tests (Hounsfield Tensometer. Specimen No. 13, 0.632 in. G.L.) are given in Table V.

Table V - Results of high temperature short time

Chemical composition Heat No.E-68/2	Temperature °C	Max. stress tons/sq.in.	% Elongation 0.632 in G.L.
Cr .. 21.5%	200	43.7	50
Mn .. 13.2%			
N .. 0.78%	300	41.3	32
C .. 0.06%			
Si .. 0.27%	450	39.3	37

Deep drawing tests

Deep drawing tests were carried out using a Swift Cupping Press of 6½" capacity. Flat bottom mandrill was used and a pressure of 80 lb/sq. in. was maintained on the pressure plate. Grease-graphite was used as a lubricant. Results of the tests are given in Table VI.

Table VI - Results of deep drawing tests carried out in Swift Cupping Press

Pressure on pressure plate - 80 lb/sq.in.
Lubricant - Grease-graphite

Nominal composition Cr:Mn:N	Blank diameter in inches	Thickness of sheet in inches	Max.drawing press. lb/sq.in	Remarks
21:14:0.7	3.8	0.035	750	Good cup
22:17:1	3.8	0.038	800	"
18:17:0.7	3.8	0.035	680	"
21:14:0.7	4	0.035	890	Good cup
22:17:1	4	0.037	970	"
18:17:0.7	4	0.035	800	"
21:14:0.7	4.1	0.036	990	Good cup
22:17:1	4.1	0.040	1020	"
18:17:0.7	4.1	0.035	830	"
21:14:0.7	4.2	0.036	1050	Good cup
22:17:1	4.2	0.038	1050	"
18:17:0.7	4.2	0.038	1000	Cracked on side

Results of corrosion tests

Corrosion tests were carried out on specimens of thesestainless steels in several media such as 65% nitric acid, vinegar, lime-juice+1% NaCl, citric, acid+NaCl. Salt spray tests were also conducted. The results of the tests are given in Table VII.

Table VII - Results of corrosion tests

Chemical composition	Medium	Temp.	Time of test hours	Loss in weight gm/dm^2 day
Heat No. A-44				
Cr .. 17.54%	65% Nitric acid	boiling	38	0.293
Mn .. 13.49%				
N .. 0.59%				
C .. 0.05%	Vinegar	41±1°C	95	0.0010
Si .. 0.25%				0.0010
	Lime-juice +1%NaCl	41±1°C	72	0.180
		41±1°C		0.197
	5% Citric acid+1% NaCl	41±1°C	64	0.0045
				0.14
	Salt spray (5%NaCl)	Room Temperature	48	0.0055
				0.0030
	5%Sulphuric acid (acrated)	41±1°C	1½	55.59
				61.56
Heat No.B_2				
Cr .. 20.9%	65% Nitric acid	boiling	48	0.1173
				0.1153
Mn .. 13.8%				
N .. 0.68%				
C .. 0.12%	Vinegar	41±1°C	72	Negligible
Si .. 0.70%				
	Lime-juice +1%NaCl	41±1°C	72	"
	5%Citric acid+1% NaCl	41±1°C	72	"
	Salt spray 5% NaCl	Room temp.	48	Negligible A few rust spots visible
	5%Sulphuric acid (aerated)	41±1°C	1½	87.76
				94.31
	1% Hydrochloric acid (aerated)	41±1°C	1	18.00

Oxidation resistance tests

Oxidation resistance tests at 900°C were carried out in a thermal balance on specimens of heat nos. B_2, B_4 and B_5 in an atmosphere of dry oxygen. Test was also conducted on a specimen of 18/8 stainless steel for the purpose of comparison. Results of the tests showed that the steels B_2 and B_4 had slightly higher oxidation rate at 900°C than 18/8 stainless steel. Steel no. B_5 containing 17.9% chromium showed much higher oxidation rate. Specimens of heat B2 and B4 showed non-uniform scaling.

CONCLUSIONS

It was established that fully stable austenitic Cr-Mn-N stainless steels can be produced on industrial and commercial scale. These steels do not have any special melting difficulties. Forging and rolling can be carried out successfully with suitable adjustment. The steels have a higher ingot to bloom conversion on account of the capping technique adopted.

The Cr-Mr-N stainless steels develop high mechanical strength and possess adequate ductility and deep drawing properties. In solution treated condition, the steels have on an average 91-94 kg/sq.mm (58-60 tons/sq in.) tensile strength that can be increased with an elongation of 50-55%; the tensile strength can be increased up to 126 kg/sq.mm (85 tons/sq.in) on 30% cold reduction. These steels have excellent deep drawing properties, and even on very deep draws, the steel does not break down to the magnetic phase. It may, however, be emphasized that heavier presses will be required for conducting such operation.

The Cr-Mn-N steels are as corrosion resistant as stainless steel AISI 304 in 65% boiling nitric acid test, salt spray and weak organic acids.

The Cr-Mr-N steels can be used in place of 18/8 Cr-Ni stainless steels for applications such as house-hold utensils, automobile and railway fittings, hospital-ware and dairy equipments etc. Proper heat treatment of the steels would result in optimum mechanical properties at both room and elevated temperatures, and these steels could be used for special applications.

ACKNOWLEDGEMENT

In updating the status of developments in the field of substitute nickel-free austenitic stainless steels during the last two decades, the author would like to express his sincere thanks to his ex-colleagues, Messrs, P.K. Gupte, S.S. Bhatnagar, J.K. Mukerjee, B.K. Guha and S.S. Dhanjal of National Metallurgical Laboratory, Jamshedpur (India) in jointly developing the family of substitute nickel-free austenitic stainless steels.

REFERENCES

1. "Manganese stainless steels" edited (1986) by The Manganese Centre, Paris - ISBN - 2-901-109-05-5
2. Recent developments of austenitic manganese steels for non-magnetic and cryogenic applications in Japan-1984 edited by Manganese Centre-Paris-ISBN-2-901-109-04-7.
3. Economic and Technological aspects of the production and uses of nickel free and low nickel stainless steels in the USSR-UN-ECE 1984-steel-Committee-steel/SEM.10/R-3
4. Influence of Nitrogen Additions to Duplex stainless steels - Akihiko-Hoshino-Tetsu-To-Hagane-No.16, vol.72-December 1986, p.2279-2286-(IssN-0021-1575)
5. Duplex stainless steels-N.Sridhar, J. Kolts, S. Srivastava and A.I. Asphahani 1983, p.481-A.S.M. - USA
6. Substitute Nickel - Free Austenitic Stainless Steels - B.R. Nijhawan, P.K.Gupte S.S. Bhatnagar, B.K. Guha and S.S. Dhanjal, J. Iron & Steel Institute, March 1967, Vol. 205, p.292-304
7. Development of substitute high temperature creep Resistant Alloys - R. Choubey, B.R. Nijhawan, P.K. Gupte, B.N. Das, S.S. Bhatnagar and K. Prasad, NML Symposium on Substitute ferrous and non-ferrous alloys, 1966, p.263-272
8. Some precipitation reactions in chromium - manganese-nitrogen stainless steels on aging and cold working - J.K. Mukerjee and B.R. Nijhawan, J. Iron and Steel Institute, January 1967 - Vol.205 p.62-69
9. Creep properties and microstructure of austenitic steels containing manganese - R. Choubey, T.B. Gibbons and K. Prasad - Metals Technology, The Metals Society, UK November 1977, p.524-529
10. Substitute high temperature creep resistant steels - R. Choubey and K. Prasad Trans. Ind. Institute of Metals, 1972, p.665
11. Precipitation reactions in chromium - manganese nitrogen stainless steels - J.K. Mukerjee and B.R. Nijhawan, J. Iron and Steel Institute, 1967-205-62
12. Indian Patent No. 796/DEL/86. A process for the production of chromium manganese nitrogen-carbon-tungsten molybednum creep resistant steels, R. Kumar, Kanhaiya Prasad and Raghbir Singh
13. English Patent No. 833.3087 (1958), B.R. Nijhawan. P.K. Gupte, S.S. Bhatnagar, B.K. Guha and S.S. Dhanjal
14. Substitute heat resistant stainless steels based on Iron, Manganese, Aluminium system, Nijhawan and Mohamed Khalaf - unpublished work.

Effect of austenitizing and tempering conditions on structure and mechanical properties of 9Cr—1Mo martensitic alloy

A Alamo, J L Boutard, M Pigoury, C Lelong and C Foucher

CEA - IRDI - SRMA - SMPA
CEN Saclay, Gif-sur-Yvette, France.
JLB is now with the Net Team, Max-Planck-Institut für Plasmaphysik, Garching, West Germany.

SYNOPSIS

The structure and mechanical properties of the 9Cr-1Mo martensitic alloy have been investigated. Phase transformation temperatures on heating and the continuous cooling transformation diagram were determined by dilatometric techniques. Results concerning the effect of solution-treatment and tempering conditions on austenitic grain size, hardness, tensile properties, creep strength and toughness impact curves are also given.

INTRODUCTION

Chromium-molybdenum martensitic steels are planned to be used as structural materials of the fuel subassembly for fast breeder reactors. For this purpose materials need to combine high strength at both ambient and elevated temperatures with adequate toughness. An important factor is the ductile-brittle transition temperature, since it is susceptible to shift toward higher temperature during irradiation.

This paper presents some results concerning a 9Cr-1Mo unstabilized alloy. We have investigated the phase transformations during heating and cooling, and the effect of solution-treatment temperatures and tempering conditions on the structure and mechanical properties.

MATERIAL AND EXPERIMENTAL TECHNIQUES

The 9Cr-1Mo martensitic alloy investigated here is a Z10 CD9 type material with the following composition: 8,8Cr - 1,1Mo - 0,1C - 0,5Mn - 0,4Si - 0,024 N - balance Fe.

Samples for heat treatments and dilatometry were obtained from a solution-treated (1 h at 1000°C) and tempered (1 h at 760°C) sheet of 4,5 mm thickness.

Heat treatments have been made under argon atmosphere.

Dilatometric studies were performed at variable heating and cooling rates on cylindrical specimens of 4 mm diameter and 20 mm length.

The effect of austenitizing or solution-treatment (Θ_A) and tempering (Θ_T) temperatures was studied in the range of $850° \leqslant \Theta_A \leqslant 1300°C$ and $100 \leqslant \Theta_T \leqslant 800°C$, respectively; all heat treatments lasted 1 hour; samples were cooled under an accelerated argon jet.

The microstructure was revealed by etching in Villela's reagent; to reveal the prior austenite grain size the method described in ref.[1] was used. Hardness was measured in a Vickers hardness testing machine under 50 N load. Values given later represent the average of at least five indentations.

Samples for mechanical testing were prepared according to the following sequence: solution-treatment, first tempering, 20% cold rolling and a second tempering. Specimens have been tested in the final condition. Table II summarizes the different heat treatments used, the identification of samples, and the mechanical tests performed for each metallurgical state.

Plate smooth specimens of 3.5 x 4 mm^2 section and a gauge length of 20 mm parallel to the rolling direction were used for tensile and creep tests. Uniaxial tensile tests were performed at temperatures ranging from 20° to 550°C with a strain rate of $8.10^{-4}s^{-1}$. Creep samples were tested at stresses in the range 160 to 250 MPa at 550°C. Charpy impact tests were conducted between -130°C and 20°C using subsize V-notch specimens of 3.5 mm thickness. Transition temperature corresponding to 50 percent shear fracture was measured by means of a planimeter on the photographs of the fracture surfaces.

STRUCTURAL TRANSFORMATION STUDIES

The studies concerning the structural transformations occurring during heating and cooling have been conducted at variable rates.

On heating, austenitization begins at about 825-845°C (AC1 point) and is terminated at about 870-920°C (AC3 point), depending on the rate. Above this temperature the structure is composed of austenite and some undissolved carbides[2]. The AC1 and AC3 temperatures are listed in Table I for the different heating rates used.

Transformation studies under continuous cooling were performed, after austenitizing 1 h at 1000°C. Dilatometric data supported by optical microscopy led us to establish the continuous cooling transformation (CCT) diagram represented in fig. 1.

Austenite (γ) is fully transformed in martensite for cooling rates higher than 100°C/h.

Cooling rates lower than 100°C/h produce increasing fractions of austenite transformed in α ferrite, a second phase having a pearlitic morphology and also rod-like carbides as shown in fig. 2.

Above a rate of 10°C/h the martensite transformation is totally suppressed.

Martensitic transformation temperature Ms is about 370°C. At lower cooling rates Ms is slightly increased simultaneously with the modifications of the austenite composition due to ferrite and carbide precipitation.

EFFECT OF SOLUTION-TREATMENT TEMPERATURES ON GRAIN SIZE AND HARDNESS

The austenitizing or solution treatments performed during 1 h from 850 to 1300°C showed that the γ-single phase field extends from 900° to 1200°C for the examined composition.

The prior austenite grain size increases with solution treatment temperatures as shown in fig. 3a. Up to 1000°C a very slow increase is observed; the austenite grain size increased rapidly beyond 1000°C. This difference of behavior is probably associated with a grain boundary pinning by undissolved $M_{23}C_6$ carbides, which is eliminated after carbide dissolution at higher temperatures as proposed in ref.[3].

Increasing fraction of δ-ferrite are detected at $\Theta_A > 1200$°C; decarburization problems at these higher temperatures have prevented measurement of the correct equilibrium fraction of δ-ferrite.

Fig. 3b shows the dependence of martensite hardness on the austenitizing temperature. The presence of δ-ferrite in the structure at $\Theta_A >$ 1200°C caused a great decrease of hardness below 300 HV.

TEMPERING CHARACTERISTICS

The tempering characteristics, i.e. isochronal hardness recovery between 100 and 800°C, after solution treatment at various temperatures, are shown in fig. 4. Except the initial hardness values, austenitizing temperatures do not modify the overall tempering characteristics. Hardness slightly decreases during tempering up to 200°C and increases up to 500°C where a secondary hardening is detected. Tempering curves for each austenitizing treatment merged after peak hardness to become identical beyond 600°C.

MECHANICAL PROPERTIES

Mechanical tests were performed on samples having different metallurgical states, i.e. prepared under different heat treatment conditions: austenitizing was made at 980°C or 1030°C, followed by tempering at temperatures ranging from 740 to 780°C during 1 or 4 hours, as summarized in Table II.

TENSILE PROPERTIES

Figs. 5 and 6 show the main results concerning the effect of heat treatments on the tensile properties.

Fig. 5 presents the 0,2% yield strength values (YS), the uniform (E_u) and total (E_t) elongations as a function of the temperature test for samples solution treated at 980°C (C and D) and 1030°C (G and H, see Table II) ; tempering was made at the same temperatures in all cases, but during one hour for C and G, and four hours for D and H.

A more stronger material is obtained for the higher austenitizing temperature and the shorter time of tempering (sample G) in the whole range of test temperatures.

Materials solution treated at higher temperatures seem to be more sensitive to the duration of tempering. Mechanical strength is decreased about 200 MPa when tempering lasted 4 hours instead of 1 hour (compare curves G and H), while this effect is much less marked for samples austenitized at 980°C (C and D). The same behavior is observed for the UTS values.

The effect of temperature and time of tempering on YS and UTS values is illustrated in fig. 6; data measured at 20°C, correspond to samples solution-treatment at 980°C and tempered 1 h and 4 hours at 740 and 780°C (samples A, B, E and F). Depending on the tempering conditions materials could present yield strength values ranging from 300 MPa to 600 MPa. Shorter times (1 h) and lower temperatures (740°C) of tempering enhance mechanical strength.

On the other hand, the ductility is not affected very much by heat treatments as shown in fig. 5. Total and uniform elongations are about 25% and 10% respectively. These values are obtained for all the examined samples and for all the temperature tests used in the present work.

CREEP PROPERTIES

Creep tests have been conducted at 550°C on C, D, G and H samples. Figs. 7a and b show respectively the dependence of the secondary creep rate and the rupture time on the applied stress.

All the examined metallurgical conditions present nearly the same behavior. Nevertheless, samples which are stronger under tensile tests, also display higher creep strength.

Compared to the ferritic-martensitic steel 9Cr - 2Mo - Nb-V (EM12)[4], the present alloy under different conditions shows creep rates one or two orders of magnitude higher than EM12, and shorter rupture life, as we can see in fig. 7.

IMPACT RESISTANCE TRANSITION CURVES

In order to study the influence of tempering conditions on the DBTT, Charpy impact tests have been conducted on specimens having the metallurgical states A, B, E and F. All of them were solution-treated at 980°C. Experimental results are show in fig. 8.

All the examined samples present the same DBTT value of about -70°C ± 10°C. Consequently, tempering at different times (1 h and 4 h) and at temperatures ranging from 740 to 780°C does not produce any change in the transition temperature values.

Nevertheless, different uppershelf energy levels of the KCV curves are obtained, ranging from 140 to 190 J/cm^2. The higher KCV values correspond to specimens F having the lower yield strength (290 MPa).

CONCLUSION

Structural transformations occurring in the 9Cr-1Mo alloy have been investigated. It has been shown that the austenitic single-phase field extends from 900 to 1200°C. Above 1200°C increasing amounts of δ-ferrite occurs with increasing temperature.

CCT diagram, determined after solution-treatment at 1000°C, shows that austenite is fully transformed in martensite below 370°C for a wide range of cooling rates, i.e. rates higher than 100°C/h. Above 10°C/h a fully ferritic material with an important fraction of precipitate carbides is obtained.

Increasing solution treatment temperatures produced increasing austenite grain size, decreasing values of martensite hardness and have no effect on the tempering characteristics, apart from affecting the general level of hardness curves on tempering. Secondary hardening occurs during tempering at 500°C.

The effect of heat treatments on mechanical properties was studied on samples austenitized at 980°C and 1030°C, and tempered from 740°C to 780°C during one or four hours.

Tensile properties are very sensitive to the performed heat-treatments. Mechanical strength (UTS and YS) increases with higher austenitizing temperature and lower tempering temperature. Both UTS and YS are strongly dependent on the tempering time ; after 4 h tempering, mechanical strength can decrease to half of the values obtained for 1 h. The ductility is not affected very much by heat treatments ; a total elongation of about 25% was obtained for all the samples.

Creep properties (secondary rate and rupture time) are nearly the same for all the examined cases.

From Charpy V tests the ductile-brittle transition temperature (DBTT) is determined. All the tested samples present the same value of the DBTT, - 70°C ± 10°C, whatever the heat treatment is. Nevertheless, the upper shelf energy of the KCV curves decreases with increasing YS values.

REFERENCES

1 - Norme Française "Détermination de la grosseur du grain ferritique ou austénitique des aciers", NF A 04-102, novembre 1969.
2 - M. Pelletier CEA-R-5131 Report, 1982.
3 - F.B. Pickering, A.D. Vassiliou, Metals Technology, p. 409, October 1980.
4 - P. Berge, J.R. Donati, F. Pellicani, M. Weisz, Proc. Int. Conf. on "Ferritic Steels for High Temperature Applications", Warren, Pennsylvania, 6-8 October 1981, Ed. A.K. Khare ASM 1983 p. 100.

Table I: Ferrite-austenite transformation temperatures (ΔA: ± 10°C)

Heating rate °C/h	AC1 °C	AC3 °C
400	845	920
50	830	925
10	825	870

Table II: Samples for mechanical testing: heat treatment conditions, identification and mechanical tests performed for each metallurgical state

Solution treatment	1 h at 980°C						1 h at 1030°	
First tempering [°C]	740 1 h	740 4 h	760 1 h	760 1 h	780 1 h	780 4 h	760 1 h	760 1 h
Cold rolling	20%							
Second tempering [°C]	740 1 h	740 4 h	740 1 h	740 4 h	780 1 h	780 4 h	740 1 h	740 4 h
Identification	A	B	C	D	E	F	G	H
Mechanical Tests*	T,V	T,V	T,C	T,C	T,V	T,V	T,C	T,C

* T : Tensile Tests ; C : Creep Tests ; V : Charpy V Tests.

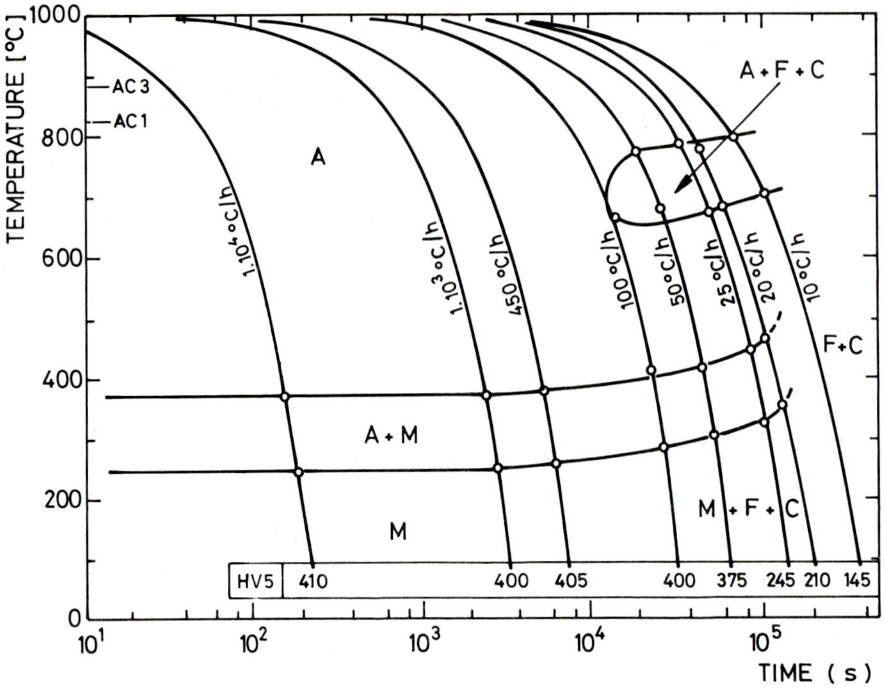

1 : Continuous cooling transformation diagram of the 9Cr-1Mo alloy obtained after austenitizing for 1 hour at 1000°C.

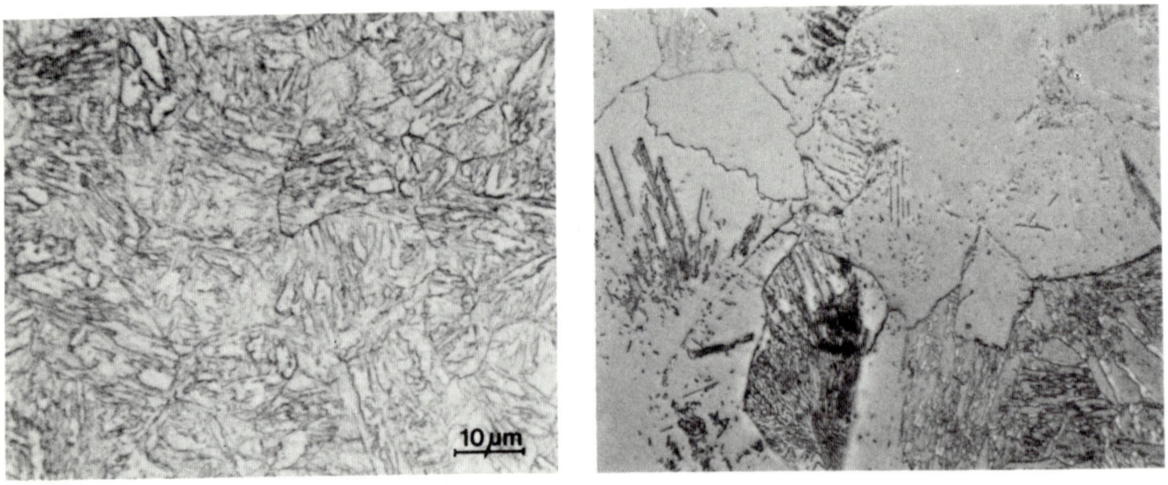

a

b

2 : Typical microstructures obtained after continuous cooling at variables rates. a) 10^4°C/h: fully martensitic material ; b) 25°C/h: ferrite, isolated carbides, pearlitic structure and martensitic regions.

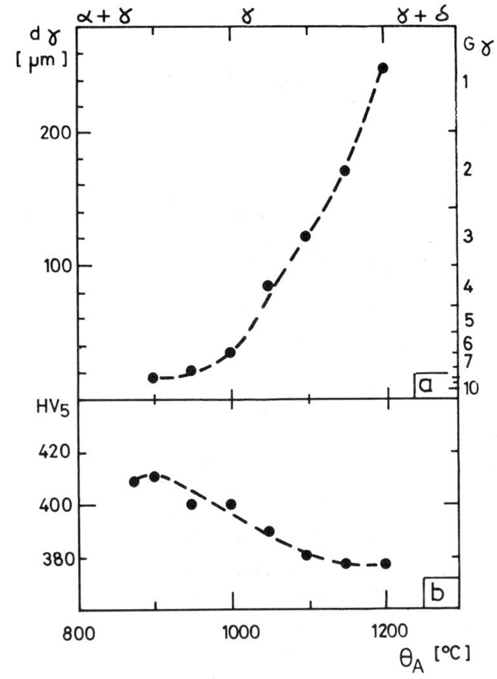

3 : Austenitic grain size and martensite hardness values as function of the solution treatment temperature.

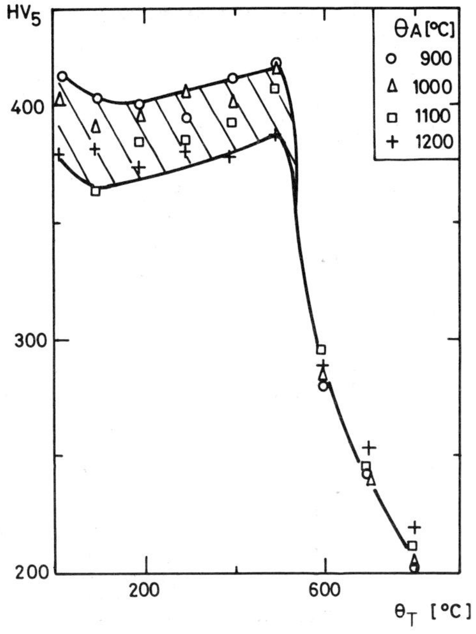

4 : Isochronal recovery of martensite hardness for various solution treatment temperatures.

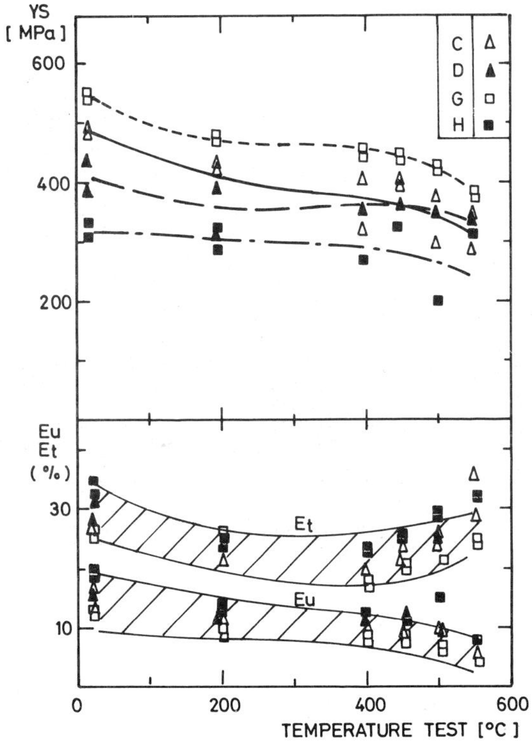

5 : Effect of austenitizing temperature and tempering time on the 0.2% yield strength, total and uniform elongations, as function of the temperature test.

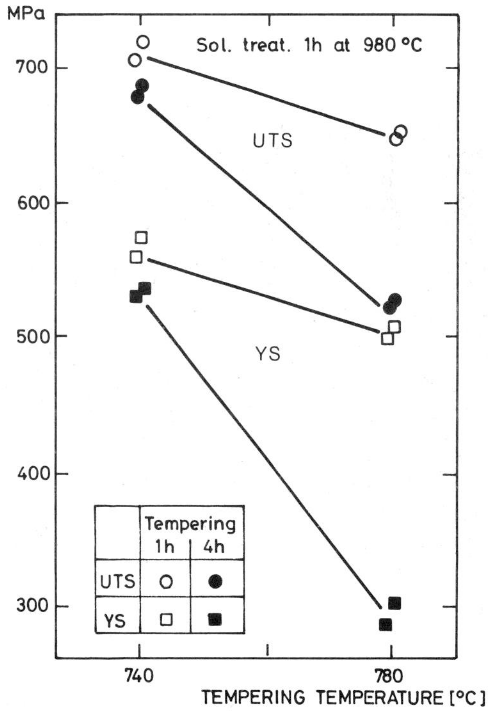

6 : Effect of the tempering conditions on the UTS and 0.2% YS values, measured at 20°C, for samples solution-treated at 980°C.

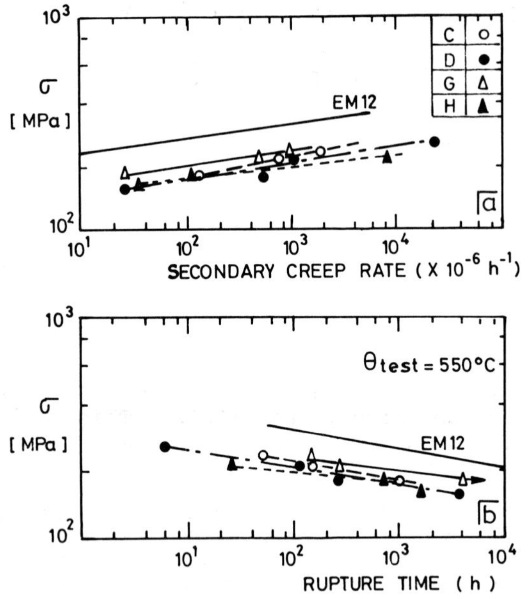

7 : Creep properties at 550°C of samples solution-treated at 980 and 1030°C and tempered at various times.

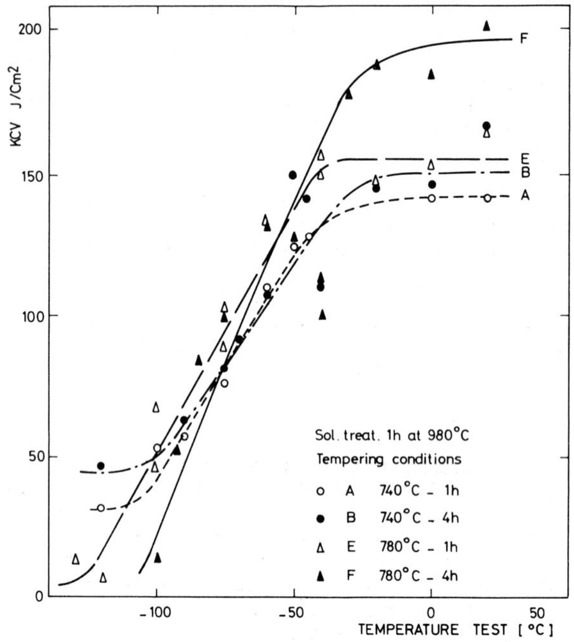

8 : Impact resistance transition curves of specimens solution-treated at 980°C and tempered at different conditions.

Elevated temperature tensile properties of a duplex low-carbon 12%Cr steel

G Leadbeater, C F Fletcher and J H Dalton

GL is a Senior Scientist, CFF a Technician and JHD Group Leader in the Alloy Development Group, Physical Metallurgy Division, Council for Mineral Technology (Mintek), Randburg, Republic of South Africa.

SYNOPSIS

Two alloys - 3CR12Ni, which contains 1.2 per cent nickel, and 3CR12, which has a nickel content of 0.6 per cent - were tested at temperatures in the range 750 to 920°C and 825 to 1000°C respectively. These temperature ranges fall within the dual-phase regions of α ferrite-austenite and δ ferrite-austenite of the phase-transformation diagram of the 3CR12 alloys.

In order that the strain-rate sensitivity of 3CR12Ni could be determined accurately, the alloy was tested in the as-received condition at various strain rates between 3×10^{-5} and 7×10^{-3} s^{-1}. It was found that, at intermediate strain rates, fracture elongations of over 300 per cent could be achieved.

In initial tests using 3CR12 in the as-received condition and intermediate strain rates, the percentage elongations produced was lower than that for 3CR12Ni. However, the application of various mechanical treatments to 3CR12 prior to testing resulted in significant increases in the elongation achieved.

INTRODUCTION

A low-carbon titanium-stabilized chromium steel, designated '3CR12' by the producers, is being marketed as a possible replacement for mild steel in mildly corrosive environments. It was developed as a modified version of AISI type 409 stainless steel, and has good weldability and toughness.

At first the alloy was produced with a nickel content of 1.2 per cent, and was known as 3CR12Ni. This alloy exhibited a dual-phase ferritic-austenitic structure in the temperature range 750 to 920°C. At present, the alloy is produced with a lower nickel content of 0.6 per cent, and is known simply as 3CR12. This alloy exhibits a dual-phase structure in the temperature range 825 to 1220°C.

It is well known that, to exhibit superplastic behaviour, an alloy must have a dual-phase structure and a fine grain size (average diameter $\leq 10\mu m$).

In the present investigation, the potential of this group of alloys as materials suitable for use in superplastic forming applications was investigated, a programme of tensile tests at elevated temperature being conducted on the material in the as-received condition and after heat and thermomechanical treatments.

SUPERPLASTICITY

Although concise reviews and accounts of the principles of superplasticity have been presented in the literature (1-3), certain points will be repeated here for clarity and emphasis with regard to the different compositions of the 3CR12 alloys investigated.

(a) *Physical and Mechanical Conditions*

Fine-structure superplasticity is usually achieved at intermediate temperatures, viz

$$\approx 0.4 - 0.7\ T_M(K),$$

and at slow to intermediate strain rates, viz

$$\approx 10^{-4} - 10^{-3}\ s^{-1}.$$

(b) *Quantitative Parameters of Superplasticity*

As stated in (a) above, one quantitative measurement that indicates superplastic behaviour is strain rate sensitivity, $m \geq 0.3$, which is given by

$$\sigma = C(\dot{\varepsilon})^m\ , \quad \ldots\ldots\ldots\ldots\ldots\ldots (1)$$

where σ is the flow stress (true stress),
C the material constant, and
$\dot{\varepsilon}$ the true strain rate.

Equation (1) reduces to

$$\frac{\Delta \log \sigma}{\Delta \log \dot{\varepsilon}} .$$

Hence m is a measure of the gradient of a plot of log σ vs log $\dot{\varepsilon}$, and this measurement is usually taken at points where

$\dot{\varepsilon} = 10^{-4} - 10^{-3}\ s^{-1}$.

In order that a value can be obtained for m, data must be accumulated over a series of tests at various values of $\dot{\varepsilon}$ and for specific temperatures. Therefore, a fairly extensive test programme needs to be conducted.

A second, more obvious and immediate, indication of super-plasticity is the percentage elongation that occurs in tensile tests at elevated temperature. An elongation greater than 200 per cent with no obvious necking can be regarded as being indicative of superplastic behaviour, particularly if this can be correlated with an applicable m value.

(c) *Mechanisms and Microstructural Requirements for Superplasticity*

It has been stated concisely (1) that the primary mechanism for superplastic deformation is the sliding and migration of grain boundaries, which is accommodated by slip processes in regions adjoining the grain boundaries. If this is the case, there are several microstructural requirements for superplasticity, as follows.

(1) The grain size must be small ($\leq 10\mu m$).

(2) A second phase, to prevent grain growth, must be present at superplastic-forming temperatures.

(3) The strength of the second phase should be of the same order as the matrix.

(4) The size and distribution of the second phase should be fine and uniform.

(5) Equiaxed grains are preferred, with grain boundaries that are high-angled and can resist tensile separation.

Of these five points, by far the most critical are the first two. A small grain size is essential, since it facilitates the movement of grains relative to one another; it also increases the density of the grain boundaries, thus making more sites available for deformation. The presence of a second phase is essential because it prevents preferential grain growth at the deformation temperature.

MATERIALS

The compositions of the materials investigated are shown in Table 1. The material was supplied in the form of hot-rolled plate, the 3CR12Ni being of 4.5 mm thickness, and the 3CR12 of two thicknesses, viz 4.5 and 12 mm. The phase diagram with respect to the the nickel content of 3CR12 material is presented in Figure 1, which shows the areas of single-phase (ferrite) and dual phase (ferrite and austenite) structure. All the alloys supplied had received their final rolling treatments at temperatures in the single-phase (α) region. Specifically, the 3CR12Ni and 3CR12 materials exhibit an $\alpha + \gamma$ structure within the temperature range 750 to 920°C and 825 to 1220°C respectively. At room temperature, of course, this dual-phase morphology is in the form of ferrite and martensite.

The microstructures of the materials in the as-received condition are shown in Figures 2 and 3.

EXPERIMENTAL PROCEDURE

Tensile tests at elevated temperature were conducted using an Instron 1175 tensile-testing machine in conjunction with a three-zone split furnace and an Instron Self-Adaptive Temperature Controller. All the tests were carried out under constant velocity in air.

(i) 3CR12Ni

At first, this material, which has a nickel content of 1.2 per cent, was tested in the as-received condition. The test temperatures covered the range 750 to 950° at 50°C intervals, and additional tests were carried out at 850 and 915°C. The strain-rate range at each test temperature was 3×10^{-3} to 7×10^{-3} s^{-1}. Collation and analysis of the test results provided sufficient data for the calculation of the strain-rate sensitivity index (m).

The results of these experiments were so encouraging that further development of the alloy was deferred until similar tests had been conducted on the 3CR12 material.

(ii) 3CR12

At first, the tensile tests at elevated temperature were carried out using the 3CR12 material in the as-received condition. The majority of specimens had been manufactured in such a way that their axes were parallel to the original rolling direction of the plate, but several specimens had been made with their axes transverse to the rolling direction so that the effects of directionality could be investigated. In the event, no significant difference in behaviour between the two types of specimens could be detected, so all the results reported here are for longitudinal specimens.

Since these tests demonstrated that the alloy in the as-received condition lacked observable highly ductile behaviour (maximum extension approximately 80 per cent), certain thermomechanical pre-treatments were applied, and the test conditions were varied. The pre-treatments consisted of cold-rolling down to various degrees of thickness, followed, in some instances, by annealing (although, in many cases, the raising of the specimen temperature for the tests served as a suitable heat treatment). The reduction in thickness varied between 15 and 50 per cent, and subsequent annealing was carried out at temperatures within the dual-phase region of the alloy. The main purpose of these treatments was to refine the grain size.

The tensile tests were generally carried out within the range 825 to 1000°C, the latter temperature being the upper limit of the split furnace. However, some tests were performed in which the temperature was raised in the course of the test through the range 775 to 840°C. This was done so that a transformation boundary would be crossed during the test, allowing any effect of this action on the tensile properties of the alloy to be observed.

The initial strain rate in all the tests was $3.3 \pm 0.05 \times 10^{-4}\ s^{-1}$, except where stated otherwise in the following section.

RESULTS

(i) 3CR12Ni

Tests on the 3CR12Ni alloy in the as-received condition showed that this material has a definite tendency to super-plastic behaviour. Data processed for the strain-rate sensitivity index show that

$$m = 0.28 - 0.38,$$

where, at all temperatures above 850°C, m was greater than 0.3. Typical elongations ranged from 185 to 325 per cent. An example of one of the specimens for which an elongation of 321 per cent was achieved, is shown in Figure 4. Figures 5 and 6 show specimens tested at other strain rates and temperatures. In Figure 5, a 'saw tooth' effect is apparent on the edges of the specimen throughout the gauge length; the failure seemingly originating from one of the valleys. A similar kind of observation was made by Engström (4), who attributed the effect to a large number of cavity-generated cracks. Further observations in the present investigation substantiated his reasoning, since cavities were frequently observed in the failed specimens.

Figure 6 shows a specimen of 3CR12Ni before and after being tested to failure at a higher temperature and a high strain rate. The failed specimen exhibits low elongation and necking, which shows that the conditions required for superplastic behaviour in this material are subject to strict limitations.

(ii) 3CR12

Tests on the 3CR12 alloy showed that it had much less of a tendency to superplastic behaviour than did the 3CR12Ni. On the material in the as-received condition, the elongations achieved were in the region of only 70 to 80 per cent, even at 1000°C. A typical specimen is shown in Figure 7, which demonstrates the obvious lack of highly ductile behaviour of this material. Because of this, fewer tests were done, and insufficient data was accumulated for the calculation of strain-rate sensitivity.

However, in the tests using specimens that had received thermomechanical pre-treatments, or in which the test parameters were varied, significant increases in ductility were observed (Figure 8). The maximum elongation achieved was approximately 180 per cent for specimens that had been reduced in thickness by about 40 per cent by cold-rolling and then tested at 925°C. Elongations of approximately 170 per cent were achieved for material in the as-received condition in tests in which the temperature was gradually increased from 775 to 840°C.

Overall, however, these elongations cannot be regarded as being indicative of superplastic behaviour and, consequently, data were not accumulated for the calculation of m-values.

DISCUSSION

The results obtained suggest that the 3CR12Ni alloy has a tendency to superplastic behaviour, but that the 3CR12 alloy does not. The microstructure of the 3CR12Ni alloy in the as-received condition does not satisfy the requirements suggested as being necessary for superplastic behaviour but, after being quenched from 880°C, the material had a much finer grain size and was dual phase in character, as can be seen in Figure 9. The etchant used - Kalling's solution - enhances the appearance of the grain boundaries, but the separate phases of the dual-phase structure do not show up clearly. It has been stated (5) that a microstructure obtained at a specific temperature under static conditions is not a true reflection of what the microstruture would be under tensile deformation. However, it can be predicted that, under dynamic conditions, strain-induced nucleation may occur, resulting in an even more refined microstructure, and that, thus, the propensity of the material for super-plastic behaviour may be enhanced.

Annealing of the 3CR12 material in the dual-phase temperature region reduced the grain size somewhat, but the refinement is not as extensive as it was for the higher-nickel alloy.

Prior cold working of the material resulted in a slight improvement in this refinement, but not to the extent required for super-plastic deformation. Even when allowance is made for any strain-induced nucleation during tensile testing, the refinement does not match that of the 3CR12Ni material.

The question of why a suitably fine grain size can be achieved in 3CR12Ni but not in 3CR12 by annealing in the dual-phase region can be answered if the role of the additional nickel content of 3CR12Ni is considered. Ball *et al.* (6) noted that, since nickel stabilizes austenite in stainless steels - an established metallurgical concept - the additional nickel content of 3CR12Ni actually creates a greater driving force for the *nucleation* of austenite in the dual phase region. The result is then a finely dispersed dual-phase structure.

The 3CR12 alloy lacks this driving force, and therefore the austenite that is formed is less finely dispersed. As a result, the overall grain structure of the material at test temperature is coarser. It is this factor that inhibits super-plastic behaviour in the alloy. However, with more severe mechanical working than was applied in the present investigation, a sufficiently fine grain structure could possibly be achieved, and superplastic behaviour might be developed. The final failure mechanism in both of the alloys was via cavitation that developed during the tensile tests. This effect can be seen clearly in Figure 10. This is not an uncommon observation in superplasticity studies on dual-phase stainless steels.

CONCLUSIONS

3CR12Ni has a propensity for super-plastic behaviour when stressed under tension at temperatures within its dual phase (α + γ) region. This occurs even without any prior grain-refining treatment.

3CR12 does not show the above tendency, even after a limited amount of grain refining treatment, although some increase in ductility at elevated temperature can be achieved by such treatments.

The difference in behaviour of the two materials is due to the role played by the additional nickel content of the 3CR12Ni alloy. This nickel gives rise to an intrinsic grain-refinement mechanism at specific temperatures in the dual-phase region temperatures owing to its stabilizing effect on austenite.

In 3CR12, the absence of this 'facility' can probably be compensated for by the application of extrinsic grain-refining techniques such as severe (>80 per cent) mechanical working and heat treatment.

ACKNOWLEDGEMENT

This paper is published by permission of the Council for Mineral Technology (Mintek).

REFERENCES

1. SHERBY, O.D., and WADSWORTH, J. Super-plasticity and superplastic forming processes. *Mater. Sci. Technol.*, vol. 1. Nov. 1985. pp. 925-935.

2. PADMANABHAN, K.A., and DAVIES, G.J. Superplasticity MRE 2. Berlin, Heidelberg, Springer-Verlag, 1980.

3. EDDINGTON, J.W. Microstructural aspects of superplasticity. *Metall. Trans. A*, vol. 13A. May 1980. pp.703-715.

4. ENGSTRÖM, E.U. Superplasticity in a microduplex stainless steel. International Conference on Duplex Stainless Steels, 26-28 Oct., 1986., The Hague.

5. BRINK, A.B., and BALL, A. Mechanical properties of a 12% chromium dual phase steel designated 3CR12. *Stainless Steels '84*. Göteburg, Sweden, Sep. 1984.

6. BALL, A., CHAUHAN, Y., and SCHAFFER, G.B. Microstructure, phase equilibria, and transformations in corrosion resistant dual phase steel designated 3CR12. *Mater. Sci. Technol.*, vol. 3. Mar. 1987. pp. 189-196.

TABLE 1

Compositions of materials used

(All values are expressed as mass percentages)

Alloy	C	N	Mn	Si	Cr	Ni	Ti	Al	Fe
3CR12Ni	0.025	0.02	1.09	0.37	12.0	1.23	0.25	<0.02	Bal
3CR12	0.03	40.014	1.24	0.4	11.3	0.63	0.24	0.14	Bal

Bal = Balance

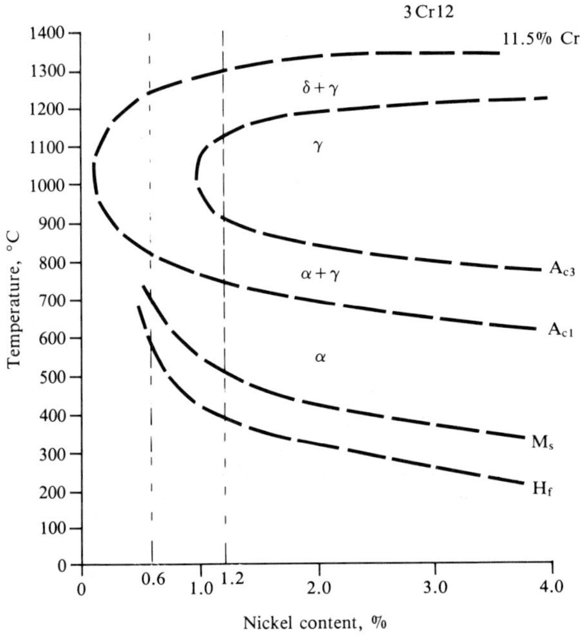

1. Phase-transformation diagram with respect to the nickel content of 3CR12, which contains 11.5 per cent chromium

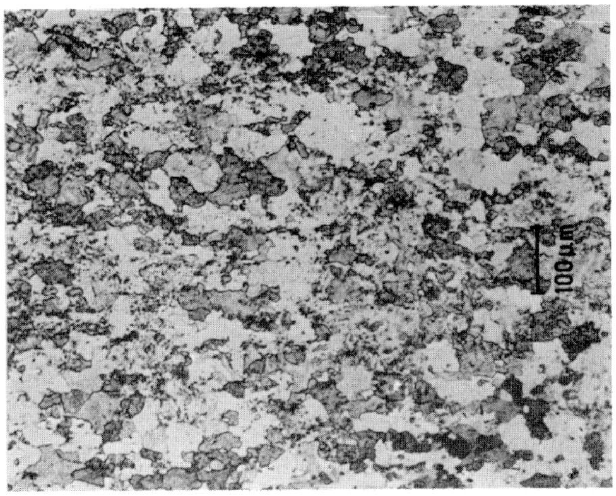

2. Microstructure of 3CR12Ni in the as-received condition. Etchant: Kalling's solution

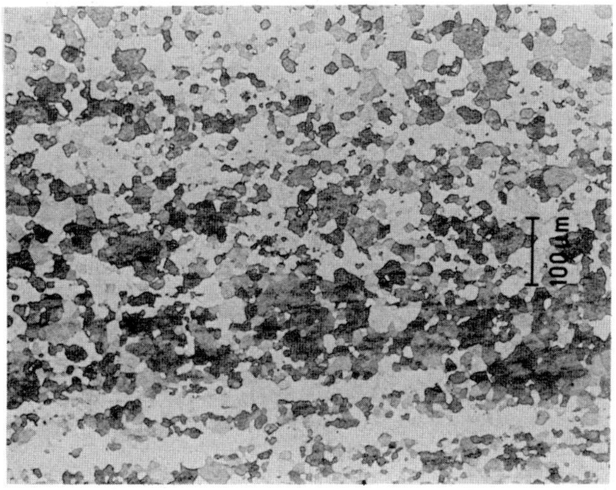

3. Microstructure of 3CR12 in the as-received condition. Etchant: Kalling's solution.

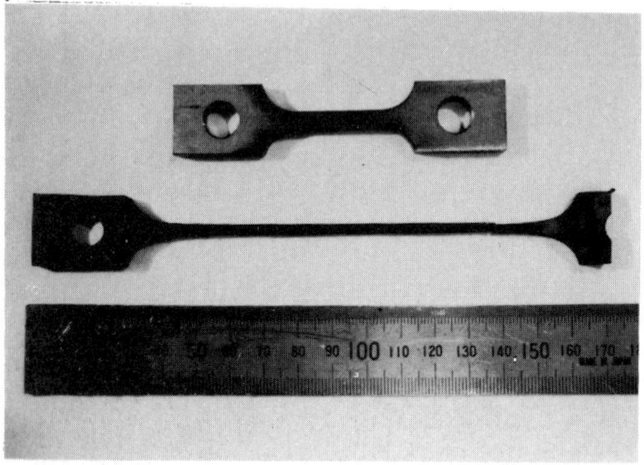

4. A 3CR12Ni specimen before and after being tested to failure at a true strain rate of 3.33×10^{-4} s^{-1} and at 880°C. Elongation 321 per cent

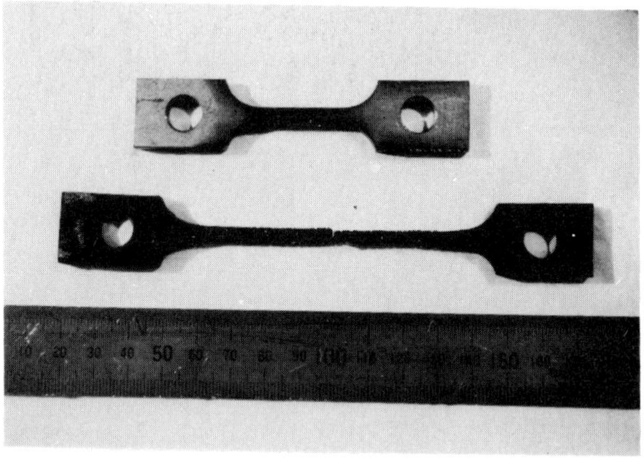

5. A 3CR12Ni specimen before and after being tested to failure at a true strain rate of 3.33×10^{-5} s^{-1} and at 880°C. Elongation 184 per cent

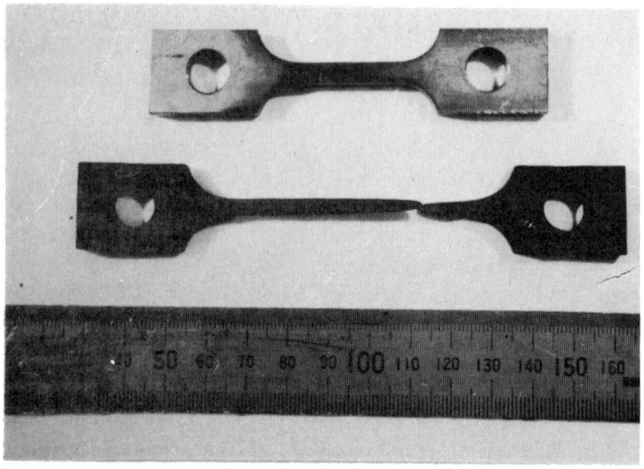

6. A 3CR12Ni specimen before and after being tested to failure at a true strain rate of 6.67×10^{-3} s^{-1} and at 915°C. Elongation 137 per cent

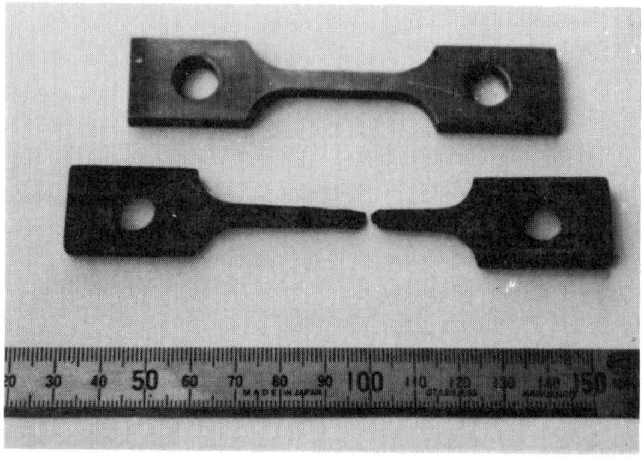

7. A 3CR12 specimen before and after being tested to failure at a true strain rate of 3.33×10^{-4} s^{-1} and at 925°C. Elongation 83 per cent

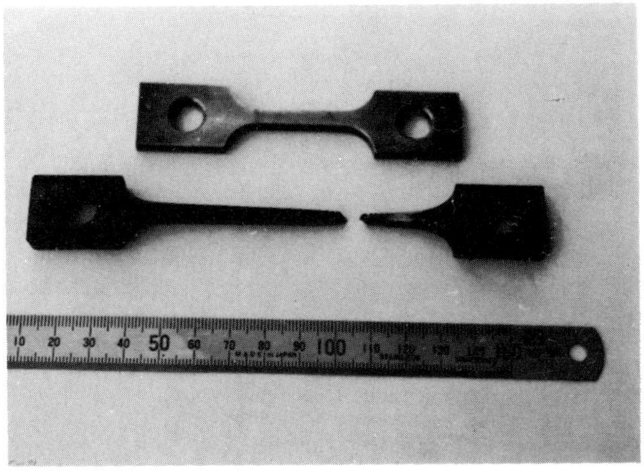

8. A 3CR12 specimen, which had been pre-reduced in thickness by 40 per cent by cold-rolling, shown before and after being tested to failure at a true strain rate of 3.30×10^{-4} s^{-1} and at $980^\circ C$. Elongation 157 per cent

9. Microstructure of 3CR12Ni alloy after annealing at $880^\circ C$ for 1 hour, followed by water-quenching. Etchant: Kalling's solution

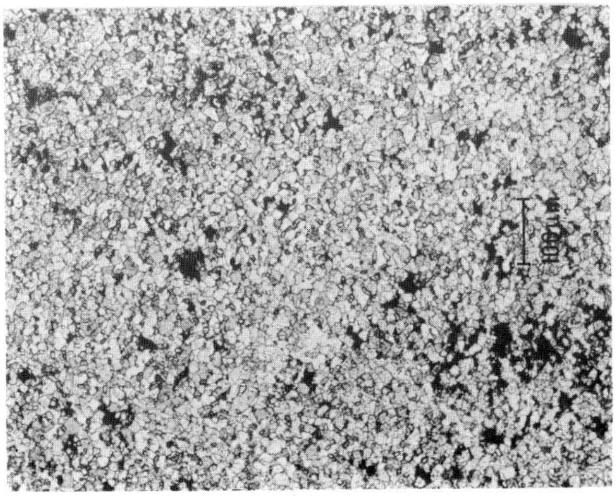

10. Microstructure of a 3CR12Ni specimen tested at a true strain rate of 3.30×10^{-5} s^{-1} and at $880^\circ C$

Phase differentiation in duplex stainless steel by potentiostatic etching

J H Dalton, M Smith and P E de Visser

JHD is Group Leader, MS an Engineer and PEdeV a Technician in the Alloy Development Group, Physical Metallurgy Division, Council for Mineral Technology (Mintek), Randburg, Republic of South Africa.

SYNOPSIS

Accurate determination of the ferrite content in duplex stainless steels would appear to be a major problem, and many measurement techniques have been tried and compared.

A convenient and reasonably rapid method for the determination of the ferrite content - and one that has often been tried - involves the use of metallographic techniques employing image analysis. The major drawback of conventional metallographic etching is the achievement of sufficient contrast to allow the austenite and ferrite phases to be distinguished, even when stain-etching techniques have been used.

Potentiostatic etching, on the other hand, has been found to give excellent colour and black-and-white contrast between the phases, which can then be determined easily and rapidly by image analysis.

This paper describes the apparatus and techniques used for the potentiostatic etching of duplex stainless steels, and compares the results obtained for ferrite content by image analysis with measurements by X-ray diffraction.

Other uses for potentiostatic etching are also outlined, including sigma-phase and carbide recognition, and the differentiation between ferrite and martensite.

INTRODUCTION

Duplex stainless steels display relatively good resistance to pitting corrosion, chloride stress-corrosion cracking, and general corrosion in acids, and their mechanical properties compare favourably with those of fully austenitic stainless steels. All these material properties are, however, strongly dependent on the ferrite-to-austenite ratio, and, therefore, strict microstructural control of both cast and wrought duplex stainless steels is of paramount importance to producers and end-users.

It is, therefore, of great importance to be able to measure accurately the ferrite content of these alloys. However, major problems arise because accurate measurement of the ferrite level in these steels is difficult. Numerous techniques, including various metallographic methods, chemical methods, and methods using magnetic permeability and magnetic attraction, have been devised, tried, and compared (1).

Metallographic methods employing image analysis appear to offer a convenient and reasonably rapid method for measurement of the ferrite content. However, the major drawback of conventional metallographic etching is that insufficient contrast is achieved to allow the austenite and ferrite phases to be distinguished by image analysis. This has led to difficulties, even when use has been made of stain-etching techniques, it being found difficult to achieve consistency in colour density within a phase, which, of course, leads to errors in measurement.

Potentiostatic etching has been found to give excellent colour and black-and-white contrast, together with consistent colour density, for determinations by image analysis.

This paper describes the apparatus used in the present investigation, and outlines the techniques and process of potentiostatic etching as applied to the measurement of the ferrite content of duplex stainless steels. Other uses for potentiostatic etching are also outlined, including sigma-phase and carbide recognition, and the differentiation between ferrite and low carbon lath martensite in a dual-phase 12 per cent chromium steel.

POTENTIOSTATIC ETCHING

The use of potentiostatic etching results in colour differences between the different phases within a metallic material. This is achieved by creating a thin transparent film on the surface of the material (2), and colours are observed as the result of interference phenomena. The incident light on the surface is reflected at the air-film and film-metal interfaces, and thickness, refractive index, and absorption coefficients of the film, and the material and the wavelength of the incident light, determine the colour observed.

Several techniques are available for the deposition of the interference layer, including heat tinting, chemical deposition, vapour deposition, and neutral and reactive sputtering (2-4). In potentiostatic etching, the interference film is formed by an electrochemical reaction between the etchant and the specimen (5). A thin, transparent, adherent layer is formed during etching, the thickness of which varies from one constituent or phase to the next.

The principles of potentiostatic etching are illustrated in Figure 1. The specimen, which is situated within an electrolytic cell, is used as the anode. The potential of the specimen is measured against the electrolyte by a calomel reference electrode. The current for the process is provided by a platinum counter electrode. During the etching process, the solution pressure (potential of the solution) is kept constant by an electronic potentiostat.

The potential applied to the specimen is of great importance. A comparison of the curves of current density versus potential for the different phases (Figure 1) allows the identification of the range of potential to reveal a specific phase selectively. If the process parameters are carefully controlled, interference films of defined thickness can be formed and, if the etching is carried out according to prescribed procedures, very consistent results can be obtained and each phase can be identified by its colour (6).

EXPERIMENTAL PROCEDURE

Specimens (in the longitudinal and transverse directions) of three types of wrought duplex stainless steel (types 2205, 2304, and 3RE60) were used to show the effect of potentiostatic etching for the measurement of their ferrite contents. Weldments of these materials were also included. In addition, specimens of type 2205 were heat-treated for extended times within the sigma-phase region to show how this technique could be used to detect the presence of sigma phase.

A specimen of a dual-phase low carbon 12 per cent chromium steel (designated 3CR12) in the heat-treated condition was included in the tests to demonstrate the differentiation between ferrite and low-carbon lath martensite. Also included, to give an exaggerated picture of the differentiation between two different carbides, was a high-chromium-iron-carbon alloy.

All specimens were prepared automatically from pre-grinding to 1 μm diamond-paste polishing using the Struer's 'ABRA' system, and finish-polished with Struer's pH-controlled OP-S solution.

The importance of the effect of preparation of the specimen on the final results obtained with this technique cannot be stressed too strongly. Because of the strong absorption of metallic materials, light cannot penetrate far into the surface. Hence, the interaction of the light with the specimen and the deposited layer occurs only in very thin surface zones, which are fully exposed to external influences such as deformation during mechanical grinding and polishing.

Grützner *et al.* (7) recommended that the final stage in the preparation of specimens should consist of electropolishing. This technique was also tried in the present investigation, and the finish obtained with OP-S solution was found to be comparable to that produced by electropolishing. Extreme cleanliness during polishing procedures is essential. Furthermore, if excellent results are to be consistently achieved, the surface of the specimen must be kept wet after final polishing and before immersion in the etchant. It is recommended that, after final polishing, the surface be washed with water and then with alcohol, after which the specimen should be immersed in alcohol before etching. The slightest contact with the air at this stage leads to erroneous results, presumably because of slight oxidation of the surface.

The most successful etchant for use with duplex stainless steels was found to be 10N NaOH solution at a pH value of 13,5. All the electrodes should be checked and cleaned thoroughly in an aqueous solution of HCl prior to etching. The etching solution must be freshly prepared and, should the solution show any sign of cloudiness during etching procedures, a fresh batch of solution should be prepared and used.

After the specimen to be etched has been connected as the anode in the cell, a period of approximately 2 minutes should be allowed to elapse prior to etching so that the sample can attain its correct corrosion potential. This change in potential can be followed by observation of the digital-voltmeter reading on the potentiostat. Etching can be commenced immediately upon immersion but, when this is done, etching times are much shorter and, as a result there is less control of the process. For the potentiostatic etching of the duplex stainless steels in NaOH solution, a potential of 1600 mV was applied. Different times are required for the etching of different alloys, and for the etching of the same alloy before and after heat treatment, owing to the difference in chemical composition of the phases. Optimum times can therefore be determined only by trial and error. After etching has been completed, the specimen should be washed in water and alcohol, and dried thoroughly. This is especially important for badly mounted or cracked specimens if staining is to be avoided.

The apparatus used for potentiostatic etching was manufactured in-house. It consists of a potentiostat of simple design and an electrolytic cell in which the specimen-holder assembly and the electrodes are mounted in a cradle. The specimen-holder assembly and the cradle, which are shown in greater detail in Figure 2, were specifically designed to give a constant distance between the electrodes and the specimen. The specimen is held against the window of the specimen-holder by an O-ring to prevent ingress of the electrolyte. This holder is screwed into the body of the assembly against another O-ring. Electrical contact is established via a spring, one end of which is in contact with a contact-plate in the body, while the other contacts the specimen through a hole drilled in the back of the bakelite mounting. The spring contact is used because it allows specimens of different lengths to be used. The front of the holder fits into a slot machined in the cradle.

The specimens were examined and photographed on an Olympus Vanox-T metallurgical microscope with halogen illumination.

Measurement of the ferrite content was carried out on a Leitz-Tas automatic image analyser. The analysis program incorporated a densitometry determination, which produces a histogram of the reflectance values of the ferrite and austenite phases. The more prominent phase (that having the lower reflectivity and greater contrast) was used for measurement of the volume fraction of ferrite. The light intensity factors were measured from the initial steep increase in density to the minimum between the two phases to ensure that grain-boundary areas would be included. The automatic image analyser was programmed to sequentially focus on and measure 50 fields on a motorized stage. Because of the good contrast available, it would have been a simple task to measure the average grain size of each of the phases separately, and to determine the average cumulative grain size, but this was not within the scope of the present investigation.

The volume fraction of austenite was measured using a Siemens D500 diffractometer using Mo $K\alpha$ radiation with a monochromator after the technique described by Jactzak(8).

RESULTS

Micrographs of the duplex stainless steel type 3RE60 are presented in Figures 3 to 5. Figures 3 and 4 show a transverse section of this alloy after etching in 10 per cent oxalic acid and by potentiostatic etching respectively. From a comparison of these two micrographs, it is clear that potentiostatic etching provides enhanced contrast and colour differentiation between austenite (yellowish blue) and ferrite (bright blue). Figure 5 is a micrograph at low magnification of a longitudinal specimen of the same alloy after potentiostatic etching. It can be seen that the boundaries between the two phases are sharp and that the colour density is consistent, which makes them suitable for image analysis.

This is also well illustrated by the micrographs of a transverse section of alloy 2205 (Figure 6) and a longitudinal section of alloy 2304 (Figure 7).

Specimens of weld metal from welded duplex alloy 2205 plate, which had been etched in 10 per cent oxalic acid and by potentiostatic etching are shown in Figures 8 and 9 respectively.

An interesting example of the effectiveness of potentiostatic etching in the differentiation of different phases is shown in Figure 10, which shows a specimen of type 2205 duplex alloy, which had been heat-treated for an extended period of time within the sigma- phase region before etching. After potentiostatic etching, all the phases are easily distinguishable because of their colour differences, the austenite being pale yellowish blue, the ferrite bright blue, and the sigma phase deep brown.

The volume percentage of low-carbon lath martensite present in the microstructure of a low-carbon 12 per cent chromium steel (3CR12) has a marked effect on the mechanical properties of this dual-phase alloy. It is difficult to measure the amount of martensite in conventionally etched material. However, this problem can be solved by the use of potentiostatic etching, which shows the ferrite as yellowish blue and the martensite as greyish brown. A specimen of 3CR12 containing martensite is shown in Figure 11.

Potentiostatic etching can also be used for the differentiation of different carbide types in the same specimen. This effect is exaggerated by the use of a specimen of a high-chromium-iron-carbon alloy. The micrograph (Figure 12) clearly shows the M_7C_3-type carbide as brown, and the $M_{23}C_6$ carbide as bright blue.

Measurements of the ferrite content by image analysis were in all cases within 2 per cent of those determined by X-ray-diffraction analysis. For example, in a determination of the average ferrite content of a transverse section of alloy 2205 by image analysis and X-ray diffraction, the values obtained were 34.72 and 35.42 per cent respectively.

CONCLUSIONS

Potentiostatic etching offers a convenient metallographic method for the differentiation of austenite and ferrite in duplex stainless-steel alloys, thus allowing the ratio of austenite to ferrite to be determined by image analysis to a good degree of accuracy. An additional advantage of potentiostatic etching in this determination is the achievement of consistent colour density within a phase.

Also, because potentionstatic etching produces marked colour differences between the phases in these alloys and a wide range of others, the phases present can be identified purely on the basis of colour, provided that care is taken in the preparation of the specimen, and control is exercised in the process.

ACKNOWLEDGEMENT

This paper is published by permission of the Council for Mineral Technology (Mintek).

REFERENCES

1. BRANTSMA, L.H., and NIJHOF, P. Ferrite measurements: an evaluation of methods and experience. Paper presented at International Conference, Duplex Stainless Steels '86 26-28 Oct., 1986, The Hague. (Late paper, not in proceedings.)

2. GAHM, H., and JEGLITSCH, F. Colour methods and their application in metallography. *Microstructural Science*, vol. 9. 1981. pp. 65-80.

3. BÜHLER, H.E., and HOUGARDY, H.P. Atlas of interference layer metallography. Deutsche Gesellschaft für Metallkunde, 1980.

4. WECK, E., and LEISTNER, E. Metallographic instructions for colour etching by immersion, Parts I-III. Deutsche Verlag für Schweisstechnik, 1983.

5. EDELEANU, M.A. The potentiostat as a metallographic tool. *J. Iron Steel Inst.* April. 1957. pp. 482-488.

6. BLÖCH, R., and LICHTENEGGER, P. The use of potentiostatic etching to reveal microstructure constituents selectively. *Practical Metallography*, vol. 12. 1975. pp. 186-193.

7. GRÜTZNER, G., *et al.* Potentiostatic colour etching of stainless steel. *Praktische Metallographic*. Jun. 1969. pp. 346-350.

8. JACTZAK, C.F., *et al.* Retained austenite and its measurement by X-ray diffraction. Warrendale, Society of Automotive Engineers, Jan. 1980.

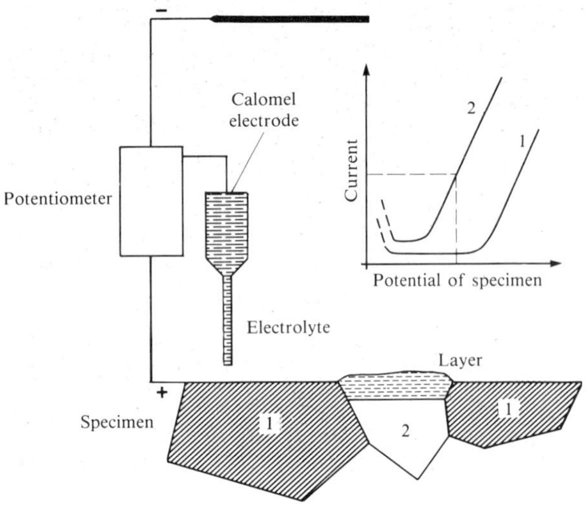

1. Principles of electrolytic potentiostatic etching (after Gahm and Jeglitsch(2))

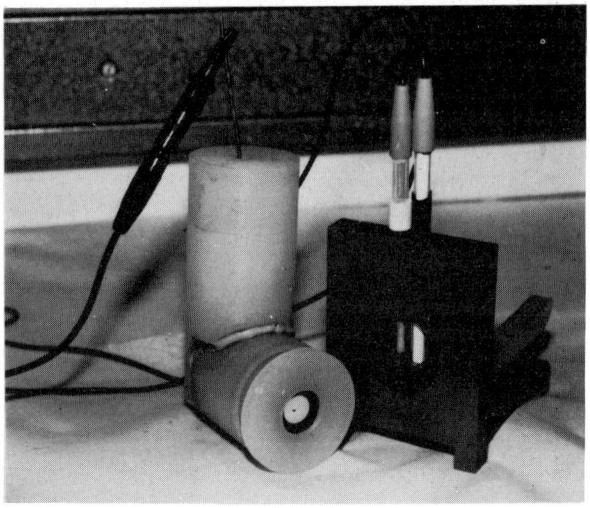

2. Detail of specimen-holder and cradle

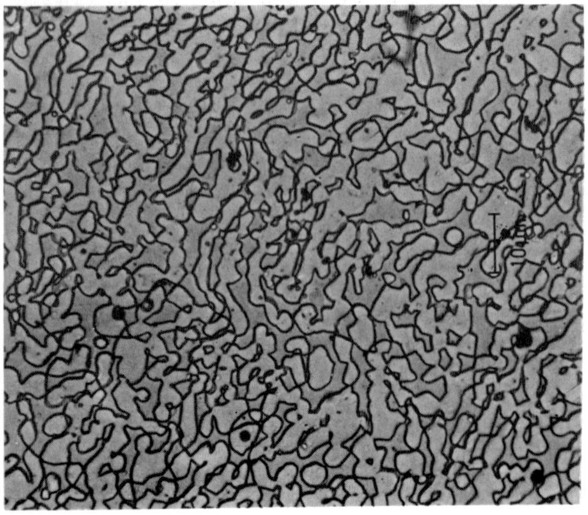

3. Alloy 3RE60 etched in 10 per cent oxalic acid (transverse section)

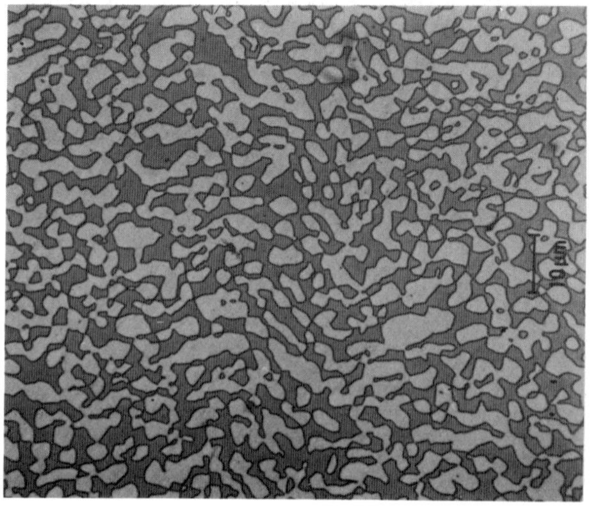

4. Alloy 3RE60 potentiostatically etched (transverse section)

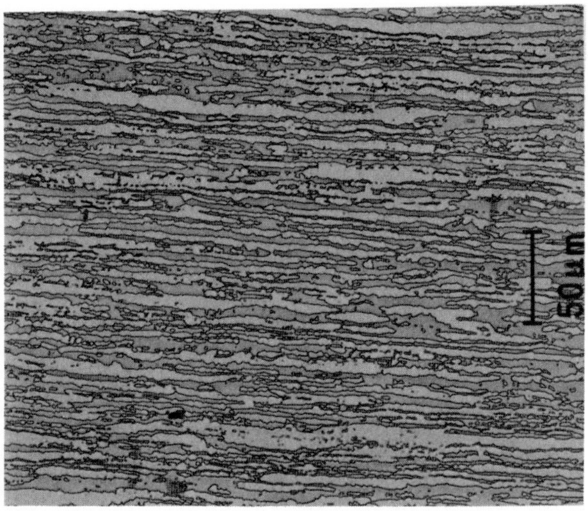

5. Low-magnification longitudinal section of 3RE60

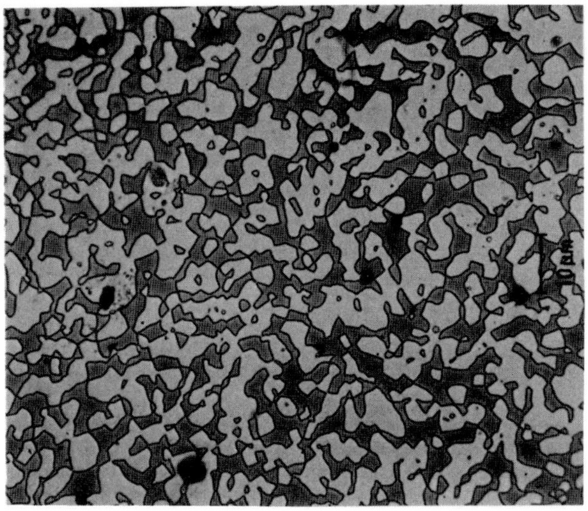

6. Alloy 2205 (transverse section)

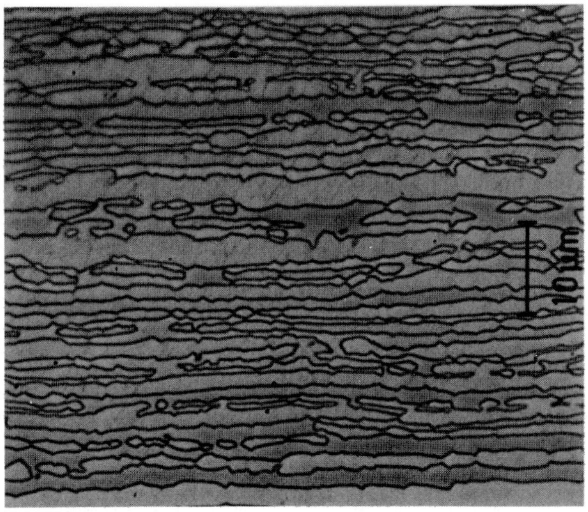

7. Alloy 2205 (longitudinal section)

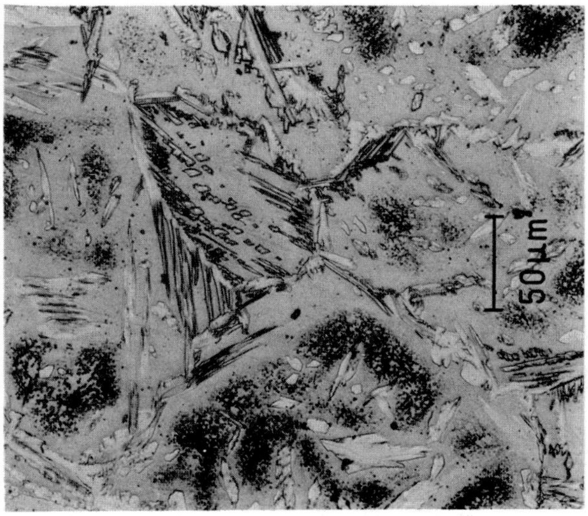

8. Alloy 2205 weld metal etched in 10 per cent oxalic acid

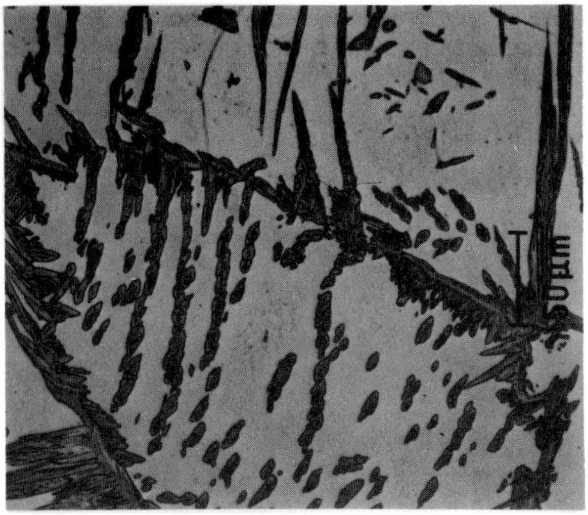

9. Alloy 2205 weld metal potentio-statically etched

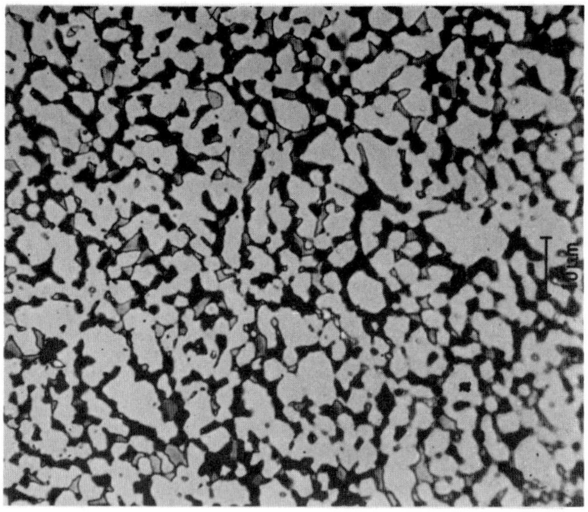

10. Heat-treated type 2205 alloy potentio-statically etched. Austenite appears pale yellowish-blue, ferrite is bright blue, and sigma phase is brown

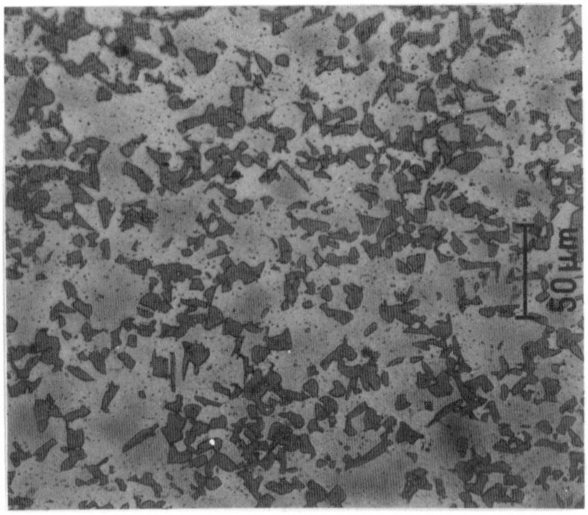

11. 3CR12 potentiostatically etched (transverse section)

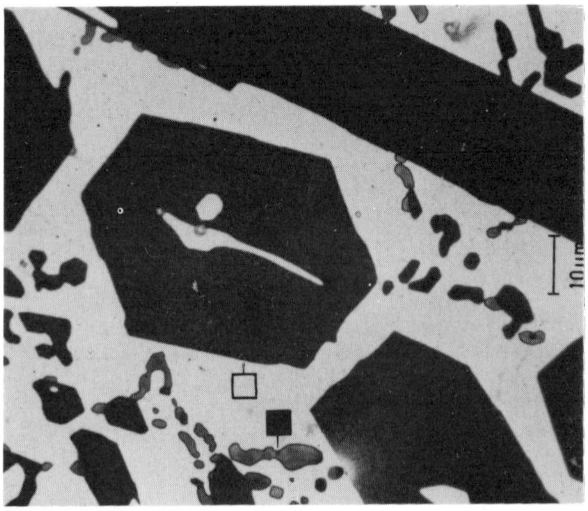

12. Potentiostatically etched high-chromium-iron-carbon alloy showing M_7C_3 carbide (brown) □ and $M_{23}C_6$ carbide (bright blue) ■

AUTHOR INDEX

SUBJECT INDEX